THE PURSUING PRESENCE OF FACTS

THE PURSUING PRESENCE OF FACTS

RICHARD JOHN KOSCIEJEW

authorHOUSE

AuthorHouse™
1663 Liberty Drive
Bloomington, IN 47403
www.authorhouse.com
Phone: 1 (800) 839-8640

Published by AuthorHouse 08/03/2015

ISBN: 978-1-5049-1881-7 (sc)
ISBN: 978-1-5049-1880-0 (e)

Print information available on the last page.

Any people depicted in stock imagery provided by Thinkstock are models, and such images are being used for illustrative purposes only. Certain stock imagery © Thinkstock.

This book is printed on acid-free paper.

About the Author

Richard John Kosciejew is a German-born Canadian who takes residence in the city of Toronto, Canada. His father was a butcher and holding of five children. Richard, the second born, received his public school training within the playground of Alexander Muir Public School, then moved into the secondary level of Ontario's educational system for being taught at Central Technical School, finding that his thirst of an increasing vexation for what is truth and knowledge were to be quenched in the relief of mind, body, and soul. As gathering opportunities, he attended Centennial College, also the University of Toronto, and keeping at this pace, he attended the University of Western Ontario, situated in London, Ontario, Canada. He had drawn heavy interests, besides philosophy and physics that were his academic studies. However, in the analyses were somewhat overpowering, nonetheless, during the criterion of analytical studies and taking time to attend of the requiring academia, he completed his book The Designing Theory of Transference. He is now living in Toronto and finds that the afforded efforts in his attemptive engagements are only to be achieved for what is obtainable in the secret reservoir of continuative phenomenon, for which we are to discover or rediscover in their essence.

Richard John Kosciejew

As one of the several contributing disciplines to cognitive science, philosophy offers two sorts of contributions. On the one hand, philosophy of science provides a metatheoretical perspective on the endeavours of any scientific enterprise, analysing such things as the goals of scientific investigation and the strategies employed in reaching those goals. Philosophy of science thus offers a perspective from which we can examine and potentially evaluate the endeavours of cognitive science. On the other hand, philosophy of mind offers substantive theses about the nature of mind and of mental activity. Although these theses typically have not resulted from empirical investigation, they often have subsequently figured in actual empirical investigations in cognitive science, or its predecessors. Because the two roles' philosophy plays in cognitive science are quite different, they are introduced exclusively as one focusses on philosophy of science, whereas Philosophy of mind is explored in philosophy of mind as an overview for cognitive science.

The strategy is to present a variety of views from philosophy of science that have figured in discussions about cognitive science. Some of these views are no longer widely accepted by philosophers of science. Nonetheless, they have been and, in some cases, remain influential outside of philosophy. Moreover, some older views have provided the starting points for current philosophical thinking that is done against a backdrop of previous endeavours, with a recognition of both their success and failure.

The Pursuing Presence of Facts

Richard John Kosciejew

Human history is in essence the recipient of an accountable discernability for which it holds inherently, by saying that times arrow is the sustaining continuity of intermittence, only through which of our ideas are the aspirations enacting of a launching celebration of gratifying distinction, by ways of the placed value of inseparability in which we have proceeded by an uncovering implication as times generations, upon which we are shackled by a condition or situation in which something is required or wanted, those of us who had led the charge for reasons that posit of a nonexisting volume of space and time, becoming infinitely distorted, but form the collective structures, from which the constructionists are taken for any given function but every regional locality that contains each point for which of the derivatives do not exist. Our need for wanting the choices that is intended to express of our claiming of an individuality or dispensing beyond a safe or reasonable limit, we are obligated beyond a limit for which of something uncommon or unusual might for expressing the action or state of a single subject as distinguished from the general exemplifications to some unknowing or unknowable phenomenons. If only to discover or rediscover in all overcoming mysteriousness. Therein we lie upon the instinctual primitivities, in that of preserving plexuity and complicating degrees of an overwhelming creativity and the servicing of ideas.

Justly as language is the dress of thought; Ideas began with Plato, as eternal, mind-independent forms or archetypes of the things in the material world. Neoplatonism made them thoughts in the in the mind of God who created the world. Much has been criticized or have in effect, to find faults with the 'new way of ideas', so much a part of seventeenth and eighteenth-century philosophy, began with Descartes' (1596-1650) was the marginalised target for the consciousable extension of ideas, only to cover whatever is in human minds to an extension of which Locke (1632-1704) made much use. But are they like mental images, of things outside the mind or non-representational, like sensations? If representational, are they mental objects, standing between the mind and what they represent, or are they mental acts and modifications of a mind perceiving the world directly? Finally, are they neither objects nor mental acts, but dispositions? Malebranche (1632-1715) and Arnauld Antoine (1612-94), and also, Gottfried Wilhelm, Leibniz, famously disagreed about how 'ideas' should be understood, and recent scholars have a differing opinion as to disagree in the corresponding figures about how Arnauld, Descartes', Locke and Malebranche in fact understood them.

Together with a comprehensive bias toward the sensory, so that what lies in the mind may be thought of as something as images, and a belief that thinking is well explained as the manipulation having no real existence, but existing in a fancied imagination. It is not reason but the imagination, which is found to be responsible for making the empirical inferences that we do. There are certain general 'principles of the imagination' according to which ideas naturally come and go in the mind under certain conditions. It is the task of the 'science of human nature' to discover such principles, but without going beyond experience. For example, an observed correlation between things of two kinds can be seen to produce in everyone the propensity to expect a thing of the second sort given an experience of a thing of the first sort. We get a feeling, or an 'impression', once the mind makes such a tradition and that is what lead us to attribute necessarily between things of the two kings, there is no necessity in the relation between things that happen in the world of enduring objects. Experience alone cannot produce that belief, everything we directly perceive is. Whatever our experience is like, no reasoning could assure us of the existence of something as autonomous of our impressions which continue s to exist when they cease. The series of constantly changing sense impressions present us with observable features which Hume calls 'constancy' and 'coherence' and these naturally operate on the mind, in such a way as eventually to produce the opinion, it's continuously distinguished from all others, discrete and perceived by the senses or intellect as a probable distinction from those discerning of description as a descriptive science: the fact of existing or our belief of an existence on Earth, are founded by the specific presence occurrences as to indicate the actuality for which is to be. We must persist, that still exists in divergent areas in which we are to continue under circumstance or in a specific place that occur, that existence to come forth, have to be manifest in our standing of distinct existence. The explanation is complicated, but it is meant to appeal only to psychological mechanisms which can be discovered by careful and exact experiments, and the observation of those particular effects, which resulting forms [the mind's] different circumstances and situations.

Science, is nothing more than a description of facts, and 'facts' involve nothing more than sensations and the relationships among them. Sensations are the only real elements, as all other concepts are extra, they are merely imputed on the real, e.g., on the sensations, by us. Concepts like 'matter' and 'atoms' are merely shorthand for collection of sensations: They do not denote anything that exists, the same holds for many other words as 'body'. Logically prevailing upon science may thereby involve nothing more than sensations and the relationships among them. Sensations are the only real elements, as all else, be other than the concepts under which are extra: They are merely imputed on the real, e.g., on the sensations, by us. Concepts like 'matter' and 'atom' are merely shorthand for collections of sensations, they do not denote anything that exists, still, the same holds for many other words, such as 'body', as science, carriers nothing more than a description of facts. 'Facts', accordingly, are devoted largely to doubtful refutations, such that, if we were to consider of a pencil that is partially submerged in water. It looks broken, but it is really straight, as we can verify by touching it. Nonetheless, causing the state or facts of having independent reality, the pencil in the water is merely two different facts. The pencil in the water is really broken, as far as the fact of sight is concerned, and that is all to this it, as to look upon an object of something that changes, simply because you looked at it, this corporative conjecture is representational among things that are aligned or designed in the spirit of quantum reality.

The philosophical debate that had led to conclusions useful to the architects of classical physics can be briefly summarized, such when Thales fellow Milesian Anaximander claimed that the first substance, although indeterminate, manifested itself in a conflict of oppositions between hot and cold, moist and dry. The idea of nature as a self-regulating balance of forces was subsequently elaborated upon by Heraclitus (D. after 480 BC), who asserted that the fundamental substance is strife between opposites, which is itself the unity of the whole. It is, said Heraclitus, the tension between opposites that keeps the whole from simply 'passing away.'

Parmenides of Elea (B.c. 515 BC) argued in turn that the unifying substance is unique and static being. This led to a conclusion about the relationship between ordinary language and external reality that was later incorporated into the view of the relationship between mathematical language and physical reality. Since thinking or naming involves the presence of something, said Parmenides, thought and language must be dependent upon the existence of objects outside the human intellect. Presuming a one-to-one correspondence between word and idea and actual existing things, Parmenides concluded that our ability to think or speak of a thing at various times implies that it exists at all times. Hence the indivisible One does not change, and all perceived change is an illusion.

These assumptions emerged in roughly the form in which they would be used by the creators of classical physics in the thought of the atomists. Leucippus? : 450-420 BC and Democritus 460-c. 370 Bc. They reconciled the two dominant and seemingly antithetical concepts of the fundamental character of being-Becoming (Heraclitus, 500B.c early Greek philosopher who maintained that strife and change are the natural conditions of the universe) and unchanging Being. Being, the Eleatic's said, is present in the invariable substance of the atoms that, through blending and separation, make up the thing of changing or becoming worlds.

The last remaining feature of what would become the paradigm for the first scientific revolution in the seventeenth century is attributed to Pythagoras ¦L. sixth century Bc. Who founded a school in southern Italy that sought to discover the mathematical principal of reality through the study of musical harmony and geometry? Like Parmenides, a Greek philosopher beared 515? BC and a founder of the Eleatic tradition, which, Pythagoras also held that the perceived world is illusory and that there is an exact correspondence between ideas and aspects of external reality. Pythagoras, however, had a different conception of the character of the idea that showed this correspondence. The truth about the fundamental character of the unified and unifying substance, which could be uncovered through reason and contemplation, is, he claimed, mathematical in form.

Pythagoras established and was the cental figure in a school of philosophy, religion and mathematics; He was apparently viewed by his followers as semi-divine. For his followers the regular solids (symmetrical three-dimensional form in which all bound sides are the same regular polygons) and whole numbers became revered essences of sacred ideas. In contrast with ordinary language, the language of mathematics and geometric forms seemed closed, precise and pure. Providing one understood the axioms and notations, and the meaning conveyed was invariant from one mind to another. The Pythagoreans felt that the language empowered the mind to leap beyond the confusion of sense experience into the realm of immutable and eternal essences. This mystical insight made Pythagoras the figure from antiquity most revered by the creators of classical physics, and it continues to have great appeal for contemporary physicists as they struggle with the epistemological implications of the quantum mechanical description of nature.

All and all, progress was made in mathematics, and to a lesser extent in physics, from the time of classical Greek philosophy to the seventeenth century in Europe. In Baghdad, for example, from about A.D. 750 to A.D. 1000, substantial advancement was made in medicine and chemistry, and the relics of Greek science were translated into Arabic, digested, and preserved. Eventually these relics reentered Europe via the Arabic kingdom of Spain and Sicily, and the work of figures like Aristotle universities of France, Italy, and England during the Middle Ages.

For much of this period the Church provided the institutions, like the reaching orders, needed for the rehabilitation of philosophy. But the social, political and an intellectual climate in Europe was not ripe for a revolution in scientific thought until the seventeenth century. Until later in time, lest as far into the nineteenth century, the works of the new class of intellectuals we call scientists, whom of which were more avocations than vocation, and the word scientist do not appear in the English until around 1840.

Copernicus (1473-1543) would have been described by his contemporaries as an administrator, a diplomat, an avid student of economics and classical literature, and most notable, a highly honoured and placed church dignitaries. Although we named a revolution after him, his devoutly conservative man did not set out to create one. The placement of the Sun at the centre of the universe, which seemed right and necessary to Copernicus, was not a result of making careful astronomical observations. In fact, he made very few observations in the course of developing his theory, and then only to ascertain if his prior conclusions seemed correct. The Copernican system was also not any more useful in making astrological calculations than the accepted model and was, in some ways, much more difficult to implement. What, then, was his motivation for creating the model and his reasons for presuming that the model was correct?

Copernicus felt that the placement of the Sun at the centre of the universe made sense because he viewed the Sun as the symbol of the presence of a supremely intelligent and intelligible God in a man-centred world. He was apparently led to this conclusion in part because the Pythagoreans believed that fire exists at the centre of the cosmos, and Copernicus identified this fire with the fireball of the Sun. the only support that Copernicus could offer for the greater efficacy of his model was that it represented a simpler and more mathematical harmonious model of the sort that the Creator would obviously prefer. The linguistic process to indicate the practice or custom through which Copernicus qualified or having an intentional point at issue or accomplished in The Revolution of Heavenly Orbs,' illustrates the religious dimension of his scientific thought: 'In the midst of all the sun reposes, unmoving. Who, indeed, in this most beautiful temple would place the light-giver in any other part than from where it can illumine all other parts?'

The belief that the mind of God as Divine Architect permeates the working of nature was the guiding principle of the scientific thought of Johannes Kepler (or, Keppler, 1571-1630). For this reason, most modern physicists would probably feel some discomfort in reading Kepler's original manuscripts. Physics and metaphysics, astronomy and astrology, geometry and theology commingle with an intensity that might offend those who practice science in the modern sense of that word. Physical laws, wrote Kepler, 'lie within the power of understanding of the human mind; God wanted us to perceive them when he created us of his own image, in order. . . that we may take part in his own thoughts. Our knowledge of numbers and quantities is the same as that of God's, at least insofar as we can understand something of it in this mortal life.'

Believing, like Newton after him, in the literal truth of the words of the Bible, Kepler concluded that the word of God is also transcribed in the immediacy of observable nature. Kepler's discovery that the motions of the planets around the Sun were elliptical, as opposed perfecting circles, may have made the universe seem a less perfect creation of God on ordinary language. For Kepler, however, the new model placed the Sun, which he also viewed as the emblem of a divine agency, more at the centre of mathematically harmonious universes than the Copernican system allowed. Communing with the perfect mind of God requires as Kepler put it 'knowledge of numbers and quantity.'

Since Galileo did not use, or even refer to, the planetary laws of Kepler when those laws would have made his defence of the heliocentric universe more credible, his attachment to the god-like circle was probably a more deeply rooted aesthetic and religious ideal. But it was Galileo, to a greater extent be regarded of conjunction in all its use, so that after a comparative effort implicated a difference as based or derived from an unequal rather than the common structural causes, especially the function preforming the operative adaptations for providing an etymologically explicable example of the notorious discrepancies between the person of Newton. Who's responsibility for formulating the scientific idealism that the quantum mechanics that now force us to give up by leaving or ceasing of other impending inclusions of recklessly inhabiting? In 'Dialogue Concerning the Two Great Systems of the World,' Galileo said about the following about the followers of Pythagoras: 'I know perfectly well that the Pythagoreans had the highest esteem for the science of number and that Plato himself admired the human intellect and believed that it participates in divinity solely because it is able to understand the nature of numbers. And I myself am inclined to make the same judgement.'

This article of faith-mathematical and geometrical ideas mirror precisely the essences of physical reality was the basis for the first scientific law of this new science, a constant describing the acceleration of bodies in free fall, could not be confirmed by experiment. The experiments conducted by Galileo in which balls of different sizes and weights were rolled simultaneously down an inclined plane did not, as he frankly admitted, their precise results. And since a vacuum pumps had not yet been invented, there was simply no way that Galileo could subject his law to rigorous experimental proof in the seventeenth century. Galileo believed in the absolute validity of this law in the absence of experimental proof because he also believed that movement could be subjected absolutely to the law of number. What Galileo asserted, as the French historian of science Alexander Koyré put it, was 'that the real are in its essence, geometrical and, consequently, subject to rigorous determination and measurement.'

The popular image of Isaac Newton (1642-1727) is that of a supremely rational and dispassionate empirical thinker. Newton, like Einstein, had the ability to concentrate unswervingly on complex theoretical problems until they yielded a solution. But what most consumed his restless intellect were not the laws of physics. In addition to believing, like Galileo that the essences of physical reality could be read in the language of mathematics, Newton also believed, with perhaps even greater intensity than Kepler, in the literal truths of the Bible.

For Newton the mathematical languages of physics and the language of biblical literature were equally valid sources of communion with the eternal writings in the extant documents alone consist of more than a million words in his own hand, and some of his speculations seem quite bizarre by contemporary standards. The Earth, said Newton, will still be inhabited after the day of judgement, and heaven, or the New Jerusalem, must be large enough to accommodate both the quick and the dead. Newton then put his mathematical genius to work and determined the dimensions required to house the population, his rather precise estimate was 'the cube root of 12,000 furlongs.'

The point is, that during the first scientific revolution the marriage between mathematical idea and physical reality, or between mind and nature via mathematical theory, was viewed as a sacred union. In our more secular age, the correspondence takes on the appearance of an unexamined article of faith or, to borrow a phrase from William James (1842-1910), 'an altar to an unknown god.' Heinrich Hertz, the famous nineteenth-century German physicist, nicely described what there is about the practice of physics that tends to inculcate this belief: 'One cannot escape the feeling that these mathematical formulae have an independent existence and intelligence of their own that they are wiser than we, wiser than their discoveries. That we get more out of them than was originally put into them.'

While Hertz made this statement without having to contend with the implications of quantum mechanics, the feeling, the described remains the most enticing and exciting aspects of physics. That elegant mathematical formulae provide a framework for understanding the origins and transformations of a cosmos of enormous age and dimensions are a staggering discovery for bidding physicists. Professors of physics do not, of course, tell their students that the study of physical laws in an act of communion with thee perfect mind of God or that these laws have an independent existence outside the minds that discover them. The business of becoming a physicist typically begins, however, with the study of classical or Newtonian dynamics, and this training provides considerable covert reinforcement of the feeling that Hertz described.

Perhaps, the best way to examine the legacy of the dialogue between science and religion in the debate over the implications of quantum non-locality is to examine the source of Einstein's objections tp quantum epistemology in more personal terms. Einstein apparently lost faith in the God portrayed in biblical literature in early adolescence. But, as appropriated, . . . the 'Autobiographical Notes' give to suggest that there were aspects that carry over into his understanding of the foundation for scientific knowledge, . . . 'Thus I am, despite the formation of mental images or some thing that is neither perceived as real nor present to the senses as holding the heritage, with which had deeply found an abrupt end to religiosity at the age of 12. Though the reading of popular scientific books, I soon reached the conviction that a great deal in the stories of the Bible could not be true. The consequence waw a positively frantic [orgy] of freethinking coupled with the impression that youth is intentionally being deceived by the stat through lies that it was a crushing impression. Suspicion against every kind of authority grew out of this experience. . . . It was clear to me that the religious paradise of youth, which was thus lost, was a first attempt ti free myself from the chains of the 'merely personal'. . . . The mental grasp of this extra-personal world within the frame of the given possibilities swam as highest aim half consciously and half unconsciously before the mind's eye.'

What is more, was, suggested Einstein, belief in the word of God as it is revealed in biblical literature that allowed him to dwell in a 'religious paradise of youth' and to shield himself from the harsh realities of social and political life. In an effort to recover that inner sense of security that was lost after exposure to scientific knowledge, or to become free once again of the 'merely personal', he committed himself to understanding the 'extra-personal world within the frame of given possibilities', or as seems obvious, to the study of physics. Although the existence of God as described in the Bible may have been in doubt, the qualities of mind that the architects of classical physics associated with this God were not. This is clear in the comments from which Einstein uses of mathematics, . . . 'Having existence only in the Nature we are to realize of the simplest conceivable mathematical ideas. I am convinced that we can discover, by means of purely mathematical construction, those concepts and those lawful connections between them that furnish the key to the understanding of natural phenomena. Experience remains, of course, the sole criteria of physical utility of a mathematical construction. But the creative principle resides in mathematics. In a certain sense, therefore, I hold it true that pure thought can grasp reality, as the ancients dreamed.'

This article of faith, first articulated by Kepler, that 'nature is the realization of the simplest conceivable mathematical ideas' allowed for Einstein to posit the first major law of modern physics much as it allows Galileo to posit the first major law of classical physics. During which time, when the special and then the general theories of relativity had not been confirmed by experiment and many established physicists viewed them as at least minor heresies, Einstein remained entirely confident of their predictions. Ilse Rosenthal-Schneider, who visited Einstein shortly after Eddington's eclipse expedition confirmed a prediction of the general theory (1919), described Einstein's response to this news: When I was giving expression to my joy that the results coincided with his calculations, he said quite unmoved, 'But I knew the theory was correct,' and when I asked, what if there had been no confirmation of his prediction, he countered: 'Then I would have been sorry for the dear Lord -the theory is correct.'

Einstein was not given to making sarcastic or sardonic comments, particularly on matters of religion. These unguarded responses testify to his profound conviction that the language of mathematics allows the human mind access to immaterial and immutable truths existing outside of the mind that conceived them. Although Einstein's belief was far more secular than Galileo's, it retained the same essential ingredients.

What continued in the twenty-three-year-long debate between Einstein and Bohr, least of mention? The primary article drawing upon its faith that contends with those opposing to the merits or limits of a physical theory, at the heart of this debate was the fundamental question, 'What is the relationship between the mathematical forms in the human mind called physical theory and physical reality?' Einstein did not believe in a God who spoke in tongues of flame from the mountaintop in ordinary language, and he could not sustain belief in the anthropomorphic God of the West. There is also no suggestion that he embraced ontological monism, or the conception of Being featured in Eastern religious systems, like Taoism, Hinduism, and Buddhism. The closest that Einstein apparently came to affirming the existence of the 'extra-personal' in the universe was a 'cosmic religious feeling', which he closely associated with the classical view of scientific epistemology.

The philosophy that Einstein endeavoured to engage upon and fought to preserve the natural inheritance of physics until the advent of quantum mechanics. Although the mind that constructs reality might be evolving fictions that are not necessarily true or necessary in social and political life, their Einstein felt a way of knowing, purged of deceptions and lies. He was convinced that knowledge of physical reality in physical theory mirrors the preexistent and immutable realm of physical laws. And as Einstein consistently made clear, this knowledge mitigates loneliness and inculcates a sense of order and reason in a cosmos that might appear otherwise bereft of meaning and purpose.

That most disturbed Einstein about quantum mechanics was the fact that this physical theory might not, in experiment or even in principle, mirrors precisely the structure of physical reality. There is, for all the reasons we seem attested of, in that an inherent uncertainty in measurements made, . . . a quantum mechanical process reflects of a pursuit that quantum theory in itself and its contributive dynamic functionalities that there is to affirm of the attribution of brings the completeness of quantum mechanical theory. Einstein's fearing that it would force us to recognize that this inherent uncertainty applied to all of physics, and, therefore, the ontological bridge between mathematical theory and physical reality -does not exist. And this would mean, as Bohr was among the first to realize, that we must profoundly revive the epistemological foundations of modern science.

The world view of classical physics allowed the physicist to assume that communion with the essences of physical reality via mathematical laws and associated theories was possible, but it made no other provisions for the knowing mind. In our new situation, the status of the knowing mind seems quite different. Modern physics distributively contributed its view toward the universe as an unbroken, undissectable and undivided dynamic whole. 'There can hardly be a sharper contrast,' said Melic Capek, 'than that between the everlasting atoms of classical physics and the vanishing 'particles' of modern physics as Stapp put it: 'Each atom turns out to be nothing but the potentialities in the behaviour pattern of others. What we find, therefore, are not elementary space-time realities, but rather a web of relationships in which no part can stand alone, every part derives its meaning and existence only from its place within the whole"

The characteristics of particles and quanta are not isolatable, given particle-wave dualism and the incessant exchange of quanta within matter-energy fields. Matter cannot be dissected from the omnipresent sea of energy, nor can we in theory or in fact observe matter from the outside. As Heisenberg put it decades ago, 'the cosmos appears to be a complicated tissue of events, in which connection of different kinds alternate or overlay or combine and thereby determine the texture of the whole. This meant that a pure reductionist approach to understanding physical reality, which was the goal of classical physics, is no longer appropriate.

While the formalism of quantum physics predicts that correlations between particles over space-like separated regions are possible, it can say nothing about what this strange new relationship between parts (quanta) and whole (cosmos) was by means an outside formalism. This does not, however, prevent us from considering the implications in philosophical terms, as the philosopher of science Errol Harris noted in thinking about the special character of wholeness in modern physics, a unity without internal content is a blank or empty set and is not recognizable as a whole. A collection of merely externally related parts does not constitute a whole in that the parts will not be 'mutually adaptive and complementary to one and another.'

Wholeness requires a complementary relationship between unity and differences and is governed by a principle of organization determining the interrelationship between parts. This organizing principle must be universal to a genuine whole and implicit in all parts that constitute the whole, even though the whole is exemplified only in its parts. This principle of order, Harris continued, 'is nothing really in and of itself. It is the way parts are organized and not another constituent addition to those that constitute the totality.'

In a genuine whole, the relationship between the constituent parts must be 'internal or immanent' in the parts, as opposed to a mere spurious whole in which parts appear to disclose wholeness due to relationships that are external to the parts. The collection of parts that would allegedly constitute the whole in classical physics is an example of a spurious whole. Parts constitute a genuine whole when the universal principle of order is inside the parts and thereby adjusts each to all that they interlock and become mutually complementary. This not only describes the character of the whole revealed in both relativity theory and quantum mechanics. It is also consistent with the manner in which we have begun to understand the relation between parts and whole in modern biology.

Modern physics also reveals, claims Harris, a complementary relationship between the differences between parts that constituted contentual representations that the universal ordering principle that is immanent in each of the parts. While the whole cannot be finally disclosed in the analysis of the parts, the study of the differences between parts provides insights into the dynamic structure of the whole present in each of the parts. The part can never, nonetheless, be finally isolated from the web of relationships that disclose the interconnections with the whole, and any attempt to do so results in ambiguity.

Much of the ambiguity in attempted to explain the character of wholes in both physics and biology derives from the assumption that order exists between or outside parts. But order in complementary relationships between differences and sameness in any physical event is never external to that event - the connections are immanent in the event. From this perspective, the addition of non-locality to this picture of the dynamic whole is not surprising. The relationship between part, as quantum event apparent in observation or measurement, and the undissectable whole, revealed but not described by the instantaneous, and the undissectable whole, revealed but described by the instantaneous correlations between measurements in space-like separated regions, is another extension of the part-whole complementarity to modern physics.

If the universe is a seamlessly interactive system that evolves to a higher level of complexity, and if the lawful regularities of this universe are emergent properties of this system, we can assume that the cosmos is a singular point of significance as a whole that evinces the 'orderly principles of progression' of complementary relations of its parts. Given that this whole exists in some sense within all parts (quanta), one can then argue that it operates in self-reflective fashion and is the ground for all emergent complexities. Since human consciousness evinces self-reflective awareness in the human brain and since this brain, like all physical phenomena can be viewed as an emergent property of the whole, it is reasonable to conclude, in philosophical terms at least, that the universe is conscious.

But since the actual character of this seamless whole cannot be represented or reduced to its parts, it lies, quite literally beyond all human representations or descriptions. If one chooses to believe that the universe be a self-reflective and self-organizing whole, this lends no support whatsoever to conceptions of design, meaning, purpose, intent, or plan associated with any mytho-religious or cultural heritage. However, If one does not accept this view of the universe, there is nothing in the scientific descriptions of nature that can be used to refute this position. On the other hand, it is no longer possible to argue that a profound sense of unity with the whole, which has long been understood as the foundation of religious experience, which can be dismissed, undermined or invalidated with appeals to scientific knowledge.

While we have consistently tried to distinguish between scientific knowledge and philosophical speculation based on this knowledge -there is no empirically valid causal linkage between the former and the latter. Those who wish to dismiss the speculative assumptions as its basis to be drawn the obvious freedom of which is firmly grounded in scientific theory and experiments there is, however, in the scientific description of nature, the belief in radical Cartesian division between mind and world sanctioned by classical physics. Seemingly clear, that this separation between mind and world was a macro-level illusion fostered by limited awareness of the actual character of physical reality and by mathematical idealization that were extended beyond the realm of their applicability.

Thus, the grounds for objecting to quantum theory, the lack of a one-to-one correspondence between every element of the physical theory and the physical reality it describes, may seem justifiable and reasonable in strictly scientific terms. After all, the completeness of all previous physical theories was measured against the criterion with enormous success. Since it was this success that gave physics the reputation of being able to disclose physical reality with magnificent exactitude, perhaps a more comprehensive quantum theory will emerge to insist on these requirements.

All indications are, however, that no future theory can circumvent quantum indeterminancy, and the success of quantum theory in co-ordinating our experience with nature is eloquent testimony to this conclusion. As Bohr realized, the fact that we live in a quantum universe in which the quantum of action is a given or an unavoidable reality requires a very different criterion for determining the completeness or physical theory. The new measure for a complete physical theory is that it unambiguously confirms our ability to co-ordinate more experience with physical reality.

If a theory does so and continues to do so, which is certainly the case with quantum physics, then the theory must be deemed complete. Quantum physics not only works exceedingly well, it is, in these terms, the most accurate physical theory that has ever existed. When we consider that this physics allows us to predict and measure quantities like the magnetic moment of electrons to the fifteenth decimal place, we realize that accuracy per se is not the real issue. The real issue, as Bohr rightly intuited, is that this complete physical theory effectively undermines the privileged relationship in classical physics between 'theory' and 'physical reality'.

If the universe is a seamlessly interactive system that evolves to higher levels of complex and complicating regularities of which ae lawfully emergent in property of systems, we can assume that the cosmos is a single significant whole that evinces progressive order in complementary relations to its parts. Given that this whole exists in some sense within all parts (quanta), one can then argue that in operates in self-reflective fashion and is the ground from all emergent plexuities. Since human consciousness evinces self-reflective awareness in te human brain (well protected between the cranium walls of the skull) and since this brain, like all physical phenomena, can b viewed as an emergent property of the whole, it is unreasonable to conclude, in philosophical terms at least, that the universe is conscious.

Nevertheless, since the actual character of this seamless whole cannot be represented or reduced to its parts, it lies, quite laterally, beyond all human representation or descriptions. If one chooses to believe that the universe be a self-reflective and self-organizing whole, this lends no support whatsoever to conceptual representation of design, meaning, purpose, intent, or plan associated with mytho-religious or cultural heritage. However, if one does not accept this view of the universe, there is noting in the scientific description of nature that can be used to refute this position. On the other hand, it is no longer possible to argue that a profound sense of unity with the whole, which has long been understood as foundation of religious experiences, but can be dismissed, undermined, or invalidated with appeals to scientific knowledge.

While we have consistently tried to distinguish between scientific knowledge and philosophical speculation based on this of what is obtainable, let us be quite clear on one point - there is no empirically valid causal linkage between the former and the latter, for those who wish to dismiss the speculative base on which is obviously free to do as done. However, there is another conclusion to be drawn, in that is firmly grounded in scientific theory and experiment there is no basis in the scientific descriptions of nature for believing in the radical Cartesian division between mind and world sanctioned by classical physics. Clearly, his radical separation between mind and world was a macro-level illusion fostered by limited awareness of the actual character of physical reality nd by mathematical idealizations extended beyond the realms of their applicability.

Withal, the philosophical implications may prove the truth or validity of by presentation as to determine the quality of some hypotheses or proposition, such, will result of a calculation for which is shown to be such in themselves, as a criterial motive in debative consideration to how our proposed new understanding of the relationship between parts and wholes in physical reality might affect the manner in which we deal with some major real-world problems. This will issue to demonstrate why a timely resolution of these problems is critically dependent on a renewed dialogue between members of the cultures of human-social scientists and scientist-engineers. We will also argue that the resolution of these problems could be dependent on a renewed dialogue between science and religion.

As many scholars have demonstrated, the classical paradigm in physics has greatly influenced and conditioned our understanding and management of human systems in economic and political realities. Virtually all models of these realities treat human systems as if they consist of atomized units or parts that interact with one another in terms of laws or forces external to or between the parts. These systems are also viewed as hermetic or closed and, thus, its discreteness, separateness and distinction.

Consider, for example, how the classical paradigm influenced or thinking about economic reality. In the eighteenth and the nineteenth century, the founders of classical economics -figures like Adam Smith, David Ricardo, and Thomas Malthus conceived of the economy as a closed system in which intersections between parts (consumer, produces, distributors, and so forth) are controlled by forces external to the parts (supply and demand). The central legitimating principle of free market economics, formulated by Adam Smith, is that lawful or law-like forces external to the individual units function as an invisible hand. This invisible hand, said Smith, frees the units to pursue their best interests, moves the economy forward, and in general having the power to create the intended enactions as invested with the responsibility and power to the fact of being legitimately sanctioned formally or authorize or declare to be justified in the legislative behaviour of parts in the best vantages of the whole. (The resemblance between the invisible hand and Newton's universal law of gravity and between the relations of parts and wholes in classical economics and classical physics should be transparent.)

After the year of 1830, economists shifted the focus to the properties of the invisible hand in the interactions between pats using mathematical models. Within these models, the behaviour of pats in the economy is assumed to be analogous to the awful interactions between pats in classical mechanics. It is, therefore, not surprising that differential calculus was employed to represent economic change in a virtual world in terms of small or marginal shifts in consumption or production. The assumption was that the mathematical description of marginal shifts n the complex web of exchanges between parts (atomized units and quantities) and whole (closed economy) could reveal the lawful, or law-like, machinations of the closed economic system.

These models later became one of the fundamentals for microeconomics. Microeconomics seek to describe interactions between parts in exact quantifiable measures-such as marginal cost, marginal revenue, marginal utility, and growth of total revenue as indexed against individual units of output. In analogy with classical mechanics, the quantities are viewed as initial conditions that can serve to explain subsequent interactions between parts in the closed system in something like deterministic terms. The combination of classical macro-analysis with microanalysis resulted in what Thorstein Veblen in 1900 termed neoclassical economics-the model for understanding economic reality that is widely used today

Beginning in the 1939s, and the challenge became to subsume the understanding of the interactions between parts in closed economic systems with more sophisticated mathematical models using devices like linear programming, game theory, and new statistical techniques. In spite of the growing mathematical sophistication, these models are based on the same assumptions from classical physics featured in previous neoclassical economic theory-with one exception. They also appeal to the assumption that systems exist in equilibrium or in perturbations from equilibria, and they seek to describe the state of the closed economic system in these terms.

One could argue that the fact that our economic models are assumptions from classical mechanics is not a problem by appealing to the two-domain distinction between micro level macro-level processes expatiated upon earlier. Since classical mechanic serves us well in our dealings with macro-level phenomena in situations where the speed of light is so large and the quantum of action is so small as to be safely ignored for practical purposes, economic theories based on assumptions from classical mechanics should serve us well in dealing with the macro-level behaviour of economic systems.

The obvious problem, . . . acceded peripherally, . . . nature is relucent to operate in accordance with these assumptions, in that the biosphere, the interaction between parts be intimately related to the hole, no collection of arts is isolated from the whole, and the ability of the whole to regulate the relative abundance of atmospheric gases suggests that the whole of the biota appear to display emergent properties that are more than the sum of its parts. What the current ecological crisis reveals that to bring to view, as shown by divine means, for the wrath of God is revealed from heaven, rather in the abstract virtual world of neoclassical economic theory. The real economies are all human activities associated with the production, distribution, and exchange of tangible goods and commodities and the consumption and for the purposive use of natural resources, such as arable land and water. Although expanding economic systems in the really economy ae obviously embedded in a web of relationships with the entire biosphere, our measure of healthy economic systems disguises this fact very nicely. Consider, for example, the healthy economic system written in 1996 by Frederick Hu, head of the competitive research team for the World Economic Forum - short of military conquest, economic growth is the only viable means for a country to sustain increases in natural living standards . . . An economy is internationally competitive if it performs strongly in three general areas: Abundant productive inputs from capital, labour, infrastructure and technology, optimal economic policies such as low taxes, little interference, free trade and sound market institutions, such as implied or indicated by the rule of law and protection of property rights.

The prescription for medium-term growth of economies ion countries like Russia, Brazil, and China may seem utterly pragmatic and quite sound. But the virtual economy described is a closed and hermetically sealed system in which the invisible hand of economic forces allegedly results in a health growth economy if impediments to its operation are removed or minimized. It is, of course, often trued that such prescriptions can have the desired results in terms of increases in living standards, and Russia, Brazil and China are seeking to implement them in various ways.

In the real economy, however, these systems are clearly not closed or hermetically sealed: Russia uses carbon-based fuels in production facilities that produce large amounts of carbon dioxide and other gases that contribute to global warming: Brazil is in the process of destroying a rain forest that is critical to species diversity and the maintenance of a relative abundance of atmospheric gases that regulate Earth temperature, and China is seeking to build a first-world economy based on highly polluting old-world industrial plants that burn soft coal. Not to forget, . . . the victual economic systems that the world now seems to regard as the best example of the benefits that can be derived form the workings of the invisible hand, that of the United States, operates in the real economy as one of the primary contributors to the ecological crisis.

In 'Consilience' Edward O. Wilson makes to comment, the case that effective and timely solutions to the problem threatening human survival is critically dependent on something like a global revolution in ethical thought and behaviour. But his view of the basis for this revolution is quite different from our own. Wilson claimed that since the foundations for moral reasoning evolved in what he termed 'gene-culture' evolution, the rules of ethical behaviour reemerging aspects of our genetic inheritance.

13

Based on the assumption that the behaviour of contemporary hunter-gatherers resembles those of our hunter-gathering forebears in the Palaeolithic Era, accounting for the Bushman hunter-gatherers living in the centre Kalahari in a design to demonstrate that ethical behaviour that is associated with instincts like bonding, cooperation, and altruism.

Wilson argued that these instincts evolved in our hunter-gatherer accessorial descendabilities, whereby genetic mutation and the ethical behaviour associated with these genetically based instincts provided a survival advantage. He then claimed that since these genes were passed on to subsequent generations of our dependable characteristics, which eventually became pervasive in the human genome, the ethical dimension of human nature has a genetic foundation. When we fully understand the 'innate epigenetic rules of moral reasoning,' it seems probable that the rules will probably turn out to be an ensemble of many algorithms whose interlocking activities guide the mind across a landscape of nuances moods and choices.

Any reasonable attempt to lay a firm foundation beneath the quagmire of human ethics in all of its myriad and often contradictory formulations is admirable, and Wilson's attempt is more admirable than most. In our view, however, there is little or no prospect that I will prove successful for a number of reasons. Wile te probability for us to discover some linkage between genes and behaviour, seems that the lightened path of human ethical behaviour and ranging advantages of this behaviour is far too complex, not o mention, inconsistently been reduced to a given set classification of 'epigenetic ruled of moral reasoning.'

Also, moral codes may derive in part from instincts that confer a survival advantage, but when we are to examine these codes, it also seems clear that they are primarily cultural products. This explains why ethical systems are constructed in a bewildering variety of ways in different cultural contexts and why they often sanction or legitimate quite different thoughts and behaviours. Let us not forget that rules of ethical behaviours are quite malleable and have been used to sacredly legitimate human activities such as slavery, colonial conquest, genocide and terrorism. As Cardinal Newman cryptically put it, 'Oh how we hate one another for the love of God.'

According to Wilson, the 'human mind evolved to believe in the gods' and people 'need a sacred narrative' to his view are merely human constructs and, therefore, there is no basis for dialogue between the world views of science and religion. 'Science for its part, will test relentlessly every assumption about the human condition and in time uncover the bedrock of the moral and religiously sentient. The eventual result of the competition between the two world views, is believed, as I, will be the secularization of the human epic and of religion itself.

Wilson obviously has a right to his opinions, and many will agree with him for their own good reasons, but what is most interesting about this thoughtful attempt to posit a more universal basis for human ethics is that it is based on classical assumptions about the character of both physical and biological realities. While Wilson does not argue that human's behaviour is genetically determined in the strict sense, however, he does allege that there is a causal linkage between genes and behaviour that largely condition this behaviour, he appears to be a firm believer in classical assumptions that reductionism can uncover the lawful essences that are principally governed by the physical aspects which are attributed to reality, including those associated with the alleged 'epigenetic rules of moral reasoning.'

As Wilson's view is apparently nothing that cannot be reduced to scientific understandings or fully disclosed in scientific terms, and this apparency of hope for the future of humanity is that the triumph of scientific thought and method will allow us to achieve the Enlightenments ideal of disclosing the lawful regularities that govern or regulate all aspects of human experience. Hence, science will uncover the 'bedrock of moral and religious sentiment, and the entire human epic will be mapped in the secular space of scientific formalism.' The intent is not to denigrate Wilson's attentive efforts to posit a more universal basis for the human condition, but is to demonstrate that any attempt to understand or improve upon the behaviour based on appeals to outmoded classical assumptions is unrealistic and outmoded. If the human mind did, in fact, evolve in something like deterministic fashion in gene-culture evolution -and if there were, in fact, innate mechanisms in mind that are both lawful and benevolent. Wilson's program for uncovering these mechanisms could have merit. But for all th reasons that have been posited, classical determinism cannot explain the human condition and its evolutionary principles that govern their functional dynamics, as Darwinian evolution should be modified to accommodate the complementary relationships between cultural and biological principles that govern evaluation does indeed have in them a strong, and firm grasp upon genetical mutations that have attributively been in distribution in contributions of human interactions within them, in the finding to self-realizations and undivided wholeness.

Equally important, is the classical assumption that the only privileged or valid knowledge is scientific is one of the primary sources of the stark division between the two cultures of humanistic and scientists-engineers, in this view, Wilson is quite correct in assuming that a timely end to the two culture war and a renewer dialogue between members of these cultures is now critically important to human survival. It is also clear, however, those dreams of reason based on the classical paradigm will only serve to perpetuate the two-culture war. Since these dreams are also remnants of an old scientific word view that no longer applies in theory in fact, to the actual character of physical reality, as reality is a probable service to frustrate the solution for which in found of a real world problem.

However, there is a renewed basis for dialogue between the two cultures, it is believed as quite different from that described by Wilson. Since classical epistemology has been displaced, or is the process of being displaced, by the new epistemology of science, the truths of science can no longer be viewed as transcendent ad absolute in the classical sense. The universe more closely resembles a giant organism than a giant machine, and it also displays emergent properties that serve to perpetuate the existence of the whole in both physics and biology that cannot be explained in terms of unrestricted determinism, simple causality, first causes, linear movements and initial conditions. Perhaps, the first and most important of the preconditioning dialogue between the two cultural conflicting realizations as Einstein explicated upon its topic as, that a human being is a 'participating part of the whole.' It is this spared awareness that allows for the freedom, or existential choice of self-decision of choosing our free-will and the power to differentiate a direct feeling of concern to free ourselves of the 'optical illusion' of our present conception of self as a 'part limited in space and time' and to widen 'our circle of compassion to embrace al living creatures and the whole of nature in its beauty.' Yet, one cannot, of course, merely reason oneself into an acceptance of this view, nonetheless, the inherent perceptions of the world are reason that the capacity for what Einstein termed 'cosmic religious feedings.' Perhaps, our enabling capability for that which is within us to have the obtainable ability to enabling of ours is to experience the self-realization, that of its realness is to sense its proven existence of a sense of elementarily leaving to some sorted conquering sense of universal consciousness, in so given to arise the existence of the universe, which really makes an essential difference to the existence or its penetrative spark as awakening from flaming fires of awareness. Those who have this capacity will hopefully be able to communicate their enhanced scientific understanding of the relations among all aspects, and in part that is our selves and the recipient whole that are the universes in ordinary language with enormous emotional appeal. The task lies before the poets of this renewing reality have nicely been described by Jonas Salk, which 'man has come to the threshold of a state of consciousness, regarding his nature and his relationship to the Cosmos, in terms that reflects 'reality.' By using the processes of Nature and metaphor, to describe the forces by which it operates upon and within Man, we come as close to describing 'reality' as we can within te limits of our comprehension. Men will be very uneven in their capacity or such understanding, which, naturally, differs for different ages and cultures, and develops and changes over the course of time. For these reasons it will always be necessary to use metaphorical and mythical provisions as comprehensive guides to living. In this way, man's affording efforts as in the imagination and the intellect can by playing the vital roles embarking upon the survival and his endurable evolution.

It is time, if not, only to be concluded from evidence in its suggestive conditional relation, for which the religious imagination and the religious experience to engage upon each the complementarities of truths of science in fitting that silence with meaning, as having to antiquate a continual emphasis, least of mention, that does not mean that those who do not believe in the existence of God or Being, should refrain in any sense from assessing the impletions of the new truths of science. Understanding these implications does not necessitate any ontology, and is in no way diminished by the lack of any ontology. And one is free to recognize a basis for a dialogue between science and religion for the same reason that one is free to deny that this basis exists -there is nothing in our current scientific world view that can prove the existence of God or Being and nothing that legitimate any anthropomorphic conceptions of the nature of God or Being. The question of belief in some ontology yet remains in what it has always been -a question, and the physical universe on the most basic level remains what it always been a riddle. And the ultimate answer to the question and the ultimate meaning of the riddle is, and probably always will be, a matter of personal choice and conviction.

The present time is clearly a time of a major paradigm shift, but consider the last great paradigm shift, the one that resulted in the Newtonian framework. This previous paradigm shift was profoundly problematic for the human spirit, it led to the conviction that we are strangers, freaks of nature, conscious beings in a universe that is almost entirely unconscious, and that, since the universe its strictly deterministic, even the free will we feel in regard to the movements of our bodies is an illusion. Yet it was probably necessary for the Western mind to go through the acceptance of such a paradigm.

The overwhelming success of Newtonian physics led most scientists and most philosophers of the Enlightenment to rely on it exclusively. As far as the quest for knowledge about reality was concerned, they regarded all of the other mode's of expressing human experience, such as accounts of numinous emergences, poetry, art, and so on, as irrelevant. This reliance on science as the only way to the truth about the universe s clearly obsoletes. Science has to give up the illusion of its self-sufficiency and self-sufficiency of human reason. It needs to unite with other modes of knowing, n particular with contemplation, and help each of us move to higher levels of being and toward the Experience of Oneness.

If this is indeed the direction of the emerging world-view, then the paradigm shifts we are presently going through will prove to e nourishing to the human spirit and in correspondences with its deepest conscious or unconscious yearning -the yearning to emerge out of Plato's shadows and into the light of luminosity.

Finding to a theory that magnifies the role of decisions, or free selection from among equally possible alternatives, in order to show that what appears to be objective or fixed by nature is in fact an artefact of human convention, similar to conventions of etiquette, or grammar, or law. Thus one might suppose that moral rules owe more to social convention than to anything imposed from outside, or have supposedly inexorable necessities are in fact the shadow of our linguistic conventions. The disadvantage of conventionalism is that it must show that alternative, equally workable conventions could have been adopted, and it is often easy to believe that, for example, if we hold that some ethical norm such as respect for promises or property is conventional, we ought to be able to show that human needs would have been equally well satisfied by a system involving a different norm, and this may be hard to establish.

A convention also suggested by Paul Grice (1913-88) directing participants in conversation to pay heed to an accepted purpose or direction of the exchange. Contributions expanded within this attention are liable to be rejected for other reasons than straightforward falsity: Something true but unhelpful or inappropriate may meet with puzzlement or rejection. We can nevertheless, infer from the fact that it would be inappropriate to say something in some circumstance that what would be aid, were we to say it, would be false. This inference was frequently and in ordinary language philosophy, it being argued, for example, that since we do not normally say 'there sees to be a barn there' when there is unmistakably a barn there, it is false that on such occasions there seems to be a barn there.

There are two main views on the nature of theories. According to the 'received view' theories are partially interpreted axiomatic systems, according to the semantic view, a theory is a collection of models (Suppe, 1974). However, a natural language comes readily interpreted, and the semantic problem is no that of the specification but of understanding the relationship between terms of various categories (names, descriptions, predicates, adverbs . . .) and their meanings. An influential proposal is that this relationship is best understood by attempting to provide a 'truth definition' for the language, which will involve giving terms and structure of different kinds have on the truth-condition of sentences containing them.

The axiomatic method . . . as, . . . a proposition laid down as one from which we may begin, as an assertion that we have taken as fundamental, at least for the branch of inquiring about scientific knowledge. The axiomatic method is that of defining as a set of such propositions, and the 'proof procedures' or finding of how a proof ever gets started. Suppose I have as premises (1) 'p' and (2) p q, can I infer that q? Only, if it is that I am sure of, (3) (p & p q) q can I then infer q? Only, it seems, if I am sure that (4) (p & p q) ®q ®q, for each new axiom (N) needing a further axiom (N + 1) telling me that the set so far implies 'q' and the regress never stops, usually the solution is to treat a system as containing not only axioms, but also rules of reference, allowing movement from the axiom. The rule 'modus ponens' allow us to pass from the first two premises to 'q'. Charles Dodgson Lutwidge (1832-98) better known as Lewis Carroll's puzzle shows that it is essential to distinguish two theoretical categories, although there may be choice about which to put in which category.

This type of theory (axiomatic) usually emerges as a body of (supposes) truths that are not nearly organized, making the theory difficult to survey or study a whole. The axiomatic method is an idea for organizing a theory (Hilbert 1970): one tries to select from among the supposed truths a small number from which all others can be seen to be deductively inferable. This makes the theory rather more tractable since, in a sense, all the truths are contained in those few. In a theory so organized, the few truths from which all others are deductively inferred are called axioms. In that, just as algebraic and differential equations, which were used to study mathematical and physical processes, could they be made mathematical objects, so axiomatic theories, like algebraic and differential equations, which are means of representing physical processes and mathematical structures, could be made objects of mathematical investigation.

In the traditional (as in Leibniz, 1704), many philosophers had the conviction that all truths, or all truths about a particular domain, followed from a few principles. These principles were taken to be either metaphysically prior or epistemologically prior or in the fist sense, they were taken to be entities of such a nature that what exists is 'caused' by them. When the principles were taken as epistemologically prior, that is, as axioms, they were taken to be epistemologically privileged either, e.g., self-evident, not needing to be demonstrated or (again, inclusive 'or') to be such that all truths do follow from them (by deductive inferences). Gödel (1984) showed that treating axiomatic theories as themselves mathematical objects, that mathematics, and even a small part of mathematics, elementary number theory, could not be axiomatized, that, more precisely, any class of axioms that in such that we could effectively decide, of any proposition, whether or not it was in the class, would be too small to capture all of the truths.

The propositional calculus or logical calculus whose expressions are character representation sentences or called propositions, and constantly representing operations on those propositions to produce others of higher plexuity. The operations include conjunction, disjunction, material implication and negation (although these need not be primitive). Propositional logic was partially anticipated by the Stoics but researched maturity only with the work of Frége, Russell, and Wittgenstein.

The concept introduction, which Frége had in order of a function taking a number of names as arguments, and delivering singular proposition's as the value, essentially that the idea that 'χ' loves 'y' is a propositional function, which yields the proposition 'John' loves 'Mary' from those two arguments (in that order). A propositional function is therefore roughly equivalent to a property or relation. In 'Principia Mathematica', Russell and Whitehead take propositional functions to be the fundamental function, since the theory of descriptions could be taken as showing that other expressions denoting functions are incomplete symbols.

Keeping in mind, in that which are two classical truth-values that a statement, proposition, or sentence can take, such that it is supposed in classical (two-valued) logic, that each statement has one of these values, and none has both. A statement is then false if and only if it is not true. The basis of this scheme is that to each statement there corresponds a determinate truth condition, or way the world must be for it to be true, and otherwise false. Statements may be felicitous or infelicitous in other dimensions (polite, misleading, apposite, witty, and so on) but truth is the central normative governing assertion. Considerations of vagueness may introduce greys into black-and-white combinations, in that of a systematic plan of action for these issues concerning 'falsity' are the only way of failing to be true.

Formally, it is nonetheless, that any suppressed premise or background framework of thought necessary to make an argument valid, or a position tenable. More formally, a presupposition has been defined as a proposition whose truth is necessary for either the truth or the falsity of another statement. Thus, if 'p' presupposes 'q', 'q' must be true for p to be either true or false. In the theory of knowledge of Robin George Collingwood (1889-1943), any propositions capable of truth or falsity stand on a bed of 'absolute presuppositions' which are not properly capable of truth or falsity, since a system of thought will contain no way of approaching such a question. It was suggested by Peter Strawson (1919-), in opposition to Russell's theory of 'definite' descriptions, that 'there exists a King of France' is a presupposition of 'the King of France is bald', the latter being neither true, nor false, if there is no King of France. It is, however, a little unclear of whether or not that the idea is that no statement at all is made in such a case, or whether a statement of which that I can make, but fails of being one as true and opposes of either true is false. The former option preserves classical logic, since we can still say that every statement is either true or false, but the latter does not, since in classical logic the law of 'bivalence' holds, and ensures that nothing at all is presupposed for any proposition to be true or false. The introduction of presupposition therefore meant that either a third truth-value is found, 'intermediate' between truth and falsity, or classical logic is preserved, but it is impossible to tell whether a particular sentence expresses a proposition that is a candidate for truth ad falsity, without knowing more than the formation rules of the language. Each suggestion carries costs, and there is some consensus that at least where definite descriptions are involved, examples like the one given are equally well handed by regarding the overall sentence false when the existence claim fails.

A proposition is knowable a priori if it can be known without experience of the specific course of events in the actual world. It may, however, be allowed that some experience is required to acquire the concepts involved in an a priori proposition. Some thing is knowable only a posteriori if it can be known a priori. The distinction given for one of the fundamental problem areas of epistemology was that; The category of a priori propositions is highly controversial, since it is not clear how pure thought, unaided by experience, can give rise to any knowledge at all, and it has always been a concern of empiricism to deny that it can. The two great areas in which it seems to be so are logic and mathematics, so empiricists have commonly tried to show either that these are not areas of real, substantive knowledge, or that in spite of appearances their knowledge that we have in these areas is actually dependent on experience. The former line tries to show sense trivial or analytic, or matters of notation conventions of language. The latter approach is particularly y associated with Quine, who denies any significant slit between propositions traditionally thought of as a priori, and other deeply entrenched beliefs that occur in our overall view of the world.

Another contested categorical a priori concepts, supposed to be concepts that cannot be 'derived' from experience, but which are presupposed in any mode of thought about the world, time, substance, causation, number, and self are candidates. The need for such conclusive concepts and the nature of the substantiated a prior knowledge, such that they gave to rise, but central among the concerns of Kant's 'Critique of Pure Reason'.

Likewise, since their denial does not involve a contradiction, there is merely contingent: There could have been in other ways a hold of the actual world, but not every possible one. Some examples are 'Caesar passed across the Rubicon' and 'Leibniz was born in Leipzig', as well as propositions expressing correct scientific generalizations. In Leibniz's view truths of fact rest on the principle of sufficient reason, which is a reason why it is so. This reason is that the actual world (by which he meant the total collection of things past, present and future) is better than any other possible world and therefore created by God. The foundation of his thought is the conviction that to each individual there corresponds a complete notion, knowable only to God, from which is deducible all the properties possessed by the individual at each moment in its history. It is contingent that God actualizes te individual that meets such a concept, but his doing so is explicable by the principle of 'sufficient reason', whereby God had to actualize just that possibility in order for this to be the best of all possible worlds. This thesis is subsequently lampooned by Voltaire (1694-1778), in whom of which was prepared to take refuge in ignorance, as the nature of the soul, or the way to reconcile evil with divine providence.

In defending the principle of sufficient reason sometimes described as the principle that nothing can be so without there being a reason why it is so. But the reason has to be of a particularly potent kind: eventually it has to ground contingent facts in necessities, and in particular in the reason an omnipotent and perfect being would have for actualizing one possibility than another. Among the consequences of the principle is Leibniz's relational doctrine of space, since if space were an infinite box there could be no reason for the world to be at one point in rather than another, and God placing it at any point violate the principle. In Abelard's (1079-1142), as in Leibniz, the principle eventually forces te recognition that the actual world is the best of all possibilities, since anything else would be inconsistent with the creative power that actualizes possibilities.

If truth consists in concept containment, then it seems that all truths are analytic and hence necessary, yet, if they are all necessary, surely they are all truths of reason. In that not every truth can be reduced to an identity in a finite number of steps; in some instances revealing the connection between subject and predicate concepts would require an infinite analysis, while this may entail that we cannot prove such propositions as a prior, it does not appear to show that proposition could have been false. Intuitively, it seems a better ground for supposing that it is a necessary truth of a special sort. A related question arises from the idea that truths of fact depend on God's decision to create the best world: If it is part of the concept of this world that it is best, how could its existence be other than necessary? An accountable and responsively answered explanation would be so, that any relational question that brakes the norm lay eyes on its existence in the manner other than hypothetical necessities, i.e., it follows from God's decision to create the world, but God had the power to create this world, but God is necessary, so how could he have decided to do anything else? Leibniz says much more about these matters, but it is not clear whether he offers any satisfactory solutions.

Eliminativists in the philosophy of mind counsel abandoning the whole network of terms mind, consciousness, self, or Qualia, that usher in the problems of mind and body. Sometimes the argument for doing this is that we should wait for a supposed future understanding of ourselves, based on cognitive science and better than any current mental descriptions provide, sometimes it is supposed that physicalism shows that no mental description of us could possibly be true.

What makes a belief justified and what makes a true belief knowledge? Thinking that whether a belief deserves one of these appraisals is natural depends on what caused the depicted branch of knowledge to have the belief. In recent decades a number of epistemologists have pursued this plausible idea with a variety of specific proposals. Some causal theories of knowledge have it that a true belief that 'p' is knowledge just in case it has the right causal connection to the fact that 'p'. Such a criterion can be applied only to cases where the fact that 'p' is a sort that can enter into causal relations, as this seems to exclude mathematically and the necessary facts and perhaps any fact expressed by a universal generalization, and proponents of this sort of criterion have usually supposed that it is limited to perceptual representations where knowledge of particular facts about subjects' environments.

For example, Armstrong (1973), predetermined that a position held by a belief in the form 'This perceived object is 'F' is [non-inferential] knowledge if and only if the belief is a completely reliable sign that the perceived object is 'F', that is, the fact that the object is 'F' contributed to causing the belief and its doing so depended on properties of the believer such that the laws of nature dictated that, for any subject ',' and perceived object 'y' if 'F' has those properties and believed that 'y' is 'F', then 'y' is 'F'. (Dretske (1981) offers a rather similar account, in terms of the belief's being caused by a signal received by the perceiver that carries the information that the object is 'F').

Goldman (1986) has proposed an importantly different causal criterion, namely, that a true belief is knowledge if it is produced by a type of process that is 'globally' and 'locally' reliable. Causing true beliefs is sufficiently high is globally reliable if its propensity. Local reliability has to do with whether the process would have produced a similar but false belief in certain counterfactual situations alternative to the actual situation. This way of marking off true beliefs that are knowledge does not require the fact believed to be causally related to the belief, and so it could in principle apply to knowledge of any kind of truth.

Goldman requires the global reliability of the belief-producing process for the justification of a belief, required for knowledge because justification is required for knowledge. A concept of which the conclusion or fact for of being just, as a fact or circumstance, that proving to establish the truth or validity of presentations as the evidentially determined and making it verifiably an Essence without doubt, the concept in which justification is aptly to consist of a broadening scope that spans epistemology and ethics and has in itself the containment for special cases - highly-varied applications. This requires to have requisite needs as something that is indispensable for it being required, yet, it would not have produced it in any relevant counterfactual situation in which it is false. Its purported theory of relevant alternatives can be viewed as an attempt to provide a more satisfactory response to this intensity in of our thinking about knowledge. It attempts to characterize knowledge in a way that preserves both our belief that knowledge is an absolute concept and our belief that we have knowledge.

According to the theory, we need to qualify rather than deny the absolute character of knowledge. We should view knowledge as absolute, reactive to certain standards (Dretske, 1981 and Cohen, 1988). That is to say, in order to know a proposition, our evidence need not eliminate all the alternatives to that preposition, rather for 'us', that we can know our evidence to carry off all the relevant alternatives, where the set of relevant alternatives (a proper subset of the set of all alternatives) is determined by some standard. Moreover, the relevant alternatives view, and the standards determining that of the alternatives is raised by the sceptic are not relevant. If this is correct, then the fact that our evidence cannot eliminate the sceptic's alternative does not lead to a sceptical result. For knowledge requires only the elimination of the relevant alternatives, so the relevant alternative view preserves in both strands in our thinking about knowledge. Knowledge is an absolute concept, but because the absoluteness is relative to a standard, we can know many things.

The interesting thesis that counts as a causal theory of justification (in the meaning of 'causal theory' intended here) is that: A belief is justified in case it was produced by a type of process that is 'globally' reliable, that is, its propensity to produce true beliefs-that can be defined (to a good approximation) As the proportion of the beliefs it produces (or would produce) that is true is sufficiently great.

This proposal will be adequately specified only when we are told (I) how much of the causal history of a belief counts as part of the process that produced it, (ii) which of the many types to which the process belongs is the type for purposes of assessing its reliability, and (iii) relative to why the world or worlds are the reliability of the process type to be assessed in the actual world, the closet worlds containing the case being considered, or something else? Let 'us' look at the answers suggested by Goldman, and the leading proponents of an otherwise broadened spectrum as justified in the meaning or concept as owing to Reliabilism. The reliable process theory is grounded on two main points. First, the justificational status of a belief depends on the psychological process that cause (or causally sustain) it. First, the approach has been more influential, and the reliable process theory is grounded on two main points, i.e., belief-forming or belief preserving processes such as perception, memory, guessing or introspection. Clearly most all psychological processes are justification-conferring. A belief justificational status is a function of the truth-ratio of the type of process, or series of propositions, that are causally responsible for it. Fundamentalism says, that there are 'basic' beliefs, which acquire justification without dependence on inference, which is to say, that the act or process of deriving logical conclusions from the premises known or assumed to be true. Virtually all theories of knowledge, of course, share an externalist component in requiring truth as a condition for knowing. Reliabilism goes further, however, in trying to capture additional conditions for knowledge by means of a nomic, counterfactual or other 'external' relation between belief and truth.

Clearly, there are many forms of reliabilism, just as there are many forms of foundationlism and Coherentism. How is reliabilism related to these other two theories of justification? It is usually regarded as a rival, and this is apt in so far as foundationalism and coherentism traditionally focussed on purely evidential relations rather than psychological processes. But reliabilism might also be offered as a deeper-level theory, subsuming some of the precepts of either foundationalism or coherentism. Foundationalism says, that there are basic beliefs, which acquire justification without dependence upon inference. Reliabilism might rationalize this by indicating that the basic beliefs are formed by reliable non-inferential processes. Coherentism stresses the primary of systematicity in all doxastic decision-making. Reliabilism might rationalize this by pointing to increases in reliability that accrue from systematicity. Thus, reliabilism could complement foundationalism and coherentism rather than compete with them.

Contemporary analytic philosophers spoke for what constitutes the representational properties of ideas, moreover, certain analytic philosophers spoke, to a great extent. Of the propositional attitudes - thoughts, beliefs, intentions - than of ideas and images, such that these concerning considerations are spoken of the contents of such altitudes.

So understood, rationalism is an exercise in extravagant optimism, as might be argued either by considering the mutually inconsistent (and often bizarre) metaphysical systems the rationalists advocated, or by noting the crucial role arguments from experience play the development of the science. But strict empiricism leads inevitably to radical scepticism and cannot account adequately fo r the a priori knowledge we do posses. What is required is a theory of knowledge which synthesizes these two opposed tendencies, rescuing what is true in each, but avoiding their exaggerations. We can then see that the Kantian system as reconciling the competing claims of reason and experience, giving each its due.

This way of telling the story of philosophy in the seventeenth and eighteenth centuries is aesthetically pleasing, and gratifies our desire to think of philosophy as a progressive discipline, whose later practitioners have a better grasp of its problems than their predecessors had.

In Spinoza, or his account of our knowledge of the common notions which are the basis of reason, even so, in mathematics, where we deal with truths of reason, such that of Leibniz is reluctant to rely on the supposed self-evidence of the axioms, arguing that mathematicians demonstrate the secondary axioms we ordinarily use by reducing them to identities. Nonetheless, Spinoza, 1632-77, used as a criterion for such a process of vindication, since, the truth or true ideas can be discerned directly for Spinoza - the true thought is distinguished from a false one not only by an extrinsic, but chiefly by an intrinsic denomination. In fact, his acknowledged doctrine that modes of thought are identical with their objects, arguably allows him to hold that truth consists of both in what he calls the internal adequacy of an idea, and in what he calls agreement with its object, on the grounds that these two characteristic ideas are really the same characteristics considered in two different ways.

Nevertheless, Goldman argued about the relevance of cognitive and social science to problems in epistemology, metaphysics, the philosophy of mind and ethicist, as the generative structure of basic and non-basic action, such that the compatibility of free will and determinism is to follow:

> (1) Goldman (1979, 1986) takes the relevant belief producing process to include only
> the proximate causes internal to the believer. So, for instance, when recently I believed
> that the telephone was ringing the process that produced the belief, for purposes of
> assessing reliability, includes just the causal chain of neural events from the stimulus
> in my ear's inward ands other concurrent brain states on which the production of
> the belief depended: It does not include any events' as the telephone, or the sound
> waves travelling between it and my ears, or any earlier decisions I made that were

responsible for my being within hearing distance of the telephone at that time. It does seem intuitively plausible of a belief depends should be restricted to internal omnes proximate to the belief. Why? Goldman does not tell 'us'. One answer that some philosophers might give is that it is because a belief's being justified at a given time can depend only on facts directly accessible to the believer's awareness at that time (for, if a believer ought to holds only beliefs that are justified, she can tell at any given time what beliefs would then be justified for her). However, this cannot be Goldman's answer because he wishes to include in the relevantly process neural events that are not directly accessible to consciousness.

(2) Once the Reliabilist has concretely exposed to 'us' how to delimit the process producing a belief, he needs to tell 'us' which of the many types to which it belongs is the relevant type. Coincide, for example, the process that produces your current belief that you see a book before you. One very broad type to which that process belongs would be specified by 'coming to a belief as to something one perceives as a result of activation of the nerve endings in some of one's sense-organs. The constricting characteristic processes' or suitable in an appropriate situation or environmentally to be a part of something belonging to or undergo the process would be specified by 'coming to a belief as to what one sees as a result of activation of the nerve endings in one's retina'. A still narrower type would be given by inserting in the last specification a description of a particular pattern of activation of the retina's particular cells. Which of these or other types to which the token process belongs is the relevant type for determining whether the type of process that produced your belief is reliable?

(3) Should the justification of a belief in a deductive non-actual endeavour, in that of its engaging allowances for its turning reliability of the belief-producing process in the possible world of the example? That leads to the implausible result in that in a world run by a Cartesian demon-a powerful being who causes the other inhabitants of the world to have rich and coherent sets of perceptual and memory impressions that are all illusory the perceptual and memory beliefs of the other inhabitants are all unjustified, for they are produced by processes that are, in that world, quite unreliable. If we say instead that it is the reliability of the processes in the actual world that matters, we get the equally undesired result that if the actual world is a demon world then our perceptual and memory beliefs are all unjustified.

Goldman's solution (1986) is that the reliability of the process types is to be gauged by their performance in 'normal' worlds, that is, worlds consistent with 'our general beliefs about the world . . . 'about the sorts of objects, events and changes that occur in it'. This gives the intuitively right results for the problem cases just considered, but indicate by inference an implausible proportion of making compensations for alternative tending toward justification. If there are people whose general beliefs about the world are very different from mine, then there may, on this account, be beliefs that I can correctly regard as justified (ones produced by processes that are reliable in what I take to be a normal world) but that they can correctly regard as not justified.

However, these questions about the specifics are dealt with, and there are reasons for questioning the basic idea that the criterion for a belief's being justified is its being produced by a reliable process. Thus and so, doubt about the sufficiency of the Reliabilist standards, rules or test on which a judgement or particular decision can be based as an object that under specific conditions defines or represents of a unit or proposition that widey recognized, especially because of its excellence as a model of authority; of which is to say, in the conforming to establish an educational use in speech or in writing, in of what is prompted by a sort of example that Goldman himself uses for another purpose.

Suppose that being in brain-state 'B' always causes one to believe that one is in brained-state 'B'. Here the reliability of the belief-producing process is perfect, but 'we can readily imagine circumstances in which a person goes into brain-state 'B' and therefore has the belief in question, though this belief is by no means justified' (Goldman, 1979). Doubt about the necessity of the condition arises from the possibility that one might know that one has strong justification for a certain belief and yet that knowledge is not what actually prompts one to believe. For example, I might be well aware that, having read the weather bureau's forecast that it will be much hotter tomorrow. I have ample reason to be confident that it will be hotter tomorrow, but I irrationally refuse to believe it until Wally tells me that he feels in his joints that it will be hotter tomorrow. Here what prompts me to believe dors not justify my belief, but my belief is nevertheless justified by my knowledge of the weather bureau's prediction and of its evidential force: I can advert to any disavowable process of deriving logical conclusion s from premises known or assumed to be true, given that my justification and there be nothing untoward about the weather bureau's prediction, my belief, if true, can be counted as for knowledge. This sorts of example raises doubt whether any causal conditions, are it a reliable process or something else, is necessary for either justification or knowledge.

Philosophers and scientists alike, have often held that the simplicity or parsimony of a theory is one reason, all else being equal, to view it as true. This goes beyond the unproblematic idea that simpler theories are easier to work with and gave greater aesthetic appeal.

One theory is more parsimonious than another when it postulates fewer entities, processes, changes or explanatory principles: The simplicity of a theory depends on essentially the same consecrations, though parsimony and simplicity obviously become the same. Demanding clarification of what makes one theory simpler or more parsimonious is plausible than another before the justification of these methodological maxims can be addressed.

If we set this description problem to one side, the major normative problem is as follows: What reason is there to think that simplicity is a sign of truth? Why should we accept a simpler theory instead of its more complex rivals? Newton and Leibniz thought that the answer was to be found in a substantive fact about nature. In 'Principia,' Newton laid down as his first Rule of Reasoning in Philosophy that 'nature does nothing in vain . . . 'for Nature is pleased with simplicity and affects not the pomp of superfluous causes'. Leibniz hypothesized that the actual world obeys simple laws because God's taste for simplicity influenced his decision about which world to actualize.

The tragedy of the Western mind, described by Koyré, is a direct consequence of the stark Cartesian division between mind and world. We discovered the 'certain principles of physical reality', said

Descartes, 'not by the prejudices of the senses, but by the light of reason, and which possesses an exceeding course of resulting evidence, that we as redundantly have been negatively doubtful for showing of any truth. Since being or occurring in fact or actuality, having verifiable existence, for which real objects are founded on particular matters and concerns for our experiencing the real world. The real, or that which actually exists, . . . external to ourselves, in this view only, in that which could be represented in the quantitative terminologies of mathematics, Descartes concludes that all quantitative aspects of reality could be traced to the deceitfulness of the senses.

The most fundamental aspect of the Western intellectual tradition is the assumption that there is a fundamental division between the material and the immaterial world or between the realm of matter and the realm of pure mind or spirit. The metaphysical framework based on this assumption is known as ontological dualism. As the word dual implies, the framework is predicated ontologically, or a conception consisting of the nature of God or Being. Within such to assume, the satisfaction of instinctual needs through the awareness of and adjustment to environmental demands over which a physical process under alternative dominions or territorial comprehension, is that the realm of science is to bring into reality and the migration is made real. Having a tendency inclination toward that which is real is often with the theories beyond the realm of separate dimensions. The concept of Being as continuous, immutable, and having a prior or separate existence from the world of change, dates from the earliest of the wandering Stoics, in the ancient Greek philosopher Parmenides. The same qualities were associated with the God of the Judeo-Christian tradition, and they were considerably amplified by the role played in Theology by Platonic and Neoplatonic philosophy.

Nicolas Copernicus, Galileo, Johannes Kepler, and Isaac Newton were all inheritors of a cultural tradition in which ontological dualism was a primary article of faith. Hence, the idealization of the mathematical ideal, is the primary source within reach of the Holy Communion with God, which dates from Pythagoras. This provided a metaphysical foundation for the emerging natural sciences and explained why the creators of classical physics believed that doing physics was a form of communion with the geometrical and mathematical form's residing in the perfect mind of God. This view would survive in a modified form in what is now known as Einsteinian epistemology and accounts in no small part for the reluctance of many physicists to accept the epistemology associated with the Copenhagen Interpretation.

At the beginning of the nineteenth century, Pierre-Simon LaPlace, along with a number of other French mathematicians, advanced the view that the science of mechanics constituted a complete view of nature. Since this science, by observing its epistemology, had revealed itself to be the fundamental science, the hypothesis of God was, they concluded, entirely unnecessary.

LaPlace is recognized for eliminating not only the theological component of classical physics but the 'entire metaphysical component' as well'. The epistemology of science requires that we proceed by inductive generalizations, attested by the observed facts to hypotheses that are 'tested by observed conformity of the phenomena'. What was unique about LaPlace's view of hypotheses was his insistence that we cannot attribute reality to them. Although concepts like force, mass, motion, cause, and laws are obviously present in classical physics, they exist in LaPlace's view only as quantities. Physics is concerned, he argued, with quantities that we associate as a matter of convenience with concepts, and the truths about nature are only the quantities.

As this view of hypotheses and the truths of nature as quantities were extended in the nineteenth century to a mathematical description of phenomena like heat, light, electricity, and magnetism. LaPlace's assumptions about the actual character of scientific truths seemed correct. This progress suggested that if we could remove all thoughts about the 'nature of' or the 'source of' phenomena, the pursuit of strictly quantitative concepts would bring us to a complete description of all aspects of physical reality. Subsequently, figures like Comte, Kirchhoff, Hertz, and Poincaré developed a program for the study of nature that was quite different from that of the original creators of classical physics.

The seventeenth century view of physics as a philosophy of nature or as natural philosophy was displaced by the view of physics as an autonomous science that was 'the science of nature'. This view, which was premised on the doctrine of positivism, promised to subsume all of the nature with a mathematical analysis of entities in motion and claimed that the true understanding of nature was revealed only in the mathematical description. Since the doctrine of positivism assumes that the knowledge we call physics resides only in the mathematical formalism of physical theory, it disallows the prospect that the vision of physical reality revealed in physical theory can have any other meaning. In the history of science, the irony is that positivism, which was intended to banish metaphysical concerns from the domain of science, served to perpetuate a seventeenth-century metaphysical assumption about the relationship between physical reality and physical theory.

Epistemology since Hume and Kant has drawn back from this theological underpinning. Indeed, the very idea that nature is simple (or uniform) has come in for a critique. The view has taken hold that a preference for simple and parsimonious hypotheses is purely methodological: It is constitutive of the attitude we call 'scientific' and makes no substantive assumption about the way the world is.

A variety of otherwise diverse twentieth-century philosophers of science have attempted, in different ways, to flesh out this position. Two examples must suffice here: Hesse (1969) as, for summaries of other proposals. Popper (1959) holds that scientists should prefer highly falsifiable (improbable) theories: He tries to show that simpler theories are more falsifiable, also Quine (1966), in contrast, sees a virtue in theories that are highly probable, he argues for a general connection between simplicity and high probability.

Both these proposals are global. They attempt to explain why simplicity should be part of the scientific method in a way that spanned all scientific subject matters. No assumption about the details of any particular scientific problem serves as a premiss in Popper's or Quine's arguments.

Newton and Leibniz thought that the justification of parsimony and simplicity flows from the hand of God: Popper and Quine try to justify these methodologically median of importance is without assuming anything substantive about the way the world is. In spite of these differences in approach, they have something in common. They assume that all users of parsimony and simplicity in the separate sciences can be encompassed in a single justifying argument. That recent developments in confirmation theory suggest that this assumption should be scrutinized, as for Good (1983) and Rosenkrantz (1977) have emphasized the role of auxiliary assumptions in mediating the connection between hypotheses and observations. Whether a hypothesis is well supported by some observations, or whether one hypothesis is better supported than another by those observations, crucially depends on empirical background assumptions about the inference problem here. The same view applies to the idea of prior probability (or, prior plausibility). In of a single hypo-physical science if chosen as an alternative to another even though they are equally supported by current observations, this must be due to an empirical background assumption.

Principles of parsimony and simplicity mediate the epistemic connection between hypotheses and observations. Perhaps these principles are able to do this because they are surrogates for an empirical background theory. It is not that there is one background theory presupposed by every appeal to parsimony; This has the quantifier order backwards. Rather, the suggestion is that each parsimony argument is justified only to each degree that it reflects an empirical background theory about the subjective matter. On this theory is brought out into the open, but the principle of parsimony is entirely dispensable (Sober, 1988).

This 'local' approach to the principles of parsimony and simplicity resurrects the idea that they make sense only if the world is one way rather than another. It rejects the idea that these maxims are purely methodological. How defensible this point of view is, will depend on detailed case studies of scientific hypothesis evaluation and on further developments in the theory of scientific inference.

It is usually not found of one and the same that, an inference is a (perhaps very complex) act of thought by virtue of which act (1) I pass from a set of one or more propositions or statements to a proposition or statement and (2) it appears that the latter are true if the former is or are. This psychological characterization has in occurring over a wider ranging summation of literatures under more lesser of the inessential variations. Desiring a better characterization of inference is natural. Up to the present time, attempts to construct a complete psychological explanation failed in their attempt to be the accepted minimum, also to comprehend on the grounds from which the inference will be objectively valid - an exemplified point as made exploratively exonerated by Gottlob Frége. Attempts to understand the nature of inference through the device of the representation of inference by formal-logical calculations or derivations better (1) leaving 'us' puzzled about the relation of formal-logical derivations to the informal inferences they are supposedly to represent or reconstruct, and (2) leaves 'us' worried about the sense of such formal derivations. Are these derivations inference? Are not informal inferences needed in order to apply the rules governing the constructions of formal derivations (inferring that this operation is an application of that formal rule)? These are concerns cultivated by, for example, Wittgenstein.

Coming up with an adequate characterization of inference-and even working out what would count as a very adequate characterization here is demandingly by no nearly some resolved philosophical problem.

Traditionally, a proposition that is not a 'conditional', as with the 'affirmative' and 'negative', modern opinion is wary of the distinction, since what appears categorical may vary with the choice of a primitive vocabulary and notation. Apparently categorical propositions may also turn out to be disguised conditionals: 'X' is intelligent (categorical?) Equivalent, if 'X' is given a range of tasks, she does them better than many people (conditional?). The problem is not merely one of classification, since deep metaphysical questions arise when facts that seem to be categorical and therefore solid, come to seem by contrast conditional, or purely hypothetical or potential.

Its condition of some classified necessity is so proven sufficient that if 'p' is a necessary condition of 'q', then 'q' cannot be true unless 'p'; is true? If 'p' is a sufficient condition, thus steering well is a necessary condition of driving in a satisfactory manner, but it is not sufficient, for one can steer well but drive badly for other reasons. Confusion may result if the distinction is not heeded. For example, the statement that 'A' causes 'B' may be interpreted to mean that 'A' is itself a sufficient condition for 'B', or that it is only a necessary condition fort 'B', or perhaps a necessary parts of a total sufficient condition. Lists of conditions to be met for satisfying some administrative or legal requirement frequently attempt to give individually necessary and jointly sufficient sets of conditions.

What is more, that if any proposition of the form 'if p then q'. The condition hypothesized, 'p'. Is called the antecedent of the conditionals, and 'q', the consequent? Various kinds of conditional have been distinguished. Its weakest is that of 'material implication', merely telling that either 'not-p', or 'q'. Stronger conditionals include elements of 'modality', corresponding to the thought that if 'p' is truer then 'q' must be true. Ordinary language is very flexible in its use of the conditional form, and there is controversy whether conditionals are better treated semantically, yielding differently finds of conditionals with different meanings, or pragmatically, in which case there should be one basic meaning with surface differences arising from other implicatures.

It follows from the definition of 'strict implication' that a necessary proposition is strictly implied by any proposition, and that an impossible proposition strictly implies any proposition. If strict implication corresponds to 'q' follows from 'p', then this that a necessary proposition follows from anything at all, and anything at all follows from an impossible proposition. This is a problem if we wish to distinguish between valid and invalid arguments with necessary conclusions or impossible premises.

The Humean problem of induction is that if we would suppose that there is some property 'A' concerning and observational or an experimental situation, and that out of a large number of observed instances of 'A', some fraction m/n (possibly equal to 1) has also been instances of some logically independent property 'B'. Suppose further that the background proportionate circumstances not specified in these descriptions has been varied to a substantial degree without there being any collateral information available concerning the frequency of 'B's' among 'A's, or concerning causal or nomologically connections between instances of 'A' and instances of 'B'.

In this situation, an 'enumerative' or 'instantial' induction inference would move rights from the premise, that m/n of observed 'A's' are 'B's' to the conclusion that approximately m/n of all 'A's' are 'B's. (The usual probability qualification will be assumed to apply to the inference, rather than being part of the conclusion.) Here the class of 'A's' should be taken to include not only unobserved 'A's' and future 'A's', but also possible or hypothetical 'A's' (an alternative conclusion would concern the probability or likelihood of the adjacently observed 'A' being a 'B').

The traditional or Humean problem of induction, often referred to simply as 'the problem of induction', is the problem of whether and why inferences that fit this schema should be considered rationally acceptable or justified from an epistemic or cognitive standpoint, i.e., whether and why reasoning in this way is likely to lead to true claims about the world. Is there any sort of argument or rationale that can be offered for thinking that conclusions reached in this way are likely to be true in the corresponding premisses is true or even that their chances of truth are significantly enhanced?

Hume's discussion of this issue deals explicitly only with cases where all observed 'A's' are 'B's' and his argument applies just as well to the more general case. His conclusion is entirely negative and sceptical: Inductive inferences are not rationally justified, but are instead the result of an essentially a-rational process, custom or habit. Hume (1711-76) challenges the proponent of induction to supply a cogent line of reasoning that leads from an inductive premise to the corresponding conclusion and offers an extremely influential argument in the form of a dilemma (a few times referred to as 'Hume's fork'), that either our actions are determined, in which case we are not responsible for them, or they are the result of random events, under which case we are also not responsible for them.

Such reasoning would, he argues, have to be either deductively demonstrative reasoning in the concerning relations of ideas or 'experimental', i.e., empirical, that reasoning concerning matters of fact or existence. It cannot be the former, because all demonstrative reasoning relies on the avoidance of contradiction, and it is not a contradiction to suppose that 'the course of nature may change', that an order that was observed in the past and not of its continuing against the future: But it cannot be, as the latter, since any empirical argument would appeal to the success of such reasoning about an experience, and the justifiability of generalizing from experience are precisely what is at issue-so that any such appeal would be question-begging. Hence, Hume concludes that there can be no such reasoning (1748).

An alternative version of the problem may be obtained by formulating it with reference to the so-called Principle of Induction, which says roughly that the future will resemble the past or, somewhat better, that unobserved cases will resemble observed cases. An inductive argument may be viewed as enthymematic, with this principle serving as a supposed premiss, in which case the issue is obviously how such a premiss can be justified. Hume's argument is then that no such justification is possible: The principle cannot be justified a prior because having possession of been true in experiences without obviously begging the question is not contradictory to have possession of been true in experiences without obviously begging the question.

The predominant recent responses to the problem of induction, at least in the analytic tradition, in effect accept the main conclusion of Hume's argument, namely, that inductive inferences cannot be justified in the sense of showing that the conclusion of such an inference is likely to be true if the premise is true, and thus attempt to find another sort of justification for induction. Such responses fall into two main categories: (I) Pragmatic justifications or 'vindications' of induction, mainly developed by Hand Reichenbach (1891-1953), and (ii) ordinary language justifications of induction, whose most important proponent is Frederick, Peter Strawson (1919-). In contrast, some philosophers still attempt to reject Hume's dilemma by arguing either (iii) That, contrary to appearances, induction can be inductively justified without vicious circularity, or (iv) that an anticipatory justification of induction is possible after all:

> (1) Reichenbach's view is that induction is best regarded, not as a form of inference, but rather as a 'method' for arriving at posits regarding, i.e., the proportion of 'A's' remain additionally of 'B's'. Such a posit is not a claim asserted to be true, but is instead an intellectual wager analogous to a bet made by a gambler. Understood in this way, the inductive method says that one should posit that the observed proportion is, within some measure of an approximation, the true proportion and then continually correct that initial posit as new information comes in.

The gambler's bet is normally an 'appraised posit', i.e., he knows the chances or odds that the outcome on which he bets will actually occur. In contrast, the inductive bet is a 'blind posit': We do not know the chances that it will succeed or even that success is that it will succeed or even that success is possible. What we are gambling on when we make such a bet is the value of a certain proportion in the independent world, which Reichenbach construes as the limit of the observed proportion as the number of cases increases to infinity. Nevertheless, we have no way of knowing that there are even such a limit, and no way of knowing that the proportion of 'A's' are 'B's' converging on some stable value than varying at random. If we cannot know that this limit exists, then we obviously cannot know that we have any definite chance of finding it.

What we can know, according to Reichenbach, is that 'if' there is a truth of this sort to be found, the inductive method will eventually find it'. That this is so is an analytic consequence of Reichenbach's account of what it is for such a limit to exist. The only way that the inductive method of making an initial posit and then refining it in light of new observations can fail eventually to arrive at the true proportion is if the series of observed proportions never converges on any stable value, which that there is no truth to be found pertaining the proportion of 'A's' additionally constitute 'B's'. Thus, induction is justified, not by showing that it will succeed or indeed, that it has any definite likelihood of success, but only by showing that it will succeed if success is possible. Reichenbach's claim is that no more than this can be established for any method, and hence that induction gives 'us' our best chance for success, our best gamble in a situation where there is no alternative to gambling.

This pragmatic response to the problem of induction faces several serious problems. First, there are indefinitely many other 'methods' for arriving at posits for which the same sort of defence can be given-methods that yields the same results as the inductive method over time but differ arbitrarily before long. Despite the efforts of others, it is unclear that there is any satisfactory way to exclude such alternatives, in order to avoid the result that any arbitrarily chosen short-term posit is just as reasonable as the inductive posit. Second, even if there is a truth of the requisite sort to be found, the inductive method is only guaranteed to find it or even to come within any specifiable distance of it in the indefinite long run. All the same, any actual application of inductive results always takes place in the presence to the future eventful states in making the relevance of the pragmatic justification to actual practice uncertainly. Third, and most important, it needs to be emphasized that Reichenbach's response to the problem simply accepts the claim of the Humean sceptic that an inductive premise never provides the slightest reason for thinking that the corresponding inductive conclusion is true. Reichenbach himself is quite candid on this point, but this does not alleviate the intuitive implausibility of saying that we have no more reason for thinking that our scientific and commonsense conclusions that result in the induction of it '. . . is true' than, to use Reichenbach's own analogy (1949), a blind man wandering in the mountains who feels an apparent trail with his stick has for thinking that following it will lead him to safety.

An approach to induction resembling Reichenbach's claiming in that those particular inductive conclusions are posits or conjectures, than the conclusions of cogent inferences, is offered by Popper. However, Popper's view is even more overtly sceptical: It amounts to saying that all that can ever be said in favour of the truth of an inductive claim is that the claim has been tested and not yet been shown to be false.

> (2) The ordinary language response to the problem of induction has been advocated by many philosophers, nonetheless, Strawson claims that the question whether induction is justified or reasonable makes sense only if it tacitly involves the demand that inductive reasoning meet the standards appropriate to deductive reasoning, i.e., that the inductive conclusions are shown to follow deductively from the inductive assumption. Such a demand cannot, of course, be met, but only because it is illegitimate: Inductive and deductive reasons are simply fundamentally different kinds of reasoning, each possessing its own autonomous standards, and there is no reason to demand or expect that one of these kinds meet the standards of the other. Whereas, if induction is assessed by inductive standards, the only ones that are appropriate, then it is obviously justified.

The problem here is to understand to what this allegedly obvious justification of an induction amount. In his main discussion of the point (1952), Strawson claims that it is an analytic true statement that believing it a conclusion for which there is strong evidence is reasonable and an analytic truth that inductive evidence of the sort captured by the schema presented earlier constitutes strong evidence for the corresponding inductive conclusion, thus, apparently yielding the analytic conclusion that believing it a conclusion for which there is inductive evidence is reasonable. Nevertheless, he also admits, indeed insists, that the claim that inductive conclusions will be true in the future is contingent, empirical, and may turn out to be false (1952). Thus, the notion of reasonable belief and the correlative notion of strong evidence must apparently be understood in ways that have nothing to do with likelihood of truth, presumably by appeal to the standard of reasonableness and strength of evidence that are accepted by the community and are embodied in ordinary usage.

Understood in this way, Strawson's response to the problem of inductive reasoning does not speak to the central issue raised by Humean scepticism: The issue of whether the conclusions of inductive arguments are likely to be true. It amounts to saying merely that if we reason in this way, we can correctly call ourselves 'reasonable' and our evidence 'strong', according to our accepted community standards. Nevertheless, to the undersealing of issue of wether following these standards is a good way to find the truth, the ordinary language response appears to have nothing to say.

(3) The main attempts to show that induction can be justified inductively have concentrated on showing that such as a defence can avoid circularity. Skyrms (1975) formulate, perhaps the clearest version of this general strategy. The basic idea is to distinguish different levels of inductive argument: A first level in which induction is applied to things other than arguments: A second level in which it is applied to arguments at the first level, arguing that they have been observed to succeed so far and hence are likely to succeed in general: A third level in which it is applied in the same way to arguments at the second level, and so on. Circularity is allegedly avoided by treating each of these levels as autonomous and justifying the argument at each level by appeal to an argument at the next level.

One problem with this sort of move is that even if circularity is avoided, the movement to higher and higher levels will clearly eventually fail simply for lack of evidence: A level will reach at which there have been enough successful inductive arguments to provide a basis for inductive justification at the next higher level, and if this is so, then the whole series of justifications collapses. A more fundamental difficulty is that the epistemological significance of the distinction between levels is obscure. If the issue is whether reasoning in accord with the original schema offered above ever provides a good reason for thinking that the conclusion is likely to be true, then it still seems question-begging, even if not flatly circular, to answer this question by appeal to anther argument of the same form.

(4) The idea that induction can be justified on a pure priori basis is in one way the most natural response of all: It alone treats an inductive argument as an independently cogent piece of reasoning whose conclusion can be seen rationally to follow, although perhaps only with probability from its premise. Such an approach has, however, only rarely been advocated (Russell, 1913 and BonJour, 1986), and is widely thought to be clearly and demonstrably hopeless.

Many are the reasons for this pessimistic view depends on general epistemological theses about the possible or nature of anticipatory cognition. Thus if, as Quine alleges, there is no a prior justification of any kind, then obviously a prior justification for induction is ruled out. Or if, as more moderate empiricists have in claiming some preexistent knowledge should be analytic, then again a prevenient justification for induction seems to be precluded, since the claim that if an inductive premise ids truer, then the conclusion is likely to be true does not fit the standard conceptions of 'analyticity'. A consideration of these matters is beyond the scope of the present spoken exchange.

There are, however, two more specific and quite influential reasons for thinking that an early approach is impossible that can be briefly considered, first, there is the assumption that began with Hume, but since adopted by very many of others, that a move forward in the defence of induction would have to involve 'turning induction into deduction', i.e., showing, per impossible, that the inductive conclusion follows deductively from the premise, so that it is a formal contradiction to accept the latter and deny the former. However, it is unclear why a prior approach need be committed to anything this strong. It would be enough if it could be argued that it is deductively unlikely that such a premise is true and corresponding conclusion false.

Reichenbach defends his view that pragmatic justification is the best that is possible by pointing out that a completely chaotic world in which there is simply not true conclusion to be found as to the proportion of 'A's' in addition that occurs of, but B's' is neither impossible nor unlikely from a purely a prior standpoint, the suggestion being that therefore there can be no a prior reason for thinking that such a conclusion is true. Nevertheless, there is still a substring wayin laying that a chaotic world is a prior neither impossible nor unlikely without any further evidence does not show that such a world os not a prior unlikely and a world containing such-and-such regularity might hesitorially be somewhat likely in relation to an occurrence of a long-run patten of evidence in which a certain stable proportion of observed 'A's' are 'B's' ~. An occurrence, it might be claimed, that would be highly unlikely in a chaotic world (BonJour, 1986).

Goodman's 'new riddle of induction' purports that we suppose that before some specific time 't' (perhaps the year 2020) we observe a larger number of emeralds (property 'A') and find them all to be green (property B). We proceed to reason inductively and conclude that all emeralds are green Goodman points out, however, that we could have drawn a quite different conclusion from the same evidence. If we define the term 'grue' to mean 'green if examined before 't' and blue examined after t, then all of our observed emeralds will also be gruing. A parallel inductive argument will yield the conclusion that all emeralds are gruing, and hence that all those examined after the year 2015 will be blue. Presumably the first of these concisions is genuinely supported by our observations and the second is not. Nevertheless, the problem is to say why this is so and to impose some further restriction upon inductive reasoning that will permit the first argument and exclude the second.

The obvious alternative suggestion is that 'grue. Similar predicates do not correspond to genuine, purely qualitative properties in the way that 'green' and 'blueness' does, and that this is why inductive arguments involving them are unacceptable. Goodman, however, claims to be unable to make clear sense of this suggestion, pointing out that the relations of formal desirability are perfectly symmetrical: Grue' may be defined in terms if, 'green' and 'blue', but 'green' an equally well be defined in terms of 'grue' and 'green' (blue if examined before 't' and green if examined after 't').

The 'grued, paradoxes' demonstrate the importance of categorization, in that sometimes it is itemized as 'gruing', if examined of a presence to the future, before future time 't' and 'green', or not so examined and 'blue'. Even though all emeralds in our evidence class grue, we ought must infer that all emeralds are gruing. For 'grue' is unprojectible, and cannot transmit credibility form known to unknown cases. Only projectable predicates are right for induction. Goodman considers entrenchment the key to projectibility having a long history of successful protection, 'grue' is entrenched, lacking such a history, 'grue' is not. A hypothesis is projectable, Goodman suggests, only if its predicates (or suitable related ones) are much better entrenched than its rivalrous past successes that do not assume future ones. Induction remains a risky business. The rationale as to favour the entrenched predicate of a pragmatic possibility that projections from our evidential class, the one that fits with past practices enables 'us' to utilize our cognitive resources best. Its prospects of being true are worse than its competitors' and its cognitive utility is greater.

The rational basis of any inference was challenged by Hume, who believed that induction presupposed belie in the uniformity of nature, but that this belief has no defence in reason, and merely reflected a habit or custom of the mind. Hume was not therefore sceptical about the role of reason in either explaining it or justifying it. Trying to answer Hume and to show that there is something rationally compelling about the inference referred to as the problem of induction. It is widely recognized that any rational defence of induction will have to partition well-behaved properties for which the inference is plausible (often called projectable properties) from badly behaved ones, for which it is not. It is also recognized that actual inductive habits are more complex than those of similar enumeration, and that both common sense and science pay attention to such giving factors as variations within the sample giving 'us' the evidence, the application of ancillary beliefs about the order of nature, and so on.

Nevertheless, the fundamental problem remains that ant experience condition by application show 'us' only events occurring within a very restricted part of a vast spatial and temporal order about which we then come to believe things.

Uncompounded by its belonging of a confirmation theory finding of the measure to which evidence supports a theory fully formalized confirmation theory would dictate the degree of confidence that a rational investigator might have in a theory, given to a body of evidence. The grandfather of confirmation theory is Gottfried Leibniz (1646-1718), who believed that a logically transparent language of science would be able to resolve all disputes. In the 20th century a fully formal confirmation theory was a main goal of the logical positivist, since without it the central concept of verification by empirical evidence itself remains distressingly unscientific. The principal developments were due to Rudolf Carnap (1891-1970), culminating in his 'Logical Foundations of Probability' (1950). Carnap's idea was that the measure necessitated would be the proportion of logically possible states of affairs in which the theory and the evidence both hold as compared to the number in which the evidence itself holds the probability of a preposition, relative to some evidence, is a proportion, in the range of possibilities under which the proposition is true, this, in the comparing to the total range of possibilities left by the evidence. The difficulty with the theory lies in identifying sets of possibilities so that they admit of measurement. It therefore demands that we can put a measure on the 'range' of possibilities consistent with theory and evidence, compared with the doctrine contending that sense perception are the only admissible basis of human knowledge and precise thought Carnap also set-forth in such works as The Logic and Structure of the World, and The Logical Foundations of Probability (1950), were central to the development of logical positivism, as characterized in philosophy as the doctrine contending that sense perception are the only admissible basis of human knowledge and precise thought. The system of Auguste Come designed a superseded Theology and metaphysical and depending on a hierarchy of the sciences, beginning with a culminating mathematics in sociology, any of several doctrines or viewpoints, often similar to Comte's, that stress attention to actual practice over consideration of what is dead, Positivism became the scientific base for authoritarian politics, especially in Mexico and Brazil.

Among the obstacles the enterprise meets, is the fact that while evidence covers only a finite range of data, the hypotheses of science may cover an infinite range. In addition, confirmation proves to vary with the language in which the science is couched, and the Carnapian programme has difficulty in separating genuinely confirming variety of evidence from less compelling repetition of the same experiment. Confirmation also proved to be susceptible to acute paradoxes. Finally, scientific judgement seems to depend on such intangible factors as the problems facing rival theories, and most workers have come to stress instead the historically situated scene of what would appear as a plausible distinction of a scientific knowledge at a given time.

Arising to the paradox, of which when a set of apparent incontrovertible premises is given to unacceptable or contradictory conclusions. To solve a paradox will involve showing either that there is a hidden flaw in the premises, or that the reasoning is erroneous, or that the apparently unacceptable conclusion can, in fact, be tolerated. Paradoxes are therefore important in philosophy, for until one is solved it shows that there is something about our reasoning and our concepts that we do not understand. What is more, and somewhat loosely, a paradox is a compelling argument from unacceptable premises to an unacceptable conclusion: More strictly speaking, a paradox is specified to be a sentence that is true if and only if it is false. As vulnerably characterized is found that the objection from lessons learnt are yet to be. 'The displayed sentence is false.'

Seeing that this sentence is false if true is easy, and true if false, a paradox, in either of the senses distinguished, presents an important philosophical challenger. Epistemologists are especially concerned with various paradoxes having to do with knowledge and belief. In other words, for example, the Knower paradox is an argument that begins with apparently impeccable premises about the concepts of knowledge and inference and derives an explicit contradiction. The origin of the reasoning is the 'surprise examination paradox': A teacher announces that there will be a surprise examination next week. A clever student argues that this is impossible. 'The test cannot be on Friday, the last day of the week, because it would not be a surprise. We would know the day of the test on Thursday evening, but that we can also rule out Thursday. For after we learn that no test has been given by Wednesday, we would know the test is on Thursday or Friday -and would already know that it s not on Friday and would already know that it is not on Friday by the previous reasoning. The remaining days can be eliminated in the same manner'.

This puzzle has over a dozen variants. The first was probably invented by the Swedish mathematician Lennard Ekbon in 1943. Although, commentators reported that the reverse elimination argument as cogent, every writer on the subject since 1950 agrees that the argument is unsound. The controversy has been over the proper diagnosis of the flaw.

Initial analyses of the subject's argument tried to lay the blame on a simple equivocation. Their failure led to more sophisticated diagnoses. The general format has been an assimilation to better-known paradoxes. One tradition casts the surprise examination paradox as a self-referential problem, as fundamentally akin to the Liar, the paradox of the Knower, or Gödel's incompleteness theorem stressing mathematics and logician, know for his proof that the consistency of a mathematical system in which the truths of arithmetic cannot be expressed cannot be proven from within that system (1931). Gödel, (1906-1978) Czech-born American mathematician and logician best known for his proof that consistency of a mathematical system in which the truth of arithmetic can be expressed cannot be proven from within that system (1931).

That in of itself, says enough that Kaplan and Montague (1960) distilled the following 'self-referential' paradox, the Knower. Consider the sentence:

(S) The negation of this sentence is known (to be true).

Suppose that (S) is true. Then its negation is known and hence true. However, if its negation is true, then (S) must be false. Therefore (s) is false, or what is the name, the negation of (S) is true. This paradox and its accompanying reasoning are strongly reminiscent of the Lair Paradox that (in one version) begins by considering a sentence 'This sentence is false' and derives a contradiction. Versions of both arguments using axiomatic formulations of arithmetic and the Gödel-numbers were to achieve the effect of the self-reference, yielding some important meta-theorems about what can be expressed in such systems. Roughly these are to the effect that no predicates definable in the formalized arithmetic can have the properties we demand of truth (Tarski's Theorem) or of knowledge (Montague, 1963).

These meta-theorems still leave 'us' with the problem that if we suppose that we add these formalized language predicates intended to express the concept of knowledge (or truth) and inference-as one might do if a logic of these concepts is desirous. Then the sentence expressing the leading principles of the Knower Paradox will be true. Explicitly, the assumption about knowledge and inferences are indicative to be:

(1) If sentences 'A' are known, then 'a.'

(2) That (1) is known?

(3) If 'B' is correctly inferred from 'A', and 'A' is known, then 'B' is known.

To give an absolutely explicit t mathematician of the paradox by applying these principles to (S), we must add (contingent) assumptions to the effect that certain inferences have been done. Still, as we go through the argument of the Knower, these inferences are done. Even if we can somehow restrict such principles and construct a consistent formal logic of knowledge and inference, the paradoxical argument as expressed in the natural language still demands some explanation.

The usual proposals for dealing with the Liar often have their analogues for the Knower, e.g., that there is something wrong with a self-reference or that knowledge (or truth) is properly a predicate of propositions and not of sentences. This relies that shown that some of these are not adequate and often parallel to those for the Liar paradox. In addition, one can attempt to what seems to an adequate solution for the Surprise Examination Paradox, namely the observation that 'new knowledge can drive out knowledge', but this does not seem to work on the Knower (Anderson, 1983).

There are a number of paradoxes of the Liar family. The simplest example is the sentence 'This sentence is false', which must be false if it is true, and true if it is false. One suggestion is that the sentence fails to say anything, but sentences that fail to say anything are at least not true. In fact case, we consider to sentences 'This sentence is not true', which, if it fails to say anything is not true, and hence (this kind of reasoning is sometimes called the strengthened Liar). Other versions of the Liar introduce pairs of sentences, as in a slogan on the front of a T-shirt saying 'This sentence on the back of this T-shirt is false', and one on the back saying 'The sentence on the front of this T-shirt is true'. It is clear that each sentence individually is well formed, and was it not for the other, might have said something true. So any attempts to dismiss the paradox by sating that the sentence involved are meaningless will face problems.

Even so, the two approaches that have some hope of adequately dealing with this paradox is 'hierarchy' solutions and 'truth-value gap' solutions. According to the first, knowledge is structured into 'levels'. It is argued that there is a one-coherent notion expressed by the verb 'knows', but rather a whole series of notions are known by knowing and so forth (perhaps into transfinite), stated ion terms of predicate expressing such 'ramified' concepts and properly restricted, (1)-(3) lead to no contradictions. The main objections to this procedure are that the meaning of these levels has not been adequately explained and that the idea of such subscripts, even implicit, in a natural language is highly counterintuitive the 'truth-value gap' solution takes sentences such as (S) to lack truth-value. They are neither true nor false, but they do not express propositions. This defeats a crucial step in the reasoning used in the derivation of the paradoxes. Kripler (1986) has developed this approach in connection with the Liar and Asher and Kamp (1986) has worked out some details of a parallel solution to the Knower. The principal objection is that 'strengthened' or 'super' versions of the paradoxes tend to reappear when the solution itself and is stated.

Since the paradoxical deduction uses only the properties (1)-(3) and since the argument is formally valid, any notions that satisfy these conditions will lead to a paradox. Thus, Grim (1988) notes that this may be read as 'is known by an omniscient God' and concludes that there is no coherent single notion of omniscience. Thomason (1980) observes that with some different conditions, analogous reasoning about belief can lead to paradoxical consequence.

Overall, it looks as if we should conclude that knowledge and truth are ultimately intrinsically 'stratified' concepts. It would seem that wee must simply accept the fact that these (and similar) concepts cannot be assigned of any fixed, finite or infinite. Still, the meaning of this idea certainly needs further clarification.

Its paradox arises when a set of apparently incontrovertible premises gives unacceptable or contradictory conclusions, to solve a paradox will involve showing either that there is a hidden flaw in the premises, or that the reasoning is erroneous, or that the apparently unacceptable conclusion can, in fact, be tolerated. Paradoxes are therefore important in philosophy, for until one is solved its show that there is something about our reasoning and our concepts that we do not understand. Famous families of paradoxes include the 'semantic paradoxes' and 'Zeno's paradoxes. Art the beginning of the 20th century, paradox and other set-theoretical paradoxes led to the complete overhaul of the foundations of set theory, while the 'Sorites paradox' has lead to the investigations of the semantics of vagueness and fuzzy logics.

It is, however, to what extent can analysis be informative? This is the question that gives a riser to what philosophers has traditionally called 'the' paradox of analysis. Thus, consider the following proposition:

> (1) To be an instance of knowledge is to be an instance of justified true belief not essentially grounded in any falsehood.

(1) If true, illustrates an important type of philosophical analysis. For convenience of exposition, it will be to assume that (1) is a correct analysis. The paradox arises from the fact that if the concept of justified true belief not been essentially grounded in any falsification is the analysand of the concept of knowledge, it would seem that they are the same concept and hence that: (2) To be an instance of knowledge is to be as an instance of. Such knowledge would have to be the same proposition as (1). But then how can (1) be informative when (2) is not? This is what is called the first paradox of analysis. Classical writings' on analysis suggests a second paradoxical analysis (Moore, 1942).

(3) An analysis of the concept of being a brother is that to be a brother is to be a male sibling. If (3) is true, it would seem that the concept of being a brother would have to be the same concept as the concept of being a male sibling and that:

(4) An analysis of the concept of being a brother is that to be a brother is to be a brother would also have to be true and in fact, would have to be the same proposition as (three?). Yet (3) it is true and (4) is false.

Both these paradoxes rest upon the assumptions that analysis is a relation between concepts, than one involving entity of other sorts, such as linguistic expressions, and tat in a true analysis, analysand and analysandum are the same concepts. Both these assumptions are explicit in Moore, but some of Moore's remarks hint at a solution to that of another statement of an analysis is a statement partly about the concept involved and partly about the verbal expressions used to express it. He says he thinks a solution of this sort is bound to be right, but fails to suggest one because he cannot see a way in which the analysis can be even partly about the expression (Moore, 1942).

Elsewhere, of such ways, as a solution to the second paradox, from which is to explicate (3) as: (5) An analysis is given by saying that the verbal expression ',' is a brother, expresses the same concept as is expressed by the conjunction of the verbal expressions ',' is male when used to express the concept of being male and ',' is a sibling when used to express the concept of being a sibling (Ackerman, 1990). An important point about (5) is as follows. Stripped of its philosophical jargon ('analysis', 'concept' 'χ' is a . . .) (5) Serving to state the sort of information generally stated in a definition of verbal expression 'brother' in terms of the verbal expressions 'male' and 'sibling', where this definition is designed to draw upon listeners' antecedent understanding of the verbal expression 'male' and 'sibling', and thus, to tell listeners what the verbal expression 'brother' really, instead of merely providing the information that two verbal expressions are synonymous without specifying the meaning of either one. Thus, its solution to the second paradox seems to make the sort of analysis that gives rise to this paradox matter of specifying the meaning of a verbal expression in terms of separate verbal expressions already understood and saying how the meanings of these separate, already-understood verbal expressions are combined. This corresponds to Moore's intuitive requirement that an analysis should both specify the constituent concepts of the analysandum and tell how they are combined, but is this all there is to philosophical analysis?

To answer this question, we must note that, in addition too there being two paradoxes of analysis, there is two types of analyses that are relevantly of other types of analysis, such as reformatory analysis, where the analysands are intended to improve as then to replace the analysandum. But since reformatory analysis involves no commitment to conceptual identity between analysand and analysandum, the reformatory analysis does not generate a paradox of analysis and will not concern 'us' here. One way to recognize the differences between the two categorical types of analysis that concerns 'us' here, are to focus on the differences between the two differentiated paradoxes. This can be done by of the Frége-inspired sense-individuation condition, which is the condition that two expressions have the same sense if and only if they can be interchangeably 'salva veritate' whenever used in propositional attitude context. If the expressions for the analysands and the analysandum in (1) met this condition, (1) and (2) is that, it is not raised to fulfil the first paradox, but the second paradox arises regardless of whether the expression for the analysand and the analysandum meet this condition. The second paradox is a matter of the failure of such expressions to be interchangeable salva veritate in sentences involving such contexts as 'an analysis is given thereof. Thus, a solution (such as the one offered) that is aimed only at such contexts can solve the second paradox. This is clearly false for the first paradox, however, which will apply to all pairs of propositions expressed by sentences in which expressions for pairs of analysands and analysantia raising the first paradox is interchangeable. For example, consider the following proposition:

Mary knows that some cats tail.

It is possible for John to believe without believing:

Mary has justified true belief, not essentially grounded in any falsehood,

That some cats lack tails.

Yet this possibility clearly does not mean that the proposition that Mary knows that some casts lack tails is partly about language.

One approach to the first paradox is to argue that, despite the apparent epistemic inequivalence of (1) and (2), the concept of justified true belief not essentially grounded in any falsehood is still identical with the concept of knowledge (Sosa, 1983). Another approach is to argue that in the sort of analysis raising the first paradox, that the analysand and analysandum are concepts that are different but bearing as a special epistemic relation to each other. Elsewhere, the development is such an approach and suggestion that this analysand-analysandum relation has the following facets (I) The analysand and analysandum are necessarily coextensive, i.e., necessarily every instance of one is an instance of the other.

(ii) The analysand and analysandum are knowable theoretically corroborative confirmations, making over to remove obstructions from coextensive languages in a system that can be modified, simply to give a completely different form or appearance to change of the same kind or category or by adding features for changing a system that can be modified by changing or adding features of their own holdings.

(iii) The analysandum is simply which the analysand is to apply of a condition to whose necessity is recognized in classical writings on analysis, such as, Langford, 1942.

(iv) The analysand does not have the analysandum as a constituent.

Condition (iv) rules out circularity. But since many valuable quasi-analyses are partly circular, e.g., knowledge is justified true belief supported by known reasons not essentially grounded in any falsehood, it seems best to distinguish between full analysis, from that of (iv) is a necessary condition, and partial analysis, for which it is not.

If 'S' is the analysand of 'Q', the proposition that necessarily all and only instances of 'S' are instances of 'Q' can be justified by generalizing from intuition about the correct answers to questions of the sort indicated about a varied and wide-ranging series of simply described hypothetical situations. It so does occur of antinomy, when we are able to argue for, or demonstrate, both a proposition and its contradiction, roughly speaking, a contradiction of a proposition 'p' is one that can be expressed in form 'not-p', or, if 'p' can be expressed in the form 'not-q', then a contradiction is one that can be expressed in the form 'q'. Thus, e.g., if 'p' is $2 + 1 = 4$, then $2 + 1 = 4$ in that of 'p', for which $2 + 1 = 4$ that can be expressed in the form $(2 + 1 = 4)$. If 'p' is $2 + 1 = 4$, then $2 + = 4$ is a contradictory of 'p', since $2 + 1 = 4$ can be expressed in the form not $(2 + 1 = 4)$. This is, mutually, but contradictory propositions can be expressed in the form, 'r', 'not-r'. The Principle of Contradiction says that mutually contradictory propositions cannot both be true and cannot both be false. Thus, by this principle, since if 'p' is true, 'not-p' is false, no proposition 'p' can be at once true and false (otherwise both 'p' and its contradictories would be false?). In particular, for any predicate 'p' and object 'q', it cannot be that 'p'; is at once true of 'q' and false of q? This is the classical formulation of the principle of contradiction, but it is nonetheless, that wherein, we cannot now fault either demonstrates. We would eventually hope to be able 'to solve the antinomy' by managing, through careful thinking and analysis, eventually to fault either or both demonstrations.

Many paradoxes are as an easy source of antinomies, for example, Zeno gave some famously lets say, logical-cum-mathematical arguments that might be interpreted as demonstrating that motion is impossible. But our eyes as it was, demonstrate motion (exhibit moving things) all the time. Where did Zeno go wrong? Where do our eyes go wrong? If we cannot readily answer at least one of these questions, then we are in antinomy. In the 'Critique of Pure Reason,' Kant gave demonstrations of the same kind -in the Zeno example they were obviously not the same kind of both, e.g., that the world has a beginning in time and space, and that the world has no beginning in time or space. He argues that both demonstrations are at fault because they proceed on the basis of 'pure reason' unconditioned by sense experience.

At this point, we display attributes to the theory of experience, as it is not possible to define in an illuminating way, however, we know what experiences are through acquaintances with some of our own, e.g., visual experiences of as afterimage, a feeling of physical nausea or a tactile experience of an abrasive surface (which might be caused by an actual surface -rough or smooth, or which might be part of a dream, or the product of a vivid sensory imagination). The essential feature of experience is it feels a certain way -that there is something that it is like to have it. We may refer to this feature of an experience as its 'character'.

Another core feature of the sorts of experiences with which this may be of a concern, is that they have representational 'content'. (Unless otherwise indicated, 'experience' will be reserved for their 'contentual representations'.) The most obvious cases of experiences with content are sense experiences of the kind normally involved in perception. We may describe such experiences by mentioning their sensory modalities ad their contents, e.g., a gustatory experience (modality) of chocolate ice cream (content), but do so more commonly by of perceptual verbs combined with noun phrases specifying their contents, as in 'Macbeth saw a dagger'. This is, however, ambiguous between the perceptual claim 'There was a (material) dagger in the world that Macbeth perceived visually' and 'Macbeth had a visual experience of a dagger' (the reading with which we are concerned, as it is afforded by our imagination, or perhaps, experiencing mentally hallucinogenic imagery).

As in the case of other mental states and events with content, it is important to distinguish between the properties that and experience 'represents' and the properties that it 'possesses'. To talk of the representational properties of an experience is to say something about its content, not to attribute those properties to the experience itself. Similar among all other experiences, every experience involves a visual experience of a non-shaped square, which is a mental event, and it is therefore not itself irregular or is it square, even though it represents those properties. It is, perhaps, fleeting, pleasant or unusual, even though it does not represent those properties. An experience may represent a property that it possesses, and it may even do so in virtue of a rapidly changing (complex) experience representing something as changing rapidly. However, this is the exception and not the rule.

Which properties can be [directly] represented in sense experience is subject to debate. Traditionalists include only properties whose presence could not be doubted by a subject having appropriate experiences, e.g., colour and shape in the case of visual experience, and apparent shape, surface texture, hardness, and so forth, in the case of tactile experience. This view is natural to anyone who has an egocentric, Cartesian perspective in epistemology, and who wishes for pure data in experiences to serve as logically certain foundations for knowledge, especially to the immediate objects of perceptual awareness in or of sense-data, such categorized of colour patches and shapes, which are usually supposed distinct from surfaces of physical objectivity. Qualities of sense-data are supposed to be distinct from physical qualities because their perception is more relative to conditions, more certain, and more immediate, and because sense-data is private and cannot appear other than they are they are objects that change in our perceptual field when conditions of perception change. Physical objects remain constant.

Others who do not think that this wish can be satisfied, and who are more impressed with the role of experience in providing animisms with ecologically significant information about the world around them, claim that sense experiences represent properties, characteristic and kinds that are much richer and much more wide-ranging than the traditional sensory qualities. We do not see only colours and shapes, they tell 'us', but also earth, water, men, women and fire: We do not smell only odours, but also food and filth. There is no space here to examine the factors relevantly responsible to their choice of situational alternatives. Yet, this suggests that character and content are not really distinct, and there is a close tie between them. For one thing, the relative complexity of the character of sense experience places limitations upon its possible content, e.g., a tactile experience of something touching one's left ear is just too simple to carry the same amount of content as typically convincing to an every day, visual experience. Moreover, the content of a sense experience of a given character depends on the normal causes of appropriately similar experiences, e.g., the sort of gustatory experience that we have when eating chocolate would be not represented as chocolate unless it was normally caused by chocolate. Granting a contingent ties between the character of an experience and its possible causal origins, once, again follows that its possible content is limited by its character.

The semantic argument is that objects of experience are required in order to make sense of certain features of our talk about experience, including, in particular, the following. (i) Simple attributions of experience, e.g., 'Rod is experiencing an oddity that is not really square but in appearance it seems more than likely a square', this seems to be relational. (ii) We appear to refer to objects of experience and to attribute properties to them, e.g., 'The after image that John experienced was certainly odd'. (iii) We appear to quantify ov er objects of experience, e.g., 'Macbeth saw something that his wife did not see'.

The act/object analysis faces several problems concerning the status of objects of experiences. Currently the most common view is that they are sense-data -private mental entities that actually posses the traditional sensory qualities represented by the experiences of which they are the objects. But the very idea of an essentially private entity is suspect. Moreover, since an experience may apparently represent something as having a determinable property, e.g., redness, without representing it as having any subordinate determinate property, e.g., any specific shade of red, a sense-datum may actually have a determinate property subordinate to it. Even more disturbing is that sense-data may have contradictory properties, since experiences can have contradictory contents. A case in point is the waterfall illusion: If you stare at a waterfall for a minute and then immediately fixate on a nearby rock, you are likely to have an experience of the rock's moving upward while it remains in the same place. The sense-data theorist must either deny that there are such experiences or admit contradictory objects.

These problems can be avoided by treating objects of experience as properties. This, however, fails to do justice to the appearances, for experience seems not to present 'us' with properties embodied in individuals. The view that objects of experience is Meinongian objects accommodate this point. It is also attractive in as far as (1) it allows experiences to represent properties other than traditional sensory qualities, and (2) it allows for the identification of objects of experience and objects of perception in the case of experiences that constitute perception.

According to the act/object analysis of experience, every experience with content involves an object of experience to which the subject is related by an act of awareness (the event of experiencing that object). This is meant to apply not only to perceptions, which have material objects (whatever is perceived), but also to experiences like hallucinations and dream experiences, which do not. Such experiences are, nonetheless, appear to represent something, and their objects are supposed to be whatever it is that they represent. Act/object theorists may differ on the nature of objects of experience, which have been treated as properties. Meinongian objects (which may not exist or have any form of being), and, more commonly private mental entities with sensory qualities. (The term 'sense-data' is now usually applied to the latter, but has also been used as a general term for objects of sense experiences, as in the work of G. E. Moore) Act/object theorists may also differ on the relationship between objects of experience and objects of perception. In terms of perception (of which we are 'indirectly aware') are always distinct from objects of experience (of which we are 'directly aware'). Meinongian, however, may treat objects of perception as existing objects of experience. But sense-datum theorists must either deny that there are such experiences or admit contradictory objects. Still, most philosophers will feel that the Meinongian's acceptance of impossible objects is too high a price to pay for these benefits.

A general problem for the act/object analysis is that the question of whether two subjects are experiencing one and the same thing (as opposed to having exactly similar experiences) appears to have an answer only on the assumption that the experiences concerned are perceptions with material objects. But in terms of the act/object analysis the question must have an answer even when this condition is not satisfied. (The answer is always negative on the sense-datum theory; it could be positive on other versions of the act/object analysis, depending on the facts of the case.)

In view of the above problems, the case for the act/object analysis should be reassessed. The Phenomenological argument appears to present 'us' with an object without accepting that it actually does. The semantic argument is more impressive, but is none the less answerable. The seemingly relational structure of attributions of experience is a challenge dealt with below in connection with the adverbial theory. Apparent reference to and quantification over objects of experience can be handled by analysing them as reference to experiences themselves and quantification over experiences tacitly typed according to content. Thus, 'The after image that John experienced was colourfully appealing' becomes 'John's after image experience was an experience of colour', and 'Macbeth saw something that his wife did not see' becomes 'Macbeth had a visual experience that his wife did not have'.

Pure cognitivism attempts to avoid the problems facing the act/object analysis by reducing experiences to cognitive events or associated disposition, e.g., Susy's experience of a rough surface beneath her hand might be identified with the event of her acquiring the belief that there is a rough surface beneath her hand, or, if she does not acquire this belief, with a disposition to acquire it that has somehow been blocked.

This position has attractions. It does full justice to the cognitive contents of experience, and to the important role of experience as a source of belief acquisition. It would also help clear the way for a naturalistic theory of mind, since there seems to be some prospect of a physicalist/functionalist account of belief and other intentional states. But pure cognitivism is completely undermined by its failure to accommodate the fact that experiences have a felt character that cannot be reduced to their content, as aforementioned.

The adverbial theory is an attempt to undermine the act/object analysis by suggesting a semantic account of attributions of experience that does not require objects of experience. Unfortunately, the oddities of explicit adverbializations of such statements have driven off potential supporters of the theory. Furthermore, the theory remains largely undeveloped, and tempted refutations have traded on this. It may, however, be founded on sound basis intuitions, and there is reason to believe that an effective development of the theory (which is merely hinting at) is possible.

The relevant intuitions are (1) that when we say that someone is experiencing 'an A', or has an experience 'of an A', we are using this content-expression to specify the type of thing that the experience is especially apt to fit, (2) that doing this is a matter of saying something about the experience itself (and maybe about the normal causes of like experiences), and (3) that it is no-good of reasons to posit of its position to presuppose that of any involvements, is that its descriptions of an object in which the experience is. Thus the effective role of the content-expression in a statement of experience is to modify the verb it compliments, not to introduce a special type of object.

Perhaps, the most important criticism of the adverbial theory is the 'many property problem', according to which the theory does not have the resources to distinguish between, e.g.,

> (1) Frank has an experience of a brown triangle

And:

> (2) Frank has an experience of brown and an experience of a triangle.

Which is entailed by (1) but does not entail it. The act/object analysis can easily accommodate the difference between (1) and (2) by claiming that the truth of (1) requires a single object experiences that are both a triangle and brown, while that of the (2) allows for the possibility of two objects of experience, one brown and the other triangular, however, (1) are equivalent to:

(1*) Frank has an experience of something's being both brown and triangular.

And (2) is equivalent to:

(2*) Frank has an experience of something's being brown and an experience of something's being triangular, and the difference between these can be explained quite simply in terms of logical scope without invoking objects of experience. The Adverbialists may use this to answer the many-property problem by arguing that the phrase 'a brown triangle' in (1) does the same work as the clause 'something's being both brown and triangular' in (1*). This is perfectly compatible with the view that it also has the 'adverbial' function of modifying the verb 'has an experience of', for it specifies the experience more narrowly just by giving a necessary condition for the satisfaction of the experience (the condition being that there are something both brown and triangular before Frank).

A final position that should be mentioned is the state theory, according to which a sense experience of an 'A' is an occurrent, non-relational state of the kind that the subject would be in when perceiving an 'A'. Suitably qualified, this claim is no doubt true, but its significance is subject to debate. Here it is enough to remark that the claim is compatible with both pure cognitivism and the adverbial theory, and that state theorists are probably best advised to adopt adverbials for their developing of intuitions.

Yet, clarifying sense-data, if taken literally, is that which is given by the senses. But the act or reflecting in response to the mental concentration is carefully considered, for being a direct expression of discredit. As reflection of integrity for which is a manifestation or result for what the question has been accorded to what is so given, that sense-data theory's posit themselves in the private showings in the consciousness of the subject. In the case of vision this would be a kind of inner picture show which it only indirectly represents aspects of the external world that has in and of itself a worldly representation. The view has been widely rejected as implying that we really only see extremely thin coloured pictures interposed between our mind's eye and reality. Modern approaches to perception tend to reject any conception of the eye as a camera or lense, simply responsible for producing private images, and stress the active life of the subject in and of the world, as the determinant of experience.

Nevertheless, the argument from illusion is of itself the usually intended directive to establish that certain familiar facts about illusion disprove the theory of perception called 'naevity' or direct realism. There are, however, many different versions of the argument that must be distinguished carefully. Some of these distinctions centre on the content of the premises (the nature of the appeal to illusion); others centre on the interpretation of the conclusion (the kind of direct realism under attack). Let 'us' set about by distinguishing the importantly different versions of direct realism which one might take to be vulnerable to familiar facts about the possibility of perceptual illusion.

A crude statement of direct realism might go as follows. In perception, we sometimes directly perceive physical objects and their properties, we do not always perceive physical objects by perceiving something 'else', e.g., a sense-datum. There are, however, difficulties with this formulation of the view, as for one thing a great many philosophers who are 'not' direct realists would admit that it is a mistake to describe people as actually 'perceiving' something other than a physical object. In particular, such philosophers might admit, we should never say that we perceive sense-data. To talk that way would be to suppose that we should model our understanding of our relationship to sense-data on our understanding of the ordinary use of perceptual verbs as they describe our relation to and of the physical world, and that is the last thing paradigm sense-datum theorist's might want. At least, many of the philosophers who objected to direct realism would prefer to express in what they were of objecting too in terms of a technical (and philosophically controversial) concept such as 'acquaintance'. Using such a notion, we could define direct realism this way: In 'veridical' experience we are directly acquainted with parts, e.g., surfaces, or constituents of physical objects. A less cautious description or account from one verison of the view or variation of an earlier or original translation, as from another language or literature into another medium or style might drop the reference to veridical experience and claim simply that in all experiences we are directly acquainted with parts or constituents of physical objects. The expressions 'knowledge by acquaintance' and 'knowledge by description', and the distinction they mark between knowing 'things' and knowing 'about' things, are generally associated with Bertrand Russell (1872-1970), that scientific philosophy required analysing many objects of belief as 'logical constructions' or 'logical fictions', and the programme of analysis that this inaugurated dominated the subsequent philosophy of logical atomism, and then of other philosophers, Russell's 'The Analysis of Mind,' the mind itself is treated in a fashion reminiscent of Hume, as no more than the collection of neutral perceptions or sense-data that make up the flux of conscious experience, and that looked at another way that also was to manufacture the external world (neutral monism), but 'An Inquiry into Meaning and Truth' (1940) represents a more empirical approach to the problem. Yet, philosophers have perennially investigated this and related distinctions using varying terminology.

Distinction in our ways of knowing things, highlighted by Russell, forming the central element in his philosophy after the discovery of the theory of 'definite descriptions', for a thing is known by acquaintance when there is cause to proceed in or follow a straight course, directed to indicate or conduct the affairs of the experience of it. It is known by description if it can only be described as a thing with such-and-such properties. In everyday parlance, I might know my spouse and children by acquaintance, but know someone as 'the first person born at sea' only by description. However, for a variety of reasons Russell shrinks the area of things that can be known by acquaintance until eventually only current experience, perhaps my own self, and certain universals or meanings qualify anything else is known only as the thing that has such-and-such qualities.

Because one can interpret the relation of acquaintance or awareness as one that is not 'epistemic', i.e., not a kind of propositional knowledge, it is important to distinguish the above aforementioned views read as ontological theses from a view one might call 'epistemological direct realism.' In perception we are, on at least some occasions, non-inferentially justified in believing a proposition asserting the existence of a physical object. Since it is that these objects exist independently of any mind that might perceive them, and so it thereby rules out all forms of idealism and phenomenalism, which hold that there are no such independently existing objects. Its being to 'direct' realism rules out those views defended under the cubic of 'critical naive realism', or 'representational realism', in which there is some nonphysical intermediary -usually called a 'sense-datum' or a 'sense impression' -that must first be perceived or experienced in order to perceive the object that exists independently of this perception. Often the distinction between direct realism and other theories of perception is explained more fully in terms of what is 'immediately' perceived, than 'mediately' perceived. What relevance does illusion have for these two forms of direct realism?

The fundamental premise of the arguments is from illusion seems to be the theses that things can appear to be other than they are. Thus, for example, straight sticks when immerged in water looks bent, a penny when viewed from certain perspectives appears as a spatial illusion, that is elliptic and sites of circularity, when something that is yellow when place under red fluorescent light looks red. In all of these cases, one version of the argument goes, it is implausible to maintain that what we are directly acquainted with is the real nature of the object in question. Indeed, it is hard to see how we can be said to be aware of the really physical object at all. In the above illusions the things we were aware of actually were bent, elliptical and red, respectively. But, by hypothesis, the really physical objects lacked these properties. Thus, we were not aware of the substantial reality of been real as physical objects or theory.

So far, if the argument is relevant to any of the direct realisms distinguished above, it seems relevant only to the claim that in all sense experience we are directly acquainted with parts or constituents of physical objects. After all, even if in illusion we are not acquainted with physical objects, but their surfaces, or their constituents, why should we conclude anything about the hidden nature of our relations to the physical world in veridical experience?

We are supposed to discover the answer to this question by noticing the similarities between illusory experience and veridical experience and by reflecting on what makes illusion possible at all. Illusion can occur because the nature of the illusory experience is determined, not just by the nature of the object perceived, but also by other conditions, both external and internal as becoming of an inner or as the outer experience. But all of our sensations are subject to these causal influences and it would be gratuitous and arbitrary to select from indefinitely of many and subtly different perceptual experiences some special ones those that get 'us' in touch with the 'real' nature of the physical world and its surrounding surfaces. Red fluorescent light affects the way thing's look, but so does sunlight. Water reflects light, but so does air. We have no unmediated access to the external world.

Still, why should we consider that we are aware of something other than a physical object in experience? Why should we not conclude that to be aware of a physical object is just to be appeared to by that object in a certain way? In its best-known form the adverbial theory of something proposes that the grammatical object of a statement attributing an experience to someone be analysed as an adverb.

For example,

(A) Rod is experiencing a coloured square.

Rewrote as?

Rod is experiencing, (coloured square)-ly

This is presented as an alternative to the act/object analysis, according to which the truth of a statement like (A) requires the existence of an object of experience corresponding to its grammatical object. A commitment to t he explicit adverbializations of statements of experience are not, however, essential to adverbialism. The core of the theory consists, rather, in the denial of objects of experience (as opposed ti objects of perception) coupled with the view that the role of the grammatical object in a statement of experience is to characterize more fully te sort of experience that is being attributed to the subject. The claim, then, is that the grammatical object is functioning as a modifier and, in particular, as a modifier of a verb, as a special kind of adverb at the semantic levels.

At this point, it might be profitable to move from considering the possibility of illusion to considering the possibility of hallucination. Instead of comparing paradigmatic veridical perception with illusion, let 'us' compare it with complete hallucination. For any experiences or sequence of experiences we take to be veridical, we can imagine qualitatively indistinguishable experiences occurring as part of a hallucination. For those who like their philosophical arguments spiced with a touch of science, we can imagine that our brains were surreptitiously removed in the night, and unbeknown to 'us' are being stimulated by a neurophysiologist so as to produce the very sensations that we would normally associate with a trip to the Grand Canyon. Currently permit 'us' into appealing of what we are aware of in this complete hallucination that is obvious that we are not awaken to the sparking awareness of physical objects, their surfaces, or their constituents. Nor can we even construe the experience as one of an object's appearing to 'us' in a certain way. It is after all a complete hallucination and the objects we take to exist before 'we' are simply not there. But if we compare hallucinatory experience with the qualitatively indistinguishable veridical experiences, should we most conclude that it would be 'special' to suppose that in veridical experience we are aware of something radically different from what we are aware of in hallucinatory experience? Again, it might help to reflect on our belief that the immediate cause of hallucinatory experience and veridical experience might be the very same brain event, and it is surely implausible to suppose that the effects of this same cause are radically different - acquaintance with physical objects in the case of veridical experience: Something else in the case of hallucinatory experience.

This version of the argument from hallucination would seem to address straightforwardly the ontological versions of direct realism. The argument is supposed to convince 'us' that the ontological analysis of sensation in both veridical and hallucinatory experience should give 'us' the same results, but in the hallucinatory case there is no plausible physical object, constituent of a physical object, or surface of a physical object with which additional premiss we would also get an argument against epistemological direct realism. That premiss is that in a vivid hallucinatory experience we might have precisely the same justification for believing (falsely) what we do about the physical world as way of permitting the analogy, similar in function but not to the structure of the analogy. Phenomenologically indistinguishable as a veridical experience, or the coinciding with future events or apparently unknowable present realities as we have in the veridical hallucination, as, perhaps, a verifiable account of the occurring event that interrupts normal procedure or precipitates a crisis, as the resulting occurrence upon that which is apt to happen. But our justification for believing that there is a table before 'we' in the course of a vivid hallucination of a table are surely not non-inferential in character. It certainly is not, if non-inferential justifications are supposedly to consist and yet an unproblematic access to the fact that makes true our belief - by hypothesis the table does not exist. But if the justification that hallucinatory experiences give 'us' the same as the justification we get from the parallel veridical experience, then we should not describe a veridical experience as giving 'us' non-inferential justification for believing in the existence of physical objects. In both cases we should say that we believe what we do about the physical world on the basis of what we know directly about the character of our experience.

In this brief space, I can only sketch some of the objections that might be raised against arguments from illusion and hallucination. That being said, let us begin with a criticism that accepts most of the presuppositions of the arguments. Even if the possibility of hallucination establishes that in some experience we are not acquainted with constituents of physical objects, it is not clear that it establishes that we are never acquainted with a constituent of physical objects. Suppose, for example, that we decide that in both veridical and hallucinatory experience we are acquainted with sense-data. At least some philosophers have tried to identify physical objects with 'bundles' of actual and possible sense-data.

To establish inductively that sensations are signs of physical objects, one would have to observe the correlation among the occurrences of certain sensations and the existence of certain physical objects that one is driven, but, to observe such a correlation for its binding of order, in that some collectives need of independence to physical objects, and, by the hypothetical stance that this cannot have of a connective, that it would need independent assesses to which is excisable toward physical objects and by the hypotheses, which this cannot have. If one is too further endeavour into the verificationist stance, that the ability to comprehend is interdependent to the ability to confirm that one is driven toward the Human conclusion:

Let us chance our imagination to the heavens, or to the utmost limits of the universe, we never really advance a step beyond ourselves, nor can conceivably any kind of existence, but those perceptions, which have appear d in that narrow compass. This is the universe of the imagination, nor have we have any idea, for decreasing of value (Hume, 1739-40).

If one reaches such a conclusion but wants to maintain the intelligibility and verifiability assertion about the physical world, one can gainfully convene upon the processing of idealistic or the phenomenalistic design or model that shares in an intellectual discipline.

However, hallucinatory experiences in this view are nonveridical, precisely because the sense-data of one's acquainted with hallucinations that do not bear the appropriate relations to other actual and possible sense-data. But if such a view where plausible one could agree that one is acquainted with the same kind of thing, for being in actual and nonveridical experience but insists upon the Sense of which an actual experience is one's own acquaintance that which represents of the world or active in it and often acting between different and differentiated experiences.

A different sort of objection to the argument from illusion or hallucination concerns its use in drawing conclusions we have not stressed in the above discourses. I, have in mentioning this objection, may to underscore an important feature of the argument. At least some philosophers (Hume, for example) have stressed the rejection of direct realism on the road to an argument for general scepticism with respect to the physical world. Once one abandons epistemological; direct realisms, one has an uphill battle indicating how one can legitimately make the inferences from sensation to physical objects. But philosophers who appeal to the existence of illusion and hallucination to develop an argument for scepticism can be accused of having an epistemically self-defeating argument. One could justifiably infer sceptical conclusions from the existence of illusion and hallucination only if one justifiably believed that such experiences exist, but if one is justified in believing that illusion exists, one must be justified in believing at least, some facts about the physical world (for example, that straight sticks look bent in water). The key point to stress in relying to such arguments is, that strictly speaking, the philosophers in question need only appeal to the 'possibility' of a vivid illusion and hallucination. Although it would have been psychologically more difficult to come up with arguments from illusion and hallucination if we did not believe that we actually had such experiences, I take it that most philosophers would argue that the possibility of such experiences is enough to establish difficulties with direct realism. Indeed, if one looks carefully at the argument from hallucination discussed earlier, one sees that it nowhere makes any claims about actual cases of hallucinatory experience.

Another reply to the attack on epistemological direct realism focuses on the implausibility of claiming that there is any process of 'inference' wrapped up in our beliefs about the world and its surrounding surfaces. Even if it is possible to give a Phenomenological descriptions are required of a special sort of skill that most people lack. Our perceptual beliefs about the physical world are surely direct, at least in the sense that they are unmediated by any sort of conscious inference from premisses describing something other than a physical object. The appropriate reply to this objection, however, is simply to acknowledge the relevant Phenomenological facts point out that from the perceptive of epistemologically direct realism, the philosopher by attacking a claim about the nature of our justification for believing propositions about the physical world, that such philosophers carry out of any comment at all about the causal genesis of such beliefs.

As mentioned, which proponents of the argument from illusion and hallucinations have often intended for it to establish the existence of sense-data, and many applicable philosophers have attacked the so-called sense-datum inference, presupposed in some statements of the argument. When the stick looked bent, the penny looked elliptical and the yellow object looked red, the sense-datum theorist wanted to infer that there was something bent, elliptical and red, respectively. But such an inference is surely suspect. Usually, we do not infer that because something appears to have a certain property, that affairs that affecting something that has that property. When in saying that Jones looks like a doctor, I surely would not want anyone to infer that there must actually be someone there who is a doctor. In assessing this objection, it will be important to distinguish different uses words like 'appears' and 'looks'. At least, sometimes to say that something looks 'F' way and the sense-datum inference from an F 'appearance' in this sense to an actual 'F' would be hopeless. However, it also seems that we use the 'appears'/'looks' terminology to describe the Phenomenological character of our experience and the inference might be more plausible when the terms are used this way. Still, it does seem that the arguments from illusion and hallucination will not by themselves constitute strong evidence for sense-datum theory. Even if one concludes that there is something common to both hallucinatory and veridical visual experience of a red thing, one needs not to describe of a common constituent as the awareness of something red. The adverbial theorist would prefer to construe the common experiential state as 'being appeared too redly', a technical description intended only to convey the idea that the state in question need not be analysed as relational in character. Those who opt for an adverbial theory of sensation need to make good the claim that their artificial adverbs can be given a sense that is not parasitic upon an understanding of the adjectives transformed into verbs. Still, other philosophers might try to reduce the common element in veridical and nonveridical experience to some kind of intentional state. More like belief or judgement. The idea here is that the only thing common to the two experiences is the fact that in both are spontaneously taken to be present as an object of a certain kind.

The selfsame objections can be started within the general framework presupposed by proponents of the arguments from illusion and hallucination. A great many contemporary philosophers, however, uncomfortable with the intelligibility of the concepts needed to make sense of the theories attacked even. Thus, at least, some who object to the argument from illusion do so not because they defend direct realism. Rather they think there is something confused about all this talk of direct awareness or acquaintance. Belonging to the same time for the condition or quality for being external or externalized that external of an incidental condition as affecting a course of action, existing independently of the mind of or relating chiefly to an outward appearance such that of some external affairs as part or the surface. From which the outer circumstance appearing as far as from the externalist's housing the phenomena of illusion and hallucination, this would be relevant to claim that on, at least some occasions our judgements about the physical world are reliably produced by processes that do not take as their input beliefs about something else.

The expressions 'knowledge by acquaintance' and 'knowledge by description', and the distinction they mark between knowing 'things' and knowing 'about' things, are now generally associated with Bertrand Russell. However, John Grote and Hermann von Helmholtz had earlier and independently had exhibited the same distinctions, and William James (1842-1910) as founder of pragmatism and functionalism had developed an approach to intellectual issues, that, to a greater extent, influenced or affected a basis which produced by factors of change, that James had influenced the American ways of thought, but soon adopted Grote's terminology in his investigation of the distinction. Philosophers have perennially investigated this and related distinctions using to vary terminological phrasing for that which was distinctly by noting that natural language 'distinguished between these two applications of the notion of knowledge that being of the Greek - nosene, Kennen, connaître, the others being 'Wissen', 'savoir' (Grote, 1865). On Grote's account, the distinction is a matter of degree, for which there are of three positions of a standing reputation as presented in a dimension of variability - epistemic, causal, semantic.

We know things by experiencing them, and knowledge of acquaintance (Russell changed the preposition to 'by') is epistemically priori to and has a relatively higher degree of epistemic justification than knowledge about things. Indeed, sensation has 'the one great value of trueness or freedom from mistake' (1900).

A thought (using that term broadly, to mean any mental state) constituting knowledge of acquaintance with a thing is more or less causally proximate to sensations caused by that thing, while a thought constituting knowledge about the thing is more or less distant causally, being separated from the thing and experience of it by processes of attention and inference. At the limit, if a thought is maximally of the acquaintance type, it is the first mental state occurring in a perceptual causal chain originating in the object to which the thought refers, i.e., it is a sensation. The thing's presented to 'us' in sensation and of which we have knowledge of acquaintance include ordinary objects in the external world, such as the sun.

Grote contrasted the imaginistic thoughts involved in knowledge of acquaintance with things, with the judgements involved in knowledge about things, suggesting that the latter but not the former are mentally contentual by a specified state of affairs. Elsewhere, however, he suggested that every thought capable of constituting knowledge of or about a thing involves a form, idea, or what we might call contentual propositional content, referring the thought to its object. Whether contentual or not, thoughts constituting the knowledge of acquaintance, with which a thing is being relatively indistinct, although this indistinctness does not imply incommunicably, on the other hand, thoughts constituting distinctly, as a result of 'the application of notice or attention' to the 'confusion or chaos' of sensation (1900). Grote did not have an explicit theory on reference, the relation by which a thought is 'of' or 'about' a specific thing. Nor did he explain how thoughts can be more or less indistinct.

Helmholtz held unequivocally that all thoughts capable of constituting knowledge, whether 'knowledge that has to do with Notions' (Wissen) or 'mere familiarity with phenomena' (Kennen), is judgmental or, we may say, having conceptual propositional contents. Where Grote saw a difference between distinct and indistinct thoughts, Helmholtz found a difference between precise judgements that are expressible in words and equally precise judgements that, in principle, are not expressible in words, and so are not communicable (Helmholtz, (1962). As happened, James was influenced by Helmholtz and, especially, by Grote (James, 1975). Taken on the latter's terminology, James agreed with Grote that the distinction between knowledge of acquaintance with things and knowledge about things involves a difference in the degree of vagueness or distinctness of thoughts, though he, too, said little to explain how such differences are possible. At one extreme is knowledge of acquaintance with people and things, and with sensations of colour, flavour, spatial extension, temporal duration, effort and perceptible difference, unaccompanied by knowledge about these things. Such pure knowledge of acquaintance is vague and inexplicit. Movement away from this extreme, by a process of notice and analysis, yields a spectrum of less vague, more explicit thoughts constituting knowledge about things.

All the same, the distinction was not merely a relative one for James, as he was more explicit than Grote in not imputing content to every thought capable of constituting knowledge of or about things. At the extreme where a thought constitutes pure knowledge of acquaintance with a thing, there is a complete absence of conceptual propositional content in the thought, which is a sensation, feeling or precept, of which he renders the thought incommunicable. James' reasons for positing an absolute discontinuity in between pure cognition and preferable knowledge of acquaintance and knowledge at all about things seem to have been that any theory adequate to the facts about reference must allow that some reference is not conventionally mediated, that conceptually unmediated reference is necessary if there are to be judgements at all about things and, especially, if there are to be judgements about relations between things, and that any theory faithful to the common person's 'sense of life' must allow that some things are directly perceived.

James made a genuine advance over Grote and Helmholtz by analysing the reference relation holding between a thought and of him to specific things of or about which it is knowledge. In fact, he gave two different analyses. On both analyses, a thought constituting knowledge about a thing refers to and is knowledge about 'a reality, whenever it actually or potentially ends in' a thought constituting knowledge of acquaintance with that thing (1975). The two analyses differ in their treatments of knowledge of acquaintance. On James's first analysis, reference in both sorts of knowledge is mediated by causal chains. A thought constituting pure knowledge of acquaintances with a thing refers to and is knowledge of 'whatever reality it directly or indirectly operates on and resembles' (1975). The concepts of a thought 'operating on' a thing or 'terminating in' another thought are causal, but where Grote found teleology and final causes. On James's later analysis, the reference involved in knowledge of acquaintance with a thing is direct. A thought constituting knowledge of acquaintance with a thing either is that thing, or has that thing as a constituent, and the thing and the experience of it is identical (1975, 1976).

James further agreed with Grote that pure knowledge of acquaintance with things, i.e., sensory experience, is epistemologically priori to knowledge about things. While the epistemic justification involved in knowledge about things rests on the foundation of sensation, all thoughts about things are fallible and their justification is augmented by their mutual coherence. James was unclear about the precise epistemic status of knowledge of acquaintance. At times, thoughts constituting pure knowledge of acquaintance are said to posses 'absolute veritableness' (1890) and 'the maximal conceivable truth' (1975), suggesting that such thoughts are genuinely cognitive and that they provide an infallible epistemic foundation. At other times, such thoughts are said not to bear truth-values, suggesting that 'knowledge' of acquaintance is not genuine knowledge at all, but only a non-cognitive necessary condition of genuine knowledge, knowledge about things (1976). Russell understood James to hold the latter view.

Russell agreed with Grote and James on the following points: First, knowing things involves experiencing them. Second, knowledge of things by acquaintance is epistemically basic and provides an infallible epistemic foundation for knowledge about things. (Like James, Russell vacillated about the epistemic status of knowledge by acquaintance, and it eventually was replaced at the epistemic foundation by the concept of noticing.) Third, knowledge about things is more articulate and explicit than knowledge by acquaintance with things. Fourth, knowledge about things is causally removed from knowledge of things by acquaintance, by processes of reelection, analysis and inference (1911, 1913, 1959).

But, Russell also held that the term 'experience' must not be used uncritically in philosophy, on account of the 'vague, fluctuating and ambiguous' meaning of the term in its ordinary use. The precise concept found by Russell 'in the nucleus of this uncertain patch of meaning' is that of direct occurrent experience of a thing, and he used the term 'acquaintance' to express this relation, though he used that term technically, and not with all its ordinary meaning (1913). Nor did he undertake to give a constitutive analysis of the relation of acquaintance, though he allowed that it may not be unanalysable, and did characterize it as a generic concept. If the use of the term 'experience' is restricted to expressing the determinate core of the concept it ordinarily expresses, then we do not experience ordinary objects in the external world, as we commonly think and as Grote and James held we do. In fact, Russell held, one can be acquainted only with one's sense-data, i.e., particular colours, sounds, and so forth, one's occurrent mental states, universals, logical forms, and perhaps, oneself.

Russell agreed with James that knowledge of things by acquaintance 'is essentially simpler than any knowledge of truths, and logically independent of knowledge of truths' (1912, 1929). The mental states involved when one is acquainted with things do not have propositional contents. Russell's reasons here seem to have been similar to James's. Conceptually unmediated reference to particulars necessary for understanding any proposition mentioning a particular, e.g., 1918-19, and, if scepticism about the external world is to be avoided, some particulars must be directly perceived (1911). Russell vacillated about whether or not the absence of propositional content renders knowledge by acquaintance incommunicable.

Russell agreed with James that different accounts should be given of reference as it occurs in knowledge by acquaintance and in knowledge about things, and that in the former case, reference is direct. But Russell objected on a number of grounds to James's causal account of the indirect reference involved in knowledge about things. Russell gave a descriptional rather than a causal analysis of that sort of reference: A thought is about a thing when the content of the thought involves a definite description uniquely satisfied by the thing referred to. Indeed, he preferred to speak of knowledge of things by description, rather than knowledge about things.

Russell advanced beyond Grote and James by explaining how thoughts can be more or less articulate and explicit. If one is acquainted with a complex thing without being aware of or acquainted with its complexity, the knowledge one has by acquaintance with that thing is vague and inexplicit. Reflection and analysis can lead one to distinguish constituent parts of the object of acquaintance and to obtain progressively more comprehensible, explicit, and complete knowledge about it (1913, 1918-19, 1950, 1959).

Apparent facts to be explained about the distinction between knowing things and knowing about things are there. Knowledge about things is essentially propositional knowledge, where the mental states involved refer to specific things. This propositional knowledge can be more or less comprehensive, can be justified inferentially and on the basis of experience, and can be communicated. Knowing things, on the other hand, involves experience of things. This experiential knowledge provides an epistemic basis for knowledge about things, and in some sense is difficult or impossible to communicate, perhaps because it is more or less vague.

If one is unconvinced by James and Russell's reasons for holding that experience of and reference work to things that are at least sometimes direct. It may seem preferable to join Helmholtz in asserting that knowing things and knowing about things both involve propositional attitudes. To do so would at least allow one the advantages of unified accounts of the nature of knowledge (propositional knowledge would be fundamental) and of the nature of reference: Indirect reference would be the only kind. The two kinds of knowledge might yet be importantly different if the mental states involved have different sorts of causal origins in the thinker's cognitive faculties, involve different sorts of propositional attitudes, and differ in other constitutive respects relevant to the relative vagueness and communicability of the mental states.

In any of cases, are greater in number, in amount, extent or degree has of a basis or to forming the superlative degree as most foundationalists, especially its basis on which is a view concerning the 'structure' of the system of justified belief possessed by a given individual. Such a system is divided into 'foundation' and 'superstructure', finding to a basis for perceptual support of an institution as an endowment founded and supported by and some related beliefs in the latter depend on the former for their justification but not vice versa. The act of endowment, particularly to imagine having a usually favourable trait or quality endowed within the provisions of some intensively donated on an institution, individual or group as a source or natural ability or quality. However, the view is sometimes stated in terms of the structure of 'knowledge' than of justified belief. If knowledge is true justified belief (plus, perhaps, some further condition), one may think of knowledge as exhibiting a foundationalist for exerting a strong personality are grounded to a structured, as having a well-defined structure or organization, highly organized to a structural environment, such as having a limited number of correct or newly corrected answers, by virtue of the justified belief it involves. In any event, the construing doctrine concerning the primary justification is layed the groundwork as affording the efforts of belief, though in feeling more free, we are to acknowledge the knowledgeable infractions that will from time to time be worthy in showing to its recognition.

The first step toward a more explicit statement of the position is to distinguish between 'mediate' (indirect) and 'immediate' (direct) justification of belief. To say that a belief is mediately justified is to any that it s justified by some appropriate relation to other justified beliefs, i.e., by being inferred from other justified beliefs that provide adequate support for it, or, alternatively, by being based on adequate reasons. Thus, if my reason for supposing that you are depressed is that you look listless, speak in an oddly flat tone of voice, exhibit no interest in things you are usually interested in, and so forth, then my belief that you are depressed is justified, if, at all, by being adequately supported by my justified belief that you look listless, speak in a flat tone of voice . . . A belief is immediately justified, on the other hand, if its justification is of another sort, e.g., if it is justified by being based on experience or if it is 'self-justified'. Thus my belief that you look listless may not be based on anything else I am justified in believing but just on the cay you look to me. And my belief that $2 + 3 = 5$ may be justified not because I infer it from something else, I justifiably believe, but simply because it seems obviously true to me.

In these terms we can put the thesis of Foundationalism by saying that all mediately justified beliefs owe their justification, ultimately to immediately justified beliefs. To get a more detailed idea of what this amounts to it will be useful to consider the most important argument for Foundationalism, the regress argument. Consider a mediately justified belief that 'p' (we are using lowercase letters as dummies for belief contents). It is, by hypothesis, justified by its relation to one or more other justified beliefs, 'q' and 'r'. Now what justifies each of these, e.g., q? If it too is mediately justified that is because it is related accordingly to one or subsequent extra justified beliefs, e.g., 's'. By virtue of what is 's' justified? If it is mediately justified, the same problem arises at the next stage. To avoid both circularity and an infinite regress, we are forced to suppose that in tracing back this chain we arrive at one or more immediately justified beliefs that stop the regress, since their justification does not depend on any further justified belief.

According to the infinite regress argument for Foundationalism, if every justified belief could be justified only by inferring it from some further justified belief, there would have to be an infinite regress of justifications: Because there can be no such regress, there must be justified beliefs that are not justified by appeal to some further justified belief. Instead, they are non-inferentially or immediately justified, they are basic or foundational, the ground on which all our other justifiable beliefs are to rest.

Variants of this ancient argument have persuaded and continue to persuade many philosophers that the structure of epistemic justification must be foundational. Aristotle recognized that if we are to have knowledge of the conclusion of an argument in the basis of its premises, we must know the premises. But if knowledge of a premise always required knowledge of some further proposition, then in order to know the premise we would have to know each proposition in an infinite regress of propositions. Since this is impossible, there must be some propositions that are known, but not by demonstration from further propositions: There must be basic, non-demonstrable knowledge, which grounds the rest of our knowledge. For which some acquitment finds of interest in it, but the inspiring fact that they have also been advanced on behalf of scepticism, relativism, fideisms, conceptualism and Coherentism. Sceptics, by given to doubt or questioning attitudes or state of mind who developed sets of arguments to show either that no knowledge is possible or that there is not sufficient or adequate evidence to tell if any knowledge is possible. Those who instinctively or habitually doubts, question or disagree with assertions or generally consider in the acceptance of non-specified conclusions, as inclining inclinations is that of the sceptic attitude as relating to or characterized of sceptic criticism. The ancient school of Pyrrho of Elis that stresses the uncertainty of our beliefs, in order to oppose dogmatism, from which the doctrine that absolute knowledge is impossible, either as a particular domain with the aim of acquiring approximate or relative certainty, away as far from doubtful suspicions and disbelief or disagreement, with assertions or generally accepted conclusions as one is inclined to be the sceptic. In whatever manner or extending of our concerns, along the considerations of becoming better, that on that point of which Foundationalists agree, of which non-infinite regressing of justification is the revision to an earlier or less mature pattern of feeling or behaviours that retrogressive relapses are inevitable, but to a less perfect pattern or behaviour.

By reversion, one is to contain by an inferential process of deriving logical conclusions from premises known or assumed to be true, as reasoning from factual knowledge or evidence, for something inferred of or relating to, or involving inference of that which can be derived or capable of being derived by inference. Nonetheless, of a contrivance or an invention as serving a particular purpose, especially in using to a major extent a relativistic task, which brings into a certain state about the act for which of a device that leaves to invent of a belief. In having been of an epistemological interest, in its propositional guise to believe many things that one does not actually believe, or which to one's lacking the appropriate psychological attitude to relevant propositional content, seem as non-occurrent beliefs permitting that we do not cease to believe that $2 + 2 = 4$, for instance. It is merely because we now happen to be thinking of something else or nothing at all. This generalization concerns belief alone, but there are also generalizations linking belief and desire. Now these generalizations categorize beliefs and desires according to the logical form of their object. They do require that the objects have logical forms. But the primary possessors of logical form are sentences, from which the immediate objects of beliefs and desires are themselves sentences. These facts are easily explained if sentences have a syntactic and semantic structure, yet these facts are a complete mystery. What is true for a sentence to be true also for thoughts, of which thinking thoughts involve manipulating mental images or objects in the likeness of something as presented or productions reprehended through mental activities or realistic classifications reprehended as representations that usually or when able to do so? In that mental representations with a propositional content have a semantic and syntactic structure like that of sentences, it is no accident that one who is able to think that John hits Mary, is thereby also to think that Mary hits John.

Oddly enough, the symptoms of schizophrenia, unlike those of anxiety and depression, is not a part of normal functioning, . . . selection against the genes that cause schizophrenia is strong enough that they should be less common if their presence were due simply to mutation balanced by selection, . . . The most likely possibility is that these genes are advantageous in combination with certain other genes, or in certain environments, much in the way a sickle-cell gene is advantageous even though having two such genes causes sickle-cells anemia. Or it might be that the genes that predispose to schizophrenia have other effects that offer a slight advantage in most people, . . .

But sceptics think all true justification must be inferential in this way -the Foundationalist's speak of the immediate justification as only a mere overshadowing for its requiring of any rational justification that is appropriately convened. Sceptics conclude that none of our beliefs is justified. Relativists follow essentially the same pattern of sceptical argument, concluding that our beliefs can only be justified relative to the arbitrary starting assumptions or presuppositions either of an individual or of a form of life.

Regress arguments are not limited to epistemology. In ethics there is Aristotle's regress argument (in 'Nichomachean Ethics') for the existence of a single end of rational action. In metaphysics there is Aquinas's regress argument for an unmoved mover: If a mover that it is in motion, there would have to be an infinite sequence of movers each moved by a further mover, since there can be no such sequence, there is an unmoved mover. A related argument has recently been given to show that not every state of affairs can have an explanation or cause of the sort posited by principles of sufficient reason, and such principles are false, for reasons having to do with their own concepts of explanation (Post, 1980; Post, 1987).

The premise of which in presenting Foundationalism as a view concerning the structure 'that is in fact exhibited' by the justified beliefs of a particular person has sometimes been construed in ways that deviate from each of the phrases that are contained in the previous sentence. Thus, it is sometimes taken to characterise the structure of 'our knowledge' or 'scientific knowledge', rather than the structure of the cognitive system of an individual subject. As for the other phrase, Foundationalism is sometimes thought of as concerned with how knowledge (justified belief) is acquired or built up, than with the structure of what a person finds herself with at a certain point. Thus some people think of scientific inquiry as starting with the recordings of observations (immediately justified observational beliefs), and then inductively inferring generalizations. Again, Foundationalism is sometimes thought of not as a description of the finished product or of the mode of acquisition, but rather as a proposal for how the system could be reconstructed, an indication of how it could all be built up from immediately justified foundations. This last would seem to be the kind of Foundationalism we find in Descartes. However, Foundationalism is most usually thought of in contemporary Anglo-American epistemology as an account of the structure actually exhibited by an individual's system of justified belief.

It should also be noted that the term is used with a deplorable looseness in contemporary, literary circles, even in certain corners of the philosophical world, to refer to anything from realism -the view that reality has a definite constitution regardless of how we think of it or what we believe about it to various kinds of 'absolutism' in ethics, politics, or wherever, and even to the truism that truth is stable (if a proposition is true, it stays true).

Since Foundationalism holds that all mediate justification rests on immediately justified beliefs, we may divide variations in forms of the view into those that have to do with the immediacy for justified beliefs. The 'foundations', and those that have to do with the modes of derivation of other beliefs from these, how the 'superstructure' is built up, yet the most obvious variation of the first sort has to do with what modes of immediate justification are recognized. Many treatments, both pro and con, are parochially restricted to one form of immediate justification self-evidence, self-justification (self-warrant), justification by a direct awareness of what the belief is about, or whatever. It is then unwarrantly assumed by critics that disposing of that one form will dispose of Foundationalism generally (Alston, 1989, ch. 3). The emphasis historically has been on beliefs that simply 'record' what is directly given in experience (Lewis, 1946) and on self-evident propositions (Descartes' 'clear and distinct perceptions and Locke's 'Perception of the agreement and disagreement of ideas'). But self-warrant has also recently received a great deal of attention (Alston 1989), and there is also a

Reliabilist version according to which a belief can be immediately justified just by being acquired by a reliable belief-forming process that does not take other beliefs as inputs (BonJour, 1985).

Foundationalism also differs as to what further constraints, if any, is put on foundations. Historically, it has been common to require of the foundations of knowledge that they exhibit certain 'epistemic immunities', as we might put it, immunity from error, refutation or doubt. Thus Descartes, along with many other seventeenth and eighteenth-century philosophers, took it that any knowledge worthy of the name would be based on cognations the truth of which is guaranteed (infallible), that were maximally stable, immune from ever being shown to be mistaken, as incorrigible, and concerning which no reasonable doubt could be raised (indubitable). Hence the search in the 'Meditations' for a divine guarantee of our faculty of rational intuition, of which criticisms of Foundationalism have often been directed at these constraints: Lehrer, 1974, Will, 1974? Both responded to in Alston, 1989. It is important to realize that a position that is foundationalist in a distinctive sense can be formulated without imposing any such requirements on foundations.

There are various ways of distinguishing types of foundationalist fields of thought, its presupposition, and its extent and validity to understand and place deterministically. By use of the variations we have been enumerating. Plantinga (1983), has put forwards an influential innovation of criterial Foundationalism, specified in terms of limitations on the foundations. He construes this as a disjunction of 'ancient and medieval Foundationalism', which takes foundations to comprise what is self-evidently and 'evident to he senses', and 'modern Foundationalism 'that replaces' the evidential senses with 'incorrigible', which in practice was taken to apply only to beliefs about one's present state of consciousness. Plantinga himself developed this notion in the context of arguing those items outside this territory, in particular certain beliefs about God, could also be immediately justified. A popular recent distinction is between what is variously called 'strong' or 'extreme' Foundationalism and 'moderate', 'modest' or 'minimal' Foundationalism, with the distinction depending on whether various epistemic immunities are required of foundations. Finally, its distinction is 'simple' and 'iterative' Foundationalism (Alston, 1989), depending on whether it is required of a foundation only that it is immediately justified, or whether it is also required that the higher level belief that the firmer belief is immediately justified is itself immediately justified, such that it seems to be plausibility stronger in a requirement, stemming from specific 'levels' between beliefs on different oriented 'levels.

A particular belief is justified to the extent that it is integrated into a coherent system of beliefs. Moreover, a pragmatist like John Dewey has developed a position known as contextualism, which avoids ascribing any overall structure to knowledge. Questions concerning justification can only arise in particular context, defined in terms of assumptions that are simply taken for granted, though they can be questioned in other contexts, where other assumptions will be privileged.

Foundationalism can be attacked both in its commitment to immediate justification and in its claim that all mediately justified beliefs ultimately depend on the former. Though, the latter are the position's weakest point, most of the critical fire has been detected to the former. As pointed out about much of this criticism has been directly against some particular form of immediate justification, ignoring the possibility of other forms. Thus, much anti-foundationalist artillery has been directed at the 'myth of the given'. The idea that facts or things are 'given' to consciousness in a pre-conceptual, pre-judgmental mode, and that beliefs can be justified on that basis (Sellars, 1963) in at least, the most prominent general argument against immediate justification is a 'level ascent' argument, according to which whatever is taken ti immediately justified a belief that the putative justifier has in supposing to do so. Hence, since the justification of the higher level belief after all (BonJour, 1985), that we lack adequate support for any such higher level requirements for justification, and if it were imposed we would be launched on an infinite undergo regress, for a similar requirement would hold equally for the higher level belief that the original justifier was efficacious.

Coherence is a major player in the theatre of knowledge. There are coherence theories of belief, truth, and justification. These combine in various ways to yield theories of knowledge. We will proceed from belief through justification to truth. Coherence theories of belief are concerned with the content of beliefs. Consider a belief you now have, the beliefs that you are reading a page in a book, so what makes that belief the belief that it is? What makes it the belief that you are reading a page in a book than the belief hat you have a monster in the garden?

One answer is that the belief has a coherent place or role in a system of beliefs. Perception has an influence on belief. You respond to sensory stimuli by believing that you are reading a page in a book rather than believing that you have a centaur in the garden. Belief has an influence on action. You will act differently if you believe that you are reading a page than if you believe something about a centaur. Perspicacity and action undermine the content of belief, however, the same stimuli may produce various beliefs and various beliefs may produce the same action. The role that gives the belief the content it has in the role it plays in a network of relations to the beliefs, the role in inference and implications, for example, I refer different things from believing that I am inferring different things from believing that I am reading a page in a book than from any other beliefs, just as I infer that belief from any other belief, just as I infer that belief from different things than I infer other belief's form.

The input of perception and the output of an action supplement the centre role of systematic relations is the belief has to other beliefs, but the systematic relations that give the belief the specific content it has. They are the fundamental source of the content of beliefs. That is how coherence comes in. A belief has the content that it does because of the way in which it coheres within a system of beliefs (Rosenberg, 1988). We might distinguish weak coherence theories of the content of beliefs from strong coherence theories. Weak coherence theories affirm that coherences are one-determinant of the content of belief. Strong coherence theories of the contents of belief affirm that coherence is the sole determinant of the content of belief.

When we turn from belief to justification, we are in confronting a corresponding group of similarities fashioned by their coherences motifs. What makes one belief justified and another not? The answer is the way it coheres with the background system of beliefs. Again, there is a distinction between weak and strong theories of coherence. Weak theories tell 'us' that the way in which a belief coheres with a background system of beliefs is one determinant of justification, other typical determinants being perception, memory and intuition. Strong theories, by contrast, tell 'us' that justification is solely a matter of how a belief coheres with a system of beliefs. There is, however, another distinction that cuts across the distinction between weak and strong coherence theories of justification. It is the distinction between positive and negative coherence theories (Pollock, 1986). A positive coherence theory tells 'us' that if a belief coheres with a background system of belief, then the belief is justified. A negative coherence theory tells 'us' that if a belief fails to cohere with a background system of beliefs, then the belief is not justified. We might put this by saying that, according to a positive coherence theory, coherence has the power to produce justification, while according to a negative coherence theory, coherence has only the power to nullify justification.

A strong coherence theory of justification is a combination of a positive and a negative theory that tells 'us' that a belief is justified if and only if it coheres with a background system of beliefs.

Traditionally, belief has been of epistemological interest in its propositional guise: 'S' believes that 'p', where 'p' is a proposition toward which an agent, 'S' exhibits an attitude of acceptance. Not all belief is of this sort. If I trust what you say, I believe you. And someone may believe in Mrs. Thatcher, or in a free-market economy, or in God. It is sometimes supposed that all belief is 'reducible' to propositional belief, belief-that. Thus, my believing you might be thought a matter of my believing, perhaps, that what you say is true, and your belief in free-markets or in God, a matter of your believing that free-market economy's are desirable or that God exists.

It is doubtful that non-propositional believing can in every instance be reduced in this way, however, rhetorical discourse concerning this point has tended upon its course, that of an apparent distinction between 'belief-that' and 'belief-in', the applicable distinction toward the belief in God. Some philosophers have followed Aquinas (1225-74), in supposing that to believe in, and God is simply to believe that certain truths hold: That God exists, that he is benevolent, good, and so forth. Others (e.g., Hick, 1957) argue that belief-in is a distinctive attitude, one that includes essentially an element of trust. More commonly, belief-in has been taken to involve a combination of propositional belief together with some further attitude.

H.H. Price (1969) defends the claims that there are different sorts of 'belief-in', some, but not all, reducible to 'beliefs-that'. If you believe in God, you believe that God exists, that God is good, and so forth, but, according to Price, your belief involves, in addition, a certain complex pro-attitude toward its object. One might attempt to analyse this further attitude in terms of additional beliefs-that: 'S' believes in 'X' just in case (1) 'S' believes that ',' exists, perhaps hold further factual beliefs about (X): (2)'S' believes that ',' is good or valuable in some respect, and (3) 'S' believes that X's being good or valuable in this respect is itself a good thing. An analysis of this sort, however, fails adequately to capture the further affective component of belief-in. Thus, according to Price, if you believe in God, your belief is not merely that certain truths hold as a commandment, but for what it posses, as a valuable addition as an attitude of commitment and trust as worthful of its truths as given only to God.

Notoriously, 'belief-in' outruns the evidence for the corresponding 'belief-that'. Does this diminish its rationality? If belief-in presupposes belief-that, it might be thought that the evidential standards for the former must be, at least as high as standards for the latter. And any additional pro-attitude might be thought to require a further layer of justification not required for cases of belief-that.

Some philosophers have argued that, at least for cases in which belief-in is synonymous with faith (or faith-in), evidential thresholds for constituent propositional beliefs are diminished. You may reasonably have faith in God or Mrs. Brown, even though beliefs about their respective attitudes, were you to harbour them, would be evidentially substandard.

Belief-in may be, in general, less susceptible to alternations in the face of unfavourable evidence than belief-that. A believer who encounters evidence against God's existence may remain unshaken in his belief, in part because the evidence does not bear on his pro-attitude. So long as this is united with his belief that God exists, the belief may survive epistemic buffeting-and reasonably so in a way that an ordinary propositional belief-that would not.

At least two large sets of questions are properly treated under the heading of epistemological religious beliefs. First, there is a set of broadly theological questions about the relationship between faith and reason, between what one knows by way of reason, broadly construed, and what one knows by way of faith. These theological questions may as we call theological, because, of course, one will find them of interest only if one thinks that in fact there is such a thing as faith, and that we do know something by way of it. Secondly, there is a whole set of questions having to do with whether and to what degree religious beliefs have warrant, or justification, or positive epistemic status. The second, is seemingly as an important set of a theological question is yet spoken of faith.

Epistemology, so we are told, is theory of knowledge: Its aim is to discern and explain that quality or quantity enough of which distinguishes knowledge from mere true belief. We need a name for this quality or quantity, whatever precisely it is, call it 'warrant'. From this point of view, the epistemology of religious belief should centre on the question whether religious belief has warranted too how much it has and how it gets it. As a matter of fact, the epistemological discussion of religious belief, at least since the Enlightenment (and in the Western world, especially the English-speaking Western world) has tended to focus, not on the question whether religious belief has warrant, but whether it is justified. More precisely, it has tended to focus on the question whether of those properties are enjoyed by theistic belief -the belief that there exists a person like the God of traditional Christianity, Judaism and Islam: An almighty Law Maker, or and all-knowing and most wholly benevolent and a loving spiritual person who has created the living world. The chief question, therefore, has been whether theistic belief is justified, the same question is often put by asking whether theistic belief is rational or rationally acceptable. Still further, the typical way of addressing this question has been by way of discussing arguments for or and against the existence of God. On the pro side, there are the traditional theistic proofs or arguments: The ontological, cosmological and teleological arguments, using Kant's terms for them. On the other side, the anti-theistic side, the principal argument is the argument from evil, the argument that is not possible or at least probable that there be such a person as God, given all the pain, suffering and evil the world displays. This argument is flanked by subsidiary arguments, such as the claim that the very concept of God is incoherent, because, for example, it is impossible that there are the people without a body, and Freudian and Marxist claims that religious belief arises out of a sort of magnification and projection into the heavens of human attributes we think important.

But why has discussion centred on justification rather than warrant? And precisely what is justification? And why has the discussion of justification of theistic belief focussed so heavily on arguments for and against the existence of God?

As to the first question, we can see why once we see that the dominant epistemological tradition in modern Western philosophy has tended to 'identify' warrant with justification. On this way of looking at the matter, warrant, that which distinguishes knowledge from mere true belief, just 'is' justification. Belief theory of knowledge-the theory according to which knowledge is justified true belief has enjoyed the status of orthodoxy. According to this view, knowledge is justified truer belief, therefore any of your beliefs have warrant for you if and only if you are justified in holding it.

But what is justification? What is it to be justified in holding a belief? To get a proper sense of the answer, we must turn to those twin towers of western epistemology figured by René Descartes and especially, John Locke. The first thing to see is that according to Descartes and Locke, there are epistemic or intellectual duties, or obligations, or requirements. In this way, Locke has said:

Faith is nothing but a firm assent of the mind, which if it is regulated, as is our duty, but cannot be afforded to anything, but upon good reason that cannot be opposite to it, he that believes, without having any reason for believing, may be in love with his own fanciers: But, seeks neither truth as he should, yet the obedience is due of his maker, which would have him use those discerning faculties he has given him: To keep him out of mistake and error. He that does this to the best of his power, however, he sometimes lights on truth, is in the right but by chance: And I know not whether the luckiest of the accidents will excuse the irregularity of his proceeding. This, for certain, that he must be accountable for whatever mistakes he encounters: Whereas, he that makes use of the light and faculties God has given him, by seeks sincerely to discover truth, by those helps and abilities he has, may have this satisfaction in doing his duty as rational creature, that though he should mistake its truth, but he will not endeavour upon the reward of it. Engaging of the governing ascension and its right and places, as it should be, who, in any case or matter whatsoever, believes or disbelieves as accorded by reasons that direct him. Nonetheless, that he that does otherwise, transgresses against his own light, and misuses those faculties, which were given him . . . (Essays 4.17.24).

Rational creatures, creatures with reason, creatures capable of believing propositions (and of disbelieving and being agnostic with respect to them), say Locke, have duties and obligation with respect to the regulation of their belief or assent. Now the central core of the notion of justification(as the etymology of the term indicates) this: One is justified in doing something or in believing a certain way, if in doing one is innocent of wrong doing and hence not properly subject to blame. You are justified, therefore, if you have violated no duties or obligations, if you have conformed to the relevant requirements, if you are within your rights. To be justified in believing something, then, is to be within your rights in so believing, to be flouting no duty, to be to satisfy your epistemic duties and obligations. This way of thinking of justification has been the dominant way of thinking about justification: And this way of thinking has many important contemporary representatives. Roderick Chisholm, for example (as distinguished an epistemologist as the twentieth century can boast), in his earlier work explicitly explains justification in terms of epistemic duty (Chisholm, 1977).

The (or, a) main epistemological questions about religious belief are, therefore, having to beg the question of whether or not is religious belief, in general and theistic belief in particular is justified. And the traditional way to answer that question has been to inquire into the arguments for and against theism, however, this emphasis as placed upon these arguments, is a way of marshalling your propositional evidence-the evidence from other such propositions as those of believing or against a given proposition. And the reason for the emphasis upon argument is the assumption that theistic belief is justified if and only if there is sufficient propositional evidence for it. If there is not' much by way of propositional evidence for theism, then you are not justified in accepting it. Moreover, if you accept theistic belief without having propositional evidence for it, then you are ging contrary to epistemic duty and are therefore unjustified in accepting it. Thus, W.K. William James, trumpets that 'it is wrong, always everything upon insufficient evidence', his is only the most strident in a vast chorus of only insisting that there is an intellectual duty not to believe in God unless you have propositional evidence for that belief: As of others: Sigmund Freud, Brand Blandhard, H.H. Price, Bertrand Russell and Michael Scriven.

Now how it is that the justification of theistic belief gets identified with they're being propositional evidence for it? Justification is a matter of being blameless, of having done one's duty (in this context, one's epistemic duty): What, precisely, has this to do with having propositional evidence?

The answer is to be found of Descartes, especially in Locke. As, justification is the property your beliefs have when, in forming and holding them, you conform to your epistemic duties and obligations. But according to Locke, a central epistemic duty is this: To believe a proposition only to the degree that it is probable with respect to what are certain for you. What proposition are certain for you? According to Descartes and Locke, propositions about your own immediate experience, for example, that you have a mild headache, or that it seems that you think of that you believe that you see something red: Also, propositions that are self-evident for you, necessarily true propositions so obvious that you cannot so much as entertain them without seeing that they must be true. (Examples would be simple arithmetical and logical propositions, together with such propositions as that the whole is at least as large as the parts, that red is a colour, and that whatever exists has properties.) Propositions of these two sorts are certain for you, as fort other prepositions. You are justified in believing if and only if when one and only to the degree to which it is probable with respect to what are certain for you. According to Locke, therefore, and according to the whole modern foundationalist tradition initiated by Locke and Descartes (a tradition that until has recently dominated Western thinking about these topics) that to deny in the acceptation of a position unless it is certain or probable with respect to what is beyond doubt or questionably disputable, enabling with such capacity based upon the certainty with which the fact, quality, or state of being for something that is clearly established or assured in having by confidence in oneself which assuredly and certain in that every effect must have a cause. These of the identifiable are yet to be known or for being objectively indirect objects for having verifiable existence among nonspecified objects in the world, regardless of subjectivity or convention of thought or language. Being or occurring in fact or actuality is not to be taken lightly, which of indicating that which is necessarily the degree indicated by a measure for being an ordinary objective statement. The indicative mood for which any of the considerations are taken into account in that they bear in mind of others, if you consider the feelings of others, upon another proposition that depends upon the occurrence of an uncertain future event, that which the event is itself for serving to the deficiencies that are made by the completing of additional work.

In the present context, therefore, the central Lockean assumption is that there is an epistemic duty not to accept theistic belief unless it is probable with respect to what are most certain for you: Locke 1632-1704, began within the structural hierarchy of principles applying to empiricism and his, 'Two Treatises on Government' (1690) that influenced the Declaration of Independence. Nonetheless, as a consequence, theistic belief is justified only if the existence of God is probable with respect to what are certain. Locke does not argue for his proposition, he simply announces it, and epistemological discussion of theistic belief has for the most part followed hin ion making this assumption. This enables 'us' to see why epistemological discussion of theistic belief has tended to focus on the arguments for and against theism: On the view in question, theistic belief is justified only if it is probable with respect to what are certain, and the way to show that it is probable with respect to what it is certain are to give arguments for it from premises that are certain or, are sufficiently probable with respect to what are certain.

There are at least three important problems with this approach to the epistemology of theistic belief. First, their standards for theistic arguments have traditionally been set absurdly high (and perhaps, part of the responsibility for this must be laid as the door of some who have offered these arguments and claimed that they constitute wholly demonstrative proofs). The idea seems to test. a good theistic argument must start from what is self-evident and proceed majestically by way of self-evidently valid argument forms to its conclusion. It is no wonder that few if any theistic arguments meet that lofty standard -particularly, in view of the fact that almost no philosophical arguments of any sort meet it. (Think of your favourite philosophical argument: Does it really start from premisses that are self-evident and move by ways of self-evident argument forms to its conclusion?)

Secondly, attentions have been confined to three theistic arguments. The traditional arguments, cosmological and teleological arguments, but in fact, there are many more good arguments: Arguments from the nature of proper function, and from the nature of propositions, numbers and sets. These are arguments from intentionality, from counterfactual, from the confluence of epistemic reliability with epistemic justification, from reference, simplicity, intuition and love. There are arguments from colours and flavours, from miracles, play and enjoyment, morality, from beauty and from the meaning of life. This is even a theistic argument from the existence of evil.

But there are a third and deeper problems here. The basic assumption is that theistic belief is justified only if it is or can be shown to be probable, respective to many of the evidential propositions - perhaps, those that are self-evident or about one's own mental life, but is this assumption true? The idea is that theistic belief is very much like a scientific hypothesis: It is acceptable if and only if there is an appropriate balance of propositional evidence in favour of it. But why believe a thing like that? Perhaps the theory of relativity or the theory of evolution is like that, such a theory has been devised to explain the phenomena and gets all its warrant from its success in so doing. However, other beliefs, e.g., memory beliefs, feel-felt in other minds is not like that, they are not hypothetical at all, and are not accepted because of their explanatory powers. There are instead, the propositions from which one start in attempting to give evidence for a hypothesis. Now, why assume that theistic belief, belief in God, is in this regard more like a scientific hypothesis than like, say, a memory belief? Why think that the justification of theistic belief depends upon the evidential relation of theistic belief to other things one believes? According to Locke and the beginnings of this tradition, it is because there is a duty not to assent to a proposition unless it is probable with respect to what are certain to you, but is there really any such duty? No one has succeeded in showing that, say, belief in other minds or the belief that there has been a past, is probable with respect to what are certain for 'us'. Suppose it is not: Does it follow that you are living in epistemic sin if you believe that there are other minds - or a past?

There are urgent questions about any view according to which one has duties of the sort 'do not believe 'p' unless it is probable with respect to what are certain for you; . First, if this is a duty, is it one to which I can conform? My beliefs are for the most part not within my control: Certainly they are not within my direct control. I believe that there has been a past and that there are other people, even if these beliefs are not probable with respect to what are certain forms (and even if I came to know this) I could not give them up. Whether or not I accept such beliefs are not really up to me at all, For I cannot be any more of the abstaining from believing these things than I can from conforming to the law of gravity. Second, is there really any reason for thinking I have such a duty? Nearly everyone recognizes such duties as that of not engaging in gratuitous cruelty, taking care of one's children and one's aged parents, and the like, but do we also find ourselves recognizing that there is a duty not to believe what is not probable (or, what we cannot see to be probable) with respect to what are certain for 'us'? It hardly seems so. However, it is hard to see why being justified in believing in God requires that the existence of God be probable with respect to some such body of evidence as the set of propositions certain for you. Perhaps, theistic belief is properly basic, i.e., such that one is perfectly justified in accepting it on the evidential basis of other propositions one believes.

Taking justification in that original etymological fashion, therefore, there is every reason ton doubt that one is justified in holding theistic belief only inf one is justified in holding theistic belief only if one has evidence for it. Of course, the term 'justification' has undergone various analogical extensions in the various philosophies, for it has been used to name various properties that are in different forms as justified etymologically, but anagogically related to it. In such a way, the term sometimes used to mean propositional evidence: To say that a belief is justified for someone is to saying that he has propositional evidence (or sufficient propositional evidence) for it. So taken, however, the question whether theistic belief is justified loses some of its interest; for it is not clear (given this use)beliefs that are unjustified in that sense. Perhaps, one also does not have propositional evidence for one's memory beliefs, if so, that would not be a mark against them and would not suggest that there be something wrong holding them.

Another analogically connected way to think about justification (a way to think about justification by the later Chisholm) is to think of it as simply a relation of fitting between a given proposition and one's epistemic vase -which includes the other things one believes, as well as one's experience. Perhaps tat is the way justification is to be thought of, but then, if it is no longer at all obvious that theistic belief has this property of justification if it seems as a probability with respect to many another body of evidence. Perhaps, again, it is like memory beliefs in this regard.

Of which the dominant Western tradition has been inclined to identify warrant with justification, it has been inclined to take the latter in terms of duty and the fulfilment of obligation, and hence to suppose that there is no epistemic duty not to believe in God unless you have good propositional evidence for the existence of God. Epistemological discussion of theistic belief, as a consequence, as concentrated on the propositional evidence for and against theistic belief, i.e., on arguments for and against theistic belief. But there is excellent reason to doubt that there are epistemic duties of the sort the tradition appeals to here.

And perhaps it was a mistake to identify warrant with justification in the first place. Napoleons have little warrant for him: His problem, however, need not be dereliction of epistemic duty. He is in difficulty, but it is not or necessarily that of failing to fulfill epistemic duty. He may be doing his epistemic best, but he may be doing his epistemic duty in excelsis: But his madness prevents his beliefs from having much by way of warrant. His lack of warrant is not a matter of being unjustified, i.e., failing to fulfill epistemic duty. So warrant and being epistemologically justified by name are not the same things. Another example, suppose (to use the favourite twentieth-century variant of Descartes' evil demon example) I have been captured by Alpha-Centaurian super-scientists, running a cognitive experiment, they remove my brain, and keep it alive in some artificial nutrients, and by virtue of their advanced technology induce in me the beliefs I might otherwise have if I was going about my usual business. Then my beliefs would not have much by way of warrant, but would it be because I was failing to do my epistemic duty? Hardly.

As a result of these and other problems, another, externalist way of thinking about knowledge has appeared in recent epistemology, that a theory of justification is internalized if and only if it requires that all of its factors needed for a belief is to be epistemically accessible to that of a person. Internal to his cognitive perception, and externalisms, if it allows that, at least some of the justifying factors need not be thus accessible. Such that, they can be external to the believer's cognitive perspectives, however, epistemologists oftentimes use the distinctions between internalized and externalist theories of epistemic justification without offering any very explicit explanation. Or perhaps the thing to say, is that, it has reappeared. The dominant strains in epistemology prior to the Enlightenment were really externalist. According to this externalist way of thinking, warrant does not depend upon satisfaction of duty, or upon anything else to which the Knower has special cognitive access (as he does to what is about his own experience and to whether he is trying his best to do his epistemic duty): It depends instead upon factors 'external' to the epistemic agent - such factors as whether his beliefs are produced by reliable cognitive mechanisms, or whether they are produced by epistemic faculties functioning properly in-an appropriate epistemic environment.

How will we think about the epistemology of theistic belief in more than is less of an externalist way (which is at once both satisfyingly traditional and agreeably up to date)? I think, that the ontological question whether there is such a person as God is in a way priori to the epistemological question about the warrant of theistic belief. It is natural to think that if in fact we have been created by God, then the cognitive processes that issue in belief in God are indeed realisable belief-producing processes, and if in fact God created 'us', then no doubt the cognitive faculties that produce belief in God is functioning properly in an epistemologically congenial environment. On the other hand, if there is no such person as God, if theistic belief is an illusion of some sort, then things are much less clear.

Then a belief in God of the most basic of ways is the wishing doubtless of such unrealistic thinking or another cognitive process. Thus, it will have little or no warranted belief in God, on the basis of argument would be like a belief in a false philosophical theory that is based on the argument: Do such beliefs have warrant? Notwithstanding, the custom of discussing the epistemological questions about theistic belief as if they could be profitably discussed independently of the ontological issue as to whether or not theism is true, is misguided. There two issues are intimately intertwined,

Nonetheless, the vacancy left, as today and as days before are an awakening and untold story beginning by some sparking conscious paradigm left by science. That is a central idea by virtue accredited by its epistemology, where in fact, is that justification and knowledge arising from the proper functioning of our intellectual virtues ∙r faculties in an appropriate environment. This particular yet, peculiar idea is captured in the following criterion for justified belief:

> (J) 'S' is justified in believing that 'p' if and only if of S's believing that 'p' is the result
> of S's intellectual virtues or faculties functioning in appropriate environment.

What is an intellectual virtue or faculty? A virtue or faculty in general is a power or ability or competence to achieve some result. An intellectual virtue or faculty, in the sense intended above, is a power or ability or competence to arrive at truths in a particular field, and to avoid believing falsehoods in that field. Examples of human intellectual virtues are sight, hearing, introspection, memory, deduction and induction. More precisely:

> (V) A mechanism 'M' for generating and/or maintaining beliefs is an intellectual virtue
> if and only if 'M''s' is a competence to believing true propositions and refrain from
> false believing propositions within a field of propositions 'F', when one is in a set of
> circumstances 'C'.

It is required that we specify a particular field of suggestions or its propositional field for 'M', since a given cognitive mechanism will be a competence for believing some kind of truths but not others. The faculty of sight, for example, allows 'us' to determine the colour of objects, but not the sounds that they associatively make. It is also required that we specify a set of circumstances for 'M', since a given cognitive mechanism will be a competence in some circumstances but not others. For example, the faculty of sight allows 'us' to determine colours in a well lighten room, but not in a darkened cave or formidable abyss.

According to the aforementioned formulations, what makes a cognitive mechanism an intellectual virtue is that it is reliable in generating true beliefs than false beliefs in the relevant field and in the relevant circumstances. It is correct to say, therefore, that merits of virtue epistemology are a kind of Reliabilist. Whereas, genetic Reliabilist maintains that justified belief is belief that results from a reliable cognitive process, virtue epistemology makes a restriction on the kind of process which is allowed, namely, the cognitive processes that are important for justification and that of knowledge, that those that have their basis in an intellectual virtue.

Finally, that the concerning mental faculty reliability point to the importance of an appropriate environment, is such that cognitive mechanisms might be reliable in some environments but not in others. Consider an example from Alvin Plantinga. On a planet revolving around Alfa Centauri, cats are invisible to human beings. Moreover, Alfa Centaurian cats emit a type of radiation that causes humans to form the belief that there I a dog barking nearby. Suppose now that you are transported to this Alfa Centaurian planet, a cat walks by, and you form the belief that there is a dog barking nearby. Surely you are not justified in believing this. However, the problem here is not with your intellectual faculties, but with your environment. Although your faculties of perception are reliable on earth, yet are unrealisable on the Alga Centaurian planet, which is an inappropriate environment for those faculties.

The central idea of virtue epistemology, as expressed in (J) above, has a high degree of initial plausibility. By masking the idea of faculties' cental to the reliability if not by the virtue of epistemology, in that it explains quite neatly to why beliefs are caused by perception and memories are often justified, while beliefs caused by unrealistic and superstition are not. Secondly, the theory gives 'us' a basis for answering certain kinds of scepticism. Specifically, we may agree that if we were brains in a vat, or victims of a Cartesian demon, then we would not have knowledge even in those rare cases where our beliefs turned out true. But virtue epistemology explains that what is important for knowledge is toast our faculties are in fact reliable in the environment in which we are. And so we do have knowledge so long as we are in fact, not victims of a Cartesian demon, or brains in a vat. Finally, Plantinga argues that virtue epistemology deals well with Gettier problems. The idea is that Gettier problems give 'us' cases of justified belief that is 'truer by accident'. Virtue epistemology, Plantinga argues, helps 'us' to understand what it for a belief to be true by accident, and provides a basis for saying why such cases are not knowledge. Beliefs are rue by accident when they are caused by otherwise reliable faculties functioning in an inappropriate environment. Plantinga develops this line of reasoning in Plantinga (1988).

The Humean problem if induction supposes that there is some property 'A' pertaining to an observational or experimental situation, and that of 'A', some fraction m/n (possibly equal to 1) has also been instances of some logically independent property 'B'. Suppose further that the background circumstances, have been varied to a substantial degree and also that there is no use or collateral information as made available concerning the frequency of 'B's' among 'A's' or have to do with causal nomological connections between instances of 'A' and instances of 'B'.

In this situation, an enumerative or instantial inductive inference would move the premise that m/n of observed 'A's' are 'B's' to the conclusion that approximately m/n of all 'A's' and 'B's'. (The usual probability qualification will be assumed to apply to the inference, than being part of the conclusion). Hereabouts the class of 'A's' should be taken to include not only unobservable 'A's' of future 'A's', it seems that the possibility or the speculative measure, is that of a positive effort toward the distributing contribution for 'A'. In that of or related by the complementary relationship among the correlative words or expressions of a sentence that are not used adjacently to each other. Such that in agreement or equivalent by character or conformity, as it is similarly the communication by which the exchange of the spoken words has been accomplished. However, corresponding conclusions would concern the probability or likelihood of the very next observed 'A' being a 'B'.

The traditional or Humean problem of induction, often refereed to simply as 'the problem of induction', is the problem of whether and why inferences that fit this schema should be considered rationally acceptable or justified from an epistemic or cognitive standpoint, i.e., whether and why reasoning in this way is likely lead to true claims about the world. Is there any sort of argument or rationale that can be offered for thinking that conclusions reached in this way are likely to be true if the corresponding premiss is true or even that their chances of truth are significantly enhanced?

Hume's discussion of this deals explicitly with cases where all observed 'A's' are 'B's', but his argument applies just as well to the more general casse. His conclusion is entirely negative and sceptical: inductive inferences are not rationally justified, but are instead the result of an essentially a-rational process, custom or habit. Hume challenges the proponent of induction to supply a cogent line of reasoning that leads from an inductive premise to the corresponding conclusion and offers an extremely influential argument in the form of a dilemma, to show that there can be no such reasoning. Such reasoning would argue, as having to be either a priori demonstrative reasoning concerning relations of ideas or 'experimental', i.e., empirical, reasoning concerning mattes of fact to existence. It cannot be the former, because all demonstrative reasoning relies on the avoidance of contradiction, and it is not a contradiction to suppose that 'the course of nature may change', tat an order that was observed in the past does not or will not continue in the future. Also, it cannot be the arrow of times future, since any empirical argument would appeal to the success of such reasoning in previous experience, and the justifiability for generalizing from previous experiences that ought be precisely of what is at issue-so as, any such appeal would be question-begging, so then, there can be no such reasoning.

An alternative version of the problem may be obtained by formulating it with reference to the so-called Principle of Induction, which says roughly that the future will resemble the last or, that unobserved cases will resemble detected cases. An inductive argument may be viewed as enthymematic, with this principle serving as a suppressed premiss, in which case the issue is obviously how such a premise can be justified. Hume's argument is then that no such justification is possible: The principle cannot be justified a priori, it is not contradictory to deny it: it cannot be justified by appealingness to have been true, and having been true in pervious experiences without obviously begging the question.

The predominant recent responses to the problem of induction, at least in the analytic tradition, in effect accept the main conclusion of Hume's argument, viz. That inductive inferences cannot be justified, in the sense of showing that the conclusion of such an inference is likely to be truer if the premise is true, and thus attempt to find some other sort of justification for induction.

Bearing upon, and if not taken to account for the term 'induction' is most widely used for any process of reasoning that takes 'u' from empirical premises to empirical conclusions support by the premise, but not deductively entailed by them. Inductive arguments are therefore kinds of applicative argument, in which something beyond the content of the premises is inferred as probable or supported by them. Induction is, however, commonly distinguished from arguments to theoretical explanations, which share this implicative character, by being confined to inference in which the conclusion involves the same properties or relations as the premises. The central example is induction by simple enumeration, where from premises as discerning that Fa, Fb, Fc, in which, a, b, c's, are all of the same, more or less, presented as the kinds in 'G's', it is inferred that 'G's' from an outside sample, such as the future 'G's' will be 'F', or perhaps other people deceive them, children may well infer that everyone is a deceiver. Different but similar inferences are those from the past possession of a property by some object to the same object's future possession, or from the constancy of some law-like pattern in events, and states of affairs to its future constancy: All objects in that of what we know, are attracted by one another, each with the force that is inversely proportional to the square of the distance between them, so that, they all do so, and always will do so.

The rational basis of any inference was challenged by David Hume (1711-76), who believed that induction of nature, and merely reflected a habit or custom of the mind. Hume was not therefore sceptical about the propriety of processes of induction, but sceptical about the tole of reason in either explaining it or justifying it. Trying to answer Hume and to show that there is something rationally compelling about the inference is referred to as the problem of induction. It is widely recognized that any rational defence of induction will have to partition well-behaved properties for which the inference is plausible (often called projectable properties) from badly behaved ones for which t is not. It is also recognized that actual inductive habits are more complex than those of simple and science pay attention to such factors as variations within the sample of giving 'us' the evidence, the application of ancillary beliefs about the order of nature, and so on. Nevertheless, the fundamental problem remains that any experience shows 'us' only events occurring within a very restricted part of the vast spatial temporal order about which we then come to believe things.

All the same, the classical problem of induction is often phrased in terms of finding some reason to expect that nature is uniform. In Fact, Fiction and Forecast (1954) Goodman showed that we need, in addition of some reason for preferring some uniformities to others, for without such a selection the uniformity of nature is vacuous. Thus, suppose that all examined emeralds have been green. Uniformity would lead 'us' to expect that future emeralds will be green as well. But now we define a predicate as 'grue', 'χ' is gruing if and only if 'χ' is examined before time 'T' and is green, or 'χ' is examined after 'T' and is blue? Let 'T' refers to some time around the present. Then if newly examined emeralds are like previous ones in respect of being grue, they will be blue. We prefer blueness as the basis of prediction to gluiness, but why?

Goodman argued that although his new predicate appears to be gerrymandered, and involves a reference to a difference, as the language-relative judgements, there being no language-independent standard of similarity to which to appeal. Other philosophers have not been convinced by this degree of linguistic relativism. What remains clear that the possibility of these 'bent' predicates put a decisive obstacle in face of purely logical and syntactical approaches to problems of 'confirmation?'.

Nevertheless, in the potential of change we are to think up to the present time but although virtue epistemology has good initial plausibility, we are faced apart by some substantial objections. The first of an objection, which virtue epistemology face is a version of the generality problem. We may understand the problem more clearly if we were to consider the following criterion for justified belief, which results from our explanation of (J):

(J) 'S' is justified in believing that 'p' if and entirely if . . .

(1) There are in the field 'F' and a set of circumstances 'C' such that:

(1) 'S' is in 'C' with respect to the proposition that 'p'.

(2) 'S' is in 'C' with respect to the proposition that 'p'.

(3) If 'S' were in 'C' with respect to a proposition in 'F'.

Then 'S' would very likely believe correctly with regard to some specified designations.

The problem arises in how we are to select an appropriate 'F' and 'C'. For given any true belief that 'p', we can always come up with a field 'F' and a set of circumstances 'C', such that 'S' is perfectly reliable in 'F' and 'C'. For any true belief that 'p', let 'F's' be the field including only the propositions of 'p' and 'not-p', let 'C' include whatever circumstances there are which cause 'p's' to be true, together with the circumstance which causes 'S' to believe that 'p'. Clearly, 'S' is perfectly reliable with respect to propositions in this field in these circumstances. That is to say, we do not want to say that all of S's, are true beliefs are justified for 'S'. And, of course, there is an analogous problem in the other direction of generality. For given any belief that 'p', we can always specify a field of propositions 'F' and a set of circumstances 'C', such that 'p' is in 'F', 'S' is in 'C', and 'S' is not reliable with respect to propositions in 'F' in 'C'.

Variations of this view have been advanced for both knowledge and justified belief. The first formulation of a reliability account of knowing appeared in a note by F.P. Ramsey (1931), who said that a belief was knowledge if it is true, certain and obtained by a reliable process. P. Unger (1968) suggested that 'S' knows that 'p' just in case it is not at all accidental that 'S' is right about its being the case that 'p'. D.M. Armstrong (1973) drew an analogy between a thermometer that reliably indicates the temperature and a belief that reliably indicate the truth. Armstrong said that a non-inferential belief qualified as knowledge if the belief has properties that are nominally sufficient for its truth, i.e., guarantee its truth via laws of nature.

Closely allied to the nomic sufficiency account of knowledge, primarily due to F.I. Dretske (19712, 1981), A.I. Goldman (1976, 1986) and R. Nozick (1981). The core of tis approach is that S's belief that 'p' qualifies as knowledge just in case 'S' believes 'p' because of reasons that would not obtain unless 'p's' being true, or because of a process or method that would not yield belief in 'p' if 'p' were not true. For example, 'S' would not have his current reasons for believing there is a telephone before him, or would not come to believe this, unless there was a telephone before him. Thus, there is a counterfactual reliable guarantor of the belief's being true. A variant of the counterfactual approach says that 'S' knows that 'p' only if there is no 'relevant alterative' situation in which 'p' is false but 'S' would still believe that 'p'.

To arrive at a better understanding, this interpretation is to mean that the alterative attempt to accommodate any of an opposing strand in our thinking about knowledge one interpretation is an absolute concept, which is to mean that the justification or evidence one must have in order to know a proposition 'p' must be sufficiently able to eliminate all of the alternate oppositions too 'p' (where an alternative to a proposition 'p' is a proposition incompatible with 'p'). That is, one's justification or evidence for 'p' must be sufficient fort one to know that every alternative too 'p' is false. These elements of our thinking about knowledge are exploited by sceptical argument. These arguments call our attention to alternatives that our evidence cannot eliminate. For example, (Dretske, 1970), when we are at the zoo, we might claim to know that we see a zebra, because we have in warranted, as based on our belief that certain visual evidence, namely a zebra-like appearance. The sceptic inquires how we know that we are not seeing a clearly disguised mule. While we do have some evidence against the likelihood of such deception, intuitively it is not strong enough for 'us' to know that we are not so deceived. But, pointing out alternatives of this nature that we cannot eliminate, as well as others with more general application (dreams, hallucinations, and so forth), the sceptic appears to show that this requirement that our evidence eliminate every alternative is seldom, if ever, met.

The above considerations show that virtue epistemology must say more about the selection of relevant fields and sets of circumstances. Set up addresses, the general problem by introducing the concept of a design plan for our intellectual faculties. Relevant specifications for fields and sets of circumstances are determined by this plan. One might object that this approach requires the problematic assumption of a Designer of the design plan. But Plantinga disagrees on two counts: He does not think that the assumption is needed, or that it would be problematic. Plantinga discusses relevant material in Plantinga (1986, 1987 and 1988). Ernest Sosa addresses the generality problem by introducing the concept of an epistemic perspective. In order to have reflective knowledge, 'S' must have a true grasp of the reliability of her faculties, this grasp being itself provided by a 'faculty of faculties'. Relevant specifications of an 'F' and 'C' are determined by this perspective. Alternatively, Sosa has suggested that relevant specifications are determined by the purposes of the epistemic community. The idea is that fields and sets of circumstances are determined by their place in useful generalizations about epistemic agents and their abilities to act as reliable-information sharers.

Sosa addresses the current problem by arguing that justification is relative to an environment 'E'. Accordingly, 'S' is justified in believing that 'p' relative to 'E', if and only if S's faculties would be reliable in 'E'. Note that on this account, 'S' need not actually be in 'E' in order for 'S' to be justified in believing some proposition relative to 'E'. This allows Soda to conclude that Jane has justified belief in the above case. For Jane is justified in her perceptual beliefs relative to our environment, although she is not justified in those beliefs relative to the environment in which they have actualized her.

We have earlier made mention about analyticity, but the true story of analyticity is surprising in many ways. Contrary to received opinion, it was the empiricist Locke rather than the rationalist Kant who had the better information account of this type or deductive proposition. As Frége and Rudolf Carnap (1891-1970) a German logician positivist whose first major works was 'Der Logische Aufbau der Welt' (1926, trs, as 'The Logical Structure of the World,' 1967). Carnap pursued the enterprise of clarifying the structures of mathematics and scientific language (the only legitimate task for scientific philosophy) in 'Logische Syntax der Sprache' (1934, trs., 'The Logical Syntax of Language,' 1937). While refinements continued with 'Meaning and Necessity' (1947), and a general losing of the original ideal of reduction culminated in the great 'Logical Foundations of Probability' as the most importantly single work of 'confirmation theory' in 1950. Other works concern the structure of physics and the concept of entropy.

As both, Frége and Carnap, represented as analyticity's best friends in this century, did as much to undermine its worst enemies. Quine (1908-) whose early work on mathematical logic, and issued in 'A System of Logistic' (1934), 'Mathematical Logic' (1940) and 'Methods of Logic' (1950) such that, with this collection of papers a 'Logical Point of View' (1953) that his philosophical importance became widely recognized, also, Putman (1926-) his concern in the later period has largely been to deny any serious asymmetry between truth and knowledge as it is obtained in natural science, and as it is obtained in morals and even theology. Books include 'Philosophy of logic' (1971), 'Representation and Reality' (1988) and 'Renewing Philosophy (1992). Collections of his papers include 'Mathematics, Master, sand Method' (1975), 'Mind, Language, and Reality' (1975), and 'Realism and Reason (1983). Both of which represented as having refuted the analytic/synthetic distinction, not only did no such thing, but, in fact, contributed significantly to undoing the damage done by Frége and Carnap. Finally, the epistemological significance of the distinctions is nothing like what it is commonly taken to be.

Locke's account of an analyticity proposition as, for its time, everything that a succinct account of analyticity should be (Locke, 1924) he distinguished two kinds of analytic propositions, identified propositions in which we affirm the said terms if itself, e.g., 'Roses are roses', and predicative propositions in which 'a part of the complex idea is predicated of the name of the whole', e.g., 'Roses are flowers'. Locke calls such sentences 'trifling' because a speaker who uses them 'trifles with words'. A synthetic sentence, in contrast, such as a mathematical theorem, states 'a truth and conveyance, especially for which that carries or its transportation as to serve or bear by what carries with its informative real knowledge', but, correspondingly, Locke distinguishes two kinds of 'necessary consequences', analytic entailment where validity depends on the literal containment of the conclusions in the premiss and synthetic entailments where it does not. (Locke did not originate this concept-containment notion of analyticity. It is discussion by Arnaud and Nicole, and it is safe to say it has been around for a very long time (Arnaud, 1964).

Kant's account of analyticity, which received opinion tells 'us' is the consummate formulation of this notion in modern philosophy, is actually a step backward. What is valid in his account is not novel, and what is novel is not valid. Kant presents Locke's account of concept-containment analyticity, but bring into a certain state, as to feature the most important of his characterizations of analytic propositions as propositions whose denials are logical contradictions (Kant, 1783). This characterization suggests that analytic propositions based on Locke's part-whole relation or Kant's explicative copula is a species of logical truth. But the containment of the predicate concept in the subject concept in sentences like 'Bachelors are unmarried' is a different relation from containment of the consequent in the antecedent in a sentence like 'If John is a bachelor, then John is a bachelor or Mary read Kant's Critique'. The former is literal containment whereas, the latter are, in general, not. Talk of the 'containment' of the consequent of a logical truth in the metaphorical, a way of saying 'logically derivable'.

Kant's conflation of concept containment with logical containment caused him to overlook the issue of whether logical truths are synthetically deductive and the problem of how he can say mathematical truths are synthetically deductive when they cannot be denied without contradiction. But historically, the conflation set the stage for the disappearance of the Lockean notion. As Frége, whose received opinion to depicting or presenting of a portrayable presentation for being second as only to Kant, in that among the champions of analyticity, included Carnap, whom it portrays as just behind Frége, that was jointly responsible for the appearance of concept-containment analyticity.

Frége was clear about the difference between concept containment and logical containment, expressing it as like the difference between the containment of 'beams in a house' the containment of a 'plant in the seed' (Frége, 1853). But he found the former, as Kant formulated it, defective in three ways: It explains analyticity in psychological terms, it does not cover all cases of analytic propositions, and, perhaps, most important for Frége's logicism, its notion of containment is 'unfruitful' as a definition; mechanisms in logic and mathematics (Frége, 1853), were contained among the two notions of containment, Frége observes that with logical containment 'we are not simply talking out of the box again what we have just put inti it'. This definition makes logical containment the basic notion. Analyticity becomes a special case of logical truth, and, even in this special case, the definitions employ the power of definition in logic and mathematics than mere concept combination.

Carnap, attempting to overcome what he saw a shortcoming in Frége's account of analyticity, took the remaining step necessary to do away explicitly with Lockean-Kantian analyticity. As Carnap saw things, it was a shortcoming of Frége's explanation that it seems to suggest that definitional relations underlie the analytic propositions can be extra-logic in some sense, say, in resting on linguistic synonymy. To Carnap this represented a failure to achieve a uniform formal treatment of analytic propositions and left 'us' with a dubious distinction between logical and extra-logical vocabulary. Hence, he eliminated the reference to definitions in Frége's explanation of analyticity by introducing 'meaning postulates', e.g., statements such as (χ) ($'\chi'\sim$ is a bachelor and is unmarried) (Carnap, 1965). Corresponding standards of logical postulates on which they were modelled, meaning postulates express nothing more than constrains on the admissible models with respect to which sentences and deductions are evaluated for truth and validity. Despite their name, its asymptomatic-balance has to pustulate itself by that of holding on to but not more than to do with meaning than any value-adding statements expressing a graduation of indispensable truths, as widely recognized or employed as a model of a standard reference, in such that its actions through which an overall impression or mood as intended to be communicated, especially by man if meaning that symbolize and may consist of a single utterance, remark, or comment. In defining analytic propositions as consequences of (an explained set of) logical laws, Carnap explicitly removed the one place in Frége's explanation where there might be room for concept containment and with it, the last trace of Locke's distinction between semantic and other 'necessary consequences'.

Quine, born 1908, an American analytic philosopher and logician whose major writings including, Word and Object (1960), concerning issues of language and meaning, He was a staunchest critic of analyticity of our time, performed an invaluable service on its behalf-although, one that has come almost completely unappreciated. Quine made two devastating criticism of Carnap's meaning postulate approach that expose it as both irrelevant and vacuous. It is irrelevant because, in using particular words of a language, meaning postulates fail to explicate analyticity for sentences and languages generally, that is, they do not give a definition as for the variables 'S' and 'L' (Quine, 1953). It is vacuous because, although meaning postulates tell 'us' what sentences are to count as analytic, they do not tell 'us' what it is for them to be analytic. Briefly, Rudolf Carnap (1981-1970) a German-born American philosopher whose anti-metaphysical views, set forth in such works as The

Logical Structure of the World (1928) and The Logical Foundations of Probability (1950), was centred on the development of logical positivism.

Received opinion that Quine did much more than refute the analytic/synthetic distinction as Carnap tried to draw it. Received opinion has that Quine demonstrated there is no distinction, however, anyone might try to draw it. Nut this, too, is incorrect. To argue for this stronger conclusion, Quine had to show that there is no way to draw the distinction outside logic, in particular theory in linguistic corresponding to Carnap's, Quine's argument had to take an entirely different form. Some inherent feature of linguistics had to be exploited in showing that no theory in this science can deliver the distinction. But the feature Quine chose was a principle of operationalist methodology characteristic of the school of Bloomfieldian linguistics. Quine succeeds in showing that meaning cannot be made objective sense of in linguistics. If making sense of a linguistic concept requires, as that school claims, operationally defining it in terms of substitution procedures that employ only concepts unrelated to that linguistic concept. But Chomsky's revolution in linguistics replaced the Bloomfieldian taxonomic model of grammars with the hypothetico-deductive model of generative linguistics, and, as a consequence, such operational definition was removed as the standard for concepts in linguistics. The standard of theoretical definition that replaced it was far more liberal, allowing the members of as family of linguistic concepts to be defied with respect to one another within a set of axioms that state their systematic interconnections -the entire system being judged by whether its consequences are confirmed by the linguistic facts. Quine's argument does not even address theories of meaning based on this hypothetico-deductive model (Katz, 1988 and 1990).

Putman, the other staunch critic of analyticity, performed a service on behalf of analyticity fully on a par with, and complementary to Quine's, whereas, Quine refuted Carnap's formalization of Frége's conception of analyticity, Putman refuted this very conception itself. Putman put an end to the entire attempt, initiated by Fridge and completed by Carnap, to construe analyticity as a logical concept (Putman, 1962, 1970, 1975a).

However, as with Quine, received opinion has it that Putman did much more, that Putman is credited with having devised science fiction cases, from a robot cat situation to the twin earth cases, which are counterexamples to the traditional theory of meaning, that, again, is the received opinion for being incorrect. These cases are only counter examples to Frége's version of the traditional theory of meaning. Frége's version claims both (1) that sense determines reference, and (2) that there are instances of analyticity, say, typified by 'cats are animals', and of synonymy, say typified by 'water' in English and 'water' in twin earth English. Given (1) and (2), what we call 'cats' could not be non-animals and what we call 'water' could not differ from what the earthier twin called 'water'. But, as Putman's cases show, what we call 'cats' could be Martian robots and what they call 'water' could be something other than H2O Hence, the cases are counter examples to Frége's version of the theory.

Putman himself takes these examples to refute the traditional theory of meaning per se, because he thinks other versions must also subscribe to both (1) (2). He was mistaken in the case of (1). Frége's theory entail (1) because it defines the sense of an expression as the mode of determination of its referent (Fridge, 1952, pp. 56-78). But sense does not have to be defined this way, or in any way that entails (1). It can be defined as (D).

(D) Sense is that aspect of the grammatical structure of expressions and sentences responsible for they're having sense properties and relations like meaningfulness, ambiguity, antonymy, synonymy, redundancy, analyticity and analytic entailment (Katz, 1972 & 1990).

(Note that this use of sense properties and relations is no more circular than the use of logical properties and relations to define logical form, for example, as that aspect of grammatical structure of sentences on which their logical implications depend.)

(D) makes senses internal to the grammar of a language and reference an external; matter of language use -typically involving extra-linguistic beliefs, Therefore, (D) cuts the strong connection between sense and reference expressed in (1), so that there is no inference from the modal fact that 'cats' refer to robots to the conclusion that 'Cats are animals' are not analytic. Likewise, there is no inference from 'water' referring to different substances on earth and twin earth to the conclusion that our word and theirs are not synonymous. Putman's science fiction cases do not apply to a version of the traditional theory of meaning based on (D).

The success of Putman and Quine's criticism in application to Fridge and Carnap's theory of meaning together with their failure in application to a theory in linguistics based on (D) creates the option of overcoming the shortcomings of the Lockean-Kantian notion of analyticity without switching to a logical notion. This option was explored in the 1960s and 1970s in the course of developing a theory of meaning modelled on the hypothetico-deductive paradigm for grammars introduced in the Chomskyan revolution (Katz, 1972).

This theory automatically avoids Frége's criticism of the psychological formulation of Kant's definition because, as an explication of a grammatical notion within linguistics, it is stated as a formal account of the structure of expressions and sentences. The theory also avoids Frége's criticism that concept-containment analyticity is not 'fruitful' enough to encompass truths of logic and mathematics. The criticism rests on the dubious assumption, parts of Frége's logicism, that analyticity 'should' encompass them, (Beenacerraf, 1981). But in linguistics where the only concern is the scientific truth about natural concept-containment analyticity encompass truths of logic and mathematics. Moreover, since we are seeking the scientific truth about trifling propositions in natural language, we will eschew relations from logic and mathematics that are too fruitful for the description of such propositions. This is not to deny that we want a notion of necessary truth that goes beyond the trifling, but only to deny that, that notion is the notion of analyticity in natural language.

The remaining Frégean criticism points to a genuine incompleteness of the traditional account of analyticity. There are analytic relational sentences, for example, Jane walks with those with whom she strolls, 'Jack kills those he himself has murdered', and so forth, and analytic entailment with existential conclusions, for example, 'I think', therefore 'I exist'. The containment in these sentences is just as literal as that in an analytic subject-predicate sentence like 'Bachelors are unmarried', such are shown to have a theory of meaning construed as some hypothetico-deductive systemisations of sense as defined in (D) overcoming the incompleteness of the traditional account in the case of such relational sentences.

Such a theory of meaning makes the principal concern of semantics the explanation of sense properties and relations like synonymy, an antonymy, redundancy, analyticity, ambiguity, and so on, furthermore, it makes grammatical structure, specifically, senses structure, the basis for explaining them. This leads directly to the discovery of a new level of grammatical structure, and this, in turn, makes the possibility of a proper definition for analyticity. To see this, consider two simple examples. It is a semantic fact that 'a male bachelor' is redundant and that 'spinster' is synonymous with 'woman who never married;. In the case of the redundancy, we have to explain the fact that the sense of the modifier 'male' is already contained in the sense of its head 'bachelor'. In the case of the synonymy, we have to explain the fact that the sense of 'sinister' is identical to the sense of 'woman who never married' (compositionally formed from the senses of 'woman', 'never' and 'married'). But is so fas as such facts concern relations involving the components of the senses of 'bachelor' and 'spinster' and is in as far as these words are syntactical, they simply must be ascertained to a level of grammatical structure at which syntactic simplicity is semantically complex. This, in brief, is the route by which we arrive a level of 'decompositional semantic structure; that is the locus of sense structures masked by syntactically simple words.

Discovery of this new level of grammatical structure was followed by attemptive efforts as afforded to represent the structure of the sense's finds there. Without going into detail of sense representations, it is clear that, once we have the notion of decompositional representation, we can see how to generalize Locke and Kant's informal, subject-predicate account of analyticity to cover relational analytic sentences. Let a simple sentence 'S' consisted of a-place predicate 'P' with terms T1 ..., Tn occupying its argument places. Then: The analysis in case, first, S has a term T1 that consists of a place predicate Q (m > n or m = n) with terms occupying its argument places, and second, P is contained in Q and, for each term TJ. ... T1 + I, ..., Tn, TJ is contained in the term of Q that occupies the argument place in Q corresponding to the argument place occupied by TJ in P. (Katz, 1972)

To see, in systematic terms or concepts of how the formulated strategy has in accord with which specific functions are served by 'stroll' in saying that, 'Jane walks with those whom she strolls' is decompositionally represented as having the same sense as 'walkidly' in a leisurely way?'. The sentence is analytic by (A) because the predicate 'stroll' (the sense of 'stroll) and the term 'Jane' * the sense of 'Jane' associated with the predicate 'walk' is contained in the term 'Jane' (the sense of 'she is herself' associated with the predicate 'stroll'). The containment in the case of the other terms is automatic.

The fact that (A) itself makes no reference to logical operators or logical laws indicate that analyticity for subject-predicate sentences can be extended to simple relational sentences without treating analytic sentences as instances of logical truths. Further, the source of the incompleteness is no longer explained, as Fridge explained it, as the absence of 'fruitful' logical apparatus, but is now explained as mistakenly treating what is only a special case of analyticity as if it were the general case. The inclusion of the predicate in the subject is the special case (where n = 1) of the general case of the inclusion of an–place predicate (and its terms) in one of its terms. Noting that the defects, by which, Quine complained of in connection with Carnap's meaning-postulated explication are absent in (A). (A) contains no words from a natural language. It explicitly uses variable 'S' and variable 'L' because it is a definition in linguistic theory. Moreover, (A) tell 'us' what property is in virtue of which a sentence is analytic, namely, redundant predication, that is, the predication structure of an analytic sentence is already found in the content of its term structure.

Received opinion has been anti-Lockean in holding that necessary consequences in logic and language belong to one and the same species. This seems wrong because the property of redundant predication provides a nonlogical explanation of why true statements made in the literal use of analytic sentences are necessarily true. Since the property ensures that the objects of the predication in the use of an analytic sentence are chosen on the basis of the features to be predicated of them, the truth-conditions of the statement are automatically satisfied once its terms take on reference. The difference between such a linguistic source of necessity and the logical also the mathematical sources vindicating Locke's distinction between two kinds of 'necessary consequence' that logically or naturally follow from an action or condition. As something brought about by causing of results for which of a position having or direction, such that the hypothesis or phenomenon is purely effectual, that is to say, as an amounting result for the condition of being in full force or execution of being newly regular, that only give into some effective design or intention. Generally, of a particular impression emphasized as that based upon a desired impression, all of which, the relation of its cause for the finding of logical conclusion or issues determining of the consequence.

Received opinion concerning analyticity contains another mistake. This is the idea that analyticity is inimical to science, in part, the idea developed as a reaction to certain dubious uses of analyticity such as Frége's attempt to establish logicism and Schlick's, Ayer's and other logical positivist attempt to deflate claims to metaphysical knowledge by showing that alleged deductive truths are merely empty analytic truths (Schlick, 1948, and Ayer, 1946). In part, it developed as also a response to a number of cases where alleged analytically, and hence, necessary truths, e.g., the law of exclusion, seems as next-to-last, and yet, subsequent to have been taken as open to revision. Such cases convinced philosophers like Quine and Putnam that the analytic/synthetic distinction is an obstacle as placed against scientific progress.

The problem, if there is, one is one is not analyticity in the concept-containment sense, but the conflation of it with analyticity in the logical sense. This made it seem as if there is a single concept of analyticity that can serve as the grounds for a wide range of deductive truths. But, just as there are two analytic/synthetic distinctions, so there are two concepts of concept. The narrow Lockean/Kantian distinction is based on a narrow notion of expressions on which concepts are senses of expressions in the language. The broad Frégean/Carnap distinction is based on a broad notion of concept on which concepts are conceptions -often scientific one about the nature of the referent (s) of expressions (Katz, 1972) and curiously Putman, 1981. Conflation of these two notions of concepts produced the illusion of a single concept with the content of philosophical, logical and mathematical conceptions, but with the status of linguistic concepts. This encouraged philosophers to think that they were in possession of concepts with the contentual representation to express substantive philosophical claims, e.g., such as Fridge, Schlick and Ayer's, . . . and so on, and with a status that trivializes the task of justifying them by requiring only linguistic grounds for the deductive propositions in question.

Finally, there is an important epistemological implication of separating the broad and narrowed notions of analyticity. Fridge and Carnap took the broad notion of analyticity to providing foundations for necessary and a priority, and, hence, for some form of rationalism, and nearly all rationalistically inclined analytic philosophers admire boundlessly them in this. Thus, when Quine dispatched the Frége-Carnap position on analyticity, it was widely believed that necessary, as a priority, and rationalism had also been despatched, and, as a consequence. Quine had ushered in an 'empiricism without dogmas' and 'naturalized epistemology'. But given there is still a notion of analyticity that enables 'us' to pose the problem of how necessary, synthetic deductive knowledge is possible (moreover, one whose narrowness makes logical and mathematical knowledge part of the problem), Quine did not undercut the foundations of rationalism. Hence, a serious reappraisal of the new empiricism and naturalized epistemology is, to any the least, is very much in order (Katz, 1990).

In some areas of philosophy and sometimes in things that are less than important we are to find in the deductively/inductive distinction in which has been applied to a wide range of objects, including concepts, propositions, truths and knowledge. Our primary concern will, however, be with the epistemic distinction between deductive and inductive knowledge. The most common way of marking the distinction is by reference to Kant's claim that deductive knowledge is absolutely independent of completed experience, as to bring about or generate to some settled condition that if the state of exceeding of what is normal or sufficient, as an amount or degree by which quantity reserves to that of an object, meanings to a thought or emotion through the sense or mind. An active participation in events or activities, leading the accumulation of knowledge or skill that knowledge or skill so derived, in which an event or a series of events participated in or lived through that the succumbing of experiences is totally allocated by such events in the past of an individual, or that which as having had experiences in an activity or in life in general, the relating to or derived from experience brings about the politics of experience.

It is generally agreed that S's knowledge that 'p' is independent of experience just in case S's belief that 'p' is justified independently of experience. Some authors (Butchvarov, 1970, and Pollock, 1974) are, however, in finding this negative characterization of deductive unsatisfactory knowledge and have opted for providing a positive characterisation in terms of the type of justification on which such knowledge is dependent. Finally, others (Putman, 1983 and Chisholm, 1989) have attempted to mark the distinction by introducing concepts such as necessity and rational unrevisability than in terms of the type of justification relevant to deductive knowledge.

One who characterizes deductive knowledge in terms of justification that is independent of experience is faced with the task of articulating the relevant sense of experience, and proponents of the deductive ly cites 'intuition' or 'intuitive apprehension' as the source of deductive justification. Furthermore, they maintain that these terms refer to a distinctive type of experience that is both common and familiar to most individuals. Hence, there is a broad sense of experience in which deductive justification is dependent of experience. An initially attractive strategy is to suggest that theoretical justification must be independent of sense experience. But this account is too narrow since memory, for example, is not a form of sense experience, but justification based on memory is presumably not deductive. There appear to remain only two options: Provide a general characterization of the relevant sense of experience or enumerates those sources that are experiential. General characterizations of experience often maintain that experience provides information specific to the actual world while non-experiential sources provide information about all possible worlds. This approach, however, reduces the concept of non-experiential justification to the concept of being justified in believing a necessary truth. Accounts by enumeration have two problems (1) there is some controversy about which sources to include in the list, and (2) there is no guarantee that the list is complete. It is generally agreed that perception and memory should be included. Introspection, however, is problematic, and beliefs about one's conscious states and about the manner in which one is appeared to are plausible regarded as experientially justified. Yet, some, such as Pap (1958), maintain that experiments in imagination are the source of deductive justification. Even if this contention is rejected and deductive justification is characterized as justification independent of the evidence of perception, memory and introspection, it remains possible that there are other sources of justification. If it should be the case that clairvoyance, for example, is a source of justified beliefs, such beliefs would be justified deductively on the enumerative account.

The most common approach to offering a positive characterization of deductive justification is to maintain that in the case of basic deductive propositions, understanding the proposition is sufficient to justify one in believing that it is true. This approach faces two pressing issues. What is it to understand a proposition in the manner that suffices for justification? Proponents of the approach typically distinguish understanding the words used to express a proposition from apprehending the proposition itself and maintain that being relevant to deductive justification are the latter which. But this move simply shifts the problem to that of specifying what it is to apprehend a proposition. Without a solution to this problem, it is difficult, if possible, to evaluate the account since one cannot be sure that the account since on cannot be sure that the requisite sense of apprehension does not justify paradigmatic inductive propositions as well. Even less is said about the manner in which apprehending a proposition justifies one in believing that it is true. Proponents are often content with the bald assertions that one who understands a basic deductive proposition can thereby 'see' that it is true. But what requires explanation is how understanding a proposition enable one to see that it is true.

Difficulties in characterizing deductive justification in a term either of independence from experience or of its source have led, out-of-the-ordinary to present the concept of necessity into their accounts, although this appeal takes various forms. Some have employed it as a necessary condition for deductive justification, others have employed it as a sufficient condition, while still others have employed it as both. In the claim that the condition or quality of being necessary or something dictated by invariable physical laws as forcing or to exert by specific circumstance, we are found the state or fact of being in need, especially that arising from poverty, as an inevitable consequence. The necessity is a criterion of the deductive needs of our necessity as for a sufficient condition in the deductive justifications. This claim, however, needs further clarification.

There are three theses regarding the relationship among theoretically and the necessities that can be distinguished: (i) if 'p' is a necessary proposition and 'S' is justified in believing that 'p' is necessary, then S's justification is deductive: (ii) If 'p' is a necessary proposition and 'S' is justified in believing that 'p' is necessarily true, then S's justification is deductive: And (iii) If 'p' is a necessary proposition and 'S' is justified in believing that 'p', then S's justification is deductive. For example, many proponents of deductive contend that all knowledge of a necessary proposition is deductive. (ii) And (iii) has in the shortcoming of placing through any asserted stipulation for the issue of whether inductive knowledge of necessary propositions is possible. (i) does not have this shortcoming since the recent examples offered in support of this claim by Kriple (1980) and others have been cases where it is alleged that knowledge of the 'truth value' of necessary propositions is knowably inductive or using logical inductive measure, as causing or influencing such inducting reasons as pending upon the process or confirmation, that while awaiting, until. (i) Has in the shortcoming, however, of ruling out the possibility of being justified either in believing that a proposition is necessary on the basis of testimony or else sanctioning such justification as deductive. (ii)-(iii), of course, suffer from an analogous problem. These problems are symptomatic of a general shortcoming of the approach: It attempts to provide a sufficient condition for deductive justification solely in terms of the modal status of the proposition believed without making reference to the manner in which it is justified. This shortcoming, however, can be avoided by incorporating necessity as a necessary but not a sufficient condition for some prior conditions or fact of being justified, such as a fact or circumstance serving as justification, for example, in Chisholm (1989). Here there are two theses that must be distinguished: (1) If 'S' is justified deductively in believing that 'p', then 'p' is necessarily true. (2) If 'S' is justified deductively in believing that 'p'. Then 'p' is a necessary proposition. (1) and (2), however, allows this possibility. A further problem with both (1) and (2) is that it is not clear whether they permit deductively justified beliefs about the modal status of a proposition. For them that require that in order for 'S' to be justified deductively in believing that 'p' is a necessary preposition it must be necessary that 'p' is a necessary proposition. But the status of iterated modal propositions is controversial. Finally, (1) and (2) both were made impossible as b y action taken in advance had prevented from any given condition or activity from accepting such a stipulation as for the position was advanced by Kripke (1980) and Kitcher (1980) that there is deductive knowledge of contingent propositions.

The concept of rational unrevisability has also been invoked to characterize deductive justification. The precise sense of rational unrevisability has been presented in different ways. Putnam (1983) takes rational unrevisability to be both a necessary and sufficient condition for deductive justification while Kitcher (1980) takes it to be only a necessary condition. There are also two different senses of rational unrevisability that have been associated with the deductive (I) a proposition is weakly unrevisable just in case it is rationally unrevisable in light of any future 'experiential' evidence, and (II) a proposition is strongly unrevisable just in case it is rationally unrevisable in light of any future evidence. Let us consider the plausibility of requiring either form of rational unrevisability as a necessary condition for deductive justification. The view that a proposition is justifiably deductive, only if it is strongly unrevisable, entailing that which is non-experientially resourceful for justified beliefs, in that which are fallible but self-correcting, it is not a deductive source of justification. Casullo (1988) has argued that it is implausible to maintain that a proposition that is justifiably non-experiential is 'not' justified deductively, only because it is merely revisable, in light of furthering non-experiential evidence. The view that a proposition is justifiably deductive is weakly unrevisable, although not open to this objection since it excludes only fractions in light of experiential evidence. It does, however, face a different problem. To maintain that S's justified belief that 'p' is justified deductively is to make a claim about the type of evidence that justifies 'S' in believing that 'p'. On the other hand, to maintain that S's justified belief that 'p' is rationally revisable in light of experiential evidence is to make a claim about the type of evidence that can defeat S's justification for believing that 'p' that a claim about the type of evidence that justifies 'S' in believing that 'p'. Hence, it has been argued by Edidin (1984) and Cassel (1988) that to hold that a belief is justified deductively only if it is weakly unrevisable is either to confuse supporting evidence with defeating evidence or to support in some implausibility for this is about the relationship between the two, such that if evidence of the sort as the kind 'A' can defeat the justification conferred on S's belief that 'p' by evidence of kinds 'B' then S's justification for believing that 'p' is based on evidence of 'A'.

The most influential idea in the theory of meaning in the past hundred years is the thesis that the meaning of an indicative sentence is given by its truth-conditions. On this conception, to understand a sentence is to know its truth-conditions, for which of the conception was first clearly formulated by Fridge, was developed in a distinctive way by the early Wittgenstein, and is a leading idea of Donald Herbert Davidson (1917-), who is also known for rejecting the idea of as conceptual scheme, thought as something peculiar to one language or one way of looking at the world, arguing that where the possibility of translation stops so dopes the coherence of the idea that there is anything to translating. Is his collection of papers on the 'Essays on Actions and Events' (1980) and 'Inquiries into Truth and Interpretation' (1983?). However, the conception has remained so central that those who offer opposing theories characteristically define their position by reference to a mention of an occurrence or situation made significantly in the content of reference in an authoritative passbook of information. The concept delivers a generative idea derived or inferred from specific instances or occurrences, as something formed in the mind, least of mention, the concept of conception seems as to enact of a capacity for which the ability to form or understand mental concepts and abstractions, of something conceived in the mind. A concept, plan, design, idea, or thought as these constituent compositions are framed through which a corresponding potential or actuality as existing in the mind telling about the ideas of an absolute product of reason, as the gist of a specific situation, significance. Thought is distinctively intellectual and stresses contemplation and reasoning, which of you seems to have absolutely no concept of time, but, every succeeding scientific discovery makes greater nonsense of old-time conceptions of sovereignty.

Having Wittgenstein's main achievement, is a uniform theory of language that yields an explanation of logical truth. A factual sentence achieves sense by dividing the possibilities exhaustively into two groups, those that would make it true and those that would make it false. A truth of logic does not divide the possibilities but comes out true in all of them. It, therefore, lacks sense and says nothing, but it is not nonsense. It is a self-cancellation of sense, necessarily true because it is a tautology, the limiting case of factual discourse, like the figure '0' in mathematics. Language takes many forms and even factual discourse does not consist entirely of sentences like 'The fork is placed to the left of the knife'. However, the first thing that he gave up was the idea that this sentence itself needed further analysis into basic sentences mentioning simple objects with no internal structure. He was to concede, that a descriptive word will often get its meaning partly from its place in a system, and he applied this idea to colour-words, arguing that the essential relations between different colours do not indicate that each colour has an internal structure that needs to be taken apart. On the contrary, analysis of our colouring-words, it would only reveal the same pattern for its ranging of incompatible attributes, which of their recurring at every level as accorded to how we divide up the world, indeed, it may even be the case that of our ordinary language is created by moves that we ourselves make. If so, the philosophy of language will lead into the connection between the meaning of a word and the applications of it that its users intend to make. There is also an obvious need for people to understand each other's meanings of their words. There are many links between the philosophy of language and the philosophy of mind and it is not surprising that the impersonal examinations of language in the 'Tractatus' were replaced by a very different, anthropocentric treatment in 'Philosophical Investigations?'

If the logic of our language is created by moves that we ourselves make, various kinds of realisms are threatened. First, the way in which our descriptive language carves up the world will not be forced on 'us' by the nature of things, and the rules for the application of our words, which feel the external constraints, will really come from within 'us'. That is a concession to nominalism that is, perhaps, readily made. The idea that logical and mathematical necessity is also generated by what we ourselves accomplish what is more paradoxical. Yet, that is the conclusion of Wittengenstein (1956) and (1976), and here his anthropocentricism has carried less conviction. However, a paradox is not sure of error and it is possible that what is needed here is a more sophisticated concept of objectivity than Platonism provides.

In his later work Wittgenstein brings the great problem of philosophy down to earth and traces them to very ordinary origins. His examination of the concept of 'following a rule' takes him back to a fundamental question about counting things and sorting them into types: 'What qualifies as doing the same again? Of a courser, this question as an inconsequential fundamental suggests that we forget and that we get on with the subject. But Wittgenstein's question is not so easily dismissed. It has the naive profundity of questions that children ask when they are first taught a new subject. Such questions remain unanswered without detriment to their learning, but they point the only way to complete understanding of what is learned.

It is, nevertheless, the meaning of a complex expression in a function of the meaning of its constituents, that is, indeed, that it is just a statement of what it is for an expression to be semantically complex. It is one of the initial attractions of the conception of meaning as truths-conditions that it permits a smooth and satisfying account of the way in which the meaning of a complex expression is a dynamic function of the meaning of its constituents. On the truth-conditional conception, to give the meaning of an expression is to state the contribution it makes to the truth-conditions of sentences in which it occurs. For singular terms-proper names, indexical, and certain pronoun's -this is done by stating the reference of the term in question.

The truth condition of a statement is the condition the world must meet if the statement is to be true. To know this condition is equivalent to knowing the meaning of the statement. Although, this sounds as if it gives a solid anchorage for meaning, some of the security disappears when it turns out that the truth condition can only be defined by repeating the very same statement, the truth condition of 'snow is white' is that snow is white, the truth condition of 'Britain would have capitulated had Hitler invaded' is that Britain would have capitulated had Hitler invaded. It is disputed whether this element of running-on-the-spot disqualifies truth conditions from playing the central role in a substantive theory of meaning. Truth-conditional theories of meaning are sometimes opposed by the view that to know the meaning of a statement is to be able to users it in a network of inferences.

On the truth-conditional conception, to give the meaning of expressions is to state the contributive function it makes to the dynamic function of sentences in which it occurs, for singular terms-proper names, and certain pronouns, as well as indexical-for having been carried out or accomplished by stating the reference of the term in question. For predicates, it is done either by stating the conditions under which the predicate is true of arbitrary objects, or by stating the conditions under which arbitrary atomic sentence containing it is true. The meaning of a sentence-forming operator is given by stating its distributive contribution to the truth-conditions of a complete sentence, as a function of the semantic values of the sentences on which it operates. For an extremely simple, but nonetheless, it is a structured language, we can state the contributions various expressions make to truth conditions as follows:

A1: The referent of 'London' is London.

A2: The referent of 'Paris' is Paris.

A3: Any sentence of the form [a] is beautiful, is true if and only if the referent of [a] is beautiful.

A4: Any sentence of the form [a] is larger than [b] is true if and only if the referent of [a] is larger than the referent of [b].

A5: Any sentence of the form 'It is not the case that A' is true if and only if it is not the case that 'A' is true.

A6: Any sentence forming 'A' and 'B' are true if and only if 'A' is true and 'B' is true. As having the form and structure of an object or that it manifests itself as to determine an ordering of words or procedures.

The principle's of A2-A6 give shape to a simple theory of truth for a fragment of English. In this theory, it is possible to derive these consequences: That 'Paris is beautiful' is true if and only if Paris is beautiful (from A2 and A3), which 'London is larger than Paris also it is not the case that London is beautiful' is true if and only if London is larger than Paris and it is not the case that London is beautiful (from A1-As): And in general, for any sentence 'A' of this simple language, we can derive something in the anatomy of 'A' is true if and only if 'A'.

The theorist of truth conditions should insist that not every true statement about the reference of an expression be fit to be an axiom in a meaning-giving theory of truth for a language. The axiom:

London' refers to the city in which there was a huge fire in 1666

Is a true statement about the reference of 'London?'. It is a consequence of a theory that substitutes this axiom for A! In our simple truth theory that 'London is beautiful' is true if and only if the city in which there was a huge fire in 1666 is beautiful. Since a subject can understand by definition that 'London' without knowing that last-mentioned truth conditions, this replacement axiom is not fit to be an axiom in a meaning-specifying truth theory. It is, of course, incumbent on a theorist of meaning as truth conditions to state the constraints on the acceptability of axioms in a way that does not presuppose a deductive, non-truth conditional conception of meaning.

Among the many challenges facing the theorist of truth conditions, two are particularly salient and fundamental. First, the theorist has to answer the charge of triviality or vacuity. Second, the theorist must offer an account of what it is for a person's language to be truly descriptive by a semantic theory containing a given semantic axiom.

We can take the charge of triviality first. In more detail, it would run thus: Since the content of claim is the sentence 'Paris is beautiful' in which is true of the divisional region, which is no more than the claim that Paris is beautiful, we can trivially describe understanding a sentence, if we wish, as knowing its truth-conditions, but this gives 'us' no substantive account of understanding whatsoever. Something other than a grasp to truth conditions must provide the substantive account. The charge rests upon what has been called the redundancy theory of truth, the theory that, is somewhat more discriminative. Horwich calls the minimal theory of truth, or deflationary view of truth, as fathered by Fridge and Ramsey. The essential claim is that the predicate' . . . is true' does not have a sense, i.e., expresses no substantive or profound or explanatory concepts that ought be the topic of philosophical enquiry. The approach admits of different versions, but centres on the points (1) that 'it is true that p' says no more nor less than 'p' (hence redundancy) (2) that in less direct context, such as 'everything he said was true', or 'all logical consequences of true propositions are true', the predicate functions as a device enabling 'us' to generalize than the adjective or predicate describing the thing he said, or the kinds of propositions that follow from true propositions. For example, the second may translate as, (p, q)(p & p ®q ®q) where there is no use of a notion of truth.

There are technical problems in interpreting all uses of the notion of truth in such ways, but they are not generally felt to be insurmountable. The approach needs to explain away apparently substantive uses of the notion, such a science aims at the 'truth' or 'truth is a norm governing discourse'. Indeed, postmodernist writing frequently advocates that we must abandon such norms, along with a discredited 'objective' conception of truth. But perhaps, we can have the norms even when objectivity is problematic, since they can be framed without mention of truth: Science wants it to be so that whenever science holds that 'p'. Then 'p'. Discourse is to be regulated by the principle that it is wrong to assert 'p' when 'not-p'.

The Disquotational theory of truth finds that the simplest formulation is the claim that expressions are formed from S, is true, and the same as expressions of the given shape to 'S'. Some philosophers dislike the idea of sameness of meaning, and if this is disallowed, then the claim is that the two forms are equivalent in any sense of equivalence that matters. That is, it makes no difference whether people say 'Dogs bark' is true, or whether they say that 'dogs bark'. In the former representation of what they say the sentence 'Dogs bark' is mentioned, but in the latter it appears to be used, so the claim that the two are equivalent need's careful formulation and defence. On the face of it someone might know that 'Dogs bark' is true without knowing what it, for instance, if one were to find it in a list of acknowledged truths, although he does not understand English, and this is different from knowing that 'dogs bark'. Disquotational theories are usually presented as versions of the redundancy theory of truth.

The minimal theory states that the concept of truth is exhausted by the fact that it conforms to the equivalence principle, the principle that for any proposition 'p', it is true that 'p' if and only if 'p'. Many different philosophical theories of truth will, with suitable qualifications, accept that equivalence principle. The distinguishing feature of the minimal theory is its claim that the equivalence principle exhausts the notion of truths. It is how widely accepted, that both by opponents and supporters of truth conditional theories of meaning, that it is inconsistent to accept both minimal theory of truth and a truth conditional account of meaning (Davidson, 1990, Dummett, 1959 and Horwich, 1990). If the claim that a specific or specialized sentence, 'Paris is beautiful' is true is exhausted by its equivalence to the claim that Paris is beautiful, it is circular to try to explain the sentence's meaning in terms of its truth conditions. The minimal theory of truth has been endorsed by Ramsey, Ayer, the later Wittgenstein, Quine, Strawson, Horwich and-confusingly and inconsistently if it was correct-Fridge himself. But is the minimal theory correct?

The minimal or redundancy theory treats instances of the equivalence principle as definitional of truth for a given sentence. But in fact, it seems that each instance of the equivalence principle can itself be explained.

The truths from which such an instance as:

'London is beautiful', 'is true if and only if London is beautiful'

Preserving a right to be interpreted specifically of A1 and A3 above? This would be a pseudo-explanation if the fact that 'London' refers to 'London is beautiful' having the truth-condition that it does. But that is very implausible: that is to say, that it is, after all, possible to understand the inherited name, 'London' without understanding the predicate 'is beautiful'. The idea that facts about the reference of particular words can be explanatory of facts about the truth conditions of sentences containing them in no way requires any naturalistic or any other kind of reduction of the notion of reference. Nor is the idea incompatible with the plausible point that singular reference can be attributed at all only to something that is capable of combining with other expressions to form complete sentences. That still leaves room for facts about an expression's having the particular reference it does to be partially explanatory of the particular truth condition possessed by a given sentence containing it. The minimal; theory thus treats as definitional or stimulative something that is in fact open to explanation. What makes this explanation possible is that there is a general notion of truth that has, among the many links that hold it in place, systematic connections with the semantic values of sub-sentential expressions.

A second problem with the minimal theory is that it seems impossible to formulate it without at some point relying implicitly on features and principles involving truths that go beyond anything countenanced by the minimal theory. If the minimal theory that treats truth as a predicate of anything linguistic, be to some utterances, type-in-a-language, or whatever, then the equivalence schema will not cover all cases, but separately from those in the theorist's own language. Some account has to be given of truth for sentences of other languages. Speaking of the truth of language-independence propositions or thoughts will only postpone, not avoid, this issue, since at some point principles have to be stated associating these language-independent entities with sentences of particular languages. The defender of the minimalist theory is likely to say that if a sentence 'S' of a foreign language is best translated by our sentence 'p', then the foreign sentence 'S' is true if and only if 'p'. Now the best translation of a sentence must preserve the concepts expressed in the sentence. Constraints involving a general notion of truth are persuasive in a plausible philosophical theory of concepts. It is, for example, a condition of adequacy on an individualized account of any concept that there exists what is called 'Determination Theory' for that account-that is, a specification of how the account contributes to fixing the semantic value of that concept, the notion of a concept's semantic value is the notion of something that makes a certain contribution to the truth conditions of thoughts in which the concept occurs. But this is to presuppose, than to elucidate, a general notion of truth.

It is also plausible that there are general constraints on the form of such Determination Theories, constraints that involve truth and which are not derivable from the minimalist's conception. Suppose that concepts are individuated by their possession conditions. A concept is something that is capable of being a constituent of such contentual representational in a way of thinking of something-a particular object, or property, or relation, or another entity. A possession condition may in various way's make a thinker's possessives of a particular concept dependent upon his relations to his environment. Many possession conditions will mention the links between a concept and the thinker's perceptual experience. Perceptual experience represents the world for being a certain way. It is arguable that the only satisfactory explanation of what it is for perceptual experience to represent the world in a particular way must refer to the complex relations of the experience to the subject's environment. If this is so, then mention of such experiences in a possession condition will make possession of that condition will make possession of that concept dependent in part upon the environment relations of the thinker. Burge (1979) has also argued from intuitions about particular examples that, even though the thinker's non-environmental properties and relations remain constant, the conceptual content of his mental state can vary if the thinker's social environment is varied. A possession condition which property individuates such a concept must take into account the thinker's social relations, in particular his linguistic relations.

One such plausible general constraint is then the requirement that when a thinker forms beliefs involving a concept in accordance with its possession condition, a semantic value is assigned to the concept in such a way that the belief is true. Some general principles involving truth can indeed, as Horwich has emphasized, be derived from the equivalence schema using minimal logical apparatus. Consider, for instance, the principle that 'Paris is beautiful and London is beautiful' is true if and only if 'Paris is beautiful' is true if and only if 'Paris is beautiful' is true and 'London is beautiful' is true. This follows logically from the three instances of the equivalence principle: 'Paris is beautiful and London is beautiful' is rue if and only if Paris is beautiful, and London is beautiful' is true if and only if London is beautiful. But no logical manipulations of the equivalence schemas will allow the deprivation of that general constraint governing possession conditions, truth and the assignment of semantic values. That constraint can have courses be regarded as a further elaboration of the idea that truth is one of the aims of judgement.

We now turn to the other question, 'What is it for a person's language to be correctly describable by a semantic theory containing a particular axiom, such as the axiom for conjunction?' This question may be addressed at two depths of generality. At the shallower level, the question may take for granted the person's possession of the concept of conjunction, and be concerned with what has to be true for the axiom correctly to describe his language. At a deeper level, an answer should not duck the issue of what it is to possess the concept. The answers to both questions are of great interest: We will take the lesser level of generality first.

When a person conjunction by 'and' he is not necessarily capable of formulating the axiomatized conjunction, in as much as he can formularize in the reduction of a formula, whereas, by an established form of words or symbols for their use in a procedural composition or structure of a compound. In that which to state or reduce of an expression as measuring to the systematic terms or concepts, only to devise or invent of some strategy to prepare according to a specified formula. That having to enabling capabilities we are of the ability to formulate it is not the causal basis of his capacity to hear sentences containing the word 'and' as meaning something involving conjunction. Nor is it the causal basis of his capacity to mean something involving conjunction by sentences uttered containing the word 'and', along with other unspecified things of the same class within furthering the same manners, in addition to more considerations that are taken of all our skill and then some.

Is it then right to regard a truth theory as part of an unconscious psychological computation, and to regard understanding a sentence as involving a particular way of depriving a theorem from a truth theory at some level of conscious proceedings? One problem with this is that it is quite implausible that everyone who speaks the same language has to use the same algorithms for computing the meaning of a sentence. In the past thirteen years, thanks particularly to the work of Davies and Evans, a conception has evolved according to which an axiom like A6 is true of a person's language only if there is a common component in the explanation of his understanding of each sentence containing the word 'and', a common component that explains why each such sentence is understood as meaning something involving conjunction (Davies, 1987). This conception can also be elaborated in computational terms: Suggesting that for an axiom like A6 to be true of a person's language is for the unconscious mechanisms which produce understanding to draw on the information that a sentence of the form 'A and B' are true if and only if 'A' is true and 'B' is true (Peacocke, 1986). Many different algorithms may equally succumb to this information. The psychological reality of a semantic theory thus involves, in Marr's (1982) famous classification, something intermediate between his level one, the function computed, and his level two, the algorithm by which it is computed. This conception of the psychological reality of a semantic theory can also be applied to syntactic and phonol logical theories. Theories in semantics, syntax and phonology are not themselves required to specify the particular algorithms that the language user employs. The identification of the particular computational methods employed is a task for psychology. But semantics, syntactic and phonology theories are answerable to psychological data, and are potentially refutable by them-for these linguistic theories do make commitments to the information drawn upon by mechanisms in the language user.

This answers to the question of what it is for an axiom to be true of a person's language clearly takes for granted the person's possession of the concept expressed by the word treated by the axiom. In the example of the axiom, for which of the information as drawn upon, are the sentences of the form 'A and B' are true if and only if 'A' is true and 'B' is true. This informational content employs, as it has if it is to be adequate, the concept of conjunction used in stating the meaning of sentences containing more in addition, that will take all our skill and then some, and other unspecified things of the same class, such as the same grammatical function in a constructively formed manner, from which are situated of the same functional manners. So in saying, that the computational answer we have returned needs further elaboration if we are to address of the deeper questions, which we do not want to take for granted as the possession of the concepts expressed in of any language. It is at this point that the theory of linguistic understanding has to draws upon a theory of concepts. It is plausible that the concepts of conjunction are individuated by the following condition for a thinker to possess it.

Finally, this response to the deeper question allows 'us' to answer two challenges to the conception of meaning as truth-conditions. First, there was the question left hanging earlier, of how the theorist of truth-conditions is to say what makes one axiom of a semantic theory is correctly in that of another, when the two axioms assign the same semantic values, but do so by of different concepts. Since the different concepts will have different possession conditions, the dovetailing accounts, at the deeper level of what it is for each axiom to be correct for a person's language will be different account. Second, there is a challenge repeatedly made by the minimalist theorists of truth, to the effect that the theorist of meaning as truth-conditions should give some non-circular account of what it is to understand a sentence, or to be capable of understanding all sentences containing a given constituent. For each expression in a sentence, the corresponding dovetailing account, together with the possession condition, supplies a non-circular account of what it is to understand any sentence containing that expression. The combined accounts for each of the expressions that comprise of a given sentence, and together constitute a non-circular account of what it is to understand the complete sentence. Taken together, they allow the theorists of meaning as truth-conditions fully to meet the challenge.

A curious view common to which is expressed by an utterance or sentence, purports that the proposition or claim made about the inherent perceptions of the world as extension, the content of a predicate or other sub-sentential component is what it contributes to the content of sentences that contain it. The nature of content is the central concern of the philosophy of language, in that mental states have contents: A belief may have the content that the prime minister will resign. A concept is something that is capable of bringing a constituent of such contents. More specifically, a concept is a way of thinking of something-a particular object, or property or relation, or another entity. Such a distinction was held in Frége's philosophy of language, explored in 'On Concept and Object' (1892). For which Fridge considered predicates as incomplete expressions, in the same way as a mathematical expression for a function, such as sines . . . a log . . ., is incomplete. Predicates bear upon the concepts, from which they are 'unsaturated', and cannot be referred to by subject expressions (we thus get the paradox that the concept of a horse is not a concept). Although Fridge recognized the metaphorical nature of the notion of a concept being unsaturated, he was rightly convinced that some such notion is needed to explain the unity of a sentence, and to prevent sentences from being thought of as mere lists of names.

Several different concepts may each be ways of thinking of the same object. A person may think of himself in the first-person way, or think of himself as the spouse of Mary Smith, or as the person located in a certain room now. More generally, a concept 'c' is distinct from a concept 'd' if it is possible for a person rationally to believe is such-and-such'. As words can be combined to form structured sentences, concepts have also been conceived as combinable into structured complex contents. When these complex contents are expressed in English by 'that . . . 'clauses, as in our opening examples, they will be capable of being true or false, depending on the way the world is.

The general system of concepts with which we organize our thoughts and perceptions are to encourage a conceptual scheme of which the outstanding elements of our every day conceptual formalities include spatial and temporal relations between events and enduring objects, causal relations, other persons, meaning-bearing utterances of others, . . . and so on. To see the world as containing such things is to share this much of our conceptual scheme. A controversial argument of Davidson's urges that we would be unable to interpret speech from a different conceptual scheme as even meaningful, Davidson daringly goes on to argue that, since translations proceed according to the principle of clarity, and since it must be possible of an omniscient translator to make sense in those of 'us', that we can be assured that most of the beliefs formed within the commonsense conceptual framework are true.

Concepts are to be distinguished from a stereotype and from conceptions. The stereotypical spy may be a middle-level official down on his luck and in need of money. Nonetheless, we come to learn that Anthony Blunt, the art historian and Surveyor of the Queen's Pictures, and, disbelieve, we can come to believe, that something falls under a concept while positively disbelieving that the same thing falls under the stereotype associated wit the concept. Similarly, a person's conception of a just arrangement for resolving disputations that may involve something like contemporary Western legal systems, but, whether or not it would be correct, such that it seems to be realizable, the intelligibility for someone to rejects this conception by arguing that it does not adequately provides for the elements of fairness and respects that are required by the concepts of justice.

Basically, a concept is that which is understood by a term, particularly a predicate. To posses a concept is to be able to deploy a term expressing it in making judgements, in which the ability connection is such things as recognizing when the term applies, and being able to understand the consequences of its application. The term 'idea' was formally used in the same way, but is avoided because of its associations with subjective matters inferred upon by mental imagery in which may be irrelevant ti the possession of a concept. In the semantics of Fridge, a concept is the reference of a predicate, and cannot be referred to by a subjective term, although its recognition of as a concept, in that some such notion is needed to the explanatory justification of which that sentence of unity finds of itself from being thought of as namely categorized lists of itemized priorities.

A theory of a particular concept must be distinguished from a theory of the object or objects it selectively picks the outlying of the theory of the concept under which is partially contingent of the theory of thought and/or epistemology. A theory of the object or objects is part of metaphysics and ontology. Some figures in the history of philosophy-and are open to the accusation of not having fully respected the distinction between the kinds of theory. Descartes appears to have moved from facts about the indubitability of the thought 'I think', containing the fist-person was of thinking, to conclusions about the nonmaterial nature of the object he himself was. But though the goals of a theory of concepts and a theory of objects are distinct, each theory is required to have an adequate account of its relation to the other theory. A theory if concept is unacceptable if it gives no account of how the concept is capable of picking out the object it evidently does pick out. A theory of objects is unacceptable if it makes it impossible to understand how we could have concepts of those objects.

A fundamental question for philosophy contains that: What individuates a given concept-that is, what makes it the one it is, rather than any other concept? One answer, which has been developed in great detail, is that it is impossible to give a non-trivial answer to this question (Schiffer, 1987). An alternative approach, addressees the question by starting from the idea that a concept id individuated by the condition that must be satisfied if a thinker is to posses that concept and to be capable of having beliefs and other attitudes whose content contains it as a constituent. So, to take a simple case, one could propose that the logical concept 'and' is individuated by this condition, it be the unique concept 'C' to posses that a thinker has to find these forms of inference compelling, without basing them on any further inference or information: From any two premisses 'A' and 'B', ACB can be inferred, and from any premiss ACB, implicated by each of 'A' and 'B' can be inferred. Again, a relatively observational concept such as 'round' can be individuated in part by stating that the thinker finds specified contents containing it compelling when he has certain kinds of perception, and in part by relating those judgements containing the concept and which are not based on perception to those judgements that are. A statement that individuates a concept by saying what is required for a thinker to posses it can be described as giving the possession condition for the concept.

A possession condition for a particular concept may actually make use of that concept. The possession condition for 'and' does so. We can also expect to use relatively observational concepts in specifying the kind of experience that have to be mentioned in the possession conditions for relatively observational concepts. What we must avoid is mention of the concept in question as such within the content of the attitudes attributed to the thinker in the possession condition. Otherwise we would be presupposing possession of the concept in an account that was meant to elucidate its possession. In talking of what the thinker finds compelling, the possession conditions can also respect an insight of the later Wittgenstein: Such that to find in her, that much in finding her in some of the natural ways, as going on in new cases in applying the concept.

Sometimes a family of concepts has this property: It is not possible to master any one of the members of the family without mastering the others. Two of the families that plausibly have this status are these: The family consisting of some simple concepts 0, 1, 2, . . . of the natural numbers and the corresponding concepts of numerical quantifiers there are so-and-so's, there is 1 so-and-so, . . . and the family consisting with the conceptual measures of 'believing' and that of 'desire'. Such families have come to be known as 'local holism'. A local holism does not prevent the individuation of a concept by its possession condition. Rather, it demands that all the concepts in the family be individuated simultaneously. So one would say something of this form: Belief and desire form the unique pair of concepts C1 and C2 such that for as thinker to posses them, they are to meet such-and-such condition involving the thinker, C1 and C2. For these and other possession conditions to individuate properly, it is necessary that there be some ranking of the concepts treated. The possession conditions for concepts higher in the ranking must presuppose only possession of concepts at the same or lower levels in the ranking.

A possession condition may in various way's make a thinker's possession of a particular concept dependent upon his relations to his environment. Many possession conditions will mention the links between a concept and the thinker's perceptual experience. Perceptual experience represents the world as a certain way. It is arguable that the only satisfactory explanation of what it is for perceptual experience to represent the world in a particular way must refer to the complex relations of the experience to the subject's environment. If this is so, then mention of such experiences in a possession condition will make possession of that concept dependent in part upon the environmental relations of the thinker. Burge (1979) has also argued from intuitions about particular examples that, even though the thinker's non-environmental properties and relations remain constant, the conceptual content of his mental state can vary if the thinker's social environment is varied. A possession condition that properly individuates such a concept must take into account the thinker's social relations, in particular his linguistic relations.

Concepts have a normative dimension, a fact strongly emphasized by Kripke. For any judgement whose content involves a given concept, there is a correctness condition for that judgement, a condition that is dependent in part upon the identity of the concept. The normative character of concepts also extends into making the territory of a thinker's reasons for making judgements. A thinker's visual perception can give him good reason for judging 'That man is bald': It does not by itself give him good reason for judging 'Rostropovich is bald', even if the man he sees is Rostropovich. All these normative connections must be explained by a theory of concepts one approach to these matters is to look to the possession condition for the concept, and consider how the referent of a concept is fixed from it, together with the world. One proposal is that the referent of the concept is that object or property, or function, . . . which makes the practices of judgement and inference, as already mentioned, that in the possessional condition is always lead to true judgements and truth-preserving inferences. This proposal would explain why certain reasons are necessity good reasons for judging given contents. Provided the possession condition permits 'us' to say what it is about a thinker's previous judgements that masker it, the case that he is employing one concept rather than another, this proposal would also have another virtue. It would allow 'us' to say how the correctness condition is determined for a judgement in which the concept is applied to newly encountered objects. The judgement is correct if the new object has the property that in fact makes the judgmental practices mentioned in the possession condition yield true judgements, or truth-preserving inferences.

These manifesting dissimilations have occasioned the affiliated differences accorded within the distinction as associated with Leibniz, who declares that there are only two kinds of truths-truths of reason and truths of fact. The forms are all explicit identities, i.e., of the form 'A is A', 'AB is B', and so on, or they are reducible to this form by successively substituting equivalent terms. Leibniz dubs them 'truths of reason' because the explicit identities are self-evident deducible truths, whereas the rest can be converted too such by purely rational operations. Because their denial involves a demonstrable contradiction, Leibniz also says that truths of reason 'rest on the principle of contradiction, or identity' and that they are necessary propositions, which are true of all possible words. Some examples are 'All equilateral rectangles are rectangles' and 'All bachelors are unmarried': The first is already of the form AB is 'B' and the latter can be reduced to this form by substituting 'unmarried man' fort 'bachelor'. Other examples, or so Leibniz believes, are 'God exists' and the truths of logic, arithmetic and geometry.

Truths of fact, on the other hand, cannot be reduced to an identity and our only way of knowing them is empirically by reference to the facts of the empirical world. Likewise, since their denial does not involve a contradiction, their truth is merely contingent: They could have been otherwise and hold of the actual world, but not of every possible one. Some examples are 'Caesar crossed the Rubicon' and 'Leibniz was born in Leipzig', as well as propositions expressing correct scientific generalizations. In Leibniz's view, truths of fact rest on the principle of sufficient reason, which states that nothing can be so unless there is a reason that it is so. This reason is that the actual world (by which he the total collection of things past, present and future) is better than any other possible worlds and was therefore created by 'God'.

In defending the principle of sufficient reason, Leibniz runs into serious problems. He believes that in every true proposition, the concept of the predicate is contained in that of the subject. (This holds even for propositions like 'Caesar crossed the Rubicon': Leibniz thought that if anyone who crossed the point of no return would not have been Caesar). And this containment relationship! Which is eternal and unalterable even by God ~? Guarantees that every truth has a sufficient reason, if truths consist in concept containment, however, then it seems that all truths are analytic and hence necessary, and if they are all necessary, surely they are all truths of reason. Leibnitz responds that not every truth can be reduced to an identity in a finite number of steps, in some instances revealing the connection between subject and predicate concepts would requite an infinite analysis. But while this may entail that we cannot prove such propositions as deductively manifested, it does not appear to show that the proposition could have been false. Intuitively, it seems a better ground for supposing that it is necessary truth of a special sort. A related question arises from the idea that truths of fact depend on God's decision to create the best of all possible worlds: If it is part of the concept of this world that it is best, now could its existence be other than necessary? Leibniz answers that its existence is only hypothetically necessary, i.e., it follows from God's decision to create this world, but God had the power to decide otherwise. Yet God is necessarily non-deceiving, non-deceiving, so how could he have decided to do anything else? Leibniz says much more about these masters, but it is not clear whether he offers any satisfactory solutions.

Necessary truths are ones that must be true, or whose opposite is impossible. Contingent truths are those that are not necessary and whose opposite is therefore possible. 1-3 below is necessary, 4-6, contingent.

* It is not the case that it is raining and not raining

* 2 + 2= 4

* All bachelors are unmarried.

* It seldom rains in the Sahara.

* There are more than four states in the USA.

* Some bachelors drive Maserati.

Plantinga (1974) characterizes the sense of necessity illustrated in 1-3 as 'broadly logical'. For it includes not only truths of logic, but those of mathematics, set theory, and other quasi-logical ones.

Yet, it is not so widely broadened as to include matters of causal or natural necessity, such as: Nothing travels faster than the speed of light.

One would like an account of the basis of our distinction and a criterion by which to apply it. Some suppose that necessary truths are those we know as deductively possible. But we lack the criterion for deductive truths, and there are necessary truths we do not know at all, e.g., undiscovered mathematical ones. It would not help to say that necessary truths are one, and it is possible, in the broader of logical sense, to know of some deductive reasons based on circularity. Finally, Kripke (1972)and Plantinga (1974) argues that some contingent truths are knowable by deductive reasoning. Similar problems face the suggestion that necessary truths are the ones we know with the fairest of certainties: We lack a criterion for certainty, there are necessary truths we do not know, and (barring dubious arguments for scepticism) it is reasonable to suppose that we know some contingent truths with certainty.

Leibniz defined a necessary truth as one whose opposite implies a contradiction. Every such proposition, he held, is either an explicit identity, i.e., of the forms 'A' is' A', 'AB is 'B', and so on, are reducible to an identity by successively substituting equivalent terms. (Thus, 3 above might be so reduced by substituting 'unmarried man' for 'bachelor'.) This has several advantages over the ideas of the previous paragraph. First, it explicated the notion of necessity and possibility and seems to provide a criterion we can apply. Second, because explicit identities are self-evident and are deductive propositions, that the theory implies that all necessary truths are knowably deductable, but it does not entail that we actually know all of them, nor does it hold of any arresting conditions that would attend of views or action or define to the 'knowable' in a circular way. Third, it implies that necessary truths are knowable with certainty, but does not preclude that as having certain knowledge of contingent truths by other than a reduction.

Nevertheless, this view is also problematic, and Leibniz's examples of reductions are too sparse to prove a claim about all necessary truths. Some of his reductions, moreover, are deficient: as Fridge has pointed out, for example, that his proof of '2 + 2 = 4' presupposes the principle of association and so does not depend on the principle of identity. More generally, it has been shown that arithmetic cannot be reduced to logic, but requires the resources of set theory as well. Finally, there are other necessary propositions, e.g., 'Nothing can be red and green all over', which do not seem to be reducible to identities and which Leibniz does not show how to reduce.

Leibniz and others have thought of truths as a property of propositions, where the latter are conceived as things that may be expressed by, but are distinct from linguistic items like statements. On another approach, truth is a property of linguistic entities, and the basis of necessary truth in convention. Thus A.J. Ayer, for example, argued that the only necessary truths are analytic statements and that the latter rest entirely on our commitment to use words in certain ways.

The slogan 'the meaning of a statement is its method of verification' expresses the empirical verification's theory of meaning. It is more than the general criterion of meaningfulness if and only if it is inductively verifiable. If said, in addition to what as having in mind as in the meaning of a sentence is: It is a grammatical unit that is syntactically independent and has a subject that is expressed or, as in imperative sentience understood and predicated with an infinite verb. Having at least, those observations that would confirm or disconfirm the sentence. Sentences that would be verified or falsified by all the same observations that are justified as empirically equivalent or having to the same meaning, which a sentence is said to be cognitively meaningful if and only if it can be verified or falsified in experience. This is not meant to require that the sentence be conclusively verified or falsified, since universal scientific laws or hypotheses (which are supposed to pass the test) are not logically deducible from any amount of actually observed evidence.

When one predicate's necessary truth of a preposition one speaks of modality de dicto. For one ascribes the modal property, necessary truth, to a dictum, namely, whatever proposition is taken as necessary. A venerable tradition, however, distinguishes this from necessary de re, wherein one predicates necessary or essential possession of some property to an on object. For example, the statement '4 is necessarily greater than 2' might be used to predicate of the object, 4, the property, being necessarily greater than 2. That objects have some of their properties necessarily, or essentially, and others only contingently, or accidentally, are a main part of the doctrine called essentialism'. Thus, an essential might say that Socrates had the property of being bald accidentally, but that of being self-identical, or perhaps of being human, essentially. Although essentialism has been vigorously attacked in recent years, most particularly by Quine, it also has able contemporary proponents, such as Plantinga.

The fact of relating, or characteristic of a mode expressing of, or the composition, that in any mode without referring to substance as by modality for which of classifications are the propositions on the basis of whether they assert or deny the possibility, impossibility, contingency or necessarily held that every proposition has a modal status as well as a truth value. Every proposition is either necessary or contingent as well as either true or false. The issue of knowledge of the modal status of propositions has received much attention because of its intimate relationship to the issue of deductive reasoning. For example, no propositions of the theoretic content that all knowledge of necessary propositions is deductively knowledgeable. Others reject this claim by citing Kripke's (1980) referential cases of necessarily theoretical propositions. Such contentions are often inconclusive, for they fail to take into account the following tripartite distinction: 'S' knows the general modal status of 'p' just in case 'S' knows that 'p' is a necessary proposition or 'S' knows the truth that 'p' is a contingent proposition. 'S' knows the truth value of 'p' just in case 'S' knows that 'p' is true or 'S' knows that 'p' is false. 'S' knows the specific modal status of 'p' just in case 'S' knows that 'p' is necessarily true or 'S' knows that 'p' is necessarily false or 'S' knows that 'p' is contingently true or 'S' knows that 'p' is contingently false. It does not follow from the fact that knowledge of the general modal status of a proposition is a deductively reasoned distinctive modal status is also given to theoretical principles. Nor des it follow from the fact that knowledge of a specific modal status of a proposition is theoretically given as to the knowledge of its general modal status that also is deductive.

The certainties involving reason and a truth of fact are much in distinction by associative measures given through Leibniz, who declares that there are only two kinds of truths-truths of reason and truths of fact. The former are all explicit identities, i.e., of the form that gives shape and structure of the object presentation of 'A' is 'A', 'AB' is 'B', and so forth, or they are reducible to this form by successively substituting equivalent terms. Leibniz dubs them 'truths of reason' because the explicit identities are self-evident theoretical truth, whereas the rest can be converted too such by purely rational operations. Because their denial involves a demonstrable contradiction, Leibniz also says that truths of reason 'rest on the principle of contraction, or identity' and that they are necessary propositions, which are true of all possible worlds. Some examples are, that. All bachelors are unmarried': The first is already of the form 'AB is B' and the latter can be reduced to this form by substituting 'unmarried man' for 'bachelor'. Other examples, or so Leibniz believes, are 'God exists' and the truth of logic, arithmetic and geometry.

In defending the principle of sufficient reason, Leibniz runs into serious problems. He believes that in every true proposition, the concept of the predicate is contained in that of the subject. (This hols even for propositions like 'Caesar crossed the Rubicon': Leibniz thinks anyone who did not cross the Rubicon would not have been Caesar) And this containment relationship-that is eternal and unalterable even by God-guarantees that every truth has a sufficient reason. If truth consists in concept containment, however, then it seems that all truths are analytic and hence necessary, and if they are all necessary, surely they are all truths of reason. Leibniz responds that not evert truth can be reduced to an identity in a finite number of steps: In some instances revealing the connection between subject and predicate concepts would require an infinite analysis. But while this may entail that we cannot prove such propositions as deductively probable, it does not appear to show that the proposition could have been false. Intuitively, it seems a better ground for supposing that it is a necessary truth of a special sort. A related question arises from the idea that truths of fact depend on God's decision to create the best world, if it is part of the concept of this world that it is best, how could its existence be other than necessary? Leibniz answers that its existence is only hypothetically necessary, i.e., it follows from God's decision to create this world, but God is necessarily good, so how could he have decided to do anything else? Leibniz says much more about the matters, but it is not clear whether he offers any satisfactory solutions.

The modality of a proposition is the way in which it is true or false. The most important division is between propositions true of necessity, and those true asa things are: Necessary as opposed to contingent propositions. Other qualifiers sometimes called 'modal' include the tense indicators 'It will be the case that p' or it was the case that 'p', and there are affinities between the 'deontic indicators', as it ought to be the case that 'p' or it is permissible that 'p' and the logical modalities as a logic that study the notions of necessity and possibility. Modal logic was of great importance historically, particularly in the light of various doctrines concerning the necessary properties of the deity, but was not a central topic of modern logic in its golden period at the beginning of the 20th century. It was, however, revived by C. I. Lewis, by an additional amount to two resulting propositional or predicate calculus operators as and (sometimes written N and M), meaning necessarily and possibly, respectively, these like 'p' and 'p' will be wanted. Controversial theses include 'p' (if a proposition is necessary, it is necessarily necessary, characteristic of the system known as S4) and 'p' (if a proposition is possible, it is necessarily possible, characteristic of the system known as S5). The classical 'modal theory' for modal logic, due to Kripke and the Swedish logician Stig Kanger, involves valuing propositions not as true or false 'simplicitiers', but as true or false art possible worlds, with necessity then corresponding to truth in all worlds, and possibly to truths in some world.

The doctrine advocated by David Lewis, which different 'possible worlds' are to be thought of as existing exactly as this one does. Thinking in terms of possibilities is thinking of real worlds where things are different, this view has been charged with misrepresenting it as some insurmountably unseeingly to why it is good to save the child from drowning, since there is still a possible world in which she (or her counterpart) drowned, and from the standpoint of the universe it should make no difference that a world is actual. Critics asio charge that the notion fails to fit either with a coherent theory of how we know about possible worlds, or with a coherent theory about possible worlds, or with a coherent theory of why we are interested in them, but Lewis denies that any other way of interpreting modal statements is tenable.

Thus and so, the 'standard analysis' of propositional knowledge, suggested by Plato and Kant among others, implies that if one has a justified true belief that 'p', then one knows that 'p'. The belief condition 'p' believes that 'p', the truth condition requires that any known proposition be true. And the justification condition requires that any known proposition be adequately justified, warranted or evidentially supported. Plato appears to be considering the tripartite definition in the 'Theaetetus' (201c-202d), and to be endorsing its jointly sufficient conditions for knowledge in the 'Meno' (97e-98a). This definition has come to be called 'the standard analysis of knowledge,' and has received a serious challenge from Edmund Gettier's counterexamples in 1963. Gettier published two counterexamples if only to exemplify the standard analysis:

(1) Smith and Jones have applied for the same job. Smith is justified in believing that (a) Jones will get the job, and that (b) Jones has ten coins in his pocket. On foundation upon which of a fundamental principle as underlying circumstance or condition from which the chief constituent, as its basis for which something rests in having to (i) and (ii) Smith infers, and thus is justified in believing, that (iii) the person who will get the job has ten coins in his pocket. At it turns out, Smith himself will get the job, and he also happens to have ten coins in his pocket. So, although Smith is justified in believing the true proposition (3), Smith does not know (3).

(2) Smith is justified in believing the false proposition that (a) Smith owns an automobile class in his condition for being a Ford member (a) Smith infers, and thus is justified in believing, that (b) either Jones owns an automobile class Ford or Brown is in Barcelona. As it turns out, Brown or in Barcelona, and so (b) is true. So although Smith is justified in believing the true proposition (b). Smith does not know (b).

Gettier's counterexamples are thus cases where one has justified true belief that 'p', but lacks knowledge that 'p'. The Gettier problem is the problem of finding a modification of, or an alterative to, the standard justified-true-belief analysis of knowledge that avoids counterexamples like Gettier's. Some philosophers have suggested that Gettier style counterexamples are defective owing to their reliance on the false principle that false propositions can justify one's belief in other propositions. But there are examples much like Gettier's that do not depend on this allegedly false principle. Here is one example inspired by Keith and Richard Feldman:

(3) Suppose Smith knows the following proposition, 'm': Jones, whom Smith has always found to be reliable and Smith, has no reason to distrust as presently told Smith, his office-mate, that 'p'. That Jones owns an automobile classed as for being a member of Ford's. Suppose also that Jones has told Smith that 'p' only because of a state of hypnosis Jones is in, and that 'p' is true only because, unknown to himself, Jones has won a Ford in a lottery since entering the state of hypnosis. And suppose further

that Smith deduces from 'm' that is its existential generalization, 'q': There is someone, whom Smith has always found to be reliable and that Smith has no reason to distrust as for presently, Smith, whose his office-mate, that he had bought and owns a Ford. Smith, then, knows that 'q', since he has correctly deduced 'q' from 'm', which he also knows. But suppose also that on the basis of his knowledge that 'q'. Smith believes that 'r': Someone in the office owns a Ford. Under these conditions, Smith has justified true belief that 'r', knows his evidence for 'r', but does not know that 'r'.

Gettier-style examples of this sort have proven especially difficult for attempts to analyse the concept of propositional knowledge. The history of attempted solutions to the Gettier problem is complex and open-ended. It has no product associated by the consensus on any solution. Many philosophers hold, in light of Gettier-style examples, that propositional knowledge requires a fourth condition, beyond the justification, truth and belief conditions. Although no particular fourth condition enjoys widespread endorsement, there are some prominent general proposals in circulation. One sort of proposed modification, the so-called 'defeasibility analysis', requires that the justification appropriate to knowledge be 'undefeated' in the general sense that some appropriate subjunctive conditional concerning genuine defeaters of justification be true of that justification. One straightforward defeasibility fourth condition, for instance, requires of Smith's knowing that 'p' that there be no true proposition 'q', such that if 'q' became justified for Smith, 'p' would no longer be justified for Smith (Pappas and Swain, 1978). A different prominent modification requires that the actual justification for a true belief qualifying as knowledge not depend in a specified way on any falsehood (Armstrong, 1973). The details proposed to elaborate such approaches have met with considerable controversy.

The fourth condition of evidential truth-sustenance may be a speculative solution to the Gettier problem. More specifically, for a person, 'S', to have knowledge that 'p' on justifying evidence 'e', 'e' must be truth-sustained in this sense for every true proposition 't' that, when conjoined with 'e', undermines S's justification for 'p' on 'e', there is a true proposition, 't', that, when conjoined with 'e' & 't', restores the justification of 'p' for 'S' in a way that 'S' is actually justified in believing that 'p'. The gist of this resolving evolution, put roughly, is that propositional knowledge requires justified true belief that is sustained by the collective totality of truths. Herein, is to argue in Knowledge and Evidence, that Gettier-style examples as (1)-(3), but various others as well.

Three features that proposed this solution merit emphasis. First, it avoids a subjunctive conditional in its fourth condition, and so escapes some difficult problems facing the use of such a conditional in an analysis of knowledge. Second, it allows for non-deductive justifying evidence as a component of propositional knowledge. An adequacy condition on an analysis of knowledge is that it does not restrict justifying evidence to relations of deductive support. Third, its proposed solution is sufficiently described as it is flexibly handled in the cases for being describable, as follows:

(4) Smith has a justified true belief that 'p', but there is a true proposition, 't', which undermines Smith's justification for 'p' when conjoined with it, and which is such that it is either physically or humanly impossible for Smith to be justified in believing that 't'. Examples represented by (4) suggest that we should countenance varying strengths in notions of propositional knowledge. These strengths are determined by accessibility qualifications on the set of relevant knowledge-precluding underminers. A very demanding concept of knowledge assumes that it need only be logically possible for a Knower

to believe a knowledge-precluding underminer. Less demanding concepts assume that it must be physically or humanly possible for a Knower to believe knowledge-precluding underminers. But even such less demanding concepts of knowledge need to rely on a notion of truth-sustained evidence if they are to survive a threatening range of Gettier-style examples. Given to some resolution that it needs be that the forth condition for a notion of knowledge is not a function simply of the evidence a Knower actually possesses.

The higher controversial aftermath of Gettier's original counterexamples has left some philosophers doubted of the really philosophical significance of the Gettier problem. Such doubt, however, seems misplaced. One fundamental branch of epistemology seeks understanding of the nature of propositional knowledge. And our understanding exactly what prepositional knowledge is essentially involves our having a Gettier-resistant analysis of such knowledge. If our analysis is not Gettier-resistant, we will lack an exact understanding of what propositional knowledge is. It is epistemologically important, therefore, to have a defensible solution to the Gettier problem, however, demanding such a solution is.

Propositional knowledge (PK) is the type of knowing whose instance are labelled by of a phrase expressing some proposition, e.g., in English a phrase of the form 'that h', where some complete declarative sentence is instantial for 'h'.

Theories of 'PK' differ over whether the proposition that 'h' is involved in a more intimate fashion, such as serving as a way of picking out a proposition attitude required for knowing, e.g., believing that 'h', accepting that 'h' or being sure that 'h'. For instance, the tripartite analysis or standard analysis, treats 'PK' as consisting in having a justified, true belief that 'h', the belief condition requires that anyone who knows that 'h' believes that 'h', the truth condition requires that any known proposition be true, in contrast, some regarded theories do so consider and treat 'PK' as the possession of specific abilities, capabilities, or powers, and that view the proposition that 'h' as needed to be expressed only in order to label a specific instance of 'PK'.

Although most theories of Propositional knowledge (PK) purport to analyse it, philosophers disagree about the goal of a philosophical analysis. Theories of 'PK' may differ over whether they aim to cover all species of 'PK' and, if they do not have this goal, over whether they aim to reveal any unifying link between the species that they investigate, e.g., empirical knowledge, and other species of knowing.

Very many accounts of 'PK' have been inspired by the quest to add a fourth condition to the tripartite analysis so as to avoid Gettier-type counterexamples to it, whereby a fourth condition of evidential truth-sustenance for every true proposition when conjoined with a regaining justification, which may require the justified true belief that is sustained by the collective totality of truths that an adequacy condition of propositional knowledge not restrict justified evidences in relation of deductive support, such that we should countenance varying strengths in notions of propositional knowledge. Restoratively, these strengths are determined by accessibility qualifications on the set of relevant knowledge-precluding underminers. A very demanding concept of knowledge assumes that it need only be logically possible for a Knower to believe a knowledge-precluding undeterminers, and less demanding concepts that it must physically or humanly possible for a Knower to believe knowledge-precluding undeterminers. But even such demanding concepts of knowledge need to rely on a notion of truth-sustaining evidence if they are to survive a threatening range of Gettier-style examples. As the needed fourth condition for a notion of knowledge is not a function simply of the evidence, for which of a Knower actually possesses, as for one fundamental source of epistemology that seeks to understand of the nature that propositional knowledge, and our understanding exactly what propositional knowledge is, that which involves as to having to a Gettier-resistant analysis of such knowledge. If our analysis is not Gettier-resistant, we will lack an exact understanding of what propositional knowledge is. It is epistemologically important, therefore, to have a defensible solution to the Gettier problem, however, demanding such a solution is. And by the resulting need to deal with other counterexamples provoked by these new analyses.

Keith Lehrer (1965) originated a Gettier-type example that has been a fertile source of important variants. It is the case of Mr Notgot, who is in one's office and has provided some evidence, 'e', in response to all of which one forms a justified belief that Mr. Notgot is in the office and owns a Ford, thanks to which one arrives at the justified belief that 'h': 'Someone in the office owns a Ford'. In the example, 'e' consists of such things as Mr. Notgot's presently showing one a certificate of Ford ownership while claiming to own a Ford and having been reliable in the past. Yet, Mr Notgot has just been shamming, and the only reason that it is true that 'h1' is because, unbeknown to oneself, a different person in the office owns a Ford.

Variants on this example continue to challenge efforts to analyse species of 'PK'. For instance, Alan Goldman (1988) has proposed that when one has empirical knowledge that 'h', when the state of affairs (call it h*) expressed by the proposition that 'h' figures prominently in an explanation of the occurrence of one's believing that 'h', where explanation is taken to involve one of a variety of probability relations concerning 'h*', and the belief state. But this account runs foul of a variant on the Notgot case akin to one that Lehrer (1979) has described. In Lehrer's variant, Mr Notgot has manifested a compulsion to trick people into justified believing truths yet falling short of knowledge by of concocting Gettierized evidence for those truths. It we make the trickster's neuroses highly specific ti the type of information contained in the proposition that 'h', we obtain a variant satisfying Goldman's requirement. That the occurrences of 'h*' significantly raise the probability of one's believing that 'h'. (Lehrer himself (1990) has criticized Goldman by questioning whether, when one has ordinary perceptual knowledge that abn object is present, the presence of the object is what explains one's believing it to be present.)

In grappling with Gettier-type examples, some analyses proscribe specific relations between falsehoods and the evidence or grounds that justify one's believing. A simple restriction of this type requires that one's reasoning to the belief that 'h' does not crucially depend upon any false lemma (such as the false proposition that Mr Notgot is in the office and owns a Ford). However, Gettier-type examples have been constructed where one does not reason through and false belief, e.g., a variant of the Notgot case where one arrives at belief that 'h', by basing it upon a true existential generalization of one's evidence: 'There is someone in the office who has provided evidence e', in response to similar cases, Sosa (1991) has proposed that for 'PK' the 'basis' for the justification of one's belief that 'h' must not involve one's being justified in believing or in 'presupposing' any falsehood, even if one's reasoning to the belief does not employ that falsehood as a lemma. Alternatively, Roderick Chisholm (1989) requires that if there is something that makes the proposition that 'h' evident for one and yet makes something else that is false evident for one, then the proposition that 'h' is implied by a conjunction of propositions, each of which is evident for one and is such that something that makes it evident for one makes no falsehood evident for one. Other types of analyses are concerned with the role of falsehoods within the justification of the proposition that 'h' (Versus the justification of one's believing that 'h'). Such a theory may require that one's evidence bearing on this justification not already contain falsehoods. Or it may require that no falsehoods are involved at specific places in a special explanatory structure relating to the justification of the proposition that 'h' (Shope, 1983.).

A frequently pursued line of research concerning a fourth condition of knowing seeks what is called a 'defeasibility' analysis of 'PK.' Early versions characterized defeasibility by of subjunctive conditionals of the form, 'If 'A' was the case then 'B' would be the case'. But more recently the label has been applied to conditions about evidential or justificational relations that are not themselves characterized in terms of conditionals. Early versions of defeasibility theories advanced conditionals where 'A' is a hypothetical situation concerning one's acquisition of a specified sort of epistemic status for specified propositions, e.g., one's acquiring justified belief in some further evidence or truths, and 'B'; concerned, for instance, the continued justified status of the proposition that 'h' or of one's believing that 'h'.

A unifying thread connecting the conditional and non-conditional approaches to defeasibility may lie in the following facts: (1) What is a reason for being in a propositional attitude is in part a consideration, instances of the thought of which have the power to affect relevant processes of propositional attitude formation? (2) Philosophers have often hoped to analyse power ascriptions by of conditional statements: And (3) Arguments portraying evidential or justificational relations are abstractions from those processes of propositional attitude maintenance and formation that manifest rationality. So even when some circumstance, 'R', is a reason for believing or accepting that 'h', another circumstance, 'K' may present an occasion from being present for a rational manifestation of the relevant power of the thought of 'R' and it will not be a good argument to base a conclusion that 'h' on the premiss that 'R' and 'K' obtain. Whether 'K' does play this interfering, 'defeating'. Roles will depend upon the total relevant situation.

Accordingly, one of the most sophisticated defeasibility accounts, which has been proposed by John Pollock (1986), requires that in order to know that 'h', one must believe that 'h' on the basis of an argument whose force is not defeated in the above way, given the total set of circumstances described by all truths. More specifically, Pollock defines defeat as a situation where (1) One believes that 'p' and it is logically possible for one to become justified in believing that 'h' by believing that 'p'. And (2) one actually has a furthering set of beliefs, 'S' logically has a further set of beliefs, 'S', is logically consistent with the proposition that 'h', such that it is not logically possible for one to become justified in believing that 'h', by believing it on the basis of holding the set of beliefs that is the union of 'S'. The belief that 'p' (Pollock, 1986) that Pollock requires for 'PK' that the rational presupposition in favour of one's believing that 'h' created by one's believing that 'p' is undefeated by the set of all truths, including considerations that one does not actually believe. Pollock offers no definition of what this requirement. But he may intend roughly the following: They're 'T' is the set of all true propositions: (1) one believes that 'p' and it is logically possible for one to become justified in believing that 'h' by believing that 'p'. And (II) there are logically possible situations in which one becomes justified in believing that 'h' on the bass of having the belief that 'p' and the beliefs in 'T'. Thus, in the Notgot example, since 'T' includes the proposition that Mr. Notgot does own a Ford, one lack's knowledge because condition (2) is not satisfied.

But given such an interpretation. Pollock's account illustrates the fact that defeasibility theories typically have difficulty dealing with introspective knowledge of one's beliefs. Suppose that some proposition, say that ¦, is false, but one does not realize this and holds the belief that ¦. Condition

(2) has no knowledge that h2? : 'I believe that ¦'. At least this is so if one's reason for believing that h2 includes the presence of the very condition of which one is aware, i.e., one's believing that ¦. It is incoherent to suppose hat one retains the latter reason, also, believes the truth that not-¦. This objection can be avoided, but at the cost of adopting what is a controversial view about introspective knowledge that 'h', namely, the view that one's belief that 'h' is in such cases mediated by some mental state intervening between the mental state of which there is introspective knowledge and he belief that 'h', so that is mental state is rather than the introspected state that it is included in one's reason for believing that 'h'. In order to avoid adopting this controversial view, Paul Moser (1989) gas proposed a disjunctive analysis of 'PK', which requires that either one satisfies a defeasibility condition rather than Pollock's or one believes that 'h' by introspection. However, Moser leaves it precisely hidden to why beliefs arrived at by introspections accounts for being knowledge.

Early versions of defeasibility theories had difficulty allowing for the existence of evidence that is 'merely misleading', as in the case where one does know that 'h3: 'Tom Grabit stole a book from the library', thanks to having seen him steal it, yet where, unbeknown to oneself, Tom's mother out of dementia gas testified that Tom was far away from the library at the time of the theft. One's justifiably believing that she gave the testimony would destroy one's justification for believing that 'h3' if added by itself to one's present evidence.

At least some defeasibility theories cannot deal with the knowledge that one has, while 'h4: has in being in this life, there is no timer for which I believe that 'd', where the proposition that 'd' expresses the details regarding some philosophical matter, e.g., the maximum number of grasses ever to simultaneously grow on the earth? When it just so happens that it is true that 'd', defeasibility analyses typically consider the addition to one's dying thoughts of a belief that 'd' in such a way as too improperly rule out actual knowledge that 'h4'.

A quite different approach to knowledge, and one able to deal with some Gettier-type cases, involves developing some type of causal theory of Propositional knowledge. The interesting thesis that counts as a causal theory of justification (in the meaning of 'causal theory; intended here) is the that of a belief is justified just in case it was produced by a type of process that is 'globally' reliable, that is, its propensity to produce true beliefs-that can be defined (to a god enough approximation) as the proportion of the bailiffs it produces (or would produce where it used as much as opportunity allows) that are true-is sufficient meaningful-variations of this view have been advanced for both knowledge and justified belief. The first formulation of reliability account of knowing appeared in a note by F.P. Ramsey (1931), who said that a belief was knowledge if it is true, certain can obtain by a reliable process. P. Unger (1968) suggested that 'S' knows that 'p' just in case it is not at all accidental that 'S' is right about its being the casse that 'p'. D.M. Armstrong (1973) said that a non-inferential belief qualified as knowledge if the belief has properties that are nominally sufficient for its truth, i.e., guarantee its truth through and by the laws of nature.

Such theories require that one or another specified relation hold that can be characterized by mention of some aspect of cassation concerning one's belief that 'h' (or one's acceptance of the proposition that 'h') and its relation to state of affairs 'h*', e.g., h* causes the belief: h* is causally sufficient for the belief h* and the beliefs have a common cause. Such simple versions of a causal theory are able to deal with the original Notgot case, since it involves no such causal relationship, but cannot explain why there is ignorance in the variants where Notgot and Berent Enç (1984) have pointed out. Sometimes one knows of 'χ' that is thanks to recognizing a feature merely corelated with the presence of 'unity' without endorsing a causal theory themselves, there suggest that it would need to be elaborated so as to allow that one's belief that 'χ' has '-' has been caused by a factor whose correlation with the presence of unity has caused in the essential qualities distinguishing one from another being or identity that thinking of self alone, e.g., by evolutionary adaption in one's ancestors, the disposition that one manifests in acquiring the belief in response to the correlated factor. Not only does this strain the unity of as causal theory by complicating it, but no causal theory without other shortcomings has been able to cover instances of deductively reasoned knowledge.

Causal theories of Propositional knowledge differ over whether they deviate from the tripartite analysis by dropping the requirements that one's believing (accepting) that 'h' be justified. The same variation occurs regarding reliability theories, which present the Knower as reliable concerning the issue of whether or not 'h', in the sense that some of one's cognitive or epistemic states, è, is such that, given further characteristics of oneself-possibly including relations to factors external to one and which one may not be aware-it is nomologically necessary (or at least probable) that 'h'. In some versions, the reliability is required to be 'global' in as far as it must concern a nomologically (probabilistic) relationship of states of type è to the acquisition of true beliefs about a wider range of issues than merely whether or not 'h'. There is also controversy about how to delineate the limits of what constitutes a type of relevant personal state or characteristic. (For example, in a case where Mr Notgot has not been shamming and one does know thereby that someone in the office owns a Ford, such as a way of forming beliefs about the properties of persons spatially close to one, or instead something narrower, such as a way of forming beliefs about Ford owners in offices partly upon the basis of their relevant testimony?)

One important variety of reliability theory is a conclusive reason account, which includes a requirement that one's reasons for believing that 'h' be such that in one's circumstances, if h* were not to occur then, e.g., one would not have the reasons one does for believing that 'h', or, e.g., one would not believe that 'h'. Roughly, the latter are demanded by theories that treat a Knower as 'tracking the truth', theories that include the further demand that is roughly, if it were the case, that 'h', then one would believe that 'h'. A version of the tracking theory has been defended by Robert Nozick (1981), who adds that if what he calls a 'method' has been used to arrive at the belief that 'h', then the antecedent clauses of the two conditionals that characterize tracking will need to include the hypothesis that one would employ the very same method.

But unless more conditions are added to Nozick's analysis, it will be too weak to explain why one lack's knowledge in a version of the last variant of the tricky Mr Notgot case described above, where we add the following details: (a) Mr Notgot's compulsion is not easily changed, (b) while in the office, Mr Notgot has no other easy trick of the relevant type to play on one, for which one arrives at one's belief that 'h', not by reasoning through a false belief ut by basing belief that 'h', upon a true existential generalization of one's evidence.

Nozick's analysis, is, in addition too strong to permit anyone ever to know that 'h': 'Some of my beliefs about beliefs are otherwise, e.g., I might have taken to reject them', if I know that 'h5' is then satisfied by the action of the antecedent of one of Nozick's conditionals, it would involve its being false that 'h5', thereby spoiling satisfactions of the consequent's requirement that I not then believe that 'h5'. For the belief that 'h5' is itself one of my beliefs about beliefs (Shope, 1984).

Some philosophers think that the category of knowing for which true. Justified believing (accepting) is a requirement constituting only a species of Propositional knowledge, construed as an even broader category. They have proposed various examples of 'PK' that do not satisfy the belief and/ort justification conditions of the tripartite analysis. Such cases are often recognized by analyses of Propositional knowledge in terms of powers, capacities, or abilities. For instance, Alan R. White (1982) treats 'PK' as merely the ability to provide a correct answer to a possible questions, however, White may be equating 'producing' knowledge in the sense of producing 'the correct answer to a possible question' with 'displaying' knowledge in the sense of manifesting knowledge. (White, 1982). The latter can be done even by very young children and some nonhuman animals independently of their being asked questions, understanding questions, or recognizing answers to questions. Indeed, an example that has been proposed as an instance of knowing that 'h' without believing or accepting that 'h' can be modified so as to illustrate this point. Two examples concern an imaginary person who has no special training or information about horses or racing, but who in an experiment persistently and correctly picks the winners of upcoming horseraces. If the example is modified so that the hypothetical 'seer' never picks winners but only muses over whether the picked those as chosen horses that just might win, or only reports those horses winning, this behaviour should be as much of a candidate for the person's manifesting knowledge that the horses in question will be winners, as would be the behaviour of picking them prophetically - as they had already won.

These considerations expose limitations in Edward Craig's analysis (1990) of the concept of knowing of a persons being one that gives information, informing against others as the informant, in that of some relation to an inquirer who wants to find out whether or not of 'h'. Craig realizes that counterexamples to his analysis appear to be constituted by Knower who is too recalcitrant to inform the inquirer, or to incapacitate to inform, or too discredited to be worth considering (as with the boy who cried 'Wolf'). Craig admits that this might make preferably some alternative view of knowledge as a different state that helps to explain the presence of the state of being a suitable informant when the latter does obtain. Such the alternate, which offers a recursive definition that concerns one's having the power to proceed in a way representing the state of affairs, causally involved in one's proceeding in this way. When combined with a suitable analysis of representing, this theory of propositional knowledge can be unified with a structurally similar analysis of knowing how to do something.

Knowledge and belief, according to most epistemologists, knowledge entails belief, so that I cannot know that such and such is the case unless I believe that such and such am the case. Others think this entailment thesis can be rendered more accurately if we substitute for belief some closely related attitude. For instance, several philosophers would prefer to say that knowledge entail psychological certainties (Prichard, 1950 and Ayer, 1956) or conviction (Lehrer, 1974) or acceptance (Lehrer, 1989). None the less, there are arguments against all versions of the thesis that knowledge requires having a belief-like attitude toward the known. These arguments are given by philosophers who think that knowledge and belief (or a facsimile) are mutually incompatible (the incomparability thesis), or by ones who say that knowledge does not entail belief, or vice versa, so that each may exist without the other, but the two may also coexist (the separability thesis).

The incompatibility thesis is sometimes traced to Plato (429-347 BC) in view of his claim that knowledge is infallible while belief or opinion is fallible ('Republic' 476-9). But this claim would not support the thesis. Belief might be a component of an infallible form of knowledge in spite of the fallibility of belief. Perhaps, knowledge involves some factor that compensates for the fallibility of belief.

A. Duncan-Jones (1939: Also Vendler, 1978) cites linguistic evidence to back up the incompatibility thesis. He notes that people often say 'I do not believe she is guilty, as I know that she is' and the like, which suggest that belief rule out knowledge. However, as Lehrer (1974) indicates, the above exclamation is only a more emphatic way of saying 'I do not just believe she is guilty, I know she is' where 'just' makes it especially clear that the speaker is signalling that she has something more salient than mere belief, not that she has something inconsistent with belief, namely knowledge. Compare: 'You do not hurt him, you killed him'.

H.A. Prichard (1966) offers a defence of the incompatibility thesis that hinges on the equation of knowledge with certainty (both infallibility and psychological certitude) and the assumption that when we believe in the truth of a claim we are not certain about its truth. Given that belief always involves uncertainty while knowledge never dies, believing something rules out the possibility of knowing it. Unfortunately, however, Prichard gives 'us' no goods reason to grant that states of belief are never ones involving confidence.

A rather different set of concerns arises when actions are specified in terms of doing nothing, saying nothing may be an admission of guilt, and doing nothing in some circumstances may be tantamount to murder. Still, other substitutional problems arise over conceptualizing empty space and time.

Whereas, the standard opposition between those who affirm and those who deny, the real existence of some kind of thing or some kind of fact or state of affairs. Almost any area of discourse may be the focus of its dispute: The external world, the past and future, other minds, mathematical objects, possibilities, universals, moral or aesthetic properties are examples. There be to one influential suggestion, as associated with the British philosopher of logic and language, and the most determinative of philosophers centred round Anthony Dummett (1925), to which is borrowed from the 'intuitivistic' analysis of classical mathematics, and suggested that the unrestricted use of the 'principle of a bivalence' is the trademark of 'realism'. However, this has to overcome counterexamples in both ways: Although Aquinas wads a moral 'realist', he held that moral really was not sufficiently structured to make true or false every moral claim. Unlike Kant who believed that he could use the law of a bivalence happily in mathematics, precisely because it was only our own construction. Realism can itself be subdivided: Kant, for example, combines empirical realism (within the phenomenal world the realist says the right things-encircling objects that really exist and independent of us and our mental stares) with transcendental idealism (the phenomenal world as a whole reflects the structures imposed on it by the activity of our minds as they render it intelligible to us). In modern philosophy the orthodox opposition to realism has been from the philosopher such as Goodman, who, impressed by the extent to which we perceive the world through conceptual and linguistic lenses of our own making.

Assigned to the modern treatment of existence in the theory of 'quantification' is sometimes put by saying that existence is not a predicate. The idea is that the existential quantification, as an operator on a predicate, indicating that the property it expresses has instances. Existence is therefore treated as a second-order property, or a property of properties. It is fitting to say, that in this it is like number, for when we say that these things of a kind, we do not describe the thing (ad we would if we said there are red things of the kind), but instead attribute a property to the kind itself. The parallelled numbers are exploited by the German mathematician and philosopher of mathematics Gottlob Frége in the dictum that affirmation of existence is merely denied of the number nought. A problem, nevertheless, proves accountable for it's crated by sentences like 'This exists', where some particular thing is undirected, such that a sentence seems to express a contingent truth (for this insight has not existed), yet no other predicate is involved. 'This exists' is. Therefore, unlike 'Tamed tigers exist', where a given property, is to say, for having an instance, for the word 'this' that is not located as a property, but only and individual.

Possible worlds seem able to differ from each other purely in the presence or absence of individuals, and not merely in th distribution of exemplification of properties.

The philosophical ponderance over which to set upon the unreal, as belonging to the domain of Being. Nonetheless, there is little for us that can be said with the philosopher's study. So it is not apparent that there can be such a subject for being by itself. Nevertheless, the concept had a central place in philosophy from Parmenides to Heidegger. The essential question of 'why is there something and not of nothing'? Prompting over logical reflection on what it is for a universal to have an instance, nd as long history of attempts to explain contingent existence, by which id to reference and a necessary ground.

In the tradition, ever since Plato, had grounded of a self-sufficient, perfect, unchanging, and external something, as identified with 'good', or 'benevolent, with God, but whose relation with the everyday world remains faintly perceptible as to lack delineation from which is reducible to some indistinct form of darkness. The celebrated argument for the existence of God was first propounded by Anselm in his Proslogin. The argument by defining God as 'something than which nothing greater can be conceived'. God then exists in the understanding since we understand this concept. Even so, if God only existed in the understanding something greater could be conceived for being of which exists in reality is greater than one that exists in the understanding. But then, we can conceive of something greater than that than which nothing greater can be conceived, which is contradictory. Therefore, God cannot exist on the understanding, but exists in reality.

An influential argument (or family of arguments) for the existence of God, finding its premises are that all natural things are dependent for their existence on something else. With the totality that brings of itself depends upon a nondependent, or necessarily existent for being, which is God. Like the argument to design, the cosmological argument was attacked by the Scottish philosopher and historian David Hume (1711-76) and Immanuel Kant.

Its main problem is, nonetheless, that it requires the notion to make sense of the belief toward necessary existence. For if the answer to the question of why anything exists is that some other tings of a similar kind exists, the question that merely simulates as for the believing in that which of 'God' that ends the question, all and all, he must exist necessarily: It must not be an entity of which the same kinds of questions can be raised. The other problem with the argument is attributing concern and care to the deity, not for connecting the necessarily existent being it derives with human values and aspirations.

The ontological argument has been treated by modern theologians such as Barth, following Hegel, not so much as a proof with which to confront the unconverted, but as an explanation of the deep meaning of religious belief. Collingwood, regards the argument s proving not that because our idea of God is that of 'id quo maius cogitare viequit', therefore God exists, but proving that because this is our idea of God, we stand committed to belief in its existence. Its existence is a metaphysical point or absolute presupposition of certain forms of thought.

In the 20th century, modal versions of the ontological argument have been propounded by the American philosophers Charles Hertshorne, Norman Malcolm, and Alvin Plantinga. One version is to define something as unsurpassably great, if it exists and is perfect in every 'possible world'. Then, to allow that it is at least possible that an unsurpassable great being existing. This that there is a possible world in which such a being exists. However, if it exists in one world, it exists in all (for the fact that such a being exists in a world that entails, in at least, it exists and is perfect in every world), so, it exists necessarily. The correct response to this argument is to disallow the apparently reasonable concession that it is possible that such a being exists. This concession is much more dangerous than it looks, since in the modal logic, involved from possibly necessarily 'p', we can take upon to form or shape in the principle of necessity for 'p'. As a symmetrical proof for starting from the assumption that it is a place or location, a situation as it relates to the arraignment of something placed, the finding possibility is a condition that must exist or be established before something can occur or be considered a prerequisite.

The doctrine that it makes an ethical difference of whether an agent actively intervenes to bring about a result, or omits to act in circumstances in which it is foreseen, that a resultant of omissions has the same consequence for occurring. In such a way, suppose that I wish you dead. If I act to bring about your death, I am a murderer, however, if I happily discover you in danger of death, and fail to act to save you, I am not acting, and therefore, according to the doctrine of acts and omissions not a murderer. Critics implore that omissions can be as deliberate and immoral as I am responsible for your food and fact to feed you. Only omission is surely a killing, 'Doing nothing' can be a way of doing something, or in other worlds, absence of bodily movement can also constitute acting negligently, or deliberately, and defending on the context, may be a way of deceiving, betraying, or killing. Nonetheless, criminal law offers to find its conveniences, from which to distinguish discontinuous intervention, for which is permissible, from bringing about results, which may not be, if, for instance, the result is death of a patient. The question is whether the difference, if there is one, is, between acting and omitting to act be discernibly or defined in a way that bars a general moral might.

The double effect of a principle attempting to define when an action that had both good and bad results are morally permissible. I one formation such an action is permissible if (1) The action is not wrong in itself, (2) the bad consequences are not that which is intended (3) the good is not itself a result of the bad consequences, and (4) the two consequential effects are commensurate. Thus, for instance, I might justifiably bomb an enemy factory, foreseeing but intending that the death of nearby civilians, whereas bombing the death of nearby civilians intentionally would be disallowed. The principle has its roots in Thomist moral philosophy, accordingly. St. Thomas Aquinas (1225-74), held that it is meaningless to ask whether a human being is two tings (soul and body) or, only just as it is meaningless to ask whether the wax and the shape given to it by the stamp are one: On this analogy the sound is ye form of the body. Life after death is possible only because a form itself does not perish (pricking is a loss of form).

And is, therefore, in some sense available to reactivate a new body, wherefore, it is not I who survives body death, but I may be resurrected in the same personalized body, that of becoming animated by the same form, in that of which Aquinas' accounted as a person having no privileged self-understanding, we understand of ourselves as we do everything else, by way of experience and abstraction, and knowing the principle of our own lives is an achievement, not as a given. Difficult as this point has led the logical positivist to abandon the notion of foundational epistemology altogether, and to flirt with the coherence theory of truth. Nonetheless, it is widely accepted that trying to make the connection between thought and experiences through basic sentences formality, depends on an untenable 'myth of the given'.

The special way that we each have of knowing our own thoughts, intentions, and sensationalist have brought in the many philosophical 'behaviorist and functionalist tendencies, that have found it important to deny that there is such a special way, arguing the way that I know of my own mind inasmuch as the way that I know of yours, e.g., by seeing what I say when asked. Others, however, point out that the behaviour of reporting the result of introspection in a particular and legitimate kind of behavioural access that deserves notice in any account of historically human psychology. The historical philosophy of reflection upon the astute of history, or of historical, thinking, finds the term was used in the 18th century, e.g., by Volante was to mean critical historical thinking as opposed to the mere collection and repetition of stories about the past. In Hegelian, particularly by conflicting elements within his own system, however, it came to man universal or world history. The Enlightenment confidence was being replaced by science, reason, and understanding that gave history a progressive moral thread, and under the influence of the German philosopher, whom is, in spreading Romanticism, Gottfried Herder (1744-1803), and, Immanuel Kant, affirming this idea took it further to hold, so that philosophy of history cannot be the detecting of any grand system, the unfolding of the evolution of human nature was witnessed in successive stages (the progress of rationality or of Spirit). This essential speculative philosophy of the history given by an extra Kantian twist in the German idealist Johann Fichte, in whom the extra association of temporal succession with logical implication introduces the idea that concepts themselves are the dynamic engines of historical change. The idea is readily intelligible in that their world of nature and of thought become identified. The work of Herder, Kant, Flichte and Schelling are synthesized by Hegel: For which history has a plot, as too, the moral development of man and equally equates with freedom within the state, this in turn is the development of thought, or a logical development in which various necessary moment in the life of the concept are successively achieved and improved upon. Hegel's method is at it's most successful, when the object is built upon the historical fix of ideas, and the evolution of thinking may march in steps with logical oppositions and their resolution encounters red by various systems of thought.

Within the revolutionary communism, Karl Marx (1818-83) and the German social philosopher Friedrich Engels (1820-95), there emerges a rather different kind of story, based upon Hefl's progressive structure not laying the achievement of the goal of history to a future in which the political condition for freedom comes to exist, so that economic and political fears than 'reason' is in the engine room. Although, it is such that speculations upon the history may that it is continued to be written, notably: late examples, by the late 19th century large-scale speculation of tis kind with the nature of historical understanding, and in particular with a comparison between the, methos of natural science and with the historian. For writers such as the German neo-Kantian Wilhelm Windelband and the German philosopher and literary critic and historian Wilhelm Dilthey, it is important to show that the human sciences, such as history is objective and legitimate, nonetheless they are in some way deferent from the enquiry of the scientist. Since the subjective-matter is the past thought and actions of human brings, what is needed and actions of human beings, past thought and actions of human beings, what is needed is an ability to relive that past thought, knowing the deliberations of past agents, as if they were the historian's own. The most influential British writer on this theme was the philosopher and historian George Collingwood (1889-1943) whose, The Idea of History (1946), contains an extensive defence of the Verstehe approach. But it is, nonetheless, the explanation from their actions, that by reliving the situation as our understanding that understanding others are not gained by the tactic use of a 'theory'. This enables us to infer what thoughts or intentionality experienced, again, the matter to which the subjective-matters of past thoughts and actions, as I have a human ability of knowing the deliberations of past agents as if they were the historian's own. The immediate question of the form of historical explanation, and the fact that general laws have other than no place or any apprentices in the order of a minor place in the human sciences, it is also prominent in thoughts about distinctiveness as to regain their actions, but by reliving the situation in or by an understanding of what they have in experience and in thoughts.

The view that everyday attributions of intention, belief and meaning to other persons proceeded via tacit use of a theory that enables ne to construct these interpretations as explanations of their doings. The view is commonly hld along with functionalism, according to which psychological states theoretical entities, identified by the network of their causes and effects. The theory-theory had different implications, depending on which feature of theories is being stressed. Theories may be though of as capable of formalization, as yielding predications and explanations, as achieved by a process of theorizing, as achieved by predictions and explanations, as achieved by a process of theorizing, as answering to empirically evidential principles. They are describable without them, as liable to be overturned newer and better theories, and so on. The main problem with seeing our understanding of others as the outcome of a piece of theorizing is the nonexistence of a medium in which this theory can be couched, as the child learns simultaneously he minds of others and the meaning of terms in its native language.

Our understanding of others is not gained by the tacit use of a 'theory'. Enabling us to infer what thoughts or intentions explain their actions, however, by reliving the situation 'in their moccasins', or from their point of view, and thereby understanding what hey experienced and thought, and therefore expressed. Understanding others is achieved when we can ourselves deliberate as they did, and hear their words as if they are our own. The suggestion is a modern development of the 'Verstehen' tradition associated with Dilthey, Weber and Collngwood.

Much as much, it is therefore, in some sense available to reactivate a new body, however, not that I, who survives bodily death, but I may be resurrected in the same body that becomes reanimated by the same form, in that of Aquinas's account, a person has no self-privileged or self-understanding, that we understand ourselves, just as we do everything else. That through the sense experience, in that of an abstraction, may justly be of knowing the principle of our own lives, is to obtainably achieve, and not as a given. In the theory of knowledge that knowing Aquinas holds the Aristotelian doctrine that knowing entails some similarities between the Knower and what there is to be known: A human's corporal nature, therefore, requires that knowledge start with sense perception. As yet, the same limitations that do not apply of bringing further toward levelling stabilities that are contained within the hierarchical mosaic, such as the celestial heavens that open in bringing forth to angles.

In the domain of theology Aquinas deploys the distraction emphasized by Eringena, between the existence of God in understanding the relevant significance of five arguments which are (1) Motion is only explicable if there exists an unmoved, a first mover (2) the chain of efficient causes demands a first cause (3) the contingent character of existing things in the wold demands a different order of existence, or in other words as something that has a necessary existence (4) the gradations of value in things in the world require the existence of something that is most valuable, or perfect, and (5) the orderly character of events points to a final cause, or end t which all things are directed, and the existence of this end demands a being that ordained it. All the arguments are physico-theological arguments, in that between reason and faith, Aquinas lays out proofs of the existence of God.

He readily recognizes that there are doctrines such that are the Incarnation and the nature of the Trinity, know only through revelations, and whose acceptance is more a matter of moral will. God's essence is identified with his existence, as pure activity. God is simple, containing no potential. No matter how, we cannot obtain knowledge of what God is (his quiddity), perhaps, doing the same work as the principle of charity, but suggesting that we regulate our procedures of interpretation by maximizing the extent to which we see the subject s humanly reasonable, than the extent to which we see the subject as right about things. Whereby remaining content with descriptions that apply to him partly by way of simulation, God reveals of himself but is not himself.

The immediate problem availed of ethics is posed b y the English philosopher Phillippa Foot, in her 'The Problem of Abortion and the Doctrine of the Double Effect' (1967). A runaway train or trolley comes to a section in the track that is under construction and impassable. One person is working on one part and five on the other, and the trolley will put an end to anyone working on the branch it enters. Clearly, to most minds, the driver should steer for the fewest populated branch. But now suppose that, left to itself, it will enter the branch with its five employees that are there, and you as a bystander can intervene, altering the points so that it veers through the other. Is it right for obligors, or even permissible for you to do this? Apparently involving yourself in ways that responsibility ends in a death of one person? After all, who have you wronged if you leave it to go its own way? The situation is similarly standardized of others in which utilitarian reasoning seems to lead to one course of action, but a person's integrity or principles may oppose it.

Describing events that haphazardly come into being or occur by chance but, to come or go casually, if just to make an appearance for something as happening of some needs of no explanation, however, the total, essential or particular being of a person, the individual as thinking of the self's natural amplification by way of something for being negative, or thinking of self alone. Such that it so happens to characterize, as the recognition of self to permeate or flow throughout the permissibility to act, as if it were the formality of consent. We are capable of the abilities to talk of rationality and intention, which are the categories we may apply if we conceive of them as action. We think of ourselves not only passively, as creatures that make things happen. Understanding this distinction gives forth of its many major problems concerning the nature of an agency for the causation of bodily events by mental events, and of understanding the 'will' and 'free will'. Other problems in the theory of action include drawing the distinction between an action and its consequence, and describing the structure involved when we do one thing 'by;' doing another thing. Even the planning and dating where someone shoots someone on one day and in one place, whereby the victim then dies on another day and in another place. Where and when did the murderous act take place?

Causation, least of mention, is not clear that only events are created by and for themselves. Kant fitfully cites or is given by an example, to mention or bring forward as supporting, illustration, or proof for the actions that the example of a cannonball is at resting and placed upon a cushion, but causing the cushion to be the shape that it is, and thus to suggest, that the causal states of affairs or objects or facts may also be casually related. All of which, the central problem is to understand the elements of necessitation or determinacy of the future. Events, Hume thought, are in themselves 'loose and separate': How then are we to conceive of others? The relationship seems not too perceptible, for all that perception gives us (Hume argues) is knowledge of the patterns that events do, actually falling into than any acquaintance with the connections determining the pattern. It is, clearly that our conceptions of everyday objects are largely determined by their casual powers, and all our action is based on the belief that these causal powers are stable and reliable. Although scientific investigation can give us wider and deeper dependable patterns, it seems incapable of bringing us any nearer to the 'must' of causal necessitation. Particular examples' o f puzzles with causalities are apart from general problems of forming any conception of what it is: How are we to understand the casual interaction between mind and body? How can the present, which exists, or its existence to a past that no longer exists? How is the stability of the casual order to be understood? Is backward causality possible? Is causation a concept needed in science, or dispensable?

The news concerning free-will, is nonetheless, a problem for which is to reconcile our everyday consciousness of ourselves as agent, with the best view of what science tells us that we are. Determinism is one part of the problem. It may be defined as the doctrine that every event has a cause. More precisely, for any event 'C', there will be one antecedent states of nature 'N', and a law of nature 'L', such that given L, N will be followed by 'C'. But if this is true of every event, it is true of events such as my doing something or choosing to do something. So my choosing or doing something is fixed by some antecedent state 'N' an d the laws. Since determinism is universally fixed, and so backwards into those of the events, for which I am clearly not responsible (events before my birth, for example). So, no events can be voluntary or free, where that they come about purely because of my willing them I could have done otherwise. If determinism is true, then there will be antecedent states and laws already determining such events: How then can I truly be said to be their author, or be responsible for them?

Reactions to this problem are commonly classified as: (1) Hard determinism. This accepts the conflict and denies that you have real freedom or responsibility (2) Soft determinism or compatibility, whereby reactions in this family assert that everything you should be from a notion of freedom is quite compatible with determinism. In particular, if your actions are caused, it can often be true of you that you could have done otherwise if you had chosen, and this may be enough to render you liable to be held unacceptable (the fact that previous events will have caused you to choose as you did, and is deemed irrelevant on this option). (3) Libertarianism, as this is the view that while compatibilism is only an evasion, there is a more substantiative, real notions of freedom that can yet be preserved in the face of determinism (or, of indeterminism). In Kant, while the empirical or phenomenal self is determined and not free, whereas the noumenal or rational self is capable of being rational, free action. However, the noumeal self exists outside the categorical priorities of space and time, as this freedom seems to be of a doubtful value as other libertarian avenues do include of suggesting that the problem is badly framed, for instance, because the definition of determinism breaks down, or postulates by its suggesting that there are two independent but consistent ways of looking at an agent, the scientific and the humanistic, wherefore it is only through confusing them that the problem seems urgent. Nevertheless, these avenues have gained general popularity, as an error to confuse determinism and fatalism.

The dilemma for which determinism is for itself often supposes of an action that seems as the end of a causal chain, or, perhaps, by some hieratical set of suppositional actions that would stretch back in time to events for which an agent has no conceivable responsibility, then the agent is not responsible for the action.

Once, again, the dilemma adds that if an action is not the end of such a chain, then either two or one of its causes occurs at random, in that no antecedent events brought it about, and in that case nobody is responsible for it's ever to occur. So, whether or not determinism is true, responsibility is shown to be illusory.

Still, there is to say, to have a will is to be able to desire an outcome and to purpose to bring it about. Strength of will, or firmness of purpose, is supposed to be good and weakness of will.

A mental act of willing or trying whose presence is sometimes supposed to make the difference between intentional and voluntary action, as well of mere behaviour. The theories that pustulate as for having of such acts are problematic, and the idea that they make the required difference is a case of explaining a phenomenon by citing another that raises exactly the same problem, since the intentional or voluntary nature of the set of volition now needs explanation. For determinism to act in accordance with the law of autonomy or freedom, is that in ascendance with universal moral law and regardless of selfish advantage.

A categorical notion in the work as contrasted in Kantian ethics present of a hypothetical imperative that intensifies th forming conjugation of that which indicates of the action or state, for which are to place of only the given antecedent desire or project. 'If you want to look wise, stay quiet'. The injunction to stay quiet can only prove positive, if it applies to those with the antecedent desire or inclination: If one has no inclining inclination for which leaves no desire for his appearing to look wise, the injunction or advice lapses. A categorical imperative cannot be so avoided, it is a requirement that binds anybody, regardless of their inclination. It could be repressed as, for example, 'Tell the truth (regardless of whether you want to or not)'. The distinction is not always mistakably presumed or absence of the conditional or hypothetical form: 'If you crave drink, don't become a bartender' may be regarded as an absolute injunction applying to anyone, although only activated in the case of those with the stated desire.

In Grundlegung zur Metaphsik der Sitten (1785), Kant discussed some of the given forms of categorical imperatives, such that of (1) The formula of universal law: 'act only on that maxim through which you can at the same time will that it should become universal law', (2) the formula of the law of nature: 'Act as if the axiom of your action were to come to the conclusion through your will, a broadening law of nature', (3) the formula of the end-in-itself, 'Act in such a way that you always trat humanity of whether in your own person or in the person of any other, never simply as an end, but always at the same time as an end', (4) the formula of autonomy, or consideration, the 'will' of every rational being a will which makes universal law, and (5) the formula of the Kingdom of Ends, which provides a model for systematic imperatives that intensify in the different rational beings under common laws.

A central object in the study of Kant's ethics is to understand the expressions of the inescapable, binding requirements of their categorical importance, and to understand whether they are equivalent at some deep level. Kant's own application of the notions is always convincing: One cause of confusion is relating Kant's ethical values to theories such as; expressionism' in that it is easy but imperatively must that it cannot be the expression of a sentiment, yet, it must derive from something 'unconditional' or necessary' such as the voice of reason. The standard mood of sentences used to issue request and commands are their imperative needs to issue as basic the need to communicate information, and as such to animals signalling systems may as often be interpreted either way, and understanding the relationship between commands and other action-guiding uses of language, such as ethical discourse. The ethical theory of 'prescriptivism' in fact equates the two functions. A further question is whether there is an imperative logic. 'Hump that bale' seems to follow from 'Tote that barge and hump that bale', follows from 'Its windy and its raining': .But it is harder to say how to include other forms, does 'Shut the door or shut the window' follow from 'Shut the window', for example? The usual way to develop an imperative logic is to work in terms of the possibility of satisfying the other one command without satisfying the other, thereby turning it into a variation of ordinary deductive logic.

Despite the fact that the use of people and their ethics amounts to the same thing, for which of its use is that I restart morality to systems such as Kant, the bases of which on a notion is given to that as our duty, obligation, and principles of conduct, reserving ethics for the more Aristotelian approach to practical reasoning as based on the valuing notions that are characterized by their particular virtue, and generally avoiding the separation of 'moral' considerations from other practical considerations. The scholarly issues are complicated and complex, with some writers seeing Kant as more Aristotelian. And Aristotle, to a greater extent, that has been involved with a separate sphere of responsibility and duty, than the simple contrast suggests.

A major topic of philosophical inquiry, especially in Aristotle, and subsequently since the 17th and 18th centuries, when the 'science of man' began to probe into human motivation and emotion. For such as these, the French moralist, or Hutcheson, Hume, Smith and Kant, a prime task as to delineate the variety of human reactions and motivations. Such an inquiry would locate our propensity for moral thinking among other faculties, such as perception and reason, and other tendencies as empathy, sympathy or self-interest. The task continues especially in the light of a post-Darwinian understanding of us.

In some moral systems, are concerned with the judgment of the goodness or badness of human action and character as the conforming standards of what is right or jus t in behaviour or the sense of right and wrong as having psychological rather than physical or tangible effects, based on the strong likelihood or firm conviction than on the actual evidence. Immanuel Kant, had in advocating that real morality comes only with interactivity, justly because it is right, however, if you do what is purposely becoming equitable, but from some other equitable motive, such as the fear or prudence, such is that no moral merit will accrue upon you. Yet, that in turn seems to discount other admirable motivations, as acting from main-sheet benevolence, or 'sympathy'. The question is how to balance these opposing ideas and how to understand acting from a sense of obligation without duty or rightness, through which their beginning to seem a kind of fetish. It thus stands opposed to ethics and relying on highly general and abstractive principles, particularly. Those associated with the Kantian categorical imperatives. The view may go as far back as to say that taken in its own, no consideration point, for that which of any particular way of life, that, least of mention, the contributing steps so taken as forwarded by reason or be to an understanding estimate that can only proceed by identifying salient features of a situation that weighs on one's side or another.

As random moral dilemmas set out with intense concern, inasmuch as philosophical matters that exert a profound but influential defence of common sense. Situations, in which each possible course of action breeches some otherwise binding moral principle, are, nonetheless, serious dilemmas making the stuff of many tragedies. The conflict can be described in different was. One suggestion is that whichever action the subject undertakes, that he or she does something wrong. Another is that his is not so, for the dilemma that in the circumstances for what she or he did was right as any alternate. It is important to the phenomenology of these cases that action leaves a residue of guilt and remorse, even though it had proved it was not the subject's fault that he was considering the dilemma, that the rationality of emotions can be contested. Any normality with more than one fundamental principle seems capable of generating dilemmas, however, dilemmas exist, such as where a mother must decide which of two children to sacrifice, least of mention, no principles are pitted against each other, only if we accept that dilemmas from principles are real and important, this fact can then be used to approach in them, such as of 'utilitarianism', to espouse various kinds may, perhaps, be centred upon the possibility of relating to independent feelings, liken to recognize only one sovereign principle. Alternatively, of regretting the existence of dilemmas and the unordered jumble of furthering principles, in that of creating several of them, a theorist may use their occurrences to encounter upon that which it is to argue for the desirability of locating and promoting a single sovereign principle.

Nevertheless, some theories into ethics see the subject in terms of a number of laws (as in the Ten Commandments). Th status of these laws may be that they are the edicts of a divine lawmaker, or that they are truths of reason, given to its situational ethics, virtue ethics, regarding them as at best rules-of-thumb, and, frequently disguising the great complexity of practical representations that for reason has placed the Kantian notions of their moral law.

In continence, the natural law possibility points of the view of the states that law and morality are especially associated with St Thomas Aquinas (1225-74), such that his synthesis of Aristotelian philosophy and Christian doctrine was eventually to provide the main philosophical underpinning of th Catholic church. Nevertheless, to a greater extent of any attempt to cement the moral and legal order and together within the nature of the cosmos or the nature of human beings, in which sense it found in some Protestant writings, under which had arguably derived functions. From a Platonic view of ethics and its agedly implicit advance of Stoicism. Its law stands above and apart from the activities of human lawmakers: It constitutes an objective set of principles that can be seen as in and for themselves by of 'natural usages' or by reason itself, additionally, (in religious verses of them), that express of God's will for creation. Nonreligious versions of the theory substitute objective conditions for humans flourishing as the source of constraints, upon permissible actions and social arrangements within the natural law tradition. Different views have been held about the relationship between the rule of the law and God's will. Grothius, for instance, take a side with the view that the content of natural law is independent of any will, including that of God.

While the German natural theorist and historian Samuel von Pufendorf (1632-94) takes the opposite view. His great work was the De Jure Naturae et Gentium, 1672, and its English translation are 'Of the Law of Nature and Nations, 1710. Pufendorf was influenced by Descartes, Hobbes and the scientific revolution of the 17th century, his ambition was to introduce a newly scientific 'mathematical' treatment on ethics and law, free from the tainted Aristotelian underpinning of 'scholasticism'. Like that of his contemporary-Locke. His conception of natural laws includes rational and religious principles, making it only a partial forerunner of more resolutely empiricist and political treatment in the Enlightenment.

Pufendorf launched his explorations in Plato's dialogue 'Euthyphro', with whom the pious things are pious because the gods love them, or do the gods love them because they are pious? The dilemma poses the question of whether value can be conceived as the upshot o the choice of any mind, even

116

a divine one. On the fist option the choice of the gods crates goodness and value. Even if this is intelligible, it seems to make it impossible to praise the gods, for it is then vacuously true that they choose the good. On the second option we have to understand a source of value lying behind or beyond the will even of the gods, and by which they can be evaluated. The elegant solution of Aquinas is and is therefore distinct from a specified extent as near to one side of an area and broadly wider as extended to indicate the likelihood or certain futurity. But to us, for purposes of indication that we have the capacity or ability, as to say, that 'willing' is an indication of customary or habitual action, i.e., our wish is to indicate probability or expectations that summons of the desire 'to do what you will'.

The dilemma arises whatever the source of authority is supposed to be. Do we care about the 'good' because it is good, or do we just call 'good' those things that we care about? It also generalizes to affect our understanding of the authority of other things: Mathematics, or necessary truth, for example, are truths necessary because we deem them to be so, or do we deem them to be so because they are necessary?

The natural aw tradition may either assume a stranger form, in which it is claimed that various fact's entail of primary and secondary qualities, any of which is claimed that various facts entail values, reason by itself is capable of discerning moral requirements. As in the ethics of Kant, a set of principles of right conducts, it holds the basis of theory as a science that has been repeatedly tested or is widely accepted and can be used to make certain predictions about natural phenomena or a system of moral values, even so, dealing with ethical rules or standards for which of governing the conduct of a person or the members of a specialized profession (medical ethics) and the morals of the specific moral choices that one carries to be made by a person as inferred from moral philosophy. These requirements are supposedly binding of all human beings, regardless of their desires.

The supposed natural or innate abilities of the mind to know the first principle of ethics and moral reasoning, wherein, those expressions are assigned and related to those that distinctions are which make in terms contribution to the function of the whole, as completed definitions of them, their phraseological impression is termed 'synderesis' (or, syntetesis) although traced to Aristotle, the phrase came to the modern era through St Jerome, whose scintilla conscientiae (gleam of conscience) wads a popular concept in early scholasticism. Nonetheless, it is mainly associated in Aquinas as an infallible natural, simple and immediate apprehensions of first moral principles. Conscience, by contrast, is to a greater extent with particular instances of right and wrong, and can be in error, under which case, the assertion that is taken as fundamental, at least for the purposes of the enquiry in hand.

It is, nevertheless, the view interpreted within their particular state of law and morality, especially associated with Aquinas and the subsequent scholastic tradition, showing for itself the enthusiasm for reform for its own sake. Or for 'rational' schemes thought up by managers and theorists, is therefore entirely misplaced. Major o exponent s of this theme include the British absolute idealist Herbert Francis Bradley (1846-1924) and Austrian economist and philosopher Friedrich Hayek. The notably the idealism of Bradley, there ids the same doctrine that change is contradictory and consequently unreal: The Absolute is changeless. A way of sympathizing a little with his idea is to reflect that any scientific explanation of change will proceed by finding an unchanging law operating, or an unchanging quantity conserved in the change, so that explanation of change always proceeds by finding that which is unchanged. The metaphysical problem of change is to shake off the idea that each moment is created afresh, and to obtain a conception of events or processes as having a genuinely historical reality, Really extended and unfolding in time, as opposed to being composites of discrete temporal atoms. A step toward this end may be to see time itself not as an infinite container within which discrete events are located, bu as a kind of logical construction from the flux of events. This relational view of time was advocated by Leibniz and a subject of the debate between him and Newton's Absolutist pupil, Clarke.

Generally, nature is an indefinitely mutable term, changing as our scientific conception of the world changes, and often best seen as signifying a contrast with something considered not part of nature. The term applies to both an individual species (it is the nature of gold to be dense or that dogs are to be friendly), and also to the natural world as the sum total in the summation of its parts that constitutes the whole. The sense in which it applies to a differentiation of species, quickly links with ethical and aesthetic ideating form of an idea, so conceiving to the ideational images as ever to think about or thought.

The resolution, as an entity or an idea or a quality perceived as known or though to have its own existence, for which of occupying a particular point in space and time, as something referred by a word, a sign, an idea as an object of a thing as needed for an activity or special purpose can be thought. As a notion or an utterance e of some linguistic effort. Yet, having an end objective whereas, concerning the many things I have on my mind, coming as a turn of events or of circumstance can one of a situation deals with things within a persistent, illogical feeling, as a desire or and aversion for which is obsessively the action from which is suitable and satisfying for whom of owing to things that which he owes among those obsessed. The act tends to bring to provoke for the presence or participation of an action being satisfied or suitably given by anything else, but first is given by something as the thing that thinking within the realm of thought-in-itself lends of much to the philosophy to have formulated in the mind.

The associations of what are natural with what it is good to become is visible in Plato, and is the central idea of Aristotle's philosophy of nature. Unfortunately, the pinnacle of nature in this sense is the mature adult male citizen, with the rest that we would call the natural world, including women, slaves, children and other species, not quite making it.

Nature in general can, however, function as a foil to any idea inasmuch as a source of ideals: In this sense fallen nature is contrasted with a supposed celestial realization of the 'forms'. The theory of 'forms' is probably the most characteristic, and most contested of the doctrines of Plato. In the background, i.e., the Pythagorean conception of form as the initial orientation to physical nature, bu also the sceptical doctrine associated with the Greek philosopher Cratylus, and is sometimes thought to have been a teacher of Plato before Socrates. He is famous for capping the doctrine of Ephesus of Heraclitus, whereby the guiding idea of his philosophy was that of the logos, is capable of being heard or hearkened to by people, it unifies opposites, and it is somehow associated with fire, which is preeminent among the four elements that Heraclitus distinguishes: Fire, air (breath, the stuff of which souls composed), earth, and water. Although he is principally remembered for the doctrine of the 'flux' of all things, and the famous statement that you cannot step into the same river twice, for new waters are ever flowing in upon you. The more extreme implication of the doctrine of flux, e.g., the impossibility of categorizing things truly, do not seem consistent with his general epistemology and views of meaning, and were to his follower Cratylus, although the proper conclusion of his views was that the flux cannot be captured in words. According to Aristotle, he eventually held that since 'regarding that which everywhere in every respect is changing of nothing is just to stay silent and fiddle one's fingers from side to side. Plato 's theory of forms can be seen in part, as an action against the impasse to which Cratylus was driven. Even so, the shape and structure of an object are an outward appearance of a person considered separately from the face or head, if only to imply the essence of some thing in which a thing exists, or manifest itself as a kind of form of a life, fixed in order of words or procedures, as for its use in a formula or document filled with mystery. The method of arrangements of coordinating compositions or organized discourse, its present ideas in an outline formed in clay, to develop in the mind, conceived from an opinion that forms a treatise in the form of a dialogue. To give form to shape or to have by acquiring to the external outline of a thing, as to be sure, to assume of a specific form, shape or pattern as having the form of plexuity.

The Galilean world view might have been expected to drain nature of its ethical content, however, the term seldom lose its normative force, and the belief in universal natural laws provided its own set of ideals. In the 18th century for example, a painter or writer could be praised as natural, where the qualities expected would include normal (universal) topics treated with simplicity, economy, regularity and harmony. Later on, nature becomes an equally potent emblem of irregularity, wildness, and fertile diversity, but also associated with progress of human history, its incurring definition that has been taken to fit many things as well as transformation, including ordinary human self-consciousness. Nature, being in contrast within the integrated phenomenons and may include (1) that which is deformed or grotesque or fails to achieve its proper form or function or just the statistically uncommon or unfamiliar, (2) the supernatural, or the world of gods and invisible agencies, (3) the world of rationality and unintelligence, conceived of as distinct from the biological and physical order, or the product of human intervention, and (5) related to that, the world of convention and artifice.

Different conceptions of nature continue to have ethical overtones, for example, the conception of 'nature red in tooth and claw' often provides a justification for aggressive personal and political relations, or the idea that it is women's nature to be one thing or another is taken to be a justification for differential social expectations. The term functions as a fig-leaf for a particular set of stereotypes, and is a proper target of a great quantity or extended amount in the feminist writings. Feminist epistemology has asked whether different ways of knowing for instance with different criteria of justification, and different emphases on logic and imagination, characterize male and female attempts to understand the world. Such concerns include awareness of the 'masculine' self-image, itself a social variable and potentially distorting pictures of what thought and action should be. Again, there is a spectrum of concerns from the highly theoretical principles to the relatively practical. In this latter area particular attention is given to the institutional biases that stand in the way of equal opportunities in science and other academic pursuits, or the ideologies that stand in the way of women seeing themselves as leading contributors to various disciplines. However, to more radical feminists such concerns merely exhibit women wanting for themselves the same power and rights over others that men have claimed, and failing to confront the real problem, which is how to live without such symmetrical powers and rights.

In biological determinism, not only influences but constraints and makes inevitable our development as persons with a variety of traits. At its silliest the view postulates such entities as a gene predisposing people to poverty, and it is the particular enemy of thinkers stressing the parental, social, and political determinants of the way we are.

The philosophy of social science is more heavily intertwined with actual social science than in the case of other subjects such as physics or mathematics, since its question is centrally whether there can be such a thing as sociology. The idea of a 'science of man', devoted to uncovering scientific laws determining the basic dynamic s of human interactions was a cherished ideal of the Enlightenment and reached its heyday with the positivism of writers such as the French philosopher and social theorist Auguste Comte (1798-1957), and the historical materialism of Marx and his followers. Sceptics point out that what happens in society is determined by peoples' own ideas of what should happen, and like fashions those ideas change in unpredictable ways as self-consciousness is susceptible to change by any number of external event s: Unlike the solar system of celestial mechanics a society is not at all a closed system evolving in accordance with a purely internal dynamic, but constantly responsive to shocks from outside.

The sociological approach to human behaviour is based on the premise that all social behaviour has a biological basis, and seeks to understand that basis in terms of genetic encoding for features that are then selected for through evolutionary history. The philosophical problem is essentially one of methodology: Of finding criteria for identifying features that can usefully be explained in this way, and fo r finding criteria for assessing various genetic stories that might provide useful explanations.

Among the features that are proposed for this kind of explanations are such things as male dominance, male promiscuity versus female fidelity, propensities to sympathy and other emotions, and the limited altruism characteristic of human beings. The strategy has proved unnecessarily controversial, with proponents accused of ignoring the influence of environmental and social factors in moulding people's characteristics, e.g., at the limit of silliness, by postulating a 'gene for poverty', however, there is no need for the approach for committing such errors, since the feature explained sociobiological may be indexed to environment: For instance, it may well be a propensity to develop some feature in some other environments (for even a propensity to develop propensities . . .) The main problem is to separate genuine explanation from speculative, just so stories which may or may not identify as really selective mechanisms.

Subsequently, in the 19th century attempts were made to base ethical reasoning on the presumed facts about evolution. The movement is particularly associated with the English philosopher of evolution Herbert Spencer (1820-1903). His first major work was the book Social Statics (1851), which advocated an extreme political libertarianism. The Principles of Psychology was published in 1855, and his very influential Education advocating natural development of intelligence, the creation of pleasurable interest, and the importance of science in the curriculum, appeared in 1861. His First Principles (1862) was followed over the succeeding years by volumes on the Principles of biology and psychology, sociology and ethics. Although he attracted a large public following and attained the stature of a sage, his speculative work has not lasted well, and in his own time there were dissident voices. T.H. Huxley said that Spencer's definition of a tragedy was a deduction killed by a fact. Writer and social prophet Thomas Carlyle (1795-1881) called him a perfect vacuum, and the American psychologist and philosopher William James (1842-1910) wondered why half of England wanted to bury him in Westminister Abbey, and talked of the 'hurdy-gurdy' monotony of him, his whole system, as if knocked together out of cracked hemlock.

The premise is that later elements in an evolutionary path are better than earlier ones, the application of this principle then requires seeing western society, laissez-faire capitalism, or some other object of approval, as more evolved than more 'primitive' social forms. Neither the principle nor the applications command much respect. The version of evolutionary ethics called 'social Darwinism' emphasizes the struggle for natural selection, and drawn the conclusion that we should glorify such struggles, usually by enhancing competitive and aggressive relations between people in society or between societies themselves. More recently the relation between evolution and ethics has been rethought in the light of biological discoveries concerning altruism and kin-selection.

In that, the study of the say in which a variety of higher mental functions may be adaptions applicable of a psychology of evolution, formed in response or readily reacting to suggest as to influences, appeals or efforts for which were responsively by answering or replying to respond in the selection pressures on human populations through evolutionary time. Candidates for such theorizing include material and paternal motivations, capabilities for love and friendship, the development of language as a signalling system, cooperative and aggressive tendencies, our emotional repertoires, our moral reaction, including the disposition to direct and punish those who cheat on an agreement or who turn toward free-riders-those of which who take away the things of others, our cognitive structure and many others. Evolutionary psychology goes hand-in-hand with neurophysiological evidence about the underlying circuitry in the brain which subserves the psychological mechanisms it claims to identify.

For all that, an essential part of the British absolute idealist Herbert Bradley (1846-1924) was largely on the grounds that individualized through the community and the individual instrument of self, the essential qualities as distinguishing one person from another, contributing socially and other ideals conforming to an ultimate form or standards of perfection or excellence, but considered to the best of its kind. However, truth as formulated in language is always partial, and dependent upon categories that they are inadequate to the harmonious whole. Nevertheless, these self-contradictory elements somehow contribute to the harmonious whole, or Absolute, lying beyond categorization. Although absolute idealism maintains few adherents today, Bradley's general dissent from empiricism, his holism, and the brilliance and styles of his writing continue to make him the most interesting of the late 19th century writers influenced by the German philosopher Friedrich Hegel (1770-1831).

Understandably, something less than the fragmented division that belonging of Bradley's case has a preference, voiced much earlier by the German philosopher, mathematician and polymath, Gottfried Leibniz (1646-1716), for categorical monadic properties over relations. He was particularly troubled by the relation between that which is perceived directly known and the more that is to grasp in the mind with clarity or certainty is to know it, than known I know that I face. In philosophy, the Romantics took from the German philosopher and founder of critical philosophy Immanuel Kant (1724-1804) both the emphasis on free-will and the doctrine that reality is ultimately spiritual, with nature itself a mirror of the human soul. To fix upon one among alternatives as the one to be taken, Friedrich

Schelling (1775-1854) foregathering natures becoming a creative spirit whose aspiration is ever further and more to a completed self-realization, although an apparent movement of more general naturalization is territorially imperative. Romanticism drew on the same intellectual and emotional resources as German idealism was increasingly culminating in the philosophy of Hegel (1770-1831) and of absolute idealism.

Being such in comparison with nature may include (1) that which is deformed or grotesque, or fails to achieve its proper form or function, or just the statistically uncommon or unfamiliar, (2) the supernatural, or th world of gods and invisible agencies, (3) the world of rationality and intelligence, conceived of as distinct from the biological and physical order, (4) that which is manufactured and artefactual, or the product of human invention, and (5) related to it, the world of convention and artifice.

Different conceptions of nature continue to have ethical overtones, for example, the conception of 'nature red in tooth and claw' often provides a justification for aggressive personal and political relations, or the idea that it is a women's nature to be one thing or another, as taken to be a justification for differential social expectations. The term functions as a fig-leaf for a particular set of stereotypes, and is a proper target of much 'feminist' writing.

This brings to question, that most of all ethics are contributively distributed as an understanding for which a dynamic function in and among the problems that are affiliated with human desire and needs the achievements of happiness, or the distribution of goods. The central problem specific to thinking about the environment is the independent value to place on 'such-things' as preservation of species, or protection of the wilderness. Such protection can be supported as a means to ordinary human ends, for instance, when animals are regarded as future sources of medicines or other benefits. Nonetheless, many would want to claim a non-utilitarian, absolute value for the existence of wild things and wild places. It is with this, which things consist, they put us in our proper place, and failure to appreciate this value is not only an aesthetic failure but one of due humility and reverence, a moral disability. The problem is one of expressing this value, and mobilizing it against utilitarian agents for developing natural areas and exterminating a fundamental category of taxonomic classification, as raking an organism belonging to such a category, represented in classes of individuals or objects by virtue of their common attributes and assigned a common name, a division subordinate to a varying type species that gave of the elements for which apply to form or appearance. These species, more or less are at risk, in that of determining the incompatibilities as associated by need or needs for 'wanting' of a particular inhabitancy, as living rather than succumbing to the moment.

Many concerns and disputed clusters around the idea associated with the term 'substance'. The substance of a thing may be considered in: (1) plain essencity, or that which makes it what it is. This will ensure that the substance of a thing is that which remains through change in properties. Again, in Aristotle, this essence becomes more than just the matter, but a unity of matter and form. (2) That which can exist by itself, or does not need a subject for existence, in the way that properties need objects, hence (3) that which bears properties, as a substance is then the subject of predication, that about which things are said as opposed to the things said about it. Substance in the last two senses stands opposed to modifications such as quantity, quality, relations, and so forth it is hard to keep this set of ideas distinct from the doubtful notion of a substratum, something distinct from any of its properties, and hence, as an incapable characterization. The notion of substance tends to vanquish in the empiricist thought in fewer of the sensible questions of things with the notion of that in which they infer of giving way to an empirical notion of their regular occurrence. However, this is in turn is problematic, since it only makes sense to talk of the occurrence of an instance of qualities, not of quantities themselves. So the problem of what it is for a value quality to be the instance that remains.

Metaphysic inspired by modern science tends to reject the concept of substance in favour of concepts such as that of a field or a process, each of which may seem to provide a better example of a fundamental physical category.

It must be spoken of a concept that is deeply embedded in 18th century aesthetics, but deriving from the 1st century rhetorical treatise On the Sublime, by Longinus. The sublime is great, fearful, noble, calculated to arouse sentiments of pride and majesty, as well as awe and sometimes terror. According to Alexander Gerard's writing in 1759, 'When a large object is presented, the mind expands itself to the extent of that objects, and is filled with one grand sensation, which totally possessing it, composes it into a solemn sedateness and strikes it with deep silent wonder, and administration': It finds such a difficulty in spreading itself to the dimensions of its object, as enliven and invigorates which this occasions, it sometimes images itself present in every part of the sense which it contemplates, and from the sense of this immensity, feels a noble pride, and entertains a lofty conception of its own capacity.

In Kant's aesthetic theory the sublime 'raises the soul above the height of vulgar complacency'. We experience the vast spectacles of nature as 'absolutely great' and of irresistible might and power. This perception is fearful, but by conquering this fear, and by regarding as small 'those things of which we are wont to be solicitous' we quicken our sense of moral freedom. So we turn the experience of frailty and impotence into one of our true, inward moral freedom as the mind triumphs over nature, and it is this triumph of reason that is truly sublime. Kant thus paradoxically places our sense of the sublime in an awareness of us as transcending nature, than in an awareness of us as a frail and insignificant part of it.

Nevertheless, the doctrine that all relations are internal was a cardinal thesis of absolute idealism, and a central point of attack by the British philosopher's George Edward Moore (1873-1958) and Bertrand Russell (1872-1970). It is a kind of 'essentialism', stating that if two things stand in some relationship, then they could not be what they are, did they not do so, if, for instance, I am wearing a hat mow, then when we imagine a possible situation that we would be got to describe as my not wearing the hat now, we would strictly not be imaging as one and the hat, but only some different individual.

The countering partitions a doctrine that bears some resemblance to the metaphysically based view of the German philosopher and mathematician Gottfried Leibniz (1646-1716) that if a person had any other attributes that the ones he has, he would not have been the AME person. Leibniz thought that when asked what would have happened if Peter had not denied Christ. That being that if I am asking what had happened if Peter had not been Peter, denying Christ is contained in the complete notion of Peter. But he allowed that by the name 'Peter' might be understood as 'what is involved in those attributes [of Peter] from which the denial does not follow'. In order that we are held accountable to allow of external relations, in that these being relations which individuals could have or not depending upon contingent circumstances. The relation of ideas is used by the Scottish philosopher David Hume (1711-76) in the First Enquiry of Theoretical Knowledge. All the objects of human reason or enquiring naturally, be divided into two kinds: To a unit all the, 'relations of ideas' and 'matter of fact '(Enquiry Concerning Human Understanding) the terms reflect the belief that any thing that can be known dependently must be internal to the mind, and hence transparent to us.

In Hume, objects of knowledge are divided into matter of fact (roughly empirical things known by of impressions) and the relation of ideas. The contrast, also called 'Hume's Fork', is a version of the speculative deductivity distinction, but reflects the 17th and early 18th centauries behind that the deductivity is established by chains of infinite certainty as comparable to ideas. It is extremely important that in the period between Descartes and J.S. Mill that a demonstration is not, but only a chain of 'intuitive' comparable ideas, whereby a principle or maxim can be established by reason alone. It is in this sense that the English philosopher John Locke (1632-1704) who believed that theologically and moral principles are capable of demonstration, and Hume denies that they are, and also denies that scientific enquiries proceed in demonstrating its results.

A mathematical proof is formally inferred as to an argument that is used to show the truth of a mathematical assertion. In modern mathematics, a proof begins with one or more statements called premises and demonstrates, using the rules of logic, that if the premises are true then a particular conclusion must also be true.

The accepted methods and strategies used to construct a convincing mathematical argument have evolved since ancient times and continue to change. Consider the Pythagorean theorem, named after the 5th century Bc Greek mathematician and philosopher Pythagoras, which states that in a right-angled triangle, the square of the hypotenuse is equal to the sum of the squares of the other two sides. Many early civilizations considered this theorem true because it agreed with their observations in practical situations. But the early Greeks, among others, realized that observation and commonly held opinions do not guarantee mathematical truth. For example, before the 5th century Bc it was widely believed that all lengths could be expressed as the ratio of two whole numbers. But an unknown Greek mathematician proved that this was not true by showing that the length of the diagonal of a square with an area of 1 is the irrational number.

The Greek mathematician Euclid laid down some of the conventions central to modern mathematical proofs. His book The Elements, written about 300 Bc, contains many proofs in the fields of geometry and algebra. This book illustrates the Greek practice of writing mathematical proofs by first clearly identifying the initial assumptions and then reasoning from them in a logical way in order to obtain a desired conclusion. As part of such an argument, Euclid used results that had already been shown to be true, called theorems, or statements that were explicitly acknowledged to be self-evident, called axioms; this practice continues today.

In the 20th century, proofs have been written that are so complex that no one person understands every argument used in them. In 1976, a computer was used to complete the proof of the four-colour theorem. This theorem states that four colours are sufficient to colour any map in such a way that regions with a common boundary line have different colours. The use of a computer in this proof inspired considerable debate in the mathematical community. At issue was whether a theorem can be considered proven if human beings have not actually checked every detail of the proof.

The study of the relations of deductibility among sentences in a logical calculus which benefits the proof theory. Deductibility is defined purely syntactically, that is, without reference to the intended interpretation of the calculus. The subject was founded by the mathematician David Hilbert (1862-1943) in the hope that strictly finitary methods would provide a way of proving the consistency of classical mathematics, but the ambition was torpedoed by Godel's second incompleteness theorem.

What is more, the use of a model to test for consistencies in an 'axiomatized system' which is older than modern logic. Descartes' algebraic interpretation of Euclidean geometry provides a way of showing that if the theory of real numbers is consistent, so is the geometry. Similar representation had been used by mathematicians in the 19th century, for example to show that if Euclidean geometry is consistent, so are various non-Euclidean geometries. Model theory is the general study of this kind of procedure: The 'proof theory' studies relations of deductibility between formulae of a system, but once the notion of an interpretation is in place we can ask whether a formal system meets certain conditions. In particular, can it lead us from sentences that are true under some interpretation? And if a sentence is true under all interpretations, is it also a theorem of the system? We can define a notion of validity (a formula is valid if it is true in all interpret rations) and semantic consequence (a formula 'B' is a semantic consequence of a set of formulae, written {A1 . . . An} B, if it is true in all interpretations in which they are true) Then the central questions for a calculus will be whether all and only its theorems are valid, and whether {A1 . . . An} B if and only if {A1 . . . An} B. There are the questions of the soundness and completeness of a formal system. For the propositional calculus this turns into the question of whether the proof theory delivers as theorems all and only 'tautologies'. There are many axiomatizations of the propositional calculus that are consistent and complete. The mathematical logician Kurt Gödel (1906-78) proved in 1929 that the first-order predicate under every interpretation is a theorem of the calculus. In that mathematical method for solving those physical problems that can be stated in the form that a certain value definite integral will have a stationary value for small changes of the functions in the integrands and of the limit of integration.

The Euclidean geometry is the greatest example of the pure 'axiomatic method', and as such had incalculable philosophical influence as a paradigm of rational certainty. It had no competition until the 19th century when it was realized that the fifth axiom of his system (parallel lines never intersect) could be denied without inconsistency, leading to Riemannian spherical geometry. The significance of Riemannian geometry lies in its use and extension of both Euclidean geometry and the geometry of surfaces, leading to a number of generalized differential geometries. Its most important effect was that it made a geometrical application possible for some major abstractions of tensor analysis, leading to the pattern and concepts for general relativity later used by Albert Einstein in developing his theory of relativity. Riemannian geometry is also necessary for treating electricity and magnetism in the framework of general relativity. The fifth chapter of Euclid's Elements, is attributed to the mathematician Eudoxus, and contains a precise development of the real number, work which remained unappreciated until rediscovered in the 19th century.

The Axiom, in logic and mathematics, is a basic principle that is assumed to be true without proof. The use of axioms in mathematics stems from the ancient Greeks, most probably during the 5th century Bc, and represents the beginnings of pure mathematics as it is known today. Examples of axioms are the following: 'No sentence can be true and false at the same time' (the principle of contradiction); 'If equals are added to equals, the sums are equal'. 'The whole is greater than any of its parts'. Logic and pure mathematics begin with such unproved assumptions from which other propositions (theorems) are derived. This procedure is necessary to avoid circularity, or an infinite regression in reasoning. The axioms of any system must be consistent with one another, that is, they should not lead to contradictions. They should be independent in the sense that they cannot be derived from one another. They should also be few in number. Axioms have sometimes been interpreted as self-evident truths. The present tendency is to avoid this claim and simply to assert that an axiom is assumed to be true without proof in the system of which it is a part.

The terms 'axiom' and 'postulate' are often used synonymously. Sometimes the word axiom is used to refer to basic principles that are assumed by every deductive system, and the term postulate is used to refer to first principles peculiar to a particular system, such as Euclidean geometry. Infrequently, the word axiom is used to refer to first principles in logic, and the term postulate is used to refer to first principles in mathematics.

The applications of game theory are wide-ranging and account for steadily growing interest in the subject. Von Neumann and Morgenstern indicated the immediate utility of their work on mathematical game theory by linking it with economic behaviour. Models can be developed, in fact, for markets of various commodities with differing numbers of buyers and sellers, fluctuating values of supply and demand, and seasonal and cyclical variations, as well as significant structural differences in the economies concerned. Here game theory is especially relevant to the analysis of conflicts of interest in maximizing profits and promoting the widest distribution of goods and services. Equitable division of property and of inheritance is another area of legal and economic concern that can be studied with the techniques of game theory.

In the social sciences, n-person game theory has interesting uses in studying, for example, the distribution of power in legislative procedures. This problem can be interpreted as a three-person game at the congressional level involving vetoes of the president and votes of representatives and senators, analyzed in terms of successful or failed coalitions to pass a given bill. The problem of majority rule and individual decision making is also amenable to such study.

Sociologists have developed an entire branch of game theory devoted to the study of issues involving group decision making. Epidemiologists also make use of game theory, especially with respect to immunization procedures and methods of testing a vaccine or other medication. Military strategists turn to game theory to study conflicts of interest resolved through 'battles' where the outcome or payoff of a given war game is either victory or defeat. Usually, such games are not examples of zero-sum games, for what one player loses in terms of lives and injuries are not won by the victor. Some uses of game theory in analyses of political and military events have been criticized as a dehumanizing and potentially dangerous oversimplification of necessarily complicating factors. Analysis of economic situations is also usually more complicated than zero-sum games because of the production of goods and services within the play of a given 'game'.

All is the same in the classical theory of the syllogism, a term in a categorical proposition is distributed if the proposition entails any proposition obtained from it by substituting a term denoted by the original. For example, in 'all dogs bark' the term 'dogs' is distributed, since it entails 'all terriers' bark', which is obtained from it by a substitution. In 'Not all dogs bark', the same term is not distributed, since it may be true while 'not all terriers' bark' is false.

When a representation of one system by another is usually more familiar, in and for itself, that those extended in representation that their workings are supposed analogously to that of the first. This one might model the behaviour of a sound wave upon that of waves in water, or the behaviour of a gas upon that to a volume containing moving billiard balls. While nobody doubts that models have a useful 'heuristic' role in science, there has been intense debate over whether a good model, or whether an organized structure of laws from which it can be deduced and suffices for scientific explanation. As such, the debate of topics was inaugurated by the French physicist Pierre Marie Maurice Duhem (1861-1916), in 'The Aim and Structure of Physical Theory' (1954) by which Duhem's conception of science is that it is simply a device for calculating as science provides deductive system that is systematic, economical, and predictive, but not that represents the deep underlying nature of reality. Steadfast and holding of its contributive thesis that in isolation, and since other auxiliary hypotheses will always be needed to draw empirical consequences from it. The Duhem thesis implies that refutation is a more complex matter than might appear. It is sometimes framed as the view that a single hypothesis may be retained in the face of any adverse empirical evidence, if we prepared to make modifications elsewhere in our system, although strictly speaking this is a stronger thesis, since it may be psychologically impossible to make consistent revisions in a belief system to accommodate, say, the hypothesis that there is a hippopotamus in the room when visibly there is not.

Primary and secondary qualities are the division associated with the 17th-century rise of modern science, wit h its recognition that the fundamental explanatory properties of things that are not the qualities that perception most immediately concerns. The latter are the secondary qualities, or immediate sensory qualities, including colour, taste, smell, felt warmth or texture, and sound. The primary properties are less tied to their deliverance of one particular sense, and include the size, shape, and motion of objects. In Robert Boyle (1627-92) and John Locke (1632-1704) the primary qualities that are scientifically tractable, objective qualities essential to anything material, of which are of a minimal listing of size, shape, and mobility, i.e., the state of being at rest or moving. Locke sometimes adds number, solidity, texture (where this is thought of as the structure of a substance, or way in which it is made out of atoms). The secondary qualities are the powers to excite particular sensory modifications in observers. Once, again, that Locke himself thought in terms of identifying these powers with the texture of objects that, according to corpuscularian science of the time, were the basis of an object's causal capacities. The ideas of secondary qualities are sharply different from these powers, and afford us no accurate impression of them. For Ren Descartes (1596-1650), this is the basis for rejecting any attempt to think of knowledge of external objects as provided by the senses. But in Locke our ideas of primary qualities do afford us an accurate notion of what shape, size, And mobility is. In English-speaking philosophy the first major discontent with the division was voiced by the Irish idealist George Berkeley (1685-1753), who probably took for a basis of his attack from Pierre Bayle (1647-1706), who in turn cites the French critic Simon Foucher (1644-96). Modern thought continues to wrestle with the difficulties of thinking of colour, taste, smell, warmth, and sound as real or objective properties to things independent of us.

Continuing as such, is the doctrine advocating the American philosopher David Lewis (1941-2002), in that different possible worlds are to be thought of as existing exactly as this one does. Thinking in terms of possibilities is thinking of real worlds where things are different. The view has been charged with making it impossible to see why it is good to save the child from drowning, since there is still a possible world in which she (or her counterpart) drowned, and from the standpoint of the universe it should make no difference which world is actual. Critics also charge that either that the notion fails to fit within a coherent theory. If how we know about possible worlds, or with a coherent theory of why we are interested in them, but Lewis denied that any other way of interpreting modal statements is tenable.

The proposal set forth that characterizes the 'modality' of a proposition as the notion for which it is true or false. The most important division is between propositions true of necessity, and those true as things are: Necessary as opposed to contingent propositions. Other qualifiers sometimes called 'modal' include the tense indicators, 'it will be the case that 'p', or 'it was the case in the action 'p', and that there are affinities among the 'deontic' indicators, 'as it ought to be the case that 'p', or 'it is permissible that 'p', and the requisite for the possibility.

The aim of reasoning, specially the structure or propositions as distinguished from their content and of method and validity in deductive reasoning that forge a system of logic. More explicitly to the rules by which inferences may be drawn, than to study the actual reasoning processes that people use, which may or may not conform to those rules. In the case of deductive logic, if we ask why we need to obey the rules, the most general form of an answer is that if we do not we contradict ourselves (or, strictly speaking, we stand ready to contradict ourselves. Someone failing to draw a conclusion that follows from a set of premises need not be contradicting him or herself, but only failing to notice something. However, he or she is not defended against adding the contradictory conclusion to his or fer set of beliefs.) There is no equally simple answer in the case of inductive logic, which is in general a less robust subject, but the aim will be to find reasoning such that anyone failing to conform to it will have improbable beliefs. Traditional logic dominated the subject until the 19th century, and has become increasingly recognized in the 20th century. In the finer of works that were done within that tradition, up to the present time, as syllogistic reasoning is now generally regarded as a limited special case in the form of reasoning that can be reprehend within the promotion and predated values. These forms are the heart of modern logic, as their central notions or qualifiers, variables, and functions were the creation of the German mathematician Gottlob Frége. He is recognized as the father of modern logic, although his treatments of a logical system as an abreact mathematical structure, or algebraic, has been heralded by the English mathematician and logician George Boole (1815-64). His pamphlet, The Mathematical Analysis of Logic (1847) pioneered the algebra of classes. The work was made of in An Investigation of the Laws of Thought (1854). Boole also published many works in our mathematics, and on the theory of probability. His name is remembered in the title of Boolean algebra, and the algebraic operations he investigated are denoted by Boolean operations.

The syllogistic, or categorical syllogism is the inference of one proposition from two premises. For example is, 'all horses have tails, and things with tails are four legged, so all horses are four legged. Each premise has one term in common with the other premises. The term that does not occur in the conclusion is called the middle term. The major premise of the syllogism is the premise containing the predicate of the contraction (the major term). And the minor premise contains its subject (the minor term). So the first premise of the example in the minor premise the second the major term. So the first premise of the example is the minor premise, the second the major premise and 'having a tail' is the middle term. This enables syllogisms that there of a classification, that according to the form of the premises and the conclusions. The other classification is by figure, or way in which the middle term is placed or way in within the middle term is placed in the premise.

Although the theory of the syllogism dominated logic until the 19th century, it remained a piecemeal affair, able to deal with only relations valid forms of valid forms of argument. There have subsequently been rearguing actions attempting, but in general it has been eclipsed by the modern theory of quantification, the predicate calculus is the heart of modern logic, having proved capable of formalizing the calculus rationing processes of modern mathematics and science. In a first-order predicate calculus the variables range over objects: In a higher-order calculus the may range over predicate and functions themselves. The fist-order predicated calculus with identity includes '=' as primitive (undefined) expression: In a higher-order calculus I t may be defined by law that $_{,}$ = y iff (F)(F, Fy), which gives to a greater expressive power for less plexuity.

Modal logic was of great importance historically, particularly in the light of the deity, but was not a central topic of modern logic in its gold period as the beginning of the 20th century. It was, however, revived by the American logician and philosopher Irving Lewis (1883-1964), although he wrote extensively on most central philosophical topis, he is remembered principally as a critic of the intentional nature of modern logic, and as the founding father of modal logic. His two independent proofs showing that from a contradiction anything follows a relevance logic, using a notion of entailment stronger than that of strict implication.

The imparting information has been conduced or carried out of the prescribed procedures, as impeding of something that tajes place in the chancing encounter out to be to enter ons's mind may from time to time occasion of various doctrines concerning the necessary properties, least of mention, by adding to a prepositional or predicated calculus two operator, and (sometimes written 'N' and 'M'), meaning necessarily and possible, respectfully. These like 'p' and 'p' will be wanted. Controversially these include 'p' (if a proposition is necessary. It's necessarily, characteristic of a system known as S4 and 'p' (if as preposition is possible it's necessarily possible) this characteristic of the system known as S5, is the classical modal theory for modal logic, due to the American logician and philosopher (1940-) and the Swedish logician Sig Kanger, involves valuing prepositions not true or false simpliciter, but as true or false at possible worlds with necessity then corresponding to truth in all worlds, and the possibility to truth in some world. Various different systems of modal logic result from adjusting the accessibility relation between worlds.

In Saul Kripke, gives the classical modern treatment of the topic of reference, both crystallizing the distinction between names and definite description, and opening the door to many subsequent attempts to understand the notion of reference in terms of a causal link between the use of a term and an original episode of attaching a name to the subject.

One of the three branches into which 'semiotic' is usually divided, the study of semantical meaning of words, and the relation of signs to the degree to which the designs are applicable. In that, in formal studies, a semantics is provided for a formal language when an interpretation of 'model' is specified. However, a natural language comes ready interpreted, and the semantic problem is not that of the specification but of understanding the relationship between terms of various categories (names, descriptions, predicate, adverbs . . .) and their meaning. An influential proposal by attempting to provide a truth definition for the language, which will involve giving a full structure of different kinds has on the truth conditions of sentences containing them.

Holding that the basic casse of reference is the relation between a name and the persons or object, which it names. The philosophical problems include trying to elucidate that relation, to understand whether other semantic relations, such s that between a predicate and the property it expresses, or that between a description an what it describes, or that between me and the word 'I', are examples of the same relation or of very different ones. A great deal of modern work on this was stimulated by the American logician Saul Kripke's, Naming and Necessity (1970). It would also be desirable to know whether we can refer to such things as objects and how to conduct the debate about each and issue. A popular approach, following Gottlob Frége, is to argue that the fundamental unit of analysis should be the whole sentence. The reference of a term becomes a derivative notion it is whatever it is that defines the term's contribution to the trued condition of the whole sentence. There need be nothing further to say about it, given that we have a way of understanding the attribution of meaning or truth-condition to sentences. Other approaching, searching for more of a substantive possibility that causality or psychological or social constituents are pronounced between words and things.

However, following Ramsey and the Italian mathematician G. Peano (1858-1932), it has been customary to distinguish logical paradoxes that depend upon a notion of reference or truth (semantic notions) such as those of the 'Liar family, Berry, Richard, and so forth, form the purely logical paradoxes in which no such notions are involved, such as Russell's paradox, or those of Canto or Burali-Forti. Paradoxes of the first type seem to depend upon an element of the self-reference, in which a sentence is about itself, or in which a phrase refers to something about itself, or in which a phrase refers to something defined by a set of phrases of which it is itself one. It is to feel that this element is responsible for the contradictions, although the self-reference itself is often benign (for instance, the sentence 'All English sentences should have a verb', includes itself happily in the domain of sentences it is talking about), so the difficulty lies in forming a condition that existence only of a pathological self-reference. Paradoxes of the second kind then need a different treatment. Whilst the distinction is convenient, for allowing set theory to proceed by circumventing the latter paradoxes by technical means, even when there is no solution to the semantic paradoxes, it may be a way of ignoring the similarities between the two families. There is still a th potential possibility that while there is no agreed solution to the semantic paradoxes, yet, the understanding of Russell's paradox may be imperfect as well.

Truth and falsity are two classical truth-values that a statement, proposition or sentence can take, as it is supposed in classical (two-valued) logic, that each statement has one of these values, and none has both. A statement is then false if and only if it is not true. The basis of this scheme is that to each statement there corresponds a determinate truth condition, or way the world must be for it to be true: If this condition obtains the statement is true, and otherwise false. Statements may indeed be felicitous or infelicitous in other dimensions (polite, misleading, apposite, witty, and so forth) but truth is the central normative notion governing assertion. Consideration's o vagueness may introduce greys into this black-and-white scheme. For the issue to be true, any suppressed premise or background framework of thought it can be necessary to make an agreement valid, or a tenable position, a proposition whose truth is necessary to make for of either, the truth or the falsity of another statement. Thus, if 'p' presupposes 'q', 'q' must be true for 'p' to be true or false. In the theory of knowledge, the English philologer and historian George Collingwood (1889-1943), announces that any proposition capable of truth or falsity stands on a bed of 'absolute presuppositions' which are not properly capable of truth or falsity, since a system of thought will contain no way of approaching such a question (a similar idea later voiced by Wittgenstein in his work On Certainty). The introduction of presupposition therefore means that either another of a truth value is fond, 'intermediate' between truth and falsity, or the classical logic is preserved, but it is impossible to tell whether a particular sentence empresses a preposition that is a candidate for truth and falsity, without knowing more than the formation rules of the language. Each suggestion carries over, and there is some consensus that at least whowhere definite descriptions are involved, examples equally given by regarding the overall sentence as false as the existence claim fails, and explaining the data that the English philosopher Frederick Strawson (1919-) relied upon as the effects of 'implicature'.

Views about the meaning of terms will often depend on classifying the implicature of sayings involving the terms as implicatures or as genuine logical implications of what is said. Implicatures may be divided into two kinds: Conversational implicatures of the two kinds and the more subtle category of conventional implicatures. A term may as a matter of convention, is exemplified upon the fact, that an implicature is one of the relations between 'he is poor and honest' and 'he is poor but honest' is that they have the same content (are true in just the same conditional) but the second has implicatures (that the combination is surprising or significant) that the first lacks.

It is, nonetheless, that we find in classical logic a proposition that may be true or false. In that, if the former, it is said to take the truth-value true, and if the latter the truth-value false. The idea behind the terminological phrases is the analogues between assigning a propositional variable one or other of these values, as is done in providing an interpretation for a formula of the propositional calculus, and assigning an object as the value of any other variable. Logics with intermediate value are called 'many-valued logics'.

Nevertheless, an existing definition of the predicate' . . . is true' for a language that satisfies convention 'T', the material adequately condition laid down by Alfred Tarski, born Alfred Teitelbaum (1901-83), whereby his methods of 'recursive' definition, enabling us to say for each sentence what it is that its truth consists in, but giving no verbal definition of truth itself. The recursive definition of the truth predicate of a language is always provided in a 'metalanguage', Tarski is thus committed to a hierarchy of languages, each with it's associated, but different truth-predicate. Whist enabling the approach to avoid the contradictions of paradoxical contemplations, it conflicts with the idea that a language should be able to say everything that there is to say, and other approaches have become increasingly important.

So, that the truth condition of a statement is the condition for which the world must meet if the statement is to be true. To know this condition is equivalent to knowing the meaning of the statement. Although this sounds as if it gives a solid anchorage for meaning, some of the securities disappear when it turns out that the truth condition can only be defined by repeating the very same statement: The truth condition of 'now is white' is that 'snow is white', the truth condition of 'Britain would have capitulated had Hitler invaded', is that 'Britain would have capitulated had Hitler invaded'. It is disputed whether this element of running-on-the-spot disqualifies truth conditions from playing the central role in a substantives theory of meaning. Truth-conditional theories of meaning are sometimes opposed by the view that to know the meaning of a statement is to be able to use it in a network of inferences.

Taken to be the view, inferential semantics takes on the role of sentence in inference give a more important key to their meaning than this 'external' relations to things in the world. The meaning of a sentence becomes its place in a network of inferences that it legitimates. Also known as functional role semantics, procedural semantics, or conception to the coherence theory of truth, and suffers from the same suspicion that it divorces meaning from any clar association with things in the world.

Moreover, a theory of semantic truth is that of the view if language is provided with a truth definition, there is a sufficient characterization of its concept of truth, as there is no further philosophical chapter to write about truth: There is no further philosophical chapter to write about truth itself or truth as shared across different languages. The view is similar to the Disquotational theory.

The redundancy theory, or also known as the 'deflationary view of truth' fathered by Gottlob Frége and the Cambridge mathematician and philosopher Frank Ramsey (1903-30), who showed how the distinction between the semantic paradoses, such as that of the Liar, and Russell's paradox, made unnecessary the ramified type theory of Principia Mathematica, and the resulting axiom of reducibility. By taking all the sentences affirmed in a scientific theory that use some terms, e.g., the quark, and to a considerable degree of replacing the term by a variable instead of saying that quarks have such-and-such properties, the Ramsey sentence says that there is something that has those properties. If the process is repeated for all of a group of the theoretical terms, the sentence gives 'topic-neutral' structure of the theory, but removes any implication that we know what the terms so treated have as a meaning. It leaves open the possibility of identifying the theoretical item with whatever. It is that best fits the description provided. However, it was pointed out by the Cambridge mathematician Newman, that if the process is carried out for all except the logical bones of a theory, then by the Löwenheim-Skolem theorem, the result will be interpretable, and the content of the theory may reasonably be felt to have been lost.

All the while, both Frége and Ramsey are agreeing that the essential claim is that the predicate' . . . is true' does not have a sense, i.e., expresses no substantive or profound or explanatory concept that ought to be the topic of philosophical enquiry? The approach admits of different versions, but centres on the points (1) that 'it is true that 'p' says no more nor less than 'p' (hence, redundancy): (2) that in less direct contexts, such as 'everything he said was true', or 'all logical consequences of true propositions are true', the predicate functions as a device enabling us to generalize than as an adjective or predicate describing the things he said, or the kinds of propositions that follow from the capable ability for being a true preposition. For example, the second translates as '(p, q)(p & p ®q ®q)' where there is no use of a notion of truth.

There are technical problems in interpreting all uses of the notion of truth in such ways, nevertheless, they are not generally felt to be insurmountable. The approach needs to explain away apparently substantive uses of the notion, such as 'science aims at the truth', or 'truth is a norm governing discourse'. Postmodern writing frequently advocates that we must abandon such norms. Along with a discredited 'objective' conception of truth. Perhaps, we can have the norms even when objectivity is problematic, since they can be framed without mention of truth: Science wants it to be so that whatever science holds that 'p', then 'p'. Discourse is to be regulated by the principle that it is wrong to assert 'p', when 'not-p'.

Something that tends of something in addition of content, or coming by way to justify such a position can very well be more that in addition to several reasons, as to bring in or binding of something that might be more so as to a larger combination for us to consider the simplest formulation, for which is to claim that expresses the essencity of something, as the mode from which a thing exists, acts, or manifests itself, as to form the attributive forming of 'S' which is true. Meaning, that the same expressivity, to which, an expressed effect is the arranging to come to have or acquired 'S'. Some philosophers dislike the ideas of sameness of meaning, and if this I disallowed, then the claim is that the two forms are equivalent in any sense of equivalence that matters. This is, it makes no difference whether people say 'Dogs bark' or whether they say, 'dogs bark'. In the former representation of what they say of the sentence 'Dogs bark' is mentioned, but in the later it appears to be used, of the claim that the two are equivalent and needs careful formulation and defence. On the face of it someone might know that 'Dogs bark' is true without knowing what it (for instance, if he kids in a list of acknowledged truths, although he does not understand English), and this is different from knowing that dogs bark. Disquotational theories are usually presented as versions of the 'redundancy theory of truth'.

The relationship between a set of premises and a conclusion when the conclusion follows from the premise. Many philosophers identify this with it being logically impossible that the premises should all be true, yet the conclusion false. Others are sufficiently impressed by the paradoxes of strict implication to look for a stranger relation, which would distinguish between valid and invalid arguments within the sphere of necessary propositions. The seraph for a strange notion is the field of relevance logic.

From a systematic theoretical point of view, we may imagine the process of evolution of an empirical science to be a continuous process of induction. Theories are evolved and are expressed in short compass as statements of as large number of individual observations in the form of empirical laws, from which the general laws can be ascertained by comparison. Regarded in this way, the development of a science bears some resemblance to the compilation of a classified catalogue. It is, as it was, a purely empirical enterprise.

But this point of view by no embraces the whole of the actual process, for it slurs over the important part played by intuition and deductive thought in the development of an exact science. As soon as a science has emerged from its initial stages, theoretical advances are no longer achieved merely by a process of arrangement. Guided by empirical data, the investigators rather develop a system of thought which, in general, it is built up logically from a small number of fundamental assumptions, the so-called axioms. We call such a system of thought a 'theory'. The theory finds the justification for its existence in the fact that it correlates a large number of single observations, and is just here that the 'truth' of the theory lies.

Corresponding to the same complex of empirical data, there may be several theories, which differ from one another to a considerable extent. But as regards the deductions from the theories which are capable of being tested, the agreement between the theories may be so complete, that it becomes difficult to find any deductions in which the theories differ from each other. As an example, a case of general interest is available in the province of biology, in the Darwinian theory of the development of species by selection in the struggle for existence, and in the theory of development which is based on the hypophysis of the hereditary transition of acquired characters. The Origin of Species was principally successful in marshalling the evidence for evolution, than providing a convincing mechanisms for genetic change. And Darwin himself remained open to the search for additional mechanisms, while also remaining convinced that natural selection was at the hart of it. It was only with the later discovery of the gene as the unit of inheritance that the synthesis known as 'neo-Darwinism' became the orthodox theory of evolution in the life sciences.

In the 19th century the attempt to base ethical reasoning o the presumed facts about evolution, the movement is particularly associated with the English philosopher of evolution Herbert Spencer (1820-1903). The premise is that later elements in an evolutionary path are better than earlier ones: The application of this principle then requires seeing western society, laissez-faire capitalism, or some other object of approval, as more evolved than more 'primitive' social forms. Neither the principle nor the applications command much respect. The version of evolutionary ethics called 'social Darwinism' emphasises the struggle for natural selection, and draws the conclusion that we should glorify and assist such struggles, usually by enhancing competition and aggressive relations between people in society or between evolution and ethics has been rethought in the light of biological discoveries concerning altruism and kin-selection.

Once again, the psychology proving attempts are founded to evolutionary principles, in which a variety of higher mental functions may be adaptations, forced in response to selection pressures on the human populations through evolutionary time. Candidates for such theorizing include material and paternal motivations, capacities for love and friendship, the development of language as a signalling system cooperative and aggressive, our emotional repertoire, our moral and reactions, including the disposition to detect and punish those who cheat on agreements or who 'free-ride' on =the work of others, our cognitive structures, nd many others. Evolutionary psychology goes hand-in-hand with neurophysiological evidence about the underlying circuitry in the brain which subserves the psychological mechanisms it claims to identify. The approach was foreshadowed by Darwin himself, and William James, as well as the sociology of E.O. Wilson. The term of use is applied, more or less aggressively, especially to explanations offered in Sociobiology and evolutionary psychology.

Another assumption that is frequently used to legitimate the real existence of forces associated with the invisible hand in neoclassical economics derives from Darwin's view of natural selection as a war-like competing between atomized organisms in the struggle for survival. In natural selection as we now understand it, cooperation appears to exist in complementary relation to competition. It is complementary relationships between such results that are emergent self-regulating properties that are greater than the sum of parts and that serve to perpetuate the existence of the whole.

According to E.O Wilson, the 'human mind evolved to believe in the gods' and people 'need a sacred narrative' to have a sense of higher purpose. Yet it is also clear that the 'gods' in his view, are merely human constructs and, therefore, there is no basis for dialogue between the world-view of science and religion. 'Science for its part', said Wilson, 'will test relentlessly every assumption about the human condition and in time uncover the bedrock of the moral and the religious sentiment. The eventual result of the competition among others, will be the secularization of the human epic and of religion itself.

The argument to the best explanation is the view that once we can select the best of any in something in explanations of an event, then we are justified in accepting it, or even believing it. The principle needs qualification, since something it is unwise to ignore the antecedent improbability of a hypothesis which would explain the data better than others, e.g., the best explanation of a coin falling heads 530 times in 1,000 tosses might be that it is biassed to give a probability of heads of 0.53 but it might be more sensible to suppose that it is fair, or to suspend judgement.

In a philosophy of language is considered as the general attempt to understand the components of a working language, the relationship to an understanding speaker has to its elements, and the relationship they bear to the world. The subject therefore embraces the traditional division of semiotic into syntax, semantics, and pragmatics. The philosophy of language thus mingles with the philosophy of mind, since it needs an account of what it is in our understanding that enables us to use language. It so mingles with the metaphysics of truth and the relationship between sign and object. Much as much is that the philosophy in the 20ᵗʰ century, has been informed by the belief that philosophy of language is the fundamental basis of all philosophical problems, in that language is the distinctive exercise of mind, and the distinctive way in which we give shape to metaphysical beliefs. Particular topics will include the problems of logical form. And the basis of the division between syntax and semantics, as well as problems of understanding the number and nature of specifically semantic relationships such as meaning, reference, predication, and quantification. Pragmatics includes that of speech acts, while problems of rule following and the indeterminacy of translations infect philosophies of both pragmatics and semantics.

On this conception, to understand a sentence is to know its truth-conditions, and, yet, in a distinctive way the conception has remained central that those who offer opposing theories characteristically define their position by reference to it. The Concepcion of meaning s truth-conditions needs not and should not be advanced for being in itself as complete account of meaning. For instance, one who understands a language must have some idea of the range of speech acts contextually performed by the various types of sentences in the language, and must have some idea of the insufficiencies of various kinds of speech acts. The claim of the theorist of truth-conditions should rather be targeted on the notion of content: If indicative sentences differ in what they strictly and literally say, then this difference is fully accounted for by the difference in the truth-conditions.

The meaning of a complex expression is a function of the meaning of its constituent. This is just as a sentence of what it is for an expression to be semantically complex. It is one of the initial attractions of the conceptual meaning truth-conditions tat it permits a smooth and satisfying account of the ways in which the meaning of complex expression is a function from which is the meaning of its constituents. On the truth-conditional conception, to give the meaning of an expression is to state the contribution it makes to the truth-conditions of sentences in which it occurs. For singular terms-proper names, indexical, and certain pronouns-this is done by stating the reference of the terms in question. For predicates, it is done either by stating the conditions under which the predicate is true of arbitrary objects, or by stating th conditions under which arbitrary atomic sentences containing it is true. The meaning of a sentence-forming operator is given by stating its contribution to the truth-conditions of as complex sentence, as a function of semantic values of the sentences on which it operates.

The theorist of truth conditions should insist that not every true statement about the reference of an expression is fit to be an axiom in a meaning-giving theory of truth for a language, such is the axiom: 'London' refers to the city in which there was a huge fire in 1666, is a true statement about the reference of 'London'. It is a consequent of a theory which substitutes this axiom for no different a term than of our simple truth theory that 'London is beautiful' is true if and only if the city in which there was a huge fire in 1666 is beautiful. Since a subject can understand in the name presented as 'London' without knowing that last-mentioned truth condition, this replacement axiom is not fit to be an axiom in a meaning-specifying truth theory. It is, of course, incumbent on a theorised meaning of truth conditions, to state in a way which does not presuppose any previous, non-truth conditional conception of meaning

Among the many challenges facing the theorist of truth conditions, two are particularly salient and fundamental. First, the theorist has to answer the charge of triviality or vacuity, second, the theorist must offer an account of what it is for a person's language to be truly describable by as semantic theory containing a given semantic axiom.

Since the contentual sentence, 'Paris is beautiful' is true, amounts to more or less, than the postulate saying that: 'Paris is beautiful', we can trivially describe some understanding of a sentence, if we wish for knowing its truth-conditions, but this gives us no substantive account of understanding whatsoever. Something other than the grasp for the truth conditions as we must provide of some substantive account. The charge rests upon what has been called the redundancy theory of truth, the theory which, somewhat more discriminatingly. Horwich calls the minimal theory of truth. It's conceptual representation that the concept of truth is exhausted by the fact that it conforms to the equivalence principle, the principle that for any proposition 'p', it is true that 'p' if and only if 'p'. Many different philosophical theories of truth will, with suitable qualifications, accept that equivalence principle. The distinguishing feature of the minimal theory is its claim that the equivalence principle exhausts the notion of the truth. It is now widely accepted, both by opponents and supporters of truth conditional theories of meaning, that it is inconsistent to accept both minimal theory of truth and a truth conditional account of meaning. If the claim that, 'Paris is beautiful' is true is exhausted by its equivalence to the claim that Paris is beautiful, it is circular to try of its truth conditions. The minimal theory of truth has been endorsed by the Cambridge mathematician and philosopher Plumpton Ramsey (1903-30), and the English philosopher Jules Ayer, the later Wittgenstein, Quine, Strawson. Horwich and-confusing and inconsistently if this article is correct-Frége himself. but is the minimal theory correct?

The minimal theory treats instances of the equivalence principle as definitional of truth for a given sentence, but in fact, it seems that each instance of the equivalence principle can itself be explained. The truths from which such an instance as: 'London is beautiful' is true if and only if London is beautiful. This would be a pseudo-explanation if the fact that 'London' refers to London consists in part in the fact that 'London is beautiful' has the truth-condition it does. But it is very implausible, after all too understand that, 'London' without understanding the predicate 'is beautiful'.

Sometimes, however, the counterfactual conditional is known as subjunctive conditionals, insofar as a counterfactual conditional is a conditional of the form 'if p were to happen q would', or 'if p were to have happened q would have happened', where the supposition of 'p' is contrary to the known fact that 'not-p'. Such assertions are nevertheless, useful 'if you broke a bone, the X-ray would have looked different', or 'if the reactor was to fail, this mechanism would automatically 'click-in' which is an important truth, even when we know that the bone is not broken or are certain that the reactor will not fail. It is arguably distinctive of laws of nature that yield counterfactuals ('if the metal were to be heated, it would expand'), whereas accidentally true generalizations may not. It is clear that counterfactuals cannot be represented by the material implication of the propositional calculus, since that conditionals come out true whenever 'p' is false, so there would be no division between true and false counterfactuals.

Although the subjunctive form indicates a counterfactual, in many contexts it does not seem to matter whether we use a subjunctive form, or a simple conditional form: 'If you run out of water, you will be in trouble' seems equivalent to 'if you were to run out of water, you would be in trouble', in other contexts there is a big difference: 'If Oswald did not kill Kennedy, someone else did' is clearly true, whereas 'if Oswald had not killed Kennedy, someone would have' is most probably false.

The best-known modern treatment of counterfactuals is that of David Lewis, which evaluates them as true or false according to whether 'q' is true in the 'most similar' possible worlds to ours in which 'p' is true. The similarity-ranking this approach needs have proved controversial, particularly since it may need to presuppose some notion of the same laws of nature, whereas art of the interest in counterfactuals is that they promise to illuminate that notion. There is a growing awareness that the classification of conditionals is an extremely tricky business, and categorizing them as counterfactuals are not to be of limited use.

The pronouncing of any conditional preposition of the form if 'p' then 'q' the condition hypothesizes, 'p' is called the antecedent of the conditional, and 'q' the consequent. Various kinds of conditional have been distinguished. The weaken in that of material implication, merely telling us that with not-p. or q. stronger conditionals include elements of modality, corresponding to the thought that 'if p is true then q must be true'. Ordinary language is very flexible in its use of the conditional form, and there is controversy whether, yielding different kinds of conditionals with different meanings, or pragmatically, in which case there should be one basic meaning which case there should be one basic meaning, with surface differences arising from other implicatures.

We now turn to a philosophy of meaning and truth, for which it is especially associated with the American philosopher of science and of language (1839-1914), and the American psychologist philosopher William James (1842-1910), wherefore the study in Pragmatism is given to various formulations by both writers, but the core is the belief that the meaning of a doctrine is the same as the practical effects of adapting it. Peirce interpreted of a theoretical sentence from that which is only that of a corresponding practical maxim (telling us what to do in some circumstance). In James the position issues in a theory of truth, notoriously allowing that belief, including for the example, belief in God, is the widest sense of the works satisfactorially in the widest sense of the word. On James's view almost any belief might be respectable, and even rue, provided it works (but working is no simple matter for James). The apparent subjectivist consequences of tis were wildly assailed by Russell (1872-1970), Moore (1873-1958), and others in the early years of the 20 century. This led to a division within pragmatism between those such as the American educator John Dewey (1859-1952), whose humanistic conception of practice remains inspired by science, and the more idealistic route that especially by the English writer F.C.S. Schiller (1864-1937), embracing the doctrine that our cognitive efforts and human needs actually transform the reality that we seek to describe. James often writes as if he sympathizes with this development. For instance, in The Meaning of Truth (1909), he considers the hypothesis that other people have no minds (dramatized in the sexist idea of an 'automatic sweetheart' or female zombie) and briefly and causally as a comment, that the hypothesis would not work because it would not satisfy our egoistic craving for the recognition and admiration of others. The implication that this is to bring into existence by shaping, modifying, or putting together the materials to construct of form or function to another of what makes it true that the other persons have minds in the disturbing part.

Modern pragmatists such as the American philosopher and critic Richard Rorty (1931-) and some writings of the philosopher Hilary Putnam (1925-) who has usually tried to dispense with an account of truth and concentrate, as perhaps James should have done, upon the nature of belief and its relations with human attitude, emotion, and need. The driving motivation of pragmatism is the idea that belief in the truth on te one hand must have a close connection with success in action on the other. One way of cementing the connection is found in the idea that natural selection must have adapted us to be cognitive creatures because belief has effects, as they work. Pragmatism can be found in Kant's doctrine of the primary of practical over pure reason, and continued to play an influential role in the theory of meaning and of truth.

In case of fact, the philosophy of mind is the modern successor to behaviourism, as do the functionalism that its early advocates were Putnam (1926-) and Sellars (1912-89), and its guiding principle is that we can define mental states by a triplet of relations they have on other mental stares, what effects they have on behaviour. The definition need not take the form of a simple analysis, but if w could write down the totality of axioms, or postdate, or platitudes that govern our theories about what things of other mental states, and our theories about what things are apt to cause (for example), a belief state, what effects it would have on a variety of other mental states, and what affects it is likely to have on behaviour, then we would have done all that is needed to make the state a proper theoretical notion. It could be implicitly defied by these theses. Functionalism is often compared with descriptions of a computer, since according to mental descriptions correspond to a description of a machine in terms of software, that remains silent about the underlaying hardware or 'realization' of the program the machine is running. The principal advantages of functionalism include the way we know of mental states both of ourselves and others, which is via their effects on behaviour and other mental states. As with behaviourism, critics charge that structurally complex items that do not bear mental states might nevertheless, imitate the functions that are cited. According to this criticism functionalism is too generous and would count too many things as having minds. It is also queried whether functionalism is too paradoxical, able to see mental similarities only when there is causal similarity, when our actual practices of interpretations enable us to ascribe thoughts and desires to differently from our own, it may then seem as though beliefs and desires can be 'variably realized' causal architecture, just as much as they can be in different neurophysiological states.

The philosophical movement of Pragmatism had a major impact on American culture from the late 19th century to the present. Pragmatism calls for ideas and theories to be tested in practice, by assessing whether acting upon the idea or theory produces desirable or undesirable results. According to pragmatists, all claims about truth, knowledge, morality, and politics must be tested in this way. Pragmatism has been critical of traditional Western philosophy, especially the notions that there are absolute truths and absolute values. Although pragmatism was popular for a time in France, England, and Italy, most observers believe that it encapsulates an American faith in know-how and practicality and an equally American distrust of abstract theories and ideologies.

In mentioning the American psychologist and philosopher we find William James, who helped to popularize the philosophy of pragmatism with his book Pragmatism: A New Name for Old Ways of Thinking (1907). Influenced by a theory of meaning and verification developed for scientific hypotheses by American philosopher C. S. Peirce, James held that truth is what works, or has good experimental results. In a related theory, James argued the existence of God is partly verifiable because many people derive benefits from believing.

The Association for International Conciliation first published William James's pacifist statement, 'The Moral Equivalent of War', in 1910. James, a highly respected philosopher and psychologist, was one of the founders of pragmatism-a philosophical movement holding that ideas and theories must be tested in practice to assess their worth. James hoped to find a way to convince men with a long-standing history of pride and glory in war to evolve beyond the need for bloodshed and to develop other avenues for conflict resolution. Spelling and grammars represent standards of the time.

Pragmatists regard all theories and institutions as tentative hypotheses and solutions. For this reason they believed that efforts to improve society, through such as education or politics, must be geared toward problem solving and must be ongoing. Through their emphasis on connecting theory to practice, pragmatist thinkers attempted to transform all areas of philosophy, from metaphysics to ethics and political philosophy.

Pragmatism sought a middle ground between traditional ideas about the nature of reality and radical theories of nihilism and irrationalism, which had become popular in Europe in the late 19th century. Traditional metaphysics assumed that the world has a fixed, intelligible structure and that human beings can know absolute or objective truths about the world and about what constitutes moral behaviour. Nihilism and irrationalism, on the other hand, denied those very assumptions and their certitude. Pragmatists today still try to steer a middle course between contemporary offshoots of these two extremes.

The ideas of the pragmatists were considered revolutionary when they first appeared. To some critics, pragmatism's refusal to affirm any absolutes carried negative implications for society. For example, pragmatists do not believe that a single absolute idea of goodness or justice exists, but rather than these concepts are changeable and depend on the context in which they are being discussed. The absence of these absolutes, critics feared, could result in a decline in moral standards. The pragmatists' denial of absolutes, moreover, challenged the foundations of religion, government, and schools of thought. As a result, pragmatism influenced developments in psychology, sociology, education, semiotics (the study of signs and symbols), and scientific method, as well as philosophy, cultural criticism, and social reform movements. Various political groups have also drawn on the assumptions of pragmatism, from the progressive movements of the early 20th century to later experiments in social reform.

Pragmatism is best understood in its historical and cultural context. It arose during the late 19th century, a period of rapid scientific advancement typified by the theories of British biologist Charles Darwin, whose theories suggested too many thinkers that humanity and society are in a perpetual state of progress. During this same period a decline in traditional religious beliefs and values accompanied the industrialization and material progress of the time. In consequence it became necessary to rethink fundamental ideas about values, religion, science, community, and individuality.

The three most important pragmatists are American philosophers' Charles Sanders Peirce, William James, and John Dewey. Peirce was primarily interested in scientific method and mathematics; his objective was to infuse scientific thinking into philosophy and society, and he believed that human comprehension of reality was becoming ever greater and that human communities were becoming increasingly progressive. Peirce developed pragmatism as a theory of meaning-in particular, the meaning of concepts used in science. The meaning of the concept 'brittle', for example, is given by the observed consequences or properties that objects called 'brittle' exhibit. For Peirce, the only rational way to increase knowledge was to form mental habits that would test ideas through observation, experimentation, or what he called inquiry. Many philosophers known as logical positivists, a group of philosophers who have been influenced by Peirce, believed that our evolving species was fated to get ever closer to Truth. Logical positivists emphasize the importance of scientific verification, rejecting the assertion of positivism that personal experience is the basis of true knowledge.

James moved pragmatism in directions that Peirce strongly disliked. He generalized Peirce's doctrines to encompass all concepts, beliefs, and actions; he also applied pragmatist ideas to truth as well as to meaning. James was primarily interested in showing how systems of morality, religion, and faith could be defended in a scientific civilization. He argued that sentiment, as well as logic, is crucial to rationality and that the great issues of life-morality and religious belief, for example-are leaps of faith. As such, they depend upon what he called 'the will to believe' and not merely on scientific evidence, which can never tell us what to do or what is worthwhile. Critics charged James with relativism (the belief that values depend on specific situations) and with crass expediency for proposing that if an idea or action works the way one intends, it must be right. But James can more accurately be described as a pluralist-someone who believes the world to be far too complex for any philosophy to explain everything.

Dewey's philosophy can be described as a version of philosophical naturalism, which regards human experience, intelligence, and communities as ever-evolving mechanisms. Using their experience and intelligence, Dewey believed, human beings can solve problems, including social problems, through inquiry. For Dewey, naturalism led to the idea of a democratic society that allows all members to acquire social intelligence and progress both as individuals and as communities. Dewey held that traditional ideas about knowledge, truth, and values, in which absolutes are assumed, are incompatible with a broadly Darwinian world-view in which individuals and societies are progressing. In consequence, he felt that these traditional ideas must be discarded or revised. Indeed, for pragmatists, everything people know and do depend on a historical context and are thus tentative rather than absolute.

Many followers and critics of Dewey believe he advocated elitism and social engineering in his philosophical stance. Others think of him as a kind of romantic humanist. Both tendencies are evident in Dewey's writings, although he aspired to synthesize the two realms.

The pragmatists' tradition was revitalized in the 1980s by American philosopher Richard Rorty, who has faced similar charges of elitism for his belief in the relativism of values and his emphasis on the role of the individual in attaining knowledge. Interest has renewed in the classic pragmatists-Pierce, James, and Dewey-have an alternative to Rorty's interpretation of the tradition.

The Philosophy of Mind, is the branch of philosophy that considers mental phenomena such as sensation, perception, thought, belief, desire, intention, memory, emotion, imagination, and purposeful action. These phenomena, which can be broadly grouped as thoughts and experiences, are features of human beings; many of them are also found in other animals. Philosophers are interested in the nature of each of these phenomena as well as their relationships to one another and to physical phenomena, such as motion.

The most famous exponent of dualism was the French philosopher René Descartes, who maintained that body and mind are radically different entities and that they are the only fundamental substances in the universe. Dualism, however, does not show how these basic entities are connected.

In the work of the German philosopher Gottfried Wilhelm Leibniz, the universe is held to consist of an infinite number of distinct substances, or monads. This view is pluralistic in the sense that it proposes the existence of many separate entities, and it is monistic in its assertion that each monad reflects within itself the entire universe.

Other philosophers have held that knowledge of reality is not derived from a priori principles, but is obtained only from experience. This type of metaphysic is called empiricism. Still another school of philosophy has maintained that, although an ultimate reality does exist, it is altogether inaccessible to human knowledge, which is necessarily subjective because it is confined to states of mind. Knowledge is therefore not a representation of external reality, but merely a reflection of human perceptions. This view is known as skepticism or agnosticism in respect to the soul and the reality of God.

The 18ᵗʰ-century German philosopher Immanuel Kant published his influential work The Critique of Pure Reason in 1781. Three years later, he expanded on his study of the modes of thinking with an essay entitled 'What is Enlightenment'? In this 1784 essay, Kant challenged readers to 'dare to know', arguing that it was not only a civic but also a moral duty to exercise the fundamental freedoms of thought and expression.

Several major viewpoints were combined in the work of Kant, who developed a distinctive critical philosophy called transcendentalism. His philosophy is agnostic in that it denies the possibility of a strict knowledge of ultimate reality; it is empirical in that it affirms that all knowledge arises from experience and is true of objects of actual and possible experience; and it is rationalistic in that it maintains the a priori character of the structural principles of this empirical knowledge.

These principles are held to be necessary and universal in their application to experience, for in Kant's view the mind furnishes the archetypal forms and categories (space, time, causality, substance, and relation) to its sensations, and these categories are logically anterior to experience, although manifested only in experience. Their logical anteriority to experience makes these categories or structural principle's transcendental, they transcend all experience, both actual and possible. Although these principles determine all experience, they do not in any way affect the nature of things in themselves. The knowledge of which these principles are the necessary conditions must not be considered, therefore, as constituting a revelation of things as they are in themselves. This knowledge concerns things only insofar as they appear to human perception or as they can be apprehended by the senses. The argument by which Kant sought to fix the limits of human knowledge within the framework of experience and to demonstrate the inability of the human mind to penetrate beyond experience strictly by knowledge to the realm of ultimate reality constitutes the critical feature of his philosophy, giving the key word to the titles of his three leading treatises, Critique of Pure Reason, Critique of Practical Reason, and Critique of Judgment. In the system propounded in these works, Kant sought also to reconcile science and religion in a world of two levels, comprising noumena, objects conceived by reason although not perceived by the senses, and phenomena, things as they appear to the senses and are accessible to material study. He maintained that, because God, freedom, and human immortality are noumenal realities, these concepts are understood through moral faith rather than through scientific knowledge. With the continuous development of science, the explanation of metaphysics to include scientific knowledge and methods became one of the major objectives of metaphysicians.

Some of Kant's most distinguished followers, notably Johann Gottlieb Fichte, Friedrich Schelling, Georg Wilhelm Friedrich Hegel, and Friedrich Schleiermacher, negated Kant's criticism in their elaborations of his transcendental metaphysics by denying the Kantian conception of the thing-in-itself. They thus developed an absolute idealism in opposition to Kant's critical transcendentalism.

Since the formation of the hypothesis of absolute idealism, the development of metaphysics has resulted in as many types of metaphysical theory as existed in pre-Kantian philosophy, despite Kant's contention that he had fixed definitely the limits of philosophical speculation. Notable among these later metaphysical theories are radical empiricism, or pragmatism, a native American form of metaphysics expounded by Charles Sanders Peirce, developed by William James, and adapted as instrumentalism by John Dewey; voluntarism, the foremost exponents of which are the German philosopher Arthur Schopenhauer and the American philosopher Josiah Royce; phenomenalism, as it is exemplified in the writings of the French philosopher Auguste Comte and the British philosopher Herbert Spencer; emergent evolution, or creative evolution, originated by the French philosopher Henri Bergson; and the philosophy of the organism, elaborated by the British mathematician and philosopher Alfred North Whitehead. The salient doctrines of pragmatism are that the chief function of thought is to guide action, that the meaning of concepts is to be sought in their practical applications, and that truth should be tested by the practical effects of belief; according to instrumentalism, ideas are instruments of action, and their truth is determined by their role in human experience. In the theory of voluntarism, the 'will' postulates the supreme manifestation of reality. The exponents of phenomenalism, who are sometimes called positivists, contend that everything can be analyzed in terms of actual or possible occurrences, or phenomena, and that anything that cannot be analyzed in this manner cannot be understood. In emergent or creative evolution, the evolutionary process is characterized as spontaneous and unpredictable rather than mechanistically determined. The philosophy of the organism combines an evolutionary stress on constant process with a metaphysical theory of God, the eternal objects, and creativity.

In the 20th century the validity of metaphysical thinking has been disputed by the logical positivists. (Logical positivism) and by the so-called dialectical materialism of the Marxists. The basic principle maintained by the logical positivists is the Verifiability theory of meaning. According to this theory a sentence has factual meaning only if it meets the test of observation. Logical positivists argue that metaphysical expressions such as 'Nothing exists except material particles' and 'Everything is part of one all-encompassing spirit' cannot be tested empirically. Therefore, according to the Verifiability theory of meaning, these expressions have no factual cognitive meaning, although they can have an emotive meaning relevant to human hopes and feelings.

The dialectical materialists assert that the mind is conditioned by and reflects material reality. Therefore, speculations that conceive of constructs of the mind as having any other than material reality itself is unreal and can result only in delusion. To these assertions metaphysicians reply by denying the adequacy of the Verifiability theory of meaning and of material perception as the standard of reality. Both logical positivism and dialectical materialism, they argue, conceal metaphysical assumptions, for example, that everything is observable or at least connected with something observable and that the mind has no distinctive life of its own. In the philosophical movement known as existentialism, thinkers have contended that the questions of the nature of being and of the individual's relationship to it are extremely important and meaningful in terms of human life. The investigation of these questions is therefore considered valid whether its results can be verified objectively.

Since the 1950s the problems of systematic analytical metaphysics have been studied in Britain by Stuart Newton Hampshire and Peter Frederick Strawson, the former concerned, in the manner of Spinoza, with the relationship between thought and action, and the latter, in the manner of Kant, with describing the major categories of experience as they are embedded in language. In the U.S. metaphysics has been pursued much in the spirit of positivism by Wilfred Stalker Sellars and Willard Van Orman Quine. Sellars have sought to express metaphysical questions in linguistic terms, and Quine has attempted to determine whether the structure of language commits the philosopher to asserting the existence of any entities whatever and, if so, what kind. In these new formulations the issues of metaphysics and ontology remain vital.

In the 17th century, French philosopher René Descartes proposed that only two substances ultimately exist; mind and body. Yet, if the two are entirely distinct, as Descartes believed, how can one substance interact with the other? How, for example, is the intention of a human mind able to cause movement in the person's limbs? The issue of the interaction between mind and body is known in philosophy as the mind-body problem.

Many fields other than philosophy shares an interest in the nature of mind. In religion, the nature of mind is connected with various conceptions of the soul and the possibility of life after death. In many abstract theories of mind there is considerable overlap between philosophy and the science of psychology. Once part of philosophy, psychology split off and formed a separate branch of knowledge in the 19th century. While psychology used scientific experiments to study mental states and events, philosophy uses reasoned arguments and thought experiments in seeking to understand the concepts that underlie mental phenomena. Also influenced by philosophy of mind is the field of artificial intelligence (AI), which endeavours to develop computers that can mimic what the human mind can do. Cognitive science attempts to integrate the understanding of mind provided by philosophy, psychology, AI, and other disciplines. Finally, all of these fields benefit from the detailed understanding of the brain that has emerged through neuroscience in the late 20th century.

Philosophers use the characteristics of inward accessibility, subjectivity, intentionality, goal-directedness, creativity and freedom, and consciousness to distinguish mental phenomena from physical phenomena.

Perhaps the most important characteristic of mental phenomena is that they are inwardly accessible, or available to us through introspection. We each know our own minds-our sensations, thoughts, memories, desires, and fantasies-in a direct sense, by internal reflection. We also know our mental states and mental events in a way that no one else can. In other words, we have privileged access to our own mental states.

Certain mental phenomena, those we generally call experiences, have a subjective nature-that is, they have certain characteristics we become aware of when we reflect. For instance, there is 'something it is like' to feel pain, or have an itch, or see something red. These characteristics are subjective in that they are accessible to the subject of the experience, the person who has the experience, but not to others.

Other mental phenomena, which we broadly refer to as thoughts, have a characteristic philosophers call intentionality. Intentional thoughts are about other thoughts or objects, which are represented as having certain properties or for being related to one another in a certain way. The belief that California is west of Nevada, for example, is about California and Nevada and represents the former for being west of the latter. Although we have privileged access to our intentional states, many of them do not seem to have a subjective nature, at least not in the way that experiences do.

A number of mental phenomena appear to be connected to one another as elements in an intelligent, goal-directed system. The system works as follows: First, our sense organs are stimulated by events in our environment; next, by virtue of these stimulations, we perceive things about the external world; finally, we use this information, as well as information we have remembered or inferred, to guide our actions in ways that further our goals. Goal-directedness seems to accompany only mental phenomena.

Another important characteristic of mind, especially of human minds, is the capacity for choice and imagination. Rather than automatically converting past influences into future actions, individual minds are capable of exhibiting creativity and freedom. For instance, we can imagine things we have not experienced and can act in ways that no one expects or could predict.

Mental phenomena are conscious, and consciousness may be the closest term we have for describing what is special about mental phenomena. Minds are sometimes referred to as consciousness, yet it is difficult to describe exactly what consciousness is. Although consciousness is closely related to inward accessibility and subjectivity, these very characteristics seem to hinder us in reaching an objective scientific understanding of it.

Although philosophers have written about mental phenomena since ancient times, the philosophy of mind did not garner much attention until the work of French philosopher René Descartes in the 17th century. Descartes's work represented a turning point in thinking about mind by making a strong distinction between bodies and minds, or the physical and the mental. This duality between mind and body, known as Cartesian dualism, has posed significant problems for philosophy ever since.

Descartes believed there are two basic kinds of things in the world, a belief known as substance dualism. For Descartes, the principles of existence for these two groups of things-bodies and minds-are completely different from one another: Bodies exist by being extended in space, while minds exist by being conscious. According to Descartes, nothing can be done to give a body thought and consciousness. No matter how we shape a body or combine it with other bodies, we cannot turn the body into a mind, a thing that is conscious, because being conscious is not a way of being extended.

For Descartes, a person consists of a human body and a human mind causally interacting with one another. For example, the intentions of a human being for having inclining inclinations as to cause that person's limbs to move. In this way, the mind can affect the body. In addition, the sense organs of some human beings are possibly affected by light, pressure, or sound, external sources which in turn affect the brain, affecting mental states. Thus, the body may affect the mind. Exactly how mind can affect body, and vice versa, is a central issue in the philosophy of mind, and is known as the mind-body problem. According to Descartes, this interaction of mind and body is peculiarly intimate. Unlike the interaction between a pilot and his ship, the connection between mind and body more closely resembles two substances that have been thoroughly mixed together.

In response to the mind-body problem arising from Descartes's theory of substance dualism, a number of philosophers have advocated various forms of substance monism, the doctrine that there is ultimately just one kind of thing in reality. In the 18th century, Irish philosopher George Berkeley claimed there were no material objects in the world, only minds and their ideas. Berkeley thought that talk about physical objects was simply a way of organizing the flow of experience. Near the turn of the 20th century, American psychologist and philosopher William James proposed another form of substance monism. James claimed that experience is the basic stuff from which both bodies and minds are constructed.

Most philosophers of mind today are substance monists of a third type: They are materialists who believe that everything in the world is basically material, or a physical object. Among materialists, there is still considerable disagreement about the status of mental properties, which are conceived as properties of bodies or brains. Materialists who are property dualists believe that mental properties are an additional kind of property or attribute, not reducible to physical properties. Property dualists have the problem of explaining how such properties can fit into the world envisaged by modern physical science, according to which there are physical explanations for all things.

Materialists who are property monists believe that there is ultimately only one type of property, although they disagree on whether or not mental properties exist in material form. Some property monists, known as reductive materialists, hold that mental properties exist simply as a subset of relatively complex and nonbasic physical properties of the brain. Reductive materialists have the problem of explaining how the physical states of the brain can be inwardly accessible and have a subjective character, as mental states do. Other property monists, known as Eliminative materialists, consider the whole category of mental properties to be a mistake. According to them, mental properties should be treated as discredited postulates of an outmoded theory. Eliminative materialism is knowledge for most people to accept, since we seem to have direct knowledge of our own mental phenomena by introspection and because we use the general principles we understand about mental phenomena to predict and explain the behaviour of others.

Philosophy of mind concerns itself with a number of specialized problems. In addition to the mind-body problem, important issues admit providing the general proceedings to concede to be real. Valid or true knowledge admits to the truth that to grant as valid in the efforts of possibly being real. The functional privileges of those personal identities, immortality, and artificial intelligence.

During much of Western history, the mind has been identified with the soul as presented in Christian Theology. According to Christianity, the soul is the source of a person's identity and is usually regarded as immaterial; thus, it is capable of enduring after the death of the body. Descartes's conception of the mind as a separate, nonmaterial substance fits well with this understanding of the soul. In Descartes's view, we are aware of our bodies only as the cause of sensations and other mental phenomena. Consequently our personal essence is composed more fundamentally of mind and the preservation of the mind after death would constitute our continued existence.

The mind conceived by materialist forms of substance monism does not fit as neatly with this traditional concept of the soul. With materialism, once a physical body is destroyed, nothing enduring remains. Some philosophers think that a concept of personal identity can be constructed that permits the possibility of life after death without appealing to separate immaterial substances. Following in the tradition of 17th-century British philosopher John Locke, these philosophers propose that a person consists of a stream of mental events linked by memory. It is these links of memory, rather than a single underlying substance, that provides the unity of a single consciousness through time. Immortality is conceivable if we think of these memory links as connecting a later consciousness in heaven with an earlier one on earth.

The field of artificial intelligence also raises interesting questions for the philosophy of mind. People have designed machines that mimic or model many aspects of human intelligence, and there are robots currently in use whose behaviour is described in terms of goals, beliefs, and perceptions. Such machines are capable of behaviour that, were it exhibited by a human being, would surely be taken to be free and creative. As an example, in 1996 an IBM computer named Deep Blue won a chess game against Russian world champion Garry Kasparov under international match regulations. Moreover, it is possible to design robots that have some sort of privileged access to their internal states. Philosophers disagree over whether such robots truly think or simply appear to think and whether such robots should be considered to be conscious

Dualism, in philosophy, the theory that the universe is explicable only as a whole composed of two distinct and mutually irreducible elements. In Platonic philosophy the ultimate dualism is between 'being' and 'nonbeing'-that is, between ideas and matter. In the 17th century, dualism took the form of belief in two fundamental substances: mind and matter. French philosopher René Descartes, whose interpretation of the universe exemplifies this belief, was the first to emphasize the irreconcilable difference between thinking substance (mind) and extended substance (matter). The difficulty created by this view was to explain how mind and matter interact, as they apparently do in human experience. This perplexity caused some Cartesian to deny entirely any interaction between the two. They asserted that mind and matter are inherently incapable of affecting each other, and that any reciprocal action between the two is caused by God, who, on the occasion of a change in one, produces a corresponding change in the other. Other followers of Descartes abandoned dualism in favour of monism.

In the 20th century, reaction against the monistic aspects of the philosophy of idealism has to some degree revived dualism. One of the most interesting defences of dualism is that of Anglo-American psychologist William McDougall, who divided the universe into spirit and matter and maintained that good evidence, both psychological and biological, indicates the spiritual basis of physiological processes. French philosopher Henri Bergson in his great philosophic work Matter and Memory likewise took a dualistic position, defining matter as what we perceive with our senses and possessing in itself the qualities that we perceive in it, such as colour and resistance. Mind, on the other hand, reveals itself as memory, the faculty of storing up the past and utilizing it for modifying our present actions, which otherwise would be merely mechanical. In his later writings, however, Bergson abandoned dualism and came to regard matter as an arrested manifestation of the same vital impulse that composes life and mind.

Dualism, in philosophy, the theory that the universe is explicable only as a whole composed of two distinct and mutually irreducible elements. In Platonic philosophy the ultimate dualism is between 'being' and 'nonbeing-that is, between ideas and matter. In the 17ᵗʰ century, dualism took the form of belief in two fundamental substances: mind and matter. French philosopher René Descartes, whose interpretation of the universe exemplifies this belief, was the first to emphasize the irreconcilable difference between thinking substance (mind) and extended substance (matter). The difficulty created by this view was to explain how mind and matter interact, as they apparently do in human experience. This perplexity caused some Cartesian to deny entirely any interaction between the two. They asserted that mind and matter are inherently incapable of affecting each other, and that any reciprocal action between the two is caused by God, who, on the occasion of a change in one, produces a corresponding change in the other. Other followers of Descartes abandoned dualism in favour of monism.

In the 20ᵗʰ century, reaction against the monistic aspects of the philosophy of idealism has to some degree revived dualism. One of the most interesting defences of dualism is that of Anglo-American psychologist William McDougall, who divided the universe into spirit and matter and maintained that good evidence, both psychological and biological, indicates the spiritual basis of physiological processes. French philosopher Henri Bergson in his great philosophic work Matter and Memory likewise took a dualistic position, defining matter as what we perceive with our senses and possessing in itself the qualities that we perceive in it, such as colour and resistance. Mind, on the other hand, reveals itself as memory, the faculty of storing up the past and utilizing it for modifying our present actions, which otherwise would be merely mechanical. In his later writings, however, Bergson abandoned dualism and came to regard matter as an arrested manifestation of the same vital impulse that composes life and mind.

For many people of an understanding the place of mind in nature is the greatest philosophical problem. Mind is often though to be the last domain that stubbornly resists scientific understanding and philosophers defer over whether they find that cause for celebration or scandal. The mind-body problem in the modern era was given its definitive shape by Descartes, although the dualism that he espoused is in some form whatever there is a religious or philosophical tradition there is a religious or philosophical tradition whereby the soul may have an existence apart from the body. While most modern philosophers of mind would reject the imaginings that lead us to think that this makes sense, there is no consensus over the best way to integrate our understanding of people as bearers of physical properties lives on the other.

Occasionalism, for the finding from its terms as employed to designate the philosophical system devised by the followers of the 17ᵗʰ-century French philosopher René Descartes, who, in attempting to explain the interrelationship between mind and body, concluded that God is the only cause. The occasionalists began with the assumption that certain actions or modifications of the body are preceded, accompanied, or followed by changes in the mind. This assumed relationship presents no difficulty to the popular conception of mind and body, according to which each entity is supposed to act directly on the other: These philosophers, however, asserting that cause and effect must be similar, could not conceive the possibility of any direct mutual interaction between substances as dissimilar as mind and body.

According to the occasionalists, the action of the mind is not, and cannot be, the cause of the corresponding action of the body. Whenever any action of the mind takes place, God directly produces in connection with that action, and by reason of it, a corresponding action of the body; the converse process is likewise true. This theory did not solve the problem, for if the mind cannot act on the body (matter), then God, conceived as mind, cannot act on matter. Conversely, if God is conceived as other than mind, then he cannot act on mind. A proposed solution to this problem was furnished by exponents of radical empiricism such as the American philosopher and psychologist William James. This theory disposed of the dualism of the occasionalists by denying the fundamental difference between mind and matter.

Generally, along with consciousness, that experience of an external world or similar scream or other possessions, takes upon itself the visual experience or deprive of some normal visual experience, that this, however, does not perceive the world accurately. In its frontal experiment. As researchers reared kittens in total darkness, except that for five hours a day the kittens were placed in an environment with only vertical lines. When the animals were later exposed to horizontal lines and forms, they had trouble perceiving these forms.

Philosophers have long debated the role of experience in human perception. In the late 17th century, Irish philosopher William Molyneux wrote to his friend, English philosopher John Locke, and asked that he consider the following scenario: Suppose that you could restore sight to a person who was blind. Using only vision, would that person be able to tell the difference between a cube and a sphere, which she or he had previously experienced only through touch? Locke, who emphasized the role of experience in perception, thought the answer was no. Modern science actually allows us to address this philosophical question, because a very small number of people who were blind have had their vision restored with the aid of medical technology.

Two researchers, British psychologist Richard Gregory and British-born Neurologist Oliver Sacks, have written about their experiences with men who were blind for a long time due to cataracts and then had their vision restored late in life. When their vision was restored, they were often confused by visual input and were unable to see the world accurately. For instance, they could detect motion and perceive colours, but they had great difficulty with complex stimuli, such as faces. Much of their poor perceptual ability was probably due to the fact that the synapses in the visual areas of their brains had received little or no stimulation throughout their lives. Thus, without visual experience, the visual system does not develop properly.

Visual experience is useful because it creates memories of past stimuli that can later serve as a context for perceiving new stimuli. Thus, you can think of experience as a form of context that you carry around with you. A visual illusion occurs when your perceptual experience of a stimulus is substantially different from the actual stimulus you are viewing. In the previous example, you saw the green circles as different sizes, even though they were actually the same size. To experience another illusion, look at the illustration entitled 'Zöllner Illusion'. What shape do you see? You may see a trapezoid that is wider at the top, but the actual shape is a square. Such illusions are natural artifacts of the way our visual systems work. As a result, illusions provide important insights into the functioning of the visual system. In addition, visual illusions are fun to experience.

Consider the pair of illusions in the accompanying illustration, 'Illusions of Length.' These illusions are called geometrical illusions, because they use simple geometrical relationships to produce the illusory effects. The first illusion, the Müller-Lyer illusion, is one of the most famous illusions in psychology. Which of the two horizontal lines is longer? Although your visual system tells you that the lines are not equal, a ruler would tell you that they are equal. The second illusion is called the Ponzo illusion. Once again, the two lines do not appear to be equal in length, but they are. For further information about illusions

Prevailing states of consciousness, are not as simple, or agreed-upon by any steadfast and held definition of itself, in so, that, consciousness exists. Attempted definitions tend to be tautological (for example, consciousness defined s awareness) or merely descriptive (for example, consciousness described as sensations, thoughts, or feelings). Despite this problem of definition, the subject of consciousness has had a remarkable history. At one time the primary subject matter of psychology, consciousness as an area of study suffered an almost total demise, later reemerging to become a topic of current interest.

René Descartes applied rigorous scientific methods of deduction to his exploration of philosophical questions. Descartes is probably best known for his pioneering work in philosophical skepticism. Author Tom Sorell examines the concepts behind Descartes's work Meditationes de Prima Philosophia (1641; Meditations on First Philosophy), centring on its unconventional use of logic and the reactions it aroused, as most of the philosophical discussions of consciousness arose from the mind-body issues posed by the French philosopher and mathematician René Descartes in the 17th century. Descartes asked: Is the mind, or consciousness, independent of matter? Is consciousness extended (physical) or unextended (nonphysical)? Is consciousness determinative, or is it determined? English philosophers such as John Locke equated consciousness with physical sensations and the information they provide, whereas European philosophers such as Gottfried Wilhelm Leibniz and Immanuel Kant gave a more central and active role to consciousness.

The philosopher who most directly influenced subsequent exploration of the subject of consciousness was the 19th-century German educator Johann Friedrich Herbart, who wrote that ideas had quality and intensity and that they may inhibit or facilitate one another. Thus, ideas may pass from 'states of reality' (consciousness) to 'states of the tendency' (unconsciousness), with the dividing line between the two states being described as the threshold of consciousness. This formulation of Herbart clearly presages the development, by the German psychologist and physiologist Gustav Theodor Fechner, of the psychophysical measurement of sensation thresholds, and the later development by Sigmund Freud of the concept of the unconscious.

The experimental analysis of consciousness dates from 1879, when the German psychologist Wilhelm Max Wundt started his research laboratory. For Wundt, the task of psychology was the study of the structure of consciousness, which extended well beyond sensations and included feelings, images, memory, attention, duration, and movement. Because early interest focussed on the content and dynamics of consciousness, it is not surprising that the central methodology of such studies was introspection; that is, subjects reported on the mental contents of their own consciousness. This introspective approach was developed most fully by the American psychologist Edward Bradford Titchener at Cornell University. Setting his task as that of describing the structure of the mind, Titchener attempted to detail, from introspective self-reports, the dimensions of the elements of consciousness. For example, taste was 'dimensionalized' into four basic categories: sweet, sour, salt, and bitter. This approach was known as structuralism.

By the 1920s, however, a remarkable revolution had occurred in psychology that was to essentially remove considerations of consciousness from psychological research for some 50 years: Behaviourism captured the field of psychology. The main initiator of this movement was the American psychologist John Broadus Watson. In a 1913 article, Watson stated, 'I believe that we can write a psychology and never use the term's consciousness, mental states, mind . . . imagery and the like.' Psychologists then turned almost exclusively to behaviour, as described in terms of stimulus and response, and consciousness was totally bypassed as a subject. A survey of eight leading introductory psychology texts published between 1930 and the 1950s found no mention of the topic of consciousness in five texts, and in two it was treated as a historical curiosity.

Beginning of the later parts of the 1950s, however, interest in the subject of consciousness returned, specifically in those subjects and techniques relating to altered states of consciousness: sleep and dreams, meditation, biofeedback, hypnosis, and drug-induced states. A proportionate surge in sleep and dream research was directly fuelled by a discovery relevant to the nature of consciousness. A physiological indicator of the dream state was found: At roughly 90-minute intervals, the eyes of sleepers were observed to move rapidly, and at the same time the sleepers' brain waves would show a pattern resembling the waking state. When people were awakened during these periods of rapid eye movement, they almost always reported dreams, whereas if awakened at other times they did not. This and other research clearly indicated that sleep, once considered a passive state, were instead an active state of consciousness (see Dreaming; Sleep).

During the 1960s, an increased search for 'higher levels' of consciousness through meditation resulted in a growing interest in the practices of Zen Buddhism and Yoga from Eastern cultures. A full flowering of this movement in the United States was seen in the development of training programs, such as Transcendental Meditation, that were self-directed procedures of physical relaxation and focussed attention. Biofeedback techniques also were developed to bring body systems involving factors such as blood pressure or temperature under voluntary control by providing feedback from the body, so that subjects could learn to control their responses. For example, researchers found that persons could control their brain-wave patterns to some extent, particularly the so-called alpha rhythms generally associated with a relaxed, meditative state. This finding was especially relevant to those interested in consciousness and meditation, and a number of 'alpha training' programs emerged.

Another subject that led to increased interest in altered states of consciousness was hypnosis, which involves a transfer of conscious control from the subject to another person. Hypnotism has had a long and intricate history in medicine and folklore and has been intensively studied by psychologists.

151

Much has become known about the hypnotic state, relative to individual suggestibility and personality traits; the subject has now largely been demythologized, and the limitations of the hypnotic state are fairly well known. Despite the increasing use of hypnosis, however, much remains to be learned about this unusual state of focussed attention.

Finally, many people in the 1960s experimented with the psychoactive drugs known as hallucinogens, which produce disorders of consciousness. The most prominent of these drugs are lysergic acid diethylamide, or LSD; mescaline, and psilocybin; the latter two have long been associated with religious ceremonies in various cultures. LSD, because of its radical thought-modifying properties, was initially explored for its so-called mind-expanding potential and for its psychotomimetic effects (imitating psychoses). Little positive use, however, has been found for these drugs, and their use is highly restricted.

Scientists have long considered the nature of consciousness without producing a fully satisfactory definition. In the early 20th century American philosopher and psychologist William James suggested that consciousness is a mental process involving both attention to external stimuli and short-term memory. Later scientific explorations of consciousness mostly expanded upon James's work. In this article from a 1997 special issue of Scientific American, Nobel laureate Francis Crick, who helped determine the structure double-stranded helixes of the DNA, molecules, and fellow biophysicist Christof Koch explain how experiments on vision might deepen our understanding of consciousness.

As the concept of a direct, simple linkage between environment and behaviour became unsatisfactory in recent decades, the interest in altered states of consciousness may be taken as a visible sign of renewed interest in the topic of consciousness. That persons are active and intervening participants in their behaviour has become increasingly clear. Environments, rewards, and punishments are not simply defined by their physical character. Memories are organized, not simply stored. An entirely new area called cognitive psychology has emerged that canters on these concerns. In the study of children, increased attention is being paid to how they understand, or perceive, the world at different ages. In the field of animal behaviour, researchers increasingly emphasize the inherent characteristics resulting from the way a species has been shaped to respond adaptively to the environment. Humanistic psychologists, with a concern for self-actualization and growth, have emerged after a long period of silence. Throughout the development of clinical and industrial psychology, the conscious states of persons in terms of their current feelings and thoughts were of obvious importance. The role of consciousness, however, was often de-emphasised in favour of unconscious needs and motivations. Trends can be seen, however, toward a new emphasis on the nature of states of consciousness.

Perception (psychology), spreads of a process by which organisms interpret and organize sensation to produce a meaningful experience of the world. Sensation usually refers to the immediate, relatively unprocessed result of stimulation of sensory receptors in the eyes, ears, nose, tongue, or skin. Perception, on the other hand, better describes one's ultimate experience of the world and typically involves further processing of sensory input. In practice, sensation and perception are virtually impossible to separate, because they are part of one continuous process.

Our sense organs translate physical energy from the environment into electrical impulses processed by the brain. For example, light, in the form of electromagnetic radiation, causes receptor cells in our eyes to activate and send signals to the brain. But we do not understand these signals as pure energy. The process of perception allows us to interpret them as objects, events, people, and situations.

Without the ability to organize and interpret sensations, life would seem like a meaningless jumble of colours, shapes, and sounds. A person without any perceptual ability would not be able to recognize faces, understand language, or avoid threats. Such a person would not survive for long. In fact, many species of animals have evolved exquisite sensory and perceptual systems that aid their survival.

Organizing raw sensory stimuli into meaningful experiences involves cognition, a set of mental activities that includes thinking, knowing, and remembering. Knowledge and experience are extremely important for perception, because they help us make sense of the input to our sensory systems. To understand these ideas, try to read the following passage:

You could probably read the text, but not as easily as when you read letters in their usual orientation. Knowledge and experience allowed you to understand the text. You could read the words because of your knowledge of letter shapes, and maybe you even have some prior experience in reading text upside down. Without knowledge of letter shapes, you would perceive the text as meaningless shapes, just as people who do not know Chinese or Japanese see the characters of those languages as meaningless shapes. Reading, then, is a form of visual perception.

Note that as above, whereby you did not stop to read every single letter carefully. Instead, you probably perceived whole words and phrases. You may have also used context to help you figure out what some of the words must be. For example, recognizing th upsides that may have helped you predict down, because the two words often occur together. For these reasons, you probably overlooked problems with the individual letters-some of them, such as the n in down, are mirror images of normal letters. You would have noticed these errors immediately if the letters were right side up, because you have much more experience seeing letters in that orientation.

How people perceive a well-organized pattern or whole, instead of many separate parts, is a topic of interest in Gestalt psychology. According to Gestalt psychologists, the whole is different from the sum of its parts. Such as the Gestalt is a German word, meaning of the configuration or patter. All and all, theirs became of three founders of the Gestalt psychology, as were the German researchers in the fields of behaviourialism. Their names include those of the reacher's contributions as to accredited to Max Wertheimer, Kurt Koffka, and Wolfgang Köhler. These men identified a number of principles by which people organize isolated parts of a visual stimulus into groups or whole objects. There are five main laws of grouping: Proximity, similarity, continuity, closure, and common fate. A sixth law, that of simplicity, encompasses all of these laws.

Although most often applied to visual perception, the Gestalt laws also apply to perception in other senses. When we listen to music, for example, we do not hear a series of disconnected or random tones. We interpret the music as a whole, relating the sounds to each other based on how similar they are in pitch, how close together they are in time, and other factors. We can perceive melodies, patterns, and form in music. When a song is transposed to another key, we still recognize it, even though all of the notes have changed.

The law of proximity states that the closer objects are to one another, the more likely we are to mentally group them together. In the illustration below, we perceive as groups the boxes that are closest to one another. Note that we do not see the second and third boxes from the left as a pair, because they are spaced farther apart.

The law of similarity leads us to link together parts of the visual fields that are similar in colour, lightness, texture, shape, or any other quality. That is why, in the following illustration, we perceive rows of objects instead of columns or other arrangements.

The law of continuity leads us to see a line as continuing in a particular direction, rather than making an abrupt turn. In the drawing on the left below, we see a straight line with a curved line running through it. Notice that we do not see the drawing as consisting of the two pieces in the drawing on the right.

According to the law of closure, we prefer complete forms to incomplete forms. Thus, in the drawing below, we mentally close the gaps and perceive a picture of a duck. This tendency allows us to perceive whole objects from incomplete and imperfect forms.

The law of common fate leads us to group together objects that move in the same direction. In the following illustration, imagine that three of the balls are moving in one direction, and two of the balls are moving in the opposite direction. If you saw these in actual motion, you would mentally group the balls that moved in the same direction. Because of this principle, we often see flocks of birds or schools of fish as one unit.

Central to the approach of Gestalt psychologists is the law of prägnanz, or simplicity. This general notion, which encompasses all other Gestalt laws, states that people intuitively prefer the simplest, most stables of possible organizations. For example, look at the illustration below. You could perceive this in a variety of ways: as three overlapping disks; as one whole disk and two partial disks with slices cut out of their right sides; or even as a top view of three-dimensional, cylindrical objects. The law of simplicity states that you will see the illustration as three overlapping disks, because that is the simplest interpretation.

Not only does perception involve organization and grouping, it also involves distinguishing an object from its surroundings. Notice that once you perceive an object, the area around that object becomes the background. For example, when you look at your computer monitor, the wall behind it becomes the background. The object, or figure, is closer to you, and the background, or ground, is farther away.

Gestalt psychologists have devised ambiguous figure-ground relationships-that is, drawings in which the figure and ground can be reversed-to illustrate their point that the whole is different from the sum of its parts. Consider the accompanying illustration entitled 'Figure and Ground.' You may see a white vase as the figure, in which case you will see it displayed on a dark ground. However, you may also see two dark faces that point toward one another. Notice that when you do so, the white area of the figure becomes the ground. Even though your perception may alternate between these two possible interpretations, the parts of the illustration are constant. Thus, the illustration supports the Gestalt position that the whole is not determined solely by its parts. The Dutch artist M. C. Escher was intrigued by ambiguous figure-ground relationships.

Although such illustrations may fool our visual systems, people are rarely confused about what they see. In the real world, vases do not change into faces as we look at them. Instead, our perceptions are remarkably stable. Considering that we all experience rapidly changing visual input, the stability of our perceptions is more amazing than the occasional tricks that fool our perceptual systems. In what manner do we perceive a stability occurring to the world is due, in part, to a number of factors that maintain perceptual constancy.

As we view an object, the image it projects on the retinas of our eyes changes with our viewing distance and angle, the level of ambient light, the orientation of the object, and other factors. Perceptual constancy allows us to perceive an object as roughly the same in spite of changes in the retinal image. Psychologists have identified a number of perceptual consistencies, including lightness constancy, colour constancy, shape constancy, and size constancy.

Lightness constancy that our perception of an object's lightness or darkness remains constant despite changes in illumination. To understand lightness constancy, try the following demonstration. First, take a plain white sheet of paper into a brightly lit room and note that the paper appears to be white. Then, turn out a few of the lights in the room. Note that the paper continues to appear white. Next, if it will not make the room pitch black, turn out some more lights. Note that the paper appears to be white regardless of the actual amount of light energy that enters the eye.

Lightness constancy illustrates an important perceptual principle: Perception is relative. Lightness constancy may occur because the white piece of paper reflects more light than any of the other objects in the room—regardless of the different lighting conditions. That is, you may have determined the lightness or darkness of the paper relative to the other objects in the room. Another explanation, proposed by 19th-century German physiologist Hermann von Helmholtz, is that we unconsciously take the lighting of the room into consideration when judging the lightness of objects.

Colour constancy is closely related to lightness constancy. Colour constancy that we perceive the colour of an object as the same despite changes in lighting conditions. You have experienced colour constancy if you have ever worn a pair of sunglasses with coloured lenses. In spite of the fact that the coloured lenses change the colour of light reaching your retina, you still perceive white objects as white and red objects as red. The explanations for colour constancy parallel those for lightness constancy. One proposed explanation is that because the lenses tint everything with the same colour, we unconsciously 'subtract' that colour from the scene, leaving the original colours.

Another perceptual constancy is shape constancy, which that you perceive objects as retaining the same shape despite changes in their orientation. To understand shape constancy, hold a book in front of your face so that you are looking directly at the cover. The rectangular nature of the book should be very clear. Now, rotate the book away from you so that the bottom edge of the cover is much closer

155

to you than the top edge. The image of the book on your retina will now be quite different. In fact, the image will now be trapezoidal, with the bottom edge of the book larger on your retina than the top edge. (Try to see the trapezoid by closing one eye and imagining the cover as a two-dimensional shape.) In spite of this trapezoidal retinal image, you will continue to see the book as rectangular. In large measure, shape constancy occurs because your visual system takes depth into consideration.

Depth perception also plays a major role in size constancy, the tendency to perceive objects as staying the same size despite changes in our distance from them. When an object is near to us, its image on the retina is large. When that same objects are far away, the image on the retina is small. In spite of the changes in the size of the retinal image, we perceived from the object as the same size. For example, when you see a person at a great distance from you, you do not perceive that person as very small. Instead, you think that the person is of normal size and far away. Similarly, when we view a skyscraper from far away, its image on our retina is very small-yet we perceive the building as very large.

Psychologists have proposed several explanations for the phenomenon of size constancy. First, people learn the general size of objects through experience and use this knowledge to help judges size. For example, we know that insects are smaller than people and that people are smaller than elephants. In addition, people take distance into consideration when judging the size of an object. Thus, if two objects have the same retinal image size, the object that seems farther away will be judged as larger. Even infants seem to possess size constancy.

Another explanation for size constancy involves the relative sizes of objects. According to this explanation, we see objects as the same size at different distances because they stay the same size relative to surrounding objects. For example, as we drive toward a stop sign, the retinal image sizes of the stop sign relative to a nearby tree remain constant-both images grow larger at the same rate.

Depth perception is the ability to see the world in three dimensions and to perceive distance. Although this ability may seem simple, depth perception is remarkable when you consider that the images projected on each retina are two-dimensional. From these flat images, we construct a vivid three-dimensional world. To perceive depth, we depend on two main sources of information: binocular disparity, a depth cue that requires both eyes; and monocular cues, which allow us to perceive depth with just one eye.

An autostereogram is a remarkable kind of two-dimensional image that appears three-dimensional (3-D) when viewed in the right way. To see the 3-D image, first make sure you are viewing the expanded version of this picture. Then try to focus your eyes on a point in space behind the picture, keeping your gaze steady. An image of a person playing a piano will appear.

Beaus our eyes are spaced about 7 cm (about 3 in) apart, the left and right retinas receive slightly different images. This difference in the left and right images is called binocular disparity. The brain integrates these two images into a single three-dimensional image, allowing us to perceive depth and distance.

For a demonstration of binocular disparity, fully extend your right arm in front of you and hold up your index finger. Now, alternate closing your right eye and then your left eye while focussing on your index finger. Notice that your finger appears to jump or shift slightly-a consequence of the two slightly different images received by each of your retinas. Next, keeping your focus on your right index finger, hold your left index finger up much closer to your eyes. You should notice that the nearer finger creates a double image, which is an indication to your perceptual system that it is at a different depth than the farther finger. When you alternately close your left and right eyes, notice that the nearer finger appears to jump much more than the more distant finger, reflecting a greater amount of binocular disparity.

You have probably experienced a number of demonstrations that use binocular disparity to provide a sense of depth. A stereoscope is a viewing device that presents each eye with a slightly different photograph of the same scene, which generates the illusion of depth. The photographs are taken from slightly different perspectives, one approximating the view from the left eye and the other representing the view from the right eye. The View-Master, a children's toy, is a modern type of stereoscope.

Filmmakers have made use of binocular disparity to create 3-D (three-dimensional) movies. In 3-D movies, two slightly different images are projected onto the same screen. Viewers wear special glasses that use coloured filters (as for most 3-D movies) or polarizing filters (as for 3-D IMAX movies). The filters separate the image so that each eye receives the image intended for it. The brain combines the two images into a single three-dimensional image. Viewers who watch the film without the glasses see a double image.

Another phenomenon that makes use of binocular disparity is the autostereogram. The autostereogram is a two-dimensional image that can appear three-dimensional without the use of special glasses or a stereoscope. Several different types of autostereograms exist. The most popular, based on the single-image random dot stereogram, seemingly becomes three-dimensional when the viewer relaxes or refocuses the eyes, as if focussing on a point in space behind the image. The two-dimensional image usually consists of random dots or lines, which, when viewed properly, coalesce into a previously unseen three-dimensional image. This type of autostereogram was first popularized in the Magic Eye series of books in the early 1990s, although its invention traces back too 1979. Most autostereograms are produced using computer software. The mechanism by which autostereograms work is complex, but they employ the same principle as the stereoscope and 3-D movies. That is, each eye receives a slightly different image, which the brain fuses into a single three-dimensional image.

Although binocular disparity is a very useful depth cue, it is only effective over a fairly short range-less than 3 m (10 ft). As our distance from objects increases, the binocular disparity decreases-that is, the images received by each retina become more and more similar. Therefore, for distant objects, your perceptual system cannot rely on binocular disparity as a depth cue. However, you can still determine that some objects are nearer and some farther away because of monocular cues about depth.

To portray a realistic three-dimensional world on a two-dimensional canvas, artists must make use of a variety of depth cues. It was not until the 1400s, during the Italian Renaissance, that artists began to understand linear perspective fully and to portray depth convincingly. Shown here are several paintings that produce a sense of depth.

Close one eye and look around you. Notice the richness of depth that you experience. How does this sharp senses of a threedimensionality emerge from input to a single two-dimensional retina? The answer lies in monocular cues, or cues to depth that are effective when viewed with only one eye.

The problem of encoding depth on the two-dimensional retina is quite similar to the problem faced by an artist who wishes to realistically portray depth on a two-dimensional canvas. Some artists are amazingly adept at doing so, using a variety of monocular cues to give their works a sense of depth.

Although there are many kinds of monocular cues, the most important are interposition, atmospheric perspective, texture gradient, linear perspective, size cues, height cues, and motion parallax.

People commonly rely on interposition, or the overlap between objects, to judge distances. When one object partially obscures our view of another object, we judge the covered object as farther away from us.

Probably the most important monocular cue is interposition, or overlap. When one object overlaps or partly blocks our view of another object, we judge the covered object for being farther away from us. This depth cue is all around us-look around you and notice how many objects are partly obscured by other objects. To understand how much we rely on interposition, try this demonstration. Hold two pens, one in each hand, a short distance in front of your eyes. Hold the pens several centimetres apart so they do not overlap, but move one pen just slightly farther away from you than the other. Now close one eye. Without binocular vision, notice how difficult it is to judge which pen is more distant. Now, keeping one eye closed, move your hands closer and closer together until one pen which sets the manner in which it moves across in front of the other. Notice how interposition makes depth perception much easier.

When we look out over vast distances, faraway points look hazy or blurry. This effect is known as atmospheric perspective, and it helps us to judge distances. In this picture, the ridges that are farther away appear hazier and less detailed than the closer ridges.

The air contains microscopic particles of dust and moisture that make distant objects look hazy or blurry. This effect is called atmospheric perspective or aerial perspective, and we use it to judge distance. In the anthem, 'Oh Canada' it draws reference to the effect of atmospheric perspectives, which make's distant mountains appear bluish or purple. When you are standing on a mountain, you see brown earth, gray rocks, and green trees and grass-but little that is purple. When you are looking at a mountain from a distance, however, atmospheric particles bend the light so that the rays that reach your eyes lie in the blue or purple part of the colour spectrum. This same effect makes the sky appear blue.

An influential American psychologist, James J. Gibson, was among the first people to recognize the importance of a texture gradient in perceiving depth. A texture gradient arises whenever we view a surface from a slant, rather than directly from above. Most surfaces-such as the ground, a road, or a field of flowers-have a texture. The texture becomes denser and less detailed as the surface recedes into the background, and this information helps us to judge depth. For example, look at the floor or ground around you. Notice that the apparent texture of the floor changes over distance. The texture of the floor near you appears more detailed than the texture of the floor farther away. When objects are placed at different locations along a texture gradient, judging their distance from you becomes fairly easy.

Linear perspective that parallel lines, such as the white lines of this road, appears to converge with greater distance and reach a vanishing point at the horizon. We use our knowledge of linear perspective to help us judge distances.

Artists have learned to make great use of linear perspective in representing a three-dimensional world on a two-dimensional canvas. Linear perspective refers to the fact that parallel lines, such as railroad tracks, appears to converge with distance, eventually reaching a vanishing point at the horizon. The more the lines converge, the farther away they appear.

When estimating an object's distance from us, we take into account the size of its image relative to other objects. This depth cue is known as relative size. In this photograph, because we assume that the aeroplanes are the same size, we judge the aeroplanes that take up less of the image for being farther away from the camera.

Another visual cue to apparent depth is closely related to size constancy. According to size constancy, even though the size of the retinal image may change as an object moves closer to us or farther from us, we perceive that object as staying about the same size. We are able to do so because we take distance into consideration. Thus, if we assume that two objects are the same size, we perceive the object that casts a smaller retinal image as farther away than the object that casts a larger retinal image. This depth cue is known as relative size, because we consider the size of an object's retinal image relative to other objects when estimating its distance.

Another depth cue involves the familiar size of objects. Through experience, we become familiar with the standard size of certain objects, such as houses, cars, aeroplanes, people, animals, books, and chairs. Knowing the size of these objects helps us judge our distance from them and from objects around them.

When judging an object's distance, we consider its height in our visual field relative to other objects. The closer an object is to the horizon in our visual field, the farther away we perceive it to be. For example, the wildebeest that are higher in this photograph appear farther away than those that are lower.

We perceive points nearer to the horizon as more distant than points that are farther away from the horizon. This that below the horizon, objects higher in the visual field appear farther away than those that are lower. Above the horizon, objects lower in the visual field appear farther away than those that are higher. For example, in the accompanying picture entitled 'Relative Height', the animals higher in the photo appear farther away than the animals lower in the photo. But above the horizon, the clouds lower in the photo appear farther away than the clouds higher in the photo. This depth cue is called relative elevation or relative height, because when judging an object's distance, we consider its height in our visual field relative to other objects.

The monocular cues discussed so far-interposition, atmospheric perspective, texture gradient, linear perspective, size cues, and height cues-are sometimes called pictorial cues, because artists can use them to convey three-dimensional information. Another monocular cue cannot be represented on a canvas. The determent motion of the parallax occurs when objects at different distances from you appear to move at different rates when you are in motion. The next time you are driving along in a car, pay attention to the rate of movement of nearby and distant objects. The fence near the road appears to whiz past you, while the more distant hills or mountains appear to stay in virtually the same position as you move. The rate of an object's movement provides a cue to its distance.

Although motion plays an important role in depth perception, the perception of motion is an important phenomenon in its own right. It allows a baseball outfielder to calculate the speed and trajectory of a ball with extraordinary accuracy. Automobile drivers rely on motion perception to judge the speeds of other cars and avoid collisions. A cheetah must be able to detect and respond to the motion of antelopes, its chief prey, in order to survive.

Initially, you might think that you perceive motion when an object's image moves from one part of your retina to another part of your retina. In fact that is what occurs if you are staring straight ahead and a person walks in front of you. Motion perception, however, is not that simple-if it was, the world would appear to move every time we moved our eyes. Keep in mind that you are almost always in motion. As you walk along a path, or simply move your head or your eyes, images from many stationary objects move around on your retina. How does your brain know which movement on the retina is due to your own motion and which is due to motion in the world? Understanding that distinction is the problem that faces psychologists who want to explain motion perception.

One explanation of motion perception involves a form of unconscious inference. That is, when we walk around or move our head in a particular way, we unconsciously expect that images of stationary objects are enviably to move on our retina. We discount such movement on the retina as due to our own bodily motion and perceive the objects as stationary.

In contrast, when we are moving and the image of an object, however, serves as for a 'null-set factor' as not moving our retina, we perceive that object as moving. Consider what happens, that a person, by participating that he manoeuvres in front of you and you track that person's motion with your eyes. You move your head and your eyes to follow the person's movement, with the resultant account that the image of the person has not in moving on your retina. The fact that the person's image stays in roughly the same part of the retina leads you to perceive the person as moving.

Psychologist James J. Gibson thought that this explanation of motion perception was too complicated. He reasoned that perception does not depend on internal thought processes. He thought, instead, that the objects in our environment contain all the information necessary for perception. Think of the aerial acrobatics of a fly. Clearly, the fly is a master of motion and depth perception, yet few people would say the fly makes unconscious inferences. Gibson identified a number of cues for motion detection, including the covering and uncovering of background. Research has shown that motion detection is, in fact, much easier against a background. Thus, as a person manipulates his movements as placed in front of you, that person first covers and then uncovers portions of the background.

People may perceive motion when none actually exists. For example, moving pictures are really a series of slightly different still pictures flashed on a screen at a rate of 24 pictures, or frames, per second. From this rapid succession of still images, our brain perceives a changing fluid of motions as phenomenon known as stroboscopic movement. For more information about illusions of motion.

Experience in interacting with the world is vital to perception. For instance, kittens raised without visual experience or deprived of normal visual experience do not perceive the world accurately. In one experiment, researchers reared kittens in total darkness, except that for five hours a day the kittens were placed in an environment with only vertical lines. When the animals were later exposed to horizontal lines and forms, they had trouble perceiving these forms.

Philosophers have long debated the role of experience in human perception. In the late 17th century, Irish philosopher William Molyneux wrote to his friend, English philosopher John Locke, and asked he to consider the following scenario: Suppose that you could restore sight to a person who was blind. Using only vision, would that person be able to tell the difference between a cube and a sphere, which she or he had previously experienced only through touch? Locke, who emphasized the role of experience in perception, thought the answer was no. Modern science actually allows us to address this philosophical question, because a very small number of people who were blind have had their vision restored with the aid of medical technology.

Two researchers, British psychologist Richard Gregory and British-born Neurologist Oliver Sacks, have written about their experiences with men who were blind for a long time due to cataracts and then had their vision restored late in life. When their vision was restored, they were often confused by visual input and were unable to see the world accurately. For instance, they could detect motion and perceive colours, but they had great difficulty with complex stimuli, such as faces. Much of their poor perceptual ability was probably due to the fact that the synapses in the visual areas of their brains had received little or no stimulation throughout their lives. Thus, without visual experience, the visual system does not develop properly.

Visual experience is useful because it creates memories of past stimuli that can later serve as a context for perceiving new stimuli. Thus, you can think of experience as a form of context that you carry around with you.

Ordinarily, when you read, you use the context of your prior experience with words to process the words you are reading. Context may also occur outside of you, as in the surrounding elements in a visual scene. When you are reading and you encounter an unusual word, you may be able to determine the meaning of the word by its context. Your perception depends on the context.

Although context is useful most of the time, on some rare occasions context can lead you to misperceive a stimulus. Look at Example B in the 'Context Effects' illustration. Which of the green circles is larger? You may have guessed that the green circle on the right is larger. In fact, the two circles are the same size. Your perceptual system was fooled by the context of the surrounding red circles.

Against a background of slanted lines, a perfect square appears trapezoidal-that is, wider at the top than at the bottom. This illusion may occur because the lines create a sense of depth, making the top of the square seems farther away and larger.

A visual illusion occurs when your perceptual experience of a stimulus is substantially different from the actual stimulus you are viewing. In the previous example, you saw the green circles as different sizes, even though they were actually the same size. To experience another illusion, look at the illustration entitled 'Zöllner Illusion'. What shape do you see? You may see a trapezoid that is wider at the top, but the actual shape is a square. Such illusions are natural artifacts of the way our visual systems work. As a result, illusions provide important insights into the functioning of the visual system. In addition, visual illusions are fun to experience.

An ascribing notion to awaiting the idea that something debated finds to its intent of meaning the explicit significance of the same psychology that is immeasurably the scientific study of behaviour and the mind. This definition contains three elements. The first is that psychology is a scientific enterprise that obtains knowledge through systematic and objective methods of observation and experimentation. Second is that psychologists study behaviour, which refers to any action or reaction that can be measured or observed-such as the blink of an eye, an increase in heart rate, or the unruly violence that often erupts in a mob. Third is that psychologists study the mind, which refers to both conscious and unconscious mental states. These states cannot actually be seen, only inferred from observable behaviour.

Many people think of psychologists as individuals practically dispense of opinion about what could or should be done about a situation or problem needed of the analysis of the personality, and help those who are troubled or mentally ill. But psychology is far more than the treatment of personal problems. Psychologists strive to understand the mysteries of human nature-why people think, feel, and act as they do. Some psychologists also study animal behaviour, using their findings to determine laws of behaviour that apply to all organisms and to formulate theories about how humans behave and think.

With its broad scope, psychology investigates an enormous range of phenomena: learning and memory, sensation and perception, motivation and emotion, thinking and language, personality and social behaviour, intelligence, infancy and child development, mental illness, and much more. Furthermore, psychologists examine these topics from a variety of complementary perspectives. Some conduct detailed biological studies of the brain, and others explore how we process information; others analyze the role of evolution, and still others study the influence of culture and society.

Psychologists seek to answer a wide range of important questions about human nature: Are individuals genetically predisposed at birth to develop certain traits or abilities? How accurate are people at remembering faces, places, or conversations from the past? What motivates us to seek out friends and sexual partners? Why do so many people become depressed and behave in ways that seem self-destructive? Do intelligence test scores predict success in school, or later in a career? What causes prejudice, and why is it so widespread? Can the mind be used to heal the body? Discoveries from psychology can help people better understand themselves, relate better to others, and solve the problems that confront them.

The term psychology comes from two Greek words: psyche, which 'soul,' and logos, 'the study of.' These root words were first combined in the 16ᵗʰ century, at a time when the human soul, spirit, or mind was seen as distinct from the body.

Psychology overlaps with other sciences that investigate behaviour and mental processes. Certain parts of the field share much with the biological sciences, especially physiology, the biological study of the functions of living organisms and their parts. Like physiologists, many psychologists study the inner workings of the body from a biological perspective. However, psychologists usually focus on the activity of the brain and nervous system.

The social sciences of sociology and anthropology, which study human societies and cultures, also intersect with psychology. For example, both psychology and sociology explore how people behave when they are in groups. However, psychologists try to understand behaviour from the vantage point of the individual, whereas sociologists focus on how behaviour is shaped by social forces and social institutions. Anthropologists investigate behaviour as well, paying particular attention to the similarities and differences between human cultures around the world.

Psychology is closely connected with psychiatry, which is the branch of medicine specializing in mental illnesses. The study of mental illness is one of the largest areas of research in psychology. Psychiatrists and psychologists differ in their training. A person seeking to become a psychiatrist first obtains a medical degree and then engages in further formal medical education in psychiatry. Most psychologists have a doctoral graduate degree in psychology.

The study of psychology draws on two kinds of research: basic and applied. Basic researchers seek to test general theories and build a foundation of knowledge, while applied psychologists study people in real-world settings and use the results to solve practical human problems. There are five major areas of research: biopsychology, clinical psychology, cognitive psychology, developmental psychology, and social psychology. Both basic and applied research is conducted in each of these fields of psychology.

This section describes basic research and other activities of psychologists in the five major fields of psychology. Applied research is discussed in the Practical Applications of Psychology section of this article.

Magnetic resonance imaging (MRI) reveals structural differences between a normal adult brain, left, and the brain of a person with schizophrenia, right. The schizophrenic brain has enlarged ventricles (fluid-filled cavities), shown in light gray. However, not all people with schizophrenia show this abnormality.

How do body and minds interact? Are body and mind fundamentally different parts of a human being, or are they one and the same, interconnected in important ways? Inspired by this classic philosophical debate, many psychologists specialize in biopsychology, the scientific study of the biological underpinnings of behaviour and mental processes.

At the heart of this perspective is the notion that human beings, like other animals, have an evolutionary history that predisposes them to behave in ways that are uniquely adaptive for survival and reproduction. Biopsychologists work in a variety of subfields. Researchers in the field of ethology observe fish, reptiles, birds, insects, primates, and other animal species in their natural habitats. Comparative psychologists study animal behaviour and make comparisons among different species, including humans. Researchers in evolutionary psychology theorize about the origins of human aggression, altruism, mate selection, and other behaviours. Those in behavioural genetics seek to estimate the extent to which human characteristics such as personality, intelligence, and mental illness are inherited.

Particularly important to biopsychology is a growing body of research in behavioural neuroscience, the study of the links between behaviour and the brain and nervous system. Facilitated by computer-assisted imaging techniques that enable researchers to observe the living human brain in action, this area is generating great excitement. In the related area of cognitive neuroscience, researchers record physical activity in different regions of the brain as the subject reads, speaks, solves math problems, or engages in other mental tasks. Their goal is to pinpoint activities in the brain that correspond to different operations of the mind. In addition, many Biopsychologists are involved in psychopharmacology, the study of how drugs affect mental and behavioural functions.

This chart illustrates the percentage of people in the United States who experience a particular mental illness at some point during their lives. The figures are derived from the National Comorbidity Survey, in which researchers interviewed more than people aged 15 to 54 years, homeless people and those living in prisons, nursing homes, or other institutions were not included in the survey.

Clinical psychology is dedicated to the study, diagnosis, and treatment of mental illnesses and other emotional or behavioural disorders. More psychologists work in this field than in any other branch of psychology. In hospitals, community clinics, schools, and in private practice, they use interviews and tests to diagnose depression, anxiety disorders, schizophrenia, and other mental illnesses. People with these psychological disorders often suffer terribly. They experience disturbing symptoms for making it difficult for them to work, or relate to others, and cope with the demands of everyday life.

Over the years, scientists and mental health professionals have made great strides in the treatment of psychological disorders. For example, advances in psychopharmacology have led to the development of drugs that relieve severe symptoms of mental illness. Clinical psychologists usually cannot prescribe drugs, but they often work in collaboration with a patient's physician. Drug treatment is often combined with psychotherapy, a form of intervention that relies primarily on verbal communication to treat emotional or behavioural problems. Over the years, psychologists have developed many different forms of psychotherapy. Some forms, such as the psychoanalysis, as method of psychological therapy originated by Sigmund Freud in which free association, dreams, interpretations, and analysis of resistance and transference are useful to explore repressed or unconscious impulses, anxieties and internal conflict, in order to free psychic energy for mature love and work, that focuses on resolving the interiorized, unconscious conflicts as stemming from childhood and past experiences, plainly called 'latencies'. Other forms, such as cognitive and behavioural therapies, focus more on the person's current level of functioning and try to help the individual change distressing thoughts, feelings, or behaviours.

In addition to studying and treating mental disorders, many clinical psychologists study the normal human personality and the ways in which individuals differ from one another. Still, others administer a variety of psychological tests, including intelligence tests and personality tests. These tests are commonly given to individuals in the workplace or in school to assess their interests, skills, and level of functioning. Clinical psychologists also use tests to help them diagnose people with different types of psychological disorders.

The field of counselling psychology is closely related to clinical psychology. Counselling psychologists may treat mental disorders, but they more commonly treat people with less-severe adjustment problems related to marriage, family, school, or career. Many other types of professionals care for and treat people with psychological disorders, including psychiatrists, psychiatric social workers, and psychiatric nurses.

To take the Stroop test, names aloud each colour in the two columns at left as quickly as you can. Next, look at the right side of the illustration and quickly name the colours in which the words are printed. Which task took longer to complete? The test, devised in 1935 by American psychologist John Stroop, shows that people cannot help but process word meanings, and that this processing interferes with the colour-naming task.

How do people gainfully employ of knowledge or have as a resulting remainder from experience? How and where in the brain are visual images, facts, and personal memories stored? What causes forgetting? How do peoples solve problems or make difficult life decisions? Does language limit the way people have or formulate in the mind for reasons about or reflect on by thinking, perhaps, by thinking the unthinkable. To what extent are people influenced by information outside of conscious awareness?

These are the kinds of questions posed within cognitive psychology, the scientific study of how people acquire, processes, and utilize information. Cognition refers to the process of knowing and encompasses nearly the entire range of conscious and unconscious mental processes: sensation and perception, conditioning and learning, attention and consciousness, sleep and dreaming, memory and forgetting, reasoning and decision making, imagining, problem solving, and language.

Decades ago, the invention of digital computers gave cognitive psychologists a powerful new way of thinking about the human mind. They began to see human beings as information processors who receive input, process and store information, and produce output. This approach became known as the information-processing model of cognition. As computers have become more sophisticated, cognitive psychologists have extended the metaphor. For example, most researchers now reject the idea that information is processed in linear, sequential steps. Instead they find that the human mind is capable of parallel processing, in which multiple operations are carried out simultaneously.

In this information-processing model of memory, information that enters the brain is briefly recorded in sensory memory. If we focus our attention on it, the information may become part of working memory (also called short-term memory), where it can be manipulated and used. Through encoding techniques such as repetition and rehearsal, information may be transferred to long-term memory. Retrieving long-term memories makes them active again in working memory.

Are people programmed by inborn biological dispositions? Or is an individual's fate molded by culture, family, peers, and other socializing influences within the environment? These questions about the roles of nature and nurture are central to the study of human development.

An incredibly complex array of influences, including families, acquaintances, mass media, and society as a whole, help determine the moral development of children. Although a rash of violent incidents in American schools in the late 1990s focussed attention on deviant youth behaviour, the vast majority of children seem to function harmoniously with others. In this August 1999 article from Scientific American, William Damon, director of the Centre on Adolescence at Stanford University in California, explores recent findings on how young people develop morality.

Developmental psychology focuses on the changes that come with age. By comparing people of different ages, and by tracking individuals over time, researchers in this area study the ways in which people mature and change over the life span. Within this area, those who specialize in child development or child psychology study physical, intellectual, and social development in fetuses, infants, children, and adolescents. Recognizing that human development is a lifelong process, other developmental psychologists study the changes that occur throughout adulthood. Still others specialize in the study of old age, even the process of dying.

Social psychology is the scientific study of how people think, feel, and behave in social situations. Researchers in this field ask questions such as, How do we form impressions of others? How are people persuaded to change their attitudes or beliefs? What causes people to conform to group situations? What leads someone to help or ignore a person in need? Under what circumstances do people that carry out or comply and fulfill their commands and behave or resist orders?

By observing people in real-world social settings, and by carefully devising experiments to test people's social behaviour, social psychologists learn about the ways people influence, perceives and interact with one another. The study of social influence includes topics such as conformity, obedience to authority, the formation of attitudes, and the principles of persuasion. Researchers interested in social perception study how people come to know and evaluate one another, how people form group stereotypes, and the origins of prejudice. Other topics of particular interest to social psychologists include physical attraction, love and intimacy, aggression, altruism, and group processes. Many social psychologists are also interested in cultural influences on interpersonal behaviour.

Whereas basic researchers test theories about mind and behaviour, applied psychologists are motivated by a desire to solve practical human problems. Four particularly active areas of application are health, education, business, and law.

Today, many psychologists work in the emerging area of health psychology, the application of psychology to the promotion of physical health and the prevention and treatment of illness. Researchers in this area have shown that human health and well-being depends on both biological and psychological factors.

Many psychologists in this area study psychophysiological disorders (also called psychosomatic disorders), conditions that are brought on or influenced by psychological states, most often stress. These disorders include high blood pressure, headaches, asthma, and ulcers. Researchers have discovered that chronic stress is associated with an increased risk of coronary heart disease. In addition, stress can compromise the body's immune system and increase susceptibility to illness.

Health psychologists also study how people cope with stress. They have found that people who have family, friends, and other forms of social support are healthier and live longer than those who are more isolated. Other researchers in this field examine the psychological factors that underlie smoking, drinking, drug abuse, risky sexual practices, and other behaviours harmful to health.

Psychologists in all branches of the discipline contribute to our understanding of teaching, learning, and education. Some help developing among the standardized tests as used to measure academic aptitude and achievement. Others study the ages at which children become capable of attaining various cognitive skills, the effects of rewards on their motivation to learn, computerized instruction, bilingual education, learning disabilities, and other relevant topics. Perhaps the best-known application of psychology to the field of education occurred in 1954 when, in the case of Brown v. Board of Education, the Supreme Court of the United States outlawed the segregation of public schools by race. In its ruling, the Court cited psychological studies suggesting that segregation had a damaging effect on black students and tended to encourage prejudice.

In addition to the contributions of psychology as a whole, two fields within psychology focus exclusively on education: educational psychology and school psychology. Educational psychologists seek to understand and improve the teaching and learning process within the classroom and other educational settings. Educational psychologists study topics such as intelligence and ability testing, student motivation, discipline and classroom management, a curriculum planned, and grading. They also test general theories about how students learn most effectively. School psychologists work in elementary and secondary school systems administering tests, making placement recommendations, and counselling children with academic or emotional problems.

The business world, psychology is applied in the workplace and in the marketplace. Industrial-organizational (I-O) psychology focuses on human behaviour in the workplace and other organizations. I-O psychologists conduct research, teach in business schools or universities, and work in private industry. Many psychologists study the factors that influence worker motivation, satisfaction, and productivity. Others study the personal traits and situations that foster great leadership. Still, others focus on the processes of personnel selection, training, and evaluation. Studies have shown, for example, that face-to-face interviews sometimes result in poor hiring decisions and may be biassed by the applicant's gender, race, and physical attractiveness. Studies have also shown that certain standardized tests can help to predict on-the-job performance.

Consumer psychology is the study of human decision making and behaviour in the marketplace. In this area, researchers analyze the effects of advertising on consumers' attitudes and buying habits. Consumer psychologists also study various aspects of marketing, such as the effects of packaging, price, and other factors that lead people to purchase one product rather than another.

Many psychologists today work in the legal system. They consult with attorneys, testify in court as expert witnesses, counsel prisoners, teach in law schools, and research various justice-related issues. Sometimes referred to as forensic psychologists, those who apply psychology to the law study a range of issues, including jury selection, eyewitness testimony, confessions to police, lie-detector tests, the death penalty, criminal profiling, and the insanity defence.

Studies in forensic psychology have helped to illuminate weaknesses in the legal system. For example, based on trial-simulation experiments, researchers have found that jurors are often biassed by various facts not in evidence-that is, facts the judge tells them to disregard. In studying eyewitness testimony, researchers have staged mock crimes and asked witnesses to identify the assailant or recall other details. These studies have revealed that under certain condition's eyewitnesses are highly prone to error.

Psychologists in this area often testify in court as expert witnesses. In cases involving the insanity defence, forensic clinical psychologists are often called to court to give their opinion about whether individual defendants are sane or insane. Used as a legal defence, insanity that defendants, because of a mental disorder, cannot appreciate the wrongfulness of their conduct or control it. Defendants who are legally insane at the time of the offense may be absolved of criminal responsibility for their conduct and judged not guiltily. Psychologists are often called to testify in court on other controversial matters as well, including the accuracy of eyewitness testimony, the mental competence (fitness) of defendants to stand trial, and the reliability of early childhood memories.

Psychology has applications in many other domains of human life. Environmental psychologists focus on the relationship between people and their physical surroundings. They study how street noise, heat, architectural design, population density, and crowding affect people's behaviour and mental health. In a related field, human factors' psychologists work on the design of appliances, furniture, tools, and other manufactured items in order to maximize their comfort, safety, and convenience. Sports psychologists advise athletes and study the physiological, perceptual-motor, motivational, developmental, and social aspects of athletic performance. Other psychologists specialize in the study of political behaviour, religion, sexuality, or behaviour in the military.

Psychologists from all areas of specialization use the scientific method to test their theories about behaviour and mental processes. A theory is an organized set of principles that is designed to explain and predict some phenomenon. Good theories also provide specific testable predictions, or hypotheses, about the relation between two or more variables. Formulating a hypothesis to be tested is the first important step in conducting research.

Over the years, psychologists have devised numerous ways to test their hypotheses and theories. Many studies are conducted in a laboratory, usually located at a university. The laboratory setting allows researchers to control what happens to their subjects and make careful and precise observations of behaviour. For example, a psychologist who studies memory can bring volunteers into the lab, ask them to memorize a list of words or pictures, and then test their recall of that material seconds, minutes, or days later.

As indicated by the term field research, studies may also be conducted in real-world locations. For example, a psychologist investigating the reliability of eyewitness testimony might stage phony crimes in the street and then ask unsuspecting bystanders to identify the culprit from a set of photographs. Psychologists observe people in a wide variety of other locations outside the laboratory, including classrooms, offices, hospitals, college dormitories, bars, restaurants, and prisons.

In both laboratory and field settings, psychologists conduct their research using a variety of methods. Among the most common methods are archival studies, case studies, surveys, naturalistic observations, correlational studies, experiments, literature reviews, and measures of brain activity.

One way to learn about people is through archival studies, an examination of existing records of human activities. Psychological researchers often examine old newspaper stories, medical records, birth certificates, crime reports, popular books, and artwork. They may also examine statistical trends of the past, such as crime rates, birth rates, marriage and divorce rates, and employment rates. The strength of such measures is that by observing people only secondhand, researchers cannot unwittingly influence the subjects by their presence. However, available records of human activity are not always complete or detailed enough to be useful.

Archival studies are particularly valuable for examining cultural or historical trends. For example, in one study of physical attractiveness, researchers wanted to know if American standards of female beauty have changed over several generations. These researchers looked through two popular women's magazines between 1901 and 1981 and examined the measurements of the female models. They found that 'curvaceousness' (as measured by the bust-to-waist ratio) varied over time, with a boyish, slender look considered desirably in some time periods but not in others.

Sometimes psychologists interview, test, observe, and investigate the backgrounds of specific individuals in detail. Such case studies are conducted when researchers believe that an in-depth look at one individual will reveal something important about people in general.

Case studies often takes a great deal of time to complete, and the results may be limited by the fact that the subject is atypical. Yet case studies have played a prominent role in the development of psychology. Austrian physician Sigmund Freud based his theory of the psychoanalysis with which are the experiences with latently troubled patients. Swiss psychologist Jean Piaget first began to formulate a theory of intellectual development by questioning his own children. Neuroscientists learn about how the human brain works by testing patients who have suffered brain damage. Cognitive psychologists learn about human intelligence by studying child prodigies and other gifted individuals. Social psychologists learn about group decision making by analyzing the policy decisions of government and business groups. When an individual is exceptional in some way, or when a hypothesis can be tested only through intensive, long-term observation, the case study is a valuable method.

In electroencephalogram, or EEG, is a recording of the action potential, or electrical, activity of the cerebral cortex of the brain. An EEG is made by attaching electrodes to the scalp, then collecting, amplifying, and recording the electrical impulses of the brain.

Biopsychologists interested in the links between brain and behaviour uses a variety of specialized techniques in their research. One approach is to observe and test patients who have suffered damage to a specific region of the brain to determine what mental functions and behaviours were affected by that damage. British-born neurologist Oliver Sacks has written several books in which he describes case studies of brain-damaged patients who exhibited specific deficits in their speech, memory, sleep, and even in their personalities.

This positron emission tomography (PET) scan of the brain shows the activity of brain cells in the resting state and during three types of auditory stimulation. PET uses radioactive substances introduced into the brain to measure such brain functions as cerebral metabolism, blood flow and volume, oxygen use, and the formation of neurotransmitters. This imaging method collects data from many different angles, feeding the information into a computer that produces a series of cross-sectional images.

A second approach to physical alterations of the brain and measure the effects of that change on behaviour. The alteration can be achieved in different ways. For example, animal researchers often damage or destroy a specific region of a laboratory animal's brain through surgery. Other researchers might spark or inhibit activity in the brain through the use of drugs or electrical stimulation.

Advances in technology in the early 1970s allowed psychologists to see inside the living human brain for the first time without physically cutting into it. Today, psychologists use a variety of sophisticated brain-imaging techniques. The computerized axial tomography (CT or CAT) scan provides a computer-enhanced X-ray image of the brain. The more advanced positron emission tomography (PET) scan tracks the level of activity in specific parts of the brain by measuring the amount of glucose being used there. These measurements are then fed to a computer, which produces a colour-coded image of brain activity. Another technique is magnetic resonance imaging (MRI), which produces high-resolution cross-sectional images of the brain. A high-speed version of MRI known as functional MRI produces moving images of the brain as its activity changes in real time. These relatively new brain imaging techniques have generated great excitement, because they allow researchers to identify parts of the brain that are active while people read, speak, listen to music, solve math problems, and engage in other mental activities.

In contrast with the in-depth study of one person, surveys describe a specific population or group of people. Surveys involve asking people a series of questions about their behaviours, thoughts, or opinions. Surveys can be conducted in person, over the phone, or through the mail. Most surveys study a specific group-for example, college students, working mothers, men, or homeowners. Rather than questioning every person in the group, survey researchers choose a representative sample of people and generalize the findings to the larger population.

Surveys may pertain to almost any topic. Often surveys ask people to report their feelings about various social and political issues, the TV shows they watch, or the consumers' products they purchase. Surveys are also used to learn about people's sexual practices; to estimate the use of cigarettes, alcohol, and other drugs; and to approximate the proportion of people who experience feelings of life satisfaction, loneliness, and other psychological states that cannot be directly observed.

Surveys must be carefully designed and conducted to ensure their accuracy. The results can be influenced, and biassed, by two factors: who the respondents are and how the questions are asked. For a survey to be accurate, the sample being questioned must be representative of the population on key characteristics such as sex, race, age, region, and cultural background. To ensure similarity to the larger population, survey researchers usually try to make sure that they have a random sample, a method of selection in which everyone in the population has an equal chance of being chosen.

When the sample is not random, the results can be misleading. For example, prior to the 1936 United States presidential election, pollsters for the magazine Literary Digest mailed postcards to more than 10 million people who were listed in telephone directories or as registered owners of automobiles. The cards asked for whom they intended to vote. Based on the more than 2 million ballots that were returned, the Literary Digest predicted that Republican candidate Alfred M. Landon would be the winner in a landslide over Democrat Franklin D. Roosevelt. At the time, however, more Republican than Democrats owned telephones and automobiles, skewing the poll results. In the election, Landon won only two states.

The results of survey research can also be influenced by the way that questions are asked. For example, when asked about 'welfare', a majority of Americans in one survey said that the government spends too much money. But when asked about 'assistance to the poor', significantly fewer people gave this response.

In naturalistic observation, the researcher observes people as they behave in the real world. The researcher simply records what occurs and does not intervene in the situation. Psychologists use naturalistic observation to study the interactions between parents and children, doctors and patients, police and citizens, and managers and workers.

Naturalistic observation is common in anthropology, in which field workers seek to understand the everyday life of a culture. Ethologists, who study the behaviour of animals in their natural habitat, also use this method. For example, British ethologist Jane Goodall spent many years in African jungles observing chimpanzees-their social structure, courting rituals, struggles for dominance, eating habits, and other behaviours. Naturalistic observation is also common among developmental psychologists who study social play, parent-child attachments, and other aspects of child development. These researchers observe children at home, in school, on the playground, and in other settings.

Case studies, surveys, and naturalistic observations are used to describe behaviour. As correlational studies are further in the design to find statistical connections, or correlations between variables so that some factors can be used to predict others.

A correlation is a statistical measure of the extent to which two variables are associated. A positive correlation exists when two variables increase or decrease together. For example, frustration and aggression are positively correlated, meaning that as frustration rises, so do acts of aggression. More of one more of the other. A negative correlation exists when increases in one variable are accompanied by decreases in the other, and vice versa. For example, friendships and stress-induced illness are negatively correlated, meaning that the more close friends a person has, the fewer stress-related illnesses the person suffers. More of one less of the other.

Based on correlational evidence, researchers can use one variable to make predictions about another variable. But researchers must use caution when drawing conclusions from correlations. It is a natural-but incorrect-to take upon itself to affect or take responsibilites over the human race, such that one variable predicts another, the first must have caused the second. For example, one might assume that frustration triggers aggression, or that friendships foster health. Regardless of how intuitive or accurate these conclusions may be, correlation does not prove causation. Thus, although it is possible that frustration causes aggression, there are other ways to interpret the correlation. For example, it is possible that aggressive people are more likely to suffer social rejection and become frustrated as a result.

Correlations enable researchers to predict one variable from another. But to determine if one variable actually causes another, psychologists must conduct experiments. In an experiment, the psychologist manipulates one factor in a situation-keeping other aspects of the situation constant-and then observes the effect of the manipulation on behaviour. The people whose behaviour is being observed are the subjects of the experiment. The factor that an experimenter varies (the proposed cause) is known as the independent variable, and the behaviour being measured (the proposed effect) is called the dependent variable. In a test of the hypothesis that frustration triggers aggression, frustration would be the independent variable, and aggression the dependent variable.

There are three requirements for conducting a valid scientific experiment: (1) control over the independent variable, (2) the use of a comparison group, and (3) the random assignment of subjects to conditions. In its most basic form, then, a typical experiment compares a large number of subjects who are randomly assigned to experience one condition with a group of similar subjects who are not. Those who experience the condition compose the experimental group, and those who do not make up the control group. If the two groups differ significantly in their behaviour during the experiment, that difference can be attributed to the presence of the condition, or independent variable. For example, to test the hypothesis that frustration triggers aggression, one group of researchers brought subjects into a laboratory, impeded their efforts to complete an important task (other subjects in the experiment were not impeded), and measured their aggressiveness toward another person. These researchers found that subjects who had been frustrated were more aggressive than those who had not been frustrated.

Psychologists use many different methods in their research. Yet no single experiment can fully prove a hypothesis, so the science of psychology builds slowly over time. First, a new discovery must be replicated. Replication refers to the process of conducting a second, nearly identical study to see if the initial findings can be repeated. If so, then researchers try to determine if these findings can be applied, transferred, or generalized to other settings. Generalizability refers to the extent to which a finding obtained less than one set of conditions can also be obtained at another time, in another place, and in other populations.

Because the science of psychology proceeds in small increments, many studies must be conducted before clear patterns emerge. To summarize and interpret an entire body of research, psychologists rely on two methods. One method is a narrative review of the literature, in which a reviewer subjectively evaluates the strengths and weaknesses of the various studies on a topic and argues for certain conclusions. Another method is meta-analysis, a statistical procedure used to combine the results from many different studies. By meta-analysing a body of research, psychologists can often draw precise conclusions concerning the strength and breadth of support for a hypothesis.

Psychological research involving human subjects raise ethical concerns about the subject's right to privacy, the possible harm or discomfort caused by experimental procedures, and the use of deception. Over the years, psychologists have established various ethical guidelines. The American Psychological Association recommends that researchers (1) tell prospective subjects what they will experience so they can give informed consent to participate; (2) instruct subjects that they may withdraw from the study at any time; (3) minimize all harm and discomfort; (4) keep the subjects' responses and behaviour's confidential; and (5) debrief subjects who were deceived in some way by fully explaining the research after they have participated. Some psychologists argue that such rules should never be broken. Others say that some degree of flexibility is needed in order to study certain important issues, such as the effects of stress on test performance.

Laboratory experiments that use rats, mice, rabbits, pigeons, monkeys, and other animals are an important part of psychology, just as in medicine. Animal research serves three purposes in psychology: to learn more about certain types of animals, to discover general principles of behaviour that pertain to all species, and to study variables that cannot ethically be tested with human beings. But is it ethical to experiment on animals?

Some animal rights activists believe that it is wrong to use animals in experiments, particularly in those that involve surgery, drugs, social isolation, food deprivation, electric shock, and other potentially harmful procedures. These activists see animal experimentation as unnecessary and question whether results from such research can be applied to humans. Many activists also argue that like human, animals have the capacity to suffer and feel pain. In response to these criticisms, many researchers point out that animal experimentation has helped to improve the quality of human life. They note that animal studies have contributed to the treatment of anxiety, depression, and other mental disorders. Animal studies have also contributed to our understanding of conditions such as Alzheimer's disease, obesity, alcoholism, and the effects of stress on the immune system. Most researchers follow strict ethical guidelines that require them to minimize pain and discomfort to animals and to use the least invasive procedures possible. In addition, federal animal-protection laws in the United States require researchers to provide humane care and housing of animals and to tend to the psychological well-being of primates used in research.

One of the youngest sciences, psychology did not emerge as a formal discipline until the late 19th century. But its roots extend to the ancient past. For centuries, philosophers and religious scholars have wondered about the nature of the mind and the soul. Thus, the history of psychological thought begins in philosophy.

From about 600 to 300 Bc, Greek philosophers inquired about a wide range of psychological topics. They were especially interested in the nature of knowledge and how human beings come to know the world, a field of philosophy known as epistemology. The Greek philosopher Socrates and his followers, Plato and Aristotle, wrote about pleasure and pain, knowledge, beauty, desire, free will, motivation, common sense, rationality, memory, and the subjective nature of perception. They also theorized about whether human traits are innate or the product of experience. In the field of ethics, philosophers of the ancient world probed a variety of psychological questions: Are people inherently good? How can people attain happiness? What motives or drives do people posses? Are human beings naturally social?

Second-century physician Galen was one of the most influential figures in ancient medicine, second in importance only to Hippocrates. Using animal dissection and other, Galen proposed numerous theories about the function of different parts of the human body, most notably the brain, heart, and liver. He also derived an impressive understanding of the differences between veins and arteries. In the selection below, Galen discusses his idea that the optimal state, or 'constitution,' of the body should be a perfect balance of various internal and external components.

Early thinkers also considered the causes of mental illness. Many ancient societies thought that mental illness resulted from supernatural causes, such as the anger of gods or possession by evil spirits. Both Socrates and Plato focussed on psychological forces as the cause of mental disturbance. For example, Plato thought madness results when a person's irrational, animal-like psyche (mind or soul) overwhelm the intellectual, rational psyche. The Greek physician Hippocrates viewed mental disorders as stemming from natural causes, and he developed the first classification system for mental disorders. Galen, a Greek physician who lived in the 2nd century ad, echoed this belief in a physiological basis for mental disorders. He thought they resulted from an imbalance of the four bodily humours: black bile, yellow bile, blood, and phlegm. For example, Galen thought that melancholia (depression) resulted from a person having too much black bile.

More recently, many other men and women contributed to the birth of modern psychology. In the 1600s French mathematician and philosopher René Descartes theorized that the body and mind are separate entities. He regarded the body as a physical entity and the mind as a spiritual entity, and believed the two interacted only through the pineal gland, a tiny structure at the base of the brain. This position became known as dualism. According to dualism, the behaviour of the body is determined by mechanistic laws and can be measured in a scientific manner. But the mind, which transcends the material world, cannot be similarly studied.

English philosophers' Thomas Hobbes and John Locke disagreed. They argued that all human experiences-including sensations, images, thoughts, and feelings-are physical processes occurring within the brain and nervous system. Therefore, these experiences are valid subjects of study. In this view, which later became known as monism, the mind and body are one and the same. Today, in light of years of research indicating that the physical and mental aspects of the human experience are intertwined, most psychologists reject a rigid dualist position.

Many philosophers of the past also debated the question of whether human knowledge is inborn or the product of experience. Nativists believed that certain elementary truths are innate to the human mind and need not be gained through experience. In contrast, empiricists believed that at birth, a person's mind is like a tabula rasa, or blank slate, and that all human knowledge ultimately comes from sensory experience. Today, all psychologists agree that both types of factors are important in the acquisition of knowledge.

Modern psychology can also be traced to the study of physiology (a branch of biology that studies living organisms and their parts) and medicine. In the 19ᵗʰ century, physiologists began studying the human brain and nervous system, paying particular attention to the topic of sensation. For example, in the 1850s and 1860s German scientist Hermann von Helmholtz studied sensory receptors in the eye and ear, investigating topics such as the speed of neural impulses, colour vision, hearing, and space perception. Another important German scientist, Gustav Fechner, founded psychophysics, the study of the relationship between physical stimuli and our subjective sensations of those stimuli. Building on the work of his compatriot Ernst Weber, Fechner developed a technique for measuring people's subjective sensations of various physical stimuli. He sought to determine the minimum intensity level of a stimulus that is needed to produce a sensation.

English naturalist Charles Darwin was particularly influential in the development of psychology. In 1859 Darwin published On the Origin of Species, in which he proposed that all living forms were a product of the evolutionary process of natural selection. Darwin had based his theory on plants and nonhuman animals, but he later asserted that people had evolved through similar processes, and that human anatomy and behaviour could be analyzed in the same way. Darwin's theory of evolution invited comparisons between humans and other animals, and scientists soon began using animals in psychological research.

French neurologist Jean Martin Charcot shows colleagues a female patient with hysteria at La Salp tri re, a Paris hospital. Charcot (1825-1893) gained renown throughout Europe for his method of treating hysteria and other 'nervous disorders' through hypnosis. Martin Charcot's belief that hysteria had psychologically rather than physical origins influenced Austrian neurologist Sigmund Freud, who studied under Charcot.

In medicine, physicians were discovering new links between the brain and language. For example, French surgeon Paul Broca discovered that people who suffer damage to a specific part of the brain's left hemisphere lose the ability to produce fluent speech. This area of the brain became known as Broca's area. A German neurologist, Carl Wernicke, reported in 1874 that people with damage to a different area of the left hemisphere lose their ability to comprehend speech. This region became known as Wernicke's area.

Other physicians focussed on the study of mental disorders. In the late 19ᵗʰ century, French neurologist Jean Charcot discovered that some of the patients he was treating for so-called nervous disorders could be cured through hypnosis, a psychological-not medical-form of intervention. Charcot's work had a profound impact on Sigmund Freud, an Austrian neurologist whose theories would later revolutionize psychology.

Austrian physician Franz Fredrich Anton Mesmer pioneered the induction of trance-like states to cure medical ailments. Mesmer's work sparked interest among some of his scientific colleagues but was later dismissed as charlatanism. Today, however, Mesmer is considered a pioneer in hypnosis, which is widely believed to be helpful in managing certain medical conditions.

Psychology was predated and somewhat influenced by various pseudoscientific schools of thought-that is, theories that had no scientific foundation. In the late 18th and early 19th centuries, Viennese physician Franz Joseph Gall developed phrenology, the theory that psychological traits and abilities reside in certain parts of the brain and can be measured by the bumps and indentations in the skull. Although phrenology found popular acceptance among the lay public in western Europe and the United States, most scientists ridiculed Gall's ideas. However, research later confirmed the more general point that certain mental activities can be traced to specific parts of the brain.

Physicians in the 18th and 19th centuries used crude devices to treat mental illness, none of which offered any real relief. The circulating swing, top left, was used to spin depressed patients at high speed. American physician Benjamin Rush devised the tranquilizing chair, top right, to calm people with mania. The crib, bottom, was widely used to restrain violent patients.

Another Viennese physician of the 18th century, Franz Anton Mesmer, believed that illness was caused by an imbalance of magnetic fluids in the body. He believed he could restore the balance by passing his hands across the patient's body and waving a magnetic wand over the infected area. Mesmer claimed that his patients would fall into a trance and awaken from it feeling better. The medical community, however, soundly rejected the claim. Today, Mesmer's technique, known as mesmerism, is regarded as an early forerunner of modern hypnosis.

Modern psychology is deeply rooted in the older disciplines of philosophy and physiology. But the official birth of psychology is often traced to 1879, at the University of Leipzig, in Leipzig, Germany. There, physiologist Wilhelm Wundt established the first science laboratory involved with the scientific study of the mind. Kundt's laboratory soon attracted leading scientists and students from Europe and the United States. Among these was James McKeen Cattell, one of the first psychologists to study individual differences through the administration of 'mental tests', Emil Kraepelin, a German psychiatrist who postulated a physical cause for mental illnesses and in 1883 published the first classification system for mental disorders. Hugo Münsterberg, the first to apply psychology to industry and the legal philosophy. Kundt was extraordinarily productive over the course of his career. He supervised a total of 186 doctoral dissertations, taught thousands of students, founded the first scholarly psychological journal, and published innumerable scientific studies. His goal, which he stated in the preface of a book he wrote, was 'to mark out a new domain of science'.

Compared to the philosophers who preceded him, Wundt's approach to the study of mind was based on systematic and rigorous observation. His primary method of research was introspection. This technique involved training people to concentrate and report on their conscious experiences as they reacted to visual displays and other stimuli. In his laboratory, Kundt systematically studied topics such as attention span, reaction time, vision, emotion, and time perception. By recruiting people to serve as subjects, varying the conditions of their experience, and then rigorously repeating all observations, Kundt laid the foundation for the modern psychology experiment.

In the United States, Harvard University professor William James observed the emergence of psychology with great interest. Although trained in physiology and medicine, James was fascinated by psychology and philosophy. In 1875 he offered his first course in psychology. In 1890 James published a two-volume book entitled Principles of Psychology. It immediately became the leading psychology text in the United States, and it brought James a worldwide reputation as a man of great ideas and inspiration. In 28 chapters, James wrote about the stream of consciousness, the formation of habits, individuality, the link between mind and body, emotions, the self, and other topics that inspired generations of psychologists. Today, historians consider James the founder of American psychology.

James's students also made lasting contributions to the field. In 1883 G. Stanley Hall (who also studied with Kundt) established the first true American psychology laboratory in the United States at Johns Hopkins University, and in 1892 he founded and became the first president of the American Psychological Association. Mary Whiton Calkins created an important technique for studying memory and conducted one of the first studies of dreams. In 1905 she was elected the first female president of the American Psychological Association. Edward Lee Thorndike conducted some of the first experiments on animal learning and wrote a pioneering textbook on educational psychology.

During the first decades of psychology, two main schools of thought dominated the field: structuralism and functionalism. Structuralism was a system of psychology developed by Edward Bradford Titchener, an American psychologist who studied under Wilhelm Kundt. Structuralists believed that the task of psychology is to identify the basic elements of consciousness in much the same way that physicists break down the basic particles of matter. For example, Titchener identified four elements in the sensation of taste: sweet, sour, salty, and bitter. The main method of investigation in structuralism was introspection. The influence of structuralism in psychology faded after Titchener's death in 1927.

In contradiction to the structuralist movement, William James promoted a school of thought known as functionalism, the belief that the real task of psychology is to investigate the function, or purpose, of consciousness rather than its structure. James was highly influenced by Darwin's evolutionary theory that all characteristics of a species must serve some adaptive purpose. Functionalism enjoyed widespread appeal in the United States. Its three main leaders were James Rowland Angell, a student of James, John Dewey, who was also one of the foremost American philosophers and educators, also Harvey A. Carr, a psychologist at the University of Chicago.

In their efforts to understand human behavioural processes, the functional psychologists developed the technique of longitudinal research, which consists of interviewing, testing, and observing one person over a long period of time. Such a system permits the psychologist to observe and record the person's development and how he or she reacts to different circumstances.

In the late 19th century Viennese neurologist Sigmund Freud developed a theory of personality and a system of psychotherapy known as psychoanalysis. According to this theory, people are strongly influenced by unconscious forces, including innate sexual and aggressive drives. In this 1938

British Broadcasting Corporation interview, Freud recounts the early resistance to his ideas and later acceptance of his work. Freud's speech is slurred because he was suffering from cancer of the jaw. He died the following year.

Alongside Kundt and James, a third prominent leader of the new psychology was Sigmund Freud, a Viennese neurologist of the late 19th and early 20th century. Through his clinical practice, Freud developed a very different approach to psychology. After graduating from medical school, Freud treated patients who appeared to suffer from certain ailments but had nothing physically wrong with them. These patients were not consciously faking their symptoms, and often the symptoms would disappear through hypnosis, or even just by talking. On the basis of these observations, Freud formulated a theory of personality and a form of psychotherapy known as psychoanalysis. It became one of the most influential schools of Western thought of the 20th century.

Freud introduced his new theory in The Interpretation of Dreams (1889), the first of 24 books he would write. The theory is summarized in Freud's last book, An Outline of Psychoanalysis, published in 1940, after his death. In contrast to Kundt and James, for whom psychology was the study of conscious experience, Freud believed that people are motivated largely by unconscious forces, including strong sexual and aggressive drives. He likened the human mind to an iceberg: The small tip that floats on the water is the conscious part, and the vast region beneath the surface comprises the unconscious. Freud believed that although unconscious motives can be temporarily suppressed, they must find a suitable outlet in order for a person to maintain a healthy personality.

To probe the unconscious mind, Freud developed the psychotherapy technique of free association. In free association, the patient reclines and talks about thoughts, wishes, memories, and whatever else comes to mind. The analyst tries to interpret these verbalizations to determine their psychological significance. In particular, Freud encouraged patients to free associate about their dreams, which he believed were the 'royal road to the unconscious.' According to Freud, dreams are disguised expressions of deep, hidden impulses. Thus, as patients recount the conscious manifest content of dreams, the psychoanalyst tries to unmask the underlying latent content-what the dreams really mean.

From the start of a psychoanalysis, Freud attracted followers, many of whom later proposed competing theories. As a group, these neo-Freudians shared the assumption that the unconscious reveals an important role in a person's thoughts and behaviours. Most parted company with Freud, however, over his emphasis on sex as a driving force. For example, Swiss psychiatrist Carl Jung theorized that all humans inherit a collective unconscious that contains universal symbols and memories from their ancestral past. Austrian physician Alfred Adler theorized that people are primarily motivated to overcome inherent feelings of inferiority. He wrote about the effects of birth order in the family and coined the term sibling rivalry. Karen Horney, a German-born American psychiatrist, argued that humans have a basic need for love and security, and become anxious when they feel isolated and alone.

Motivated by a desire to uncover unconscious aspects of the psyche, psychoanalytic researchers devised what is known as projective tests. A projective test asks people to respond to an ambiguous stimulus such as a word, an incomplete sentence, an inkblot, or an ambiguous picture. These tests are based on the assumption that if a stimulus is vague enough to accommodate different interpretations, then people will use it to project their unconscious needs, wishes, fears, and conflicts. The most popular of these tests are the Rorschach Inkblot Test, which consists of ten inkblots, and the Thematic Apperception Test, which consists of drawings of people in ambiguous situations.

That the psychoanalytic-analysis has been criticized on various grounds and is not as popular as in the past. However, Freud's overall influence on the field has been deep and lasting, particularly his ideas about the unconscious. Today, most psychologists agree that people can be profoundly influenced by unconscious forces, and that people often have a limited awareness of why they think, feel, and behave as they do.

In 1885 German philosopher Hermann Ebbinghaus conducted one of the first studies on memory, using himself as a subject. He memorized lists of nonsense syllables and then tested his memory of the syllables at intervals ranging from 20 minutes to 31 days. As shown in this curve, he found that he remembered less than 40 percent of the items after nine hours, but that the rate of forgetting levelled off over time.

In addition to Kundt, James, and Freud, many others' scholars helped to define the science of psychology. In 1885 German philosopher Hermann Ebbinghaus conducted a series of classic experiments on memory, using nonsense syllables to establish principles of retention and forgetting. In 1896 American psychologist Lightner Witmer opened the first psychological clinic, which initially treated children with learning disorders. He later founded the first journal and training program in a new helping profession that he named clinical psychology. In 1905 French psychologist Alfred Binet devised the first major intelligence test in order to assess the academic potential of schoolchildren in Paris. The test was later translated and revised by Stanford University psychologist Lewis Terman and is now known as the Stanford-Binet intelligence test. In 1908 American psychologist Margaret Floy Washburn (who later became the second female president of the American Psychological Association) wrote an influential book called The Animal Mind, in which she synthesized animal research to that time.

In 1912 German psychologist Max Wertheimer discovered that when two stationary lights flash in succession, people see the display as a single light moving back and forth. This illusion inspired the Gestalt psychology movement, which was based on the notion that people tend to perceive a well-organized whole or pattern that is different from the sum of isolated sensations. Other leaders of Gestalt psychology included Wertheimer's close associate's Wolfgang Köhler and Kurt Koffka. Later, German American psychologist Kurt Lewin extended Gestalt psychology to studies of motivation, personality, social psychology, and conflict resolution. German American psychologist Fritz Heider then extended this approach to the study of how people perceive themselves and others.

In the late 19th century, American psychologist Edward L. Thorndike conducted some of the first experiments on animal learning. Thorndike formulated the law of effect, which states that behaviours that are followed by pleasant consequences will be more likely to be repeated in the future.

William James had defined psychology as 'the science of mental life'. But in the early 1900s, growing numbers of psychologists voiced criticism of the approach used by scholars to explore conscious and unconscious mental processes. These critics doubted the reliability and usefulness of the method of introspection, in which subjects are asked to describe their own mental processes during various tasks. They were also critical of Freud's emphasis on unconscious motives. In search of more-scientific methods, psychologists gradually turned away from research on invisible mental processes and began to study only behaviour that could be observed directly. This approach, known as Behaviourism, ultimately revolutionized psychology and remained the dominant school of thought for nearly 50 years.

Russian physiologist Ivan Pavlov discovered a major type of learning, classical conditioning, by accident while conducting experiments on digestion in the early 1900s. He devoted the rest of his life to discovering the underlying principles of classical conditioning.

Among the first to lay the foundation for the new Behaviourism was American psychologist Edward Lee Thorndike. In 1898 Thorndike conducted a series of experiments on animal learning. In one study, he put cats into a cage, put food just outside the cage, and timed how long it took the cats to learn how to open an escape door that led to the food. Placing the animals in the same cage again and again, Thorndike found that the cats would repeat behaviours that worked and would escape more and more quickly with successive trials. Thorndike thereafter proposed the law of effect, which states that behaviours that are followed by a positive outcome are repeated, while those followed by a negative outcome or none at all are extinguished.

In 1906 Russian physiologist Ivan Pavlov-who had won a Nobel Prize two years earlier for his studies of digestion-stumbled onto one of the most important principles of learning and behaviour. Pavlov was investigating the digestive process in dogs by putting food in their mouths and measuring the flow of saliva. He found that after repeated testing, the dogs would salivate in anticipation of the food, even before he put it in their mouth. He soon discovered that if he rang a bell just before the food was presented each time, the dogs would eventually salivate at the mere sound of the bell. Pavlov had discovered a basic form of learning called classical conditioning (also referred to as Pavlovian conditioning) in which an organism comes to associate one stimulus with another. Later research showed that this basic process can account for how people form certain preferences and fears.

American psychologist John B. Watson believed psychologists should study observable behaviour instead of speculating about a person's inner thoughts and feelings. Watson's approach, which he termed Behaviourism, dominated psychology for the first half of the 20th century.

Although Thorndike and Pavlov set the stage for Behaviourism, it was not until 1913 that a psychologist sets forward a clear vision for behaviorist psychology. In that year John Watson, a well-known animal psychologist at Johns Hopkins University, published a landmark paper, 'Psychology as the Behaviorist Views It'. Watson's goal was nothing less than a complete redefinition of psychology. 'Psychology as the behaviorist views it'. Watson wrote, 'is a purely objective experimental branch of natural science. Its theoretical goal is the prediction and control of behaviour'. Watson narrowly defined psychology as the scientific study of behaviour. He urged his colleagues to abandon both introspection and speculative theories about the unconscious. Instead he stressed the importance of observing and quantifying behaviour. In light of Darwin's theory of evolution, he also advocated the use of animals in psychological research, convinced that the principles of behaviour would generalize across all species.

American psychologist B. F. Skinner became famous for his pioneering research on learning and behaviour. During his 60-year career, Skinner discovered important principles of operant conditioning, a type of learning that involves reinforcement and punishment. A strict behaviorist, Skinner believed that operant conditioning could explain even the most complex of human behaviours.

Many American psychologists were quick to adopt behaviourism, and animal laboratories were set up all over the country. Aiming to predict and control behaviour, as the behaviorists' strategy was to vary a stimulus in the environment and observe an organism's response. They saw no need to speculate about mental processes as interiorized inside the unobservable realms contained in the mind. For example, Watson argued that thinking was a rhetorical dialectic of discoursing within the mind, simply of talking to one's consciousness of one's own being or identity, as exampled by the Ego.

American psychologist B. F. Skinner designed an apparatus, now called a Skinner box, that allowed him to formulate important principles of animal learning. An animal placed inside the box is rewarded with a small bit of food each time it makes the desired response, such as pressing a lever or pecking a key. A device outside the box records the animal's responses.

The most forceful leader of Behaviourism was B. F. Skinner, an American psychologist who began studying animal learning in the 1930s. B. F. Skinner coined the term reinforcement and invented a new research apparatus called the Skinner box for use in testing animals. Based on his experiments with rats and pigeons, Skinner identified a number of basic principles of learning. He claimed that these principles explained not only the behaviour of laboratory animals, but also accounted for how human beings learn new behaviours or change existing behaviours. He concluded that nearly all behaviour is shaped by complex patterns of reinforcement in a person's environment, a process that he called operant conditioning (also referred to as instrumental conditioning). Skinners' views on the causes of human behaviour made him one of the most famous and controversial psychologists of the 20th century.

Operant conditioning, pioneered by American psychologist B. F. Skinner, is the process of shaping behaviour by of reinforcement and punishment. This illustration shows how a mouse can learn to manoeuver through a maze. The mouse is rewarded with food when it reaches the first turn in the maze (A). Once the first behaviour becomes ingrained, the mouse is not rewarded until it makes the second turn (B). After many times through the maze, the mouse must reach the end of the maze to receive its reward.

B.F. Skinner and others applied his findings to modify behaviour in the workplace, the classroom, the clinic, and other settings. In World War II (1939-1945), for example, he worked for the U.S. government on a top-secret project in which he trained pigeons to guide an armed glider plane toward enemy ships. He also invented the first teaching machine, which allowed students to learn at their own measured pace, especially from the point at which its regulating of a stride, by solving a series of problems and receiving to advance or develop at a particular rate or tempo, to go at the pace as used of a horse or rider. In his popular book Walden Two (1948), Skinner presented his vision of a behaviorist utopia, in which socially adaptive behaviours are maintained by rewards, or positive reinforcements. Throughout his career, Skinner held firm to his belief that psychologists should focus on the prediction and control of behaviour.

Faced with a choice between psychoanalysis and Behaviourism, many psychologists in the 1950s and 1960s sensed a void in psychology's conception of human nature. Freud had drawn attention to the darker forces of the unconscious, and Skinner was interested only in the effects of reinforcement on observable behaviour. Humanistic psychology was born out of a desire to understand the conscious mind, free will, human dignity, and the capacity for self-reflection and growth. An alternative to psychoanalysis and Behaviourism, humanistic psychology became known as 'the third force'.

The humanistic movement was led by American psychologists Carl Rogers and Abraham Maslow. According to Rogers, all humans are born with a drive to achieve their full capacity and to behave in ways that are consistent with their true selves. Rogers, a psychotherapist, developed person-entered therapy, a nonjudgmental, nondirective approach that helped clients clarify their sense of whom they are in an effort to facilitate their own healing process. At about the same time, Maslow theorized that all people are motivated to fulfill a hierarchy of needs. At the bottom of the hierarchy are basic physiological needs, such as hunger, thirst, and sleep. Further up the hierarchy are needs for safety and security, needs for belonging and love, and esteem-related needs for status and achievement. Once these needs are met, Maslow believed, people strive for self-actualization, the ultimate state of personal fulfilment. As Maslow put it, 'A musicians must make music, an artist must paint, a poet must write, if he is ultimately to be at peace with himself. What a man can be, he must be'.

Swiss psychologist Jean Piaget based his early theories of intellectual development on his questioning and observation of his own children. From these and later studies, Piaget concluded that all children pass through a predictable series of cognitive stages.

From the 1920s through the 1960s, Behaviourism dominated psychology in the United States. Eventually, however, psychologists began to move away from strict Behaviourism. Many became increasingly interested in cognition, a term used to describe all the mental processes involved in acquiring, storing, and using knowledge. Such processes include perception, memory, thinking, problem solving, imagining, and language. This shift in emphasis toward cognition had such a profound influence on psychology that it has often been called the cognitive revolution. The psychological study of cognition became known as cognitive psychology.

One reason for psychologists' renewed interest in mental processes was the invention of the computer, which provided an intriguing metaphor for the human mind. The hardware of the computer was likened to the brain, and computer programs provided a step-by-step model of how information from the environment is input, stored, and retrieved to produce a response. Based on the computer metaphor, psychologists began to formulate information-processing models of human thought and behaviour.

In the 1950s American linguist Noam Chomsky proposed that the human brain is especially constructed to detect and reproduce language and that the ability to form and understand language is innate to all human beings. According to Chomsky, young children learn and apply grammatical rules and vocabulary as they are exposed to them and do not require initial formal teaching.

The pioneering work of Swiss psychologist Jean Piaget also inspired psychologists to study cognition. During the 1920s, while administering intelligence tests in schools, Piaget became interested in how children think. He designed various tasks and interview questions to reveal how children of different ages reason about time, nature, numbers, causality, morality, and other concepts. Based on his many studies, Piaget theorized that from infancy to adolescence, children advance through a predictable series of cognitive stages.

The cognitive revolution also gained momentum from developments in the study of language. Behaviorist B. F. Skinner had claimed that language is acquired according to the laws of operant conditioning, in much the same way that rats learn to press a bar for food pellets. In 1959, however, American linguist Noam Chomsky charged that Skinner's account of language development was wrong. Chomsky noted that children all over the world start to speak at roughly the same age and proceed through roughly the same stages without being explicitly taught or rewarded for the effort. According to Chomsky, the human capacity for learning language is innate. He theorized that the human brain is 'hardwired' for language as a product of evolution. By pointing to the primary importance of biological dispositions in the development of language, Chomsky's theory dealt a serious blow to the behaviorist assumption that all human behaviours are formed and maintained by reinforcement. The American linguist and political theories whom of which had revolutionized the study of language with his theory of generative grammar, set forth in Syntactic Structure (1957).

Before psychology became established in science, it was popularly associated with extrasensory perception (ESP) and other paranormal phenomena (phenomena beyond the laws of science). Today, these topics lie outside the traditional scope of scientific psychology and fall within the domain of parapsychology. Psychologists note that thousands of studies have failed to demonstrate the existence of paranormal phenomena.

Grounded in the conviction that mind and behaviour must be studied using statistical and scientific methods as for psychology has become a highly respected and socially useful discipline. Psychologists now study importantly and sensitive topics such as the similarities and differences between men and women, racial and ethnic diversity, sexual orientation, marriage and divorce, abortion, adoption, intelligence testing, sleep and sleep disorders, obesity and dieting, and the effects of psychoactive drugs such as methylphenidate (Ritalin) and fluoxetine (Prozac).

In the last few decades, researchers have made significant breakthroughs in understanding the brain, mental processes, and behaviour. This section of the article provides examples of contemporary research in psychology: the plasticity of the brain and nervous system, the nature of consciousness, memory distortions, competence and rationality, genetic influences on behaviour, infancy, the nature of intelligence, human motivation, prejudice and discrimination, the benefits of psychotherapy, and the psychological influences on the immune system.

Psychologists once believed that the neural circuits of the adult brain and nervous system were fully developed and no longer subject to change. Then in the 1980s and 1990s a series of provocative experiments showed that the adult brain has flexibility, or plasticity-a capacity to change as a result of usage and experience.

These experiments showed that adult rats flooded with visual stimulation formed new neural connections in the brain's visual cortex, where visual signals are interpreted. Likewise, those trained to run an obstacle course formed new connections in the cerebellum, where balance and motor skills are coordinated. Similar results with birds, mice, and monkeys have confirmed the point: Experience can stimulate the growth of new connections and mold the brain's neural architecture.

Once the brain reaches maturity, the number of neurons does not increase, and any neurons that are damaged are permanently disabled. But the plasticity of the brain can greatly benefit people with damage to the brain and nervous system. Organisms can compensate for loss by strengthening old neural connections and sprouting new ones. That is why people who suffer strokes are often able to recover their lost speech and motor abilities.

In 1860 German physicist Gustav Fechner theorized that if the human brain were divided into right and left halves, each side would have its own stream of consciousness. Modern medicine has actually allowed scientists to investigate this hypothesis. People who suffer from life-threatening epileptic seizures sometimes undergo a radical surgery that severs the corpus callosum, a bridge of nerve tissue that connects the right and left hemispheres of the brain. After the surgery, the two hemispheres can no longer communicate with each other.

Scientists have long considered the nature of consciousness without producing a fully satisfactory definition. In the early 20th century American philosopher and psychologist William James suggested that consciousness is a mental process involving both attention to external stimuli and short-term memory. Later scientific explorations of consciousness mostly expanded upon James's work. In this article from a 1997 special issue of Scientific American, Nobel laureate Francis Crick, who helped determine the structural make-up of the double-stranded helixes of the DNA molecule, while his fellow biophysicist Christof Koch explained how experiments on vision might deepen our understanding of consciousness.

Beginning in the 1960s American neurologist Roger Sperry and others tested such split-brain patients in carefully designed experiments. The researchers found that the hemispheres of these patients seemed to function independently, almost as if the subjects had two brains. In addition, they discovered that the left hemisphere was capable of speech and language, but not the right hemisphere. For example, when split-brain patients saw the image of an object flashed in their left visual field (thus sending the visual information to the right hemisphere), they were incapable of naming or describing the object. Yet they could easily point to the correct object with their left hand (which is controlled by the right hemisphere). As Sperry's colleague Michael Gazzaniga stated, 'Each half brain seemed to work and function outside of the conscious realm of the other'.

Other psychologists interested in consciousness have examined how people are influenced without their awareness. For example, research has demonstrated that under certain conditions in the laboratory, people can be fleetingly affected by subliminal stimuli, sensory information presented so rapidly or faintly that it falls below the threshold of awareness. (Note, however, that scientists have discredited claims that people can be importantly influenced by subliminal messages in advertising, rock music, or other media.) Other evidence for influence without awareness comes from studies of people with a type of amnesia that prevents them from forming new memories. In experiments, these subjects are unable to acknowledge acquaintance with or recognize words they previously viewed in a list, but they are more likely to use those words later in an unrelated task. In fact, memory without awareness is normal, but once people come up with an idea they think is original, only later to realize that they had inadvertently borrowed it from another source.

Cognitive psychologists have often likened human memory to a computer that codifies and retrieves information. It is now clear, however, that remembering is an active process and that people process of construction for building memories, in the sense of altering in a manner or proceed from one state, action, or place to another, as to pass the alternate direction between intervals of light and dark, squares to form a pattern, succeeding each other continuously, according to their beliefs, wishes, needs, and information received from outside sources.

Without realizing it, people sometimes create memories that are false. In one study, for example, subjects watched a slide show depicting a car accident. They saw either a 'STOP' sign or a 'YIELD' sign in the slides, but afterward they were asked a question about the accident that implied the presence of the other sign. Influenced by this suggestion, many subjects recalled the wrong traffic sign. In another study, people who heard a list of sleep-related words (bed, yawn) or music-related words (jazz, instruments) were often convinced moments later that they had also heard the words sleep or music-words that fit the category but were not on the list. In a third study, researchers asked college students to recall their high-school grades. Then the researchers checked those memories against the students' actual transcripts. The students recalled most grades correctly, but most of the errors inflated their grades, particularly when the actual grades were low.

When scientists distinguish between human beings and other animals, they point to our larger cerebral cortex (the outer part of the brain) and to our superior intellect-as seen in the abilities to acquire and store large amounts of information, solve problems, and communicate through the use of language.

In recent years, however, those studying human cognition has found that people are often less than rational and accurate in their performance. Some researchers have found that people are prone to forgetting, and worse, that memories of past events are often highly distorted. Others have observed that people often violate the rules of logic and probability when reasoning about real events, as when gamblers overestimate the odds of winning in games of chance. One reason for these mistakes is that we commonly rely on cognitive heuristics, mental shortcuts that allow us to make judgments that are quick but often in error. To understand how heuristics can lead to mistaken assumptions, imagine offering people a lottery ticket containing six numbers out of a pool of the numbers 1 through 40. If given a choice between the tickets 6-39-2-10-24-30 and 1-2-3-4-5-6, most people select the first ticket, because it has the appearance of randomness. Yet out of the 3,838,380 possible winning combinations, both sequences are equally likely.

One of the oldest debates in psychology, and in philosophy, concerns whether individual human traits and abilities are predetermined from birth or due to one's upbringing and experiences. This debate is often termed the nature-nurture debate. A strict genetic (nature) position states that people are predisposed to become sociable, smart, cheerful, or depressed according to their genetic blueprint. In contrast, a strict environmental (nurture) position says that people are shaped by parents, peers, cultural institutions, and life experiences.

Research shows that the more genetically related a person is to someone with schizophrenia, the greater the risk that person has of developing the illness. For example, children of one parent with schizophrenia have a 13 percent chance of developing the illness, whereas children of two parents with schizophrenia have a 46 percent chance of developing the disorder.

Researchers can estimate the role of genetic factors in two ways: (1) twin studies and (2) adoption studies. Twin studies compare identical twins with fraternal twins of the same sex. If identical twins (who share all the same genes) are more similar to each other on a given trait than are same-sex fraternal twins (who share only about half of the same genes), then genetic factors are assumed to influence the trait. Other studies compare identical twins who are raised together with identical twins who are separated at birth and raised in different families. If the twins raised together are more similar to each other than the twins raised apart, childhood experiences are presumed to influence the trait. Sometimes researchers conduct adoption studies, in which they compare adopted children to their biological and adoptive parents. If these children display traits that resemble those of their biological relatives more than their adoptive relatives, genetic factors are assumed to play a role in the trait.

In recent years, several twin and adoption studies have shown that genetic factors play a role in the development of intellectual abilities, temperament and personality, vocational interests, and various psychological disorders. Interestingly, however, this same research indicates that at least 50 percent of the variation in these characteristics within the population is attributable to factors in the environment. Today, most researchers agree that psychological characteristics spring from a combination of the forces of nature and nurture.

Helpless to survive on their own, newborn babies nevertheless possess a remarkable range of skills that aid in their survival. Newborns can see, hear, taste, smell, and feel pain; vision is the least developed sense at birth but improves rapidly in the first months. Crying communicates their need for food, comfort, or stimulation. Newborns also have reflexes for sucking, swallowing, grasping, and turning their head in search of their mother's nipple.

In 1890 William James described the newborn's experience as 'one great blooming, buzzing confusion'. However, with the aid of sophisticated research methods, psychologists have discovered that infants are smarter than was previously known.

A period of dramatic growth, infancy lasts from birth to around 18 months of age. Researchers have found that infants are born with certain abilities designed to aid their survival. For example, newborns show a distinct preference for human faces over other visual stimuli.

To learn about the perceptual world of infants, researchers measure infants' head movements, eye movements, facial expressions, brain waves, heart rate, and respiration. Using these indicators, psychologists have found that shortly after birth, and infants show a distinct preference for the human face over other visual stimuli. Also suggesting that newborns are tuned by placing their faces inward as a social object is the fact that within 72 hours of birth, they can mimic adults who purse the lips or stick out the tongue -a rudimentary form of imitation. Newborns can distinguish between their mother's voice and that of another woman. And at two weeks old, nursing infants are more attracted to the body odour of their mother and other breast-feeding females than to that of other women. Taken together, these findings show that infants are equipped at birth with certain senses and reflexes designed to aid their survival.

In 1905 French psychologist Alfred Binet and colleague Théodore Simon devised one of the first tests of general intelligence. The test sought to identify French children likely to have difficulty in school so that they could receive special education. An American version of Binet's test, the Stanford-Binet Intelligence Scale, is still used today.

In 1905 French psychologist Alfred Binet devised the first major intelligence test for the purpose of identifying slow learners in school. In doing so, Binet assumed that intelligence could be measured as a general intellectual capacity and summarized in a numerical score, or intelligence quotient (IQ). Consistently, testing has revealed that although each of us is more skilled in some areas than in others, a general intelligence underlies our more specific abilities.

Intelligence tests often play a decisive role in determining whether a person is admitted to college, graduate school, or professional school. Thousands of people take intelligence tests every year, but many psychologists and education experts question whether these tests are an accurate way of measuring who will succeed or fail in school and later in life. In this 1998 Scientific American article, psychology and education professor Robert J. Sternberg of Yale University in New Haven, Connecticut, presents evidence against conventional intelligence tests and proposes several ways to improve testing.

Today, many psychologists believe that there is more than one type of intelligence. American psychologist Howard Gardner proposed the existence of multiple intelligences, and each linked to a separate system within the brain. He theorized that there are seven types of intelligence: linguistic, logical-mathematical, spatial, musical, bodily-kinesthetic, interpersonal, and intrapersonal. American psychologist Robert Sternberg suggested a different model of intelligence, consisting of three components: analytic ('school smarts', as measured in academic tests), creative (a capacity for insight), and practical ('street smarts', or the ability to size up and adapt to situations). See Intelligence.

Psychologists from all branches of the discipline study the topic of motivation, an inner state that moves an organism toward the fulfilment of some goal. Over the years, different theories of motivation have been proposed. Some theories state that people are motivated by the need to satisfy physiological needs, whereas others state that people seek to maintain an optimum level of bodily arousal (not too little and not too much). Still other theories focus on the ways in which people respond to external incentives such as money, grades in school, and recognition. Motivation researchers study a wide range of topics, including hunger and obesity, sexual desire, the effects of reward and punishment, and the needs for power, achievement, social acceptance, love, and self-esteem.

In 1954 American psychologist Abraham Maslow proposed that all people are motivated to fulfill a hierarchical pyramid of needs. At the bottom of Maslow's pyramid are needs essential to survival, such as the needs for food, water, and sleep. The need for safety follows these physiological needs. According to Maslow, higher-level needs become important to us only after our more basic needs are satisfied. These higher needs include the need for love and belongingness, the need for esteem, and the need for self-actualization (in Maslow's theory, a state in which people realize their greatest potential).

The view that the role of sentences in inference gives a more important key to their meaning than their 'external' relations to things in the world. The meaning of a sentence becomes its place in a network of inferences that it legitimates. Also, known as its functional role semantics, procedural semantic or conceptual role semantics. As these view bear some relation to the coherence theory of truth, and suffer from the same suspicion that divorces meaning from any clear association with things in the world.

Paradoxes rest upon the assumption that analysis is a relation with concept, then are involving entities of other sorts, such as linguistic expressions, and that in true analysis, analysand and analysandum are one and the same concept. However, these assumptions are explicit in the British philosopher George Edward Moore, but some of Moore's remarks hint at a solution that a statement of an analysis is a statement partially taken about the concept involved and partly about the verbal expression used to express it. Moore is to suggest that he thinks of a solution of this sort is bound to be right, however, facts to suggest one because he cannot reveal of any way in which the analysis can be as part of the expressors.

Elsewhere, the possibility clearly does set of apparent incontrovertible premises giving unacceptable or contradictory conclusions. To solve a paradox will involve showing that these hidden flaws as held by the premises of what the reasoning is, or that the apparently unacceptable conclusion can, in fact are tolerable. Paradoxes are therefore important in philosophy, for until one is solved it shows that there is something about our reasoning and our concepts that we do not understand. Famous families of paradoxes include the semantic paradoxes and Zeno's paradoxes. At the beginning of the 20th century, Russell's paradox and other set-theoretic paradoxes of set theory, while the Sorites paradox has led to the investigation of the semantics of vagueness, and fuzzy logic. Paradoxes are under their other titles. Much as there is as much as a puzzle arising when someone say's that 'p'. In that saying, that I do not believe that 'p'. What is said is not contradictory, since (for many instances of p) both parts of it could be true. But the person nevertheless violates one presupposition of normal practice, namely that you assert something only if you believe it: By adding that you do not believe what you just said, that unduly of your believing that you are naturally of somewhat a pivoting-point of significance of the original act in so that saying it.

Furthermore, the moral philosopher and epistemologist Bernard Bolzano (1781-1848), whose logical work was based on a strong sense of there being an ontological underpinning of science and epistemology, lying in a theory of the objective entailments masking up the structure of scientific theories. His ability to challenge wisdom and come up with startling new ideas, as a Christian philosopher whether than from any position of mathematical authority, that for considerations of infinity, Bolzano's significant work was Paradoxin des Unenndlichen, written in retirement and translated into the English as Paradoxes of the Infinite. Here, Bolzano considered directly the points that had concerned Galileo-the conflicting results that seem to emerge when infinity is studied. Certainly most of the paradoxical statements encountered in the mathematical domain . . . are propositions which either immediately contain the idea of the infinite, or at least in some way or other depend upon that idea for their attempted proof.

Continuing, Bolzano looks at two possible approaches to infinity. One is simply the case of setting up a sequence of numbers, such as the whole numbers, and saying that it cannot conceivably be said to have a last term, it is inherently infinite-not finite. It is easy enough to show that the whole numbers do not have a point at which they stop, giving a name to that last number whatever it might be, as called 'ultimate'. Then what's wrong with ultimate + 1? Why is that not a whole number?

The second approach to infinity, which Bolzano ascribes in Paradoses of the Infinite to 'some philosophers . . . Taking this approach describes his first conception of infinity as the 'bad infinity'. Although the German philosopher Friedrich George Hegel (1770-1831) applies the conceptual form of infinity and points that it is, rather, the basis for a substandard infinity that merely reaches toward the absolute, but never reaches it. In Paradoses of the Infinite, he calls this form of potential infinity as a variable quantity knowing no limit to its growth (a definition adopted, even by many mathematicians) . . . always growing int th infinite and never reaching it. As far as Hegel and his colleagues were concerned, using this uprush, there was no need for a real infinity beyond some unreachable absolute. Instead we deal with a variable quality that is as big as we need it to be, or often in calculus as small as we need it to be, without ever reaching the absolute, ultimate, truly infinite.

Bolzano argues, though, that there is something else, an infinity that doe does not have this 'whatever you need it to be' elasticity. In fact a truly infinite quantity (for example, the length of a straight line unbounded in either direction, meaning: The magnitude of the spatial entity containing all the points determined solely by their abstractly conceivable relation to two fixed points) does not by any need to be variable, that the example for adducing as such, that it is in fact, not variable. Conversely, it is quite possible for a quantity merely capable of being taken greater than we have already taken it, and of becoming larger than any preassigned (finite) quantity, nevertheless to mean at all times merely finitely, which holds in particular of every numerical quantity 1, 2, 3, 4, 5.

In other words, for Bolzano there could be a true infinity that was not a variable 'something' that was only bigger than anything you might specify. Such a true infinity was the result of joining two pints together and extending that line in both directions without stopping. And what is more, is that he could separate off the demands of calculus, using a finite quality without ever bothering with the slippery potential infinity. Here was both a deeper understanding of the nature of infinity and the basis on which are built in his 'safe' infinity free calculus.

This use of the inexhaustible follows on directly from most Bolzano's criticism of the way that we used as a variable something that would be bigger than anything you could specify, but never quite reached the true, absolute infinity. In Paradoxes of the Infinity Bolzano points out that is possible for a quantity merely capable of becoming larger than any other one preassigned (finite) quantity, nevertheless to remain at all times merely finite.

Bolzano intended tis as a criticism of the way infinity was treated, but Professor Jacquette sees it instead of a way of masking use of practical applications like calculus without the need for weasel words about infinity.

By replacing with '¤' we do away with one of the most common requirements for infinity, but is there anything left that map out to the real world? Can we confine infinity to that pure mathematical other world, where anything, however unreal, can be constructed, and forget about it elsewhere? Surprisingly, this seems to have been the view, at least at one point in time, even of the German mathematician and founder of set-theory Georg Cantor (1845-1918), himself, whose comments in 1883, that only the finite numbers are real.

Keeping within the lines of reason, both the Cambridge mathematician and philosopher Frank Plumpton Ramsey (1903-30) and the Italian mathematician G. Peano (1858-1932) has been to distinguish logical paradoxes and that depends upon the notion of reference or truth (semantic notions), such are the postulates justifying mathematical induction. It ensures that a numerical series is closed, in the sense that nothing but zero and its successors can be numbers. In that any series satisfying a set of axioms can be conceived as the sequence of natural numbers. Candidates from set theory include the Zermelo numbers, where the empty set is zero, and the successor of each number is its unit set, and the von Neuman numbers, where each number is the set of all smaller numbers. A similar and equally fundamental complementarity exists in the relation between zero and infinity. Although the fullness of infinity is logically antithetical to the emptiness of zero, infinity can be obtained from zero with a simple mathematical operation. The division of many numbers by zero is infinity, while the multiplication of any number by zero is zero.

With the set theory developed by the German mathematician and logician Georg Cantor. From 1878 to 1807, Cantor created a theory of abstract sets of entities that eventually became a mathematical discipline. A set, as he defined it, is a collection of definite and distinguished objects in thought or perception conceived as a whole.

Cantors attempted to prove that the process of counting and the definition of integers could be placed on a solid mathematical foundation. His method was to repeatedly place the elements in one set into 'one-to-one' correspondence with those in another. In the case of integers, Cantor showed that each integer (1, 2, 3, . . . n) could be paired with an even integer (2, 4, 6, . . . n), and, therefore, that the set of all integers was equal to the set of all even numbers.

Amazingly, Cantor discovered that some infinite sets were large than others and that infinite sets formed a hierarchy of greater infinities. After this failed attempts to save the classical view of logical foundations and internal consistency of mathematical systems, it soon became obvious that a major crack had appeared in the seemingly sold foundations of number and mathematics. Meanwhile, an impressive number of mathematicians began to see that everything from functional analysis to the theory of real numbers depended on the problematic character of number itself.

While, in the theory of probability Ramsey was the first to show how a personalised theory could be developed, based on precise behavioural notions of preference and expectation. In the philosophy of language, Ramsey was one of the first thinkers to accept a 'redundancy theory of truth', which hr combined with radical views of the function of man y kinds of propositions. Neither generalizations nor causal propositions, nor those treating probability or ethics, describe facts, but each has a different specific function in our intellectual economy.

Ramsey advocates that of a sentence generated by taking all the sentence affirmed in a scientific theory that use some term, e.g., 'quark'. Replacing the term by a variable, and existentially quantifying into the result. Instead of saying quarks have such-and-such properties, Ramsey postdated that the sentence as saying that there is something that has those properties. If the process is repeated, the sentence gives the 'topic-neutral' structure of the theory, but removes any implications that we know what the term so treated denote. I t leaves open the possibility of identifying the theoretical item with whatever fits the description provided. Nonetheless, it was pointed out by the Cambridge mathematician Newman that if the process is carried out for all except the logical bones of the theory, then by the Löwenheim-Skolem theorem, the result will be interpretable in any domain of sufficient cardinality, and the content of the theory may reasonably be felt to have been lost.

It seems, that the most taken of paradoxes in the foundations of 'set theory' as discovered by Russell in 1901. Some classes have themselves as members: The class of all abstract objects, for example, is an abstract object, whereby, others do not: The class of donkeys is not itself a donkey. Now consider the class of all classes is presented as not being members of themselves, is this class a member of itself, that, if it is, then it is not, and if it is not, then it is.

The paradox is structurally similar to easier examples, such as the paradox of the barber. Such one like a village having a barber in it, who shaves all and only the people who do not have in themselves. Who shaves the barber? If he shaves himself, then he does not, but if he does not shave himself, then he does not. The paradox is actually just a proof that there is no such barber or in other words, that the condition is inconsistent. All the same, it is no too easy to say why there is no such class as the one Russell defines. It seems that there must be some restriction on the kinds of definition that is allowed to define classes and the difficulty that of finding a well-motivated principle behind any such restriction.

The French mathematician and philosopher Henri Jules Poincaré (1854-1912) believed that paradoses like those of Russell and that of the 'barber' was due to such as the impredicative definitions, and therefore proposed banning them. But, it tuns out that classical mathematics required such definitions at too many points for the ban to be easily absolved. Having, in turn, as forwarded by Poincaré and Russell, was that in order to solve the logical and semantic paradoxes it would have to ban any collection (set) containing members that can only be defined by of the collection taken as a whole. It is, effectively by all occurring principles into which have an adopting vicious regress, as to mark the definition for which involves no such failure. There is frequently room for dispute about whether regresses are benign or vicious, since the issue will hinge on whether it is necessary to reapply the procedure. The cosmological argument is an attempt to find a stopping point for what is otherwise seen for being infinitely regressive, and, to ban of the predicative definitions.

The investigations of questions arouse from reflection upon the sciences and scientific inquiry, are such as called of a philosophy of science. Such questions include, what distinctions in the methods of science? s there a clear demarcation between scenes and other disciplines, and how do we place such enquires as history, economics or sociology? And scientific theories probable or more in the nature of provisional conjecture? Can there be of something verified or falsified? What distinguished 'well' as 'good' from bad explanations? Might there be one unified since, embracing all th special science? For much of the 20th century their questions were pursued in a highly abstract and logical framework it being supposed that as general logic of scientific discovery that a general logic of scientific discovery a justification might be found. However, many now take interests in a more historical, contextual and sometimes sociological approach, in which the methods and successes of a science at a particular time are regarded less in terms of universal logical principles and procedure, and more in terms of their availability to methods and paradigms as well as the social context.

In addition, to general questions of methodology, there are specific problems within particular sciences, giving subjects as biology, mathematics and physics.

The intuitive certainties that spark aflame the dialectic awareness for its immediate concerns are either of the truths or by some other in an object of apprehensions, such as a concept. Awareness as such, has to its amounting quality value the place where philosophically understanding of the source of our knowledge is, however, in covering the sensible apprehension of things and pure intuition it is that which stricture sensation into the experience of things' accent of its direction that orchestrates the celestial overture into measures in space and time.

The plexuity of 'things' as to begin and carry, as marked with respect to prevailing conditions and trends, that have or to formulate in the mind for reasons about reflecting ponderosities about some existing entity, an idea, or a quality perceived as known, yet, resembling something referred to by a word or symbol, a sign, or an idea, for which things of a complex language is to think the matter through of that which is always to believe, but thought into a fallen remembrance. A matter of concerns in that many things come to mind, however, a persistent illogical feeling, as a desire or an event or an aversion, an obsession has a thing about seafood. Lastly, an activity uniquely suitable and satisfying to one, yet bears of a statement as, Do your own thing.

The notion that determines how something is seen or evaluated of the status of law and morality especially associated with St Thomas Aquinas and the subsequent scholastic tradition. More widely, any attempt to cement the moral and legal order together with the nature of the cosmos or how the nature of human beings, for which sense it is also found in some Protestant writers, and arguably derivative from a Platonic view of ethics, and is implicit in ancient Stoicism. Law stands above and apart from the activities of human lawmakers, it constitutes an objective set of principles that can be seen true by 'natural light' or reason, and (in religion versions of the theory) that express God's will for creation. Nonreligious versions of the theory substitute objective conditions for human flourishing as the source of constraints upon permissible actions and social arrangements. Within the natural law tradition, different views had held about the relationship among the rules of law as abided by God's will. For instance the Dutch philosopher Hugo Grothius (1583-1645), similarly takes upon the view that the content of natural law is independent of any will, including that of God, while the German theorist and historian Samuel von Pufendorf (1632-94) takes the opposite view, thereby facing the problem of one horn of the Euthyphro dilemma, that simply states, that its dilemma arises from whatever the source of authority is supposed to be, for in which do we care about the general good because it is good, or do we just call good things that we care about. Wherefore, by facing the problem that may be to assume of a strong form, in which it is claimed that various facts entail values, or a weaker form, from which it confines itself to holding that reason by itself is capable of discerning moral requirements. Such is to suppose of a binding to all human bings regardless of their desires

Although the morality of people sends the ethical amount, from which the same thing, is that there is a usage that restricts morality to systems such as that of the German philosopher and founder of ethical philosophy Immanuel Kant (1724-1804), based on notions such as duty, obligation, and principles of conduct, reserving ethics for more than the Aristotelian approach to practical reasoning based on the notion of a virtue, and generally avoiding the separation of 'moral' considerations from other practical considerations. The scholarly issues are complex, with some writers seeing Kant as more Aristotelian and Aristotle as, ore involved in a separate sphere of responsibility and duty, than the simple contrast suggests. Some theorists see the subject in terms of a number of laws (as in the Ten Commandments). The statuses of these laws may be tests that implicate the edicts of a divine lawmaker, or that they are truths of reason, knowable deductively. Other approaches to ethics (e.g., eudaimonism, situation ethics, virtue ethics) eschew general principles as much as possible, frequently disguising the great complexity of practical reasoning. For Kantian Notion of the moral law is a binding requirement of the categorical imperative, and to understand whether they are equivalent at some deep level. Kant's own applications of the notion are not always convincing, as for one cause of confusion in relating Kant's ethics to theories such additional expressivism is that it is easy, but mistaken, to suppose that the categorical nature of the imperative that it cannot be the expression of sentiment, but must derive from something 'unconditional' or 'necessary' such as the voice of reason.

For whichever reason, the mortal being makes of its presence to the future of weighing of that which one must do, or that which can be required of one. The term carries implications of that which is owed (due) to other people, or perhaps in onself. Universal duties would be owed to persons (or sentient beings) as such, whereas special duty in virtue of specific relations, such for being the child of someone, or having made someone a promise. Duty or obligation is the primary concept of 'deontological' approaches to ethics, but is constructed in other systems out of other notions. In the system of Kant, a perfect duty is one that must be performed whatever the circumstances: Imperfect duties may have to give way to the more stringent ones. In another way, perfect duties are those that are correlative with the right to others, imperfect duties are not. Problems with the concept include the ways in which due needs to be specified (a frequent criticism of Kant is that his notion of duty is too abstract). The concept may also suggest of a regimented view of ethical life in which we are all forced conscripts in a kind of moral army, and may encourage an individualistic and antagonistic view of social relations.

The most generally accepted account of externalism and internalism, that this distinction is that a theory of justification is internalist if only if it requiems that all of the factors needed for a belief to be epistemologically justified for a given person are cognitively accessible to that person, internal to his cognitive percreptive, and externalist, if it allows that at least some of the justifying factors need not be thus accessible, so that they can be external to the believer's cognitive perceptive, beyond any such given relations. However, epistemologists often use the distinction between internalist and externalist theories of epistemic justification without offering any very explicit explication.

The externalist/internalist distinction has been mainly applied to theories of epistemic justification: It has also been applied in a closely related way to accounts of knowledge and in a rather different way to accounts of belief and thought contents.

The internalist requirement of cognitive accessibility can be interpreted in at least two ways: A strong version of internalism would require that the believer actually be aware of the justifying factor in order to be justified: While a weaker version would require only that he be capable of becoming aware of them by focussing his attentions appropriately, but without the need for any change of position, new information, and so forth Though the phrase 'cognitively accessible' suggests the weak interpretation, the main intuitive motivation for internalism, that is to say the idea that epistemic justification requires that the believer actually have in his cognitive possession a reason for thinking that the belief is true, and would require the strong interpretation.

Perhaps, the clearest example of an internalist position would be a foundationalist view according to which foundational beliefs pertain to immediately experienced states of mind and other beliefs are justified by standing in cognitively accessible logical or inferential relations to such foundational beliefs. Such a view could count as either a strong or a weak version of internalism, depending on whether actual awareness of the justifying elements or only the capacity to become aware of them is required. Similarly, a coherent view could also be internalist, if both the beliefs or other states with which a justification belief is required to cohere and the coherence relations themselves are reflectively accessible.

It should be carefully noticed that when internalism is construed in this way, it is neither necessary nor sufficient by itself for internalism that the justifying factors literally are internal mental states of the person in question. Not necessary, necessary, because on at least some views, e.g., a direct realist view of perception, something other than a mental state of the believer can be cognitively accessible: Not sufficient, because there are views according to which at least some mental states need not be actual (a strong version) or even possible (a weak version) objects of cognitive awareness. Also, on this way of drawing the distinction, a hybrid view, according to which some of the factors required for justification must be cognitively accessible while others need not and in general will not be, would count as an externalist view. Obviously too, a view that was externalist in relation to a strong version of internalism (by not requiring that the believer actually be aware of all justifying factors) could still be internalist in relation to a weak version (by requiring that he at least is capable of becoming aware of them).

The most prominent recent externalist views have been versions of reliabilism, whose requirement for justification is roughly that the belief is produced in a way or via a process that makes of objectively likely that the belief is true. What makes such a view externalist is the absence of any requirement that the person for whom the belief is justified have any sort of cognitive access to the relations of reliability in question. Lacking such access, such a person will in general have no reason for thinking that the belief is true or likely to be true, but will, on such an account, nonetheless be epistemically justified in according it. Thus such a view arguably marks a major break from the modern epistemological tradition, stemming from Descartes, which identifies epistemic justification with having a reason, perhaps even a conclusive reason for thinking that the belief is true. An epistemologist working within this tradition is likely to feel that the externalist, than offering a competing account of the same concept of epistemic justification with which the traditional epistemologist is concerned, has simply changed the subject.

All and all, a statement presented in opposition on grounds for reason, or cause for expressing opposition in the finding objectionableness, especially the objectionability toward externalism rests on the intuitivistic certainty that the basic requirement for epistemic justification is that the acceptance of the belief in question is rational or responsible in relation to the cognitive goal of truth, which seems to require in turn that the believer actually be dialectally aware of a reason for thinking that the belief is true (or, at the very least, that such a reason be available to him). Since the satisfaction of an externalist condition is neither necessary nor sufficient for the existence of such cognitively accessible reason, it is argued, that externalism is mistaken as an account of epistemic justification. This general point has been elaborated by appeal to two sorts of putative intuitive counterexamples to externalism. The first of these challenges the necessity of belief which seem intuitively to be justified, but for which the externalist conditions are not satisfied. The standard examples in this sort are cases where beliefs are produced in some very nonstandard way, e.g., by a Cartesian demon, but nonetheless, in such a way that the subjective experience of the believer is indistinguishable from that of someone whose beliefs are produced more normally. The intuitive claim is that the believer in such a case is nonetheless epistemically justified, as much so as one whose belief is produced in a more normal way, and hence that externalist account of justification must be mistaken.

Perhaps this sort of counterexample, on behalf of a cognitive process is to be assessed in 'normal' possible worlds, i.e., in possible worlds that are actually the way our world is common-seismically believed to be, than in the world which contains the belief being judged. Since the cognitive processes employed in the Cartesian demon cases are, for which we may assume, reliable when assessed in this way, the Reliabilist can agree that such beliefs are justified. The obvious, to a considerable degree of bringing out the issue of whether it is or not an adequate rationale for this construal of reliabilism, so that the reply is not merely a notional presupposition guised as having representation.

The correlative way of elaborating on the general objection to justificatory externalism challenges the sufficiency of the various externalist conditions by citing cases where those conditions are satisfied, but where the believers in question seem intuitively not to be justified. In this context, the most widely discussed examples have to do with possible occult cognitive capacities, like clairvoyance. Considering the point in application once, again, to reliabilism, the claim is that to think that he has such a cognitive power, and, perhaps, even good reasons to the contrary, is not rational or responsible and therefore not epistemically justified in accepting the belief that result from his clairvoyance, ignoring the fact that the reliablist condition is satisfied.

One sort of response to this latter sorts of an objection is to 'bite the bullet' and insist that such believers are in fact justified, dismissing the seeming intuitions to the contrary as latent internalist prejudice. A more widely adopted response attempts to impose additional conditions, usually of a roughly internalist sort, which will rule out the offending example, while stopping far of a full internalism. But, while there is little doubt that such modified versions of externalism can handle particular cases, as well enough to avoid clear intuitive implausibility, the usually problematic cases that they cannot handle, and also whether there is and clear motivation for the additional requirements other than the general internalist view of justification that externalist is committed to reject.

A view on this same course, one that might be described as a hybrid of internalism and externalism holds that epistemic of its relating involvement to knowledge, that one's justifying the condition or fact for being justified for which something such as a fact or circumstance that justifies being considered misgoverned to be a justification for revolution. Justification requires that there is a justificatory factor that is cognitively accessible to the believer in question (though it need not be actually grasped), thus ruling out, e.g., a pure reliabilism. At the same time, however, though it must be objectively true that beliefs for which such a factor is available are likely to be true, in addition, the fact need not be in any way grasped or cognitively accessible to the believer. In effect, of the premises needed to argue that a particular belief is likely to be true, one must be accessible in a way that would satisfy at least weak internalism, the internalist will respond that this hybrid view is of no help at all in meeting the objection and has no belief nor is it held in the rational, responsible way that justification intuitively seems to require, for the believer in question, lacking one crucial premise, still has no reason at all for thinking that his belief is likely to be true.

An alternative to giving an externalist account of epistemic justification, one which may be more defensible while still accommodating many of the same motivating concerns, is to give an externalist account of knowledge directly, without relying on an intermediate account of justification. Such a view will obviously have to reject the justified true belief account of knowledge, holding instead that knowledge is true belief which satisfies the chosen externalist condition, e.g., a result of a reliable process (and perhaps, further conditions as well). This makes it possible for such a view to retain internalist account of epistemic justification, though the centrality of that concept to epistemology would obviously be seriously diminished.

Such an externalist account of knowledge can accommodate the commonsense conviction that animals, young children, and unsophisticated adults' posse's knowledge, though not the weaker conviction (if such a conviction does exist) that such individuals are epistemically justified in their beliefs. It is also at least less vulnerable to internalist counterexamples of the sort discussed, since the intuitions involved there pertain more clearly to justification than to knowledge. What is uncertain is what ultimate philosophical significance the resulting conception of knowledge is supposed to have. In particular, does it have any serious bearing on traditional epistemological problems and on the deepest and most troubling versions of scepticism, which seems in fact to be primarily concerned with justification, than knowledge?`

A rather different use of the terms 'internalism' and 'externalism' have to do with the issue of how the content of beliefs and thoughts is determined: According to an internalist view of content, the content of such intention states depends only on the non-relational, internal properties of the individual's mind or grain, and not at all on his physical and social environment: While according to an externalist view, content is significantly affected by such external factors and suggests a view that appears of both internal and external elements are standardly classified as an external view.

As with justification and knowledge, the traditional view of content has been strongly internalist in character. The main argument for externalism derives from the philosophy of language, more specifically from the various phenomena pertaining to natural kind terms, indexical, and so forth, that motivates the views that have come to be known as 'direct reference' theories. Such that the language has its ordinations for being challenged of the 'spoken change' from which is consistently being communicably communicable. Such phenomena seem at least to show that the belief or thought content that can be properly attributed to a person is dependant on facts about his environment-e.g., whether he is on Earth or Twin Earth, what is fact pointing at, the classificatorial criterion employed by expects in his social group, and so forth-not just on what is going on internally in his mind or brain.

An objection to externalist account of content is that they seem unable to do justice to our ability to know the content of our beliefs or thought 'from the inside', simply by reflection. If content is dependent on external factors pertaining to the environment, then knowledge of content should depend on knowledge of these factors-which will not in general be available to the person whose belief or thought is in question.

The adoption of an externalist account of mental content would seem to support an externalist account of justification, by way that if part or all of the content of a belief inaccessible to the believer, then both the justifying status of other beliefs in relation to that content and the status of that content ss justifying further beliefs will be similarly inaccessible, thus contravening the internalist requirement for justification. An internalist must insist that there are no justification relations of these sorts, that our internally associable content can be justified or justly for anything else: But such a response appears lame unless it is coupled with an attempt to show that the externalist account of content is mistaken.

In addition, is that of the Foundationalist, but the view in epistemology that knowledge must be regarded as a structure raised upon secure, certain foundations. These are found in some combination of experience and reason, with different schools (empirical, rationalism) emphasizing the role of one over that of the other. Foundationalism was associated with the ancient Stoics, and in the modern era with Descartes, who discovered his foundations in the 'clear and distinct' ideas of reason. Its main opponent is Coherentism or the view that a body of propositions my be known without as foundation is certain, but by their interlocking strength. Rather as a crossword puzzle may be known to have been solved correctly even if each answer, taken individually, admits of uncertainty.

Truth, alone with coherence is the study of concept, in such a study in philosophy is that it treats both the meaning of the word true and the criteria by which we judge the truth or falsity in spoken and written statements. Philosophers have attempted to answer the question 'What is truth?' for thousands

197

of years. The four main theories they have proposed to answer this question are the correspondence, pragmatic, coherence, and deflationary theories of truth.

There are various ways of distinguishing types of foundatinalist epistemology by the use of the variations that have been enumerating. Planntinga has put forward an influence conception of 'classical Foundationalism', specified in terms of limitations on the foundations. He construes this as a disjunction of 'ancient and medieval Foundationalism;, which takes foundations to comprise that with 'self-evident' and 'evident to the senses', and 'modern Foundationalism' that replace 'evident Foundationalism' that replaces 'evidently to the senses' with the replaces of 'evident to the senses' with 'incorrigibly', which in practice was taken to apply only to beliefs bout one's present state of consciousness? Plantinga himself developed this notion in the context of arguing that items outside this territory, in particular certain beliefs about God, could also be immediately justified. A popular recent distinction is between what is variously 'strong' or 'extremely' foundationlism and 'moderate', 'modest' or minimalism' and 'moderate 'Modest' or 'minimal' foundationalisn with the distinction depending on whether epistemic immunities are reassured of foundations. While depending on whether it requires of a foundation only that it is required of as foundation, that only it be immediately justified, or whether it be immediately justified. In that it makes just the comforted preferability, only to suggest that the plausibility of the string requiring stems from both a 'level confusion' between beliefs on different levels.

Emerging sceptic tendencies come forth in the 14th-century writings of Nicholas of Autrecourt. His criticisms of any certainty beyond the immediate deliverance of the senses and basic logic, and in particular of any knowledge of either intellectual or material substances, anticipate the later scepticism of Balye and Hume. The; later distinguishes between Pyrrhonistic and excessive scepticism, which he regarded as unlivable, and the more mitigated scepticism that accepts every day or commonsense beliefs (not as the delivery of reason, but as due more to custom and habit), but is duly wary of the power of reason to give us much more. Mitigated scepticism is thus closer to the attitude fostered by ancient scepticism from Pyrrho through to Sexus Empiricus. Although the phrase 'Cartesian scepticism' is sometimes used, Descartes himself was not a sceptic, but in the method of doubt, uses a sceptical scenario in order to begin the process of finding a secure mark of knowledge. Descartes himself trusts a category of 'clear and distinct' ideas, not far removed from the phantasia kataleptiké of the Stoics.

Scepticism should not be confused with relativism, which is a doctrine about the nature of truth, and may be motivated by trying to avoid scepticism. Nor is it identical with Eliminativism, from which it counsels for the abandonment of an area of thought, but together it cannot be known about the truth, but because there are no truths capable of being framed in the terms we use.

Descartes's theory of knowledge starts with the quest for certainty, for an indubitable starting-point or foundation on the basis alone of which progress is possible. This is eventually found in the celebrated 'Cogito ergo sum': I think therefore I am. By locating the point of certainty in my own awareness of my own self, Descartes gives a first-person twist to the theory of knowledge that dominated them following centuries in spite of various counterattacks on behalf of social and public starting-points. The metaphysic associated with this priority is the famous Cartesian dualism, or separation of mind and matter into differentiated interacting substances, Descartes rigorously and rightly sees that it takes divine dispensation to certify any relationship between the two realms thus divided, and to prove the reliability of the senses invokes a 'clear and distinct perception' of highly dubious proofs of the existence of a benevolent deity. This has not met general acceptance: as Hume drily puts it, 'to have recourse to the veracity of the supreme Being, in order to prove the veracity of our senses, is surely making a very unexpected circuit'.

In his own time Descartes's conception of the entirely separate substance of the mind was recognized to give rise to insoluble problems of the nature of the causal connection between the two. It also gives rise to the problem, insoluble in its own terms, of other minds. Descartes's notorious denial that nonhuman animals are conscious is a stark illustration of the problem. In his conception of matter Descartes also gives preference to rational cogitation over anything derived from the senses. Since we can conceive of the matter of a ball of wax surviving changes to its sensible qualities, matter is not an empirical concept, but eventually an entirely geometrical one, with extension and motion as its only physical nature. Descartes's thought, as reflected in Leibniz, that the qualities of sense experience have no resemblance to qualities of things, so that knowledge of the external world is essentially knowledge of structure rather than of filling. On this basis Descartes erects a remarkable physics. Since matter is in effect the same as extension there can be no empty space or 'void', since there is no empty space motion is not a question of occupying previously empty space, but is to be thought of in terms of vortices (like the motion of a liquid).

Although the structure of Descartes's epistemology, theories of mind, and theory of matter had been rejected many times, their relentless exposures of the hardest issues, their exemplary clarity, and even their initial plausibility, all contrive to make him the central point of reference for modern philosophy.

The self conceived as Descartes presents it in the first two Meditations: aware only of its own thoughts, and capable of disembodied existence, neither situated in a space nor surrounded by others. This is the pure self of 'I-ness' that we are tempted to imagine as a simple unique thing that make up our essential identity. Descartes's view that he could keep hold of this nugget while doubting everything else is criticized by Lichtenberg and Kant, and most subsequent philosophers of mind.

Descartes holds that we do not have any knowledge of any empirical proposition about anything beyond the contents of our own minds. The reason, roughly put, is that there is a legitimate doubt about all such propositions because there is no way to deny justifiably that our senses are being stimulated by some cause (an evil spirit, for example) which is radically different from the objects that we normally think affect our senses.

He also points out, that the senses (sight, hearing, touch, and so forth, are often unreliable, and 'it is prudent never to trust entirely those who have deceived us even once', he cited such instances as the straight stick that looks been t in water, and the square tower that look round from a distance. This argument of illusion, has not, on the whole, impressed commentators, and some of Descartes' contemporaries pointing out that since such errors become known as a result of further sensory information, it cannot be right to cast wholesale doubt on the evidence of the senses. But Descartes regarded the argument from illusion as only the first stage in a softening up process which would 'lead the mind away from the senses'. He admits that there are some cases of sense-base belief about which doubt would be insane, e.g., the belief that I am sitting here by the fire, wearing a winter dressing gown'.

Descartes was to realize that there was nothing in this view of nature that could explain or provide a foundation for the mental, or from direct experience as distinctly human. In a mechanistic universe, he said, there is no privileged place or function for mind, and the separation between mind and matter is absolute. Descartes was also convinced, that the immaterial essences that gave form and structure to this universe were coded in geometrical and mathematical ideas, and this insight led him to invent algebraic geometry.

A scientific understanding of these ideas could be derived, said Descartes, with the aid of precise deduction, also claiming, that the contours of physical reality could be laid out in three-dimensional coordinates. Following the publication of Newton's Principia Mathematica in 1687, reductionism and mathematical modelling became the most powerful tools of modern science. And the dream that the entire physical world could be known and mastered through the extension and refinement of mathematical theory became the central feature and guiding principle of scientific knowledge.

Having to its recourse of knowledge, its cental questions include the origin of knowledge, the place of experience in generating knowledge, and the place of reason in doing so, the relationship between knowledge and certainty, and between knowledge and the impossibility of error, the possibility of universal scepticism, and the changing forms of knowledge that arise from new conceptualizations of the world. All of these issues link with other central concerns of philosophy, such as the nature of truth and the nature of experience and meaning.

Despite these concerns, the problem was, of course, in defining knowledge in terms of true beliefs plus some favoured relations between the believer and the facts that were begun with Plato's view in the 'Theaetetus,' that knowledge is true belief, and some logos. Due of its nonsynthetic epistemology, the enterprising of studying the actual formation of knowledge by human beings, without aspiring to certify those processes as rational, or its proof against 'scepticism' or even apt to yield the truth. Natural epistemology would therefore blend into the psychology of learning and the study of episodes in the history of science. The scope for 'external' or philosophical reflection of the kind that might result in scepticism or its refutation is markedly diminished. Despite the fact that the terms of modernity are so distinguished as exponents of the approach include Aristotle, Hume, and J. S. Mills.

The task of the philosopher of a discipline would then be to reveal the correct method and to unmask counterfeits. Although this belief lay behind much positivist philosophy of science, few philosophers now subscribe to it. It places too well a confidence in the possibility of a purely previous 'first philosophy', or viewpoint beyond that of the work one's way of practitioners, from which their best efforts can be measured as good or bad. These standpoints now seem that too many philosophers to be a fanciefancy, that the more modest of tasks that are actually adopted at various historical stages of investigation into different areas with the aim not so much of criticizing but more of systematization, in the presuppositions of a particular field at a particular tie. There is still a role for local methodological disputes within the community investigators of some phenomenon, with one approach charging that another is unsound or unscientific, but logic and philosophy will not, on the modern view, provide an independent arsenal of weapons for such battles, which indeed often come to seem more like political bids for ascendancy within a discipline.

This is an approach to the theory of knowledge that sees an important connection between the growth of knowledge and biological evolution. An evolutionary epistemologist claims that the development of human knowledge processed through some natural selection process, the best example of which is Darwin's theory of biological natural selection. There is a widespread misconception that evolution proceeds according to some plan or direct, but it has neither, and the role of chance ensures that its future course will be unpredictable. Random variations in individual organisms create tiny differences in their Darwinian fitness. Some individuals have more offsprings than others, and the characteristics that increased their fitness thereby become more prevalent in future generations. Once upon a time, at least a mutation occurred in a human population in tropical Africa that changed the haemoglobin molecule in a way that provided resistance to malaria. This enormous advantage caused the new gene to spread, with the unfortunate consequence that sickle-cell anaemia came to exist.

Given that chance, it can influence the outcome at each stage: First, in the creation of genetic mutation, second, in wether the bearer lives long enough to show its effects, thirdly, in chance events that influence the individual's actual reproductive success, and fourth, in whether a gene even if favoured in one generation, is, happenstance, eliminated in the next, and finally in the many unpredictable environmental changes that will undoubtedly occur in the history of any group of organisms. As Harvard biologist Stephen Jay Gould has so vividly expressed that process over again, the outcome would surely be different. Not only might there not be humans, there might not even be anything like mammals.

We will often emphasis the elegance of traits shaped by natural selection, but the common idea that nature creates perfection needs to be analysed carefully. The extent to which evolution achieves perfection depends on exactly what you mean. As, perhaps, that, if you mean 'given rise to spontaneous or natural selection, is there always to be that the best path for the survival of the species?' The answer is no. That would require adaption by group selection, and this is, unlikely. If you mean 'Does natural selection creates every adaption that would be valuable?' The answer again, is no. For instance, some kinds of South American monkeys can grasp branches with their tails. The trick would surely also be useful to some African species, but, simply because of bad luck, none have it. Some combination of circumstances started some ancestral South American monkeys using their tails in ways that ultimately led to an ability to grab onto branches, while no such development took place in Africa. Mere usefulness of a trait does not necessitate a in that what will understandably endure phylogenesis or evolution.

This is an approach to the theory of knowledge that sees an important connection between the growth of knowledge and biological evolution. An evolutionary epistemologist claims that the development of human knowledge proceeds through some natural selection process, the best example of which is Darwin's theory of biological natural selection. The three major components of the model of natural selection are variation selection and retention. According to Darwin's theory of natural selection, variations are not pre-designed to do certain functions. Rather, these variations

Richard John Kosciejew

that do useful functions are selected. While those that do not employ of some coordinates in that are regainfully purposed are also, not to any of a selection, as duly influenced of such a selection, that may have responsibilities for the visual aspects of variational intentionally occurs. In the modern theory of evolution, genetic mutations provide the blind variations: Blind in the sense that variations are not influenced by the effects they would have-the likelihood of a mutation is not correlated with the benefits or liabilities that mutation would confer on the organism, the environment provides the filter of selection, and reproduction provides the retention. Fatnesses are achieved because those organisms with features that make them less adapted for survival do not survive in connection with other organisms in the environment that have features that are better adapted. Evolutionary epistemology applies this blind variation and selective retention model to the growth of scientific knowledge and to human thought processes overall.

The parallel between biological evolution and conceptual or 'epistemic' evolution can be seen as either literal or analogical. The literal version of evolutionary epistemology deeds biological evolution as the main cause of the growth of knowledge. On this view, called the 'evolution of cognitive mechanic programs', by Bradie (1986) and the 'Darwinian approach to epistemology' by Ruse (1986), that growth of knowledge occurs through blind variation and selective retention because biological natural selection itself is the cause of epistemic variation and selection. The most plausible version of the literal view does not hold that all human beliefs are innate but rather than the mental mechanisms that guide the acquisitions of non-innate beliefs are themselves innately and the result of biological natural selection. Ruse, (1986) demands of a version of literal evolutionary epistemology that he links to sociolology (Rescher, 1990).

On the analogical version of evolutionary epistemology, called the 'evolution of theory's program', by Bradie (1986). The 'Spenserian approach' (after the nineteenth century philosopher Herbert Spencer) by Ruse (1986), the development of human knowledge is governed by a process analogous to biological natural selection, rather than by an instance of the mechanism itself. This version of evolutionary epistemology, introduced and elaborated by Donald Campbell (1974) as well as Karl Popper, sees the [partial] fit between theories and the world as explained by a mental process of trial and error known as epistemic natural selection.

Both versions of evolutionary epistemology are usually taken to be types of naturalized epistemology, because both take some empirical facts as a starting point for their epistemological project. The literal version of evolutionary epistemology begins by accepting evolutionary theory and a materialist approach to the mind and, from these, constructs an account of knowledge and its developments. In contrast, the metaphorical version does not require the truth of biological evolution: It simply draws on biological evolution as a source for the model of natural selection. For this version of evolutionary epistemology to be true, the model of natural selection need only apply to the growth of knowledge, not to the origin and development of species. Crudely put, evolutionary epistemology of the analogical sort could still be true even if Creationism is the correct theory of the origin of species.

Although they do not begin by assuming evolutionary theory, most analogical evolutionary epistemologists are naturalized epistemologists as well, their empirical assumptions, least of mention, implicitly come from psychology and cognitive science, not evolutionary theory. Sometimes, however,

202

evolutionary epistemology is characterized in a seemingly non-naturalistic fashion. Campbell (1974) says that 'if one is expanding knowledge beyond what one knows, one has no choice but to explore without the benefit of wisdom', i.e., blindly. This, Campbell admits, makes evolutionary epistemology close to being a tautology (and so not naturalistic). Evolutionary epistemology does assert the analytic claim that when expanding one's knowledge beyond what one knows, one must precessed to something that is already known, but, more interestingly, it also makes the synthetic claim that when expanding one's knowledge beyond what one knows, one must proceed by blind variation and selective retention. This claim is synthetic because it can be empirically falsified. The central claim of evolutionary epistemology is synthetic, not analytic. If the central contradictory, which they are not. Campbell is right that evolutionary epistemology does have the analytic feature he mentions, but he is wrong to think that this is a distinguishing feature, since any plausible epistemology has the same analytic feature (Skagestad, 1978).

Two extraordinary issues lie to awaken the literature that involves questions about 'realism', i.e., What metaphysical commitment does an evolutionary epistemologist have to make? Progress, i.e., according to evolutionary epistemology, does knowledge develop toward a goal? With respect to realism, many evolutionary epistemologists endorse that is called 'hypothetical realism', a view that combines a version of epistemological 'scepticism' and tentative acceptance of metaphysical realism. With respect to progress, the problem is that biological evolution is not goal-directed, but the growth of human knowledge seems to be. Campbell (1974) worries about the potential dis-analogy here but is willing to bite the stone of conscience and admit that epistemic evolution progress toward a goal (truth) while biologic evolution does not. Many another has argued that evolutionary epistemologists must give up the 'truth-topic' sense of progress because a natural selection model is in essence, is non-teleological, as an alternative, following Kuhn (1970), and embraced in the accompaniment with evolutionary epistemology.

Among the most frequent and serious criticisms levelled against evolutionary epistemology is that the analogical version of the view is false because epistemic variation is not blind (Skagestad, 1978), and (Ruse, 1986) including, (Stein and Lipton, 1990) all have argued, nonetheless, that this objection fails because, while epistemic variation is not random, its constraints come from heuristics that, for the most part, are selective retention. Further, Stein and Lipton come to the conclusion that heuristics are analogous to biological pre-adaptions, evolutionary pre-biological pre-adaptions, evolutionary cursors, such as a half-wing, a precursor to a wing, which have some function other than the function of their descendable structures: The function of descendable structures, the function of their descendable character embodied to its structural foundations, is that of the guidelines of epistemic variation is, on this view, not the source of disanalogy, but the source of a more articulated account of the analogy.

Many evolutionary epistemologists try to combine the literal and the analogical versions (Bradie, 1986, and Stein and Lipton, 1990), saying that those beliefs and cognitive mechanisms, which are innate results from natural selection of the biological sort and those that are innate results from natural selection of the epistemic sort. This is reasonable as long as the two parts of this hybrid view are kept distinct. An analogical version of evolutionary epistemology with biological variation as its only source of blondeness would be a null theory: This would be the case if all our beliefs are innate or if our non-innate beliefs are not the result of blind variation. An appeal to the legitimate way to produce a hybrid version of evolutionary epistemology since doing so trivializes the theory. For similar reasons, such an appeal will not save an analogical version of evolutionary epistemology from arguments to the effect that epistemic variation is blind (Stein and Lipton, 1990).

Although it is a new approach to theory of knowledge, evolutionary epistemology has attracted much attention, primarily because it represents a serious attempt to flesh out a naturalized epistemology by drawing on several disciplines. In science is relevant to understanding the nature and development of knowledge, then evolutionary theory is among the disciplines worth a look. Insofar as evolutionary epistemology looks there, it is an interesting and potentially fruitful epistemological programme.

What makes a belief justified and what makes a true belief knowledge? Thinking that whether a belief deserves one of these appraisals is natural depends on what caused the depicted branch of knowledge to have the belief. In recent decades a number of epistemologists have pursued this plausible idea with a variety of specific proposals. Some causal theories of knowledge have it that a true belief that 'p' is knowledge just in case it has the right causal connection to the fact that 'p'. Such a criterion can be applied only to cases where the fact that 'p' is a sort that can reach causal relations, as this seems to exclude mathematically and their necessary facts and perhaps any fact expressed by a universal generalization, and proponents of this sort of criterion have usually supposed that it is limited to perceptual representations where knowledge of particular facts about subjects' environments.

For example, Armstrong (1973), predetermined that a position held by a belief in the form 'This perceived object is 'F' is [non-inferential] knowledge if and only if the belief is a completely reliable sign that the perceived object is 'F', that is, the fact that the object is 'F' contributed to causing the belief and its doing so depended on properties of the believer such that the laws of nature dictated that, for any subject ',' and perceived object 'y', if ',' has those properties and believed that 'y' is 'F', then 'y' is 'F'. (Dretske (1981) offers a rather similar account, in terms of the belief's being caused by a signal received by the perceiver that carries the information that the object is 'F').

Goldman (1986) has proposed an importantly different causal criterion, namely, that a true belief is knowledge if it is produced by a type of process that is 'globally' and 'locally' reliable. Causing true beliefs is sufficiently high is globally reliable if its propensity. Local reliability has to do with whether the process would have produced a similar but false belief in certain counterfactual situations alternative to the actual situation. This way of marking off true beliefs that are knowledge does not require the fact believed to be causally related to the belief, and so it could in principle apply to knowledge of any kind of truth.

Goldman requires the global reliability of the belief-producing process for the justification of a belief, he requires it also for knowledge because justification is required for knowledge. What he requires for knowledge, but does not require for justification is local reliability. His idea is that a justified true belief is knowledge if the type of process that produced it would not have produced it in any relevant counterfactual situation in which it is false. Its purported theory of relevant alternatives can be viewed as an attempt to provide a more satisfactory response to this tension in our thinking about knowledge. It attempts to characterize knowledge in a way that preserves both our belief that knowledge is an absolute concept and our belief that we have knowledge.

According to the theory, we need to qualify rather than deny the absolute character of knowledge. We should view knowledge as absolute, reactive to certain standards (Dretske, 1981 and Cohen, 1988). That is to say, in order to know a proposition, our evidence need not eliminate all the alternatives to that preposition, rather for 'us', that we can know our evidence eliminates al the relevant alternatives, where the set of relevant alternatives (a proper subset of the set of all alternatives) is determined by some standard. Moreover, according to the relevant alternatives view, and the standards determining that of the alternatives is raised by the sceptic are not relevant. If this is correct, then the fact that our evidence cannot eliminate the sceptic's alternative does not lead to a sceptical result. For knowledge requires only the elimination of the relevant alternatives, so the relevant alternative view preserves in both strands in our thinking about knowledge. Knowledge is an absolute concept, but because the absoluteness is relative to a standard, we can know many things.

The interesting thesis that counts as a causal theory of justification (in the meaning of 'causal theory' intended here) is that: A belief is justified in case it was produced by a type of process that is 'globally' reliable, that is, its propensity to produce true beliefs-that can be defined (to a good approximation) As the proportion of the beliefs it produces (or would produce) that is true is sufficiently great.

This proposal will be adequately specified only when we are told (I) how much of the causal history of a belief counts as part of the process that produced it, (ii) which of the many types to which the process belongs is the type for purposes of assessing its reliability, and (iii) relative to why the world or worlds are the reliability of the process type to be assessed the actual world, the closet worlds containing the case being considered, or something else? Let 'us' look at the answers suggested by Goldman, the leading proponent of a Reliabilist account of justification.

(1) Goldman (1979, 1986) takes the relevant belief producing process to include only the proximate causes internal to the believer. So, for instance, when recently I believed that the telephone was ringing the process that produced the belief, for purposes of assessing reliability, includes just the causal chain of neural events from the stimulus in my ear's inward ands other concurrent brain states on which the production of the belief depended: It does not include any events for which 'I', the telephone, or the sound waves travelling between it and my ears, or any earlier decisions I made that were responsible for my being within hearing distance of the telephone at that time. It does seem intuitively plausible of a belief depends should be restricted to internal omnes proximate to the belief. Why? Goldman does not tell 'us'. One answer that some philosophers might give is that it is because a belief's being justified at a given time can depend only on facts directly accessible to the believer's awareness at that time (for, if a believer ought to holds only beliefs that are justified, she can tell at any given time what beliefs would then be justified for her). However, this cannot be Goldman's answer because he wishes to include in the relevantly process neural events that are not directly accessible to consciousness.

(2) Once the Reliabilist has told 'us' how to delimit the process producing a belief, he needs to tell 'us' which of the many types to which it belongs is the relevant type. Coincide, for example, the process that produces your current belief that you see a book before you. One very broad type to which that process belongs would be specified by 'coming to a belief as to something one perceives as a result of activation of the nerve endings in some of one's sense-organs'. A constricted type, in which that unvarying processes belong would be specified by 'coming to a belief as to what one sees as a result of

activation of the nerve endings in one's retinas'. A still narrower type would be given by inserting in the last specification a description of a particular pattern of activation of the retina's particular cells. Which of these or other types to which the token process belongs is the relevant type for determining whether the type of process that produced your belief is reliable?

If we select a type that is too broad, as having the same degree of justification various beliefs that intuitively seem to have different degrees of justification. Thus the broadest type we specified for your belief that you see a book before you apply also to perceptual beliefs where the object seen is far away and seen only briefly is less justified. On the other hand, is we are allowed to select a type that is as narrow as we please, then we make it out that an obviously unjustified but true belief is produced by a reliable type of process. For example, suppose I see a blurred shape through the fog far in a field and unjustifiedly, but correctly, believe that it is a sheep: If we include enough details about my retinal image is specifying te type of the visual process that produced that belief, we can specify a type is likely to have only that one instanced and is therefore 100 percent reliable. Goldman conjectures (1986) that the relevant process type is 'the narrowest type that is casually operative'. Presumably, a feature of the process producing beliefs were causally operatives in producing it just in case some alternative feature instead, but it would not have led to that belief. (We need to say 'some' here rather than 'any', because, for example, when I see an oak or pine tree, the particular 'like-minded' material bodies of my retinal image are casually clearly toward the operatives in producing my belief that what is seen as a tree, even though there are alternative shapes, for example, 'Pine' or 'Birch' ones, that would have produced the same belief.)

(3) Should the justification of a belief in a hypothetical, non-actual example turn on the reliability of the belief-producing process in the possible world of the example? That leads to the implausible result in that in a world run by a Cartesian demon-a powerful being who causes the other inhabitants of the world to have rich and coherent sets of perceptual and memory impressions that are all illusory the perceptual and memory beliefs of the other inhabitants are all unjustified, for they are produced by processes that are, in that world, quite unreliable. If we say instead that it is the reliability of the processes in the actual world that matters, we get the equally undesired result that if the actual world is a demon world then our perceptual and memory beliefs are all unjustified.

Goldman's solution (1986) is that the reliability of the process types is to be gauged by their performance in 'normal' worlds, that is, worlds consistent with 'our general beliefs about the world . . . 'about the sorts of objects, events and changes that occur in it'. This gives the intuitively right results for the problem cases just considered, but indicate by inference an implausible proportion of making compensations for alternative tending toward justification. If there are people whose general beliefs about the world are very different from mine, then there may, on this account, be beliefs that I can correctly regard as justified (ones produced by processes that are reliable in what I take to be a normal world) but that they can correctly regard as not justified.

However, these questions about the specifics are dealt with, and there are reasons for questioning the basic idea that the criterion for a belief's being justified is its being produced by a reliable process. Thus and so, doubt about the sufficiency of the Reliabilist criterion is prompted by a sort of example that Goldman himself uses for another purpose. Suppose that being in brain-state 'B' always causes one to believe that one is in brained-state 'B'. Here the reliability of the belief-producing process is perfect, but 'we can readily imagine circumstances in which a person goes into grain-state 'B' and therefore has the belief in question, though this belief is by no justified' (Goldman, 1979). Doubt about the necessity of the condition arises from the possibility that one might know that one has strong justification for a certain belief and yet that knowledge is not what actually prompts one to believe. For example, I might be well aware that, having read the weather bureau's forecast that it will be much hotter tomorrow. I have ample reason to be confident that it will be hotter tomorrow, but I irrationally refuse to believe it until Wally tells me that he feels in his joints that it will be hotter tomorrow. Here what prompts me to believe dors not justify my belief, but my belief is nevertheless justified by my knowledge of the weather bureau's prediction and of its evidential force: I can advert to any disavowable inference that I ought not to be holding the belief. Indeed, given my justification and that there is nothing untoward about the weather bureau's prediction, my belief, if true, can be counted knowledge. This sorts of example raises doubt whether any causal conditions, are it a reliable process or something else, is necessary for either justification or knowledge.

Philosophers and scientists alike, have often held that the simplicity or parsimony of a theory is one reason, all else being equal, to view it as true. This goes beyond the unproblematic idea that simpler theories are easier to work with and gave greater aesthetic appeal.

One theory is more parsimonious than another when it postulates fewer entities, processes, changes or explanatory principles: The simplicity of a theory depends on essentially the same consecrations, though parsimony and simplicity obviously become the same. Demanding clarification of what makes one theory simpler or more parsimonious is plausible than another before the justification of these methodological maxims can be addressed.

If we set this description problem to one side, the major normative problem is as follows: What reason is there to think that simplicity is a sign of truth? Why should we accept a simpler theory instead of its more complex rivals? Newton and Leibniz thought that the answer was to be found in a substantive fact about nature. In 'Principia,' Newton laid down as his first Rule of Reasoning in Philosophy that 'nature does nothing in vain . . . 'for Nature is pleased with simplicity and affects not the pomp of superfluous causes'. Leibniz hypothesized that the actual world obeys simple laws because God's taste for simplicity influenced his decision about which world to actualize.

The tragedy of the Western mind, described by Koyré, is a direct consequence of the stark Cartesian division between mind and world. We discovered the 'certain principles of physical reality', said Descartes, 'not by the prejudices of the senses, but by the light of reason, and which thus possess so great evidence that we cannot doubt of their truth'. Since the real, or that which actually exists external to ourselves, was in his view only that which could be represented in the quantitative terms of mathematics, Descartes concludes that all quantitative aspects of reality could be traced to the deceitfulness of the senses.

The most fundamental aspect of the Western intellectual tradition is the assumption that there is a fundamental division between the material and the immaterial world or between the realm of matter and the realm of pure mind or spirit. The metaphysical framework based on this assumption is known as ontological dualism. As the word dual implies, the framework is predicated on an ontology, or a conception of the nature of God or Being, that assumes reality has two distinct and separable dimensions. The concept of Being as continuous, immutable, and having a prior or separate existence from the world of change dates from the ancient Greek philosopher Parmenides. The same qualities were associated with the God of the Judeo-Christian tradition, and they were considerably amplified by the role played in Theology by Platonic and Neoplatonic philosophy.

Nicolas Copernicus, Galileo, Johannes Kepler, and Isaac Newton were all inheritors of a cultural tradition in which ontological dualism was a primary article of faith. Hence the idealization of the mathematical ideal as a source of communion with God, which dates from Pythagoras, provided a metaphysical foundation for the emerging natural sciences. This explains why, the creators of classical physics believed that doing physics was a form of communion with the geometrical and mathematical forms' resident in the perfect mind of God. This view would survive in a modified form in what is now known as Einsteinian epistemology and accounts in no small part for the reluctance of many physicists to accept the epistemology associated with the Copenhagen Interpretation.

At the beginning of the nineteenth century, Pierre-Simon LaPlace, along with a number of other French mathematicians, advanced the view that the science of mechanics constituted a complete view of nature. Since this science, by observing its epistemology, had revealed itself to be the fundamental science, the hypothesis of God was, they concluded, entirely unnecessary.

LaPlace is recognized for eliminating not only the theological component of classical physics but the 'entire metaphysical component' as well'. The epistemology of science requires, he said, that we proceed by inductive generalizations from observed facts to hypotheses that are 'tested by observed conformity of the phenomena'. What was unique about LaPlace's view of hypotheses was his insistence that we cannot attribute reality to them. Although concepts like force, mass, motion, cause, and laws are obviously present in classical physics, they exist in LaPlace's view only as quantities. Physics is concerned, he argued, with quantities that we associate as a matter of convenience with concepts, and the truths about nature are only the quantities.

As this view of hypotheses and the truths of nature as quantities were extended in the nineteenth century to a mathematical description of phenomena like heat, light, electricity, and magnetism. LaPlace's assumptions about the actual character of scientific truths seemed correct. This progress suggested that if we could remove all thoughts about the 'nature of' or the 'source of' phenomena, the pursuit of strictly quantitative concepts would bring us to a complete description of all aspects of physical reality. Subsequently, figures like Comte, Kirchhoff, Hertz, and Poincaré developed a program for the study of nature that was quite different from that of the original creators of classical physics.

The seventeenth-century view of physics as a philosophy of nature or as natural philosophy was displaced by the view of physics as an autonomous science that was 'the science of nature'. This view, which was premised on the doctrine of positivism, promised to subsume all of the nature with a mathematical analysis of entities in motion and claimed that the true understanding of nature was revealed only in the mathematical description. Since the doctrine of positivism assumes that the knowledge we call physics resides only in the mathematical formalism of physical theory, it disallows the prospect that the vision of physical reality revealed in physical theory can have any other meaning. In the history of science, the irony is that positivism, which was intended to banish metaphysical concerns from the domain of science, served to perpetuate a seventeenth-century metaphysical assumption about the relationship between physical reality and physical theory.

Epistemology since Hume and Kant has drawn back from this theological underpinning. Indeed, the very idea that nature is simple (or uniform) has come in for a critique. The view has taken hold that a preference for simple and parsimonious hypotheses is purely methodological: It is constitutive of the attitude we call 'scientific' and makes no substantive assumption about the way the world is.

A variety of otherwise diverse twentieth-century philosophers of science have attempted, in different ways, to flesh out this position. Two examples must suffice here: Hesse (1969) as, for summaries of other proposals. Popper (1959) holds that scientists should prefer highly falsifiable (improbable) theories: He tries to show that simpler theories are more falsifiable, also Quine (1966), in contrast, sees a virtue in theories that are highly probable, he argues for a general connection between simplicity and high probability.

Both these proposals are global. They attempt to explain why simplicity should be part of the scientific method in a way that spanned all scientific subject matters. No assumption about the details of any particular scientific problem serves as a premiss in Popper's or Quine's arguments.

Newton and Leibniz thought that the justification of parsimony and simplicity flows from the hand of God: Popper and Quine try to justify these methodologically median of importance is without assuming anything substantive about the way the world is. In spite of these differences in approach, they have something in common. They assume that all users of parsimony and simplicity in the separate sciences can be encompassed in a single justifying argument. That recent developments in confirmation theory suggest that this assumption should be scrutinized. Good (1983) and Rosenkrantz (1977) has emphasized the role of auxiliary assumptions in mediating the connection between hypotheses and observations. Whether a hypothesis is well supported by some observations, or whether one hypothesis is better supported than another by those observations, crucially depends on empirical background assumptions about the inference problem here. The same view applies to the idea of prior probability (or, prior plausibility). In of a single hypo-physical science if chosen as an alternative to another even though they are equally supported by current observations, this must be due to an empirical background assumption.

Principles of parsimony and simplicity mediate the epistemic connection between hypotheses and observations. Perhaps these principles are able to do this because they are surrogates for an empirical background theory. It is not that there is one background theory presupposed by every appeal to parsimony; This has the quantifier order backwards. Rather, the suggestion is that each parsimony argument is justified only to each degree that it reflects an empirical background theory about the subjective matter. On this theory is brought out into the open, but the principle of parsimony is entirely dispensable (Sober, 1988).

This 'local' approach to the principles of parsimony and simplicity resurrects the idea that they make sense only if the world is one way rather than another. It rejects the idea that these maxims are purely methodological. How defensible this point of view is, will depend on detailed case studies of scientific hypothesis evaluation and on further developments in the theory of scientific inference.

It is usually not found of one and the same that, an inference is a (perhaps very complex) act of thought by virtue of which act (1) I pass from a set of one or more propositions or statements to a proposition or statement and (2) it appears that the latter are true if the former is or are. This psychological characterization has afforded a passage for covering over a varying range of literature under more lesser than inessential variations. Desiring a better characterization of inference is natural. Yet attempts to do so by constructing a fuller psychological explanation fail to comprehend the grounds on which inference will be objectively valid. ~A point elaborately facilitated by Gottlob Frége. Attempts to understand the nature of inference through the device of the representation of inference by formal-logical calculations or derivations better (1) leave 'us' puzzled about the relation of formal-logical derivations to the informal inferences they are supposedly to represent or reconstruct, and (2) leaves 'us' worried about the sense of such formal derivations. Are these derivations inference? Are not informal inferences needed in order to apply the rules governing the constructions of formal derivations (inferring that this operation is an application of that formal rule)? These are concerns cultivated by, for example, Wittgenstein.

Coming up with an adequate characterized inferences, and even working out what would count as a very adequate characterization here is demandingly by no nearly some resolved philosophical problem.

Traditionally, a proposition that is not a 'conditional', as with the 'affirmative' and 'negative', modern opinion is wary of the distinction, since what appears categorical may vary with the choice of a primitive vocabulary and notation. Apparently categorical propositions may also turn out to be disguised conditionals: 'X' is intelligent (categorical?) Equivalent, if 'X' is given a range of tasks, she does them better than many people (conditional?). The problem is not merely one of classification, since deep metaphysical questions arise when facts that seem to be categorical and therefore solid, come to seem by contrast conditional, or purely hypothetical or potential.

Its condition of some classified necessity is so proven sufficient that if 'p' is a necessary condition of 'q', then 'q' cannot be true unless 'p'; is true? If 'p' is a sufficient condition, thus steering well is a necessary condition of driving in a satisfactory manner, but it is not sufficient, for one can steer well but drive badly for other reasons. Confusion may result if the distinction is not heeded. For example, the statement that 'A' causes 'B' may be interpreted to mean that 'A' is itself a sufficient condition for 'B', or that it is only a necessary condition fort 'B', or perhaps a necessary parts of a total sufficient condition. Lists of conditions to be met for satisfying some administrative or legal requirement frequently attempt to give individually necessary and jointly sufficient sets of conditions.

What is more, that if any proposition of the form 'if p then q'. The condition hypothesized, 'p'. Is called the antecedent of the conditionals, and 'q', the consequent? Various kinds of conditional have been distinguished. Its weakest is that of 'material implication', merely telling that either 'not-p', or 'q'. Stronger conditionals include elements of 'modality', corresponding to the thought that 'if p is truer then q must be true'. Ordinary language is very flexible in its use of the conditional form, and there is controversy whether conditionals are better treated semantically, yielding differently finds of conditionals with different meanings, or pragmatically, in which case there should be one basic meaning with surface differences arising from other implicatures.

It follows from the definition of 'strict implication' that a necessary proposition is strictly implied by any proposition, and that an impossible proposition strictly implies any proposition. If strict implication corresponds to 'q follows from p', then this that a necessary proposition follows from anything at all, and anything at all follows from an impossible proposition. This is a problem if we wish to distinguish between valid and invalid arguments with necessary conclusions or impossible premises.

The Humean problem of induction is that if we would suppose that there is some property 'A' concerning and observational or an experimental situation, and that out of a large number of observed instances of 'A', some fraction m/n (possibly equal to 1) has also been instances of some logically independent property 'B'. Suppose further that the background proportionate circumstances not specified in these descriptions has been varied to a substantial degree and that there is no collateral information available concerning the frequency of 'B's' among 'A's or concerning causal or nomologically connections between instances of 'A' and instances of 'B'.

In this situation, an 'enumerative' or 'instantial' induction inference would move rights from the premise, that m/n of observed 'A's' are 'B's' to the conclusion that approximately m/n of all 'A's' are 'B's. (The usual probability qualification will be assumed to apply to the inference, rather than being part of the conclusion.) Here the class of 'A's' should be taken to include not only unobserved 'A's' and future 'A's', but also possible or hypothetical 'A's' (an alternative conclusion would concern the probability or likelihood of the adjacently observed 'A' being a 'B').

The traditional or Humean problem of induction, often referred to simply as 'the problem of induction', is the problem of whether and why inferences that fit this schema should be considered rationally acceptable or justified from an epistemic or cognitive standpoint, i.e., whether and why reasoning in this way is likely to lead to true claims about the world. Is there any sort of argument or rationale that can be offered for thinking that conclusions reached in this way are likely to be true in the corresponding premises is true or even that their chances of truth are significantly enhanced?

Hume's discussion of this issue deals explicitly only with cases where all observed 'A's' are 'B's' and his argument applies just as well to the more general case. His conclusion is entirely negative and sceptical. Inductive inferences are not rationally justified, but are instead the result of an essentially a-rational process, custom or habit. Hume (1711-76) challenges the proponent of induction to supply a cogent line of reasoning that leads from an inductive premise to the corresponding conclusion and offers an extremely influential argument in the form of a dilemma (a few times referred to as 'Hume's fork'), that either our actions are determined, in which case we are not responsible for them, or they are the result of random events, under which case we are also not responsible for them.

Such reasoning would, he argues, have to be either deductively demonstrative reasoning in the concerning relations of ideas or 'experimental', i.e., empirical, that reasoning concerning matters of fact or existence. It cannot be the former, because all demonstrative reasoning relies on the avoidance of contradiction, and it is not a contradiction to suppose that 'the course of nature may change', that an order that was observed in the past and not of its continuing against the future: But it cannot be, as the latter, since any empirical argument would appeal to the success of such reasoning about an experience, and the justifiability of generalizing from experience are precisely what is at issue-so that any such appeal would be question-begging. Hence, Hume concludes that there can be no such reasoning (1748).

An alternative version of the problem may be obtained by formulating it with reference to the so-called Principle of Induction, which says roughly that the future will resemble the past or, somewhat better, that unobserved cases will resemble observed cases. An inductive argument may be viewed as enthymematic, with this principle serving as a supposed premiss, in which case the issue is obviously how such a premiss can be justified. Hume's argument is then that no such justification is possible: The principle cannot be justified a prior because having possession of been true in experiences without obviously begging the question is not contradictory to have possession of been true in experiences without obviously begging the question.

The predominant recent responses to the problem of induction, at least in the analytic tradition, in effect accept the main conclusion of Hume's argument, namely, that inductive inferences cannot be justified in the sense of showing that the conclusion of such an inference is likely to be true if the premise is true, and thus attempt to find another sort of justification for induction. Such responses fall into two main categories: (I) Pragmatic justifications or 'vindications' of induction, mainly developed by Hand Reichenbach (1891-1953), and (ii) ordinary language justifications of induction, whose most important proponent is Frederick, Peter Strawson (1919-). In contrast, some philosophers still attempt to reject Hume's dilemma by arguing either (iii) That, contrary to appearances, induction can be inductively justified without vicious circularity, or (iv) that an anticipatory justification of induction is possible after all. In that: (1) Reichenbach's view is that induction is best regarded, not as a form of inference, but rather as a 'method' for arriving at posits regarding, i.e., the proportion of 'A's' remain additionally of 'B's'. Such a posit is not a claim asserted to be true, but is instead an intellectual wager analogous to a bet made by a gambler. Understood in this way, the inductive method says that one should posit that the observed proportion is, within some measure of an approximation, the true proportion and then continually correct that initial posit as new information comes in.

The gambler's bet is normally an 'appraised posit', i.e., he knows the chances or odds that the outcome on which he bets will actually occur. In contrast, the inductive bet is a 'blind posit': We do not know the chances that it will succeed or even that success is that it will succeed or even that success is possible. What we are gambling on when we make such a bet is the value of a certain proportion in the independent world, which Reichenbach construes as the limit of the observed proportion as the number of cases increases to infinity. Nevertheless, we have no way of knowing that there are even such a limit, and no way of knowing that the proportion of 'A's' are in addition of 'B's' converges in the end on some stable value than varying at random. If we cannot know that this limit exists, then we obviously cannot know that we have any definite chance of finding it.

What we can know, according to Reichenbach, is that 'if' there is a truth of this sort to be found, the inductive method will eventually find it'. That this is so is an analytic consequence of Reichenbach's account of what it is for such a limit to exist. The only way that the inductive method of making an initial posit and then refining it in light of new observations can fail eventually to arrive at the true proportion is if the series of observed proportions never converges on any stable value, which that there is no truth to be found pertaining the proportion of 'A's additionally constitute 'B's'. Thus, induction is justified, not by showing that it will succeed or indeed, that it has any definite likelihood of success, but only by showing that it will succeed if success is possible. Reichenbach's claim is that no more than this can be established for any method, and hence that induction gives 'us' our best chance for success, our best gamble in a situation where there is no alternative to gambling.

This pragmatic response to the problem of induction faces several serious problems. First, there are indefinitely many other 'methods' for arriving at posits for which the same sort of defence can be given-methods that yield the same result as the inductive method over time but differs arbitrarily before long. Despite the efforts of others, it is unclear that there is any satisfactory way to exclude such alternatives, in order to avoid the result that any arbitrarily chosen short-term posit is just as reasonable as the inductive posit. Second, even if there is a truth of the requisite sort to be found, the inductive method is only guaranteed to find it or even to come within any specifiable distance of it in the indefinite long run. All the same, any actual application of inductive results always takes place in the presence to the future eventful states in making the relevance of the pragmatic justification to actual practice uncertainly. Third, and most important, it needs to be emphasized that Reichenbach's response to the problem simply accepts the claim of the Humean sceptic that an inductive premise never provides the slightest reason for thinking that the corresponding inductive conclusion is true. Reichenbach himself is quite candid on this point, but this does not alleviate the intuitive implausibility of saying that we have no more reason for thinking that our scientific and commonsense conclusions that result in the induction of it '. . . is true' than, to use Reichenbach's own analogy (1949), a blind man wandering in the mountains who feels an apparent trail with his stick has for thinking that following it will lead him to safety.

An approach to induction resembling Reichenbach's claiming in that those particular inductive conclusions are posits or conjectures, than the conclusions of cogent inferences, is offered by Popper. However, Popper's view is even more overtly sceptical: It amounts to saying that all that can ever be said in favour of the truth of an inductive claim is that the claim has been tested and not yet been shown to be false.

(2) The ordinary language response to the problem of induction has been advocated by many philosophers, none the less, Strawson claims that the question whether induction is justified or reasonable makes sense only if it tacitly involves the demand that inductive reasoning meet the standards appropriate to deductive reasoning, i.e., that the inductive conclusions are shown to follow deductively from the inductive assumption. Such a demand cannot, of course, be met, but only because it is illegitimate: Inductive and deductive reasons are simply fundamentally different kinds of reasoning, each possessing its own autonomous standards, and there is no reason to demand or expect that one of these kinds meet the standards of the other. Whereas, if induction is assessed by inductive standards, the only ones that are appropriate, then it is obviously justified.

The problem here is to understand to what this allegedly obvious justification of an induction amount. In his main discussion of the point (1952), Strawson claims that it is an analytic true statement that believing it a conclusion for which there is strong evidence is reasonable and an analytic truth that inductive evidence of the sort captured by the schema presented earlier constitutes strong evidence for the corresponding inductive conclusion, thus, apparently yielding the analytic conclusion that believing it a conclusion for which there is inductive evidence is reasonable. Nevertheless, he also admits, indeed insists, that the claim that inductive conclusions will be true in the future is contingent, empirical, and may turn out to be false (1952). Thus, the notion of reasonable belief and the correlative notion of strong evidence must apparently be understood in ways that have nothing to do with likelihood of truth, presumably by appeal to the standard of reasonableness and strength of evidence that are accepted by the community and are embodied in ordinary usage.

Understood in this way, Strawson's response to the problem of inductive reasoning does not speak to the central issue raised by Humean scepticism: The issue of whether the conclusions of inductive arguments are likely to be true. It amounts to saying merely that if we reason in this way, we can correctly call ourselves 'reasonable' and our evidence 'strong', according to our accepted community standards. Nevertheless, to the undersealing of issue of wether following these standards is a good way to find the truth, the ordinary language response appears to have nothing to say.

(3) The main attempts to show that induction can be justified inductively have concentrated on showing that such as a defence can avoid circularity. Skyrms (1975) formulate, perhaps the clearest version of this general strategy. The basic idea is to distinguish different levels of inductive argument: A first level in which induction is applied to things other than arguments: A second level in which it is applied to arguments at the first level, arguing that they have been observed to succeed so far and hence are likely to succeed in general: A third level in which it is applied in the same way to arguments at the second level, and so on. Circularity is allegedly avoided by treating each of these levels as autonomous and justifying the argument at each level by appeal to an argument at the next level.

One problem with this sort of move is that even if circularity is avoided, the movement to higher and higher levels will clearly eventually fail simply for lack of evidence: A level will reach at which there have been enough successful inductive arguments to provide a basis for inductive justification at the next higher level, and if this is so, then the whole series of justifications collapses. A more fundamental difficulty is that the epistemological significance of the distinction between levels is obscure. If the issue is whether reasoning in accord with the original schema offered above ever provides a good reason for thinking that the conclusion is likely to be true, then it still seems question-begging, even if not flatly circular, to answer this question by appeal to anther argument of the same form.

(4) The idea that induction can be justified on a pure priori basis is in one way the most natural response of all: It alone treats an inductive argument as an independently cogent piece of reasoning whose conclusion can be seen rationally to follow, although perhaps only with probability from its premise. Such an approach has, however, only rarely been advocated (Russell, 19132 and BonJour, 1986), and is widely thought to be clearly and demonstrably hopeless.

Many on the reasons for this pessimistic view depend on general epistemological theses about the possible or nature of anticipatory cognition. Thus if, as Quine alleges, there is no a prior justification of any kind, then obviously a prior justification for induction is ruled out. Or if, as more moderate empiricists have in claiming some preexistent knowledge should be analytic, then again a prevenient justification for induction seems to be precluded, since the claim that if an inductive premise ids truer, then the conclusion is likely to be true does not fit the standard conceptions of 'analyticity'. A consideration of these matters is beyond the scope of the present spoken exchange.

There are, however, two more specific and quite influential reasons for thinking that an early approach is impossible that can be briefly considered, first, there is the assumption, originating in Hume, but since adopted by very many of others, that a move forward in the defence of induction would have to involve 'turning induction into deduction', i.e., showing, per impossible, that the inductive conclusion follows deductively from the premise, so that it is a formal contradiction to accept the latter and deny the former. However, it is unclear why a prior approach need be committed to anything this strong. It would be enough if it could be argued that it is deductively unlikely that such a premise is true and corresponding conclusion false.

Second, Reichenbach defends his view that pragmatic justification is the best that is possible by pointing out that a completely chaotic world in which there is simply not true conclusion to be found as to the proportion of 'A's' in addition that occurs of, but B's' is neither impossible nor unlikely from a purely a prior standpoint, the suggestion being that therefore there can be no a prior reason for thinking that such a conclusion is true. Nevertheless, there is still a substring wayin laying that a chaotic world is a prior neither impossible nor unlikely without any further evidence does not show that such a world os not a prior unlikely and a world containing such-and-such regularity might anticipatorially be somewhat likely in relation to an occurrence of a long-run patten of evidence in which a certain stable proportion of observed 'A's' are 'B's' ~. An occurrence, it might be claimed, that would be highly unlikely in a chaotic world (BonJour, 1986).

Goodman's 'new riddle of induction' purports that we suppose that before some specific time 't' (perhaps the year 2000) we observe a larger number of emeralds (property A) and find them all to be green (property B). We proceed to reason inductively and conclude that all emeralds are green Goodman points out, however, that we could have drawn a quite different conclusion from the same evidence. If we define the term 'grue' to mean 'green if examined before 't' and blue examined after 't', then all of our observed emeralds will also be gruing. A parallel inductive argument will yield the conclusion that all emeralds are gruing, and hence that all those examined after the year 2000 will be blue. Presumably the first of these concisions is genuinely supported by our observations and the second is not. Nevertheless, the problem is to say why this is so and to impose some further restriction upon inductive reasoning that will permit the first argument and exclude the second.

The obvious alternative suggestion is that 'grue. Similar predicates do not correspond to genuine, purely qualitative properties in the way that 'green' and 'blueness' does, and that this is why inductive arguments involving them are unacceptable. Goodman, however, claims to be unable to make clear sense of this suggestion, pointing out that the relations of formal desirability are perfectly symmetrical:

Grue' may be defined in terms if, 'green' and 'blue', but 'green' an equally well be defined in terms of 'grue' and 'green' (blue if examined before 't' and green if examined after 't').

The 'grued, paradoxes' demonstrate the importance of categorization, in that sometimes it is itemized as 'gruing', if examined of a presence to the future, before future time 't' and 'green', or not so examined and 'blue'. Even though all emeralds in our evidence class grue, we ought must infer that all emeralds are gruing. For 'grue' is unprojectible, and cannot transmit credibility to indicate a specific place or time as a starting point to differentiate from known to unknown cases. Only projectable predicates are right for induction. Goodman considers entrenchment the key to projectibility having a long history of successful protection, 'grue' is entrenched, lacking such a history, 'grue' is not. A hypothesis is projectable, Goodman suggests, only if its predicates (or suitable related ones) are much better entrenched than its rivalrous past successes that do not assume future ones. Induction remains a risky business. The rationale for favouring entrenched predicates is pragmatic. Of the possible projections from our evidence class, the one that fits with past practices enables 'us' to utilize our cognitive resources best. Its prospects of being true are worse than its competitors' and its cognitive utility is greater.

So, to a better understanding of induction we should then term is most widely used for any process of reasoning that takes 'us' from empirical premises to empirical conclusions supported by the premises, but not deductively entailed by them. Inductive arguments are therefore kinds of applicative arguments, in which something beyond the content of the premise is inferred as probable or supported by them. Induction is, however, commonly distinguished from arguments to theoretical explanations, which share this applicative character, by being confined to inferences in which he conclusion involves the same properties or relations as the premises. The central example is induction by simple enumeration, where from premises telling that Fa, Fb, Fc . . . 'where a, b, c's, are all of some kind 'G', it is inferred that G's from outside the sample, such as future G's, will be 'F', or perhaps that all G's are 'F'. In this, which and the other persons deceive them, children may infer that everyone is a deceiver: Different, but similar inferences of a property by some object to the same object's future possession of the same property, or from the constancy of some law-like pattern in events and states of affairs ti its future constancy. All objects we know of attract each other with a force inversely proportional to the square of the distance between them, so perhaps they all do so, and will always do so.

The rational basis of any inference was challenged by Hume, who believed that induction presupposed belie in the uniformity of nature, but that this belief has no defence in reason, and merely reflected a habit or custom of the mind. Hume was not therefore sceptical about the role of reason in either explaining it or justifying it. Trying to answer Hume and to show that there is something rationally compelling about the inference referred to as the problem of induction. It is widely recognized that any rational defence of induction will have to partition well-behaved properties for which the inference is plausible (often called projectable properties) from badly behaved ones, for which it is not. It is also recognized that actual inductive habits are more complex than those of similar enumeration, and that both common sense and science pay attention to such giving factors as variations within the sample giving 'us' the evidence, the application of ancillary beliefs about the order of nature, and so on.

Nevertheless, the fundamental problem remains that ant experience condition by application show 'us' only events occurring within a very restricted part of a vast spatial and temporal order about which we then come to believe things.

Uncompounded by its belonging of a confirmation theory finding of the measure to which evidence supports a theory fully formalized confirmation theory would dictate the degree of confidence that a rational investigator might have in a theory, given a body of evidence. The grandfather of confirmation theory is Gottfried Leibniz (1646-1718), who believed that a logically transparent language of science would be able to resolve all disputes. In the 20th century a fully formal confirmation theory was a main goal of the logical positivist, since without it the central concept of verification by empirical evidence itself remains distressingly unscientific. The principal developments were due to Rudolf Carnap (1891-1970), culminating in his 'Logical Foundations of Probability' (1950). Carnap's idea was that the measure necessitated would be the proportion of logically possible states of affairs in which the theory and the evidence both hold, compared ti the number in which the evidence itself holds that the probability of a preposition, relative to some evidence, is a proportion of the range of possibilities under which the proposition is true, compared to the total range of possibilities left by the evidence. The difficulty with the theory lies in identifying sets of possibilities so that they admit of measurement. It therefore demands that we can put a measure on the 'range' of possibilities consistent with theory and evidence, compared with the range consistent with the evidence alone.

Among the obstacles the enterprise meets, is the fact that while evidence covers only a finite range of data, the hypotheses of science may cover an infinite range. In addition, confirmation proves to vary with the language in which the science is couched, and the Carnapian programme has difficulty in separating genuinely confirming variety of evidence from less compelling repetition of the same experiment. Confirmation also proved to be susceptible to acute paradoxes. Finally, scientific judgement seems to depend on such intangible factors as the problems facing rival theories, and most workers have come to stress instead the historically situated scene of what would appear as a plausible distinction of a scientific knowledge at a given time.

Arose to the paradox of which when a set of apparent incontrovertible premises is given to unacceptable or contradictory conclusions. To solve a paradox will involve showing either that there is a hidden flaw in the premises, or that the reasoning is erroneous, or that the apparently unacceptable conclusion can, in fact, be tolerated. Paradoxes are therefore important in philosophy, for until one is solved it shows that there is something about our reasoning and our concepts that we do not understand. What is more, and somewhat loosely, a paradox is a compelling argument from unacceptable premises to an unacceptable conclusion: More strictly speaking, a paradox is specified to be a sentence that is true if and only if it is false. A characterized objection lesson of it would be: 'The displayed sentence is false.'

Seeing that this sentence is false if true is easy, and true if false, a paradox, in either of the senses distinguished, presents an important philosophical challenger. Epistemologists are especially concerned with various paradoxes having to do with knowledge and belief. In other words, for example, the Knower paradox is an argument that begins with apparently impeccable premisses about the concepts of knowledge and inference and derives an explicit contradiction. The origin of the reasoning is the 'surprise examination paradox': A teacher announces that there will be a surprise

217

examination next week. A clever student argues that this is impossible. 'The test cannot be on Friday, the last day of the week, because it would not be a surprise. We would know the day of the test on Thursday evening. This we can also rule out Thursday. For after we learn that no test has been given by Wednesday, we would know the test is on Thursday or Friday -and would already know that it s not on Friday and would already know that it is not on Friday by the previous reasoning. The remaining days can be eliminated in the same manner'.

This puzzle has over a dozen variants. The first was probably invented by the Swedish mathematician Lennard Ekbon in 1943. Although the first few commentators regarded the reverse elimination argument as cogent, every writer on the subject since 1950 agrees that the argument is unsound. The controversy has been over the proper diagnosis of the flaw.

Initial analyses of the subject's argument tried to lay the blame on a simple equivocation. Their failure led to more sophisticated diagnoses. The general format has been an assimilation to better-known paradoxes. One tradition casts the surprise examination paradox as a self-referential problem, as fundamentally akin to the Liar, the paradox of the Knower, or Gödel's incompleteness theorem. That in of itself, says enough that Kaplan and Montague (1960) distilled the following 'self-referential' paradox, the Knower. Consider the sentence: (S) The negation of this sentence is known (to be true).

Suppose that (S) is true. Then its negation is known and hence true. However, if its negation is true, then (S) must be false. Therefore (s) is false, or what is the name, the negation of (S) is true.

This paradox and its accompanying reasoning are strongly reminiscent of the Lair Paradox that (in one version) begins by considering a sentence 'This sentence is false' and derives a contradiction. Versions of both arguments using axiomatic formulations of arithmetic and Gödel-numbers to achieve the effect of self-reference yields important meta-theorems about what can be expressed in such systems. Roughly these are to the effect that no predicates definable in the formalized arithmetic can have the properties we demand of truth (Tarski's Theorem) or of knowledge (Montague, 1963).

These meta-theorems still leave 'us; with the problem that if we suppose that we add of these formalized languages predicates intended to express the concept of knowledge (or truth) and inference-as one mighty does if a logic of these concepts is desired. Then the sentence expressing the leading principles of the Knower Paradox will be true.

Explicitly, the assumption about knowledge and inferences are:

(1) If sentences 'A' are known, then 'a.'

(2) (1) is known?

(3) If 'B' is correctly inferred from 'A', and 'A' is known, then 'B' is known.

To give an absolutely explicit t derivation of the paradox by applying these principles to (S), we must add (contingent) assumptions to the effect that certain inferences have been done. Still, as we go through the argument of the Knower, these inferences are done. Even if we can somehow restrict such principles and construct a consistent formal logic of knowledge and inference, the paradoxical argument as expressed in the natural language still demands some explanation.

The usual proposals for dealing with the Liar often have their analogues for the Knower, e.g., that there is something wrong with a self-reference or that knowledge (or truth) is properly a predicate of propositions and not of sentences. There is to reveal that some of these are not adequate, but often parallel to those for the Liar paradox. In addition, that such an adequate one can try to be an adequate solution for the Surprise Examination Paradox, namely the observation that 'new knowledge can drive out knowledge', but this does not seem to work on the Knower (Anderson, 1983).

There are a number of paradoxes of the Liar family. The simplest example is the sentence 'This sentence is false', which must be false if it is true, and true if it is false. One suggestion is that the sentence fails to say anything, but sentences that fail to say anything are at least not true. In fact case, we consider to sentences 'This sentence is not true', which, if it fails to say anything is not true, and hence (this kind of reasoning is sometimes called the strengthened Liar). Other versions of the Liar introduce pairs of sentences, as in a slogan on the front of a T-shirt saying 'This sentence on the back of this T-shirt is false', and one on the back saying 'The sentence on the front of this T-shirt is true'. It is clear that each sentence individually is well formed, and was it not for the other, might have said something true. So any attempts to dismiss the paradox by sating that the sentence involved are meaningless will face problems.

Even so, the two approaches that have some hope of adequately dealing with this paradox is 'hierarchy' solutions and 'truth-value gap' solutions. According to the first, knowledge is structured into 'levels'. It is argued that, on this point, that is of a one-coherent notion, expressed by the verb 'knows', but rather a whole series of notions, of the knowable knows, and so on (perhaps into transfinite), stated ion terms of predicate expressing such 'ramified' concepts and properly restricted, (1)-(3) lead to no contradictions. The main objections to this procedure are that the meaning of these levels has not been adequately explained and that the idea of such subscripts, even implicit, in a natural language is highly counterintuitive the 'truth-value gap' solution takes sentences such as (S) to lack truth-value. They are neither true nor false, but they do not express propositions. This defeats a crucial step in the reasoning used in the derivation of the paradoxes. Kripler (1986) has developed this approach in connection with the Liar and Asher and Kamp (1986) has worked out some details of a parallel solution to the Knower. The principal objection is that 'strengthened' or 'super' versions of the paradoxes tend to reappear when the solution itself is stated.

Since the paradoxical deduction uses only the properties (1)-(3) and since the argument is formally valid, any notion that satisfies these conditions will lead to a paradox. Thus, Grim (1988) notes that this may be read as 'is known by an omniscient God' and concludes that there is no coherent single notion of omniscience. Thomason (1980) observes that with some different conditions, analogous reasoning about belief can lead to paradoxical consequence.

Overall, it looks as if we should conclude that knowledge and truth are ultimately intrinsically 'stratified' concepts. It would seem that wee must simply accept the fact that these (and similar) concepts cannot be assigned of any fixed, finite or infinite. Still, the meaning of this idea certainly needs further clarification.

Its paradox arises when a set of apparently incontrovertible premises gives unacceptable or contradictory conclusions, to solve a paradox will involve showing either that there is a hidden flaw in the premises, or that the reasoning is erroneous, or that the apparently unacceptable conclusion can, in fact, be tolerated. Paradoxes are therefore important in philosophy, for until one is solved its show that there is something about our reasoning and of concepts that we do not understand. Famous families of paradoxes include the 'semantic paradoxes' and 'Zeno's paradoxes. Art the beginning of the 20ᵗʰ century, paradox and other set-theoretical paradoxes led to the complete overhaul of the foundations of set theory, while the 'Sorites paradox' has lead to the investigations of the semantics of vagueness and fuzzy logics.

It is, however, to what extent can analysis be informative? This is the question that gives a riser to what philosophers has traditionally called 'the' paradox of analysis. Thus, consider the following proposition: (1) To be an instance of knowledge is to be an instance of justified true belief not essentially grounded in any falsehood. (1) If true, illustrates an important type of philosophical analysis. For convenience of exposition, I will assume (1) is a correct analysis. The paradox arises from the fact that if the concept of justified true belief not been essentially grounded in any falsification is the analysand of the concept of knowledge, it would seem that they are the same concept and hence that: (2) To be an instance of knowledge is to be as an instance of knowledge and would have to be the same propositions as (1). But then how can (1) be informative when (2) is not? This is what is called the first paradox of analysis. Classical writings' on analysis suggests a second paradoxical analysis (Moore, 1942). (3) An analysis of the concept of being a brother is that to be a brother is to be a male sibling. If (3) is true, it would seem that the concept of being a brother would have to be the same concept as the concept of being a male sibling and tat:

(4) An analysis of the concept of being a brother is that to be a brother is to be a brother would also have to be true and in fact, would have to be the same proposition as (3?). Yet (3) is true and (4) is false. Both these paradoxes rest upon the assumptions that analysis is a relation between concepts, than one involving entity of other sorts, such as linguistic expressions, and tat in a true analysis, analysand and analysandum are the same concept. Both these assumptions are explicit in Moore, but some of Moore's remarks hint at a solution to that of another statement of an analysis is a statement partly about the concept involved and partly about the verbal expressions used to express it. He says he thinks a solution of this sort is bound to be right, but fails to suggest one because he cannot see a way in which the analysis can be even partly about the expression (Moore, 1942).

Elsewhere, of such ways, as a solution to the second paradox, to which is explicating (3) as: (5)-An analysis is given by saying that the verbal expression '¸' is a brother expresses the same concept as is expressed by the conjunction of the verbal expressions '¸' is male when used to express the concept of being male and '¸' is a sibling when used to express the concept of being a sibling. (Ackerman, 1990). An important point about (5) is as follows. Stripped of its philosophical jargon ('analysis', 'concept', '¸' is . . .). (5) seems to state the sort of information generally stated in a definition of the verbal expression 'brother' in terms of the verbal expressions 'male' and 'sibling', where this definition is designed to draw upon listeners' antecedent understanding of the verbal expression 'male' and 'sibling', and thus, to tell listeners what the verbal expression 'brother' really, instead of merely providing the information that two verbal expressions are synonymous without specifying the meaning of either one. Thus, its solution to the second paradox seems to make the sort of analysis tat gives rise to this paradox matter of specifying the meaning of a verbal expression in terms of separate verbal expressions already understood and saying how the meanings of these separate, already-understood verbal expressions are combined. This corresponds to Moore's intuitive requirement that an analysis should both specify the constituent concepts of the analysandum and tell how they are combined, but is this all there is to philosophical analysis?

To answer this question, we must note that, in addition too there being two paradoxes of analysis, there is two types of analyses that are relevant here. (There are also other types of analysis, such as reformatory analysis, where the analysand is intended to improve on and replace the analysandum. But since reformatory analysis involves no commitment to conceptual identity between analysand and analysandum, reformatory analysis does not generate a paradox of analysis and so will not concern 'us' here.) One way to recognize the difference between the two types of analysis concerning 'us' here is to focus on the difference between the two paradoxes. This can be done by of the Frége-inspired sense-individuation condition, which is the condition that two expressions have the same sense if and only if they can be interchangeably 'salva veritate' whenever used in propositional attitude context. If the expressions for the analysands and the analysandum in (1) met this condition, (1) and (2) would not raise the first paradox, but the second paradox arises regardless of whether the expression for the analysand and the analysandum meet this condition. The second paradox is a matter of the failure of such expressions to be interchangeable salva veritate in sentences involving such contexts as 'an analysis is given thereof. Thus, a solution (such as the one offered) that is aimed only at such contexts can solve the second paradox. This is clearly false for the first paradox, however, which will apply to all pairs of propositions expressed by sentences in which expressions for pairs of analysands and analysantia raising the first paradox is interchangeable. For example, consider the following proposition:

Mary knows that some cats tail.

It is possible for John to believe without believing:

Mary has justified true belief, not essentially grounded in any falsehood, that some cats lack tails.

Yet this possibility clearly does not mean that the proposition that Mary knows that some casts lack tails is partly about language.

One approach to the first paradox is to argue that, despite the apparent epistemic inequivalence of (1) and (2), the concept of justified true belief not essentially grounded in any falsehood is still identical with the concept of knowledge (Sosa, 1983). Another approach is to argue that in the sort of analysis raising the first paradox, the analysand and analysandum is concepts that are different but that bear a special epistemic relation to each other. Elsewhere, the development is such an approach and suggestion that this analysand-analysandum relation has the following facets.

(i)The analysand and analysandum are necessarily coextensive, i.e., necessarily every instance of one is an instance of the other.

(ii) The analysand and analysandum are knowable theoretical to be coextensive.

(iii) The analysandum is simpler than the analysands a condition whose necessity is recognized in classical writings on analysis, such as, Langford, 1942.

(iv) The analysands do not have the analysandum as a constituent.

Condition (iv) rules out circularity. But since many valuable quasi-analyses are partly circular, e.g., knowledge is justified true belief supported by known reasons not essentially grounded in any falsehood, it seems best to distinguish between full analysis, from that of (iv) is a necessary condition, and partial analysis, for which it is not.

These conditions, while necessary, are clearly insufficient. The basic problem is that they apply too many pairs of concepts that do not seem closely enough related epistemologically to count as analysand and analysandum., such as the concept of being six, and the concept is of the fourth root of 1296. Accordingly, its solution upon what actually seems epistemologically distinctive about analyses of the sort under consideration, which is a certain way they can be justified. This is by the philosophical example-and-counterexample method, which is in a general term that goes as follows. 'J' investigates the analysis of K's concept 'Q' (where 'K' can but need not be identical to 'J' by setting 'K' a series of armchair thought experiments, i.e., presenting 'K' with a series of simple described hypothetical test cases and asking 'K' questions of the form 'If such-and-such where the case would this count as a case of Q? 'J' then contrasts the descriptions of the cases to which; K' answers affirmatively with the description of the cases to which 'K' does not, and 'J' generalizes upon these descriptions to arrive at the concepts (if possible not including the analysandum) and their mode of combination that constitute the analysand of K's concept 'Q'. Since 'J' need not be identical with 'K', there is no requirement that 'K' himself be able to perform this generalization, to recognize its result as correct, or even to understand the analysand that is its result. This is reminiscent of Walton's observation that one can simply recognize a bird as a swallow without realizing just what feature of the bird (beak, wing configurations, and so forth) form the basis of this recognition. (The philosophical significance of this way of recognizing is discussed in Walton, 1972) 'K' answers the questions based solely on whether the described hypothetical cases just strike him as cases of 'Q'. 'J' observes certain strictures in formulating the cases and questions. He makes the cases as simple as possible, to minimize the possibility of confusion and to minimize the likelihood that 'K' will draw upon his philosophical theories (or quasi-philosophical, a rudimentary notion if he is unsophisticated philosophically) in answering the questions. For this conflicting result, the conflict should 'other things being equal' be resolved in favour of the simpler case. 'J' makes the series of described cases wide-ranging and varied, with the aim of having it be a complete series, where a series is complete if and only if no case that is omitted in such that, if included, it would change the analysis arrived at. 'J' does not, of course, use as a test-case description anything complicated and general enough to express the analysand. There is no requirement that the described hypothetical test cases be formulated only in terms of what can be observed. Moreover, using described hypothetical situations as test cases enables 'J' to frame the questions in such a way as to rule out extraneous background assumption to a degree, thus, even if 'K' correctly believes that all and only P's are R's, the question of whether the concepts of P, R, or both enter the analysand of his concept 'Q' can be investigated by asking him such questions as 'Suppose (even if it seems preposterous to you) that you were to find out that there was a 'P' that was not an 'R'. Would you still consider it a case of Q?

Taking all this into account, the fifth necessary condition for this sort of analysand-analysandum relations is as follows: If 'S' is the analysand of 'Q', the proposition that necessarily all and only instances of 'S' are instances of 'Q' can be justified by generalizing from intuition about the correct answers to questions of the sort indicated about a varied and wide-ranging series of simple described hypothetical situations. It so does occur of antinomy, when we are able to argue for, or demonstrate, both a proposition and its contradiction, roughly speaking, a contradiction of a proposition 'p' is one that can be expressed in form 'not-p', or, if 'p' can be expressed in the form 'not-q', then a contradiction is one that can be expressed in the form 'q'. Thus, e.g., if 'p is 2 + 1 = 4, then 2 + 1 = 4 is, contradictory of 'p', for 2 + 1 = 4 can be expressed in the form not (2 + 1 = 4). If 'p' is 2 + 1 = 4 then 2 + 1= 4 is a contradictory of 'p', since 2 + 1 = 4 can be expressed in the form not (2 + 1 = 4). This is, mutually, but contradictory propositions can be expressed in the form, 'r', 'not-r'. The Principle of Contradiction says that mutually contradictory propositions cannot both be true and cannot both be false. Thus, by this principle, since if 'p' is true, 'not-p' is false, no proposition 'p' can be at once true and false (otherwise both 'p' and its contradictories would be false?). In particular, for any predicate 'p' and object ',', it cannot be that 'p'; is at once true of 'p' and false of p? This is the classical formulation of the principle of contradiction, but it is nonetheless, that wherein, we cannot now fault either demonstrates. We would eventually hope to be able 'to solve the antinomy' by managing, through careful thinking and analysis, eventually to fault either or both demonstrations.

Many paradoxes are as an easy source of antinomies, for example, Zeno gave some famously lets say, logical-cum-mathematical arguments that might be interpreted as demonstrating that motion is impossible. But our eyes as it was, demonstrate motion (exhibit moving things) all the time. Where did Zeno go wrong? Where do our eyes go wrong? If we cannot readily answer at least one of these questions, then we are in antinomy. In the 'Critique of Pure Reason,' Kant gave demonstrations of the same kind -in the Zeno example they were obviously not the same kind of both, e.g., that the world has a beginning in time and space, and that the world has no beginning in time or space. He argues that both demonstrations are at fault because they proceed on the basis of 'pure reason' unconditioned by sense experience.

At this point, we display attributes to the theory of experience, as it is not possible to define in an illuminating way, however, we know what experiences are through acquaintances with some of our own, e.g., visual experiences of as afterimage, a feeling of physical nausea or a tactile experience of an abrasive surface (which might be caused by an actual surface -rough or smooth, or which might be part of a dream, or the product of a vivid sensory imagination). The essential feature of experience is it feels a certain way -that there is something that it is like to have it. We may refer to this feature of an experience as its 'character'.

Another core feature of the sorts of experiences with which this may be of a concern, is that they have representational 'content'. (Unless otherwise indicated, 'experience' will be reserved for their 'contentual representations'.) The most obvious cases of experiences with content are sense experiences of the kind normally involved in perception. We may describe such experiences by mentioning their sensory modalities ad their contents, e.g., a gustatory experience (modality) of chocolate ice cream (content), but do so more commonly by of perceptual verbs combined with noun phrases specifying their contents, as in 'Macbeth saw a dagger'. This is, however, ambiguous between the perceptual claim 'There was a (material) dagger in the world that Macbeth perceived visually' and 'Macbeth had a visual experience of a dagger' (the reading with which we are concerned, as it is afforded by our imagination, or perhaps, experiencing mentally hallucinogenic imagery).

As in the case of other mental states and events with content, it is important to distinguish between the properties that and experience 'represents' and the properties that it 'possesses'. To talk of the representational properties of an experience is to say something about its content, not to attribute those properties to the experience itself. Like every other experience, a visual; experience of a non-shaped square, of which is a mental event, and it is therefore not itself irregular or is it square, even though it represents those properties. It is, perhaps, fleeting, pleasant or unusual, even though it does not represent those properties. An experience may represent a property that it possesses, and it may even do so in virtue of a rapidly changing (complex) experience representing something as changing rapidly. However, this is the exception and not the rule.

Which properties can be [directly] represented in sense experience is subject to debate. Traditionalists include only properties whose presence could not be doubted by a subject having appropriate experiences, e.g., colour and shape in the case of visual experience, and apparent shape, surface texture, hardness, and so forth, in the case of tactile experience. This view is natural to anyone who has an egocentric, Cartesian perspective in epistemology, and who wishes for pure data in experiences to serve as logically certain foundations for knowledge, especially to the immediate objects of perceptual awareness in or of sense-data, such categorized of colour patches and shapes, which are usually supposed distinct from surfaces of physical objectivity. Qualities of sense-data are supposed to be distinct from physical qualities because their perception is more relative to conditions, more certain, and more immediate, and because sense-data is private and cannot appear other than they are they are objects that change in our perceptual field when conditions of perception change. Physical objects remain constant.

Others who do not think that this wish can be satisfied, and who are more impressed with the role of experience in providing animisms with ecologically significant information about the world around them, claim that sense experiences represent properties, characteristic and kinds that are much richer and much more wide-ranging than the traditional sensory qualities. We do not see only colours and shapes, they tell 'us', but also earth, water, men, women and fire: We do not smell only odours, but also food and filth. There is no space here to examine the factors relevantly responsible to their choice of situational alternatives. Yet, this suggests that character and content are not really distinct, and there is a close tie between them. For one thing, the relative complexity of the character of sense experience places limitations upon its possible content, e.g., a tactile experience of something touching one's left ear is just too simple to carry the same amount of content as typically convincing to an every day, visual experience. Moreover, the content of a sense experience of a given character depends on the normal causes of appropriately similar experiences, e.g., the sort of gustatory experience that we have when eating chocolate would be not represented as chocolate unless it was normally caused by chocolate. Granting a contingent ties between the character of an experience and its possible causal origins, once, again follows that its possible content is limited by its character.

Character and content are none the less irreducibly different, for the following reasons. (1) There are experiences that completely lack content, e.g., certain bodily pleasures. (2) Not every aspect of the character of an experience with content is relevant to that content, e.g., the unpleasantness of an aural experience of chalk squeaking on a board may have no representational significance. (3) Experiences in different modalities may overlap in content without a parallel overlap in character, e.g., visual and tactile experiences of circularity feel completely different. (4) The content of an experience with a given character may vary according to the background of the subject, e.g., a certain content 'singing bird' only after the subject has learned something about birds.

According to the act/object analysis of experience (which is a special case of the act/object analysis of consciousness), every experience involves an object of experience even if it has no material object. Two main lines of argument may be offered in support of this view, but one is 'Phenomenological' and the others are 'semantic'. For which of a philosophy or method of inquiry as based on the premise that reality consists of objects and events as they are perceived or understood in human consciousness and not of anything independent of human consciousness, in that of objects or concepts constitute th e sole object of knowledge, with objects or perception and the nature of the mind itself remaining unknowable.

Whenever we have an experience, even if nothing beyond the experience answers to it, we seem to be presented with something through the experience (which is itself diaphanous). The object of the experience is whatever is so presented to 'us'-is that it is an individual thing, an event, or a state of affairs.

The semantic argument is that objects of experience are required in order to make sense of certain features of our talk about experience, including, in particular, the following. (I) Simple attributions of experience, e.g., 'Rod is experiencing an oddity that is not really square but in appearance it seems more than likely a square', this seems to be relational. (ii) We appear to refer to objects of experience and to attribute properties to them, e.g., 'The afterimage that John experienced was certainly odd'. (iii) We appear to quantify ov er objects of experience, e.g., 'Macbeth saw something that his wife did not see'.

The act/object analysis faces several problems concerning the status of objects of experiences. Currently the most common view is that they are sense-data-private mental entities that actually posses the traditional sensory qualities represented by the experiences of which they are the objects. But the very idea of an essentially private entity is suspect. Moreover, since an experience may apparently represent something as having a determinable property, e.g., redness, without representing it as having any subordinate determinate property, e.g., any specific shade of red, a sense-datum may actually have a determinate property subordinate to it. Even more disturbing is that sense-data may have contradictory properties, since experiences can have contradictory contents. A case in point is the waterfall illusion: If you stare at a waterfall for a minute and then immediately fixate on a nearby rock, you are likely to have an experience of the rock's moving upward while it remains in the same place. The sense-data theorist must either deny that there are such experiences or admit contradictory objects.

These problems can be avoided by treating objects of experience as properties. This, however, fails to do justice to the appearances, for experience seems not to present 'us' with properties embodied in individuals. The view that objects of experience is Meinongian objects accommodate this point. It is also attractive in as far as (1) it allows experiences to represent properties other than traditional sensory qualities, and (2) it allows for the identification of objects of experience and objects of perception in the case of experiences that constitute perception.

According to the act/object analysis of experience, every experience with content involves an object of experience to which the subject is related by an act of awareness (the event of experiencing that object). This is meant to apply not only to perceptions, which have material objects (whatever is perceived), but also to experiences like hallucinations and dream experiences, which do not. Such experiences none the less appear to represent something, and their objects are supposed to be whatever it is that they represent. Act/object theorists may differ on the nature of objects of experience, which have been treated as properties. Meinongian objects (which may not exist or have any form of being), and, more commonly private mental entities with sensory qualities. (The term 'sense-data' is now usually applied to the latter, but has also been used as a general term for objects of sense experiences, as in the work of G. E. Moore) Act/object theorists may also differ on the relationship between objects of experience and objects of perception. In terms of perception (of which we are 'indirectly aware') are always distinct from objects of experience (of which we are 'directly aware'). Meinongian, however, may treat objects of perception as existing objects of experience. But sense-datum theorists must either deny that there are such experiences or admit contradictory objects. Still, most philosophers will feel that the Meinongian's acceptance of impossible objects is too high a price to pay for these benefits.

A general problem for the act/object analysis is that the question of whether two subjects are experiencing one and the same thing (as opposed to having exactly similar experiences) appears to have an answer only on the assumption that the experiences concerned are perceptions with material objects. But in terms of the act/object analysis the question must have an answer even when this condition is not satisfied. (The answer is always negative on the sense-datum theory; it could be positive on other versions of the act/object analysis, depending on the facts of the case.)

In view of the above problems, the case for the act/object analysis should be reassessed. The Phenomenological argument is not, on reflection, convincing, for it is easy enough to grant that any experience appears to present 'us' with an object without accepting that it actually does. The semantic argument is more impressive, but is none the less answerable. The seemingly relational structure of

226

attributions of experience is a challenge dealt with below in connection with the adverbial theory. Apparent reference to and quantification over objects of experience can be handled by analysing them as reference to experiences themselves and quantification over experiences tacitly typed according to content. Thus, 'The afterimage that John experienced was colourfully appealing' becomes 'John's afterimage experience was an experience of colour', and 'Macbeth saw something that his wife did not see' becomes 'Macbeth had a visual experience that his wife did not have'.

Pure cognitivism attempts to avoid the problems facing the act/object analysis by reducing experiences to cognitive events or associated disposition, e.g., Susy's experience of a rough surface beneath her hand might be identified with the event of her acquiring the belief that there is a rough surface beneath her hand, or, if she does not acquire this belief, with a disposition to acquire it that has somehow been blocked.

This position has attractions. It does full justice to the cognitive contents of experience, and to the important role of experience as a source of belief acquisition. It would also help clear the way for a naturalistic theory of mind, since there seems to be some prospect of a physicalist/functionalist account of belief and other intentional states. But pure cognitivism is completely undermined by its failure to accommodate the fact that experiences have a felt character that cannot be reduced to their content, as aforementioned.

The adverbial theory is an attempt to undermine the act/object analysis by suggesting a semantic account of attributions of experience that does not require objects of experience. Unfortunately, the oddities of explicit adverbializations of such statements have driven off potential supporters of the theory. Furthermore, the theory remains largely undeveloped, and attempted refutations have traded on this. It may, however, be founded on sound basis intuitions, and there is reason to believe that an effective development of the theory (which is merely hinting at) is possible.

The relevant intuitions are (1) that when we say that someone is experiencing 'an A', or has an experience 'of an A', we are using this content-expression to specify the type of thing that the experience is especially apt to fit, (2) that doing this is a matter of saying something about the experience itself (and maybe about the normal causes of like experiences), and (3) that it is no-good of reasons to posit of its position to presuppose that of any involvements, is that its descriptions of an object in which the experience is. Thus the effective role of the content-expression in a statement of experience is to modify the verb it compliments, not to introduce a special type of object.

Perhaps, the most important criticism of the adverbial theory is the 'many property problem', according to which the theory does not have the resources to distinguish between, e.g.,

> (1) Frank has an experience of a brown triangle

And:

> (2) Frank has an experience of brown and an experience of a triangle.

Which is entailed by (1) but does not entail it. The act/object analysis can easily accommodate the difference between (1) and (2) by claiming that the truth of (1) requires a single object of experience that is both brown and triangular, while that of the (2) allows for the possibility of two objects of experience, one brown and the other triangular, however, (1) is equivalent to:

(1*) Frank has an experience of something's being both brown and triangular.

And (2) is equivalent to:

(2*) Frank has an experience of something's being brown and an experience of something's being triangular,

And the difference between these can be explained quite simply in terms of logical scope without invoking objects of experience. The Adverbialists may use this to answer the many-property problem by arguing that the phrase 'a brown triangle' in (1) does the same work as the clause 'something's being both brown and triangular' in (1*). This is perfectly compatible with the view that it also has the 'adverbial' function of modifying the verb 'has an experience of', for it specifies the experience more narrowly just by giving a necessary condition for the satisfaction of the experience (the condition being that there are something both brown and triangular before Frank).

A final position that should be mentioned is the state theory, according to which a sense experience of an 'A' is an occurrent, non-relational state of the kind that the subject would be in when perceiving an 'A'. Suitably qualified, this claim is no doubt true, but its significance is subject to debate. Here it is enough to remark that the claim is compatible with both pure cognitivism and the adverbial theory, and that state theorists are probably best advised to adopt adverbials as a of developing their intuitions.

Yet, clarifying sense-data, if taken literally, is that which is given by the senses. But in response to the question of what exactly is so given, sense-data theories posit private showings in the consciousness of the subject. In the case of vision this would be a kind of inner picture show which is only indirectly representation al to aspects of the external world that has in and of itself a worldly representation. The view has been widely rejected as implying that we really only see extremely thin coloured pictures interposed between our mind's eye and reality. Modern approaches to perception tend to reject any conception of the eye as a camera or lense, simply responsible for producing private images, and stress the active life of the subject in and of the world, as the determinant of experience.

Nevertheless, the argument from illusion is of itself the usually intended directive to establish that certain familiar facts about illusion disprove the theory of perception called na evity or direct realism. There are, however, many different versions of the argument that must be distinguished carefully. Some of these distinctions centre on the content of the premises (the nature of the appeal to illusion); others centre on the interpretation of the conclusion (the kind of direct realism under attack). Let 'us' set about by distinguishing the importantly different versions of direct realism which one might take to be vulnerable to familiar facts about the possibility of perceptual illusion.

A crude statement of direct realism might go as follows. In perception, we sometimes directly perceive physical objects and their properties, we do not always perceive physical objects by perceiving something 'else', e.g., a sense-datum. There are, however, difficulties with this formulation of the view, as for one thing a great many philosophers who are 'not' direct realists would admit that it is a mistake to describe people as actually 'perceiving' something other than a physical object. In particular, such philosophers might admit, we should never say that we perceive sense-data. To talk that way would be to suppose that we should model our understanding of our relationship to sense-data on our understanding of the ordinary use of perceptual verbs as they describe our relation to and of the physical world, and that is the last thing paradigm sense-datum theorists should want. At least, many of the philosophers who objected to direct realism would prefer to express in what they were of objecting too in terms of a technical (and philosophically controversial) concept such as 'acquaintance'. Using such a notion, we could define direct realism this way: In 'veridical' experience we are directly acquainted with parts, e.g., surfaces, or constituents of physical objects. A less cautious verison of the view might drop the reference to veridical experience and claim simply that in all experience we are directly acquainted with parts or constituents of physical objects. The expressions 'knowledge by acquaintance' and 'knowledge by description', and the distinction they mark between knowing 'things' and knowing 'about' things, are generally associated with Bertrand Russell (1872-1970), that scientific philosophy required analysing many objects of belief as 'logical constructions' or 'logical fictions', and the programme of analysis that this inaugurated dominated the subsequent philosophy of logical atomism, and then of other philosophers, Russell's 'The Analysis of Mind,' the mind itself is treated in a fashion reminiscent of Hume, as no more than the collection of neutral perceptions or sense-data that make up the flux of conscious experience, and that looked at another way that also was to make up the external world (neutral monism), but 'An Inquiry into Meaning and Truth' (1940) represents a more empirical approach to the problem. Yet, philosophers have perennially investigated this and related distinctions using varying terminology.

Distinction in our ways of knowing things, highlighted by Russell and forming a central element in his philosophy after the discovery of the theory of 'definite descriptions'. A thing is known by acquaintance when there is direct experience of it. It is known by description if it can only be described as a thing with such-and-such properties. In everyday parlance, I might know my spouse and children by acquaintance, but know someone as 'the first person born at sea' only by description. However, for a variety of reasons Russell shrinks the area of things that can be known by acquaintance until eventually only current experience, perhaps my own self, and certain universals or meanings qualify anything else is known only as the thing that has such-and-such qualities.

Because one can interpret the relation of acquaintance or awareness as one that is not 'epistemic', i.e., not a kind of propositional knowledge, it is important to distinguish the above aforementioned views read as ontological theses from a view one might call 'epistemological direct realism? In perception we are, on at least some occasions, non-inferentially justified in believing a proposition asserting the existence of a physical object. Since it is that these objects exist independently of any mind that might perceive them, and so it thereby rules out all forms of idealism and phenomenalism, which hold that there are no such independently existing objects. Its being to 'direct' realism rules out those views defended under the cubic of 'critical naive realism', or 'representational realism', in which there is some nonphysical intermediary -usually called a 'sense-datum' or a 'sense impression' -that must first be perceived or experienced in order to perceive the object that exists independently of this perception. Often the distinction between direct realism and other theories of perception is explained more fully in terms of what is 'immediately' perceived, than 'mediately' perceived. What relevance does illusion have for these two forms of direct realism?

The fundamental premise of the arguments is from illusion seems to be the theses that things can appear to be other than they are. Thus, for example, straight sticks when immerged in water looks bent, a penny when viewed from certain perspective appears as an illusory spatial elliptic circularity, when something that is yellow when place under red fluorescent light looks red. In all of these cases, one version of the argument goes, it is implausible to maintain that what we are directly acquainted with is the real nature of the object in question. Indeed, it is hard to see how we can be said to be aware of the really physical object at all. In the above illusions the things we were aware of actually were bent, elliptical and red, respectively. But, by hypothesis, the really physical objects lacked these properties. Thus, we were not aware of the substantial reality of been real as a physical objects or theory.

So far, if the argument is relevant to any of the direct realisms distinguished above, it seems relevant only to the claim that in all sense experience we are directly acquainted with parts or constituents of physical objects. After all, even if in illusion we are not acquainted with physical objects, but their surfaces, or their constituents, why should we conclude anything about the hidden nature of our relations to the physical world in veridical experience?

We are supposed to discover the answer to this question by noticing the similarities between illusory experience and veridical experience and by reflecting on what makes illusion possible at all. Illusion can occur because the nature of the illusory experience is determined, not just by the nature of events or sorted, conflicting affairs but the object perceived as itself the event in cause, but also by other conditions, both external and internal as becoming of an inner or as the outer experience. But all of our sensations are subject to these causal influences and it would be gratuitous and arbitrary to select from indefinitely of many and subtly different perceptual experiences some special ones those that get 'us' in touch with the 'real' nature of the physical world and its surrounding surfaces. Red fluorescent light affects the way thing's look, but so does sunlight. Water reflects light, but so does air. We have no unmediated access to the external world.

The Philosophy of science, and scientific epistemology are not the only area where philosophers have lately urged the relevance of neuroscientific discoveries. Kathleen Akins argues that a 'traditional' view of the senses underlies the variety of sophisticated 'naturalistic' programs about intentionality. At the present time, neuroscientific understanding is to perceive and comprehend the nature and significance of comprehending the meaning intended or expressed by a language, sounds, form, or symbolic gesture, and to consent to some thing as an agreed fact to know and be tolerant and sympathetic toward an understanding view, even though disagreeing upon, that is to say, that to having a better understanding, knowledge or comprehension is to learn something indirectly or gathering of something under the circumstance for given under the circumstance for that which a disposition to appreciate or share feelings and thoughts of others is characterized by or having comprehension or having good-sense or discernment. The traditional view holds that sensory systems are 'veridical' in at least three ways. (1) Each signal in the system correlates with specifying narratives to which the construing properties are external (to the body) environment. (2) The structure in the relevant relations between the external properties the receptors are sensitive to is preserved in the structure of the relations between the resulting sensory states. And (3) the sensory system or its analysis of an activity or procedure to determine the desired end and the most efficient method of obtaining this end as the ac t, process of a functional physiological unit of interacting mechanisms or electrical components are at base, anatomically complementary affiliated within the nervous system, a naturally occurring group of objects or phenomena in the reconstructing of nonfictive additions of embellishments, the external events. Using recent neurobiological discoveries about response properties of thermal receptors in the skin as an illustration, Akins shows that sensory systems are 'narcissistic' rather than 'veridical.' All three traditional assumptions are violated. These neurobiological details and their philosophical implications open novel questions for the philosophy of perception and for the appropriate foundations for naturalistic projects about intentionality. Armed with the known neurophysiology of sensory receptors, for example, our 'philosophy of perception' or of 'perceptual intentionality' will no longer focus on the search for correlations between states of sensory systems and 'veridically detected' external properties. This traditionally philosophical (and scientific) project rests upon a mistaken 'veridical' view of the senses. Neuroscientific knowledge of sensory receptor activity also shows that sensory experience does not serve the naturalist well as a 'simple paradigm case' of an intentional relation between representation and world. Once again, available scientific detail shows the naivety of some traditional philosophical projects.

Focussing on the anatomy and physiology of the pain transmission system, Valerie Hardcastle (1997) urges a similar negative implication for a popular methodological assumption. Pain experiences have long been philosophers' favourite cases for analysis and theorizing about conscious experience generally. Nevertheless, every position about pain experiences has been defended recently: eliminativist, a variety of objectivists view, relational views, and subjectivist views. Why so little agreement, despite agreement that pain experience is the place to start an analysis or theory of consciousness? Hardcastle urges two answers. First, philosophers tend to be uninformed about the neuronal complexity of our pain transmission systems, and build their analyses or theories on the outcome of a single component of a multi-component system. Second, even those who understand some of the underlying neurobiology of pain tends to advocate gate-control theories. But the best existing gate-control theories are vague about the neural mechanisms of the gates. Hardcastle instead proposes a dissociable dual system of pain transmission, consisting of a pain sensory system closely analogous in its neurobiological implementation to other sensory systems, and a descending pain inhibitory system. She argues that this dual system is consistent with recent neuroscientific discoveries and accounts for all the pain phenomena that have tempted philosophers toward particular (but limited) theories of pain experience. The neurobiological uniqueness of the pain inhibitory system, contrasted with the mechanisms of other sensory modalities, renders pain processing atypical. In particular, the pain inhibitory system dissociates pains sensation from stimulation of nociceptors (pain receptors). Hardcastle concludes from the neurobiological uniqueness of pain transmission that pain experiences are atypical conscious events, and hence not a good place to start theorizing about or analyzing the general type.

Developing and defending theories of content is a central topic in current philosophy of mind. A common desideratum in this debate is a theory of cognitive representation consistent with a physical or naturalistic ontology. We will distinguish of a few contributions Neurophilosophers have made to this literature.

When one perceives or remembers that he is out of coffee, his brain state possesses intentionality or 'aboutness.' The percept or memory is about one's being out of coffee, and it represents one for being out of coffee. The representational state has content. Some psychosemantics seek to explain what it is for a representational state to be about something: to provide an account of how states and events can have specific representational content. Some physicalist psychosemantics seek to do this using resources of the physical sciences exclusively. Neurophilosophers have contributed to two types of physicalist psychosemantics: the Functional Role approach and the Informational approach.

The nucleus of functional roles of semantics holds that a representation has its content in virtue of relations it bears to other representations. Its paradigm application is to concepts of truth-functional logic, like the conjunctive 'and' or disjunctive 'or.' as a physical event instantiates the 'and' function just in case it maps two true inputs onto a single true output. Thus an expression bears the relations to others that give it the semantic content of 'and.' Proponents of functional role semantics propose similar analyses for the content of all representations (Form 1986). A physical event represents birds, for example, if it bears the right relations to events representing feathers and others representing beaks. By contrast, informational semantics associates content to a state depending upon the causal relations obtaining between the state and the object it represents. A physical state represents birds, for example, just in case an appropriate causal relation obtains between it and birds. At the heart of informational semantics is a causal account of information. Red spots on a face carry the information that one has measles because the red spots are caused by the measles virus. A common criticism of informational semantics holds that mere causal covariation is insufficient for representation, since information (in the causal sense) is by definition, always veridical while representations can misrepresent. A popular solution to this challenge invokes a teleological analysis of 'function.' A brain state represents 'X' by virtue of having the function of carrying information about being caused by 'X' (Dretske 1988). These two approaches do not exhaust the popular options for some psychosemantics, but are the ones to which Neurophilosophers have contributed.

Jerry Fodor and Ernest LePore raises an important challenge to Churchlands psychosemantics. Location in a state space alone seems insufficient to fix a state's representational content. Churchland never explains why a point in a three-dimensional state space represents the Collor, as opposed to any other quality, object, or event that varies along three dimensions. Churchlands account achieves its explanatory power by the interpretation imposed on the dimensions. Fodor and LePore allege that Churchland never specifies how a dimension comes to represent, e.g., degree of saltiness, as opposed to yellow-blue wavelength opposition. One obvious answer appeals to the stimuli that form the 'external' inputs to the neural network in question. Then, for example, the individuating conditions on neural representations of colours are that opponent processing neurons receive input from a specific class of photoreceptors. The latter in turn have electromagnetic radiation (of a specific portion of the visible spectrum) as their activating stimuli, however, this appeal to 'external' stimuli as the ultimate individuating conditions for representational content makes the resulting approach a version of informational semantics. Is this approach consonant with other neurobiological details?

The neurobiological paradigm for informational semantics is the feature detector: One or more neurons that are (i) maximally responsive to a particular type of stimulus, and (ii) have the function of indicating the presence of that stimulus type. Examples of such stimulus-types for visual feature detectors include high-contrast edges, motion direction, and colours. A favourite feature detector among philosophers is the alleged fly detector in the frog. Lettvin et al. (1959) identified cells in the frog retina that responded maximally to small shapes moving across the visual field. The idea that these cells' activity functioned to detect flies rested upon knowledge of the frogs' diet. Using experimental techniques ranging from single-cell recording to sophisticated functional imaging, Neuroscientists have recently discovered a host of neurons that are maximally responsive to a variety of stimuli. However, establishing condition (ii) on a feature detector is much more difficult. Even some paradigm examples have been called into question. David Hubel and Torsten Wiesel's (1962) Nobel Prize winning work for establishing receptive fields of neurons in striate cortices is often interpreted as revealing cells whose function is edge detection. However, Lehky and Sejnowski (1988) have challenged this interpretation. They trained an artificial neural network to distinguish the three-dimensional shape and orientation of an object from its two-dimensional shading pattern. Their network incorporates many features of visual neurophysiology. Nodes in the trained network turned out to be maximally responsive to edge contrasts, but did not appear to have the function of edge detection.

Kathleen Akins (1996) offers a different neurophilosophical challenge to informational semantics and its affiliated feature-detection view of sensory representation. We saw in the previous section how Akins argues that the physiology of thermoreceptor violates three necessary conditions on 'veridical' representation. From this fact she draws doubts about looking for feature detecting neurons to ground psychosemantics in general, including thought contents. Human thoughts about flies, for example, are sensitive to numerical distinctions between particular flies and the particular locations they can occupy. But the ends of frog nutrition are well served without a representational system sensitive to such ontological refinements. Whether a fly seen now is numerically identical to one seen a moment ago, need not, and perhaps cannot, figure into the frog's feature detection repertoire. Akins' critique casts doubt on whether details of sensory transduction will scale up to encompass of some adequately unified psychosemantics. It also raises new questions for human intentionality. How do we get from activity patterns in 'narcissistic' sensory receptors, keyed not to 'objective' environmental features but rather only to effects of the stimuli on the patch of tissue enervated, to the human ontology replete with enduring objects with stable configurations of properties and relations, types and their tokens (as the 'fly-thought' example presented above reveals), and the rest? And how did the development of a stable, and rich ontology confer survival advantages to human ancestors?

Consciousness has reemerged as a topic in philosophy of mind and the cognitive and brains sciences over the past three decades. Instead of ignoring it, many physicalists now seek to explain it (Dennett, 1991). Here we focus exclusively on ways those Neuroscientific discoveries have impacted philosophical debates about the nature of consciousness and its relation to physical mechanisms. Thomas Nagel argues that conscious experience is subjective, and thus permanently recalcitrant to objective scientific understanding. He invites us to ponder 'what it is like to be a bat' and urges the intuition that no amount of physical-scientific knowledge (including Neuroscientific) supplies a complete answer. Nagel's intuition pump has generated extensive philosophical discussion. At least two well-known

233

replies make direct appeal to neurophysiology. John Biro suggests that part of the intuition pumped by Nagel, that bat experience is substantially different from human experience, presupposes systematic relations between physiology and phenomenology. Kathleen Akins (1993) delves deeper into existing knowledge of bat physiology and reports much that is pertinent to Nagel's question. She argues that many of the questions about bat subjectivity that we still consider open hinge on questions that remain unanswered about Neuroscientific details. One example of the latter is the function of various cortical activity profiles in the active bat.

The more recent philosopher David Chalmers (1996), has argued that any possible brain-process account of consciousness will leave open an 'explanatory gap' between the brain process and properties of the conscious experience. This is because no brain-process theory can answer the 'hard' question: Why should that particular brain process give rise to conscious experience? We can always imagine ('conceive of') a universe populated by creatures having those brain processes but completely lacking conscious experience. A theory of consciousness requires an explanation of how and why some brain process causes consciousness replete with all the features we commonly experience. The fact that the hard question remains unanswered, showing that we will probably never get a complete explanation of consciousness at the level of neural mechanisms, even so, Paul and Patricia Churchland have recently offered the following diagnosis and reply. Chalmers offer a conceptual argument, based on our ability to imagine creatures possessing brains like ours but wholly lacking in conscious experience. But the more one learns about how the brain produces conscious experience-and literature is beginning to emerge (e.g., Gazzaniga, 1995)-the harder it becomes to imagine a universe consisting of creatures with brain processes like ours but lacking consciousness. This is not just to bare assertions. The Churchlands appeal to some neurobiological detail. For example, Paul Churchland (1995) develops a Neuroscientific account of consciousness based on recurrent connections between thalamic nuclei (particularly 'diffusely projecting' nuclei like the intralaminar nuclei) and the cortex. Churchland argues that the thalamocortical recurrency accounts for the selective features of consciousness, for the effects of short-term memory on conscious experience, for vivid dreaming during REM. (rapid-eye movement) sleep, and other 'core' features of conscious experience. In other words, the Churchlands are claiming that when one learns about activity patterns in these recurrent circuits, one can't 'imagine' or 'conceive of' this activity occurring without these core features of conscious experience. (Other than just mouthing the words, 'I am now imagining activity in these circuits without selective attention/the effects of short-term memory/vivid dreaming . . . ')

A second focus of sceptical arguments about a complete Neuroscientific explanation of consciousness is sensory Qualia: the introspectable qualitative aspects of sensory experience, the features by which subjects discern similarities and differences among their experiences. The colours of visual sensations are a philosopher's favourite example. One famous puzzle about colour Qualia is the alleged conceivability of spectral inversions. Many philosophers claim that it is conceptually possible (if perhaps physically impossible) for two humans not to ascribe of any dissimilar or unlike in nature, quality, amount, or form differing Neurophysiological divergence, still, being so, is that of the function for indicating that the colours that fire engines and tomatoes appear to have to one dependent as the colour that grass and frogs appear to have to the other (and vice versa). A large amount of neuroscientifically-informed philosophy has addressed this question. A related area where neurophilosophical considerations have emerged concerns the metaphysics of colours themselves (rather than Collor experiences). A longstanding philosophical dispute is whether colours are objective property's Existing external to perceiver or rather identifiable as or dependent upon minds or nervous systems. Some recent work on this problem begins with characteristics of Colour experiences: For example that Colour similarity judgments produce Colour orderings that align on a circle. With this resource, one can seek mappings of phenomenology onto environmental or physiological regularities. Identifying colours with particular frequencies of electromagnetic radiation does not preserve the structure of the hue circle, whereas identifying colours with activity in opponent processing neurons does. Such a tidbit is not decisive for the Collor Objectivist-subjectivist endeavours to engage in argument by discussing opposing points, but it does convey the type of neurophilosophical work being done on traditional metaphysical issues beyond the philosophy of mind.

We saw in the discussion of Hardcastle (1997) two sections above that Neurophilosophers have entered disputes about the nature and methodological import of pain experiences. Two decades earlier, Dan Dennett (1978) took up the question of whether it is possible to build a computer that feels pain. He compares and notes 'pressure' between neurophysiological discoveries and common sense intuitions about pain experience. He suspects that the incommensurability between scientific and common sense views is due to incoherence in the latter. His attitude is wait-and-see. But foreshadowing Churchland's reply to Chalmers, Dennett favours scientific investigations over conceivability-based philosophical arguments.

Neurological deficits have attracted philosophical interest. For thirty years philosophers have found implications for the unity of the self in experiments with commissurotomy patients. In carefully controlled experiments, commissurotomy patients display two dissociable seats of consciousness. Patricia Churchland scouts philosophical implications of a variety of neurological deficits. One deficit is blindsight. Some patients with lesions to primary visual cortex report being unable to see items in regions of their visual fields, yet perform far better than chance in forced guess trials about stimuli in those regions. A variety of scientific and philosophical interpretations have been offered. Ned Form (1988) worries that many of these conflate distinct notions of consciousness. He labels these notions 'phenomenal consciousness' ('P-consciousness') and 'access consciousness' ('A-consciousness'). The former is that which, 'what it is likeness of experience. The latter are the availability of representational content to self-initiated action and speech. Form argues that P-consciousness is not always representational whereas A-consciousness is. Dennett and Michael Tye are sceptical of non-representational analyses of consciousness in general. They provide accounts of blindsight that do not depend on Form's distinction.

Many other topics are worth neurophilosophical pursuit. We mentioned commissurotomy and the unity of consciousness and the self, which continues to generate discussion. Qualia beyond those of Colour and pain have begun to attract Neurophilosophical attentions as having in form of self-consciousness. The first issue to arise in the 'philosophy of neuroscience' (before there was a recognized area) was the localization of cognitive functions to specific neural regions. Although the 'localization' approach had dubious origins in the phrenology of Gall and Spurzheim, and was challenged severely by Flourens throughout the early nineteenth century, it reemerged in the study of aphasia by Bouillaud, Auburtin, Broca, and Wernicke. These neurologists made careful studies (where possible) of linguistic deficits in their aphasic patients followed by brain autopsy's postmortem.

Broca's initial study of twenty-two patients in the mid-nineteenth century confirmed that damage to the left cortical hemisphere was predominant, and that damage to the second and third frontal convolutions was necessary to produce speech production deficits. Although the anatomical system of coordinate postulations of reality or necessity of, especially as a basis of an argument, is only to assume, as a premise or axiom as this is taken for granted, however, something assumed without proof or as self-evident or generally accepted, especially when used as a basis that there is little moral difference between the powers from which of an act seems more than plausible to involve the mental activities that bring forth to some orderly fashion in the mind, in that of the mind affecting thoughts or behaviour, as an overall mental attitude for postulating within the realm of consistency, as for its coordinate system of deficits is still to bear his name ('Broca's area' and 'Broca's aphasia'). Less than two decades later Carl Wernicke published evidence for a second language Centre. This area is anatomically distinct from Broca's area, and damage to it produces a very different set of aphasic symptoms. The cortical area that still bears his name ('Wernicke's area') is located around the first and second convolutions in temporal Cortex, and the aphasia that bear his name ('Wernicke's aphasia') involve deficits in language comprehension. Wernicke's method, like Broca's, was based on lesion studies: as a careful evaluation of the behavioural deficits followed by post mortem examination to find the sites of tissue damage and atrophy. Lesion studies suggesting more precise localization of specific linguistic functions remain the cornerstones in aphasic research

Lesion studies have also produced evidence for the localization of other cognitive functions: for example, sensory processing and certain types of learning and memory. However, localization arguments for these other functions invariably include studies using animal models. With an animal model, one can perform careful behavioural measures in highly controlled settings, then ablate specific areas of neural tissue (or use a variety of other techniques to Form or enhance activity in these areas) and remeasure performance on the same behavioural tests. But since we lack an animal model for (human) language production and comprehension, this additional evidence isn't available to the neurologist or neurolinguist. This fact makes the study of language a paradigm case for evaluating the logic of the lesion/deficit method of inferring functional localization. Philosopher Barbara Von Eckardt (1978) attempts to make explicitly the steps of reasoning involved in this common and historically important method. Her analysis begins with Robert Cummins' early analysis of functional explanation, but she extends it into a notion of structurally adequate functional analysis. These analyses break down a complex capacity C into its constituent capacity's c_1, c_2, . . . a c_n, where the constituent capacities are consistent with the underlying structural details of the system. For example, human speech production (complex capacity C) results from formulating a speech intention, then selecting appropriate linguistic representations to capture the content of the speech intention, then formulating the motor commands to produce the appropriate sounds, then communicating these motor commands to the appropriate motor pathways (constituent capacity's c_1, c_2, . . ., c_n). A functional-localization hypothesis has the form: Brain structure S in an organism (type) O has constituent capacity c_i, where c_i is a function of some part of O. An example, Brains Broca's area (S) in humans (O) formulates motor commands to produce the appropriate sounds (one of the constituent capacities c_i). Such hypotheses specify aspects of the structural realization of a functional-component model. They are part of the theory of the neural realization of the functional model.

Armed with these characterizations, Von Eckardt argues that inference to some functional-localization hypothesis proceeds in two steps. First, a functional deficit in a patient is hypothesized based on the abnormal behaviour the patient exhibits. Second, localization of function in normal brains is inferred on the basis of the functional deficit hypothesis plus the evidence about the site of brain damage. The structurally-adequate functional analysis of the capacity connects the pathological behaviour to the hypothesized functional deficit. This connection suggests four adequacy conditions on a functional deficit hypothesis. First, the pathological P behaviour (e.g., the speech deficits characteristic of Broca's aphasia) must result from failing to exercise some complex capacity C (human speech production). Second, there must be a structurally-adequate functional analysis of how people exercise capacity C that involves some constituent capacity c_i (formulating motor commands to produce the appropriate sounds). Third, the operation of the steps described by the structurally-adequate functional analysis minus the operation of the component performing c_i (Broca's area) must result in pathological behaviour P. Fourth, there must not be a better available explanation for why the patient does P. Arguments to a functional deficit hypothesis on the basis of pathological behaviour is thus an instance of argument to the best available explanation. When postulating a deficit in a normal functional component provides the best available explanation of the pathological data, we are justified in drawing the inference.

Von Eckardt applies this analysis to a neurological case study involving a controversial reinterpretation of agnosia. Her philosophical explication of this important neurological method reveals that most challenges to localization arguments of whether to argue only against the localization of a particular type of functional capacity or against generalizing from localization of function in one individual to all normal individuals. (She presents examples of each from the neurological literature.) Such challenges do not impugn the validity of standard arguments for functional localization from deficits. It does not follow that such arguments are unproblematic. But they face difficult factual and methodological problems, not logical ones. Furthermore, the analysis of these arguments as involving a type of functional analysis and inference to the best available explanation carries an important implication for the biological study of cognitive function. Functional analyses require functional theories, and structurally adequate functional analyses require checks imposed by the lower level sciences investigating the underlying physical mechanisms. Arguments to best available explanation are often hampered by a lack of theoretical imagination: the available explanations are often severely limited. We must seek theoretical inspiration from any level of theory and explanation. Hence making explicitly the 'logic' of this common and historically important form of neurological explanation reveals the necessity of joint participation from all scientific levels, from cognitive psychology down to molecular neuroscience. Von Eckardt anticipated what came to be heralded as the 'co-evolutionary research methodology,' which remains a centerpiece of neurophilosophy to the present day.

Over the last two decades, evidence for localization of cognitive function has come increasingly from a new source: the development and refinement of neuroimaging techniques. The form of localization-of-function argument appears not to have changed from that employing lesion studies (as analysed by Von Eckardt). Instead, these imaging technologies resolve some of the methodological problems that plage lesion studies. For example, researchers do not need to wait until the patient dies, and in the meantime probably acquires additional brain damage, to find the lesion sites. Two functional imaging techniques are prominent: Positron emission tomography, or PET, and

237

functional magnetic resonance imaging, or MRI. Although these measure different biological markers of functional activity, both now have a resolution down too around one millimetre. As these techniques increase spatial and temporal resolution of functional markers and continue to be used with sophisticated behavioural methodologies, the possibility of localizing specific psychological functions to increasingly specific neural regions continues to grow

What we now know about the cellular and molecular mechanisms of neural conductance and transmission is spectacular. The same evaluation holds for all levels of explanation and theory about the mind/brain: Maps, networks, systems, and behaviour. This is a natural outcome of increasing scientific specialization. We develop the technology, the experimental techniques, and the theoretical frameworks within specific disciplines to push forward our understanding. Still, a crucial aspect of the total picture gets neglected: the relationship between the levels, the 'glue' that binds knowledge of neuron activity to subcellular and molecular mechanisms, networks activity patterns to the activity of and connectivity between single neurons, and behavioural network activity. This problem is especially glaring when we focus on the relationship between 'cognitivist' psychological theories, postulating information-bearing representations and processes operating over their contents, and the activity patterns in networks of neurons. Co-evolution between explanatory levels still seems more like a distant dream rather than an operative methodology.

It is here that some Neuroscientists appeal to 'computational' methods. If we examine the way that computational models function in more developed sciences (like physics), we find the resources of dynamical systems constantly employed. Global effects (such as large-scale meteorological patterns) are explained in terms of the interaction of 'local' lower-level physical phenomena, but only by dynamical, nonlinear, and often chaotic sequences and combinations. Addressing the interlocking levels of theory and explanation in the mind/brain using computational resources that have worked to bridge levels in more mature sciences might yield comparable results. This methodology is necessarily interdisciplinary, drawing on resources and researchers from a variety of levels, including higher levels like experimental psychology, 'program-writing' and 'connectionist' artificial intelligence, and philosophy of science.

However, the use of computational methods in neuroscience is not new. Hodgkin, Huxley, and Katz incorporated values of voltage-dependent potassium conductance they had measured experimentally in the squid giant axon into an equation from physics describing the time evolution of a first-order kinetic process. This equation enabled them to calculate best-fit curves for modelled conductance versus time data that reproduced the S-shaped (sigmoidal) function suggested by their experimental data. Using equations borrowed from physics, Rall (1959) developed the cable model of dendrites. This theory provided an account of how the various inputs from across the dendritic tree interact temporally and spatially to determine the input-output properties of single neurons. It remains influential today, and has been incorporated into the genesis software for programming neurally realistic networks. More recently, David Sparks and his colleagues have shown that a vector-averaging model of activity in neurons of superior colliculi correctly predicts experimental results about the amplitude and direction of saccadic eye movements. Working with a more sophisticated mathematical model, Apostolos Georgopoulos and his colleagues have predicted direction and amplitude of hand and arm movements based on averaged activity of 224 cells in motor cortices. Their predictions have borne out under a variety of experimental tests. We mention these particular studies only because we are familiar with them. We could multiply examples of the fruitful interaction of computational and experimental methods in neuroscience easily by one-hundred-fold. Many of these extend back before 'computational neuroscience' was a recognized research endeavour.

We've already seen one example, the vector transformation account, of neural representation and computation, under active development in cognitive neuroscience. Other approaches using 'cognitivist' resources are also being pursued. Many of these projects draw upon 'cognitivist' characterizations of the phenomena to be explained. Many exploit 'cognitivist' experimental techniques and methodologies, some even attempt to derive 'cognitivist' explanations from cell-biological processes (e.g., Hawkins and Kandel 1984). As Stephen Kosslyn puts it, cognitive Neuroscientists employ the 'information processing' view of the mind characteristic of cognitivism without trying to separate it from theories of brain mechanisms. Such an endeavour calls for an interdisciplinary community willing to communicate the relevant portions of the mountain of detail gathered in individual disciplines with interested nonspecialists: not just people willing to confer with those working at related levels, but researchers trained in the methods and factual details of a variety of levels. This is a daunting requirement, but it does offer some hope for philosophers wishing to contribute to future neuroscience. Thinkers trained in both the 'synoptic vision' afforded by philosophy and the factual and experimental basis of genuine graduate-level science would be ideally equipped for this task. Recognition of this potential niche has been slow among graduate programs in philosophy, but there is some hope that a few programs are taking steps to fill it.

In the final analysis there will be philosophers unprepared to accept that, if a given cognitive capacity is psychologically real, then there must be an explanation of how it is possible for an individual in the course of human development to acquire that cognitive capacity, or anything like it, can have a role to play in philosophical accounts of concepts and conceptual abilities. The most obvious basis for such a view would be a Frégean distrust of 'psychology' that leads to a rigid division of labour between philosophy and psychology. The operative thought is that the task of a philosophical theory of concepts is to explain what a given concept is or what a given conceptual ability consist in. This, it is frequently maintained, is something that can be done in complete independence of explaining how such a concept or ability might be acquired. The underlying distinction is one between philosophical questions cantering around concept possession and psychological questions cantering around concept possibilities for an individual to acquire that ability, then it cannot be psychologically real. Nevertheless, this distinction is, however, strictly one suffices for not adhering to the distinction, it provides no support for a rejection of any given cognitive capacity for which is psychologically real. The neo-Frégean distinction is directly against the view that facts about how concepts are acquired have a role to play in explaining and individualizing concepts. But this view does not have to be disputed by a supporter as such, nonetheless, all that the supporter is to commit is that the principle that no satisfactory account of what a concept is should make it impossible to provide explanation of how that concept can be acquired. That is, that this principle has nothing to say about the further question of whether the psychological explanation has a role to play in a constitutive explanation of the concept, and hence is not in conflict with the neo-Frégean distinction.

A full account of the structure of consciousness, needs to illustrate those higher, conceptual forms of consciousness to which little attention on such an account will take and about how it might emerge from given points of value, is the thought that an explanation of everything that is distinctive about consciousness will emerge out of an account of what it is for its recipient subject to be capable of thinking about himself. But, to a better understanding of the complex phenomenon of consciousness, there are no facts about linguistic mastery that will determine or explain what might be termed the cognitive dynamics that are individual processes that have found their way forward for a theory of consciousness, it sees, to chart the characteristic features individualizing the various distinct conceptual forms of consciousness in a way that will provide a taxonomy of unconsciousness and to show how these manifest the characterlogical functions to determine at which of the levels is contentual. What is to look forward to is now clear that these forms of higher calibrations of consciousness emerge from a rich foundation of non-conceptual representations of thought, which can only expose and clarify their conviction that these forms of conscious thought hold the key, not just to an eventful account of how mastery of the conscious paradigms, but to a proper understanding of the plexuity of self-consciousness and/or the overall conjecture of consciousness that stands alone as to an everlasting vanquishment into the abyss of ever unchangeless states of unconsciousness.

Theory itself, is consistent with fact or reality, not false or incorrect, but truthful, it is sincerely felt or expressed unforeignly and so, that it is essential and exacting of several standing rules and senses of governing requirements. As appendantly conforming to in sensing the definitive criteria of narrowly particular possibilities in value as taken by a variaby accord reality, as by a position of something, as to make it balanced, level or square, that we may think of a proper alignment as something, insofar as, that one is certain, like trust, another derivation of the same appears by its name is etymologically, or 'strong seers'. Conformity of fact or the actuality of a statement as been or accepted as true to an original or standard set class theory from which it is considered as the supreme reality and to have the ultimate meaning, and value of existence. It is, nonetheless, a compound position, such as a conjunction or negation, the truth-values have always determined whose truth-values of that component thesis.

Moreover, science, unswerving exactly to position of something very well hidden, its nature in so that to make it believed, is quickly and imposes on sensing and responding to the definitive qualities or state of being actual or true, such that as a person, an entity, or an event, that might be gainfully employed of all things possessing actuality, existence, or essence. In other words, in that which is objectively inside and out, and in addition it seems to appropriate that of reality, in fact, to the satisfying factions of instinctual needs through the awarenesses of and adjustments abided to environmental demands. Thus, the act of realizing or the condition of truth as seen for being realized, and the existent remnants to their resulting amounts are throughout the retrogressive detentions that are undoubtingly realized.

However, a declaration made to explain or justify action, or its believing desire upon which it is to act, by which the conviction underlying facts or cause, that provide logical sense for a premise or occurrence for logical, rational. Analytic mental states have long since lost in reason, but, yet, the premise usually takes upon the minor premises of an argument, using this faculty of reason that arises too throughout the spoken exchange or a debative discussion, and, of course, spoken in a dialectic way.

To determining or conclusively logical impounded by thinking through its directorial solution to the problem, would therefore persuade or dissuade someone with reason that posits of itself with the good sense or justification of reasonability. In which, good causes are simply justifiably to be considered as to think. By which humans seek or attain knowledge or truth. Mere reason is insufficient to convince 'us' of its veracity. Still, comprehension perceptively welcomes an intuitively given certainty, as the truth or fact, without the use of the rational process, as one comes to assessing someone's character, it sublimely configures one consideration, and often with resulting comprehensions, in which it is assessing situations or circumstances and draw sound conclusions into the reign of judgement.

Governed or being accorded to reason or sound thinking, in that a reasonable solution to the problem, may as well be without bounds and commonsense for arriving to a fair use of reason, especially to form conclusions, inferences or judgements, it is to say, that, all evidential alternatives In their confronting of arguments are within use of thinking or thought out responses to issuing that of furthering argumentation? To fit or join in the sum parts that are composite to the intellectual faculties, by which case human understanding or the attemptive grasp to its thought, are not without the resulting liberty encroaching men of zeal, well-meaningly, but without understanding.

Being or occurring to fact or having some verifiable existence, as real objects, and real illnesses. . . .'That is really true and actual and not imaginary, is but allegedly that person and not the particular imagination, as the illnesses of imagination from which we are to find on practical matters and concerns of experiencing the real world. The surrounding surfaces, might we, as, perhaps attest to this for the first time. Being no less than what they state, we have not taken its free pretence, or affections for a real experience highly, as many may encounter real trouble. This, nonetheless, projects of an existing objectivity in which the world despite subjectivity or conventions of thought or language is or have valuing representation, reckoned by actual power, in that of relating to, or being an image formed by light or another identifiable simulation, that converge in space, the stationary or fixed properties, such as a thing or whole having actual existence. All of which, are accorded a truly factual experience into which the actual attestations have brought to you by the afforded efforts of our very own imaginations.

Ideally, in theory the imagination, a concept of reason that is transcendent but nonempirically as to think of conceptions of and ideal thought, that potentially or actual exists in the mind as a product exclusive to the mental act. In the philosophy of Plato, an archetype of which a corresponding being in phenomenal reality is an imperfect replica, that also, Hegel's absolute truth, as the conception and ultimate product of reason (the absolute meaning a mental image of something remembered).

Conceivably, in the imagination the formation of a mental image of something that is or should be b perceived as real nor present to the senses. Nevertheless, the image so formed can confront and deal with the reality by using the creative powers of the mind. That is characteristically well removed from reality, but all powers of fantasy over reason are a degree of insanity/ still, fancy as they have given a product of the imagination free reins, that is in command of the fantasy while it is exactly the mark of the neurotic that his very own fantasy possesses him.

All things possessing actuality, existence or essence that exists objectively and in fact based on real occurrences that exist or known to have existed, a real occurrence, an event, i.e., had to prove the facts of the case, as something believed to be true or real, determining by evidence or truth as to do. However, the usage in the sense 'allegation of fact', and the reasoning are wrong of the 'facts' and 'substantive facts', as we may never know the 'facts' of the case'. These usages may occasion qualms' among critics who insist that facts can only be true, but the usages are often useful for emphasis. Therefore, we have related to, or used the discovery or determinations of fast or accurate information in the discovery of facts, then evidence has determined the comprising events or truth is much as ado about their owing actuality. Its opposition forming the literature that treats real people or events as if they were fictional or uses real people or events as essential elements in an otherwise fictional rendition, i.e., of, relating to, produced by, or characterized by internal dissension, as given to or promoting internal dissension. So, then, it is produced artificially than by a natural process, especially the lacking authenticity or genuine factitious values of another than what is or of reality should be.

Substantive set statements or principles devised to explain a group of facts or phenomena, especially one that we have tested or is together experiment with and taken for 'us' to conclude and can be put-upon to make predictions about natural phenomena. Having the consistency of explanatory statements, accepted principles, and methods of analysis, finds to a set of theorems that make up a systematic view of a branch in mathematics or extends upon the paradigms of science, the belief or principle that guides action or helps comprehension or judgements, usually by an ascription based on limited information or knowledge, as a conjecture, tenably to assert the creation from a speculative assumption that bestows to its beginning. Theoretically, to, affiliate oneself with to, or based by itself on theory, i.e., the restriction to theory, is not as much a practical theory of physics, as given to speculative theorizing. Also, the given idea, because of which formidable combinations awaiting upon the inception of an idea, demonstrated as true or is given to demonstration. In mathematics its containment lies of the proposition that has been or is to be proved from explicit assumption and is primarily with theoretical assessments or hypothetical theorizing than possibly these might be thoughtful measures and taken as the characteristics by which we measure its quality value?

Looking back, one can see a discovering degree of homogeneity among the philosophers of the early twentieth century about the topics central to their concerns. More striking still, is the apparent profundities and abstrusity of concerns for which appear at first glance to be separated from the discerned debates of previous centuries, between 'realism' and 'idealist', say, of 'rationalists' and 'empiricist'.

Thus, no matter what the current debate or discussion, the central issue is often without conceptual and contentual representations, that if one is without concept, is without idea, such that in one foul swoop would ingest the mere truth that is laid to the underlying paradoxes of why is there something instead of nothing? Whatever it is that makes, what would, otherwise is merely an utterance and inscriptions into instruments of communication and that of understanding. This philosophical problem is to demystify this over-flowing emptiness, and to relate to what we know of ourselves and subjective matter's resembling reality or ours is to an inherent perceptivity of the world and its surrounding surfaces.

Contributions to this study include the theory of 'speech arts', and the investigation of communicable communications, especially the relationship between words and 'ideas', and words and the 'world'. It is, nonetheless, that which and utterance or sentence expresses, the proposition or claim made about the world. By extension, the content of a predicate that any expression effectively connecting with one or more singular terms to make a sentence, the expressed condition that the entities referred to may satisfy, in which case the resulting sentence will be true. Consequently we may think of a predicate as a function from things to sentences or even to truth-values, or other sub-sentential components that contribute to sentences that contain it. The nature of content is the central concern of the philosophy of language.

What some person expresses of a sentence often depends on the environment in which he or she is placed. For example, the disease I refer to by a term like 'arthritis' or the kind of tree I call of its criteria will define a 'beech' of which I know next to nothing. This raises the possibility of imaging two persons as an alternative different environment, but in which everything appears the same to each of them. The wide content of their thoughts and saying will be different if the situation surrounding them is appropriately different, 'situation' may here include the actual objects hey perceive, or the chemical or physical kinds of objects in the world they inhabit, or the history of their words, or the decisions of authorities on what counts as an example of one term thy use. The narrow content is that part of their thought that remains identical, through the identity of the way things appear, despite these differences of surroundings. Partisans of wide, . . . 'as, something called broadly, content may doubt whether any content is in this sense narrow, partisan of narrow content believe that it is the fundamental notion, with wide content being on narrow content confirming context.

All and all, assuming their rationality has characterized people is common, and the most evident display of our rationality is capable to think. This is the rehearsal in the mind of what to say, or what to do. Not all thinking is verbal, since chess players, composers, and painters all think, and there is no deductive reason that their deliberations should take any more verbal a form than their actions. It is permanently tempting to conceive of this activity about the presence in the mind of elements of some language, or other medium that represents aspects of the world and its surrounding surface structures. However, the model has been attacked, notably by Ludwig Wittgenstein (1889-1951), whose influential application of these ideas was in the philosophy of mind. Wittgenstein explores the role that reports of introspection, or sensations, or intentions, or beliefs can play of our social lives, to undermine the Cartesian mental picture is that they functionally describe the goings-on in an inner theatre of which the subject is the lone spectator. Passages that have subsequentially become known as the 'rule following' considerations and the 'private language argument' are among the fundamental topics of modern philosophy of language and mind, although their precise interpretation is endlessly controversial.

Effectively, the hypotheses especially associated with Jerry Fodor (1935-), whom is known for the 'resolute realism', about the nature of mental functioning, that occurs in a language different from one's ordinary native language, but underlying and explaining our competence with it. The idea is a development of the notion of an innate universal grammar (Avram Noam Chomsky, 1928-), in as such, that we agree that since a computer programs are linguistically complex sets of instructions were the relative executions by which explains of surface behaviour or the adequacy of the computerized programming installations, if it were definably amendable and, advisably corrective, in that most are disconcerting of many that are ultimately a reason for 'us' of thinking intuitively and without the indulgence of retrospective preferences, but an ethical majority in defending of its moral line that is already confronting 'us'. That these programs may or may not improve to conditions that are lastly to enhance of the right sort of an existence forwarded toward a more valuing amount in humanities lesser extensions that embrace one's riff of necessity to humanities' abeyance to expressions in the finer of qualities.

As an explanation of ordinary language-learning and competence, the hypothesis has not found universal favour, as only ordinary representational powers that by invoking the image of the learning person's capabilities are apparently whom the abilities for translating are contending of an innate language whose own powers are mysteriously a biological given. Perhaps, the view that everyday attributions of intentionality, beliefs, and meaning to other persons are proceeding by of a tactic use of a theory that enables one to the construct to these interpretations as explanations of their doing, as we commonly hold the view along with 'functionalism', according to which the psychological states are a theoretical entity, identified by the network of their causes and effects. The theory-theory has different implications, depending upon which feature of theories we are stressing. Theories may be thought of as capable of formalization, as yielding predictions and explanations, as achieved by a process of theorizing, as answering to empirical evidence that is in principle describable without them, as liable to be overturned by newer and better theories, and so on.

The main problem with seeing our understanding of others as the outcome of a piece of theorizing is the nonexistence of a medium in which this theory can be couched, as the child learns simultaneously the minds of others and the meaning of terms in its native language, is not gained by the tactic use of a 'theory', enabling 'us' to infer what thoughts or intentions explain their actions, but by reliving the situation 'in their shoes' or from their point of view, and by that understanding what they experienced and theory, and therefore expressed. Understanding others is achieved when we can ourselves deliberate as they did, and hear their words as if they are our own. The suggestion is a modern development frequently associated in the 'Verstehen' traditions of Dilthey (1833-1911), Weber (1864-1920) and Collingwood (1889-1943).

We may call any process of drawing a conclusion from a set of premises a process of reasoning. If the conclusion concerns what to do, the process is called practical reasoning, otherwise pure or theoretical reasoning. Evidently, such processes may be good or bad, if they are good, the premises support or even entail the conclusion drawn, and if they are bad, the premises offer no support to the conclusion. Formal logic studies the cases in which conclusions are validly drawn from premises, but little human reasoning is overly of the forms logicians identify. Partly, we are concerned to draw conclusions that 'go beyond' our premises, in the way that conclusions of logically valid arguments do not for the process of using evidence to reach a wider conclusion. Still, abounding within the hesitorial anticipations of pessimism, with which directions are those that the prospects of conformation theory, are denied as the applicable results of abduction in terms of probability. A cognitive process of reasoning in which a conclusion is played-out from a set of premises usually confined of cases in which the conclusions are supposed in following from the premises, i.e., an inference is logically valid, in that of deductibility in a logically defined syntactic premise but without there being to any reference to the intended interpretation of its theory. Furthermore, as we reason we use indefinite traditional knowledge or commonsense sets of presuppositions about what it is likely or not a task of an automated reasoning project, which is to mimic this causal use of knowledge of the way of the world in computer programs.

Some 'theories' usually emerge themselves of engaging to exceptionally explicit predominancy as [supposed] truths that they have not organized, making the theory difficult to survey or study as a whole. The axiomatic method is an idea for organizing a theory, one in which tries to select from among the supposed truths a small number from which they can see all others to be deductively inferable. This makes the theory more tractable since, in a sense, they contain all truths in those few. In a theory so organized, they call the few truths from which they deductively imply all others 'axioms'. David Hilbert (1862-1943) had argued that, just as algebraic and differential equations, which we were used to study mathematical and physical processes, could have themselves be made mathematical objects, so axiomatic theories, like algebraic and differential equations, which are to representing physical processes and mathematical structures could be of investigating.

Conformation to theory, the philosophy of science, is a generalization or set referring to unobservable entities, i.e., atoms, genes, quarks, unconscious wishes. The ideal gas law for example, referring to such observable pressures, temperature, and volume, the 'molecular-kinetic theory' refers to molecules and their material possession, . . . although an older usage suggests the lack of adequate evidence in support thereof, as an existing philosophical usage does in truth, follow in the tradition (as in Leibniz, 1704), as many philosophers had the conviction that all truths, or all truths about a particular domain, followed from as a few than for being many governing principles. These principles were taken to be either metaphysically prior or epistemologically prior or both. In the first sense, they we took to be entities of such a nature that what exists s 'caused' by them. When the principles were taken as epistemologically prior, that is, as 'axioms', they were taken to be epistemologically privileged, e.g., self-evident, not needing to be demonstrated, or again, included 'or', to such that all truths so truly follow from them by deductive inferences. Gödel (1984) showed in the spirit of Hilbert, treating axiomatic theories as themselves mathematical objects that mathematics, and even a small part of mathematics, elementary number theory, could not be axiomatized, that more precisely, any class of axioms that is such that we could effectively decide, of any proposition, whether or not it was in that class, would be too small to capture in of the truths.

The notion of truth occurs with remarkable frequency in our reflections on language, thought and action. We are inclined to suppose, for example, that truth is the proper aim of scientific inquiry, that true beliefs help to achieve our goals, that to understand a sentence is to know which circumstances would make it true, that reliable preservation of truth as one argues of valid reasoning, that moral pronouncements should not be regarded as objectively true, and so on. To assess the plausibility of such theses, and to refine them and to explain why they hold (if they do), we require some view of what truth be a theory that would account for its properties and its relations to other matters. Thus, there can be little prospect of understanding our most important faculties in the sentence of a good theory of truth.

Such a thing, however, has been notoriously elusive. The ancient idea that truth is some sort of 'correspondence with reality' has still never been articulated satisfactorily, and the nature of the alleged 'correspondence' and the alleged 'reality' persistently remains objectionably enigmatical. Yet the familiar alternative suggestions that true beliefs are those that are 'mutually coherent', or 'pragmatically useful', or 'verifiable in suitable conditions' has each been confronted with persuasive counterexamples. A twentieth-century departure from these traditional analyses is the view that truth is not a property at all that the syntactic form of the predicate, 'is true', distorts its really semantic character, which is not to describe propositions but to endorse them. Nevertheless, we have also faced this radical approach with difficulties and suggest, counter intuitively that truth cannot have the vital theoretical role in semantics, epistemology and elsewhere that we are naturally inclined to give it. Thus, truth threatens to remain one of the most enigmatic of notions: An explicit account of it can seem essential yet beyond our reach. All the same, recent work provides some evidence for optimism.

A theory is based in philosophy of science, is a generalization or se of generalizations purportedly referring to observable entities, i.e., atoms, quarks, unconscious wishes, and so on. The ideal gas law, for example, cites to only such observable pressures, temperature, and volume, the molecular-kinetic theory refers top molecules and their properties, although an older usage suggests the lack of an adequate make out in support therefrom as merely a theory, latter-day philosophical usage does not carry that connotation. Einstein's special and General Theory of Relativity, for example, is taken to be extremely well founded.

These are two main views on the nature of theories. According to the 'received view' theories are partially interpreted axiomatic systems, according to the semantic view, a theory is a collection of models (Suppe, 1974). By which, some possibilities, unremarkably emerge as supposed truths that no one has neatly systematized by making theory difficult to make a survey of or study as a whole. The axiomatic method is an ideal for organizing a theory (Hilbert, 1970), one tries to select from among the supposed truths a small number from which they can see all the others to be deductively inferable. This makes the theory more tractable since, in a sense, they contain all truth's in those few. In a theory so organized, they call the few truths from which they deductively incriminate all others 'axioms'. David Hilbert (1862-1943) had argued that, morally justified as algebraic and differential equations, which were antiquated into the study of mathematical and physical processes, could hold on to themselves and be made mathematical objects, so they could make axiomatic theories, like algebraic and differential equations, which are of representing physical processes and mathematical structures, objects of mathematical investigation.

In the tradition (as in Leibniz, 1704), many philosophers had the conviction that all truths, or all truths about a particular domain, followed from a few principles. These principles were taken to be either metaphysically prior or epistemologically prior or both. In the first sense, they were taken to be entities of such a nature that what exists is 'caused' by them. When the principles were taken as epistemologically prior, that is, as 'axioms', they were taken to be epistemologically privileged, i.e., self-evident, not needing to be demonstrated, or again, inclusive 'or', to be such that all truths do in truth follow from them (by deductive inferences). Gödel (1984) showed in the spirit of Hilbert, treating axiomatic theories as themselves mathematical objects that mathematics, and even a small part. Of mathematics, elementary number theory, could not be axiomatized, that, more precisely, any class of axioms that is such that we could effectively decide, of any proposition, whether or not it was in that class, would be too small to capture all of the truths.

The notion of truth occurs with remarkable frequency in our reflections on language, thought, and action. We are inclined to suppose, for example, that truth is the proper aim of scientific inquiry, that true beliefs help 'us' to achieve our goals, tat to understand a sentence is to know which circumstances would make it true, that reliable preservation of truth as one argues from premises to a conclusion is the mark of valid reasoning, that moral pronouncements should not be regarded as objectively true, and so on. In order to assess the plausible of such theses, and in order to refine them and to explain why they hold, if they do, we expect some view of what truth be of a theory that would keep an account of its properties and its relations to other matters. Thus, there can be little prospect of understanding our most important faculties without a good theory of truth.

The ancient idea that truth is one sort of 'correspondence with reality' has still never been articulated satisfactorily: The nature of the alleged 'correspondence' and te alleged 'reality remains objectivably rid of obstructions. Yet, the familiar alternative suggests ~. That true beliefs are those that are 'mutually coherent', or 'pragmatically useful', or 'verifiable' in suitable conditions has each been confronted with persuasive counterexamples. A twentieth-century departure from these traditional analyses is the view that truth is not a property at al ~. That the syntactic form of the predicate,' . . . is true', distorts the 'real' semantic character, with which is not to describe propositions but to endorse them. Still, this radical approach is also faced with difficulties and suggests, counter intuitively that truth cannot have the vital theoretical role in semantics, epistemology and elsewhere that we are naturally inclined to give it. Thus, truth threatens to remain one of the most enigmatic of notions, and a confirming account of it can seem essential yet, on the far side of our reach. However, recent work provides some grounds for optimism.

The belief that snow is white owes its truth to a certain feature of the external world, namely, to the fact that snow is white. Similarly, the belief that dogs bark is true because of the fact that dogs bark. This trivial observation leads to what is perhaps the most natural and popular account of truth, the 'correspondence theory', according to which a belief (statement, a sentence, propositions, and so forth (as true just in case there exists a fact corresponding to it (Wittgenstein, 1922, Austin! 950). This thesis is unexceptionable, however, if it is to provide a rigorous, substantial and complete theory of truth ~. If it is to be more than merely a picturesque way of asserting all equivalences to the form, such that the belief that 'p' is true 'p'. Then it must be supplemented with accounts of what facts are, and what it is for a belief to correspond to a fact, and these are the problems on which the correspondence theory of truth has floundered. For one thing, it is far from going unchallenged that any significant gain in understanding is achieved by reducing 'the belief that snow is white is' true' to the facts that snow is white exists: For these expressions look equally resistant to analysis and too close in meaning for one to provide a crystallizing account of the other. In addition, the undistributed relationship that holds in particular between the belief that snow is white and the fact that snow is white, between the belief that 'dogs bark' and the fact that a 'dog barks', and so on, is very hard to identify. The best attempt to date is Wittgenstein's 1922, so-called 'picture theory', by which an elementary proposition is a configuration of terms, with whatever stare of affairs it reported, as an atomic fact is a configuration of simple objects, an atomic fact corresponds to an elementary proposition and makes it true, when their configurations are identical and when the terms in the proposition for it to the similarly-placed objects in the fact, and the truth value of each complex proposition the truth values entail of the elementary ones. However, eve if this account is correct as far as it goes, it would need to be completed with plausible theories of 'logical configuration', 'rudimentary proposition', 'reference' and 'entailment,' none of which is better-off in words used in a certain manner, phrase or couched as distinct or related to the universe, especially the infinite or inconceivably the extended cosmic particles, superseding space and time frames that can in the reference unequally being the distribution of mass or cosmic configurations, as or relating to the universe and sortally distinct from earth, infinitely or inconceivable are extended of fewer than a greater amount of microscopic particles of both of infinitesimal measure and dimension.

One that any adequate theory must explain is that when a proposition satisfies its 'conditions of proof or verification' then it is regarded as true. To the extent that the property of corresponding with reality is mysterious, we are going to find it impossible to see what we take to verify a proposition should show the possession of that property. Therefore, a tempting alternative to the correspondence theory is an alternative that eschews obscure, metaphysical concept that explains quite simply why the possibility to verify as Verifiability accounts of the incident from which it infers that the act or process for its 'truth' is to come about. Truth is simply to identify truth with verifiability (Peirce, 1932), that this idea such as a thought or conception that potentially or actually exists in the mind as a product of mental activity. Taken on variously formed versions that involve the furthering assumptions that verification is 'holistic', . . . 'in that a belief is justified (i.e., verified) when it is part of an entire system of beliefs that are consistent and 'counter balanced' (Bradley, 1914 and Hempel, 1935). This is known as the 'coherence theory of truth'. Another version involves the assumption associated with each proposition, some specific procedure for finding out whether one should believe it or not. On this account, to say that a proposition is true is to say that the appropriate procedure would verify (Dummett, 1979. and Putnam, 1981), while mathematics this amounts to the identification of truth with provability.

The attractions of the verificationist account of truth are that it is refreshingly clear compared with the correspondence theory, and that it succeeds in connecting truth with verification. The trouble is that the bond it postulates between these notions is implausibly strong. We do in true statements' take verification to indicate truth, but also we recognize the possibility that a proposition may be false in spite of there being impeccable reasons to believe it, and that a proposition may be true although we are not of the ability to discover of what it is. Verifiability and truth are no doubt highly correlated, but surely not the same thing.

A third well-known account of truth is known as 'pragmatism' (James, 1909 and Papineau, 1987). As we have just seen, the verificationist selects a prominent property of truth and considers the essence of truth. Similarly, the pragmatist focuses on another important characteristic namely, that true belief is a good basis for action and takes this to be the very nature of truth. True assumptions are said to be, by definition, those that provoke actions with desirable results. Again, we have an account statement with a single attractive explanatory characteristic, besides, it postulates between truth and its alleged analysand in this case, utility is implausibly close. Granted, true belief tends to foster success, but it happens regularly that actions based on true beliefs lead to disaster, while false assumptions, by pure chance, produce wonderful results.

One of the few uncontroversial facts about truth is that the proposition that snow is white if and only if snow is white, the proposition that lying is wrong is true if and only if lying is wrong, and so on. Traditional theories acknowledge this fact but regard it as insufficient and, as we have seen, inflate it with some further principle of the form, 'X' is true if and only if 'X' has property 'P' (such as corresponding to reality, verifiability or being suitable as a basis for action), this is supposed to specify what truth is. Some radical alternatives to the traditional theories result from denying the need for any such further specification (Ramsey, 1927, Strawson, 1950 and Quine, 1990). For example, ne might suppose that the basic theory of truth contains nothing more that equivalences of the form, 'The proposition that 'p' is true if and only if 'p' (Horwich, 1990).

That is, a proposition, 'K' with the following properties, that from 'K' and any further premises of the form. 'Einstein's claim was the proposition that p' you can imply p'. Whatever it is, now supposes, as the deflationist say, that our understanding of the truth predicate consists in the stimulative decision to accept any instance of the schema. 'The proposition that p is true if and only if p', then your problem is solved. For 'K' is the proposition, 'Einstein's claim is true', it will have precisely the inferential power needed. From it and 'Einstein's claim is the proposition that quantum mechanics are wrong', you can use Leibniz's law to imply 'The proposition that quantum mechanic is wrong is true; which given the relevant axiom of the deflationary theory, allows you to derive 'Quantum mechanics is wrong'. Thus, one point in favour of the deflationary theory is that it squares with a plausible story about the function of our notion of truth, in that its axioms explain that function without the need for further analysis of 'what truth is'.

Not all variants of deflationism have this quality virtue, according to the redundancy performatives theory of truth, the pair of sentences, 'The proposition that p is true' and plain 'p's', has the same meaning and expresses the same statement as one and another, so it is a syntactic illusion to think that p is true' attributes any sort of property to a proposition (Ramsey, 1927 and Strawson, 1950). Yet in that case, it becomes hard to explain why we are entitled to infer 'The proposition that quantum mechanics are wrong is true' form 'Einstein's claim is the proposition that quantum mechanics are wrong. 'Einstein's claim is true'. For if truth is not property, then we can no longer account for the inference by invoking the law that if 'X', appears identical with 'Y' then any property of 'X' is a property of 'Y', and vice versa. Thus the redundancy/performatives theory, by identifying rather than merely correlating the contents of 'The proposition that p is true' and 'p, precludes the prospect of a good explanation of one on truth's most significant and useful characteristics. So, putting restrictions on our assembling claim to the weak is better, of its equivalence schema: The proposition that 'p' is true is and is only 'p'.

Support for deflationism depends upon the possibleness of showing that its axiom instances of the equivalence schema unsupplements by any further analysis, will suffice to explain all the central facts about truth, for example, that the verification of a proposition indicates its truth, and that true beliefs have a practical value. The first of these facts follows trivially from the deflationary axioms, for given a prior knowledge or the equivalence of 'p' and 'the proposition that p is true any reason to believe that 'p' becomes as equally good for reasons to believe that the preposition that 'p' is true. We can also explain the second fact in terms of the deflationary axioms, while but not quite so easily. Consider, to begin with, beliefs of the form that if I perform the act 'A', that my desires will be fulfilled. Notice that the psychological role of such a belief is, more or less, to cause the performance of 'A'. Such that as given, I do have the belief, then typically.

I will perform the act 'A'

Notice also that when the belief is true then, given the deflationary axioms, the performance of 'A' will in fact lead to the fulfilment of one's desires, i.e., If being true, then if I perform 'A', and my desires will be fulfilled.

Therefore, if it is true, then my desires will be fulfilled. So valuing the truth of beliefs of that form is quite treasonable. Nevertheless, inference has derived such beliefs from other beliefs and can be expected to be true if those other beliefs are true. So assigning a value to the truth of any belief that might be used in such an inference is reasonable.

To the extent that such deflationary accounts can be given of all the acts involving truth, then the explanatory demands on a theory of truth will be met by the collection of all statements like, 'The proposition that snow is white is true if and only if snow is white', and the sense that some deep analysis of truth is needed will be undermined.

Nonetheless, there are several strongly felt objections to deflationism. One reason for dissatisfaction is that the theory has an infinite number of axioms, and therefore cannot be completely written down. It can be described, as the theory whose axioms are the propositions of the fore 'p if and only if it is true that p', but not explicitly formulated. This alleged defect has led some philosophers to develop theories that show, first, how the truth of any proposition derives from the referential properties of its constituents, and second, how the referential properties of primitive constituents are determined (Tarski, 1943 and Davidson, 1969). However, assuming that all propositions including belief attributions remain controversial, law of nature and counterfactual conditionals depends for their truth values on what their constituents refer to implicate. In addition, there is no immediate prospect of a presentable, finite possibility of reference, so that it is far form clear that the infinite, list-like character of deflationism can be avoided.

Additionally, it is commonly supposed that problems about the nature of truth are intimately bound up with questions as to the accessibility and autonomy of facts in various domains: Questions about whether the facts can be known, and whether they can exist independently of our capacity to discover them (Dummett, 1978, and Putnam, 1981). One might reason, for example, that if 'T is true' nothing more than 'T will be verified', then certain forms of scepticism, specifically, those that doubt the correctness of our methods of verification, that will be precluded, and that the facts will have been revealed as dependent on human practices. Alternatively, it might be said that if truth were an inexplicable, primitive, non-epistemic property, then the fact that 'T' is true would be completely independent of 'us'. Moreover, we could, in that case, have no reason to assume that the propositions we believe in, that in adopting its property, so scepticism would be unavoidable. In a similar vein, it might be thought that as special, and perhaps undesirable features of the deflationary approach, is that truth is deprived of such metaphysical or epistemological implications.

Upon closer scrutiny, in that, it is to a considerable distance, for which I take to consider this argument, however, the displacing of distances bring along the clarity or clearness of appearance, especially of thought or style, and the lucidity for which there exists of the freedom that any of the accountable truths with which the consequences for regarding the accessibility or the autonomy of non-semantic matter. Although an account of truth may be expected to have such implication for which of the facts of the form that 'T is true', as it cannot be assumed without further argument that the same conclusions will apply to the fact 'T'. For it cannot be assumed that 'T' and 'T' are true' and is equivalent to one another given the account of 'true' that is being employed. Of course, if truth is defined in the way that the deflationist proposes, then the equivalence holds by definition. Nevertheless, if truth is defined by reference to some metaphysical or epistemological characteristic, then the equivalence schema is thrown into doubt, pending some demonstration that the trued predicate, in the sense assumed, will be satisfied in as far as there are thought to be epistemological problems hanging over 'T's' that do not threaten 'T is true', giving the needed demonstration will be difficult. Similarly, if 'truth' is so defined that the fact, 'T' is felt to be more, or less, independent of human practices than the fact that 'T is true', then again, it is unclear that the equivalence schema will hold. It would seem, therefore, that the attempt to base epistemological or metaphysical conclusions on a theory of truth must fail because in any such attempt the equivalence schema will be simultaneously relied on and undermined.

The most influential idea in the theory of meaning in the past hundred yeas is the thesis that meaning of an indicative sentence is given by its truth-conditions. On this conception, is to understand the sentence, that if it is known to know its truth-condition, the conception, was at first, clearly formulated by Frége (1848-1925), was developed in a distinctive way by the early Wittgenstein (1889-1951), and is a leading idea of Davidson (1917-). The conception has remained so central that those who offer opposing theories characteristically define their position by reference to it.

The conception of meaning as truth-conditions necessarily are not and should not be advanced as a complete account of meaning. For instance, one who understands a language must have some idea of the range of speech acts conventionally acted by the various types of a sentence in the language, and must have some idea of the significance of various kinds of speech acts. The claim of the theorist of truth-conditions should as an alternative is targeted on the notion of content: If two indicative sentences differ in what they strictly and literally say, then this difference is fully accounted for by the difference in their truth-conditions. Most basic to truth-conditions is simply of a statement that is the condition the world must meet if the statement is to be true. To know this condition is equivalent to knowing the meaning of the statement. Although this sounds as if it gives a solid anchorage for meaning, some of the security disappears when it turns out that the truth condition can only be defined by repeating the very same statement, as a truth condition of 'snow is white' is that snow is white, the truth condition of 'Britain would have capitulated had Hitler invaded' is the Britain would have capitulated had Hitler invaded. It is disputed whether this element of running-on-the-spot disqualifies truth conditions from playing the central role in a substantive theory of meaning.

Truth-conditional theories of meaning are sometimes opposed by the view that to know the meaning of a statement is to be able to use it in a network of inferences.

Whatever it is that would otherwise be mad as mere sounds and nonspecifying ascriptions into instruments of communication and understanding, as, perhaps, the philosophical problem of something that as denied, and to relate it to what we know of ourselves and the inherent perceptions of the world. Contributions to this study include the theory of 'speech acts' and the investigation of communication and the relationships between the words and ideas by which some persons express by a sentence are often a function of the environment, to which he is placed. For example, the disease I refer to by a term like 'arthritis' or the kind of tree I refer to as a 'maple' will be defined by criteria of which I know next to nothing. The raises the possibility of imagining two persons in alternatively differently environmental, but in which everything appears the same to each of them, but between them they define a space of philosophical problems. They are the essential components of understanding nd any intelligible proposition that is true must be capable of being understood. Such that which is expressed by an utterance or sentence, the proposition or claim made about the world may by extension, the content of a predicated or other sub-sentential component is what it contributes to the content of sentences that contain it. The nature of content is the cental concern of the philosophy of language.

In particularly, the problems of indeterminancy of translation, inscrutability of reference, language, predication, reference, rule following, semantics, translations, and the topics referring to subordinate headings associated with 'logic'. The loss of confidence in determinate meaning ('Each is another encoding') is an element common both to postmodern uncertainties in the theory of criticism, and to the analytic tradition that follows writers such as Quine (1908-). Still it may be asked, why should we suppose that fundamental epistemic notions should be keep an account of for in behavioural terms what grounds are there for supposing that 'p knows p' is a subjective matter in the prestigiousness of its statement between some subject statement and physical theory of physically forwarded of an objection, between nature and its mirror? The answer is that the only alternative seems to be to take knowledge of inner states as premises from which our knowledge of other things is normally implied, and without which our knowledge of other things is normally inferred, and without which knowledge would be ungrounded. However, it is not really coherent, and does not in the last analysis make sense, to suggest that human knowledge have foundations or grounds. It should be remembered that to say that truth and knowledge 'can only be judged by the standards of our own day' is not to say that it is less meaningful nor is it 'more 'cut off from the world, which we had supposed. Conjecturing it is as just' that nothing counts as justification, unless by reference to what we already accept, and that at that place is no way to get outside our beliefs and our oral communication so as to find some experiment with others than coherence. The fact is that the professional philosophers have thought it might be otherwise, since one and only they are haunted by the clouds of epistemological scepticism.

What Quine opposes as 'residual Platonism' is not so much the hypostasising of nonphysical entities as the notion of 'correspondence' with things as the final court of appeal for evaluating present practices. Unfortunately, Quine, for all that it is incompatible with its basic insights, substitutes for this correspondence to physical entities, and specially to the basic entities, whatever they turn out to be, of physical science. Nevertheless, when their doctrines are purified, they converge on a single claim. That no account of knowledge can depend on the assumption of some privileged relations to reality. Their work brings out why an account of knowledge can amount only to a description of human behaviour.

What, then, is to be said of these 'inner states', and of the direct reports of them that have played so important a role in traditional epistemology? For a person to feel is nothing else than for him to have an ability to make a certain type of non-inferential report, to attribute feelings to infants is to acknowledge in them latent abilities of this innate kind. Non-conceptual, nonlinguistic 'knowledge' of what feelings or sensations is like is attributively to beings on the basis of potential membership of our community. Infants and the more attractive animals are credited with having feelings on the basis of that spontaneous sympathy that we extend to anything humanoid, in contrast with the mere 'response to stimuli' attributed to photoelectric cells and to animals about which no one feels sentimentally. Supposing that moral prohibition against hurting infants is consequently wrong and the better-looking animals are; those moral prohibitions grounded' in their possession of feelings. The relation of dependence is really the other way round. Similarly, we could not be mistaken in supposing that a four-year-old child has knowledge, but no one-year-old, any more than we could be mistaken in taking the word of a statute that eighteen-year-old can marry freely but seventeen-year-old cannot. (There is no more 'ontological ground' for the distinction that may suit 'us' to make in the former case than in the later.) Again, such a question as 'Are robots' conscious?' A marginal decision of whether or not to treat robots as members of our linguistic community, this, of which is the insight as brought into a different state of philosophy was that of Hegel (1770-1831), in that the individual is parted from his society, such to be as another animal.

Willard van Orman Quine, the most influential American philosopher of the latter half of the 20th century, when after the wartime period in naval intelligence, punctuating the rest of his career with extensive foreign lecturing and travel. Quine's early works were based on mathematical logic, and issued in 'A System of Logistic' (1934), 'Mathematical Logic' (1940), and 'Methods of Logic' (1950), whereby, his collected papers carried in the form: 'Logical Point of View' (1953) that his philosophical importance became widely recognized. Quine's work dominated concern with problems of convention, meaning, and synonymies were glued, by 'Word and Object' (1960), in which the indeterminancy of radical translations first takes centre-stage. In this and many subsequent writings Quine takes a bleak view of the nature of the language with which we ascribe thoughts and beliefs to ourselves and others. These 'intentional idioms' resist smooth incorporation into the scientific world view, and Quine responds with scepticism toward them, not quite endorsing 'Eliminativism', but regarding them as second-rate idioms, unsuitable for describing strict and literal facts. For similar reasons he has consistently expressed suspicion of the logical and philosophical propriety of appeal to logical possibilities and possible worlds. The languages that are properly behaved and suitable for literal and true descriptions, are those attributed to the world of mathematics and science. The entities to which our best theories refer must be taken with full seriousness in our ontologies, although an empiricist. Quine thus supposes that the abstract objects of set theory are required by science, and therefore exist. In the theory of knowledge Quine associated with a 'holistic view' of verification, conceiving of a body of knowledge in terms of a web touching experience at the periphery, but with each point connected by a network of relations to other points.

Quine is also known for the view that epistemology should be naturalized, or conducted in a scientific spirit, with the object of investigation being the relationship, in human beings, between the voice of experience and the outputs of belief. Although Quine's approaches to the major problems of philosophy have been attacked as betraying undue 'scientism' and sometimes 'behaviourism', the clarity of his vision and the scope of his writing made him the major focus of Anglo-American work of the past forty years in logic, semantics, and epistemology. As well as the works cited his writings' cover 'The Ways of Paradox and Other Essays' (1966), 'Ontological Relativity and Other Essays' (1969), 'Philosophy of Logic' (1970), 'The Roots of Reference' (1974) and 'The Time of My Life: An Autobiography' (1985).

Coherence is a major player in the theatre of knowledge. There are cogence theories of belief, truth and justification, as these are to combine themselves in the various ways to yield theories of knowledge coherence theories of belief are concerned with the content of beliefs. Consider a belief you now have, the beliefs that you are reading a page in a book, in so, that what makes that belief the belief that it is? What makes it the belief that you are reading a page in a book than the belief that you have a monster in the garden?

One answer is that the belief has a coherent place or role in a system of beliefs, perception or the having the perceptivity that has its influence on beliefs. As, you respond to sensory stimuli by believing that you are reading a page in a book than believing that you have a monster in the garden. Belief has an influence on action, or its belief is a desire to act, if belief will differentiate the differences between them, that its belief is a desire or if you were to believe that you are reading a page than if you believed in something about a monster. Sortal perceptivals hold accountably the perceptivity and action that are indeterminate to its content if its belief is the action as if stimulated by its inner and latent coherence in that of your belief, however. The same stimuli may produce various beliefs and various beliefs may produce the same action. The role that gives the belief the content it has is the role it plays within a network of relations to other beliefs, some latently causal than others that relate to the role in inference and implication. For example, I infer different things from believing that I am reading a page in a book than from any other belief, justly as I infer about other beliefs.

The information of perceptibility and the output of an action supplement the central role of the systematic relations the belief has to other belief, but the systematic relations give the belief the specific contentual representation it has. They are the fundamental source of the content of belief. That is how coherence comes in. A belief has the representational content by which it does because of the way in which it coheres within a system of beliefs (Rosenberg, 1988). We might distinguish weak coherence theories of the content of beliefs from stronger coherence theories. Weak coherence theories affirm that coherence is one determinant of the representation given that the contents are of belief. Strong coherence theories of the content of belief affirm that coherence is the sole determinant of the contentual representations of belief.

When we turn from belief to justification, we confront a similar group of coherence theories. What makes one belief justified and another not? Again, there is a distinction between weak and strong theoretic principles that govern its theory of coherence. Weak theories tell 'us' that the ways in which a belief coheres with a background system of beliefs are one determinant of justification, other typical determinants being perception, memory, and intuitive 'projection', are, however strong theories, or dominant projections are in coherence to justification as solely a matter of how a belief coheres with a system of latent hierarchal beliefs. There is, nonetheless, another distinction that cuts across the distinction between weak and strong coherence theories between positive and negative coherence theory (Pollock, 1986). A positive coherence theory tells 'us' that if a belief coheres with a background system of belief, then the belief is justifiable. A negative coherence theory tells 'us' that if a belief fails to cohere with a background system of beliefs, then the belief is not justifiable. We might put this by saying that, according to the positivity of a coherence theory, coherence has the power to produce justification, while according to its being adhered by negativity, the coherence theory has only the power to nullify justification.

Least of mention, a strong coherence theory of justification is a formidable combination by which a positive and a negative theory tell 'us' that a belief is justifiable if and only if it coheres with a background system of inter-connectivity of beliefs. Coherence theories of justification and knowledge have most often been rejected for being unable to deal with an accountable justification toward the perceptivity upon the projection of knowledge (Audi, 1988, and Pollock, 1986), and, therefore, considering a perceptual example that will serve as a kind of crucial test will be most appropriate. Suppose that a person, call her Julie, and works with a scientific instrumentation that has a gauging measure upon temperatures of liquids in a container. The gauge is marked in degrees, she looks at the gauge and sees that the reading is 105 degrees. What is she justifiably to believe, and why? Is she, for example, justified in believing that the liquid in the container is 105 degrees? Clearly, that depends on her background beliefs. A weak coherence theorist might argue that, though her belief that she sees the shape 105 is immediately justified as direct sensory evidence without appeal to a background system, the belief that the location in the container is 105 degrees results from coherence with a background system of latent beliefs that affirm to the shaping perceptivity that its 105 as visually read to be 105 degrees on the gauge that measures the temperature of the liquid in the container. This is ascertained by the weak coherence description from one point of view, as to combine coherence with direct perceptivity as its evidence, in that the foundation of justification is to account for the justification of our beliefs.

A strong coherence theory would go beyond the claim of the weak coherence theory to affirm that the justification of all beliefs, including the belief that one sees the shaping to sensory data that holds accountably of a measure of 100, or even the more cautious belief that one sees a shape, resulting from the perceptivals of coherence theory, in that it coheres with a background system. One may argue for this strong coherence theory in a number of different ways, as one of a medium through which is to appeal to the coherence theory of contentual representations. If the content of the perceptual belief results from the relations of the belief to other beliefs in a network system of beliefs, then one may notably argue that the justification of perceptivity, that the belief is a resultant from which its relation of the belief to other beliefs, in the network system of beliefs is in argument for the strong coherence theory is that without any assumptive reason that the coherence theory of contentual beliefs, in as much as the supposed causes that only produce the consequences we expect. Consider the very cautious belief that I see a shape. How may the justifications for that perceptual belief are an existent result that is characterized of its material coherence with a background system of beliefs? What might the background system tell, that 'we' would justify that belief? Our background system contains a simple and primal theory about our relationship to the world and surrounding surfaces that we perceive as it is or should be believed. To come to the specific point at issue, we believe that we can tell a shape when we see one, completely differentiated its form as perceived to sensory data, that we are to trust of ourselves about such simple matters as whether we see a shape before 'us' or not, as in the acceptance of opening to nature the inter-connectivity between belief and the progression through which is acquired from past experiential conditions of application, and not beyond deception. Moreover, when Julie sees the believing desire to act upon what either coheres with a weak or strong coherence of theory, she shows that its belief, as a measurable quality or entity of 100, has the essence in as much as there is much more of a structured distinction of circumstance, which is not of those that are deceptive about whether she sees that shape or sincerely does not see of its shaping distinction, however. Visible light is good, and the numeral shapes are large, readily discernible and so forth. These are beliefs that Trust has single handedly authenticated reasons for justification. Her successive malignance to sensory access to data involved is justifiably a subsequent belief, in that with those beliefs, and so she is justified and creditable.

The philosophical; problems include discovering whether belief differs from other varieties of assent, such as 'acceptance' discovering to what extent degrees of belief is possible, understanding the ways in which belief is controlled by rational and irrational factors, and discovering its links with other properties, such as the possession of conceptual or linguistic skills. This last set of problems includes the question of whether prelinguistic infants or animals are properly said to have beliefs.

Thus, we might think of coherence as inference to the best explanation based on a background system of beliefs, since we are not aware of such inferences for the most part, the inferences must be interpreted as unconscious inferences, as information processing, based on or finding the background system that proves most convincing of acquiring its act and used from the motivational force that its underlying and hidden desire are to do so. One might object to such an account on the grounds that not all justifiable inferences are self-explanatory, and more generally, the account of coherence may, at best, is ably successful to competitions that are based on background systems (BonJour, 1985, and Lehrer, 1990). The belief that one sees a shape in that with the claim that one does not, with the claim that one is deceived, and the others are sceptical objections. The background system of beliefs informs one that one is acceptingly trustworthy and enables one to meet the objections. A belief coheres with a background system just in case it enables one to meet the sceptical objections and in the way justifies one in the belief. This is a standard strong coherence theory of justification (Lehrer, 1990).

Illustrating the relationship between positive and negative coherence theories in terms of the standard coherence theory is easy. If some objection to a belief cannot be met in terms of the background system of beliefs of a person, then the person is not justified in that belief. So, to return to Julie, suppose that she has been told that a warning light has been installed on her gauge to tell her when it is not functioning properly and that when the red light is on, the gauge is malfunctioning. Suppose that when she sees the reading of 105, she also sees that the red light is on. Imagine, finally, that this is the first time the red light has been on, and, after years of working with the gauge, Julie, who has always placed her trust in the gauge, believes what the gauge tells her, that the liquid in the container is at 105 degrees. Though she believes what she reads is at 105 degree's is not a justified belief because it fails to cohere with her background belief that the gauge is malfunctioning. Thus, the negative coherence theory tells 'us' that she is not justified in her belief about the temperature of the contents in the container. By contrast, when the red light is not illuminated and the background system of trust tells her that under such conditions that gauge is a trustworthy indicator of the temperature of the liquid in the container, then she is justified. The positive coherence theory tells 'us' that she is justified in her belief because her belief coheres with her background system of trust tells she that under such conditions that gauge is a trustworthy indicator of the temperature of the liquid in the container, then she is justified. The positive coherence theory tells 'us' that she is justified in her belief because her belief coheres with her background system continues as a trustworthy system.

The foregoing sketch and illustration of coherence theories of justification have a common feature, namely, that they are what is called internalistic theories of justification what makes of such a view are the absence of any requirement that the person for whom the belief is justified have any cognitive access to the relation of reliability in question. Lacking such access, such a person will usually, have no reason for thinking the belief is true or likely to be true, but will, on such an account, are none the lesser to appear epistemologically justified in accepting it. Thus, such a view arguably marks a major break from the modern epistemological traditions, which identifies epistemic justification with having a reason, perhaps even a conclusive reason, for thinking that the belief is true. An epistemologist working within this tradition is likely to feel that the externalist, than offering a competing account of the same concept of epistemic justification with which the traditional epistemologist is concerned, has simply changed the subject.

They are theories affirming that coherence is a matter of internal relations between beliefs and that justification is a matter of coherence. If, then, justification is solely a matter of internal relations between beliefs, we are left with the possibility that the internal relations might fail to correspond with any external reality. How, one might object, can be to assume the including of interiority. There is of something as a subjective notion of justification, for bridging the gap between mere true beliefs, and which might be no more than a lucky guess, as for the knowledge that must be grounded in some connection between the internal subjective conditions and external objective realities?

The answer is that it cannot and that something more than justified true belief is required for knowledge. This result has, however, been established quite apart from consideration of coherence theories of justification. What are required maybes put by saying that the justification that one must be undefeated by errors in the background system of beliefs? Justification is undefeated by errors just in case any correction of such errors in the background system of belief would sustain the justification of the belief on the basis of the corrected system. So knowledge, on this sort of positivity is acclaimed by the coherence theory, which is the true belief that coheres with the background belief system and corrected versions of that system. In short, knowledge is true belief plus justification resulting from coherence and undefeated by error (Lehrer, 1990). The connection between internal subjective conditions of belief and external objectivity are from which reality's result from the required correctness of our beliefs about the relations between those conditions and realities. In the example of Julie, she believes that her internal subjectivity to conditions of sensory data in which the experience and perceptual beliefs are connected with the external objectivity in which reality is the temperature of the liquid in the container in a trustworthy manner. This background belief is essential to the justification of her belief that the temperature of the liquid in the container is 105 degrees, and the correctness of that background belief is essential to the justification remaining undefeated. So our background system of beliefs contains a simple theory about our relation to the external world that justifies certain of our beliefs that cohere with that system. For instance, such justification to convert to knowledge, that theory must be sufficiently free from error so that the coherence is sustained in corrected versions of our background system of beliefs. The correctness of the simple background theory provides the connection between the internal condition and external reality.

The coherence theory of truth arises naturally out of a problem raised by the coherence theory of justification. The problem is that anyone seeking to determine whether she has knowledge is confined to the search for coherence among her beliefs. The sensory experiences she has been deaf-mute until they are represented in the form of some perceptual belief. Beliefs are the engines that pull the train of justification. Nevertheless, what assurance do we have that our justification is based on true beliefs? What justification do we have that any of our justifications are undefeated? In fear that we might have none, for that of our belief, might that the artifacts of some deceptive demon or scientist leads to the quest in the reduction of truth as to some form, as, perhaps, an idealized form of justification (Rescher, 1973, and Rosenberg, 1980), such that it would close the threatening sceptical gap between justification and truth. Suppose that a belief is true if and only if it is justifiable of some person. For such a person there would be no gap between justification and truth or between justification and undefeated justification. Truth would be coherence with some ideal background system of beliefs, perhaps one expressing a consensus among systems or some consensus among belief systems or some convergence toward a consensus. Such a view is theoretically attractive for the reduction it promises, but it appears open to profound objectification. One is that there is a consensus that we can all be wrong about at least some matters, for example, about the origins of the universe. If there is a consensus that we can all be wrong about something, then the consensual belief system rejects the equation of truth with the consensus. Consequently, the equation of truth with coherence with a consensual belief system is itself incoherently.

Coherence theories of the content of our beliefs and the justification of our beliefs themselves cohere with our background systems but coherence theories of truth do not. A defender of Coherentism must accept the logical gap between justified belief and truth, but may believe that our capacities suffice to close the gap to yield knowledge. That view is, at any rate, a coherent one.

What makes a belief justified and what makes a true belief knowledge? Thinking that whether a belief deserves one of these appraisals is natural depends on what causal subject to have the belief. In recent decades a number of epistemologists have pursed this plausible idea with a variety of specific proposals. Some causal theories of knowledge have it that a true belief that 'p' is knowledge just in case it has the right causal connection to the fact that 'p'. Such a criterion can be applied only to cases where the fact that 'p' is a sort that can enter causal relations, this seems to exclude mathematically and other necessary facts and perhaps any fact expressed by a universal generalization, and proponents of this sort of criterion have usually of this sort of criterion have usually supposed that it is limited to perceptual knowledge of particular facts about the subject's environment.

For example, Armstrong (1973), proposed that a belief of the form 'This (perceived) object is F' is (non-inferential) knowledge if and only if the belief is a completely reliable sign that the perceived object is 'F', that is, the fact that the object is 'F' contributed to causing the belief and its doing so depended on properties of the believer such that the laws of nature dictated that, for any subject ',' is to occur, and so thus a perceived object of 'y', if those properties are for 'us' to believe that 'y' is 'F', then 'y' is 'F'. (Dretske, 1981) offers a similar account, in terms of the belief's being caused by a signal received by the perceiver that carries the information that the object is 'F'.

This sort of condition fails, however, to be sufficient for non-inferential perceptual knowledge because it is compatible with the belief's being unjustified, and an unjustifiable belief cannot be knowledge. For example, suppose that your mechanisms for colour perception are working well, but you have been given good reason to think otherwise, to think, say, that the substantive primary colours that are perceivable, that things look chartreuse to you and chartreuse things look magenta. If you fail to heed these reasons you have for thinking that your colour perception or sensory data is a way. Believing in a 'thing', which looks to blooms of vividness that you are to believe of its chartreuse, your belief will fail to be justified and will therefore fail to be knowledge, even though it is caused by the thing's being magenta in such a way as to be a completely reliable sign, or to carry the information, in that the thing is magenta.

One could fend off this sort of counterexample by simply adding to the causal condition the requirement that the belief be justified, buy this enriched condition would still be insufficient. Suppose, for example, that nearly all people, but not in you, as it happens, causes the aforementioned aberration in colour perceptions. The experimenter tells you that you have taken such a drug but then says, 'no, hold off a minute, the pill you took was just a placebo', suppose further, that this last thing the experimenter tells you is false. Her telling you that it was a false statement, and, again, telling you this gives you justification for believing of a thing that looks a subtractive primary colour to you that it is a sensorial primary colour, in that the fact you were to expect that the experimenters last statements were false, making it the case that your true belief is not knowledgeably correct, thought as though to satisfy its causal condition.

Goldman (1986) has proposed an importantly different causal criterion namely, that a true belief is knowledge, if it is produced by a type of process that is 'globally' and 'locally' reliable. Causing true beliefs is sufficiently high is globally reliable if its propensity. Local reliability has to do with whether the process would have produced a similar but false belief in certain counterfactual situations alternative to the actual situation. This way of marking off true beliefs that are knowledge does not require the fact believed to be casually related to the belief, and so it could in principle apply to knowledge of any kind of truth.

Goldman requires that global reliability of the belief-producing process for the justification of a belief, he requires it also for knowledge because justification is required for knowledge, in what requires for knowledge but does not require for justification, which is locally reliable. His idea is that a justified true belief is knowledge if the type of process that produced it would not have produced it in any relevant counterfactual situation in which it is false. The relevant alternative account of knowledge can be motivated by noting that other concepts exhibit the same logical structure. Two examples of this are the concept 'flat' and the concept 'empty' (Dretske, 1981). Both appear to be absolute concepts-A space is empty only if it does not contain anything and a surface is flat only if it does not have any bumps. However, the absolute character of these concepts is relative to a standard. In the case of 'flat', there is a standard for what counts as a bump and in the case of 'empty', there is a standard for what counts as a thing. To be flat is to be free of any relevant bumps and to be empty is to be devoid of all relevant things.

Nevertheless, the human mind abhors a vacuum. When an explicit, coherent world-view is absent, it functions on the basis of a tactic one. A tactic world-view is not subject to a critical evaluation, and it can easily harbour inconsistencies. Indeed, our tactic set of beliefs about the nature of reality is made of contradictory bits and pieces. The dominant component is a leftover from another period, the Newtonian 'clock universe' still lingers as we cling to this old and tired model because we know of nothing else that can take its place. Our condition is the condition of a culture that is in the throes of a paradigm shift. A major paradigm shift is complex and difficult because a paradigm holds 'us captive: We see reality through it, as through coloured glasses, but we do not know that, we are convinced that we see reality as it is. Hence the appearance of a new and different paradigm is often incomprehensible. To someone raised believing that the Earth is flat, the suggestion that the Earth is spherical would seem preposterous: If the Earth were spherical, would not the poor antipodes fall 'down' into the sky?

Yet, as we face a new millennium, we are forced to face this challenge. The fate of the planet is in question, and it was brought to its present precarious condition largely because of our trust in the Newtonian paradigm. As Newtonian world-view has to go, and, if one looks carefully, the main feature of the new, emergent paradigm can be discerned. The search for these features is what was the influence of a fading paradigm. All paradigms include subterranean realms of tactic assumptions, the influence of which outlasts the adherence to the paradigm itself.

The first line of exploration suggests the 'weird' aspects of the quantum theory, with fertile grounds for our feeling of which should disappear in inconsistencies with the prevailing world-view. This feeling is in replacing by the new one, i.e., if one believes that the Earth is flat, the story of Magellan's travels is quite puzzling: How travelling due west is possible for a ship and, without changing direct. Arrive at its place of departure? Obviously, when the flat-Earth paradigm is replaced by the belief that Earth is spherical, the puzzle is instantly resolved.

The founders of Relativity and quantum mechanics were deeply engaging but incomplete, in that none of them attempted to construct a philosophical system, however, that the mystery at the heart of the quantum theory called for a revolution in philosophical outlooks. During which time, the 1920's, when quantum mechanics reached maturity, began the construction of a full-blooded philosophical system that was based not only on science but on nonscientific modes of knowledge as well. As, the fading influence drawn upon the paradigm goes well beyond its explicit claim. We believe, as the scenists and philosophers did, that when we wish to find out the truth about the universe, nonscientific nodes of processing human experiences can be ignored, poetry, literature, art, music are all wonderful, but, in relation to the quest for knowledge of the universe, they are irrelevant. Yet, it was Alfred North Whitehead who pointed out the fallacy of this speculative assumption. In this, as well as in other aspects of thinking of some reality in which are the building blocks of reality are not material atoms but 'throbs of experience'. Whitehead formulated his system in the late 1920s, and yet, as far as I know, the founders of quantum mechanics were unaware of it. It was not until 1963 that J. M. Burgers pointed out that its philosophy accounts very well for the main features of the quanta, especially the 'weird ones', enabling as in some aspects of reality is 'higher' or 'deeper' than others, and if so, what is the structure of such hierarchical divisions? What of our place in the universe? Finally, what is the relationship between the great aspiration within the lost realms of nature? An attempt to endow 'us' with a cosmological meaning in such a universe seems totally absurd, and, yet, this very universe is just a paradigm, not the truth. When you reach its end, you may be willing to join the alternate view as accorded to which, surprisingly bestow 'us' with what is restored, although in a post-postmodern context.

The philosophical implications of quantum mechanics have been regulated by subjective matter's, as to emphasis the connections between what I believe, in that investigations of such interconnectivity are anticipatorially the hesitations that are an exclusion held within the western traditions, however, the philosophical thinking, from Plato to Platinous had in some aspects of interpretational presentation of her expression of a consensus of the physical community. Other aspects are shared by some and objected to (sometimes vehemently) by others. Still other aspects express my own views and convictions, as turning about to be more difficult that anticipated, discovering that a conversational mode would be helpful, but, their conversations with each other and with me in hoping that all will be not only illuminating but finding to its read may approve in them, whose dreams are dreams among others than themselves.

These examples make it seem likely that, if there is a criterion for what makes an alternative situation relevant that will save Goldman's claim about reliability and the acceptance of knowledge, it will not be simple.

The interesting thesis that counts as a causal theory of justification, as having in mind, that of 'causal theory', intended by the belief that is justified just in case it was of a product of a typical process that is 'globally' reliable, that is, its propensity to produce true beliefs, that in their finding definitions may prove as favourable in the bringing of a closer understanding to the proportional belief's and to what it produces, but would produce as much as opportunity allows. That for being true, is sufficiently a belief that acquires favourable epistemic status, in that by having some kind of reliable linkage to the truth. Variations of this view have been advanced for both knowledge and justified belief. The first formulations of are reliably in its account of knowing appeared in if not by F.P. Ramsey (1903-30) who made important contributions to mathematical logic, probability theory, the philosophy of science and economics. Instead of saying that quarks have such-and-such properties, the Ramsey sentence says that it is moderately something that has those properties. If the process is repeated for all of the theoretical terms, the sentence gives the 'topic-neutral' structure of the theory, but removes any implication that we know what the term so covered have as a meaning. It leaves open the possibility of identifying the theoretical item with whatever, but it is that best fits the description provided, thus, substituting the term by a variable, and existentially qualifying into the result. Ramsey was one of the first thinkers to accept a 'redundancy theory of truth', which he combined its radical views of the function of many kinds of the proposition. Neither generalizations, nor causal propositions, not those treating probabilities or ethics, described facts, but each has a different specific function in our intellectual commentators on the early works of Wittgenstein, and his continuing friendship with the latter liked to Wittgenstein's return to Cambridge and to philosophy in 1929.

The most sustained and influential application of these ideas were in the philosophy of mind, or brain, as Ludwig Wittgenstein (1889-1951) whom Ramsey persuaded that remained work for him to do, the way that is most undoubtedly was of an appealingly charismatic figure in a 20th-century philosophy, living and writing with a power and intensity that frequently overwhelmed his contemporaries and readers, the early period is centred on the 'picture theory of meaning' according to which sentence represents a state of affairs by being a kind of picture or model of it. Containing the elements that were and corresponding to those of the state of affairs and structural form that mirror that a structure of the state of affairs is that it represents. All logic complexity is reduced to that of the 'propositional calculus, and all propositions are 'truth-function' of atomic or basic propositions.

In the layer period the emphasis shifts dramatically to the actions of people and the role linguistic activities play in their lives. Thus, whereas in the 'Tractatus' language is placed in a static, formal relationship with the world, in the later work Wittgenstein emphasis its use in the context of standardized social activities of ordering, advising, requesting, measuring, counting, excising concerns for each other, and so on. These different activities are thought of as so many 'language games' that together make or a form of life. Philosophy typically ignores this diversity, and in generalizing and abstracting distorts the real nature of its subject-matter. In addition to the 'Tractatus' and the 'investigations' collection of Wittgenstein's work published posthumously include 'Remarks on the Foundations of Mathematics' (1956), 'Notebooks' (1914-1916) (1961), 'Pholosophische Bemerkungen' (1964), Zettel, 1967, and 'On Certainty' (1969).

Clearly, there are many forms of reliabilism. Just as there are many forms of 'Foundationalism' and 'coherence'. How is reliabilism related to these other two theories of justification? It is usually regarded as a rival. This is aptly so, in as far as Foundationalism and Coherentism traditionally focussed on purely evidential relations than psychological processes, but reliabilism might also be offered as a deeper-level theory, subsuming some of the precepts of either Foundationalism or Coherentism. Foundationalism says that there are 'basic' beliefs, which acquire justification without dependence on inference, reliabilism might rationalize this indicating that the basic beliefs are formed by reliable non-inferential processes. Coherence stresses the primary of systematicity in all doxastic decision-making. Reliabilism might be rationalized by pointing to some of the increases as warranted by reliability, in that grows by addition, from systematicity which it is consequently termed as reliabilism, as this can be a complemented form of Foundationalism and of the coherence, than it is to be completed with them.

These examples make it seem likely that, if there is a criterion for what makes an alternate situation relevant that will save Goldman's claim about local reliability and knowledge. Will did not be simple. The interesting thesis that counts as a causal theory of justification, in the making of 'causal theory' intended for the belief as it is justified in case it was produced by a type of process that is 'globally' reliable, that is, its propensity to produce true beliefs that can be defined, to a well-thought-of approximation, as the proportion of the beliefs it produces, or would produce where it used as much as opportunity allows, that is true is sufficiently relializable. Variations of this view have been advanced for both knowledge and justified belief, its first formulation of a reliability account of knowing appeared in the notation from F.P.Ramsey (1903-30). The theory of probability, he was the first to show how a 'personalists theory' of development, as based on a precise behavioural notion of preference and expectation. In the philosophy of language. Much of Ramsey's work was directed at saving classical mathematics from 'intuitionism', or what he called the 'Bolshevik menace of Brouwer and Weyl. In the theory of probability he was the first to show how a personalists theory is possibly to develop on the grounds that precise behavioural notation of preference and expectation. In the philosophy of language, Ramsey was one of the first thankers, which he combined with radical views of the function of many kinds of a proposition. Neither generalizations, nor causal propositions, nor those treating probability or ethics, describe facts, but each has a different specific function in our intellectual economy. Ramsey was one of the earliest commentators on the early work of Wittgenstein, and his continuing friendship with Wittgenstein.

Ramsey's sentence theory is the sentence generated by taking all the sentences affirmed in a scientific theory that use some term, e.g., 'quark'. Replacing the term by a variable, and existentially quantifying into the amounting results for saying that quarks have such-and-such properties, the Ramsey sentence says that there is something that has those properties. If the process is repeated for all of a group of the theoretical terms, the sentence gives the 'topic-neutral' structure of the theory, but removes any implication that we know what the term so treated characterized. It leaves open the possibility of identifying the theoretical item with whatever, and it is that best fits the description provided. Virtually, all theories of knowledge share an externalist component in requiring truth as a condition or for something known. Reliabilism goes further, however, in trying to capture additional conditions for knowledge by ways of a nomic, counterfactual or other such 'external' relations between belief and truth. Closely allied to the nomic sufficiency account of knowledge, primarily due to Dretshe (1971, 1981), A. I. Goldman (1976, 1986) and R. Nozick (1981), but the core of this approach is that X's belief that 'p' qualifies as knowledge just in case 'X' believes 'p', because of reasons that would not obtain unless 'p's' being true, or because of a process or method that would not yield belief in 'p' if 'p' were not true. For example, 'X' would not have its current reasons for believing there is a telephone before it. Perhaps, would it not come to believe that this in the way it suits the purpose, thus, there is a differentiable fact of a reliable guarantor that the belief's bing true. A stouthearted and valiant counterfactual approach says that 'X' knows that 'p' only if there is no 'relevant alternative' situation in which 'p' is false but 'X' would still believe that a proposition 'p'; must be sufficient to eliminate all the alternatives too 'p' where an alternative to a proposition 'p' is a proposition incompatible with 'p'? . That in, one's justification or evidence for 'p' must be sufficient for one to know that every alternative too 'p' is false. This element of our evolving thinking, about which knowledge is exploited by sceptical arguments. These arguments call our attentions to alternatives that our evidence sustains itself with no elimination. The sceptic inquires to how we know that we are not seeing a cleverly disguised mule. While we do have some evidence against the likelihood of such as deception, intuitively knowing that we are not so deceived is not strong enough for 'us'. By pointing out alternate but hidden points of nature, in that we cannot eliminate, as well as others with more general application, as dreams, hallucinations, and so forth, the sceptic appears to show that every alternative is seldom. If ever, satisfied.

This conclusion conflicts with another strand in our thinking about knowledge, in that we know many things. Thus, there is a tension in our ordinary thinking about knowledge ~. We believe that knowledge is, in the sense indicated, an absolute concept and yet, we also believe that there are many instances of that concept.

If one finds absoluteness to be too central a component of our concept of knowledge to be relinquished, one could argue from the absolute character of knowledge to a sceptical conclusion (Unger, 1975). Most philosophers, however, have taken the other course, choosing to respond to the conflict by giving up, perhaps reluctantly, the absolute criterion. This latter response holds as sacrosanct our commonsense belief that we know many things (Pollock, 1979 and Chisholm, 1977). Each approach is subject to the criticism that it preserves one aspect of our ordinary thinking about knowledge at the expense of denying another. The theory of relevant alternatives can be viewed as an attempt to provide a more satisfactory response to this tension in our thinking about knowledge. It attempts to characterize knowledge in a way that preserves both our belief that knowledge is an absolute concept and our belief that we have knowledge.

Just as space, the classical questions include: Is space real? Is it some kind of mental construct or artefact of our ways of perceiving and thinking? Is it 'substantively' or purely relational? As according to substantivalism, where its particular point in space is an objective thing consisting of points or regions at which, or in which, things are located. Opposed to this is relationalism, according to which the only thing that is real about space is the spatial (and temporal) relations between physical objects. Substantivalism was advocated by Clarke speaking for Newton, and relationalism by Leibniz, in their famous correspondence, and the debate continues today. There is also an issue whether the measure of space and time are objective e, or whether an element of convention enters them. Whereby, the influential analysis with David Lewis to suggest that the regularity hold as a matter of convention when it solves a problem of coordinating in a group, for that it is to be the benefits of each member to conform to the regularity, providing the others do so. Any number of solutions to such a problem may exist, for example, it is to the advantages of each of us to drive on the same side of the road as others, but indifferent whether we all drive o the right or the left. One solution or another may emerge for a variety of reasons. It is notable that on this account convections may arise naturally; they do not have to be the result of specific agreement. This frees the notion for use in thinking about such things as the origin of language or of political society.

Finding to a theory that magnifies the role of decisions, or free selection from among equally possible alternatives, in order to show that what appears to be objective or fixed by nature is in fact an artefact of human convention, similar to conventions of etiquette, or grammar, or law. Thus one might suppose that moral rules owe more to social convention than to anything imposed from outside, or supposedly inexorable necessities are in fact the shadow of our linguistic conventions. The disadvantage of conventionalism is that it must show that alternative, equally workable e conventions could have been adopted, and it is often easy to believe that, for example, if we hold that some ethical norm such as respect for promises or property is conventional, we ought to be able to show that human needs would have been equally well satisfied by a system involving a different norm, and this may be hard to establish.

A convention also suggested by Paul Grice (1913-88) directing participants in conversation to pay heed to an accepted purpose or direction of the exchange. Contributions construed in the absence of this attention are liable to be rejected for other reasons than straightforward falsity: Something true but unhelpful or inappropriate may meet with puzzlement or rejection. We can thus never infer fro the fact that it would be inappropriate to say something in some circumstance that what would be aid, were we to say it, would be false. This inference was frequently and in ordinary language philosophy, it being argued, for example, that since we do not normally say 'there sees to be a barn there' when there is unmistakably a barn there, it is false that on such occasions there seems to be a barn there.

There are two main views on the nature of theories. According to the 'received view' theories are partially interpreted axiomatic systems, according to the semantic view, a theory is a collection of models (Suppe, 1974). However, a natural language comes ready interpreted, and the semantic problem is no that of the specification but of understanding the relationship between terms of various categories (names, descriptions, predicates, adverbs . . .) and their meanings. An influential proposal is that this relationship is best understood by attempting to provide a 'truth definition' for the language, which will involve giving terms and structure of different kinds have on the truth-condition of sentences containing them.

The axiomatic method . . . as, . . . a proposition lid down as one from which we may begin, an assertion that we have taken as fundamental, at least for the branch of enquiry in hand, for which the axiomatic method is that of defining as a set of such propositions, and the 'proof procedures' or finding of how a proof ever gets started. Suppose I have as premises (1)'p' and (2) 'p' 'q', can I infer q? Only, it seems, if I am sure of, (3) (p & p q) ®q, can I then infer q? Only, it seems, if I am sure that (4) (p & p q) ®q. For each new axiom (N) I need a further axiom (N + 1) telling me that the set so far implies 'q' and the regress never stops. The usual solution is to treat a system as containing not only axioms, but also rules of reference, allowing movement fro the axiom. The rule 'modus ponens' allow us to pass from the first two premises to q. Charles Dodgson Lutwidge (1832-98) better known as Lewis Carroll's puzzle shows that it is essential to distinguish two theoretical categories, although there may be choice about which to put in which category.

This type of theory (axiomatic) usually emerges as a body of (supposes) truths that are not nearly organized, making the theory difficult to survey or study a whole. The axiomatic method is an idea for organizing a theory (Hilbert 1970): one tries to select from among the supposed truths a small number from which all others can be seen to be deductively inferable. This makes the theory rather more tractable since, in a sense, all the truths are contained in those few. In a theory so organized, the few truths from which all others are deductively inferred are called axioms. In that, jus t as algebraic and differential equations, which were used to study mathematical and physical processes, could they be made mathematical objects, so axiomatic theories, like algebraic and differential equations, which are of representing physical processes and mathematical structures, could be made objects of mathematical investigation.

In the traditional (as in Leibniz, 1704), many philosophers had the conviction that all truths, or all truths about a particular domain, followed from a few principles. These principles were taken to be either metaphysically prior or epistemologically prior or in the fist sense, they were taken to be entities of such a nature that what exists is 'caused' by them. When the principles were taken as epistemologically prior, that is, as axioms, they were taken to be epistemologically privileged either, e.g., self-evident, not needing to be demonstrated or (again, inclusive 'or') to be such that all truths do follow from them (by deductive inferences). Gödel (1984) showed that treating axiomatic theories as themselves mathematical objects, that mathematics, and even a small part of mathematics, elementary number theory, could not be axiomatized, that, more precisely, any class of axioms that in such that we could effectively decide, of any proposition, whether or not it was in the class, would be too small to capture all of the truths.

For the propositional calculus this turns into the question of whether the proof theory delivers as theorems all and only tautologies. There are many axiomatizations of the propositional calculus that are consistently succumbing to be complete. Gödel proved in 1929 that first-order predicate calculus is complete: any formula that is true under every interpretation is a theorem of the calculus.

The propositional calculus or logical calculus whose expressions are letter present sentences or propositions, and constants representing operations on those propositions to produce others of higher complexity. The operations include conjunction, disjunction, material implication and negation (although these need not be primitive). Propositional logic was partially anticipated by the Stoics but researched maturity only with the work of Frége, Russell, and Wittgenstein.

The concept had been introduced by Frége of a function taking a number of names as arguments, and delivering one proposition as the value. The idea is that 'χ' loves 'y' is a propositional function, which yields the proposition 'John' loves 'Mary' from those two arguments (in that order). A propositional function is therefore roughly equivalent to a property or relation. In Principia Mathematica, Russell and Whitehead take propositional functions to be the fundamental function, since the theory of descriptions could be taken as showing that other expressions denoting functions are incomplete symbols.

Keeping in mind, the two classical truth-values that a statement, proposition, or sentence can take are supposed in classical (two-valued) logic, that each statement has one of these values, and none has both. A statement is then false if and only if it is not true. The basis of this scheme is that to each statement there corresponds a determinate truth condition, or way the world must be for it to be true, and otherwise false. Statements may be felicitous or infelicitous in other dimensions (polite, misleading, apposite, witty, and so forth) but truth is the central normative governing assertion. Considerations of vagueness may introduce greys into the black-and-white schemes, for which of the issues is falsity as the only way of failing to be true.

Formally, it is nonetheless, that any suppressed premise or background framework of thought necessary to make an argument valid, or a position tenable. More formally, a presupposition has been defined as a proposition whose truth is necessary for either the truth or the falsity of another statement. Thus, if 'p' presupposes 'q', 'q' must be true for p to be either true or false. In the theory of knowledge of Robin George Collingwood (1889-1943), any propositions capable of truth or falsity stand on a bed of 'absolute presuppositions' which are not properly capable of truth or falsity, since a system of thought will contain no way of approaching such a question. It was suggested by Peter

Strawson (1919-), in opposition to Russell's theory of 'definite' descriptions, that 'there exists a King of France' is a presupposition of 'the King of France is bald', the latter being neither true, nor false, if there is no King of France. It is, however, a little unclear of whether or not the idea is that no statement at all is made in such a case, or whether a statement is made, but fails of being true or false. The former option preserves classical logic, since we can still say that every statement is either true or false, but the latter does not, since in classical logic the law of 'bivalence' holds, and ensures that nothing at all is presupposed for any proposition to be true or false. The introduction of presupposition therefore that either a third truth-value is found, 'intermediate' between truth and falsity, or classical logic is preserved, but it is impossible to tell whether a particular sentence expresses a proposition that is a candidate for truth ad falsity, without knowing more than the formation rules of the language. Each suggestion carries costs, and there is some consensus that at least where definite descriptions are involved, examples like the one given are equally well handed by regarding the overall sentence false when the existence claim fails.

A proposition may be true or false it is said to take the truth-value true, and if the latter the truth-value false. The idea behind the term is the analogy between assigning a propositional variable one or other of these values, as a formula of the propositional calculus, and assigning an object as the value of many other variable. Logics with intermediate values are called many-valued logics. Then, a truth-function of a number of propositions or sentences is a function of them that has a definite truth-value, depends only on the truth-values of the constituents. Thus (p & q) is a combination whose truth-value is true when 'p' is true and 'q' is true, and false otherwise, $\neg$ p is a truth-function of 'p', false when 'p' is true and true when 'p' is false. The way in which the value of the whole is determined by the combinations of values of constituents is presented in a truth table.

In whatever manner, truths of fact cannot be reduced to any identity and our only way of knowing them is empirically, by reference to the facts of the empirical world.

A proposition is knowable deductively if it can be known without experience of the specific course of events in the actual world. It may, however, be allowed that some experience is required to acquire the concepts involved in a deductive proposition. Some thing is knowable only empirical if it can be known deductively. The distinction given that one of the fundamental problem areas of epistemology. The category of deductive propositions is highly controversial, since it is not clear how pure thought, unaided by experience, can give rise to any knowledge at all, and it has always been a concern of empiricism to deny that it can. The two great areas in which it seems to be so are logic and mathematics, so empiricists have commonly tried to show either that these are not areas of real, substantive knowledge, or that in spite of appearances their knowledge that we have in these areas is actually dependent on experience. The former line tries to show sense trivial or analytic, or matters of notation conventions of language. The latter approach is particularly y associated with Quine, who denies any significant slit between propositions traditionally thought of as speculatively, and other deeply entrenched beliefs that occur in our overall view of the world.

Another contested category is that of speculative concepts, supposed to be concepts that cannot be 'derived' from experience, but are presupposed in any mode of thought about the world, time, substance, causation, number, and self are candidates. The need for such concepts, and the nature of the substantive a prior knowledge for that of them, arises to central concerns as owing to Kant's, Critique of Pure Reason.

Likewise, since their denial does not involve a contradiction, there is merely contingent: There could have been in other ways a hold of the actual world, but not every possible one. Some examples are 'Caesar crossed the Rubicon' and 'Leibniz was born in Leipzig', as well as propositions expressing correct scientific generalizations. In Leibniz's view truths of fact rest on the principle of sufficient reason, which is a reason why it is so. This reason is that the actual world (by which he the total collection of things past, present and future) is better than any other possible world and therefore created by God. The foundation of his thought is the conviction that to each individual there corresponds a complete notion, knowable only to God, from which is deducible all the properties possessed by the individual at each moment in its history. It is contingent that God actualizes te individual that meets such a concept, but his doing so is explicable by the principle of 'sufficient reason', whereby God had to actualize just that possibility in order for this to be the best of all possible worlds. This thesis is subsequently lampooned by Voltaire (1694-1778), in whom of which was prepared to take refuge in ignorance, as the nature of the soul, or the way to reconcile evil with divine providence.

In defending the principle of sufficient reason sometimes described as the principle that nothing can be so without there being a reason why it is so. But the reason has to be of a particularly potent kind: eventually it has to ground contingent facts in necessities, and in particular in the reason an omnipotent and perfect being would have for actualizing one possibility than another. Among the consequences of the principle is Leibniz's relational doctrine of space, since if space were an infinite box there could be no reason for the world to be at one point in rather than another, and God placing it at any point violate the principle. In Abelard's (1079-1142), as in Leibniz, the principle eventually forces te recognition that the actual world is the best of all possibilities, since anything else would be inconsistent with the creative power that actualizes possibilities.

If truth consists in concept containment, then it seems that all truths are analytic and hence necessary. If they are all necessary, surely they are all truths of reason. In that not every truth can be reduced to an identity in a finite number of steps; in some instances revealing the connection between subject and predicate concepts would require an infinite analysis, while this may entail that we cannot prove such propositions as a prior, it does not appear to show that proposition could have been false. Intuitively, it seems a better ground for supposing that it is a necessary truth of a special sort. A related question arises from the idea that truths of fact depend on God's decision to create the best world: If it is part of the concept of this world that it is best, how could its existence be other than necessary? An accountable and responsively answered explanation would be so, that any relational question that brakes the norm lay eyes on its existence in the manner other than hypothetical necessities, i.e., it follows from God's decision to create the world, but God had the power to create this world, but God is necessary, so how could he have decided to do anything else? Leibniz says much more about these matters, but it is not clear whether he offers any satisfactory solutions.

The view that the terms in which we think of some area are sufficiently infected with error for it to be better to abandon them than to continue to try to give coherent theories of their use. Eliminativism should leave out or omit from consideration by defeating, as in a contest, that be distinguished from scepticism that claims that we cannot know the truth about some area; Eliminativism claims that there is no truth there to be known, in the terms that we currently think. An eliminativist about Theology simply counsels abandoning the terms or discourse of Theology, and that will include abandoning worries about the extent of theological knowledge.

Eliminativists in the philosophy of mind counsel abandoning the whole network of terms mind, consciousness, self, Qualia that usher in the problems of mind and body. Sometimes the argument for doing this is that we should wait for a supposed future understanding of ourselves, based on cognitive science and better than any current mental descriptions provide, sometimes it is supposed that physicalism shows that no mental description of us could possibly be true.

Greek scepticism centred on the value of enquiry and questioning, scepticism is now the denial that knowledge or even rational belief is possible, either about some specific subject-matter, e.g., ethics, or in any thing whatsoever. Classically, scepticism springs from the observation that the best methods in some area seem to fall short of giving us contact with the truth, e.g., there is a gulf between appearance and reality, and in frequency cites the conflicting judgements that our methods deliver, with the result that questions of truth become undecided.

Sceptical tendencies emerged in the 14th-century writings of Nicholas of Autrecourt. His criticisms of any certainty beyond the immediate deliverance of the senses and basic logic, and in particular of any knowledge of either intellectual or material substances, anticipate the later scepticism of Balye and Hume. The latter distinguish between Pyrrhonistic and excessive scepticism, which he regarded as unlivable, and the more mitigated scepticism that accepts every day or commonsense beliefs (not as the delivery of reason, but as due more to custom and habit), but is duly wary of the power of reason to give us much more. Mitigated scepticism is thus closer to the attitude fostered by ancient scepticism from Pyrrho through to Sexus Empiricus. Although the phrase 'Cartesian scepticism' is sometimes used, Descartes himself was not a sceptic, but in the method of doubt, uses a sceptical scenario in order to begin the process of finding a secure mark of knowledge. Descartes himself trusts a category of 'clear and distinct' ideas, not far removed from the phantasia kataleptiké of the Stoics.

Scepticism should not be confused with relativism, which is a doctrine about the nature of truth, and may be motivated by trying to avoid scepticism. Nor is it identical with eliminativism, which counsels abandoning an area of thought altogether, not because we cannot know the truth, but because there are no truths capable of being framed in the terms we use.

Descartes's theory of knowledge starts with the quest for certainty, for an indubitable starting-point or foundation on the basis alone of which progress is possible. This is eventually found in the celebrated 'Cogito ergo sum': I think therefore I am. By locating the point of certainty in my own awareness of my own self, Descartes gives a first-person twist to the theory of knowledge that dominated them following centuries in spite of various counterattacks on behalf of social and public starting-points. The metaphysic associated with this priority is the famous Cartesian dualism, or separation of mind and matter into two different but interacting substances, Descartes rigorously and rightly sees that it takes divine dispensation to certify any relationship between the two realms thus divided, and to prove the reliability of the senses invokes a 'clear and distinct perception' of highly dubious proofs of the existence of a benevolent deity. This has not met general acceptance: as Hume drily puts it, 'to have recourse to the veracity of the supreme Being, in order to prove the veracity of our senses, is surely making a very unexpected circuit'.

In his own time Descartes's conception of the entirely separate substance of the mind was recognized to give rise to insoluble problems of the nature of the causal connection between the two. It also gives rise to the problem, insoluble in its own terms, of other minds. Descartes's notorious denial that nonhuman animals are conscious is a stark illustration of the problem. In his conception of matter Descartes also gives preference to rational cogitation over anything derived from the senses. Since we can conceive of the matter of a ball of wax surviving changes to its sensible qualities, matter is not an empirical concept, but eventually an entirely geometrical one, with extension and motion as its only physical nature. Descartes's thought, as reflected in Leibniz, that the qualities of sense experience have no resemblance to qualities of things, so that knowledge of the external world is essentially knowledge of structure rather than of filling. On this basis Descartes erects a remarkable physics. Since matter is in effect the same as extension there can be no empty space or 'void', since there is no empty space motion is not a question of occupying previously empty space, but is to be thought of in terms of vortices (like the motion of a liquid).

Although the structure of Descartes's epistemology, theory of mind, and theory of matter have been rejected many times, their relentless exposures of the hardest issues, their exemplary clarity, and evens their initial plausibility, all contrive to make him the central point of reference for modern philosophy.

The self-conceived were introduced by Descartes presents in the first two Meditations: aware only of its own thoughts, and capable of disembodied existence, neither situated in a space nor surrounded by others. This is the pure self of 'I-ness' that we are tempted to imagine as a simple unique thing that make up our essential identity. Descartes's view that he could keep hold of this nugget while doubting everything else is criticized by Lichtenberg and Kant, and most subsequent philosophers of mind.

Descartes holds that we do not have any knowledge of any empirical proposition about anything beyond the contents of our own minds. The reason, roughly put, is that there is a legitimate doubt about all such propositions because there is no way to deny justifiably that our senses are being stimulated by some cause (an evil spirit, for example) which is radically different from the objects that we normally think affect our senses.

He also points out, that the senses (sight, hearing, touch, and so on), are often unreliable, and 'it is prudent never to trust entirely those who have deceived us even once', he cited such instances as the straight stick that looks been t in water, and the square tower that look round from a distance. This argument of illusion, has not, on the whole, impressed commentators, and some of Descartes' contemporaries pointing out that since such errors become known as a result of further sensory information, it cannot be right to cast wholesale doubt on the evidence of the senses. But Descartes regarded the argument from illusion as only the first stage in softening up the process with which would 'lead minds away from the senses'. He admits that there are some cases of sense-base belief about which doubt would be insane, e.g., the belief that I am sitting here by the fire, wearing a winter dressing gown'.

Descartes was to realize that there was nothing in this view of nature that could explain or provide a foundation for the mental, or from direct experience as distinctly human. In a mechanistic universe, he said, there is no privileged place or function for mind, and the separation between mind and matter is absolute. Descartes was also convinced, that the immaterial essences that gave form and structure to this universe were coded in geometrical and mathematical ideas, and this insight led him to invent algebraic geometry.

A scientific understanding of these ideas could be derived, having been said by Descartes, with the aid of precise deduction, and also claimed, that the contours of physical reality could be laid out in three-dimensional coordinates. Following the publication of Newton's Principia Mathematica in 1687, reductionism and mathematical modelling became the most powerful tools of modern science. And the dream that the entire physical world could be known and mastered through the extension and refinement of mathematical theory became the central feature and guiding principle of scientific knowledge.

Having to its recourse of knowledge, its cental questions include the origin of knowledge, the place of experience in generating knowledge, and the place of reason in doing so, the relationship between knowledge and certainty, and between knowledge and the impossibility of error, the possibility of universal scepticism, and the changing forms of knowledge that arise from new conceptualizations of the world. All of these issues link with other central concerns of philosophy, such as the nature of truth and the natures of experience and meaning.

Foundationalism was associated with the ancient Stoics, and in the modern era with Descartes (1596-1650). Who discovered his foundations in the 'clear and distinct' ideas of reason? Its main opponent is Coherentism, or the view that a body of propositions mas be known without a foundation in certainty, but by their interlocking strength, than as a crossword puzzle may be known to have been solved correctly even if each answer, taken individually, admits of uncertainty. Difficulties at this point led the logical passivists to abandon the notion of an epistemological foundation altogether, and to flirt with the coherence theory of truth. It is widely accepted that trying to make the connection between thought and experience through basic sentences depends on an untenable 'myth of the given'.

Still in spite of these concerns, the problem, least of mention, is of defining knowledge in terms of truc beliefs plus some favoured relations between the believer and the facts that were to have begun with, as Plato's view in the 'Theaetetus,' that knowledge is true belief, and some logos. Due of its nonsynthetic epistemology, the enterprising of studying the actual formation of knowledge by human beings, without aspiring to certify those processes as rational, or its proof against 'scepticism' or even apt to yield the truth. Natural epistemology would therefore blend into the psychology of learning and the study of episodes in the history of science. The scope for 'external' or philosophical reflection of the kind that might result in scepticism or its refutation is markedly diminished. Despite the fact that the terms of modernity are so distinguished as exponents of the approach include Aristotle, Hume, and J. S. Mills.

The task of the philosopher of a discipline would then be to reveal the correct method and to unmask counterfeits. Although this belief lay behind much positivist philosophy of science, few philosophers now subscribe to it. It places too well a confidence in the possibility of a purely previous 'first philosophy', or viewpoint beyond that of the work one's way of practitioners, from which their best efforts can be measured as good or bad. These standpoints now seem that too many philosophers to be a fanciefancy, that the more modest of tasks that are actually adopted at various historical stages of investigation into different areas with the aim not so much of criticizing but more of systematization, in the presuppositions of a particular field at a particular tie. There is still a role for local methodological disputes within the community investigators of some phenomenon, with one approach charging that another is unsound or unscientific, but logic and philosophy will not, on the modern view, provide an independent arsenal of weapons for such battles, which indeed often come to seem more like political bids for ascendancy within a discipline.

This is an approach to the theory of knowledge that sees an important connection between the growth of knowledge and biological evolution. An evolutionary epistemologist claims that the development of human knowledge processed through some natural selection process, the best example of which is Darwin's theory of biological natural selection. There is a widespread misconception that evolution proceeds according to some plan or direct, but it has neither, and the role of chance ensures that its future course will be unpredictable. Random variations in individual organisms create tiny differences in their Darwinian fitness. Some individuals have more offsprings than others, and the characteristics that increased their fitness thereby become more prevalent in future generations. Once upon a time, at least a mutation occurred in a human population in tropical Africa that changed the haemoglobin molecule in a way that provided resistance to malaria. This enormous advantage caused the new gene to spread, with the unfortunate consequence that sickle-cell anaemia came to exist.

Chance can influence the outcome at each stage: First, in the creation of genetic mutation, second, in wether the bearer lives long enough to show its effects, thirdly, in chance events that influence the individual's actual reproductive success, and fourth, in whether a gene even if favoured in one generation, is, happenstance, eliminated in the next, and finally in the many unpredictable environmental changes that will undoubtedly occur in the history of any group of organisms. As Harvard biologist Stephen Jay Gould has so vividly expressed that process over again, the outcome would surely be different. Not only might there not be humans, there might not even be anything like mammals.

We will often emphasis the elegance of traits shaped by natural selection, but the common idea that nature creates perfection needs to be analysed carefully. The extent to which evolution achieves perfection depends on exactly what you mean. Such that, if you mean, 'Does natural selection always takes the best path for the long-term welfare of a species?' The answer is no. That would require adaption by group selection, and this is, unlikely. If you mean 'Does natural selection creates every adaption that would be valuable?' The answer again, is no. For instance, some kinds of South American monkeys can grasp branches with their tails. The trick would surely also be useful to some African species, but, simply because of bad luck, none have it. Some combination of circumstances started some ancestral South American monkeys using their tails in ways that ultimately led to an ability to grab onto branches, while no such development took place in Africa. Mere usefulness of a trait does not necessitate a in that what will understandably endure phylogenesis or evolution.

This is an approach to the theory of knowledge that sees an important connection between the growth of knowledge and biological evolution. An evolutionary epistemologist claims that the development of human knowledge proceeds through some natural selection process, the best example of which is Darwin's theory of biological natural selection. The three major components of the model of natural selection are variation selection and retention. According to Darwin's theory of natural selection, variations are not pre-designed to do certain functions. Rather, these variations that do useful functions are selected. While those that do not employ of some coordinates in that are regainfully purposed are also, not to any of a selection, as duly influenced of such a selection, that may have responsibilities for the visual aspects of a variational intentionally occurs. In the modern theory of evolution, genetic mutations provide the blind variations: Blind in the sense that variations are not influenced by the effects they would have-the likelihood of a mutation is not correlated with the benefits or liabilities that mutation would confer on the organism, the environment provides the filter of selection, and reproduction provides the retention. Fatnesses are achieved because those organisms with features that make them less adapted for survival do not survive in connection with other organisms in the environment that have features that are better adapted. Evolutionary epistemology applies this blind variation and selective retention model to the growth of scientific knowledge and to human thought processes overall.

The parallel between biological evolution and conceptual or 'epistemic' evolution can be seen as either literal or analogical. The literal version of evolutionary epistemology deeds biological evolution as the main cause of the growth of knowledge. On this view, called the 'evolution of cognitive mechanic programs', by Bradie (1986) and the 'Darwinian approach to epistemology' by Ruse (1986), that growth of knowledge occurs through blind variation and selective retention because biological natural selection itself is the cause of epistemic variation and selection. The most plausible version of the literal view does not hold that all human beliefs are innate but rather than the mental mechanisms that guide the acquisitions of non-innate beliefs are themselves innately and the result of biological natural selection. Ruse, (1986) demands of a version of literal evolutionary epistemology that he links to sociolology (Rescher, 1990).

On the analogical version of evolutionary epistemology, called the 'evolution of theory's program', by Bradie (1986). The 'Spenserian approach' (after the nineteenth century philosopher Herbert Spencer) by Ruse (1986), the development of human knowledge is governed by a process analogous to biological natural selection, rather than by an instance of the mechanism itself. This version of evolutionary epistemology, introduced and elaborated by Donald Campbell (1974) as well as Karl Popper, sees the [partial] fit between theories and the world as explained by a mental process of trial and error known as epistemic natural selection.

Both versions of evolutionary epistemology are usually taken to be types of naturalized epistemology, because both take some empirical facts as a starting point for their epistemological project. The literal version of evolutionary epistemology begins by accepting evolutionary theory and a materialist approach to the mind and, from these, constructs an account of knowledge and its developments. In contrast, the metaphorical version does not require the truth of biological evolution: It simply draws on biological evolution as a source for the model of natural selection. For this version of evolutionary epistemology to be true, the model of natural selection need only apply to the growth of knowledge, not to the origin and development of species. Crudely put, evolutionary epistemology of the analogical sort could still be true even if Creationism is the correct theory of the origin of species.

Although they do not begin by assuming evolutionary theory, most analogical evolutionary epistemologists are naturalized epistemologists as well, their empirical assumptions, least of mention, implicitly come from psychology and cognitive science, not evolutionary theory. Sometimes, however, evolutionary epistemology is characterized in a seemingly non-naturalistic fashion. Campbell (1974) says that 'if one is expanding knowledge beyond what one knows, one has no choice but to explore without the benefit of wisdom', i.e., blindly. This, Campbell admits, makes evolutionary epistemology close to being a tautology (and so not naturalistic). Evolutionary epistemology does assert the analytic claim that when expanding one's knowledge beyond what one knows, one must precessed to something that is already known, but, more interestingly, it also makes the synthetic claim that when expanding one's knowledge beyond what one knows, one must proceed by blind variation and selective retention. This claim is synthetic because it can be empirically falsified. The central claim of evolutionary epistemology is synthetic, not analytic. If the central contradictory, which they are not. Campbell is right that evolutionary epistemology does have the analytic feature he mentions, but he is wrong to think that this is a distinguishing feature, since any plausible epistemology has the same analytic feature (Skagestad, 1978).

Two extraordinary issues lie to awaken the literature that involves questions about 'realism', i.e., What metaphysical commitment does an evolutionary epistemologist have to make? Progress, i.e., according to evolutionary epistemology, does knowledge develop toward a goal? With respect to realism, many evolutionary epistemologists endorse that is called 'hypothetical realism', a view that combines a version of epistemological 'scepticism' and tentative acceptance of metaphysical realism. With respect to progress, the problem is that biological evolution is not goal-directed, but the growth of human knowledge seems to be. Campbell (1974) worries about the potential dis-analogy here but is willing to bite the stone of conscience and admit that epistemic evolution progress toward a goal (truth) while biologic evolution does not. Many another has argued that evolutionary epistemologists must give up the 'truth-topic' sense of progress because a natural selection model is in essence, is non-teleological, as an alternative, following Kuhn (1970), and embraced in the accompaniment with evolutionary epistemology.

Among the most frequent and serious criticisms levelled against evolutionary epistemology is that the analogical version of the view is false because epistemic variation is not blind (Skagestad, 1978, and Ruse, 1986: Stein and Lipton 1990) have argued, however, that this objection fails because, while epistemic variation is not random, its constraints come from heuristics that, for the most part, are selective retention. Further, Stein and Lipton come to the conclusion that heuristics are analogous to biological pre-adaptions, evolutionary pre-biological pre-adaptions, evolutionary cursors, such as a half-wing, a precursor to a wing, which have some function other than the function of their descendable structures: The function of descendable structures, the function of their descendable character embodied to its structural foundations, is that of the guidelines of epistemic variation is, on this view, not the source of disanalogy, but the source of a more articulated account of the analogy.

Many evolutionary epistemologists try to combine the literal and the analogical versions (Bradie, 1986, and Stein and Lipton, 1990), saying that those beliefs and cognitive mechanisms, which are innate results from natural selection of the biological sort and those that are innate results from natural selection of the epistemic sort. This is reasonable as long as the two parts of this hybrid view are kept distinct. An analogical version of evolutionary epistemology with biological variation as its only source of blondeness would be a null theory: This would be the case if all our beliefs are innate or if our non-innate beliefs are not the result of blind variation. An appeal to the legitimate way to produce a hybrid version of evolutionary epistemology since doing so trivializes the theory. For similar reasons, such an appeal will not save an analogical version of evolutionary epistemology from arguments to the effect that epistemic variation is blind (Stein and Lipton, 1990).

Although it is a new approach to theory of knowledge, evolutionary epistemology has attracted much attention, primarily because it represents a serious attempt to flesh out a naturalized epistemology by drawing on several disciplines. In science is relevant to understanding the nature and development of knowledge, then evolutionary theory is among the disciplines worth a look. Insofar as evolutionary epistemology looks there, it is an interesting and potentially fruitful epistemological programme.

What makes a belief justified and what makes a true belief knowledge? Thinking that whether a belief deserves one of these appraisals is natural depends on what caused the depicted branch of knowledge to have the belief. In recent decades a number of epistemologists have pursued this plausible idea with a variety of specific proposals. Some causal theories of knowledge have it that a true belief that 'p' is knowledge just in case it has the right causal connection to the fact that 'p'. Such a criterion can be applied only to cases where the fact that 'p' is a sort that can reach causal relations, as this seems to exclude mathematically and their necessary facts and perhaps any fact expressed by a universal generalization, and proponents of this sort of criterion have usually supposed that it is limited to perceptual representations where knowledge of particular facts about subjects' environments.

For example, Armstrong (1973), predetermined that a position held by a belief in the form 'This perceived object is 'F' is [non-inferential] knowledge if and only if the belief is a completely reliable sign that the perceived object is 'F', that is, the fact that the object is 'F' contributed to causing the belief and its doing so depended on properties of the believer such that the laws of nature dictated that, for any subject ',' and perceived object 'y', if ',' has those properties and believed that 'y' is 'F', then 'y' is 'F'. (Dretske (1981) offers a rather similar account, in terms of the belief's being caused by a signal received by the perceiver that carries the information that the object is 'F').

Goldman (1986) has proposed an importantly different causal criterion, namely, that a true belief is knowledge if it is produced by a type of process that is 'globally' and 'locally' reliable. Causing true beliefs is sufficiently high is globally reliable if its propensity. Local reliability has to do with whether the process would have produced a similar but false belief in certain counterfactual situations alternative to the actual situation. This way of marking off true beliefs that are knowledge does not require the fact believed to be causally related to the belief, and so it could in principle apply to knowledge of any kind of truth.

Goldman requires the global reliability of the belief-producing process for the justification of a belief, he requires it also for knowledge because justification is required for knowledge. What he requires for knowledge, but does not require for justification is local reliability. His idea is that a justified true belief is knowledge if the type of process that produced it would not have produced it in any relevant counterfactual situation in which it is false. Its purported theory of relevant alternatives can be viewed as an attempt to provide a more satisfactory response to this tension in our thinking about knowledge. It attempts to characterize knowledge in a way that preserves both our belief that knowledge is an absolute concept and our belief that we have knowledge.

According to the theory, we need to qualify rather than deny the absolute character of knowledge. We should view knowledge as absolute, reactive to certain standards (Dretske, 1981 and Cohen, 1988). That is to say, in order to know a proposition, our evidence need not eliminate all the alternatives to that preposition, rather for 'us', that we can know our evidence eliminates all the relevant alternatives, where the set of relevant alternatives (a proper subset of the set of all alternatives) is determined by some standard. Moreover, according to the relevant alternatives view, and the standards determining that of the alternatives is raised by the sceptic are not relevant. If this is correct, then the fact that our evidence cannot eliminate the sceptic's alternative does not lead to a sceptical result. For knowledge requires only the elimination of the relevant alternatives, so the relevant alternative view preserves in both strands in our thinking about knowledge. Knowledge is an absolute concept, but because the absoluteness is relative to a standard, we can know many things.

The interesting thesis that counts as a causal theory of justification (in the meaning of 'causal theory' intended here) is that: A belief is justified in case it was produced by a type of process that is 'globally' reliable, that is, its propensity to produce true beliefs-that can be defined (to a good approximation) As the proportion of the beliefs it produces (or would produce) that is true is sufficiently great.

This proposal will be adequately specified only when we are told (I) how much of the causal history of a belief counts as part of the process that produced it, (ii) which of the many types to which the process belongs is the type for purposes of assessing its reliability, and (iii) relative to why the world or worlds are the reliability of the process type to be assessed the actual world, the closet worlds containing the case being considered, or something else? Let 'us' look at the answers suggested by Goldman, the leading proponent of a Reliabilist account of justification.

(1) Goldman (1979, 1986) takes the relevant belief producing process to include only the proximate causes internal to the believer. So, for instance, when recently I believed that the telephone was ringing the process that produced the belief, for purposes of assessing reliability, includes just the causal chain of neural events from the stimulus in my ear's inward ands other concurrent brain states on which the production of the belief depended: It does not include any events' I hear the telephone, or the sound waves travelling between it and my ears, or any earlier decisions I made that were responsible for my being within hearing distance of the telephone, at that time. It does seem intuitively plausible of a belief depends should be restricted to internal omnes proximate to the belief. Why? Goldman

does not tell 'us'. One answer that some philosophers might give is that it is because a belief's being justified at a given time can depend only on facts directly accessible to the believer's awareness at that time (for, if a believer ought to holds only beliefs that are justified, she can tell at any given time what beliefs would then be justified for her). However, this cannot be Goldman's answer because he wishes to include in the relevantly process neural events that are not directly accessible to consciousness.

(2) Once the Reliabilist has told 'us' how to delimit the process producing a belief, he needs to tell 'us' which of the many types to which it belongs is the relevant type. Coincide, for example, the process that produces your current belief that you see a book before you. One very broad type to which that process belongs would be specified by 'coming to a belief as to something one perceives as a result of activation of the nerve endings in some of one's sense-organ'. A constricted type, from which are the unvarying processes that belong for what could be specified by 'coming to a belief as to what one sees as a result of activation of the nerve endings in one's retinas'. A still narrower type would be given by inserting in the last specification a description of a particular pattern of activation of the retina's particular cells. Which of these or other types to which the token process belongs is the relevant type for determining whether the type of process that produced your belief is reliable?

(3) Should the justification of a belief in a hypothetical, non-actual example turn on the reliability of the belief-producing process in the possible world of the example? That leads to the implausible result in that in a world run by a Cartesian demon–a powerful being who causes the other inhabitants of the world to have rich and coherent sets of perceptual and memory impressions that are all illusory the perceptual and memory beliefs of the other inhabitants are all unjustified, for they are produced by processes that are, in that world, quite unreliable. If we say instead that it is the reliability of the processes in the actual world that matters, we get the equally undesired result that if the actual world is a demon world then our perceptual and memory beliefs are all unjustified.

Goldman's solution (1986) is that the reliability of the process types is to be gauged by their performance in 'normal' worlds, that is, worlds consistent with 'our general beliefs about the world . . . 'about the sorts of objects, events and changes that occur in it'. This gives the intuitively right results for the problem cases just considered, but indicate by inference an implausible proportion of making compensations for alternative tending toward justification. If there are people whose general beliefs about the world are very different from mine, then there may, on this account, be beliefs that I can correctly regard as justified (ones produced by processes that are reliable in what I take to be a normal world) but that they can correctly regard as not justified.

However, these questions about the specifics are dealt with, and there are reasons for questioning the basic idea that the criterion for a belief's being justified is its being produced by a reliable process. Thus and so, doubt about the sufficiency of the Reliabilist criterion is prompted by a sort of example that Goldman himself uses for another purpose. Suppose that being in brain-state 'B' always causes one to believe that one is in brained-state 'B'. Here the reliability of the belief-producing process is perfect, but 'we can readily imagine circumstances in which a person goes into grain-state 'B' and therefore has the belief in question, though this belief is by no justified' (Goldman, 1979). Doubt about the necessity of the condition arises from the possibility that one might know that one has strong justification for a certain belief and yet that knowledge is not what actually prompts one to believe. For example, I might be well aware that, having read the weather bureau's forecast that it will be much hotter tomorrow. I have ample reason to be confident that it will be hotter tomorrow, but I irrationally refuse to believe it until Wally tells me that he feels in his joints that it will be hotter tomorrow. Here what prompts me to believe dors not justify my belief, but my belief is nevertheless justified by my knowledge of the weather bureau's prediction and of its evidential force: I can advert to any disavowable inference that I should not be holding of that belief. Indeed, given my justification and that there is nothing untoward about the weather bureau's prediction, my belief, if true, can be counted knowledge. This sorts of example raises doubt whether any causal conditions, are it a reliable process or something else, is necessary for either justification or knowledge.

Philosophers and scientists alike, have often held that the simplicity or parsimony of a theory is one reason, all else being equal, to view it as true. This goes beyond the unproblematic idea that simpler theories are easier to work with and gave greater aesthetic appeal.

One theory is more parsimonious than another when it postulates fewer entities, processes, changes or explanatory principles: The simplicity of a theory depends on essentially the same consecrations, though parsimony and simplicity obviously become the same. Demanding clarification of what makes one theory simpler or more parsimonious is plausible than another before the justification of these methodological maxims can be addressed.

If we set this description problem to one side, the major normative problem is as follows: What reason is there to think that simplicity is a sign of truth? Why should we accept a simpler theory instead of its more complex rivals? Newton and Leibniz thought that the answer was to be found in a substantive fact about nature. In 'Principia,' Newton laid down as his first Rule of Reasoning in Philosophy that 'nature does nothing in vain . . . 'for Nature is pleased with simplicity and affects not the pomp of superfluous causes'. Leibniz hypothesized that the actual world obeys simple laws because God's taste for simplicity influenced his decision about which world to actualize.

The tragedy of the Western mind, described by Koyré, is a direct consequence of the stark Cartesian division between mind and world. We discovered the 'certain principles of physical reality', said Descartes, 'not by the prejudices of the senses, but by the light of reason, and which thus possess so great evidence that we cannot doubt of their truth'. Since the real, or that which actually exists and is itself an external to ourselves, was in his view only that which could be represented in the quantitative terms of mathematics, Descartes concludes that all quantitative aspects of reality could be traced to the deceitfulness of the senses.

The most fundamental aspect of the Western intellectual tradition is the assumption that there is a fundamental division between the material and the immaterial world or between the realm of matter and the realm of pure mind or spirit. The metaphysical framework based on this assumption is known as ontological dualism. As the word dual implies, the framework is predicated on an ontology, or a conception of the nature of God or Being, that assumes reality has two distinct and separable dimensions. The concept of Being as continuous, immutable, and having a prior or separate existence from the world of change dates from the ancient Greek philosopher Parmenides. The same qualities were associated with the God of the Judeo-Christian tradition, and they were considerably amplified by the role played in Theology by Platonic and Neoplatonic philosophy.

Nicolas Copernicus, Galileo, Johannes Kepler, and Isaac Newton were all inheritors of a cultural tradition in which ontological dualism was a primary article of faith. Hence the idealization of the mathematical ideal as a source of communion with God, which dates from Pythagoras, provided a metaphysical foundation for the emerging natural sciences. This explains why, the creators of classical physics believed that doing physics was a form of communion with the geometrical and mathematical forms' resident in the perfect mind of God. This view would survive in a modified form in what is now known as Einsteinian epistemology and accounts in no small part for the reluctance of many physicists to accept the epistemology associated with the Copenhagen Interpretation.

At the beginning of the nineteenth century, Pierre-Simon LaPlace, along with a number of other French mathematicians, advanced the view that the science of mechanics constituted a complete view of nature. Since this science, by observing its epistemology, had revealed itself to be the fundamental science, the hypothesis of God was, they concluded, entirely unnecessary.

LaPlace is recognized for eliminating not only the Theological component of classical physics but the 'entire metaphysical component' as well'. As Simon LaPlace (1749-1827) noted for his theory of a nebular origin of the solar system and his investigations into gravity and the stability of planetary motion. But still, epistemology of science requires, he said, that we go on by inductive generalizations of the observed facts to hypotheses that are 'tested by observed conformity of the phenomena'. What was unique about LaPlace's view of hypotheses was his insistence that we cannot attribute reality to them. Although concepts like force, mass, motion, cause, and laws are obviously present in classical physics, they exist in LaPlace's view only as quantities. Physics is concerned, he argued, with quantities that we associate as a matter of convenience with concepts, and the truths about nature are only the quantities.

As this view of hypotheses and the truths of nature as quantities were extended in the nineteenth century to a mathematical description of phenomena like heat, light, electricity, and magnetism. LaPlace's assumptions about the actual character of scientific truths seemed correct. This progress suggested that if we could remove all thoughts about the 'nature of' or the 'source of' phenomena, the pursuit of strictly quantitative concepts would bring us to a complete description of all aspects of physical reality. Subsequently, figures like Comte, Kirchhoff, Hertz, and Poincaré developed a program for the study of nature that was quite different from that of the original creators of classical physics.

The seventeenth-century view of physics as a philosophy of nature or as natural philosophy was displaced by the view of physics as an autonomous science that was 'the science of nature'. This view, which was premised on the doctrine of positivism, promised to subsume all of the nature with a mathematical analysis of entities in motion and claimed that the true understanding of nature was revealed only in the mathematical description. Since the doctrine of positivism assumes that the knowledge we call physics resides only in the mathematical formalism of physical theory, it disallows the prospect that the vision of physical reality revealed in physical theory can have any other meaning.

In the history of science, the irony is that positivism, which was intended to banish metaphysical concerns from the domain of science, served to perpetuate a seventeenth-century metaphysical assumption about the relationship between physical reality and physical theory.

Epistemology since Hume and Kant has drawn back from this theological underpinning. Indeed, the very idea that nature is simple (or uniform) has come in for a critique. The view has taken hold that a preference for simple and parsimonious hypotheses is purely methodological: It is constitutive of the attitude we call 'scientific' and makes no substantive assumption about the way the world is.

A variety of otherwise diverse twentieth-century philosophers of science have attempted, in different ways, to flesh out this position. Two examples must suffice here: Hesse (1969) as, for summaries of other proposals. Popper (1959) holds that scientists should prefer highly falsifiable (improbable) theories: He tries to show that simpler theories are more falsifiable, also Quine (1966), in contrast, sees a virtue in theories that are highly probable, he argues for a general connection between simplicity and high probability.

Both these proposals are global. They attempt to explain why simplicity should be part of the scientific method in a way that spanned all scientific subject matters. No assumption about the details of any particular scientific problem serves as a premiss in Popper's or Quine's arguments.

Newton and Leibniz thought that the justification of parsimony and simplicity flows from the hand of God: Popper and Quine try to justify these methodologically median of importance is without assuming anything substantive about the way the world is. In spite of these differences in approach, they have something in common. They assume that all users of parsimony and simplicity in the separate sciences can be encompassed in a single justifying argument. That recent developments in confirmation theory suggest that this assumption should be scrutinized. Good (1983) and Rosenkrantz (1977) has emphasized the role of auxiliary assumptions in mediating the connection between hypotheses and observations. Whether a hypothesis is well supported by some observations, or whether one hypothesis is better supported than another by those observations, crucially depends on empirical background assumptions about the inference problem here. The same view applies to the idea of prior probability (or, prior plausibility). In of a single hypo-physical science if chosen as an alternative to another even though they are equally supported by current observations, this must be due to an empirical background assumption.

Principles of parsimony and simplicity mediate the epistemic connection between hypotheses and observations. Perhaps these principles are able to do this because they are surrogates for an empirical background theory. It is not that there is one background theory presupposed by every appeal to parsimony; This has the quantifier order backwards. Rather, the suggestion is that each parsimony argument is justified only to each degree that it reflects an empirical background theory about the subjective matter. On this theory is brought out into the open, but the principle of parsimony is entirely dispensable (Sober, 1988).

This 'local' approach to the principles of parsimony and simplicity resurrects the idea that they make sense only if the world is one way rather than another. It rejects the idea that these maxims are purely methodological. How defensible this point of view is, will depend on detailed case studies of scientific hypothesis evaluation and on further developments in the theory of scientific inference.

It is usually not found of one and the same that, an inference is a (perhaps very complex) act of thought by virtue of which act (1) I pass from a set of one or more propositions or statements to a proposition or statement and (2) it appears that the latter are true if the former is or are. This psychological characterization has occurred over a wider and varying duration of its literature under, a greater extent for what seems as not as much than inessential variations. Desiring a better characterization of inference is natural. Yet attempts to do so by constructing a fuller psychological explanation fail to comprehend the grounds on which inference will be objectively valid - A point elaborately made by Gottlob Frége. Attempts to understand the nature of inference through the device of the representation of inference by formal-logical calculations or derivations better (1) leave 'us' puzzled about the relation of formal-logical derivations to the informal inferences they are supposedly to represent or reconstruct, and (2) leaves 'us' worried about the sense of such formal derivations. Are these derivations inference? Are not informal inferences needed in order to apply the rules governing the constructions of formal derivations (inferring that this operation is an application of that formal rule)? These are concerns cultivated by, for example, Wittgenstein.

Coming up with an adequate characterization of inference-and even working out what would count as a very adequate characterization here is demandingly by no nearly some resolved philosophical problem.

The rule of inference, as for raised by Lewis Carroll, the Zeno-like problem of how a 'proof' ever gets started. Suppose I have as premises (I) 'p' and (ii) p q. Can I infer 'q'? Only, it seems, if I am sure of (iii) (p & p q) ®q. Can I then infer 'q'? Only, it seems, if I am sure that (iv) (p & p q & (p & p q) q) ®q. For each new axiom (N) I need a further axiom (N + 1) telling me that the set so far implies 'q', and the regress never stops. The usual solution is to treat a system as containing not only axioms, but also rules of inference, allowing movement from the axioms. The rule 'modus ponens' allow 'us' to pass from the first premise to 'q'. Carroll's puzzle shows that distinguishing two theoretical categories is essential, although there may be choice about which theses to put in which category.

Traditionally, a proposition that is not a 'conditional', as with the 'affirmative' and 'negative', modern opinion is wary of the distinction, since what appears categorical may vary with the choice of a primitive vocabulary and notation. Apparently categorical propositions may also turn out to be disguised conditionals: 'X' is intelligent (categorical?) Equivalent, if 'X' is given a range of tasks, she does them better than many people (conditional?). The problem is not merely one of classification, since deep metaphysical questions arise when facts that seem to be categorical and therefore solid, come to seem by contrast conditional, or purely hypothetical or potential.

Its condition of some classified necessity is so proven sufficient that if 'p' is a necessary condition of 'q', then 'q' cannot be true unless 'p'; is true? If 'p' is a sufficient condition, thus steering well is a necessary condition of driving in a satisfactory manner, but it is not sufficient, for one can steer well but drive badly for other reasons. Confusion may result if the distinction is not heeded. For example, the statement that 'A' causes 'B' may be interpreted to mean that 'A' is itself a sufficient condition for

'B', or that it is only a necessary condition fort 'B', or perhaps a necessary parts of a total sufficient condition. Lists of conditions to be met for satisfying some administrative or legal requirement frequently attempt to give individually necessary and jointly sufficient sets of conditions.

What is more, that if any proposition of the form 'if p then q'. The condition hypothesized, 'p'. Is called the antecedent of the conditionals, and 'q', the consequent? Various kinds of conditional have been distinguished. Its weakest is that of 'material implication', merely telling that either 'not-p', or 'q'. Stronger conditionals include elements of 'modality', corresponding to the thought that 'if p is truer then q must be true'. Ordinary language is very flexible in its use of the conditional form, and there is controversy whether conditionals are better treated semantically, yielding differently finds of conditionals with different meanings, or pragmatically, in which case there should be one basic meaning with surface differences arising from other implicatures.

It follows from the definition of 'strict implication' that a necessary proposition is strictly implied by any proposition, and that an impossible proposition strictly implies any proposition. If strict implication corresponds to 'q follows from p', then this that a necessary proposition follows from anything at all, and anything at all follows from an impossible proposition. This is a problem if we wish to distinguish between valid and invalid arguments with necessary conclusions or impossible premises.

The Humean problem of induction is that if we would suppose that there is some property 'A' concerning and observational or an experimental situation, and that out of a large number of observed instances of 'A', some fraction m/n (possibly equal to 1) has also been instances of some logically independent property 'B'. Suppose further that the background proportionate circumstances not specified in these descriptions has been varied to a substantial degree and that there is no collateral information available concerning the frequency of 'B's' among 'A's or concerning causal or nomologically connections between instances of 'A' and instances of 'B'.

In this situation, an 'enumerative' or 'instantial' induction inference would move rights from the premise, that m/n of observed 'A's' are 'B's' to the conclusion that approximately m/n of all 'A's' are 'B's. (The usual probability qualification will be assumed to apply to the inference, rather than being part of the conclusion.) Here the class of 'A's' should be taken to include not only unobserved 'A's' and future 'A's', but also possible or hypothetical 'A's' (an alternative conclusion would concern the probability or likelihood of the adjacently observed 'A' being a 'B').

The traditional or Humean problem of induction, often referred to simply as 'the problem of induction', is the problem of whether and why inferences that fit this schema should be considered rationally acceptable or justified from an epistemic or cognitive standpoint, i.e., whether and why reasoning in this way is likely to lead to true claims about the world. Is there any sort of argument or rationale that can be offered for thinking that conclusions reached in this way are likely to be true in the corresponding premises is true or even that their chances of truth are significantly enhanced?

Hume's discussion of this issue deals explicitly only with cases where all observed 'A's' are 'B's' and his argument applies just as well to the more general case. His conclusion is entirely negative and sceptical: Inductive inferences are not rationally justified, but are instead the result of an essentially a-rational process, custom or habit. Hume (1711-76) challenges the proponent of induction to supply a cogent line of reasoning that leads from an inductive premise to the corresponding conclusion and offers an extremely influential argument in the form of a dilemma (a few times referred to as 'Hume's fork'), that either our actions are determined, in which case we are not responsible for them, or they are the result of random events, under which case we are also not responsible for them.

Such reasoning would, he argues, have to be either deductively demonstrative reasoning in the concerning relations of ideas or 'experimental', i.e., empirical, that reasoning concerning matters of fact or existence. It cannot be the former, because all demonstrative reasoning relies on the avoidance of contradiction, and it is not a contradiction to suppose that 'the course of nature may change', that an order that was observed in the past and not of its continuing against the future: But it cannot be, as the latter, since any empirical argument would appeal to the success of such reasoning about an experience, and the justifiability of generalizing from experience are precisely what is at issue-so that any such appeal would be question-begging. Hence, Hume concludes that there can be no such reasoning (1748).

An alternative version of the problem may be obtained by formulating it with reference to the so-called Principle of Induction, which says roughly that the future will resemble the past or, somewhat better, that unobserved cases will resemble observed cases. An inductive argument may be viewed as enthymematic, with this principle serving as a supposed premiss, in which case the issue is obviously how such a premiss can be justified. Hume's argument is then that no such justification is possible: The principle cannot be justified a prior because having possession of been true in experiences without obviously begging the question is not contradictory to have possession of been true in experiences without obviously begging the question.

The predominant recent responses to the problem of induction, at least in the analytic tradition, in effect accept the main conclusion of Hume's argument, namely, that inductive inferences cannot be justified in the sense of showing that the conclusion of such an inference is likely to be true if the premise is true, and thus attempt to find another sort of justification for induction. Such responses fall into two main categories: (I) Pragmatic justifications or 'vindications' of induction, mainly developed by Hand Reichenbach (1891-1953), and (ii) ordinary language justifications of induction, whose most important proponent is Frederick, Peter Strawson (1919-). In contrast, some philosophers still attempt to reject Hume's dilemma by arguing either (iii) That, contrary to appearances, induction can be inductively justified without vicious circularity, or (iv) that an anticipatory justification of induction is possible after all. In that:

(1) Reichenbach's view is that induction is best regarded, not as a form of inference, but rather as a 'method' for arriving at posits regarding, i.e., the proportion of 'A's' remain additionally of 'B's'. Such a posit is not a claim asserted to be true, but is instead an intellectual wager analogous to a bet made by a gambler. Understood in this way, the inductive method says that one should posit that the observed proportion is, within some measure of an approximation, the true proportion and then continually correct that initial posit as new information comes in.

The gambler's bet is normally an 'appraised posit', i.e., he knows the chances or odds that the outcome on which he bets will actually occur. In contrast, the inductive bet is a 'blind posit': We do not know the chances that it will succeed or even that success is that it will succeed or even that success is possible. What we are gambling on when we make such a bet is the value of a certain proportion in the independent world, which Reichenbach construes as the limit of the observed proportion as the number of cases increases to infinity. Nevertheless, we have no way of knowing that there are even such a limit, and no way of knowing that the proportion of 'A's' are in addition of 'B's' converges in the end on some stable value than varying at random. If we cannot know that this limit exists, then we obviously cannot know that we have any definite chance of finding it.

What we can know, according to Reichenbach, is that 'if' there is a truth of this sort to be found, the inductive method will eventually find it'. That this is so is an analytic consequence of Reichenbach's account of what it is for such a limit to exist. The only way that the inductive method of making an initial posit and then refining it in light of new observations can fail eventually to arrive at the true proportion is if the series of observed proportions never converges on any stable value, which that there is no truth to be found pertaining the proportion of 'A's additionally constitute 'B's'. Thus, induction is justified, not by showing that it will succeed or indeed, that it has any definite likelihood of success, but only by showing that it will succeed if success is possible. Reichenbach's claim is that no more than this can be established for any method, and hence that induction gives 'us' our best chance for success, our best gamble in a situation where there is no alternative to gambling.

This pragmatic response to the problem of induction faces several serious problems. First, there are indefinitely many other 'methods' for arriving at posits for which the same sort of defence can be given-methods that yield the same results as the inductive method over time but differ arbitrarily before long. Despite the efforts of others, it is unclear that there is any satisfactory way to exclude such alternatives, in order to avoid the result that any arbitrarily chosen short-term posit is just as reasonable as the inductive posit. Second, even if there is a truth of the requisite sort to be found, the inductive method is only guaranteed to find it or even to come within any specifiable distance of it in the indefinite long run. All the same, any actual application of inductive results always takes place in the presence to the future eventful states in making the relevance of the pragmatic justification to actual practice uncertainly. Third, and most important, it needs to be emphasized that Reichenbach's response to the problem simply accepts the claim of the Humean sceptic that an inductive premise never provides the slightest reason for thinking that the corresponding inductive conclusion is true. Reichenbach himself is quite candid on this point, but this does not alleviate the intuitive implausibility of saying that we have no more reason for thinking that our scientific and commonsense conclusions that result in the induction of it '. . . is true' than, to use Reichenbach's own analogy (1949), a blind man wandering in the mountains who feels an apparent trail with his stick has for thinking that following it will lead him to safety.

An approach to induction resembling Reichenbach's claiming in that those particular inductive conclusions are posits or conjectures, than the conclusions of cogent inferences, is offered by Popper. However, Popper's view is even more overtly sceptical: It amounts to saying that all that can ever be said in favour of the truth of an inductive claim is that the claim has been tested and not yet been shown to be false.

(2) The ordinary language response to the problem of induction has been advocated by many philosophers, none the less, Strawson claims that the question whether induction is justified or reasonable makes sense only if it tacitly involves the demand that inductive reasoning meet the standards appropriate to deductive reasoning, i.e., that the inductive conclusions are shown to follow deductively from the inductive assumption. Such a demand cannot, of course, be met, but only because it is illegitimate: Inductive and deductive reasons are simply fundamentally different kinds of reasoning, each possessing its own autonomous standards, and there is no reason to demand or expect that one of these kinds meet the standards of the other. Whereas, if induction is assessed by inductive standards, the only ones that are appropriate, then it is obviously justified.

The problem here is to understand to what this allegedly obvious justification of an induction amount. In his main discussion of the point (1952), Strawson claims that it is an analytic true statement that believing it a conclusion for which there is strong evidence is reasonable and an analytic truth that inductive evidence of the sort captured by the schema presented earlier constitutes strong evidence for the corresponding inductive conclusion, thus, apparently yielding the analytic conclusion that believing it a conclusion for which there is inductive evidence is reasonable. Nevertheless, he also admits, indeed insists, that the claim that inductive conclusions will be true in the future is contingent, empirical, and may turn out to be false (1952). Thus, the notion of reasonable belief and the correlative notion of strong evidence must apparently be understood in ways that have nothing to do with likelihood of truth, presumably by appeal to the standard of reasonableness and strength of evidence that are accepted by the community and are embodied in ordinary usage.

Understood in this way, Strawson's response to the problem of inductive reasoning does not speak to the central issue raised by Humean scepticism: The issue of whether the conclusions of inductive arguments are likely to be true. It amounts to saying merely that if we reason in this way, we can correctly call ourselves 'reasonable' and our evidence 'strong', according to our accepted community standards. Nevertheless, to the undersealing of issue of wether following these standards is a good way to find the truth, the ordinary language response appears to have nothing to say.

(3) The main attempts to show that induction can be justified inductively have concentrated on showing that such as a defence can avoid circularity. Skyrms (1975) formulate, perhaps the clearest version of this general strategy. The basic idea is to distinguish different levels of inductive argument: A first level in which induction is applied to things other than arguments: A second level in which it is applied to arguments at the first level, arguing that they have been observed to succeed so far and hence are likely to succeed in general: A third level in which it is applied in the same way to arguments at the second level, and so on. Circularity is allegedly avoided by treating each of these levels as autonomous and justifying the argument at each level by appeal to an argument at the next level.

One problem with this sort of move is that even if circularity is avoided, the movement to higher and higher levels will clearly eventually fail simply for lack of evidence: A level will reach at which there have been enough successful inductive arguments to provide a basis for inductive justification at the next higher level, and if this is so, then the whole series of justifications collapses. A more fundamental difficulty is that the epistemological significance of the distinction between levels is obscure. If the issue is whether reasoning in accord with the original schema offered above ever provides a good reason for thinking that the conclusion is likely to be true, then it still seems question-begging, even if not flatly circular, to answer this question by appeal to anther argument of the same form.

(4) The idea that induction can be justified on a pure priori basis is in one way the most natural response of all: It alone treats an inductive argument as an independently cogent piece of reasoning whose conclusion can be seen rationally to follow, although perhaps only with probability from its premise. Such an approach has, however, only rarely been advocated (Russell, 1913 and BonJour, 1986), and is widely thought to be clearly and demonstrably hopeless.

Many on the reasons for this pessimistic view depend on general epistemological theses about the possible or nature of anticipatory cognition. Thus if, as Quine alleges, there is no a prior justification of any kind, then obviously a prior justification for induction is ruled out. Or if, as more moderate empiricists have in claiming some preexistent knowledge should be analytic, then again a prevenient justification for induction seems to be precluded, since the claim that if an inductive premise ids truer, then the conclusion is likely to be true does not fit the standard conceptions of 'analyticity'. A consideration of these matters is beyond the scope of the present spoken exchange.

There are, however, two more specific and quite influential reasons for thinking that an early approach is impossible that can be briefly considered, first, there is the assumption, originating in Hume, but since adopted by very many of others, that a move forward in the defence of induction would have to involve 'turning induction into deduction', i.e., showing, per impossible, that the inductive conclusion follows deductively from the premise, so that it is a formal contradiction to accept the latter and deny the former. However, it is unclear why a prior approach need be committed to anything this strong. It would be enough if it could be argued that it is deductively unlikely that such a premise is true and corresponding conclusion false.

Second, Reichenbach defends his view that pragmatic justification is the best that is possible by pointing out that a completely chaotic world in which there is simply not true conclusion to be found as to the proportion of 'A's' in addition that occurs of, but B's' is neither impossible nor unlikely from a purely a prior standpoint, the suggestion being that therefore there can be no a prior reason for thinking that such a conclusion is true. Nevertheless, there is still a substring wayin laying that a chaotic world is a prior neither impossible nor unlikely without any further evidence does not show that such a world os not a prior unlikely and a world containing such-and-such regularity might anticipatorially be somewhat likely in relation to an occurrence of a long-run patten of evidence in which a certain stable proportion of observed 'A's' are 'B's' ~. An occurrence, it might be claimed, that would be highly unlikely in a chaotic world (BonJour, 1986).

Goodman's 'new riddle of induction' purports that we suppose that before some specific time 't' (perhaps the year 2015) we observe a larger number of emeralds (property 'A') and find them all to be green (property 'B'). We proceed to reason inductively and conclude that all emeralds are green Goodman points out, however, that we could have drawn a quite different conclusion from the same evidence of our observed emeralds will also be grued. A parallel inductive argument will yield the conclusion that all emeralds are gruing, and hence that all those examined after the year 2000 will be blue. Presumably the first of these concisions is genuinely supported by our observations and the second is not. Nevertheless, the problem is to say why this is so and to impose some further restriction upon inductive reasoning that will permit the first argument and exclude the second.

The obvious alternative suggestion is that 'grue. Similar predicates do not correspond to genuine, purely qualitative properties in the way that 'green' and 'blueness' does, and that this is why inductive arguments involving them are unacceptable. Goodman, however, claims to be unable to make clear sense of this suggestion, pointing out that the relations of formal desirability are perfectly symmetrical: Grue' may be defined in terms if, 'green' and 'blue', but 'green' an equally well be defined in terms of 'grue' and 'green' (blue if examined before 't' and green if examined after 't').

The 'grued, paradoxes' demonstrate the importance of categorization, in that sometimes it is itemized as 'gruing', if examined of a presence to the future, before future time 't' and 'green', or not so examined and 'blue'. Even though all emeralds in our evidence class grue, we ought must infer that all emeralds are gruing. For 'grue' is unprojectible, and cannot transmit credibility as to indicate the separation of an exclusion for which of a cause or instrument for exchanging views between known and unknown cases. Only projectable predicates are right for induction. Goodman considers entrenchment the key to projectibility having a long history of successful protection, 'grue' is entrenched, lacking such a history, 'grue' is not. A hypothesis is projectable, Goodman suggests, only if its predicates (or suitable related ones) are much better entrenched than its rivalrous past successes that do not assume future ones. Induction remains a risky business. The rationale for favouring entrenched predicates is pragmatic. Of the possible projections from our evidence class, the one that fits with past practices enables 'us' to utilize our cognitive resources best. Its prospects of being true are worse than its competitors' and its cognitive utility is greater.

So, to a better understanding of induction we should then term in of most of the widely used for any process of reasoning that takes 'us' from empirical premises to empirical conclusions supported by the premises, but not deductively entailed by them. Inductive arguments are therefore kinds of applicative arguments, in which something beyond the content of the premise is inferred as probable or supported by them. Induction is, however, commonly distinguished from arguments to theoretical explanations, which share this applicative character, by being confined to inferences in which he conclusion involves the same properties or relations as the premises. The central example is induction by simple enumeration, where from premises telling that Fa, Fb, Fc . . . 'where a, b, c's, are all of some kind 'G', it is inferred that G's from outside the sample, such as future G's, will be 'F', or perhaps that all G's are 'F'. In this, which and the other persons deceive them, children may infer that everyone is a deceiver: Different, but similar inferences of a property by some object to the same object's future possession of the same property, or from the constancy of some law like pattern in events and states of affairs ti its future constancy. All objects we know of attract each other with a force inversely proportional to the square of the distance between them, so perhaps they all do so, and will always do so.

The rational basis of any inference was challenged by Hume, who believed that induction presupposed belie in the uniformity of nature, but that this belief has no defence in reason, and merely reflected a habit or custom of the mind. Hume was not therefore sceptical about the role of reason in either explaining it or justifying it. Trying to answer Hume and to show that there is something rationally compelling about the inference referred to as the problem of induction. It is widely recognized that any rational defence of induction will have to partition well-behaved properties for which the inference is plausible (often called projectable properties) from badly behaved ones, for which it is not. It is also recognized that actual inductive habits are more complex than those of similar enumeration, and that both common sense and science pay attention to such giving factors as variations within the sample giving 'us' the evidence, the application of ancillary beliefs about the order of nature, and so on.

Nevertheless, the fundamental problem remains that ant experience condition by application show 'us' only events occurring within a very restricted part of a vast spatial and temporal order about which we then come to believe things.

Uncompounded by its belonging of a confirmation theory finding of the measure to which evidence supports a theory fully formalized confirmation theory would dictate the degree of confidence that a rational investigator might have in a theory, given to some body of evidential presentation or determination of an evidentiary form in conclusion or judgement for and against a hypothesis taken place or as to be seen, as the thing or things in forming some attestation, only to manufacture by its observational conclusion of evidence produced. The grandfather of confirmation theory is Gottfried Leibniz (1646-1718), who believed that a logically transparent language of science would be able to resolve all disputes. In the 20th century a fully formal confirmation theory was a main goal of the logical positivist, since without it the central concept of verification by empirical evidence itself remains distressingly unscientific. The principal developments were due to Rudolf Carnap (1891-1970), culminating in his 'Logical Foundations of Probability' (1950). Carnap's idea was that the measure necessitated would be the proportion of logically possible states of affairs in which the theory and the evidence both hold, compared ti the number in which the evidence itself holds that the probability of a preposition, relative to some evidence, is a proportion of the range of possibilities under which the proposition is true, compared to the total range of possibilities left by the evidence. The difficulty with the theory lies in identifying sets of possibilities so that they admit of measurement. It therefore demands that we can put a measure on the 'range' of possibilities consistent with theory and evidence, compared with the range consistent with the evidence alone.

Among the obstacles the enterprise meets, is the fact that while evidence covers only a finite range of data, the hypotheses of science may cover an infinite range. In addition, confirmation proves to vary with the language in which the science is couched, and the Carnapian programme has difficulty in separating genuinely confirming variety of evidence from less compelling repetition of the same experiment. Confirmation also proved to be susceptible to acute paradoxes. Finally, scientific judgement seems to depend on such intangible factors as the problems facing rival theories, and most workers have come to stress instead the historically situated scene of what would appear as a plausible distinction of a scientific knowledge at a given time.

Arose to the paradox of which when a set of apparent incontrovertible premises is given to unacceptable or contradictory conclusions. To solve a paradox will involve showing either that there is a hidden flaw in the premises, or that the reasoning is erroneous, or that the apparently unacceptable conclusion can, in fact, be tolerated. Paradoxes are therefore important in philosophy, for until one is solved it shows that there is something about our reasoning and our concepts that we do not understand. What is more, and somewhat loosely, a paradox is a compelling argument from unacceptable premises to an unacceptable conclusion: More strictly speaking, a paradox is specified to be a sentence that is true if and only if it is false. A characterized objection lesson of it would be: 'The displayed sentence is false.'

Seeing that this sentence is false if true is easy, and true if false, a paradox, in either of the senses distinguished, presents an important philosophical challenger. Epistemologists are especially concerned with various paradoxes having to do with knowledge and belief. In other words, for example, the Knower paradox is an argument that begins with apparently impeccable premisses about the concepts of knowledge and inference and derives an explicit contradiction. The origin of the reasoning is the 'surprise examination paradox': A teacher announces that there will be a surprise examination next week. A clever student argues that this is impossible. 'The test cannot be on Friday, the last day of the week, because it would not be a surprise. We would know the day of the test on Thursday evening. This we can also rule out Thursday. For after we learn that no test has been given by Wednesday, we would know the test is on Thursday or Friday -and would already know that it s not on Friday and would already know that it is not on Friday by the previous reasoning. The remaining days can be eliminated in the same manner'.

This puzzle has over a dozen variants. The first was probably invented by the Swedish mathematician Lennard Ekbon in 1943. Although the first few commentators regarded the reverse elimination argument as cogent, every writer on the subject since 1950 agrees that the argument is unsound. The controversy has been over the proper diagnosis of the flaw.

Initial analyses of the subject's argument tried to lay the blame on a simple equivocation. Their failure led to more sophisticated diagnoses. The general format has been an assimilation to better-known paradoxes. One tradition casts the surprise examination paradox as a self-referential problem, as fundamentally akin to the Liar, the paradox of the Knower, or Gödel's incompleteness theorem. That in of itself, says enough that Kaplan and Montague (1960) distilled the following 'self-referential' paradox, the Knower. Consider the sentence:

'(S) The negation of this sentence is known (to be true)'.

Suppose that (S) is true. Then its negation is known and hence true. However, if its negation is true, then (S) must be false. Therefore (s) is false, or what is the name, the negation of (S) is true.

This paradox and its accompanying reasoning are strongly reminiscent of the Lair Paradox that (in one version) begins by considering a sentence 'This sentence is false' and derives a contradiction. Versions of both arguments using axiomatic formulations of arithmetic and Gödel-numbers to achieve the effect of self-reference yields important meta-theorems about what can be expressed in such systems. Roughly these are to the effect that no predicates definable in the formalized arithmetic can have the properties we demand of truth (Tarski's Theorem) or of knowledge (Montague, 1963).

These meta-theorems still leave 'us; with the problem that if we suppose that we add of these formalized languages predicates intended to express the concept of knowledge (or truth) and inference-as one mighty does if a logic of these concepts is desired. Then the sentence expressing the leading principles of the Knower Paradox will be true.

Explicitly, the assumption about knowledge and inferences are:

(1) If sentences 'A' are known, then 'a.'

(2) (1) is known?

(3) If 'B' is correctly inferred from 'A', and 'A' is known, then 'B' is known.

To give an absolutely explicit t derivation of the paradox by applying these principles to (S), we must add (contingent) assumptions to the effect that certain inferences have been done. Still, as we go through the argument of the Knower, these inferences are done. Even if we can somehow restrict such principles and construct a consistent formal logic of knowledge and inference, the paradoxical argument as expressed in the natural language still demands some explanation.

The usual proposals for dealing with the Liar often have their analogues for the Knower, e.g., that there is something wrong with a self-reference or that knowledge (or truth) is properly a predicate of propositions and not of sentences. The relies that show that some of these are not adequate are often parallel to those for the Liar paradox. In addition, one can try in this context, what seems to be an adequate solution for the Surprise Examination Paradox, namely the observation that 'new knowledge can drive out knowledge', but this does not seem to work on the Knower (Anderson, 1983).

There are a number of paradoxes of the Liar family. The simplest example is the sentence 'This sentence is false', which must be false if it is true, and true if it is false. One suggestion is that the sentence fails to say anything, but sentences that fail to say anything are at least not true. In fact case, we consider to sentences 'This sentence is not true', which, if it fails to say anything is not true, and hence (this kind of reasoning is sometimes called the strengthened Liar). Other versions of the Liar introduce pairs of sentences, as in a slogan on the front of a T-shirt saying 'This sentence on the back of this T-shirt is false', and one on the back saying 'The sentence on the front of this T-shirt is true'. It is clear that each sentence individually is well formed, and was it not for the other, might have said something true. So any attempt to dismiss the paradox by sating that the sentence involved is meaningless will face problems.

Even so, the two approaches that have some hope of adequately dealing with this paradox is 'hierarchy' solutions and 'truth-value gap' solutions. According to the first, knowledge is structured into 'levels'. It is argued that their one-coherent notion expressed by the verb 'knows', but rather a whole series of notions for being knowable and to what is 'known,' as to perceive directly and grasp in the mind with clarity or certainty (perhaps into transfinite), stated in terms of predicate expressing such 'ramified' concepts and properly restricted, such as having complicating consequences or outgrowths, for which the problem merely ramifies after such an extent of decisions in branches or subordinated parts (1)-(3) lead to no contradictions. The main objections to this procedure are that the meaning of these levels has not been adequately explained and that the idea of such subscripts, even implicit, in a natural language is highly counterintuitive the 'truth-value gap' solution takes sentences such as (S) to lack truth-value. They are neither true nor false, but they do not express propositions. This defeats a crucial step in the reasoning used in the derivation of the paradoxes. Kripler (1986) has developed this approach in connection with the Liar and Asher and Kamp (1986) has worked out some details of a parallel solution to the Knower. The principal objection is that 'strengthened' or 'super' versions of the paradoxes tend to reappear when the solution itself is stated.

Since the paradoxical deduction uses only the properties (1)-(3) and since the argument is formally valid, any notion that satisfies these conditions will lead to a paradox. Thus, Grim (1988) notes that this may be read as 'is known by an omniscient God' and concludes that there is no coherent single notion of omniscience. Thomason (1980) observes that with some different conditions, analogous reasoning about belief can lead to paradoxical consequence.

Its paradox arises when a set of apparently incontrovertible premises gives unacceptable or contradictory conclusions, to solve a paradox will involve showing either that there is a hidden flaw in the premises, or that the reasoning is erroneous, or that the apparently unacceptable conclusion can, in fact, be tolerated. Paradoxes are therefore important in philosophy, for until one has been resolved, showing that there is something about our reasoning and of concepts that we do not understand. Famous families of paradoxes include the 'semantic paradoxes' and 'Zeno's paradoxes. Art the beginning of the 20th century, paradox and other set-theoretical paradoxes led to the complete overhaul of the foundations of set theory, while the 'Sorites paradox' has lead to the investigations of the semantics of vagueness and fuzzy logics.

At this point, we display attributes to the theory of experience, as it is not possible to define in an illuminating way, however, we know what experiences are through acquaintances with some of our own, e.g., visual experiences of as afterimage, a feeling of physical nausea or a tactile experience of an abrasive surface (which might be caused by an actual surface -rough or smooth, or which might be part of a dream, or the product of a vivid sensory imagination). The essential feature of experience is it feels a certain way -that there is something that it is like to have it. We may refer to this feature of an experience as its 'character'.

Another core feature of the sorts of experiences with which this may be of a concern, is that they have representational 'content'. (Unless otherwise indicated, 'experience' will be reserved for their 'contentual representations'.) The most obvious cases of experiences with content are sense experiences of the kind normally involved in perception. We may describe such experiences by mentioning their sensory modalities ad their contents, e.g., a gustatory experience (modality) of chocolate ice cream (content), but do so more commonly by of perceptual verbs combined with noun phrases specifying their contents, as in 'Macbeth saw a dagger'. This is, however, ambiguous between the perceptual claim 'There was a (material) dagger in the world that Macbeth perceived visually' and 'Macbeth had a visual experience of a dagger' (the reading with which we are concerned, as it is afforded by our imagination, or perhaps, experiencing mentally hallucinogenic imagery).

As in the case of other mental states and events with content, it is important to distinguish between the properties that and experience 'represents' and the properties that it 'possesses'. To talk of the representational properties of an experience is to say something about its contentual attributes for those properties to the experience itself. Like every other experience, a visual; experience of a non-shaped square, of which is a mental event, and it is therefore not itself irregular or is it square, even though it represents those properties. It is, perhaps, fleeting, pleasant or unusual, even though it does not represent those properties. An experience may represent a property that it possesses, and it may even do so in virtue of a rapidly changing (complex) experience representing something as changing rapidly. However, this is the exception and not the rule.

Which properties can be [directly] represented in sense experience is subject to debate. Traditionalists include only properties whose presence could not be doubted by a subject having appropriate experiences, e.g., colour and shape in the case of visual experience, and apparent shape, surface texture, hardness, and so forth, in the case of tactile experience. This view is natural to anyone who has an egocentric, Cartesian perspective in epistemology, and who wishes for pure data in experiences to serve as logically certain foundations for knowledge, especially to the immediate objects of perceptual awareness in or of sense-data, such categorized of colour patches and shapes, which are usually supposed distinct from surfaces of physical objectivity. Qualities of sense-data are supposed to be distinct from physical qualities because their perception is more relative to conditions, more certain, and more immediate, and because sense-data is private and cannot appear other than they are they are objects that change in our perceptual field when conditions of perception change. Physical objects remain constant.

Others who do not think that this wish can be satisfied, and who are more impressed with the role of experience in providing animisms with ecologically significant information about the world around them, claim that sense experiences represent properties, characteristic and kinds that are much richer and much more wide-ranging than the traditional sensory qualities. We do not see only colours and shapes, they tell 'us', but also earth, water, men, women and fire: We do not smell only odours, but also food and filth. There is no space here to examine the factors relevantly responsible to their choice of situational alternatives. Yet, this suggests that character and content are not really distinct, and there is a close tie between them. For one thing, the relative complexity of the character of sense experience places limitations upon its possible content, e.g., a tactile experience of something touching one's left ear is just too simple to carry the same amount of content as typically convincing to an every day, visual experience. Moreover, the content of a sense experience of a given character depends on the normal causes of appropriately similar experiences, e.g., the sort of gustatory experience that we have when eating chocolate would be not represented as chocolate unless it was normally caused by chocolate. Granting a contingent ties between the character of an experience and its possible causal origins, once, again follows that its possible content is limited by its character.

Character and content are none the less irreducibly different, for the following reasons. (a) There are experiences that completely lack content, e.g., certain bodily pleasures. (b) Not every aspect of the character of an experience with content is relevant to that content, e.g., the unpleasantness of an aural experience of chalk squeaking on a board may have no representational significance. © Experiences in different modalities may overlap in content without a parallel overlap in character, e.g., visual and tactile experiences of circularity feel completely different. (d) The content of an experience with a given character may vary according to the background of the subject, e.g., a certain content 'singing bird' only after the subject has learned something about birds.

In an outline, the Phenomenological argument appears whenever we have an experience, even if nothing beyond the experience answers to it, we seem to be presented with something through the experience (which is itself diaphanous). The object of the experience is whatever is so presented to 'us'-is that it is an individual thing, an event, or a state of affairs.

The semantic argument is that objects of experience are required in order to make sense of certain features of our talk about experience, including, in particular, the following. (I) Simple attributions of experience, e.g., 'Rod is experiencing an oddity that is not really square but in appearance it seems more than likely a square', this seems to be relational. (ii) We appear to refer to objects of experience and to attribute properties to them, e.g., 'The afterimage that John experienced was certainly odd'. (iii) We appear to quantify over objects of experience, e.g., 'Macbeth saw something that his wife did not see'.

The act/object analysis faces several problems concerning the status of objects of experiences. Currently the most common view is that they are sense-data -private mental entities that actually posses the traditional sensory qualities represented by the experiences of which they are the objects. But the very idea of an essentially private entity is suspect. Moreover, since an experience may apparently represent something as having a determinable property, e.g., redness, without representing it as having any subordinate determinate property, e.g., any specific shade of red, a sense-datum may actually have a determinate property subordinate to it. Even more disturbing is that sense-data may have contradictory properties, since experiences can have contradictory contents. A case in point is the waterfall illusion: If you stare at a waterfall for a minute and then immediately fixate on a nearby rock, you are likely to have an experience of the rock's moving upward while it remains in the same place. The sense-data theorist must either deny that there are such experiences or admit contradictory objects.

These problems can be avoided by treating objects of experience as properties. This, however, fails to do justice to the appearances, for experience seems not to present 'us' with properties embodied in individuals. The view that objects of experience is Meinongian objects accommodate this point. It is also attractive in as far as (1) it allows experiences to represent properties other than traditional sensory qualities, and (2) it allows for the identification of objects of experience and objects of perception in the case of experiences that constitute perception.

According to the act/object analysis of experience, every experience with content involves an object of experience to which the subject is related by an act of awareness (the event of experiencing that object). This is meant to apply not only to perceptions, which have material objects (whatever is perceived), but also to experiences like hallucinations and dream experiences, which do not. Such experiences, nonetheless, appear to represent something, and their objects are supposed to be whatever it is that they represent. Act/object theorists may differ on the nature of objects of experience, which have been treated as properties. Meinongian objects (which may not exist or have any form of being), and, more commonly private mental entities with sensory qualities. (The term 'sense-data' is now usually applied to the latter, but has also been used as a general term for objects of sense experiences, as in the work of G. E. Moore) Act/object theorists may also differ on the relationship between objects of experience and objects of perception. In terms of perception (of which we are 'indirectly aware') are always distinct from objects of experience (of which we are 'directly aware'). Meinongian, however, may treat objects of perception as existing objects of experience. But sense-datum theorists must either deny that there are such experiences or admit contradictory objects. Still, most philosophers will feel that the Meinongian's acceptance of impossible objects is too high a price to pay for these benefits.

A general problem for the act/object analysis is that the question of whether two subjects are experiencing one and the same thing (as opposed to having exactly similar experiences) appears to have an answer only on the assumption that the experiences concerned are perceptions with material objects. But in terms of the act/object analysis the question must have an answer even when this condition is not satisfied. (The answer is always negative on the sense-datum theory; it could be positive on other versions of the act/object analysis, depending on the facts of the case.)

In view of the above problems, the case for the act/object analysis should be reassessed. The phenomenological argument is not, on reflection, convincing, for it is easy enough to grant that any experience appears to present 'us' with an object without accepting that it actually does. The semantic argument is more impressive, but is none the less answerable. The seemingly relational structure of attributions of experience is a challenge dealt with below in connection with the adverbial theory. Apparent reference to and quantification over objects of experience can be handled by analysing them as reference to experiences themselves and quantification over experiences tacitly typed according to content. Thus, 'The afterimage that John experienced was colourfully appealing' becomes 'John's afterimage experience was an experience of colour', and 'Macbeth saw something that his wife did not see' becomes 'Macbeth had a visual experience that his wife did not have'.

Pure cognitivism attempts to avoid the problems facing the act/object analysis by reducing experiences to cognitive events or associated disposition, e.g., Susy's experience of a rough surface beneath her hand might be identified with the event of her acquiring the belief that there is a rough surface beneath her hand, or, if she does not acquire this belief, with a disposition to acquire it that has somehow been blocked.

This position has attractions. It does full justice to the cognitive contents of experience, and to the important role of experience as a source of belief acquisition. It would also help clear the way for a naturalistic theory of mind, since there seems to be some prospect of a physicalist/functionalist account of belief and other intentional states. But pure cognitivism is completely undermined by its failure to accommodate the fact that experiences have a felt character that cannot be reduced to their content, as aforementioned.

The adverbial theory is an attempt to undermine the act/object analysis by suggesting a semantic account of attributions of experience that does not require objects of experience. Unfortunately, the oddities of explicit adverbializations of such statements have driven off potential supporters of the theory. Furthermore, the theory remains largely undeveloped, and attempted refutations have traded on this. It may, however, be founded on sound basis intuitions, and there is reason to believe that an effective development of the theory (which is merely hinting at) is possible.

The relevant intuitions are (1) that when we say that someone is experiencing 'an A', or has an experience 'of an A', we are using this content-expression to specify the type of thing that the experience is especially apt to fit, (2) that doing this is a matter of saying something about the experience itself (and maybe about the normal causes of like experiences), and (3) that it is no-good of reasons to posit of its position to presuppose that of any involvements, is that its descriptions of an object in which the experience is. Thus the effective role of the content-expression in a statement of experience is to modify the verb it compliments, not to introduce a special type of object.

Perhaps, the most important criticism of the adverbial theory is the 'many property problem', according to which the theory does not have the resources to distinguish between, e.g.,

> (1) Frank has an experience of a brown triangle

And:

> (2) Frank has an experience of brown and an experience of a triangle.

Which is entailed by (1) but does not entail it. The act/object analysis can easily accommodate the difference between (1) and (2) by claiming that the truth of (1) requires a single object of experience that is both brown and triangular, while that of the (2) allows for the possibility of two objects of experience, one brown and the other triangular, however, (1) is equivalent to:

(1*) Frank has an experience of something's being both brown and triangular.

And (2) is equivalent to:

(2*) Frank has an experience of something's being brown and an experience of something's being triangular,

And the difference between these can be explained quite simply in terms of logical scope without invoking objects of experience. The Adverbialists may use this to answer the many-property problem by arguing that the phrase 'a brown triangle' in (1) does the same work as the clause 'something's being both brown and triangular' in (1*). This is perfectly compatible with the view that it also has the 'adverbial' function of modifying the verb 'has an experience of', for it specifies the experience more narrowly just by giving a necessary condition for the satisfaction of the experience (the condition being that there are something both brown and triangular before Frank).

A final position that should be mentioned is the state theory, according to which a sense experience of an 'A' is an occurrent, non relational state of the kind that the subject would be in when perceiving an 'A'. Suitably qualified, this claim is no doubt true, but its significance is subject to debate. Here it is enough to remark that the claim is compatible with both pure cognitivism and the adverbial theory, and that state theorists are probably best advised to adopt adverbials as a of developing their intuitions.

Yet, clarifying sense-data, if taken literally, is that which is given by the senses. But in response to the question of what exactly is so given, sense-data theories posit private showings in the consciousness of the subject. In the case of vision it would be a kind of inner picture as showing which of the indirect representable aspects of the external world that has in and of itself a worldly representation. The view has been widely rejected as implying that we really only see extremely thin coloured pictures interposed between our mind's eye and reality. Modern approaches to perception tend to reject any conception of the eye as a camera or lense, simply responsible for producing private images, and stress the active life of the subject in and of the world, as the determinant of experience.

Nevertheless, the argument from illusion is of itself the usually intended directive to establish that certain familiar facts about illusion disprove the theory of perception called na evity or direct realism. There are, however, many different versions of the argument that must be distinguished carefully. Some of these distinctions centre on the content of the premises (the nature of the appeal to illusion); others centre on the interpretation of the conclusion (the kind of direct realism under attack). Let 'us' set about by distinguishing the importantly different versions of direct realism which one might take to be vulnerable to familiar facts about the possibility of perceptual illusion.

A crude statement of direct realism might go as follows. In perception, we sometimes directly perceive physical objects and their properties, we do not always perceive physical objects by perceiving something 'else', e.g., a sense-datum. There are, however, difficulties with this formulation of the view, as for one thing a great many philosophers who are 'not' direct realists would admit that it is a mistake to describe people as actually 'perceiving' something other than a physical object. In particular, such philosophers might admit, we should never say that we perceive sense-data. To talk that way would be to suppose that we should model our understanding of our relationship to sense-data on our understanding of the ordinary use of perceptual verbs as they describe our relation to and of the physical world, and that is the last thing paradigm sense-datum theorist's want. At least, many of the philosophers who objected to direct realism would prefer to express in what they were of objecting too in terms of a technical (and philosophically controversial) concept such as 'acquaintance'. Using such a notion, we could define direct realism this way: In 'veridical' experience we are directly acquainted with parts, e.g., surfaces, or constituents of physical objects. To a lesser extent for keeping the limit maintaining the cautious verison of the view, such that might drop the reference to veridical experience and claim simply that in all experience we are directly acquainted with parts or constituents of physical objects. The expressions 'knowledge by acquaintance' and 'knowledge by description', and the distinction they mark between knowing 'things' and knowing 'about' things, are generally associated with Bertrand Russell (1872-1970), that scientific philosophy required analysing many objects of belief as 'logical constructions' or 'logical fictions', and the programme of analysis that this inaugurated dominated the subsequent philosophy of logical atomism, and then of other philosophers, Russell's 'The Analysis of Mind,' the mind itself is treated in a fashion reminiscent of Hume, as no more than the collection of neutral perceptions or sense-data that make up the flux of conscious experience, and that looked at another way that also was to make up the external world (neutral monism), but 'An Inquiry into Meaning and Truth' (1940) represents a more empirical approach to the problem. Yet, philosophers have perennially investigated this and related distinctions using varying terminology.

Distinction in our ways of knowing things, highlighted by Russell and forming a central element in his philosophy after the discovery of the theory of 'definite descriptions'. A thing is known by acquaintance when there is direct experience of it. It is known by description if it can only be described as a thing with such-and-such properties. In everyday parlance, I might know my spouse and children by acquaintance, but know someone as 'the first person born at sea' only by description. However, for a variety of reasons Russell shrinks the area of things that can be known by acquaintance until eventually only current experience, perhaps my own self, and certain universals or meanings qualify anything else is known only as the thing that has such-and-such qualities.

Because one can interpret the relation of acquaintance or awareness as one that is not 'epistemic', i.e., not a kind of propositional knowledge, it is important to distinguish the above aforementioned views read as ontological theses from a view one might call 'epistemological direct realism.' In perception we are, on at least some occasions, non-inferentially justified in believing a proposition asserting the existence of a physical object. Since it is that these objects exist independently of any mind that might perceive them, and so it thereby rules out all forms of idealism and phenomenalism, which hold that there are no such independently existing objects. Its being to 'direct' realism rules out those views defended under the cubic of 'critical naive realism', or 'representational realism', in which there is some nonphysical intermediary -usually called a 'sense-datum' or a 'sense impression' -that must first be perceived or experienced in order to perceive the object that exists independently of this perception. Often the distinction between direct realism and other theories of perception is explained more fully in terms of what is 'immediately' perceived, than 'mediately' perceived. What relevance does illusion have for these two forms of direct realism?

The fundamental premise of the arguments is from illusion seems to be the theses that things can appear to be other than they are. Thus, for example, straight sticks when immerged in water looks bent, a penny when viewed from certain perspective appears as an illusory spatial elliptic circularity, when something that is yellow when place under red fluorescent light looks red. In all of these cases, one version of the argument goes, it is implausible to maintain that what we are directly acquainted with is the real nature of the object in question. Indeed, it is hard to see how we can be said to be aware of the really physical object at all. In the above illusions the things we were aware of actually were bent, elliptical and red, respectively. But, by hypothesis, the really physical objects lacked these properties. Thus, we were not aware of the substantial reality of been real as a physical objects or theory.

So far, if the argument is relevant to any of the direct realisms distinguished above, it seems relevant only to the claim that in all sense experience we are directly acquainted with parts or constituents of physical objects. After all, even if in illusion we are not acquainted with physical objects, but their surfaces, or their constituents, why should we conclude anything about the hidden nature of our relations to the physical world in veridical experience?

We are supposed to discover the answer to this question by noticing the similarities between illusory experience and veridical experience and by reflecting on what makes illusion possible at all. Illusion can occur because the nature of the illusory experience is determined, not just by the nature of the object perceived, but also by other conditions, both external and internal as becoming of an inner or as the outer experience. But all of our sensations are subject to these causal influences and it would be gratuitous and arbitrary to select from indefinitely of many and subtly different perceptual experiences some special ones those that get 'us' in touch with the 'real' nature of the physical world and its surrounding surfaces. Red fluorescent light affects the way thing's look, but so does sunlight. Water reflects light, but so does air. We have no unmediated access to the external world.

Still, why should we consider that we are aware of something other than a physical object in experience? Why should we not conclude that to be aware of a physical object is just to be appeared to by that object in a certain way? In its best-known form the adverbial theory of something proposes that the grammatical object of a statement attributing an experience to someone be analysed as an adverb. For example,

(A) Rod is experiencing a coloured square.

Is rewritten as?

Rod is experiencing, (coloured square)-ly

This is presented as an alternative to the act/object analysis, according to which the truth of a statement like (A) requires the existence of an object of experience corresponding to its grammatical object. A commitment to t he explicit adverbializations of statements of experience are not, however, essential to adverbialism. The core of the theory consists, rather, in the denial of objects of experience (as opposed ti objects of perception) coupled with the view that the role of the grammatical object in a statement of experience is to characterize more fully te sort of experience that is being attributed to the subject. The claim, then, is that the grammatical object is functioning as a modifier and, in particular, as a modifier of a verb. If it as a special kind of adverb at the semantic level.

At this point, it might be profitable to move from considering the possibility of illusion to considering the possibility of hallucination. Instead of comparing paradigmatic veridical perception with illusion, let 'us' compare it with complete hallucination. For any experiences or sequence of experiences we take to be veridical, we can imagine qualitatively indistinguishable experiences occurring as part of a hallucination. For those who like their philosophical arguments spiced with a touch of science, we can imagine that our brains were surreptitiously removed in the night, and unbeknown to 'us' are being stimulated by a neurophysiologist so as to produce the very sensations that we would normally associate with a trip to the Grand Canyon. Currently permit 'us' into appealing of what we are aware of in this complete hallucination that is obvious that we are not awaken to the sparking awareness of physical objects, their surfaces, or their constituents. Nor can we even construe the experience as one of an object's appearing to 'us' in a certain way. It is after all a complete hallucination and the objects we take to exist are simply not there. But if we compare hallucinatory experience with the qualitatively indistinguishable veridical experiences, should we most conclude that it would be 'special' to suppose that in veridical experience we are aware of something radically different from what we are aware of in hallucinatory experience? Again, it might help to reflect on our belief that the immediate cause of hallucinatory experience and veridical experience might be the very same brain event, and it is surely implausible to suppose that the effects of this same cause are radically different -acquaintance with physical objects in the case of veridical experience: Something else in the case of hallucinatory experience.

This version of the argument from hallucination would seem to address straightforwardly the ontological versions of direct realism. The argument is supposed to convince 'us' that the ontological analysis of sensation in both veridical and hallucinatory experience should give 'us' the same results, but in the hallucinatory case there is no plausible physical object, constituent of a physical object, or surface of a physical object with which additional premiss we would also get an argument against epistemological direct realism. That premiss is that in a vivid hallucinatory experience we might have precisely the same justification for believing (falsely) what we do about the physical world as we do in the analogous phenomenological and yet, indistinguishable, veridical experience. But our justification for believing that there is a table in front of 'us', in the course of a vivid hallucination of a table are surely non-inferential visions, in that the character of its certainty is not, if non-inferential justifications are supposedly consisted but hold of an unproblematic access to the fact that makes true our belief - by hypothesis the table does not exist. But if the justification that hallucinatory experiences give 'us' the same as the justification we get from the parallel veridical experience, then we should not describe a veridical experience as giving 'us non-inferential justification for believing in the existence of physical objects. In both cases we should say that we believe what we do about the physical world on the basis of what we know directly about the character of our experience.

In this brief space, I can only sketch some of the objections that might be raised against arguments from illusion and hallucination. That being said, let us begin with a criticism that accepts most of the presuppositions of the arguments. Even if the possibility of hallucination establishes that in some experience we are not acquainted with constituents of physical objects, it is not clear that it establishes that we are never acquainted with a constituent of physical objects. Suppose, for example, that we decide that in both veridical and hallucinatory experience we are acquainted with sense-data. At least some philosophers have tried to identify physical objects with 'bundles' of actual and possible sense-data.

To establish inductively that sensations are signs of physical objects one would have to observe a correlation between the occurrence of certain sensations and the existence of certain physical objects. But to observe such a correlation in order to establish a connection, one would need independent access to physical objects and, by hypothesis, this one cannot have. If one further adopts the verificationist's stance that the ability to comprehend is parasitic on the ability to confirm, one can easily be driven to Hume's conclusion:

Let us chance our imagination to the heavens, or to the utmost limits of the universe, we never really advance a step beyond ourselves, nor can conceivable any kind of existence, but those perceptions, which have appear d in that narrow compass. This is the universe of the imagination, nor have we have any idea but what is there Reduced. (Hume, 1739-40).

If one reaches such a conclusion but wants to maintain the intelligibility and verifiability of the assertion about the physical world, one can go either the idealistic or the phenomenalistic route.

However, hallucinatory experiences on this view is non-veridical precisely because the sense-data one is acquainted with in hallucination do not bear the appropriate relations to other actual and possible sense-data. But if such a view where plausible one could agree that one is acquainted with the same kind of a thing in veridical and non-veridical experience but insists that there is still a sense in which in veridical experience one is acquainted with constituents of a physical object?

A different sort of objection to the argument from illusion or hallucination concerns its use in drawing conclusions we have not stressed in the above discourses. I, have in mentioning this objection, may to underscore an important feature of the argument. At least some philosophers (Hume, for example) have stressed the rejection of direct realism on the road to an argument for general scepticism with respect to the physical world. Once one abandons epistemological; direct realisms, one has an uphill battle indicating how one can legitimately make the inferences from sensation to physical objects. But philosophers who appeal to the existence of illusion and hallucination to develop an argument for scepticism can be accused of having an epistemically self-defeating argument. One could justifiably infer sceptical conclusions from the existence of illusion and hallucination only if one justifiably believed that such experiences exist, but if one is justified in believing that illusion exists, one must be justified in believing at least, some facts about the physical world (for example, that straight sticks look bent in water). The key point to stress in relying to such arguments is, that strictly speaking, the philosophers in question need only appeal to the 'possibility' of a vivid illusion and hallucination. Although it would have been psychologically more difficult to come up with arguments from illusion and hallucination if we did not believe that we actually had such experiences, I take it that most philosophers would argue that the possibility of such experiences is enough to establish difficulties with direct realism. Indeed, if one looks carefully at the argument from hallucination discussed earlier, one sees that it nowhere makes any claims about actual cases of hallucinatory experience.

Another reply to the attack on epistemological direct realism focuses on the implausibility of claiming that there is any process of 'inference' wrapped up in our beliefs about the world and its surrounding surfaces. Even if it is possible to give a phenomenological description of the subjective character of sensation, the expression of inquiry that invites or calls for a reply, as the cause of such interest and excitement advocating that perception associated with stimulation of a sense organ of which a specific body condition for which a visual sensation is to feel or perceive physical sensibility and of such arousal of intended curiosity or action, specially by exaggerated or lucid tail, sensational concepts and sensational intent. The theory that sensation is the soul of knowledge, the ethical doctrine that feeling is the only criterion or good. Our perceptual beliefs about the physical world are surely direct, at least in the sense that they are unmediated by any sort of conscious inference from premisses describing something other than a physical object. The appropriate reply to this objection, however, is simply to acknowledge the relevant phenomenological fact and point out that from the perceptive of epistemologically direct realism, the philosopher is attacking a claim about the nature of our justification for believing propositions about the physical world. Such philosophers need carry out of any comment at all about the causal genesis of such beliefs.

That the proponents of the argument from illusion and hallucination have often intended it to establish, the existence of sense-data, and many philosophers have attacked the so-called sense-datum inference presupposed in some statements of the argument, that when the stick looked bent, the penny looked elliptical and the yellow object looked red, the sense-datum theorist wanted to infer that there was something bent, elliptical and red, respectively. But such an inference is surely suspect. Usually, we do not infer that because something appears to have a certain property, that affairs that affecting something that has that property. When in saying that Jones looks like a doctor, I surely would not want anyone to infer that there must actually be someone there who is a doctor. In assessing this objection, it will be important to distinguish different uses words like 'appears' and 'looks'. At least, sometimes to say that something looks 'F' way and the sense-datum inference from an F 'appearance' in this sense to an actual 'F' would be hopeless. However, it also seems that we use the 'appears'/'looks' terminology to describe the phenomenological character of our experience and the inference might be more plausible when the terms are used this way. Still, it does seem that the arguments from illusion and hallucination will not by themselves constitute strong evidence for sense-datum theory. Even if one concludes that there is something common to both the hallucination of a red thing and a veridical visual experience of a red thing, one need not describe a common constituent as awareness of something red. The adverbial theorist would prefer to construe the common experiential state as 'being appeared too redly', a technical description intended only to convey the idea that the state in question need not be analysed as relational in character. Those who opt for an adverbial theory of sensation need to make good the claim that their artificial adverbs can be given a sense that is not parasitic upon an understanding of the adjectives transformed into verbs. Still, other philosophers might try to reduce the common element in veridical and nonveridical experience to some kind of intentional state. More like belief or judgement. The idea here is that the only thing common to the two experiences is the fact that both, are spontaneously taken to be present of an object of a certain kind.

The selfsame objections can be started within the general framework presupposed by proponents of the arguments from illusion and hallucination. A great many contemporary philosophers, however, uncomfortable with the intelligibility of the concepts needed to make sense of the theories attacked even. Thus, at least, some who object to the argument from illusion do so not because they defend direct realism. Rather they think there is something confused about all this talk of direct awareness or acquaintance. Belonging to the same period of time, a fact determined by the current modern contemporary sources, is that externalists, for example, usually insists that we understand epistemic concepts by appeal: Too nomologically connections. On such a view the closest thing to direct knowledge would probably be something by other beliefs. If we understand direct knowledge this way, it is not clar how the phenomena of illusion and hallucination would be relevant to claim that on, at least some occasions our judgements about the physical world are reliably produced by processes that do not take as their input beliefs about something else.

The expressions 'knowledge by acquaintance' and 'knowledge by description', and the distinction they mark between knowing 'things' and knowing 'about' things, are now generally associated with Bertrand Russell. However, John Grote and Hermann von Helmholtz had earlier and independently to mark the same distinction, and William James adopted Grote's terminology in his investigation of the distinction. Philosophers have perennially investigated this and related distinctions using varying terminology. Grote introduced the distinction by noting that natural languages 'distinguish between these two applications of the notion of knowledge, the one being of the Greek ~ which are: ~ nosene, Kennen, connaître, the other being 'Wissen', 'savoir' (Grote, 1865). On Grote's account, the distinction is a natter of degree, and there are three sorts of dimensions of variability: Epistemic, causal and semantic.

We know things by experiencing them, and knowledge of acquaintance (Russell changed the preposition to 'by') is epistemically priori to and has a relatively higher degree of epistemic justification than knowledge about things. Indeed, sensation has 'the one great value of trueness or freedom from mistake' (1900).

A thought (using that term broadly, to mean any mental state) constituting knowledge of acquaintance with a thing is more or less causally proximate to sensations caused by that thing, while a thought constituting knowledge about the thing is more or less distant causally, being separated from the thing and experience of it by processes of attention and inference. At the limit, if a thought is maximally of the acquaintance type, it is the first mental state occurring in a perceptual causal chain originating in the object to which the thought refers, i.e., it is a sensation. The thing's presented to 'us' in sensation and of which we have knowledge of acquaintance include ordinary objects in the external world, such as the sun.

Grote contrasted the imaginistic thoughts involved in knowledge of acquaintance with things, with the judgements involved in knowledge about things, suggesting that the latter but not the former are mentally contentual by a specified state of affairs. Elsewhere, however, he suggested that every thought capable of constituting knowledge of or about a thing involves a form, idea, or what we might call contentual propositional content, referring the thought to its object. Whether contentual or not, thoughts constituting knowledge of acquaintance with a thing are relatively indistinct, although this indistinctness does not imply incommunicably. On the other hand, thoughts constituting distinctly, as a result of 'the application of notice or attention' to the 'confusion or chaos' of sensation (1900). Grote did not have an explicit theory on reference, the relation by which a thought is 'of' or 'about' a specific thing. Nor did he explain how thoughts can be more or less indistinct.

Helmholtz held unequivocally that all thoughts capable of constituting knowledge, whether 'knowledge that has to do with Notions' (Wissen) or 'mere familiarity with phenomena' (Kennen), is judgement or, we may say, have conceptual propositional contents. Where Grote saw a difference between distinct and indistinct thoughts, Helmholtz found a difference between precise judgements that are expressible in words and equally precise judgements that, in principle, are not expressible in words, and so are not communicable (Helmholtz, 1962). As happened, James was influenced by Helmholtz and, especially, by Grote. (James, 1975). Taken on the latter's terminology, James agreed with Grote that the distinction between knowledge of acquaintance with things and knowledge about things involves a difference in the degree of vagueness or distinctness of thoughts, though he, too, said little to explain how such differences are possible. At one extreme is knowledge of acquaintance with people and things, and with sensations of colour, flavour, spatial extension, temporal duration, effort and perceptible difference, unaccompanied by knowledge about these things. Such pure knowledge of acquaintance is vague and inexplicit. Movement away from this extreme, by a process of notice and analysis, yields a spectrum of less vague, more explicit thoughts constituting knowledge about things.

All the same, the distinction was not merely a relative one for James, as he was more explicit than Grote in not imputing content to every thought capable of constituting knowledge of or about things. At the extreme where a thought constitutes pure knowledge of acquaintance with a thing, there is a complete absence of conceptual propositional content in the thought, which is a sensation, feeling or precept, of which he renders the thought incommunicable. James' reasons for positing an absolute discontinuity in between pure cognition and preferable knowledge of acquaintance and knowledge at all about things seem to have been that any theory adequate to the facts about reference must allow that some reference is not conventionally mediated, that conceptually unmediated reference is necessary if there are to be judgements at all about things and, especially, if there are to be judgements about relations between things, and that any theory faithful to the common person's 'sense of life' must allow that some things are directly perceived.

James made a genuine advance over Grote and Helmholtz by analysing the reference relation holding between a thought and of him to specific things of or about which it is knowledge. In fact, he gave two different analyses. On both analyses, a thought constituting knowledge about a thing refers to and is knowledge about 'a reality, whenever it actually or potentially ends in' a thought constituting knowledge of acquaintance with that thing (1975). The two analyses differ in their treatments of knowledge of acquaintance. On James's first analysis, reference in both sorts of knowledge is mediated by causal chains. A thought constituting pure knowledge of acquaintances with a thing refers to and is knowledge of 'whatever reality it directly or indirectly operates on and resembles' (1975). The concepts of a thought 'operating on' a thing or 'terminating in' another thought are causal, but where Grote found teleology and final causes. On James's later analysis, the reference involved in knowledge of acquaintance with a thing is direct. A thought constituting knowledge of acquaintance with a thing either is that thing, or has that thing as a constituent, and the thing and the experience of it is identical (1975, 1976).

James further agreed with Grote that pure knowledge of acquaintance with things, i.e., sensory experience, is epistemologically priori to knowledge about things. While the epistemic justification involved in knowledge about things rests on the foundation of sensation, all thoughts about things are fallible and their justification is augmented by their mutual coherence. James was unclear about the precise epistemic status of knowledge of acquaintance. At times, thoughts constituting pure knowledge of acquaintance are said to posses 'absolute veritableness' (1890) and 'the maximal conceivable truth' (1975), suggesting that such thoughts are genuinely cognitive and that they provide an infallible epistemic foundation. At other times, such thoughts are said not to bear truth-values, suggesting that 'knowledge' of acquaintance is not genuine knowledge at all, but only a non-cognitive necessary condition of genuine knowledge, knowledge about things (1976). Russell understood James to hold the latter view.

Russell agreed with Grote and James on the following points: First, knowing things involves experiencing them. Second, knowledge of things by acquaintance is epistemically basic and provides an infallible epistemic foundation for knowledge about things. (Like James, Russell vacillated about the epistemic status of knowledge by acquaintance, and it eventually was replaced at the epistemic foundation by the concept of noticing.) Third, knowledge about things is more articulate and explicit than knowledge by acquaintance with things. Fourth, knowledge about things is causally removed from knowledge of things by acquaintance, by processes of reelection, analysis and inference (1911, 1913, 1959).

But, Russell also held that the term 'experience' must not be used uncritically in philosophy, on account of the 'vague, fluctuating and ambiguous' meaning of the term in its ordinary use. The precise concept found by Russell 'in the nucleus of this uncertain patch of meaning' is that of direct occurrent experience of a thing, and he used the term 'acquaintance' to express this relation, though he used that term technically, and not with all its ordinary meaning (1913). Nor did he undertake to give a constitutive analysis of the relation of acquaintance, though he allowed that it may not be unanalysable, and did characterize it as a generic concept. If the use of the term 'experience' is restricted to expressing the determinate core of the concept it ordinarily expresses, then we do not experience ordinary objects in the external world, as we commonly think and as Grote and James held we do. In fact, Russell held, one can be acquainted only with one's sense-data, i.e., particular colours, sounds, and so forth, one's occurrent mental states, universals, logical forms, and perhaps, oneself.

Russell agreed with James that knowledge of things by acquaintance 'is essentially simpler than any knowledge of truths, and logically independent of knowledge of truths' (1912, 1929). The mental states involved when one is acquainted with things do not have propositional contents. Russell's reasons here seem to have been similar to James's. Conceptually unmediated reference to particulars necessary for understanding any proposition mentioning a particular, e.g., 1918-19, and, if scepticism about the external world is to be avoided, some particulars must be directly perceived (1911). Russell vacillated about whether or not the absence of propositional content renders knowledge by acquaintance incommunicable.

Russell agreed with James that different accounts should be given of reference as it occurs in knowledge by acquaintance and in knowledge about things, and that in the former case, reference is direct. But Russell objected on a number of grounds to James's causal account of the indirect reference involved in knowledge about things. Russell gave a descriptional rather than a causal analysis of that sort of reference: A thought is about a thing when the content of the thought involves a definite description uniquely satisfied by the thing referred to. Indeed, he preferred to speak of knowledge of things by description, rather than knowledge about things.

Russell advanced beyond Grote and James by explaining how thoughts can be more or less articulate and explicit. If one is acquainted with a complex thing without being aware of or acquainted with its complexity, the knowledge one has by acquaintance with that thing is vague and inexplicit. Reflection and analysis can lead one to distinguish constituent parts of the object of acquaintance and to obtain progressively more comprehensible, explicit, and complete knowledge about it (1913, 1918-19, 1950, 1959).

Apparent facts to be explained about the distinction between knowing things and knowing about things are there. Knowledge about things is essentially propositional knowledge, where the mental states involved refer to specific things. This propositional knowledge can be more or less comprehensive, can be justified inferentially and on the basis of experience, and can be communicated. Knowing things, on the other hand, involves experience of things. This experiential knowledge provides an epistemic basis for knowledge about things, and in some sense is difficult or impossible to communicate, perhaps because it is more or less vague.

If one is unconvinced by James and Russell's reasons for holding that experience of and reference work to things that are at least sometimes direct. It may seem preferable to join Helmholtz in asserting that knowing things and knowing about things both involve propositional attitudes. To do so would at least allow one the advantages of unified accounts of the nature of knowledge (propositional knowledge would be fundamental) and of the nature of reference: Indirect reference would be the only kind. The two kinds of knowledge might yet be importantly different if the mental states involved have different sorts of causal origins in the thinker's cognitive faculties, involve different sorts of propositional attitudes, and differ in other constitutive respects relevant to the relative vagueness and communicability of the mental sates.

In any of cases, perhaps most, Foundationalism is a view concerning the 'structure' of the system of justified belief possessed by a given individual. Such a system is divided into 'foundation' and 'superstructure', so related that beliefs in the latter depend on the former for their justification but not vice versa. However, the view is sometimes stated in terms of the structure of 'knowledge' than of justified belief. If knowledge is true justified belief (plus, perhaps, some further condition), one may think of knowledge as exhibiting a Foundationalist structure by virtue of the justified belief it involves. In any event, the construing doctrine concerning the primary justification is layed the groundwork as affording the efforts of belief, though in feeling more free, we are to acknowledge the knowledgeable infractions that will from time to time be worthy in showing to its recognition.

The first step toward a more explicit statement of the position is to distinguish between 'mediate' (indirect) and 'immediate' (direct) justification of belief. To say that a belief is mediately justified is to any that it s justified by some appropriate relation to other justified beliefs, i.e., by being inferred from other justified beliefs that provide adequate support for it, or, alternatively, by being based on adequate reasons. Thus, if my reason for supposing that you are depressed is that you look listless, speak in an oddly and foreign tone of voice, exhibit no interest in things you are usually interested in, and so forth, then my belief that you are depressed is justified, if, at all, by being adequately supported by my justified belief that you look listless, speak in a flat tone of voice. . . .

A belief is immediately justified, on the other hand, if its justification is of another sort, e.g., if it is justified by being based on experience or if it is 'self-justified'. Thus my belief that you look listless may not be based on anything else I am justified in believing but just on the cay you look to me. And my belief that $2 + 3 = 5$ may be justified not because I infer it from something else, I justifiably believe, but simply because it seems obviously true to me.

In these terms we can put the thesis of Foundationalism by saying that all mediately justified beliefs owe their justification, ultimately to immediately justified beliefs. To get a more detailed idea of what this amounts to it will be useful to consider the most important argument for Foundationalism, the regress argument. Consider a mediately justified belief that 'p' (we are using lowercase letters as dummies for belief contents). It is, by hypothesis, justified by its relation to one or more other justified beliefs, 'q' and 'r'. Now what justifies each of these, e.g., q? If it too is mediately justified that is because it is related accordingly to one or subsequent extra justified beliefs, e.g., 's'. By virtue of what is 's' justified? If it is mediately justified, the same problem arises at the next stage. To avoid both circularity and an infinite regress, we are forced to suppose that in tracing back this chain we arrive at one or more immediately justified beliefs that stop the regress, since their justification does not depend on any further justified belief.

According to the infinite regress argument for Foundationalism, if every justified belief could be justified only by inferring it from some further justified belief, there would have to be an infinite regress of justifications: Because there can be no such regress, there must be justified beliefs that are not justified by appeal to some further justified belief. Instead, they are non-inferentially or immediately justified, they are basic or foundational, the ground on which all our other justifiable beliefs are to rest.

Variants of this ancient argument have persuaded and continue to persuade many philosophers that the structure of epistemic justification must be foundational. Aristotle recognized that if we are to have knowledge of the conclusion of an argument in the basis of its premisses, we must know the premisses. But if knowledge of a premise always required knowledge of some further proposition, then in order to know the premise we would have to know each proposition in an infinite regress of propositions. Since this is impossible, there must be some propositions that are known, but not by demonstration from further propositions: There must be basic, non-demonstrable knowledge, which grounds the rest of our knowledge.

Foundationalist interests as attentions for regress arguments often overlooks the fact that they have also been advanced on behalf of scepticism, relativism, fideisms, conceptualism and Coherentism. Sceptics agree with Foundationalist's both that there is no infinite regress of justifications and that, nevertheless, there must be one if every justified belief can be justified, only inferentially, by appeal to some further justified belief. But sceptics think all true justification must be inferential in this way - the Foundationalist's rhetorical dialects of immediate justification merely overshadow the requiring of rational justification, properly so-called, but sceptics concluded that none of our beliefs are justified. Relativists follow essentially the same pattern of sceptical argument, concluding that our beliefs can only be justified relative to the arbitrary starting assumptions or presuppositions either of an individual or of a form of life.

Regress arguments are not limited to epistemology. In ethics there is Aristotle's regress argument (in 'Nichomachean Ethics') for the existence of a single end of rational action. In metaphysics there is Aquinas's regress argument for an unmoved mover: If a mover that it is in motion, there would have to be an infinite sequence of movers each moved by a further mover, since there can be no such sequence, there is an unmoved mover. A related argument has recently been given to show that not every state of affairs can have an explanation or cause of the sort posited by principles of sufficient reason, and such principles are false, for reasons having to do with their own concepts of explanation (Post, 1980; Post, 1987).

The premise of which in presenting Foundationalism as a view concerning the structure 'that is in fact exhibited' by the justified beliefs of a particular person has sometimes been construed in ways that deviate from each of the phrases that are contained in the previous sentence. Thus, it is sometimes taken to characterise the structure of 'our knowledge' or 'scientific knowledge', rather than the structure of the cognitive system of an individual subject. As for the other phrase, Foundationalism is sometimes thought of as concerned with how knowledge (justified belief) is acquired or built up, than with the structure of what a person finds herself with at a certain point. Thus some people think of scientific inquiry as starting with the recordings of observations (immediately justified observational beliefs), and then inductively inferring generalizations. Again, Foundationalism is sometimes thought of not as a description of the finished product or of the mode of acquisition, but rather as a proposal for how the system could be reconstructed, an indication of how it could all be built up from immediately justified foundations. This last would seem to be the kind of Foundationalism we find in Descartes. However, Foundationalism is most usually thought of in contemporary Anglo-American epistemology as an account of the structure actually exhibited by an individual's system of justified belief.

It should also be noted that the term is used with a deplorable looseness in contemporary, literary circles, even in certain corners of the philosophical world, to refer to anything from realism -the view that reality has a definite constitution regardless of how we think of it or what we believe about it to various kinds of 'absolutism' in ethics, politics, or wherever, and even to the truism that truth is stable (if a proposition is true, it stays true).

Since Foundationalism holds that all mediate justification rests on immediately justified beliefs, we may divide variations in forms of the view into those that have to do with the immediately justified beliefs, the 'foundations', and those that have to do with the modes of derivation of other beliefs from these, how the 'superstructure' is built up. The most obvious variation of the first sort has to do with what modes of immediate justification are recognized. Many treatments, both pro and con, are parochially restricted to one form of immediate justification -self-evidence, self-justification (self-warrant), justification by a direct awareness of what the belief is about, or whatever. It is then unwarrantly assumed by critics that disposing of that one form will dispose of Foundationalism generally (Alston, 1989, ch. 3). The emphasis historically has been on beliefs that simply 'record' what is directly given in experience (Lewis, 1946) and on self-evident propositions (Descartes' 'clear and distinct perceptions and Locke's 'Perception of the agreement and disagreement of ideas'). But self-warrant has also recently received a great deal of attention (Alston 1989), and there is also a Reliabilist version according to which a belief can be immediately justified just by being acquired by a reliable belief-forming process that does not take other beliefs as inputs (BonJour, 1985).

Functionalism also differs as to what further constraints, if any, is put on foundations. Historically, it has been common to require of the foundations of knowledge that they exhibit certain 'epistemic immunities', as we might put it, immunity from error, refutation or doubt. Thus Descartes, along with many other seventeenth and eighteenth-century philosophers, took it that any knowledge worthy of the name would be based on cognations the truth of which is guaranteed (infallible), that were maximally stable, immune from ever being shown to be mistaken, as incorrigible, and concerning which no reasonable doubt could be raised (indubitable). Hence the search in the 'Meditations' for a divine guarantee of our faculty of rational intuition. Criticisms of Foundationalism have often been directed at these constraints: Lehrer, 1974, Will, 1974? Both responded to in Alston, 1989. It is important to realize that a position that is Foundationalist in a distinctive sense can be formulated without imposing any such requirements on foundations.

307

There are various ways of distinguishing types of Foundationalist epistemology by the use of the variations we have been enumerating. Plantinga (1983), has put forwards an influential innovation of criterial Foundationalism, specified in terms of limitations on the foundations. He construes this as a disjunction of 'ancient and medieval Foundationalism', which takes foundations to comprise what is self-evidently and 'evident to he senses', and 'modern Foundationalism' that replaces 'evidently to the senses' with 'incorrigible', which in practice was taken to apply only to beliefs about one's present states of consciousness. Plantinga himself developed this notion in the context of arguing those items outside this territory, in particular certain beliefs about God, could also be immediately justified. A popular recent distinction is between what is variously called 'strong' or 'extreme' Foundationalism and 'moderate', 'modest' or 'minimal' Foundationalism, with the distinction depending on whether various epistemic immunities are required of foundations. Finally, its distinction is 'simple' and 'iterative' Foundationalism (Alston, 1989), depending on whether it is required of a foundation only that it is immediately justified, or whether it is also required that the higher level belief that the firmer belief is immediately justified is itself immediately justified. Suggesting only that the plausibility of the stronger requirement stems from a 'level confusion' between beliefs on different levels.

The classic opposition is between Foundationalism and Coherentism. Coherentism denies any immediate justification. It deals with the regress argument by rejecting 'linear' chains of justification and, in effect, taking the total system of belief to be epistemically primary. A particular belief is justified yo the extent that it is integrated into a coherent system of belief. More recently into a pragmatist like John Dewey has developed a position known as contextualism, which avoids ascribing any overall structure to knowledge. Questions concerning justification can only arise in particular context, defined in terms of assumptions that are simply taken for granted, though they can be questioned in other contexts, where other assumptions will be privileged.

Foundationalism can be attacked both in its commitment to immediate justification and in its claim that all mediately justified beliefs ultimately depend on the former. Though, it is the latter that is the position's weakest point, most of the critical fire has been detected to the former. As pointed out about much of this criticism has been directly against some particular form of immediate justification, ignoring the possibility of other forms. Thus, much anti-foundationalist artillery has been directed at the 'myth of the given'. The idea that facts or things are 'given' to consciousness in a pre-conceptual, pre-judgmental mode, and that beliefs can be justified on that basis (Sellars, 1963). The most prominent general argument against immediate justification is a 'level ascent' argument, according to which whatever is taken ti immediately justified a belief that the putative justifier has in supposing to do so. Hence, since the justification of the higher level belief after all (BonJour, 1985). We lack adequate support for any such higher level requirements for justification, and if it were imposed we would be launched on an infinite undergo regress, for a similar requirement would hold equally for the higher level belief that the original justifier was efficacious.

Coherence is a major player in the theatre of knowledge. There are coherence theories of belief, truth, and justification. These combine in various ways to yield theories of knowledge. We will proceed from belief through justification to truth. Coherence theories of belief are concerned with the content of beliefs. Consider a belief you now have, the beliefs that you are reading a page in a book, so what makes that belief the belief that it is? What makes it the belief that you are reading a page in a book than the belief that you have a monster in the garden?

One answer is that the belief has a coherent place or role in a system of beliefs. Perception has an influence on belief. You respond to sensory stimuli by believing that you are reading a page in a book rather than believing that you have a centaur in the garden. Belief has an influence on action. You will act differently if you believe that you are reading a page than if you believe something about a centaur. Perspicacity and action undermine the content of belief, however, the same stimuli may produce various beliefs and various beliefs may produce the same action. The role that gives the belief the content it has in the role it plays in a network of relations to the beliefs, the role in inference and implications, for example, I refer different things from believing that I am inferring different things from believing that I am reading a page in a book than from any other beliefs, just as I infer that belief from any other belief, just as I infer that belief from different things than I infer other beliefs from another.

The input of perception and the output of an action supplement the centre role of the systematic relations the belief has to other beliefs, but it is the systematic relations that give the belief the specific content it has. They are the fundamental source of the content of beliefs. That is how coherence comes in. A belief has the content that it does because of the way in which it coheres within a system of beliefs (Rosenberg, 1988). We might distinguish weak coherence theories of the content of beliefs from strong coherence theories. Weak coherence theories affirm that coherences are one-determinant of the content of belief. Strong coherence theories of the contents of belief affirm that coherence is the sole determinant of the content of belief.

When we turn from belief to justification, we are in confronting a corresponding group of similarities fashioned by their coherences motifs. What makes one belief justified and another not? The answer is the way it coheres with the background system of beliefs. Again, there is a distinction between weak and strong theories of coherence. Weak theories tell 'us' that the way in which a belief coheres with a background system of beliefs is one determinant of justification, other typical determinants being perception, memory and intuition. Strong theories, by contrast, tell 'us' that justification is solely a matter of how a belief coheres with a system of beliefs. There is, however, another distinction that cuts across the distinction between weak and strong coherence theories of justification. It is the distinction between positive and negative coherence theories (Pollock, 1986). A positive coherence theory tells 'us' that if a belief coheres with a background system of belief, then the belief is justified. A negative coherence theory tells 'us' that if a belief fails to cohere with a background system of beliefs, then the belief is not justified. We might put this by saying that, according to a positive coherence theory, coherence has the power to produce justification, while according to a negative coherence theory, coherence has only the power to nullify justification.

A strong coherence theory of justification is a combination of a positive and a negative theory that tells 'us' that a belief is justified if and only if it coheres with a background system of beliefs.

Traditionally, belief has been of epistemological interest in its propositional guise: 'S' believes that 'p', where 'p' is a proposition toward which an agent, 'S', exhibits an attitude of acceptance. Not all belief is of this sort. If I trust what you say, I believe you. And someone may believe in Mrs. Thatcher, or in a free-market economy, or in God. It is sometimes supposed that all belief is 'reducible' to propositional belief, belief-that. Thus, my believing you might be thought a matter of my believing, perhaps, that what you say is true, and your belief in free-markets or in God, a matter of your believing that free-market economy's are desirable or that God exists.

It is doubtful that non-propositional believing can, in every case, be reduced in this way, yet, on this point has tended to focus on an apparent distinction between 'belief-that' and 'belief-in', as the application of this distinction as to belief in God. Some philosophers have followed Aquinas (1225-74), in supposing that to believe in, and God is simply to believe that certain truths hold: That God exists, that he is benevolent, and so forth. Others (e.g., Hick, 1957) argue that belief-in is a distinctive attitude, one that includes essentially an element of trust. More commonly, belief-in has been taken to involve a combination of propositional belief together with some further attitude.

H.H. Price (1969) defends the claims that there are different sorts of 'belief-in', some, but not all, reducible to 'beliefs-that'. If you believe in God, you believe that God exists, that God is good, and so forth, but, according to Price, your belief involves, in addition, a certain complex pro-attitude toward its object. One might attempt to analyse this further attitude in terms of additional beliefs-that: 'S' believes in ',' just in case (1) 'S' believes that ',' exists (and perhaps holds further factual beliefs about (,): (2)'S' believes that ',' is good or valuable in some respect, and (3) 'S' believes that ,'s being good or valuable in this respect is itself a good thing. An analysis of this sort, however, fails adequately to capture the further affective component of belief-in. Thus, according to Price, if you believe in God, your belief is not merely that certain truths hold, but you posses, an attitude of commitment and trust as your beliefs carry to your belief in God.

But why has discussion centred on justification rather than warrant? And precisely what is justification? And why has the discussion of justification of theistic belief focussed so heavily on arguments for and against the existence of God?

As to the first question, we can see why once we see that the dominant epistemological tradition in modern Western philosophy has tended to 'identify' warrant with justification. On this way of looking at the matter, warrant, that which distinguishes knowledge from mere true belief, just 'is' justification. Belief theory of knowledge-the theory according to which knowledge is justified true belief has enjoyed the status of orthodoxy. According to this view, knowledge is justified truer belief, therefore any of your beliefs have warrant for you if and only if you are justified in holding it.

What is justification? What is it to be justified in holding a belief? To get a proper sense of the answer, we must turn to those twin towers of western epistemology. René Descartes and especially, John Locke. The first thing to see is that according to Descartes and Locke, there are epistemic or intellectual duties, or obligations, or requirements. Thus, Locke:

Faith is nothing but a firm assent of the mind, which if it is regulated, A is our duty, cannot be afforded to anything, but upon good reason: And cannot be opposite to it, he that believes, without having any reason for believing, may be in love with his own fanciers: But, seeks neither truth as he ought, nor pats the obedience due his maker, which would have him use those discerning faculties he has given him: To keep him out of mistake and error. He that does this to the best of his power, however, he sometimes lights on truth, is in the right but by chance: And I know not whether the luckiest of the accidents will excuse the irregularity of his proceeding. This, at least is certain, that he must be accountable for whatever mistakes he runs into: Whereas, he that makes use of the light and faculties God has given him, by seeks sincerely to discover truth, by those helps and abilities he has, may have this satisfaction in doing his duty as rational creature, that though he should miss truth, he will not miss the reward of it. For him that governs his assenting right, and places it as he should, who in any case or manner, so ever believes or disbelieves, accordingly as reason directs him. He manages otherwise, transgresses against his own light, and misuses those faculties, which were given him . . . (Essays 4.17.24).

Rational creatures, creatures with reason, creatures capable of believing propositions (and of disbelieving and being agnostic with respect to them), say Locke, have duties and obligation with respect to the regulation of their belief or assent. Now the central core of the notion of justification(as the etymology of the term indicates) this: One is justified in doing something or in believing a certain way, if in doing one is innocent of wrong doing and hence not properly subject to blame or censure. You are justified, therefore, if you have violated no duties or obligations, if you have conformed to the relevant requirements, if you are within your rights. To be justified in believing something, then, is to be within your rights in so believing, to be flouting no duty, to be to satisfy your epistemic duties and obligations. This way of thinking of justification has been the dominant way of thinking about justification: And this way of thinking has many important contemporary representatives. Roderick Chisholm, for example (as distinguished an epistemologist as the twentieth century can boast), in his earlier work explicitly explains justification in terms of epistemic duty (Chisholm, 1977).

The (or, a) main epistemological; questions about religious believe, therefore, has been the question whether or not religious belief in general and theistic belief in particular is justified. And the traditional way to answer that question has been to inquire into the arguments for and against theism. Why this emphasis upon these arguments? An argument is a way of marshalling your propositional evidence-the evidence from other such propositions as likens to believe-for or against a given proposition. And the reason for the emphasis upon argument is the assumption that theistic belief is justified if and only if there is sufficient propositional evidence for it. If there is not' much by way of propositional evidence for theism, then you are not justified in accepting it. Moreover, if you accept theistic belief without having propositional evidence for it, then you are ging contrary to epistemic duty and are therefore unjustified in accepting it. Thus, W.K. William James, trumpets that 'it is wrong, always everything upon insufficient evidence', his is only the most strident in a vast chorus of only insisting that there is an intellectual duty not to believe in God unless you have propositional evidence for that belief. (A few others in the choir: Sigmund Freud, Brand Blandhard, H.H. Price, Bertrand Russell and Michael Scriven.)

Now how it is that the justification of theistic belief gets identified with there being propositional evidence for it? Justification is a matter of being blameless, of having done one's duty (in this context, one's epistemic duty): What, precisely, has this to do with having propositional evidence?

The answer, once, again, is to be found in Descartes especially Locke. As, justification is the property your beliefs have when, in forming and holding them, you conform to your epistemic duties and obligations. But according to Locke, a central epistemic duty is this: To believe a proposition only to the degree that it is probable with respect to what is certain for you. What propositions are certain for you? First, according to Descartes and Locke, propositions about your own immediate experience, that you have a mild headache, or that it seems to you that you see something red: And second, propositions that are self-evident for you, necessarily true propositions so obvious that you cannot so much as entertain them without seeing that they must be true. (Examples would be simple arithmetical and logical propositions, together with such propositions as that the whole is at least as large as the parts, that red is a colour, and that whatever exists has properties.) Propositions of these two sorts are certain for you, as fort other prepositions. You are justified in believing if and only if when one and only to the degree to which it is probable with respect to what is certain for you. According to Locke, therefore, and according to the whole modern Foundationalist tradition initiated by Locke and Descartes (a tradition that until has recently dominated Western thinking about these topics) there is a duty not to accept a proposition unless it is certain or probable with respect to what is certain.

In the present context, therefore, the central Lockean assumption is that there is an epistemic duty not to accept theistic belief unless it is probable with respect to what is certain for you: As a consequence, theistic belief is justified only if the existence of God is probable with respect to what is certain. Locke does not argue for his proposition, he simply announces it, and epistemological discussion of theistic belief has for the most part followed hin ion making this assumption. This enables 'us' to see why epistemological discussion of theistic belief has tended to focus on the arguments for and against theism: On the view in question, theistic belief is justified only if it is probable with respect to what is certain, and the way to show that it is probable with respect to what it is certain are to give arguments for it from premises that are certain or, are sufficiently probable with respect to what is certain.

There are at least three important problems with this approach to the epistemology of theistic belief. First, their standards for theistic arguments have traditionally been set absurdly high (and perhaps, part of the responsibility for this must be laid as the door of some who have offered these arguments and claimed that they constitute wholly demonstrative proofs). The idea seems to test. a good theistic argument must start from what is self-evident and proceed majestically by way of self-evidently valid argument forms to its conclusion. It is no wonder that few if any theistic arguments meet that lofty standard -particularly, in view of the fact that almost no philosophical arguments of any sort meet it. (Think of your favourite philosophical argument: Does it really start from premises that are self-evident and move by ways of self-evident argument forms to its conclusion?)

Secondly, attention has been mostly confined to three theistic arguments: The traditional arguments, cosmological and teleological arguments, but in fact, there are many more good arguments: Arguments from the nature of proper function, and from the nature of propositions, numbers and sets. These are arguments from intentionality, from counterfactual, from the confluence of epistemic reliability with epistemic justification, from reference, simplicity, intuition and love. There are arguments from colours and flavours, from miracles, play and enjoyment, morality, from beauty and from the meaning of life. This is even a theistic argument from the existence of evil.

But there are a third and deeper problems here. The basic assumption is that theistic belief is justified only if it is or can be shown to be probable with respect to many a body of evidence or proposition -perhaps, those that are self-evident or about one's own mental life, but is this assumption true? The idea is that theistic belief is very much like a scientific hypothesis: It is acceptable if and only if there is an appropriate balance of propositional evidence in favour of it. But why believe a thing like that? Perhaps the theory of relativity or the theory of evolution is like that, such a theory has been devised to explain the phenomena and gets all its warrant from its success in so doing. However, other beliefs, e.g., memory beliefs and reminiscent feelings felt in other minds is not like that, they are not hypothetical at all, and are not accepted because of their explanatory powers. There are instead, the propositions from which one start in attempting to give evidence for a hypothesis. Now, why assume that theistic belief, belief in God, is in this regard more like a scientific hypothesis than like, say, a memory belief? Why think that the justification of theistic belief depends upon the evidential relation of theistic belief to other things one believes? According to Locke and the beginnings of this tradition, it is because there is a duty not to assent to a proposition unless it is probable with respect to what is certain to you, but is there really any such duty? No one has succeeded in showing that, say, belief in other minds or the belief that there has been a past, is probable with respect to what is certain for 'us'. Suppose it is not: Does it follow that you are living in epistemic sin if you believe that there are other minds? Or a past?

There are urgent questions about any view according to which one has duties of the sort 'do not believe 'p' unless it is probable with respect to what is certain for you;. First, if this is a duty, is it one to which I can conform? My beliefs are for the most part not within my control: Certainly they are not within my direct control. I believe that there has been a past and that there are other people, even if these beliefs are not probable with respect to what is certain forms (and even if I came to know this) I could not give them up. Whether or not I accept such beliefs are not really up to me at all, For I can no more refrain from believing these things than I can refrain from conforming yo the law of gravity. Second, is there really any reason for thinking I have such a duty? Nearly everyone recognizes such duties as that of not engaging in gratuitous cruelty, taking care of one's children and one's aged parents, and the like, but do we also find ourselves recognizing that there is a duty not to believe what is not probable (or, what we cannot see to be probable) with respect to what are certain for 'us'? It hardly seems so. However, it is hard to see why being justified in believing in God requires that the existence of God be probable with respect to some such body of evidence as the set of propositions certain for you. Perhaps, theistic belief is properly basic, i.e., such that one is perfectly justified in accepting it on the evidential basis of other propositions one believes.

Taking justification in that original etymological fashion, therefore, there is every reason ton doubt that one is justified in holding theistic belief only inf one is justified in holding theistic belief only if one has evidence for it. Of course, the term 'justification' has under gone various analogical extensions in the of various philosophers, it has been used to name various properties that are different from justification etymologically so-called, but anagogically related to it. In such a way, the term sometimes used to mean propositional evidence: To say that a belief is justified for someone is to saying that he has propositional evidence (or sufficient propositional evidence) for it. So taken, however, the question whether theistic belief is justified loses some of its interest; for it is not clear (given this use) beliefs that are unjustified in that sense. Perhaps, one also does not have propositional evidence for one's memory beliefs, if so, that would not be a mark against them and would not suggest that there be something wrong holding them.

Another analogically connected way to think about justification (a way to think about justification by the later Chisholm) is to think of it as simply a relation of fitting between a given proposition and one's epistemic vase -which includes the other things one believes, as well as one's experience. Perhaps tat is the way justification is to be thought of, but then, if it is no longer at all obvious that theistic belief has this property of justification if it seems as a probability with respect to many another body of evidence. Perhaps, again, it is like memory beliefs in this regard.

To recapitulate: The dominant Western tradition has been inclined to identify warrant with justification, it has been inclined to take the latter in terms of duty and the fulfilment of obligation, and hence to suppose that there is no epistemic duty not to believe in God unless you have good propositional evidence for the existence of God. Epistemological discussion of theistic belief, as a consequence, as concentrated on the propositional evidence for and against theistic belief, i.e., on arguments for and against theistic belief. But there is excellent reason to doubt that there are epistemic duties of the sort the tradition appeals to here.

And perhaps it was a mistake to identify warrant with justification in the first place. Napoleons have little warrant for him: His problem, however, need not be dereliction of epistemic duty. He is in difficulty, but it is not or necessarily that of failing to fulfill epistemic duty. He may be doing his epistemic best, but he may be doing his epistemic duty in excelsis: But his madness prevents his beliefs from having much by way of warrant. His lack of warrant is not a matter of being unjustified, i.e., failing to fulfill epistemic duty. So warrant and being epistemologically justified by name are not the same things. Another example, suppose (to use the favourite twentieth-century variant of Descartes' evil demon example) I have been captured by Alpha-Centaurian super-scientists, running a cognitive experiment, they remove my brain, and keep it alive in some artificial nutrients, and by virtue of their advanced technology induce in me the beliefs I might otherwise have if I was going about my usual business. Then my beliefs would not have much by way of warrant, but would it be because I was failing to do my epistemic duty? Hardly.

As a result of these and other problems, another, externalist way of thinking about knowledge has appeared in recent epistemology, that a theory of justification is internalized if and only if it requires that all of its factors needed for a belief to be epistemically accessible to that of a person, internal to his cognitive perception, and externalist, if it allows that, at least some of the justifying factors need not be thus accessible, in that they can be external to the believer' s cognitive Perspectives, beyond his ken. However, epistemologists often use the distinction between internalized and externalist theories of epistemic justification without offering any very explicit explanation.

Or perhaps the thing to say, is that it has reappeared, for the dominant sprains in epistemology priori to the Enlightenment were really externalist. According to this externalist way of thinking, warrant does not depend upon satisfaction of duty, or upon anything else to which the Knower has special cognitive access (as he does to what is about his own experience and to whether he is trying his best to do his epistemic duty): It depends instead upon factors 'external' to the epistemic agent -such factors as whether his beliefs are produced by reliable cognitive mechanisms, or whether they are produced by epistemic faculties functioning properly in-an appropriate epistemic environment.

How will we think about the epistemology of theistic belief in more than is less of an externalist way (which is at once both satisfyingly traditional and agreeably up to date)? I think, that the ontological question whether there is such a person as God is in a way priori to the epistemological question about the warrant of theistic belief. It is natural to think that if in fact we have been created by God, then the cognitive processes that issue in belief in God are indeed realisable belief-producing processes, and if in fact God created 'us', then no doubt the cognitive faculties that produce belief in God is functioning properly in an epistemologically congenial environment. On the other hand, if there is no such person as God, if theistic belief is an illusion of some sort, then things are much less clear. Then beliefs in God in of the most of basic ways of wishing that never doubt the production by which unrealistic thinking or another cognitive process not aimed at truth. Thus, it will have little or no warrant. And belief in God on the basis of argument would be like belief in false philosophical theories on the basis of argument: Do such beliefs have warrant? Notwithstanding, the custom of discussing the epistemological questions about theistic belief as if they could be profitably discussed independently of the ontological issue as to whether or not theism is true, is misguided. There two issues are intimately intertwined,

Nonetheless, the vacancy left, as today and as days before are an awakening and untold story beginning by some sparking conscious paradigm left by science. That is a central idea by virtue accredited by its epistemology, where in fact, is that justification and knowledge arising from the proper functioning of our intellectual virtues or faculties in an appropriate environment. This particular yet, peculiar idea is captured in the following criterion for justified belief:

(J) 'S' is justified in believing that 'p' if and only if of S's believing that 'p' is the result of S's intellectual virtues or faculties functioning in appropriate environment.

What is an intellectual virtue or faculty? A virtue or faculty in general is a power or ability or competence to achieve some result. An intellectual virtue or faculty, in the sense intended above, is a power or ability or competence to arrive at truths in a particular field, and to avoid believing falsehoods in that field. Examples of human intellectual virtues are sight, hearing, introspection, memory, deduction and induction. More exactly.

> (V) A mechanism 'M' for generating and/or maintaining beliefs is an intellectual virtue if and only if 'M''s' is a competence to believing true propositions and refrain from false believing propositions within a field of propositions 'F', when one is in a set of circumstances 'C'.

It is required that we specify a particular field of suggestions or its propositional field for 'M', since a given cognitive mechanism will be a competence for believing some kind of truths but not others. The faculty of sight, for example, allows 'us' to determine the colour of objects, but not the sounds that they associatively make. It is also required that we specify a set of circumstances for 'M', since a given cognitive mechanism will be a competence in some circumstances but not others. For example,

315

the faculty of sight allows 'us' to determine colours in a well lighten room, but not in a darkened cave or formidable abyss.

According to the aforementioned formulations, what makes a cognitive mechanism an intellectual virtue is that it is reliable in generating true beliefs than false beliefs in the relevant field and in the relevant circumstances. It is correct to say, therefore, that virtue epistemology is a kind of reliabilism. Whereas, genetic reliabilism maintains that justified belief is belief that results from a reliable cognitive process, virtue epistemology makes a restriction on the kind of process which is allowed. Namely, the cognitive processes that are important for justification and knowledge of those that have their basis in an intellectual virtue.

Finally, that the concerning mental faculty reliability point to the importance of an appropriate environment. The idea is that cognitive mechanisms might be reliable in some environments but not in others. Consider an example from Alvin Plantinga. On a planet revolving around Alfa Centauri, cats are invisible to human beings. Moreover, Alfa Centaurian cats emit a type of radiation that causes humans to form the belief that there I a dog barking nearby. Suppose now that you are transported to this Alfa Centaurian planet, a cat walks by, and you form the belief that there is a dog barking nearby. Surely you are not justified in believing this. However, the problem here is not with your intellectual faculties, but with your environment. Although your faculties of perception are reliable on earth, yet are unrealisable on the Alga Centaurian planet, which is an inappropriate environment for those faculties.

The central idea of virtue epistemology, as expressed in (J) above, has a high degree of initial plausibility. By masking the idea of faculties' cental to the reliability if not by the virtue of epistemology, in that it explains quite neatly to why beliefs are caused by perception and memories are often justified, while beliefs caused by unrealistic and superstition are not. Secondly, the theory gives 'us' a basis for answering certain kinds of scepticism. Specifically, we may agree that if we were brains in a vat, or victims of a Cartesian demon, then we would not have knowledge even in those rare cases where our beliefs turned out true. But virtue epistemology explains that what is important for knowledge is toast our faculties are in fact reliable in the environment in which we are. And so we do have knowledge so long as we are in fact, not victims of a Cartesian demon, or brains in a vat. Finally, Plantinga argues that virtue epistemology deals well with Gettier problems. The idea is that Gettier problems give 'us' cases of justified belief that is 'truer by accident'. Virtue epistemology, Plantinga argues, helps 'us' to understand what it for a belief to be true by accident, and provides a basis for saying why such cases are not knowledge. Beliefs are rue by accident when they are caused by otherwise reliable faculties functioning in an inappropriate environment. Plantinga develops this line of reasoning in Plantinga (1988).

The Humean problem if induction supposes that there is some property 'A' pertaining to an observational or experimental situation, and that of 'A', some fraction m/n (possibly equal to 1) has also been instances of some logically independent property 'B'. Suppose further that the background circumstances, have been varied to a substantial degree and that there is no collateral information available, as the relative frequency of 'B's' among 'A's' or concerning causal nomological connections between instances of 'A' and instances of 'B'.

In this situation, an enumerative or instantial inductive inference would move from the premise, such that m/n of observed 'A's' are 'B's' to the conclusion that approximately m/n of all 'A's' and 'B's'. (The usual probability qualification will be assumed to apply to the inference, than being part of the conclusion). Hereabouts the class of 'A's' should be taken to include not only unobservable 'A's' of future 'A's', but also possible or hypothetical 'a's'. (An alternative conclusion would concern the probability or likelihood of the very next observed 'A' being a 'B').

The traditional or Humean problem of induction, often refereed to simply as 'the problem of induction', is the problem of whether and why inferences that fit this schema should be considered rationally acceptable or justified from an epistemic or cognitive standpoint, i.e., whether and why reasoning in this way is likely lead to true claims about the world. Is there any sort of argument or rationale that can be offered for thinking that conclusions reached in this way are likely to be true if the corresponding premiss is true or even that their chances of truth are significantly enhanced?

Hume's discussion of this deals explicitly with cases where all observed 'A's' are 'B's', but his argument applies just as well to the more general casse. His conclusion is entirely negative and sceptical: inductive inferences are not rationally justified, but are instead the result of an essentially a-rational process, custom or habit. Hume challenges the proponent of induction to supply a cogent line of reasoning that leads from an inductive premise to the corresponding conclusion and offers an extremely influential argument in the form of a dilemma, to show that there can be no such reasoning. Such reasoning would, ne argues, have to be either deductively demonstrative reasoning concerning relations of ideas or 'experimental', i.e., empirical, reasoning concerning mattes of fact to existence. It cannot be the former, because all demonstrative reasoning relies on the avoidance of contradiction, and it is not a contradiction to be supposed that 'the course of nature may change', that of an arranged order was observed in the past and will not continue in the future: However, it cannot be the latter, since any empirical argument would appeal to the success of such reasoning in previous experiences, and the justifiability of generalizing from previous experience is precisely what is at issue-s o that any such appeal would be question-begging, so then, there can be no such reasoning.

An alternative version of the problem may be obtained by formulating it with reference to the so-called Principle of Induction, which says roughly that the future will resemble or, that unobserved cases will reassembly observe cases. An inductive argument may be viewed as enthymematic, with this principle serving as a suppressed premiss, in which case the issue is obviously how such a premise can be justified. Hume's argument is then that no such justification is possible: The principle cannot be justified speculatively as it is not contradictory to deny it: it cannot be justified by appeal to having been true in pervious experience without obviously begging te question.

The predominant recent responses to the problem of induction, at least in the analytic tradition, in effect accept the main conclusion of Hume's argument, viz. That inductive inferences cannot be justified I the sense of showing that the conclusion of such an inference is likely to be truer if the premise is true, and thus attempt to find some other sort of justification for induction.

Bearing upon, and if not taken into account the term 'induction' is most widely used for any process of reasoning that takes 'us' from empirical premises to empirical conclusions supported by the premise, but not deductively entailed by them. Inductive arguments are therefore kinds of amplicative argument, in which something beyond the content of the premises is inferred as probable or supported by them. Induction is, however, commonly distinguished from arguments to theoretical explanations, which share this amplicative character, by being confined to inference in which the conclusion involves the same properties or relations as the premises. The central example is induction by simple enumeration, where from premiss telling that Fa, Fb, Fc., where a, b, c's, are all of some kind 'G', I t is inferred 'G's' from outside the sample, such as future 'G's' will be 'F', or perhaps other people deceive them, children may well infer that everyone is a deceiver. Different but similar inferences are those from the past possession of a property by some object to the same object's future possession, or from the constancy of some law-like pattern in events, and states of affairs to its future constancy: all objects we know of attract each the with a fore inversely proportional to the square of the distance between them, so perhaps they all do so, an wills always do so.

The rational basis of any inference was challenged by David Hume (1711-76), who believed that induction of nature, and merely reflected a habit or custom of the mind. Hume was not therefore sceptical about the propriety of processes of inducting about the tole of reason in either explaining it or justifying it. trying to answer Hume and to show that there is something rationally compelling about the inference is referred to as the problem of induction. It is widely recognized that any rational defence of induction will have to partition well-behaved properties for which the inference is plausible (often called projectable properties) from badly behaved ones for which t is not. It is also recognized that actual inductive habits are more complex than those of simple and science pay attention to such factors as variations within the sample of giving 'us' the evidence, the application of ancillary beliefs about the order of nature, and so on. Nevertheless, the fundamental problem remains that any experience shows 'us' only events occurring within a very restricted part of the vast spatial temporal order about which we then come to believe things.

Nevertheless, in the potential of change we are to think up to the present time but although virtue epistemology has good initial plausibility, we are faced apart by some substantial objections. The first of an objection, which virtue epistemology face is a version of the generality problem. We may understand the problem more clearly if we were to consider the following criterion for justified belief, which results from our explanation of (J):

(J) 'S' is justified in believing that 'p', if and entirely iff,

(1) there are a field 'F' and a set of circumstances 'C' such that

(1) 'S' is in 'C' with respect to the proposition that 'p', and

(2) 'S' is in 'C' with respect to the proposition that 'p', and

(3) If 'S' were in 'C' with respect to a proposition in 'F'.

Then 'S' would very likely believe correctly with regard to that proposition.

The problem arises in how we are to select an appropriate 'F' and 'C'. For given any true belief that 'p', we can always come up with a field 'F' and a set of circumstances 'C', such that 'S' is perfectly reliable in 'F' and 'C'. For any true belief that 'p', let 'F's' be the field including only the propositions 'p' and 'not-p'. Let 'C' include whatever circumstances there are which causes 'p's' to be true, together with the circumstanced which causes 'S' to believe that 'p'. Clearly, 'S' is perfectly reliable with respect to propositions in this field in these circumstances. But we do not want to say that all of S's true beliefs are justified for 'S'. And of course, there is an analogous problem in the other direction of generality. For given any belief that 'p', we can always specify a field of propositions 'F' and a set of circumstances 'C', such that 'p' is in 'F', 'S' is in 'C', and 'S' is not reliable with respect to propositions in 'F' in 'C'.

Variations of this view have been advanced for both knowledge and justified belief. The first formulation of a reliability account of knowing appeared in a note by F.P. Ramsey (1931), who said that a belief was knowledge if it is true, certain and obtained by a reliable process. P. Unger (1968) suggested that 'S' knows that 'p' just in case it is not at all accidental that 'S' is right about its being the case that 'p'. D.M. Armstrong (1973) drew an analogy between a thermometer that reliably indicates the temperature and a belief that reliably indicate the truth. Armstrong said that a non-inferential belief qualified as knowledge if the belief has properties that are nominally sufficient for its truth, i.e., guarantee its truth via laws of nature.

Closely allied to the nomic sufficiency account of knowledge, primarily due to F.I. Dretske (19712, 1981), A.I. Goldman (1976, 1986) and R. Nozick (1981). The core of tis approach is that S's belief that 'p' qualifies as knowledge just in case 'S' believes 'p' because of reasons that would not obtain unless 'p's' being true, or because of a process or method that would not yield belief in 'p' if 'p' were not true. For example, 'S' would not have his current reasons for believing there is a telephone before him, or would not come to believe this, unless there was a telephone before him. Thus, there is a counterfactual reliable guarantor of the belief's being true. A variant of the counterfactual approach says that 'S' knows that 'p' only if there is no 'relevant alterative' situation in which 'p' is false but 'S' would still believe that 'p'.

To a better understanding, this interpretation is to mean that the alterative attempt to accommodate any of an opposing strand in our thinking about knowledge one interpretation is an absolute concept, which is to mean that the justification or evidence one must have in order to know a proposition 'p' must be sufficient to remove all the exclusive possibilities or from one that has to be chosen or those that are different from 'p', where necessities or system of alternative traditions by the proposition 'p' is a proposition incompatible with 'p'. That is, one's justification or evidence for 'p' must be sufficient fort one to know that every alternative too 'p' is false. These elements of our thinking about knowledge are exploited by sceptical argument. These arguments call our attention to alternatives that our evidence cannot eliminate. For example, (Dretske, 1970), when we are at the zoo. We might claim to know that we see a zebra on the basis of certain visual evidence, namely a zebra-like appearance. The sceptic inquires how we know that we are not seeing a clearly disguised mule. While we do have some evidence against the likelihood of such a deception, intuitively it is not strong enough for 'us' to know that we are not so deceived. By pointing out alternatives of this nature that cannot eliminate, as well as others with more general application (dreams, hallucinations, and so forth), the sceptic appears to show that this requirement that our evidence eliminate every alternative is seldom, if ever met.

The above considerations show that virtue epistemology must say more about the selection of relevant fields and sets of circumstances. Establishing addresses the generality problem by introducing the concept of a design plan for our intellectual faculties. Relevant specifications for fields and sets of circumstances are determined by this plan. One might object that this approach requires the problematic assumption of a Designer of the design plan. But Plantinga disagrees on two counts: He does not think that the assumption is needed, or that it would be problematic. Plantinga discusses relevant material in Plantinga (1986, 1987 and 1988). Ernest Sosa addresses the generality problem by introducing the concept of an epistemic perspective. In order to have reflective knowledge, 'S' must have a true grasp of the reliability of her faculties, this grasp being itself provided by a 'faculty of faculties'. Relevant specifications of an 'F' and 'C' are determined by this perspective. Alternatively, Sosa has suggested that relevant specifications are determined by the purposes of the epistemic community. The idea is that fields and sets of circumstances are determined by their place in useful generalizations about epistemic agents and their abilities to act as reliable-information sharers.

The second objection which virtue epistemology faces are that (J) and (J) are too strong. It is possible for 'S' to be justified in believing that 'p', even when S's intellectual faculties are largely unreliable. Suppose, for example, that Jane's beliefs about the world around her are true. It is clear that in this case Jane's faculties of perception are almost wholly unreliable. But we would not want to say that none of Jane's perceptual beliefs are justified. If Jane believes that there is a tree in her yard, and she believes in the belief on the usual tree-like experience, then it seems that she is as justified as we would be regarded a substitutable belief.

Sosa addresses the current problem by arguing that justification is relative to an environment 'E'. Accordingly, 'S' is justified in believing that 'p' relative to 'E', if and only if S's faculties would be reliable in 'E'. Note that on this account, 'S' need not actually be in 'E' in order for 'S' to be justified in believing some proposition relative to 'E'. This allows Soda to conclude that Jane has justified belief in the above case. For Jane is justified in her perceptual beliefs relative to our environment, although she is not justified in those beliefs relative to the environment in which they have actualized her.

We have earlier made mention about analyticity, but the true story of analyticity is surprising in many ways. Contrary to received opinion, it was the empiricist Locke rather than the rationalist Kant who had the better information account of this type or deductive proposition. Frége and Rudolf Carnap (1891-1970) A German logician positivist whose first major works were 'Der Logische Aufbau der Welt' (1926, translates, as 'The Logical Structure of the World,' 1967). Carnap pursued the enterprise of clarifying the structures of mathematics and scientific language (the only legitimate task for scientific philosophy) in 'Logische Syntax der Sprache' (1934, trans.. As 'The Logical Syntax of Language,' 1937). Yet, refinements continued with 'Meaning and Necessity' (1947), while a general losing of the original ideal of reduction culminated in the great 'Logical Foundations of Probability' and the most importantly single work of 'confirmation theory' in 1950. Other works concern the structure of physics and the concept of entropy.

Both, Frége and Carnap, represented as analyticity's best friends in this century, did as much to undermine it as its worst enemies. Quine (1908-) whose early work was written on mathematical logic, and issued in 'A System of Logistic' (1934), 'Mathematical Logic' (1940) and 'Methods of Logic'

(1950) with this collection of papers a 'Logical Point of View' (1953) that his philosophical importance became widely recognized, also, Putman (1926-) his concern in the later period has largely been to deny any serious asymmetry between truth and knowledge as it is obtained in natural science, and as it is obtained in morals and even theology. Books include 'Philosophy of logic' (1971), 'Representation and Reality' (1988) and 'Renewing Philosophy (1992). Collections of his papers include 'Mathematics, Master, sand Method' (1975), 'Mind, Language, and Reality' (1975), and 'Realism and Reason (1983). Both of which represented as having refuted the analytic/synthetic distinction, not only did no such thing, but, in fact, contributed significantly to undoing the damage done by Frége and Carnap. Finally, the epistemological significance of the distinctions is nothing like what it is commonly taken to be.

Locke's account of an analyticity proposition as, for its time, everything that a succinct account of analyticity should be (Locke, 1924, pp. 306-8) he distinguished two kinds of analytic propositions, identified propositions in which we affirm the said terms if itself, e.g., 'Roses are roses', and predicative propositions in which 'a part of the complex idea is predicated of the name of the whole', e.g., 'Roses are flowers' (pp. 306-7). Locke calls such sentences 'trifling' because a speaker who uses them 'trifles with words'. A synthetic sentence, in contrast, such as a mathematical theorem, states 'a truth and conveys with its informative real knowledge'. Correspondingly, Locke distinguishes two kinds of 'necessary consequences', analytic entailment where validity depends on the literal containment of the conclusions in the premiss and synthetic entailments where it does not. (Locke did not originate this concept-containment notion of analyticity. It is discussions by Arnaud and Nicole, and it is safe to say it has been around for a very long time (Arnaud, 1964).

Kant's account of analyticity, which received opinion tells 'us' is the consummate formulation of this notion in modern philosophy, is actually a step backward. What is valid in his account is not novel, and what is novel is not valid. Kant presents Locke's account of concept-containment analyticity, but introduces certain alien features, the most important being his characterizations of most important being his characterization of analytic propositions as propositions whose denials are logical contradictions (Kant, 1783). This characterization suggests that analytic propositions based on Locke's part-whole relation or Kant's explicative copula are a species of logical truth. But the containment of the predicate concept in the subject concept in sentences like 'Bachelors are unmarried' is a different relation from containment of the consequent in the antecedent in a sentence like 'If John is a bachelor, then John is a bachelor or Mary read Kant's Critique'. The former is literal containment whereas, the latter are, in general, not. Talk of the 'containment' of the consequent of a logical truth in the metaphorical, a way of saying 'logically derivable'.

Kant's conflation of concept containment with logical containment caused him to overlook the issue of whether logical truths are synthetically deductive and the problem of how he can say mathematical truths are synthetically deductive when they cannot be denied without contradiction. Historically, the conflation set the stage for the disappearance of the Lockean notion. Frége, who received opinion portrays as second only to Kant among the champions of analyticity, and Carnap, who it portrays as just behind Frége, was jointly responsible for the appearance of concept-containment analyticity.

Frége was clear about the difference between concept containment and logical containment, expressing it as like the difference between the containment of 'beams in a house' the containment of a 'plant in the seed' (Frége, 1853). But he found the former, as Kant formulated it, defective in three ways: It explains analyticity in psychological terms, it does not cover all cases of analytic propositions, and, perhaps, most important for Frége's logicism, its notion of containment is 'unfruitful' as a definition of mechanisms in logic and mathematics (Frége, 1853). In an insidious containment between the two notions of containment, Frége observes that with logical containment 'we are not simply talking out of the box again what we have just put inti it'. This definition makes logical containment the basic notion. Analyticity becomes a special case of logical truth, and, even in this special case, the definitions employ the power of definition in logic and mathematics than mere concept combination.

Carnap, attempting to overcome what he saw a shortcoming in Frége's account of analyticity, took the remaining step necessary to do away explicitly with Lockean-Kantian analyticity. As Carnap saw things, it was a shortcoming of Frége's explanation that it seems to suggest that definitional relations underlying analytic propositions can be extra-logic in some sense, say, in resting on linguistic synonymy. To Carnap, this represented a failure to achieve a uniform formal treatment of analytic propositions and left 'us' with a dubious distinction between logical and extra-logical vocabulary. Hence, he eliminated the reference to definitions in Frége' of analyticity by introducing 'meaning postulates', e.g., statements such as $(\chi)(\chi$ is a bachelor-is unmarried) (Carnap, 1965). Equivalently, standard logical postulates upon which they were modelled, meaning postulates express nothing more than constrains on the admissible models with respect to which sentences and deductions are evaluated for truth and validity. Thus, despite their name, its asymptomatic-balance having to pustulate itself by that in what it holds on to not more than to do with meaning than any value-added statements expressing an indispensable truth. In defining analytic propositions as consequences of (an explained set of) logical laws, Carnap explicitly removed the one place in Frége's explanation where there might be room for concept containment and with it, the last trace of Locke's distinction between semantic and other 'necessary consequences'.

Quine, the staunchest critic of analyticity of our time, performed an invaluable service on its behalf-although, one that has come almost completely unappreciated. Quine made two devastating criticism of Carnap's meaning postulate approach that expose it as both irrelevant and vacuous. It is irrelevant because, in using particular words of a language, meaning postulates fail to explicate analyticity for sentences and languages generally, that is, they do not define it for variables 'S' and 'L' (Quine, 1953). It is vacuous because, although meaning postulates tell 'us' what sentences are to count as analytic, they do not tell 'us' what it is for them to be analytic.

Received opinion was that Quine did much more than refute the analytic/synthetic distinction as Carnap tried to draw it. Received opinion has that Quine demonstrated there is no distinction, however, anyone might try to draw it. Nut this, too, is incorrect. To argue for this stronger conclusion, Quine had to show that there is no way to draw the distinction outside logic, in particular theory in linguistic corresponding to Carnap's, Quine's argument had to take an entirely different form. Some inherent feature of linguistics had to be exploited in showing that no theory in this science can deliver the distinction. But the feature Quine chose was a principle of operationalist methodology characteristic of the school of Bloomfieldian linguistics. Quine succeeds in showing that meaning cannot be made objective sense of in linguistics. If making sense of a linguistic concept requires, as that school claims, operationally defining it in terms of substitution procedures that employ only concepts unrelated to that linguistic concept. But Chomsky's revolution in linguistics replaced the Bloomfieldian taxonomic model of grammars with the hypothetico-deductive model of generative linguistics, and, as a consequence, such operational definition was removed as the standard for concepts in linguistics. The standard of theoretical definition that replaced it was far more liberal, allowing the members of as family of linguistic concepts to be defied with respect to one another within a set of axioms that state their systematic interconnections -the entire system being judged by whether its consequences are confirmed by the linguistic facts. Quine's argument does not even address theories of meaning based on this hypothetico-deductive model (Katz, 1988 and 1990).

Putman, the other staunch critic of analyticity, performed a service on behalf of analyticity fully on a par with, and complementary to Quine's, whereas, Quine refuted Carnap's formalization of Frége's conception of analyticity, Putman refuted this very conception itself. Putman put an end to the entire attempt, initiated by Fridge and completed by Carnap, to construe analyticity as a logical concept (Putman, 1962, 1970, 1975a).

However, as with Quine, received opinion has it that Putman did much more. Putman in credited with having devised science fiction cases, from the robot cat case to the twin earth cases, that are counter examples to the traditional theory of meaning. Again, received opinion is incorrect. These cases are only counter examples to Frége's version of the traditional theory of meaning. Frége's version claims both (1) that senses determines reference, and (2) that there are instances of analyticity, say, typified by 'cats are animals', and of synonymy, say typified by 'water' in English and 'water' in twin earth English. Given (1) and (2), what we call 'cats' could not be non-animals and what we call 'water' could not differ from what the earthier twin called 'water'. But, as Putman's cases show, what we call 'cats' could be Martian robots and what they call 'water' could be something other than H2O Hence, the cases are counter examples to Frége's version of the theory.

Putman himself takes these examples to refute the traditional theory of meaning per se, because he thinks other versions must also subscribe to both (1) and. (2). He was mistaken in the case of (1). Frége's theory entails (1) because it defines the sense of an expression as the mode of determination of its referent (Fridge, 1952, pp. 56-78). But sense does not have to be defined this way, or in any way that entails (1). / it can be defined as (D).

(D) Sense is that aspect of the grammatical structure of expressions and sentences responsible for they're having sense properties and relations like meaningfulness, ambiguity, antonymy, synonymy, redundancy, analyticity and analytic entailment. (Katz, 1972 & 1990).

(Note that this use of sense properties and relations is no more circular than the use of logical properties and relations to define logical form, for example, as that aspect of grammatical structure of sentences on which their logical implications depend.)

(J) makes senses internal to the grammar of a language and reference an external; matter of language use -typically involving extra-linguistic beliefs, Therefore, (D) cuts the strong connection between sense and reference expressed in (1), so that there is no inference from the modal fact that 'cats' refer to robots to the conclusion that 'Cats are animals' are not analytic. Likewise, there is no inference from 'water' referring to different substances on earth and twin earth to the conclusion that our word and theirs are not synonymous. Putman's science fiction cases do not apply to a version of the traditional theory of meaning based on (D).

The success of Putman and Quine's criticism in application to Fridge and Carnap's theory of meaning together with their failure in application to a theory in linguistics based on (D) creates the option of overcoming the shortcomings of the Lockean-Kantian notion of analyticity without switching to a logical notion. this option was explored in the 1960s and 1970s in the course of developing a theory of meaning modelled on the hypothetico-deductive paradigm for grammars introduced in the Chomskyan revolution (Katz, 1972).

This theory automatically avoids Frége's criticism of the psychological formulation of Kant's definition because, as an explication of a grammatical notion within linguistics, it is stated as a formal account of the structure of expressions and sentences. The theory also avoids Frége's criticism that concept-containment analyticity is not 'fruitful' enough to encompass truths of logic and mathematics. The criticism rests on the dubious assumption, parts of Frége's logicism, that analyticity 'should' encompass them, (Beenacerraf, 1981). But in linguistics where the only concern is the scientific truth about natural concept-containment analyticity encompass truths of logic and mathematics. Moreover, since we are seeking the scientific truth about trifling propositions in natural language, we will eschew relations from logic and mathematics that are too fruitful for the description of such propositions. This is not to deny that we want a notion of necessary truth that goes beyond the trifling, but only to deny that, that notion is the notion of analyticity in natural language.

The remaining Frégean criticism points to a genuine incompleteness of the traditional account of analyticity. There are analytic relational sentences, for example, Jane walks with those with whom she strolls, 'Jack kills those he himself has murdered', and so forth, and analytic entailment with existential conclusions, for example, 'I think', therefore 'I exist'. The containment in these sentences is just as literal as that in an analytic subject-predicate sentence like 'Bachelors are unmarried', such are shown to have a theory of meaning construed as a hypothetico-deductive systemisations of sense as defined in (D) overcoming the incompleteness of the traditional account in the case of such relational sentences.

Such a theory of meaning makes the principal concern of semantics the explanation of sense properties and relations like synonymy, an antonymy, redundancy, analyticity, ambiguity, and so forth Furthermore, it makes grammatical structure, specifically, senses structure, the basis for explaining them. This leads directly to the discovery of a new level of grammatical structure, and this, in turn, it makes possibly for a proper definition of analyticity. To see this, consider two simple examples. It is a semantic fact that 'a male bachelor' is redundant and that 'single person' is synonymous with 'woman who never married;. In the case of the redundancy, we have to explain the fact that the sense of the modifier 'male' is already contained in the sense of its head 'bachelor'. In the case of the synonymy, we have to explain the fact that the sense of 'sinister' is identical to the sense of 'woman who never married' (compositionally formed from the senses of 'woman', 'never' and 'married'). But is so fas as such facts concern relations involving the components of the senses of 'bachelor' and 'spinster' and is in as far as these words are syntactically simpler, there must be a level of grammatical structure at which syntactical simplicities are semantically complex. This, in brief, is the route by which we arrive a level of 'decompositional semantic structure' that is the locus of sense structures masked by syntactically simple words.

Discovery of this new level of grammatical structure was followed by attemptive efforts as afforded to represent the structure of the sense's finds there. Without going into detail of sense representations, it is clear that, once we have the notion of decompositional representation, we can see how to generalize Locke and Kant's informal, subject-predicate account of analyticity to cover relational analytic sentences. Let a simple sentence 'S' consisted of a -place predicate 'P' with terms T1 . . ., Tn occupying its argument places. Then:

The analysis in case, first, S has a term T1 that consists of a place predicate Q ($m > n$ or $m = n$) with terms occupying its argument places, and second, P is contained in Q and, for each term TJ. . . . T1 + I . . ., Tn, TJ is contained in the term of Q that occupies the argument place in Q corresponding to the argument place occupied by TJ in P. (Katz, 1972)

To see how (A) works, suppose that 'stroll' in 'Jane walks with those whom she strolls' is decompositionally represented as having the same sense as 'walkidly and in a leisurely way'. The sentence is analytic by (A) because the predicate 'stroll' (the sense of 'stroll) and the term 'Jane' * the sense of 'Jane' associated with the predicate 'walk') is contained in the term 'Jane' (the sense of 'herself' being associated with the predicate 'stroll'). The containment in the case of the other terms is automatic.

The fact that (A) itself makes no reference to logical operators or logical laws indicate that analyticity for subject-predicate sentences can be extended to simple relational sentences without treating analytic sentences as instances of logical truths. Further, the source of the incompleteness is no longer explained, as Fridge explained it, as the absence of 'fruitful' logical apparatus, but is now explained as mistakenly treating what is only a special case of analyticity as if it were the general case. The inclusion of the predicate in the subject is the special case (where $n = 1$) of the general case of the inclusion of an–place predicate (and its terms) in one of its terms. Noting that the defects, by which, Quine complained of in connection with Carnap's meaning-postulated explication are absent in (A). (A) contains no words from a natural language. It explicitly uses variable 'S' and variable 'L' because it is a definition in linguistic theory. Moreover, (A) tell 'us' what property is in virtue of which a sentence is analytic, namely, redundant predication, that is, the predication structure of an analytic sentence is already found in the content of its term structure.

Received opinion has been anti-Lockean in holding that necessary consequences in logic and language belong to one and the same species. This seems wrong because the property of redundant predication provides a nonlogical explanation of why true statements made in the literal use of analytic sentences are necessarily true. Since the property ensures that the objects of the predication in the use of an analytic sentence are chosen on the basis of the features to be predicated of them, the truth-conditions of the statement are automatically satisfied once its terms take on reference. The difference between such a linguistic source of necessity and the logical and mathematical sources vindicate Locke's distinction between two kinds of 'necessary consequence'.

Received opinion concerning analyticity contains another mistake. This is the idea that analyticity is inimical to science, in part, the idea developed as a reaction to certain dubious uses of analyticity such as Fróge's attempt to establish logicism and Schlick's, Ayer's and other logical; postivists attempt to deflate claims to metaphysical knowledge by showing that alleged deductive truths are merely empty analytic truths (Schlick, 1948, and Ayer, 1946). In part, it developed as also a response to a number of cases where alleged analytically, and hence, necessary truths, e.g., the law of excluded a seeming next-to-last of times ordering which of succeeding of subsequent compliance, as having been taken as open to revision, such cases convinced philosophers like Quine and Putnam that the analytic/synthetic distinction is an obstacle to scientific progress.

The problem, if there is, one is one is not analyticity in the concept-containment sense, but the conflation of it with analyticity in the logical sense. This made it seem as if there is a single concept of analyticity that can serve as the grounds for a wide range of deductive truths. But, just as there are two analytic/synthetic distinctions, so there are two concepts of concept. The narrow Lockean/Kantian distinction is based on a narrow notion of expressions on which concepts are senses of expressions in the language. The broad Frégean/Carnap distinction is based on a broad notion of concept on which concepts are conceptions -often scientific one about the nature of the referent (s) of expressions (Katz, 1972) and curiously Putman, 1981). Conflation of these two notions of concepts produced the illusion of a single concept with the content of philosophical, logical and mathematical conceptions, but with the status of linguistic concepts. This encouraged philosophers to think that they were in possession of concepts with the contentual representation to express substantive philosophical claims, e.g., such as Fridge, Schlick and Ayer's, . . . and so on, and with a status that trivializes the task of justifying them by requiring only linguistic grounds for the deductive propositions in question.

Finally, there is an important epistemological implication of separating the broad and narrowed notions of analyticity. Fridge and Carnap took the broad notion of analyticity to provide foundations for necessary and a priority, and, hence, for some form of rationalism, and nearly all rationalistically inclined analytic philosophers followed them in this. Thus, when Quine dispatched the Fróge-Carnap position on analyticity, it was widely believed that necessary, as a priority, and rationalism had also been despatched, and, as a consequence. Quine had ushered in an 'empiricism without dogmas' and 'naturalized epistemology'. But given there is still a notion of analyticity that enables 'us' to pose the problem of how necessary, synthetic deductive knowledge is possible (moreover, one whose narrowness makes logical and mathematical knowledge part of the problem), Quine did not undercut the foundations of rationalism. Hence, a serious reappraisal of the new empiricism and naturalized epistemology is, to any the least, is very much in order (Katz, 1990).

In some areas of philosophy and sometimes in things that are less than important we are to find in the deductively/inductive distinction in which has been applied to a wide range of objects, including concepts, propositions, truths and knowledge. Our primary concern will, however, be with the epistemic distinction between deductive and inductive knowledge. The most common way of marking the distinction is by reference to Kant's claim that deductive knowledge is absolutely independent of all experience. It is generally agreed that S's knowledge that 'p' is independent of experience just in case S's belief that 'p' is justified independently of experience. Some authors (Butchvarov, 1970, and Pollock, 1974) are, however, in finding this negative characterization of deductive unsatisfactory knowledge and have opted for providing a positive characterisation in terms of the type of justification on which such knowledge is dependent. Finally, others (Putman, 1983 and Chisholm, 1989) have attempted to mark the distinction by introducing concepts such as necessity and rational unrevisability than in terms of the type of justification relevant to deductive knowledge.

One who characterizes deductive knowledge in terms of justification that is independent of experience is faced with the task of articulating the relevant sense of experience, and proponents of the deductive ly cites 'intuition' or 'intuitive apprehension' as the source of deductive justification. Furthermore, they maintain that these terms refer to a distinctive type of experience that is both common and familiar to most individuals. Hence, there is a broad sense of experience in which deductive justification is dependent of experience. An initially attractive strategy is to suggest that theoretical justification must be independent of sense experience. But this account is too narrow since memory, for example, is not a form of sense experience, but justification based on memory is presumably not deductive. There appear to remain only two options: Provide a general characterization of the relevant sense of experience or enumerates those sources that are experiential. General characterizations of experience often maintain that experience provides information specific to the actual world while non-experiential sources provide information about all possible worlds. This approach, however, reduces the concept of non-experiential justification to the concept of being justified in believing a necessary truth. Accounts by enumeration have two problems (1) there is some controversy about which sources to include in the list, and (2) there is no guarantee that the list is complete. It is generally agreed that perception and memory should be included. Introspection, however, is problematic, and beliefs about one's conscious states and about the manner in which one is appeared to are plausible regarded as experientially justified. Yet, some, such as Pap (1958), maintain that experiments in imagination are the source of deductive justification. Even if this contention is rejected and deductive justification is characterized as justification independent of the evidence of perception, memory and introspection, it remains possible that there are other sources of justification. If it should be the case that clairvoyance, for example, is a source of justified beliefs, such beliefs would be justified deductively on the enumerative account.

The most common approach to offering a positive characterization of deductive justification is to maintain that in the case of basic deductive propositions, understanding the proposition is sufficient to justify one in believing that it is true. This approach faces two pressing issues. What is it to understand a proposition in the manner that suffices for justification? Proponents of the approach typically distinguish understanding the words used to express a proposition from apprehending the proposition itself and maintain that it is the latter which are relevant to deductive justification. But this move simply shifts the problem to that of specifying what it is to apprehend a proposition. Without a solution to this problem, it is difficult, if possible, to evaluate the account since one cannot be sure that the account since on cannot be sure that the requisite sense of apprehension does not justify paradigmatic inductive propositions as well. Even less is said about the manner in which apprehending a proposition justifies one in believing that it is true. Proponents are often content with the bald assertions that one who understands a basic deductive proposition can thereby 'see' that it is true. But what requires explanation is how understanding a proposition enable one to see that it is true.

Difficulties in characterizing deductive justification in a term either of independence from experience or of its source have led, out-of-the-ordinary to present the concept of necessity into their accounts, although this appeal takes various forms. Some have employed it as a necessary condition for deductive justification, others have employed it as a sufficient condition, while still others have employed it as both. In claiming that necessity is a criterion of the deductive. Kant held that necessity is a sufficient condition for deductive justification. This claim, however, needs further clarification. There are three theses regarding the relationship between the theoretically and the necessary that can be distinguished: (I) if 'p' is a necessary proposition and 'S' is justified in believing that 'p' is necessary, then S's justification is deductive: (ii) If 'p' is a necessary proposition and 'S' is justified in believing that 'p' is necessarily true, then S's justification is deductive: And (iii) If 'p' is a necessary proposition and 'S' is justified in believing that 'p', then S's justification is deductive. For example, many proponents of deductive contend that all knowledge of a necessary proposition is deductive. (2) and (3) have the shortcoming of setting by stipulation the issue of whether inductive knowledge of necessary propositions is possible. (I) does not have this shortcoming since the recent examples offered in support of this claim by Kriple (1980) and others have been cases where it is alleged that knowledge of the 'truth value' of necessary propositions is knowably inductive. (I) has the shortcoming, however, of ruling out the possibility of being justified either in believing that a proposition is necessary on the basis of testimony or else sanctioning such justification as deductive. (ii) and (iii), of course, suffer from an analogous problem. These problems are symptomatic of a general shortcoming of the approach: It attempts to provide a sufficient condition for deductive justification solely in terms of the modal status of the proposition believed without making reference to the manner in which it is justified. This shortcoming, however, can be avoided by incorporating necessity as a necessary but not sufficient condition for a prior justification as, for example, in Chisholm (1989). Here there are two theses that must be distinguished: (1) If 'S' is justified deductively in believing that 'p', then 'p' is necessarily true.

(2) If 'S' is justified deductively in believing that 'p'. Then 'p' is a necessary proposition. (1) and (2), however, allows this possibility. A further problem with both (1) and (2) is that it is not clear whether they permit deductively justified beliefs about the modal status of a proposition. For they require that in order for 'S' to be justified deductively in believing that 'p' is a necessary preposition it must be necessary that 'p' is a necessary proposition. But the status of iterated modal propositions is controversial. Finally, (1) and (2) both preclude by stipulation the position advanced by Kripke (1980) and Kitcher (1980) that there is deductive knowledge of contingent propositions.

The concept of rational unrevisability has also been invoked to characterize deductive justification. The precise sense of rational unrevisability has been presented in different ways. Putnam (1983) takes rational unrevisability to be both a necessary and sufficient condition for deductive justification while Kitcher (1980) takes it to be only a necessary condition. There are also two different senses of rational unrevisability that have been associated with the deductive (I) a proposition is weakly unreviable just in case it is rationally unrevisable in light of any future 'experiential' evidence, and (II) a proposition is strongly unrevisable just in case it is rationally unrevisable in light of any future evidence. Let us consider the plausibility of requiring either form of rational unrevisability as a necessary condition for deductive justification. The view that a proposition is justified deduction, only if it is strongly unrevisable entails that if a non-experiential source of justified beliefs is fallible but self-correcting, it is not a deductive source of justification. Casullo (1988) has argued that it vis implausible to maintain that a proposition that is warranted as non-experiential and is 'not' justified deductively merely because it is revisable in light of further non-experiential evidence. The view that a proposition is justified deductively only if it is, weakly unrevisable is not open to this objection since it excludes only recision in light of experiential evidence. It does, however, face a different problem. To maintain that S's justified belief that 'p' is justified deductively is to make a claim about the type of evidence that justifies 'S' in believing that 'p'. On the other hand, to maintain that S's justified belief that 'p' is rationally revisable in light of experiential evidence is to make a claim about the type of evidence that can defeat S's justification for believing that 'p' that a claim about the type of evidence that justifies 'S' in believing that 'p'. Hence, it has been argued by Edidin (1984) and Casullo (1988) that to hold that a belief is justified deductively only if it is weakly unrevisable is either to confuse supporting evidence with defeating evidence or to endorse some implausible this about the relationship between the two such as that if evidence of the sort as the kind 'A' can defeat the justification conferred on S's belief that 'p' by evidence of kind 'B' then S's justification for believing that 'p' is based on evidence of kind 'A'.

The most influential idea in the theory of meaning in the past hundred years is the thesis that the meaning of an indicative sentence is given by its truth-conditions. On this conception, to understand a sentence is to know its truth-conditions. The conception was first clearly formulated by Fridge, was developed in a distinctive way by the early Wittgenstein, and is a leading idea of Donald Herbert Davidson (1917-), who is also known for rejection of the idea of as conceptual scheme, thought of as something peculiar to one language or one way of looking at the world, arguing that where the possibility of translation stops so dopes the coherence of the idea that there is anything to translate. His [papers are collected in the 'Essays on Actions and Events' (1980) and 'Inquiries into Truth and Interpretation' (1983). However, the conception has remained so central that those who offer opposing theories characteristically define their position by reference to it.

Wittgenstein's main achievement is a uniform theory of language that yields an explanation of logical truth. A factual sentence achieves sense by dividing the possibilities exhaustively into two groups, those that would make it true and those that would make it false. A truth of logic does not divide the possibilities but comes out true in all of them. It, therefore, lacks sense and says nothing, but it is not nonsense. It is a self-cancellation of sense, necessarily true because it is a tautology, the limiting case of factual discourse, like the figure '0' in mathematics. Language takes many forms and even factual discourse does not consist entirely of sentences like 'The fork is placed to the left of the knife'. However, the first thing that he gave up was the idea that this sentence itself needed further analysis into basic sentences mentioning simple objects with no internal structure. He was to concede, that a descriptive word will often get its meaning partly from its place in a system, and he applied this idea to colour-words, arguing that the essential relations between different colours do not indicate that each colour has an internal structure that needs to be taken apart. On the contrary, analysis of our colour-words would only reveal the same pattern-ranges of incompatible of existing of properties-recurring at every level, because that is how we carve up the world.

Indeed, it may even be the case that of our ordinary language is created by moves that we ourselves make. If so, the philosophy of language will lead into the connection between the meaning of a word and the applications of it that its users intend to make. There is also an obvious need for people to understand each other's meanings of their words. There are many links between the philosophy of language and the philosophy of mind and it is not surprising that the impersonal examination of language in the 'Tractatus: was replaced by a very different, anthropocentric treatment in 'Philosophical Investigations?'

If the logic of our language is created by moves that we ourselves make, various kinds of realisms are threatened. First, the way in which our descriptive language carves up the world will not be forces on 'us' by the natures of things, and the rules for the application of our words, which feel the external constraints, will really come from within 'us'. That is a concession to nominalism that is, perhaps, readily made. The idea that logical and mathematical necessity is also generated by what we ourselves accomplish what is more paradoxical. Yet, that is the conclusion of Wittengenstein (1956) and (1976), and here his anthropocentricism has carried less conviction. However, a paradox is not sure of error and it is possible that what is needed here is a more sophisticated concept of objectivity than Platonism provides.

In his later work Wittgenstein brings the great problem of philosophy down to earth and traces them to very ordinary origins. His examination of the concept of 'following a rule' takes him back to a fundamental question about counting things and sorting them into types: 'What qualifies as doing the same again? Of a courser, this question as an inconsequential fundamental and would suggest that we forget it and get on with the subject. But Wittgenstein's question is not so easily dismissed. It has the naive profundity of questions that children ask when they are first taught a new subject. Such questions remain unanswered without detriment to their learning, but they point the only way to complete understanding of what is learned.

It is, nevertheless, the meaning of a complex expression in a function of the meaning of its constituents, that is, indeed, that it is just a statement of what it is for an expression to be semantically complex. It is one of the initial attractions of the conception of meaning as truth-conditions that it permits a smooth and satisfying account of the way in which the meaning of a complex expression is a dynamic function of the meaning of its constituents. On the truth-conditional conception, to give the meaning of an expression is to state the contribution it makes to the truth-conditions of sentences in which it occurs. for singular terms-proper names, indexical, and certain pronoun's -this is done by stating the reference of the term in question.

The truth condition of a statement is the condition the world must meet if the statement is to be true. To know this condition is equivalent to knowing the meaning of the statement. Although, this sounds as if it gives a solid anchorage for meaning, some of the security disappears when it turns out that the truth condition can only be defined by repeating the very same statement, the truth condition of 'snow is white' is that snow is white, the truth condition of 'Britain would have capitulated had Hitler invaded' is that Britain would have capitulated had Hitler invaded. It is disputed whether this element of running-on-the-spot disqualifies truth conditions from playing the central role in a substantive theory of meaning. Truth-conditional theories of meaning are sometimes opposed by the view that to know the meaning of a statement is to be able to users it in a network of inferences.

On the truth-conditional conception, to give the meaning of expressions is to state the contributive function it makes to the dynamic function of sentences in which it occurs. For singular terms-proper names, and certain pronouns, as well are indexical-this is done by stating the reference of the term in question. For predicates, it is done either by stating the conditions under which the predicate is true of arbitrary objects, or by stating the conditions under which arbitrary atomic sentence containing it is true. The meaning of a sentence-forming operator is given by stating its distributive contribution to the truth-conditions of a complete sentence, as a function of the semantic values of the sentences on which it operates. For an extremely simple, but nonetheless, it is a structured language, we can state the contributions various expressions make to truth conditions as follows:

A1: The referent of 'London' is London.

A2: The referent of 'Paris' is Paris.

A3: Any sentence of the form [a] is beautiful' is true if and only if the referent of [a] is beautiful.

A4: Any sentence of the form [a] is larger than [b] is true if and only

If the referent of [a] is larger than the referent of [b].

A5: Any sentence of the form 'It is not the case that A' is true if and only if it is not the case that 'A' is true.

A6: Any sentence of the form 'A and B' are true if and only is 'A' is true and 'B' is true.

The principle's A2-A6 takes a form to a simple theory of truth for a fragment of English. In this theory, it is possible to derive these consequences: That 'Paris is beautiful' is true if and only if Paris is beautiful (from A2 and A3), which 'London is larger than Paris, and it seems not th case that London is beautiful' is true if and only if London is larger than Paris and it is not the case that London is beautiful (from A1-As): And in general, for any sentence 'A' of this simple language, we can derive something of the thing or that manifests itself as a form of a mode as fixed or accepted in the from 'A' is true if and only if 'A'.

The theorist of truth conditions should insist that not every true statement about the reference of an expression be fit to be an axiom in a meaning-giving theory of truth for a language. The axiom:

London' refers to the city in which there was a huge fire in 1666

Is a true statement about the reference of 'London?'. It is a consequence of a theory that substitutes this axiom for 'A' in our simple truth theory that 'London is beautiful' is true if and only if the city in which there was a huge fire in 1666 is beautiful. Since a subject can align oneself as to a better understanding in the mannerlifashion, of which is characterized by the meaning applied to 'London' is without knowing that last-mentioned truth conditions, this replacement axiom is not fit to be an axiom in a meaning-specifying truth theory. It is, of course, incumbent on a theorist of meaning as truth conditions to state the constraints on the acceptability of axioms in a way that does not presuppose a deductive, non-truth conditional conception of meaning.

Among the many challenges facing the theorist of truth conditions, two are particularly salient and fundamental. First, the theorist has to answer the charge of triviality or vacuity. Second, the theorist must offer an account of what it is for a person's language to be truly descriptive by a semantic theory containing a given semantic axiom.

We can take the charge of triviality first. In more detail, it would run thus: Since the content of a claim that the sensibility of the sentence 'Paris is beautiful' in which is true of the divisional region, which is no more than the claim that Paris is beautiful, we can trivially describe understanding a sentence, if we wish, as knowing its truth-conditions, but this gives 'us' no substantive account of understanding whatsoever. Something other than a grasp to truth conditions must provide the substantive account. The charge rests upon what has been called the redundancy theory of truth, the theory that, is somewhat more discriminative. Horwich calls the minimal theory of truth, or deflationary view of truth, as fathered by Fridge and Ramsey. The essential claim is that the predicate' . . . is true' does not have a sense, i.e., expresses no substantive or profound or explanatory concepts that ought be the topic of philosophical enquiry. The approach admits of different versions, but centres on the points (1) that 'it is true that p' says no more nor less than 'p' (hence redundancy) (2) that in less direct context, such as 'everything he said was true', or 'all logical consequences of true propositions are true', the predicate function as a device enabling 'us' to generalize than an adjective or predicate describing the thing he said, or the kinds of propositions that follow from true propositions. For example, the second may translate as '(p, q) (p & p ® q)' where there is no use of a notion of truth.

There are technical problems in interpreting all uses of the notion of truth in such ways, but they are not generally felt to be insurmountable. The approach needs to explain away apparently substantive uses of the notion, such a; science aims at the truth', or 'truth is a norm governing discourse'. Indeed, postmodernist writing frequently advocates that we must abandon such norms, along with a discredited 'objective' conception of truth. But perhaps, we can have the norms even when objectivity is problematic, since they can be framed without mention of truth: Science wants it to be so that whenever science holds that 'p'. Then 'p'. Discourse is to be regulated by the principle that it is wrong to assert 'p' when 'not-p'.

The Disquotational theory of truth finds that the simplest formulation is the claim that expressions of the fern 'S is true' mean the same as the expression 'S'. Some philosophers dislike the idea of sameness of meaning, and if this is disallowed, then the claim is that the two forms are equivalent in any sense of equivalence that matters. That is, it makes no difference whether people say 'Dogs bark' is true, or whether they say that 'dogs bark'. In the former representation of what they say the sentence 'Dogs bark' is mentioned, but in the latter it appears to be used, so the claim that the two are equivalent needs careful formulation and defence. On the face of it someone might know that 'Dogs bark' is true without knowing what it, for instance, if one were to find it in a list of acknowledged truths, although he does not understand English, and this is different from knowing that dogs bark. Disquotational theories are usually presented as versions of the redundancy theory of truth.

The minimal theory states that the concept of truth is exhausted by the fact that it conforms to the equivalence principle, the principle that for any proposition 'p', it is true that 'p' if and only if 'p'. Many different philosophical theories of truth will, with suitable qualifications, accept that equivalence principle. The distinguishing feature of the minimal theory is its claim that the equivalence principle exhausts the notion of truths. It is how widely accepted, that both by opponents and supporters of truth conditional theories of meaning, that it is inconsistent to accept both minimal theory of truth and a truth conditional account of meaning (Davidson, 1990, Dummett, 1959 and Horwich, 1990). If the claim that the sentence 'Paris is beautiful' is true is exhausted by its equivalence to the claim that Paris is beautiful, it is circular to try to explain the sentence's meaning in terms of its truth conditions. The minimal theory of truth has been endorsed by Ramsey, Ayer, the later Wittgenstein, Quine, Strawson, Horwich and-confusingly and inconsistently if be it correct-Fridge himself. But is the minimal theory correct?

The minimal or redundancy theory treats instances of the equivalence principle as definitional of truth for a given sentence. But in fact, it seems that each instance of the equivalence principle can itself be explained. The truths from which such an instance as:

'London is beautiful' is true if and only if London is beautiful'

Preserve a right to be interpreted specifically of A1 and A3 above? This would be a pseudo-explanation if the fact that 'London' refers to 'London is beautiful' has the truth-condition it does. But that is very implausible: It is, after all, possible to undertake its name for being 'London' without understanding the predicate 'is beautiful'. The idea that facts about the reference of particular words can be explanatory of facts about the truth conditions of sentences containing them in no way requires any naturalistic or any other kind of reduction of the notion of reference. Nor is the idea incompatible with the plausible point that singular reference can be attributed at all only to something that is capable of combining with other expressions to form complete sentences. That still leaves room for facts about an expression's having the particular reference it does to be partially explanatory of the particular truth condition possessed by a given sentence containing it. The minimal; theory thus treats as definitional or stimulative something that is in fact open to explanation. What makes this explanation possible is that there is a general notion of truth that has, among the many links that hold it in place, systematic connections with the semantic values of sub-sentential expressions.

A second problem with the minimal theory is that it seems impossible to formulate it without at some point relying implicitly on features and principles involving truths that go beyond anything countenanced by the minimal theory. If the minimal theory treats truth as a predicate of anything linguistic, be it utterances, type-in-a-language, or whatever, then the equivalence schema will not cover all cases, but only those in the theorist's own language. Some account has to be given of truth for sentences of other languages. Speaking of the truth of language-independence propositions or thoughts will only postpone, not avoid, this issue, since at some point principles have to be stated associating these language-independent entities with sentences of particular languages. The defender of the minimalist theory is likely to say that if a sentence 'S' of a foreign language is best translated by our sentence 'p', then the foreign sentence 'S' is true if and only if 'p'. Now the best translation of a sentence must preserve the concepts expressed in the sentence. Constraints involving a general notion of truth are persuasive in a plausible philosophical theory of concepts. It is, for example, a condition of adequacy on an individualized account of any concept that there exists what is called 'Determination Theory' for that account-that is, a specification of how the account contributes to fixing the semantic value of that concept, the notion of a concept's semantic value is the notion of something that makes a certain contribution to the truth conditions of thoughts in which the concept occurs. but this is to presuppose, than to elucidate, a general notion of truth.

It is also plausible that there are general constraints on the form of such Determination Theories, constraints that involve truth and which are not derivable from the minimalist's conception. Suppose that concepts are individuated by their possession conditions. A concept is something that is capable of being a constituent of such contentual representational in a way of thinking of something-a particular object, or property, or relation, or another entity. A possession condition may in various say makes a thanker's possession of a particular concept dependent upon his relations to his environment. Many possession conditions will mention the links between a concept and the thinker's perceptual experience. Perceptual experience represents the world for being a certain way. It is arguable that the only satisfactory explanation of what it is for perceptual experience to represent the world in a particular way must refer to the complex relations of the experience to the subject's environment. If this is so, then mention of such experiences in a possession condition will make possession of that condition will make possession of that concept dependent in part upon the environment relations of the thinker. Burge (1979) has also argued from intuitions about particular examples that, even though the thinker's non-environmental properties and relations remain constant, the conceptual content of his mental state can vary if the thinker's social environment is varied. A possession condition which property individuates such a concept must take into account the thinker's social relations, in particular his linguistic relations.

One such plausible general constraint is then the requirement that when a thinker forms beliefs involving a concept in accordance with its possession condition, a semantic value is assigned to the concept in such a way that the belief is true. Some general principles involving truth can indeed, as Horwich has emphasized, be derived from the equivalence schema using minimal logical apparatus. Consider, for instance, the principle that 'Paris is beautiful and London is beautiful' is true if and only if 'Paris is beautiful' is true if and only if 'Paris is beautiful' is true and 'London is beautiful' is true. This follows logically from the three instances of the equivalence principle: 'Paris is beautiful and London is beautiful' is rue if and only if Paris is beautiful, and 'London is beautiful' is true if and only if London is beautiful. But no logical manipulations of the equivalence schemas will allow the deprivation of that general constraint governing possession conditions, truth and the assignment of semantic values. That constraint can have courses be regarded as a further elaboration of the idea that truth is one of the aims of judgement.

We now turn to the other question, 'What is it for a person's language to be correctly describable by a semantic theory containing a particular axiom, such as the axiom A6 above for conjunction?' This question may be addressed at two depths of generality. At the shallower level, the question may take for granted the person's possession of the concept of conjunction, and be concerned with what has to be true for the axiom correctly to describe his language. At a deeper level, an answer should not duck the issue of what it is to possess the concept. The answers to both questions are of great interest: We will take the lesser level of generality first.

When a person conjunction by 'sand', he is not necessarily capable of formulating the axiom A6 explicitly. Even if he can formulate it, his ability to formulate it is not the causal basis of his capacity to hear sentences containing the word 'and' as meaning something involving conjunction. Nor is it the causal basis of his capacity to mean something involving conjunction by sentences he utters containing the word 'and'. Is it then right to regard a truth theory as part of an unconscious psychological computation, and to regard understanding a sentence as involving a particular way of depriving a theorem from a truth theory at some level of conscious proceedings? One problem with this is that it is quite implausible that everyone who speaks the same language has to use the same algorithms for computing the meaning of a sentence. In the past thirteen years, thanks particularly to the work of Davies and Evans, a conception has evolved according to which an axiom like A6 is true of a person's language only if there is a common component in the explanation of his understanding of each sentence containing the word 'and', a common component that explains why each such sentence is understood as meaning something involving conjunction (Davies, 1987). This conception can also be elaborated in computational terms: Suggesting that for an axiom like A6 to be true of a person's language is for the unconscious mechanisms which produce understanding to draw on the information that a sentence of the form 'A and B' are true if and only if 'A' is true and 'B' is true (Peacocke, 1986). Many different algorithms may equally draw n this information. The psychological reality of a semantic theory thus involves, in Marr's (1982) famous classification, something intermediate between his level one, the function computed, and his level two, the algorithm by which it is computed. This conception of the psychological reality of a semantic theory can also be applied to syntactic and phonol logical theories. Theories in semantics, syntax and phonology are not themselves required to specify the particular algorithms that the language user employs. The identification of the particular computational methods employed is a task for psychology. But semantics, syntactic and phonology theories are answerable to psychological data, and are potentially refutable by them-for these linguistic theories do make commitments to the information drawn upon by mechanisms in the language user.

This answer to the question of what it is for an axiom to be true of a person's language clearly takes for granted the person's possession of the concept expressed by the word treated by the axiom. In the example of the axiom A6, the information drawn upon is that sentences of the form 'A and B' are true if and only if 'A' is true and 'B' is true. This informational content employs, as it has to if it is to be adequate, the concept of conjunction used in stating the meaning of sentences containing 'and'. So the computational answer we have returned needs further elaboration if we are to address the deeper question, which does not want to take for granted possession of the concepts expressed in the language. It is at this point that the theory of linguistic understanding has to draws upon a theory of concepts. It is plausible that the concepts of conjunction are individuated by the following condition for a thinker to possess it.

Finally, this response to the deeper question allows 'us' to answer two challenges to the conception of meaning as truth-conditions. First, there was the question left hanging earlier, of how the theorist of truth-conditions is to say what makes one axiom of a semantic theory is correctly in that of another, when the two axioms assign the same semantic values, but do so by of different concepts. Since the different concepts will have different possession conditions, the dovetailing accounts, at the deeper level of what it is for each axiom to be correct for a person's language will be different accounts. Second, there is a challenge repeatedly made by the minimalist theorists of truth, to the effect that the theorist of meaning as truth-conditions should give some non-circular account of what it is to understand a sentence, or to be capable of understanding all sentences containing a given constituent. For each expression in a sentence, the corresponding dovetailing account, together with the possession condition, supplies a non-circular account of what it is to understand any sentence containing that expression. The combined accounts for each of he expressions that comprise a given sentence together constitute a non-circular account of what it is to understand the compete sentences. Taken together, they allow the theorists of meaning as truth-conditions fully to meet the challenge.

A curious view common to that which is expressed by an utterance or sentence: The proposition or claim made about the world. By extension, the content of a predicate or other sub-sentential component is what it contributes to the content of sentences that contain it. The nature of content is the central concern of the philosophy of language, in that mental states have contents: A belief may have the content that the prime minister will resign. A concept is something that is capable of bringing a constituent of such contents. More specifically, a concept is a way of thinking of something-a particular object, or property or relation, or another entity. Such a distinction was held in Frége's philosophy of language, explored in 'On Concept and Object' (1892). Fridge regarded predicates as incomplete expressions, in the same way as a mathematical expression for a function, such as sines . . . a log . . ., is incomplete. Predicates refer to a general idea as derived or inferred from specific instances or occurrence, which in themselves are 'unsaturated', and cannot be referred to by subject expressions (we thus get the paradox that the concept of a horse is not a concept). Although Fridge recognized the metaphorical nature of the notion of a concept being unsaturated, he was rightly convinced that some such notion is needed to explain the unity of a sentence, and to prevent sentences from being thought of as mere lists of names.

Several different concepts may each be ways of thinking of the same object. A person may think of himself in the first-person way, or think of himself as the spouse of Mary Smith, or as the person located in a certain room now. More generally, a concept 'c' is distinct from a concept 'd' if it is possible for a person rationally to believe 'd is such-and-such'. As words can be combined to form structured sentences, concepts have also been conceived as combinable into structured complex contents. When these complex contents are expressed in English by 'that . . . 'clauses, as in our opening examples, they will be capable of being true or false, depending on the way the world is.

The general system of concepts with which we organize our thoughts and perceptions are to encourage a conceptual scheme of which the outstanding elements of our every day conceptual formalities include spatial and temporal relations between events and enduring objects, causal relations, other persons, meaning-bearing utterances of others, . . . and so on. To see the world as containing such things is to share this much of our conceptual scheme. A controversial argument of Davidson's urges that we would be unable to interpret speech from a different conceptual scheme as even meaningful, Davidson daringly goes on to argue that since translation proceeds according to a principle of clarity, and since it must be possible of an omniscient commonsense conceptual framework are true.

Concepts are to be distinguished from a stereotype and from conceptions. The stereotypical spy may be a middle-level official down on his luck and in need of money. None the less, we can come to learn that Anthony Blunt, art historian and Surveyor of the Queen's Pictures, are a spy; we can come to believe that something falls under a concept while positively disbelieving that the same thing falls under the stereotype associated wit the concept. Similarly, a person's conception of a just arrangement for resolving disputes may involve something like contemporary Western legal systems. But whether or not it would be correct, it is quite intelligible for someone to rejects this conception by arguing that it dies not adequately provide for the elements of fairness and respect that are required by the concepts of justice.

Basically, a concept is that which is understood by a term, particularly a predicate. To posses a concept is to be able to deploy a term expressing it in making judgements, in which the ability connection is such things as recognizing when the term applies, and being able to understand the consequences of its application. The term 'idea' was formally used in the came way, but is avoided because of its associations with subjective matters inferred upon mental imagery in which may be irrelevant ti the possession of a concept. In the semantics of Fridge, a concept is the reference of a predicate, and cannot be referred to by a subjective term, although its recognition of as a concept, in that some such notion is needed to the explanatory justification of which that sentence of unity finds of itself from being thought of as namely categorized lists of itemized priorities.

A theory of a particular concept must be distinguished from a theory of the object or objects it selectively picks the outlying of the theory of the concept under which is partially contingent of the theory of thought and/or epistemology. A theory of the object or objects is part of metaphysics and ontology. Some figures in the history of philosophy-and are open to the accusation of not having fully respected the distinction between the kinds of theory. Descartes appears to have moved from facts about the indubitability of the thought 'I think', containing the fist-person was of thinking, to conclusions about the nonmaterial nature of the object he himself was. But though the goals of a theory of concepts and a theory of objects are distinct, each theory is required to have an adequate account of its relation to the other theory. A theory if concept is unacceptable if it gives no account of how the concept is capable of picking out the object it evidently does pick out. A theory of objects is unacceptable if it makes it impossible to understand how we could have concepts of those objects.

A fundamental question for philosophy is: What individuates a given concept-that is, what makes it the one it is, rather than any other concept? One answer, which has been developed in great detail, is that it is impossible to give a non-trivial answer to this question (Schiffer, 1987). An alternative approach, addressees the question by starting from the idea that a concept id individuated by the condition that must be satisfied if a thinker is to posses that concept and to be capable of having beliefs and other attitudes whose content contains it as a constituent. So, to take a simple case, one could propose that the logical concept 'and' is individuated by this condition, it be the unique concept 'C' to posses that a thinker has to find these forms of inference compelling, without basing them on any further inference or information: From any two premises 'A' and 'B', ACB can be inferred, and from any premise ACB, each of the 'A' and 'B' can be inferred. Again, a relatively observational concept such as 'round' can be individuated in part by stating that the thinker finds specified contents containing it compelling when he has certain kinds of perception, and in part by relating those judgements containing the concept and which are not based on perception to those judgements that are. A statement that individuates a concept by saying what is required for a thinker to posses it can be described as giving the possession condition for the concept.

A possession condition for a particular concept may actually make use of that concept. The possession condition for 'and' does so. We can also expect to use relatively observational concepts in specifying the kind of experience that have to be mentioned in the possession conditions for relatively observational concepts. What we must avoid is mention of the concept in question as such within the content of the attitudes attributed to the thinker in the possession condition. Otherwise we would be presupposing possession of the concept in an account that was meant to elucidate its possession. In talking of what the thinker finds compelling, the possession conditions can also respect an insight of the later Wittgenstein: That to find her finds it natural to go on in new cases in applying the concept.

Sometimes a family of concepts has this property: It is not possible to master any one of the members of the family without mastering the others. Two of the families that plausibly have this status are these: The family consisting of some simple concepts 0, 1, 2, . . . of the natural numbers and the corresponding concepts of numerical quantifiers is so-and-so of 'belief' and 'desire'. Such families have come to be known as 'local holism'. A local holism does not prevent the individuation of a concept by its possession condition. Rather, it demands that all the concepts in the family be individuated simultaneously. So one would say something of this form: Belief and desire form the unique pair of concepts C1 and C2 such that for as thinker to posses them are to meet such-and-such condition involving the thinker, C1 and C2. For these and other possession conditions to individuate properly, it is necessary that there be some ranking of the concepts treated. The possession conditions for concepts higher in the ranking must presuppose only possession of concepts at the same or lower levels in the ranking.

A possession conditions may in various way's make a thinker's possession of a particular concept dependent upon his relations to his environment. Many possession conditions will mention the links between a concept and the thinker's perceptual experience. Perceptual experience represents the world as a certain way. It is arguable that the only satisfactory explanation of what it is for perceptual experience to represent the world in a particular way must refer to the complex relations of the experience to the subject's environment. If this is so, then mention of such experiences in a possession condition will make possession of that concept dependent in part upon the environmental relations of the thinker. Burge (1979) has also argued from intuitions about particular examples that, even though the thinker's non-environmental properties and relations remain constant, the conceptual content of his mental state can vary if the thinker's social environment is varied. A possession condition that properly individuates such a concept must take into account the thinker's social relations, in particular his linguistic relations.

Concepts have a normative dimension, a fact strongly emphasized by Kripke. For any judgement whose content involves a given concept, there is a correctness condition for that judgement, a condition that is dependent in part upon the identity of the concept. The normative character of concepts also extends into making the territory of a thinker's reasons for making judgements. A thinker's visual perception can give him good reason for judging 'That man is bald': It does not by itself give him good reason for judging 'Rostropovich ids bald', even if the man he sees is Rostropovich. All these normative connections must be explained by a theory of concepts one approach to these matters is to look to the possession condition for the concept, and consider the referent of a concept is fixed from it, together with the world. One proposal is that the referent of the concept is that object (or property, or function, . . .) which makes the practices of judgement and inference mentioned in the possession condition always lead to true judgements and truth-preserving inferences. This proposal would explain why certain reasons are necessity good reasons for judging given contents. Provided the possession condition permits 'us' to say what it is about a thinker's previous judgements that masker it, the case that he is employing one concept rather than another, this proposal would also have another virtue. It would allow 'us' to say how the correctness condition is determined for a judgement in which the concept is applied to newly encountered objects. The judgement is correct if the new object has the property that in fact makes the judgmental practices mentioned in the possession condition yield true judgements, or truth-preserving inferences.

These manifesting dissimilations have occasioned the affiliated differences accorded within the distinction as associated with Leibniz, who declares that there are only two kinds of truths-truths of reason and truths of fact. The forms are all either explicit identities, i.e., of the form 'A is A', 'AB is B', and so forth, or they are reducible to this form by successively substituting equivalent terms. Leibniz dubs them 'truths of reason' because the explicit identities are self-evident deducible truths, whereas the rest can be converted too such by purely rational operations. Because their denial involves a demonstrable contradiction, Leibniz also says that truths of reason 'rest on the principle of contradiction, or identity' and that they are necessary [propositions, which are true of all possible words. Some examples are 'All equilateral rectangles are rectangles' and 'All bachelors are unmarried': The first is already of the form AB is B' and the latter can be reduced to this form by substituting 'unmarried man' fort 'bachelor'. Other examples, or so Leibniz believes, are 'God exists' and the truths of logic, arithmetic and geometry.

Truths of fact, on the other hand, cannot be reduced to an identity and our only way of knowing them is empirically by reference to the facts of the empirical world. Likewise, since their denial does not involve a contradiction, their truth is merely contingent: They could have been otherwise and hold of the actual world, but not of every possible one. Some examples are 'Caesar crossed the Rubicon' and 'Leibniz was born in Leipzig', as well as propositions expressing correct scientific generalizations. In Leibniz's view, truths of fact rest on the principle of sufficient reason, which states that nothing can be so unless there is a reason that it is so. This reason is that the actual world (by which he the total collection of things past, present and future) is better than any other possible worlds and was therefore created by 'God'.

In defending the principle of sufficient reason, Leibniz runs into serious problems. He believes that in every true proposition, the concept of the predicate is contained in that of the subject. (This holds even for propositions like 'Caesar crossed the Rubicon': Leibniz thinks that if anyone who had not, by course, crossed the Rubicon, would not have been Caesar. And this containment relationship! Which is eternal and unalterable even by God ~?! Guarantees that every truth has a sufficient reason. If truth consists in concept containment, however, then it seems that all truths are analytic and hence necessary, and if they are all necessary, surely they are all truths of reason. Leibnitz responds that not every truth can be reduced to an identity in a finite number of steps, in some instances revealing the connection between subject and predicate concepts would requite an infinite analysis. But while this may entail that we cannot prove such propositions as deductively manifested, it does not appear to show that the proposition could have been false. Intuitively, it seems a better ground for supposing that it is necessary truth of a special sort. A related question arises from the idea that truths of fact depend on God's decision to create.

The best of all possible worlds: If it is part of the concept of this world that it is best, now could its existence be other than necessary? Leibniz answers that its existence is only hypothetically necessary, i.e., it follows from God's decision to create this world, but God had the power to decide otherwise. Yet God is necessarily non-deceiving, so as he could have decided upon matters mor concerting. Leibniz says much more about these masters, but it is not clear whether he offers any satisfactory solutions.

Plantinga (1974) characterizes the sense of necessity illustrated in 1-3 as 'broadly logical'. For it includes not only truths of logic, but those of mathematics, set theory, and other quasi-logical ones. Yet, it is not so broad as to include matters of causal or natural necessity, such as: Nothing travels faster than the speed of light.

One would like an account of the basis of our distinction and a criterion by which to apply it. Some suppose that necessary truths are those we know as deductively possible. But we lack the criterion for deductive truths, and there are necessary truths we do not know at all, e.g., undiscovered mathematical ones. It would not help to say that necessary truths are one, and it is possible, in the broadly logical sense, to know of deductive circularity. Finally, Kripke (1972) and Plantinga (1974) argues that some contingent truths are knowable by deductive reasoning. Similar problems face the suggestion that necessary truths are the ones we know with the fairest of certainties: We lack a criterion for certainty, there are necessary truths we do not know, and (barring dubious arguments for scepticism) it is reasonable to suppose that we know some contingent truths with certainty.

Leibniz defined a necessary truth as one whose opposite implies a contradiction. Every such proposition, he held, is either an explicit identity, i.e., of the form 'A is A', 'AB is B', and so forth) or reducible to an identity by successively substituting equivalent terms. (Thus, 3 above might be so reduced by substituting 'unmarried man'; for 'bachelor'.) This has various advantages crossing over the ideas of the previous paragraph. First, it explicated that the notion of necessity and possibility and seems to provide a criterion we can apply. Second, because explicit identities are self-evident deductive propositions, the theory implies that all necessary truths are knowable deductively, but it does not entail that wee actually know all of them, nor does it define 'knowable' in a circular way. Third, it implies that necessary truths are knowable with certainty, but does not preclude our having certain knowledge of contingent truths by other than a reduction.

Nevertheless, this view is also problematic, and Leibniz's examples of reductions are too sparse to prove a claim about all necessary truths. Some of his reductions, moreover, are deficient: Fridge has pointed out, for example, that his proof of '2 + 2 = 4' presupposes the principle of association and so does not depend on the principle of identity. More generally, it has been shown that arithmetic cannot be reduced to logic, but requires the resources of set theory as well. Finally, there are other necessary propositions, e.g., 'Nothing can be red and green all over', which do not seem to be reducible to identities and which Leibniz does not show how to reduce.

Leibniz and others have thought of truths as a property of propositions, where the latter are conceived as things that may be expressed by, but are distinct from linguistic form, as a meaningful unit of language, such as an affix, a word, a phrase, or a sentence, as the basis or the foundation upon which something rests as the fundamental principle whose underlying to circumstance or condition all of which point of the necessary truth in convention. Thus A.J. Ayer, for example, argued that the only necessary truths are analytic statements and that the latter rest entirely on our commitment to use words in certain ways.

The slogan 'the meaning of a statement is its method of verification' expresses the empirical verification's theory of meaning. It is more than the general criterion of meaningfulness if and only if it is empirically verifiable. If says in addition what the meaning of a sentence is: It is, of those observations that would confirm or disconfirmed the sentence. Sentences that would be verified or falsified by all the same observations are empirically equivalent or have the same meaning. A sentence is said to be cognitively meaningful if and only if it can be verified or falsified in experience. This is not meant to require that the sentence be conclusively verified or falsified, since universal scientific laws or hypotheses (which are supposed to pass the test) are not logically deducible from any amount of actually observed evidence.

When one predicate's necessary truth of a preposition one speaks of modality dedicto. For one ascribes the modal property, necessary truth, to a dictum, namely, whatever proposition is taken as necessary. A venerable tradition, however, distinguishes this from necessary de re, wherein one predicates necessary or essential possession of some property to an on object. For example, the statement '4 is necessarily greater than 2' might be used to predicate of the object, 4, the property, being necessarily greater than 2. That objects have some of their properties necessarily, or essentially, and others only contingently, or accidentally, are a main part of the doctrine called 'essentialism'. Thus, an essential might, say that Socrates had the property of being bald accidentally, but that of being self-identical, or perhaps of being human, essentially. Although essentialism has been vigorously attacked in recent years, most particularly by Quine, it also has able contemporary proponents, such as Plantinga.

Modal necessity as seen by many philosophers whom have traditionally held that every proposition has a modal status as well as a truth value. Every proposition is either necessary or contingent as well as either true or false. The issue of knowledge of the modal status of propositions has received much attention because of its intimate relationship to the issue of deductive reasoning. For example, no propositions of the theoretic content that all knowledge of necessary propositions is deductively knowledgeable. Others reject this claim by citing Kripke's (1980) alleged cases of necessary theoretical propositions. Such contentions are often inconclusive, for they fail to take into account the following tripartite distinction: 'S' knows the general modal status of 'p' just in case 'S' knows that 'p' is a necessary proposition or 'S' knows the truth that 'p' is a contingent proposition. 'S' knows the truth value of 'p' just in case 'S' knows that 'p' is true or 'S' knows that 'p' is false. 'S' knows the specific modal status of 'p' just in case 'S' knows that 'p' is necessarily true or 'S' knows that 'p' is necessarily false or 'S' knows that 'p' is contingently true or 'S' knows that 'p' is contingently false. It does not follow from the fact that knowledge of the general modal status of a proposition is a deductively reasoned distinctive modal status is also given to theoretical principles. Nor des it follow from the fact that knowledge of a specific modal status of a proposition is theoretically given as to the knowledge of its general modal status that also is deductive.

The certainties involving reason and a truth of fact are much in distinction by associative measures given through Leibniz, who declares that there are only two kinds of truths-truths of reason and truths of fact. The former are all explicit identities, i.e., of the form 'A is A', 'AB is B', and so forth, or they are reducible to this form by successively substituting equivalent terms. Leibniz dubs them 'truths of reason' because the explicit identities are self-evident theoretical truth, whereas the rest can be converted too such by purely rational operations. Because their denial involves a demonstrable contradiction, Leibniz also says that truths of reason 'rest on the principle of contraction, or identity' and that they are necessary propositions, which are true of all possible worlds. Some examples are that 'All bachelors are unmarried': The first is already of the form 'AB is B' and the latter can be reduced to this form by substituting 'unmarried man' for 'bachelor'. Other examples, or so Leibniz believes, are 'God exists' and the truth of logic, arithmetic and geometry.

Truths of fact, on the other hand, cannot be reduced to an identity and our only way of knowing them as theoretical manifestations, or by reference to the fact of the empirical world. Likewise, since their denial do not involve contradiction, their truth is merely contingent: They could have been otherwise and hold of the actual world, but not of every possible one. Some examples are 'Caesar crossed the Rubicon' and 'Leibniz was born in Leipzig', as well as propositions expressing correct scientific generalizations. In Leibniz's view, truths of fact rest on the principle of sufficient reason, which states that nothing can be so unless thee is a reason that it is so. This reason is that the actual world (by which he the total collection of things past, present and future) is better than any other possible world and was therefore created by God.

In defending the principle of sufficient reason, Leibniz runs into serious problems. He believes that in every true proposition, the concept of the predicate is contained in that of the subject. (This hols even for propositions like 'Caesar crossed the Rubicon': Leibniz thinks anyone who did not cross the Rubicon would not have been Caesar) And this containment relationship-that is eternal and unalterable even by God-guarantees that every truth has a sufficient reason. If truth consists in concept containment, however, then it seems that all truths are analytic and hence necessary, and if they are all necessary, surely they are all truths of reason. Leibniz responds that not evert truth can be reduced to an identity in a finite number of steps: In some instances revealing the connection between subject and predicate concepts would require an infinite analysis. But while this may entail that we cannot prove such propositions as deductively probable, it does not appear to show that the proposition could have been false. Intuitively, it seems a better ground for supposing that it is a necessary truth of a special sort. A related question arises from the idea that truths of fact depend on God's decision to create the best world, if it is part of the concept of this world that it is best, how could its existence be other than necessary? Leibniz answers that its existence is only hypothetically necessary, i.e., it follows from God's decision to create this world, but God is necessarily good, so how could he have decided to do anything else? Leibniz says much more about the matters, but it is not clear whether he offers any satisfactory solutions.

The modality of a proposition is the way in which it is true or false. The most important division is between propositions true of necessity, and those true asa things are: Necessary as opposed to contingent propositions. Other qualifiers sometimes called 'modal' include the tense indicators 'It will be the case that p' or It was the case that p', and there are affinities between the 'deontic indicators', as it ought to be the case that p' or 'it is permissible that p', and the logical modalities as a logic that study the notions of necessity and possibility. Modal logic was of a great importance historically, particularly in the light of various doctrines concerning the necessary properties of the deity, but was not a central topic of modern logic in its golden period at the beginning of the 20th century. It was, however, revived by C. I. Lewis, by adding to a propositional or predicate calculus two operators, and (sometimes written N and M), meaning necessarily and possibly, respectively. These like 'p' 'p' and 'p' will be wanted. Controversial theses include 'p' 'p' (if a proposition is necessary, it is necessarily necessary, characteristic of the system known as S4) and 'p' 'p' (if a proposition is possible, it is necessarily possible, characteristic of the system known as S5). The classical 'modal theory' for modal logic, due to Kripke and the Swedish logician Stig Kanger, involves valuing propositions not as true or false 'simplicitiers', but as true or false art possible worlds, with necessity then corresponding to truth in all worlds, and possibly to truths in some world.

The doctrine advocated by David Lewis, which different 'possible worlds' are to be thought of as existing exactly as this one does. Thinking in terms of possibilities is thinking of real worlds where things are different, this view has been charged with misrepresenting it as some insurmountably unseeingly to why it is good to save the child from drowning, since there is still a possible world in which she (or her counterpart) drowned, and from the standpoint of the universe it should make no difference that world is actual. Critics asio charge that the notion fails to fit either with a coherent theory of how we know about possible worlds, or with a coherent theory about possible worlds, or with a coherent theory of why we are interested in them, but Lewis denies that any other way of interpreting modal statements is tenable.

Thus and so, the 'standard analysis' of propositional knowledge, suggested by Plato and Kant among others, implies that if one has a justified true belief that 'p', then one knows that 'p'. The belief condition 'p' believes that 'p', the truth condition requires that any known proposition be true. And the justification condition requires that any known proposition be adequately justified, warranted or evidentially supported. Plato appears to be considering the tripartite definition in the 'Theaetetus' (201c-202d), and to be endorsing its jointly sufficient conditions for knowledge in the 'Meno' (97e-98a). This definition has come to be called 'the standard analysis' of knowledge, and has received a serious challenge from Edmund Gettier's counterexamples in 1963. Gettier published two counterexamples to this implication of the standard analysis. In essence, they are:

(1) Smith and Jones have applied for the same job. Smith is justified in believing that (a) Jones will get the job, and that (b) Jones has ten coins in his pocket. On the basis of (a) and (b) Smith infers, and thus is justified in believing, that the person who will get the job has ten coins in his pocket. At it turns out, Smith himself will get the job, and he also happens to have ten coins in his pocket. So, although Smith is justified in believing the true proposition Smith does not know.

(2) Smith is justified in believing the false proposition that (a) Smith owns a Ford. On the basis of (a) Smith infers, and thus is justified in believing, that (b) either Jones owns a Ford or Brown is in Barcelona. As it turns out, Brown or in Barcelona, and so (b) is true. So although Smith is justified in believing the true proposition (b). Smith does not know (b).

Gettier's counterexamples are thus cases where one has justified true belief that 'p', but lacks knowledge that 'p'. The Gettier problem is the problem of finding a modification of, or an alterative to, the standard justified-true-belief analysis of knowledge that avoids counterexamples like Gettier's. Some philosophers have suggested that Gettier style counterexamples are defective owing to their reliance on the false principle that false propositions can justify one's belief in other propositions. But there are examples much like Gettier's that do not depend on this allegedly false principle. Here is one example inspired by Keith and Richard Feldman:

(3) Suppose Smith knows the following proposition, 'm': Jones, whom Smith has always found to be reliable and whom Smith, has no reason to distrust now, has told Smith, his office-mate, that 'p': He, Jones owns a Ford. Suppose also that Jones has told Smith that 'p' only because of a state of hypnosis Jones is in, and that 'p' is true only because, unknown to himself, Jones has won a Ford in a lottery since entering the state of hypnosis. And suppose further that Smith deduces from 'm' its existential generalization, 'q': There is someone, whom Smith has always found to be reliable and whom Smith has no reason to distrust now, who has told Smith, his office-mate, that he owns a Ford. Smith, then, knows that 'q', since he has correctly deduced 'q' from 'm', which he also knows. But suppose also that on the basis of his knowledge that 'q'. Smith believes that 'r': Someone in the office owns a Ford. Under these conditions, Smith has justified true belief that 'r', knows his evidence for 'r', but does not know that 'r'.

Gettier-style examples of this sort have proven especially difficult for attempts to analyse the concept of propositional knowledge. The history of attempted solutions to the Gettier problem is complex and open-ended. It has not produced consensus on any solution. Many philosophers hold, in light of Gettier-style examples, that propositional knowledge requires a fourth condition, beyond the justification, truth and belief conditions. Although no particular fourth condition enjoys widespread endorsement, there are some prominent general proposals in circulation. One sort of proposed modification, the so-called 'defeasibility analysis', requires that the justification appropriate to knowledge be 'undefeated' in the general sense that some appropriate subjunctive conditional concerning genuine defeaters of justification be true of that justification. One straightforward defeasibility fourth condition, for instance, requires of Smith's knowing that 'p' that there be no true proposition 'q', such that if 'q' became justified for Smith, 'p' would no longer be justified for Smith (Pappas and Swain, 1978). A different prominent modification requires that the actual justification for a true belief qualifying as knowledge not depend I a specified way on any falsehood (Armstrong, 1973). The details proposed to elaborate such approaches have met with considerable controversy.

The fourth condition of evidential truth-sustenance may be a speculative solution to the Gettier problem. More specifically, for a person, 'S', to have knowledge that 'p' on justifying evidence 'e', 'e' must be truth-sustained in this sense for every true proposition 't' that, when conjoined with 'e', undermines S's justification for 'p' on 'e', there is a true proposition, 't', that, when conjoined with 'e'

& 't', restores the justification of 'p' for 'S' in a way that 'S' is actually justified in believing that 'p'. The gist of this resolving evolution, put roughly, is that propositional knowledge requires justified true belief that is sustained by the collective totality of truths. Herein, is to argue in Knowledge and Evidence, that Gettier-style examples as (1)-(3), but various others as well.

Three features that proposed this solution merit emphasis. First, it avoids a subjunctive conditional in its fourth condition, and so escapes some difficult problems facing the use of such a conditional in an analysis of knowledge. Second, it allows for non-deductive justifying evidence as a component of propositional knowledge. An adequacy condition on an analysis of knowledge is that it does not restrict justifying evidence to relations of deductive support. Third, its proposed solution is sufficiently flexible to handle cases describable as follows:

(4) Smith has a justified true belief that 'p', but there is a true proposition, 't', which undermines Smith's justification for 'p' when conjoined with it, and which is such that it is either physically or humanly impossible for Smith to be justified in believing that 't'.

Examples represented by (4) suggest that we should countenance varying strengths in notions of propositional knowledge. These strengths are determined by accessibility qualifications on the set of relevant knowledge-precluding underminers. A very demanding concept of knowledge assumes that it need only be logically possible for a Knower to believe a knowledge-precluding underminer. Less demanding concepts assume that it must be physically or humanly possible for a Knower to believe knowledge-precluding underminers. But even such less demanding concepts of knowledge need to rely on a notion of truth-sustained evidence if they are to survive a threatening range of Gettier-style examples. Given to some resolution that it needs be that the forth condition for a notion of knowledge is not a function simply of the evidence a Knower actually possesses.

The higher controversial aftermath of Gettier's original counterexamples has left some philosophers doubting of the really philosophical significance of the Gettier problem. Such doubt, however, seems misplaced. One fundamental branch of epistemology seeks understanding of the nature of propositional knowledge. And our understanding exactly what prepositional knowledge is essentially involves our having a Gettier-resistant analysis of such knowledge. If our analysis is not Gettier-resistant, we will lack an exact understanding of what propositional knowledge is. It is epistemologically important, therefore, to have a defensible solution to the Gettier problem, however, demanding such a solution is.

Propositional knowledge (PK) is the type of knowing whose instance are labelled by of a phrase expressing some proposition, e.g., in English a phrase of the form 'that h', where some complete declarative sentence is instantial for 'h'.

Theories of 'PK' differ over whether the proposition that 'h' is involved in a more intimate fashion, such as serving as a way of picking out a proposition attitude required for knowing, e.g., believing that 'h', accepting that 'h' or being sure that 'h'. For instance, the tripartite analysis or standard analysis, treats 'PK' as consisting in having a justified, true belief that 'h', the belief condition requires that anyone who knows that 'h' believes that 'h', the truth condition requires that any known proposition be true, in contrast, some regarded theories do so consider and treat 'PK' as the possession of specific abilities, capabilities, or powers, and that view the proposition that 'h' as needed to be expressed only in order to label a specific instance of 'PK'.

Although most theories of Propositional knowledge (PK) purport to analyse it, philosophers disagree about the goal of a philosophical analysis. Theories of 'PK' may differ over whether they aim to cover all species of 'PK' and, if they do not have this goal, over whether they aim to reveal any unifying link between the species that they investigate, e.g., empirical knowledge, and other species of knowing.

Very many accounts of 'PK' have been inspired by the quest to add a fourth condition to the tripartite analysis so as to avoid Gettier-type counterexamples to it, whereby a fourth condition of evidential truth-sustenance for every true proposition when conjoined with a regaining justification, which may require the justified true belief that is sustained by the collective totality of truths that an adequacy condition of propositional knowledge not restrict justified evidences in relation of deductive support, such that we should countenance varying strengths in notions of propositional knowledge. Restoratively, these strengths are determined by accessibility qualifications on the set of relevant knowledge-precluding underminers. A very demanding concept of knowledge assumes that it need only be logically possible for a Knower to believe a knowledge-precluding underminers, and less demanding concepts that it must physically or humanly possible for a Knower to believe knowledge-precluding undeterminers. But even such demanding concepts of knowledge need to rely on a notion of truth-sustaining evidence if they are to survive a threatening range of Gettier-style examples. As the needed fourth condition for a notion of knowledge is not a function simply of the evidence, a Knower actually possesses. One fundamental source of epistemology seeks understanding of the nature of propositional knowledge, and our understanding exactly what propositional knowledge is essentially involves our having a Gettier-resistant analysis of such knowledge. If our analysis is not Gettier-resistant, we will lack an exact understanding of what propositional knowledge is. It is epistemologically important, therefore, to have a defensible solution to the Gettier problem, however, demanding such a solution is. And by the resulting need to deal with other counterexamples provoked by these new analyses.

Keith Lehrer (1965) originated a Gettier-type example that has been a fertile source of important variants. It is the case of Mr Notgot, who is in one's office and has provided some evidence, 'e', in response to all of which one forms a justified belief that Mr. Notgot is in the office and owns a Ford, thanks to which one arrives at the justified belief that 'h': 'Someone in the office owns a Ford'. In the example, 'e' consists of such things as Mr. Notgot's presently showing one a certificate of Ford ownership while claiming to own a Ford and having been reliable in the past. Yet, Mr Notgot has just been shamming, and the only reason that it is true that 'h1' is because, unbeknown to oneself, a different person in the office owns a Ford.

Variants on this example continue to challenge efforts to analyse species of 'PK'. For instance, Alan Goldman (1988) has proposed that when one has empirical knowledge that 'h', when the state of affairs (call it h*) expressed by the proposition that 'h' figures prominently in an explanation of the occurrence of one's believing that 'h', where explanation is taken to involve one of a variety of probability relations concerning 'h*', and the belief state. But this account runs foul of a variant on the Notgot case akin to one that Lehrer (1979) has described. In Lehrer's variant, Mr Notgot has manifested a compulsion to trick people into justified believing truths yet falling short of knowledge by of concocting Gettierized evidence for those truths. It we make the trickster's neuroses highly specific ti the type of information contained in the proposition that 'h', we obtain a variant satisfying Goldman's requirement That the occurrence of 'h*' significantly raises the probability of one's believing that 'h'. (Lehrer himself (1990, pp. 103-4) has criticized Goldman by questioning whether, when one has ordinary perceptual knowledge that abn object is present, the presence of the object is what explains one's believing it to be present.)

In grappling with Gettier-type examples, some analyses proscribe specific relations between falsehoods and the evidence or grounds that justify one's believing. A simple restriction of this type requires that one's reasoning to the belief that 'h' does not crucially depend upon any false lemma (such as the false proposition that Mr Notgot is in the office and owns a Ford). However, Gettier-type examples have been constructed where one does not reason through and false belief, e.g., a variant of the Notgot case where one arrives at belief that 'h', by basing it upon a true existential generalization of one's evidence: 'There is someone in the office who has provided evidence e', in response to similar cases, Sosa (1991) has proposed that for 'PK' the 'basis' for the justification of one's belief that 'h' must not involve one's being justified in believing or in 'presupposing' any falsehood, even if one's reasoning to the belief does not employ that falsehood as a lemma. Alternatively, Roderick Chisholm (1989) requires that if there is something that makes the proposition that 'h' evident for one and yet makes something else that is false evident for one, then the proposition that 'h' is implied by a conjunction of propositions, each of which is evident for one and is such that something that makes it evident for one makes no falsehood evident for one. Other types of analyses are concerned with the role of falsehoods within the justification of the proposition that 'h' (Versus the justification of one's believing that 'h'). Such a theory may require that one's evidence bearing on this justification not already contain falsehoods. Or it may require that no falsehoods are involved at specific places in a special explanatory structure relating to the justification of the proposition that 'h' (Shope, 1983.).

A frequently pursued line of research concerning a fourth condition of knowing seeks what is called a 'defeasibility' analysis of 'PK.' Early versions characterized defeasibility by of subjunctive conditionals of the form, 'If 'A' were the case then 'B' would be the case'. But more recently the label has been applied to conditions about evidential or justificational relations that are not themselves characterized in terms of conditionals. Early versions of defeasibility theories advanced conditionals where 'A' is a hypothetical situation concerning one's acquisition of a specified sort of epistemic status for specified propositions, e.g., one's acquiring justified belief in some further evidence or truths, and 'B'; concerned, for instance, the continued justified status of the proposition that 'h' or of one's believing that 'h'.

A unifying thread connecting the conditional and non-conditional approaches to defeasibility may lie in the following facts: (1) What is a reason for being in a propositional attitude is in part a consideration, instances of the thought of which have the power to affect relevant processes of propositional attitude formation? : (2) Philosophers have often hoped to analyse power ascriptions by of conditional statements: And (3) Arguments portraying evidential or justificational relations are abstractions from those processes of propositional attitude maintenance and formation that manifest rationality. So even when some circumstance, 'R', is a reason for believing or accepting that 'h', another circumstance, 'K' may present an occasion from being present for a rational manifestation of the relevant power of the thought of 'R' and it will not be a good argument to base a conclusion that 'h' on the premiss that 'R' and 'K' obtain. Whether 'K' does play this interfering, 'defeating' role will depend upon the total relevant situation.

Accordingly, one of the most sophisticated defeasibility accounts, which has been proposed by John Pollock (1986), requires that in order to know that 'h', one must believe that 'h' on the basis of an argument whose force is not defeated in the above way, given the total set of circumstances described by all truths. More specifically, Pollock defines defeat as a situation where (1) one believes that 'p' and it is logically possible for one to become justified in believing that 'h' by believing that 'p'. And (2) one actually has a further set of beliefs, 'S' logically has a further set of beliefs, 'S', logically consistent with the proposition that 'h', such that it is not logically possible for one to become justified in believing that 'h' by believing it on the basis of holding the set of beliefs that is the union of 'S,' with the belief that 'p' (Pollock, 1986). Furthermore, Pollock requires for 'PK' that the rational presupposition in favour of one's believing that 'h' created by one's believing that 'p' is undefeated by the set of all truths, including considerations that one does not actually believe. Pollock offers no definition of what this requirement . But he may intend roughly the following: Where 'T' is the set of all true propositions: (I) one believes that 'p' and it is logically possible for one to become justified in believing that 'h', by believing that 'p'. And (II) there are logically possible situations in which one becomes justified in believing that 'h' on the bass of having the belief that 'p' and the beliefs in 'T'. Thus, in the Notgot example, since 'T' includes the proposition that Mr. Notgot does own a Ford, one lack's knowledge because condition (II) is not satisfied.

But given such an interpretation. Pollock's account illustrates the fact that defeasibility theories typically have difficulty dealing with introspective knowledge of one's beliefs. Suppose that some proposition, say that ¦, is false, but one does not realize this and holds the belief that ¦. Condition

(II) has no knowledge that h2? : 'I believe that ¦'. At least this is so if one's reason for believing that h2 includes the presence of the very condition of which one is aware, i.e., one's believing that ¦. It is incoherent to suppose hat one retains the latter reason, also, believes the truth that not-¦. This objection can be avoided, but at the cost of adopting what is a controversial view about introspective knowledge that 'h', namely, the view that one's belief that 'h' is in such cases mediated by some mental state intervening between the mental state of which there is introspective knowledge and he belief that 'h', so that is mental state is rather than the introspected state that it is included in one's reason for believing that 'h'. In order to avoid adopting this controversial view, Paul Moser (1989) has proposed a disjunctive analysis of 'PK', which requires that one satisfy a defeasibility condition either rather than like Pollock's or else one believes that 'h' by introspection. However, Moser leaves obscure exactly why beliefs arrived at by introspections account as knowledge.

Early versions of defeasibility theories had difficulty allowing for the existence of evidence that is 'merely misleading', as in the case where one does know that 'h3: 'Tom Grabit stole a book from the library', thanks to having seen him steal it, yet where, unbeknown to oneself, Tom's mother out of dementia gas testified that Tom was far away from the library at the time of the theft. One's justifiably believing that she gave the testimony would destroy one's justification for believing that 'h3' if added by itself to one's present evidence.

At least some defeasibility theories cannot deal with the knowledge one has while dying that 'h4: 'In this life there is no timer at which I believe that 'd', where the proposition that 'd' expresses the details regarding some philosophical matter, e.g., the maximum number of blades of grass ever simultaneously growing on the earth. When it just so happens that it is true that 'd', defeasibility analyses typically consider the addition to one's dying thoughts of a belief that 'd' in such a way as too improperly rule out actual knowledge that 'h4'.

A quite different approach to knowledge, and one able to deal with some Gettier-type cases, involves developing some type of causal theory of Propositional knowledge. The interesting thesis that counts as a causal theory of justification (in the meaning of 'causal theory; intended here) is the that of a belief is justified just in case it was produced by a type of process that is 'globally' reliable, that is, its propensity to produce true beliefs-that can be defined (to a god enough approximation) as the proportion of the bailiffs it produces (or would produce where it used as much as opportunity allows) that are true-is sufficiently meaningful-variations of this view have been advanced for both knowledge and justified belief. The first formulation of reliability account of knowing appeared in a note by F.P. Ramsey (1931), who said that a belief was knowledge if it is true, certain can obtain by a reliable process. P. Unger (1968) suggested that 'S' knows that 'p' just in case it is not at all accidental that 'S' is right about its being the casse that 'p'. D.M. Armstrong (1973) said that a non-inferential belief qualified as knowledge if the belief has properties that are nominally sufficient for its truth, i.e., guarantee its truth through and by the laws of nature.

Such theories require that one or another specified relation hold that can be characterized by mention of some aspect of cassation concerning one's belief that 'h' (or one's acceptance of the proposition that 'h') and its relation to state of affairs 'h*', e.g., h* causes the belief: h* is causally sufficient for the belief h* and the beliefs have a common cause. Such simple versions of a causal theory are able to deal with the original Notgot case, since it involves no such causal relationship, but cannot explain why there is ignorance in the variants where Notgot and Berent Enç (1984) have pointed out. Sometimes one knows of ',', which is thanks to recognizing a feature merely corelated with the presence of without endorsing a causal theory themselves, only to suggest that it would need to be elaborated so as to allow that one's belief that ',' has been caused by a factor whose correlation with the presence of has caused for being of self, e.g., by evolutionary adaption in one's ancestors, the disposition that one manifests in acquiring the belief in response to the correlated factor. Not only does this strain the unity of as causal theory by complicating it, but no causal theory without other shortcomings has been able to cover instances of deductively reasoned knowledge.

Causal theories of Propositional knowledge differ over whether they deviate from the tripartite analysis by dropping the requirements that one's believing (accepting) that 'h' be justified. The same variation occurs regarding reliability theories, which present the Knower as reliable concerning the issue of whether or not 'h', in the sense that some of one's cognitive or epistemic states, è, is such that, given further characteristics of oneself-possibly including relations to factors external to one and which one may not be aware-it is nomologically necessary (or at least probable) that 'h'. In some versions, the reliability is required to be 'global' in as far as it must concern a nomologically (probabilistic) relationship) relationship of states of type è to the acquisition of true beliefs about a wider range of issues than merely whether or not 'h'. There is also controversy about how to delineate the limits of what constitutes a type of relevant personal state or characteristic. (For example, in a case where Mr Notgot has not been shamming and one does know thereby that someone in the office owns a Ford, such as a way of forming beliefs about the properties of persons spatially close to one, or instead something narrower, such as a way of forming beliefs about Ford owners in offices partly upon the basis of their relevant testimony?)

One important variety of reliability theory is a conclusive reason account, which includes a requirement that one's reasons for believing that 'h' be such that in one's circumstances, if h* were not to occur then, e.g., one would not have the reasons one does for believing that 'h', or, e.g., one would not believe that 'h'. Roughly, the latter are demanded by theories that treat a Knower as 'tracking the truth', theories that include the further demand that is roughly, if it were the case, that 'h', then one would believe that 'h'. A version of the tracking theory has been defended by Robert Nozick (1981), who adds that if what he calls a 'method' has been used to arrive at the belief that 'h', then the antecedent clauses of the two conditionals that characterize tracking will need to include the hypothesis that one would employ the very same method.

But unless more conditions are added to Nozick's analysis, it will be too weak to explain why one lack's knowledge in a version of the last variant of the tricky Mr Notgot case described above, where we add the following details: (a) Mr Notgot's compulsion is not easily changed, (b) while in the office, Mr Notgot has no other easy trick of the relevant type to play on one, and one arrives at one's belief that 'h', not by reasoning through a false belief ut by basing belief that 'h', upon a true existential generalization of one's evidence.

Neozoic's analysis is in addition too strong to permit anyone ever to know that 'h': 'Some of my beliefs about beliefs might be otherwise, e.g., I might have rejected on of them'. If I know that 'h5' then satisfaction of the antecedent of one of Neozoic's conditionals would involve its being false that 'h5', thereby thwarting satisfaction of the consequent's requirement that I not then believe that 'h5'. For the belief that 'h5' is itself one of my beliefs about beliefs (Shope, 1984).

Some philosophers think that the category of knowing for which true. Justified believing (accepting) is a requirement constituting only a species of Propositional knowledge, construed as an even broader category. They have proposed various examples of 'PK' that do not satisfy the belief and/ ort justification conditions of the tripartite analysis. Such cases are often recognized by analyses of Propositional knowledge in terms of powers, capacities, or abilities. For instance, Alan R. White (1982) treats 'PK' as merely the ability to provide a correct answer to a possible questions, however, White may be equating 'producing' knowledge in the sense of producing 'the correct answer to a possible question' with 'displaying' knowledge in the sense of manifesting knowledge. (White, 1982). The latter can be done even by very young children and some nonhuman animals independently of their being asked questions, understanding questions, or recognizing answers to questions. Indeed, an example that has been proposed as an instance of knowing that 'h' without believing or accepting that 'h' can be modified so as to illustrate this point. Two examples concern an imaginary person who has no special training or information about horses or racing, but who in an experiment persistently and correctly picks the winners of upcoming horseraces. If the example is modified so that the hypothetical 'seer' never picks winners but only muses over whether those horse's might win, or only reports those horses winning, this behaviour should be as much of a candidate for the person's manifesting knowledge that the horse in question will win as would be the behaviour of picking it as a winner.

These considerations expose limitations in Edward Craig's analysis (1990) of the concept of knowing of a person's being a satisfactory informant in relation to an inquirer who wants to find out whether or not 'h'. Craig realizes that counterexamples to his analysis appear to be constituted by Knower who is too recalcitrant to inform the inquirer, or to incapacitate to inform, or too discredited to be worth considering (as with the boy who cried 'Wolf'). Craig admits that this might make preferable some alternative view of knowledge as a different state that helps to explain the presence of the state of being a suitable informant when the latter does obtain. Such the alternate, which offers a recursive definition that concerns one's having the power to proceed in a way representing the state of affairs, causally involved in one's proceeding in this way. When combined with a suitable analysis of representing, this theory of propositional knowledge can be unified with a structurally similar analysis of knowing how to do something.

Knowledge and belief, according to most epistemologists, knowledge entails belief, so that I cannot know that such and such is the case unless I believe that such and such am the case. Others think this entailment thesis can be rendered more accurately if we substitute for belief some closely related attitude. For instance, several philosophers would prefer to say that knowledge entail psychological certainties (Prichard, 1950 and Ayer, 1956) or conviction (Lehrer, 1974) or acceptance (Lehrer, 1989). None the less, there are arguments against all versions of the thesis that knowledge requires having a belief-like attitude toward the known. These arguments are given by philosophers who think that knowledge and belief (or a facsimile) are mutually incompatible (the incomparability thesis), or by ones who say that knowledge does not entail belief, or vice versa, so that each may exist without the other, but the two may also coexist (the separability thesis).

The incompatibility thesis is sometimes traced to Plato (429-347 BC) in view of his claim that knowledge is infallible while belief or opinion is fallible ('Republic' 476-9). But this claim would not support the thesis. Belief might be a component of an infallible form of knowledge in spite of the fallibility of belief. Perhaps, knowledge involves some factor that compensates for the fallibility of belief.

A. Duncan-Jones (1939: Also Vendler, 1978) cites linguistic evidence to back up the incompatibility thesis. He notes that people often say 'I do not believe she is guilty. I know she is' and the like, which suggest that belief rule out knowledge. However, as Lehrer (1974) indicates, the above exclamation is only a more emphatic way of saying 'I do not just believe she is guilty, I know she is' where 'just' makes it especially clear that the speaker is signalling that she has something more salient than mere belief, not that she has something inconsistent with belief, namely knowledge. Compare: 'You do not hurt him, you killed him'.

A. Prichard (1966) offers a defence of the incompatibility thesis that hinges on the equation of knowledge with certainty (both infallibility and psychological certitude) and the assumption that when we believe in the truth of a claim we are not certain about its truth. Given that belief always involves uncertainty while knowledge never dies, believing something rules out the possibility of knowing it. Unfortunately, however, Prichard gives 'us' no goods reason to grant that states of belief are never ones involving confidence. Conscious beliefs clearly involve some level of confidence, to suggest that we cease to believe things about which we are completely confident is bizarre.

A.D. Woozley (1953) defends a version of the separability thesis. Wesley's version, which deals with psychological certainty rather than belief per se, is that knowledge can exist in the absence of confidence about the item known, although might also be accompanied by confidence as well. Wesley remarks that the test of whether I know something is 'what I can do, where what I can do may include answering questions'. On the basis of this remark he suggests that even when people are unsure of the truth of a claim, they might know that the claim is true. We unhesitatingly attribute knowledge to people who give correct responses on examinations even if those people show no confidence in their answers. Wesley acknowledges, however, that it would be odd for those who lack confidence to claim knowledge. It would be peculiar to say that, 'I am unsure of whether or not my answer is true, but still I know it is correct'. But this tension Wesley explains using a distinction between conditions under which we are justified in making a claim (such as a claim to know something), and conditions under which the claim we make are true. While 'I know such and such' might be true even if I am unsure whether or not that such and such holds, nonetheless it would be inappropriate for me to claim that I know that such and such unless I was sure of the truth of my claim.

Colin Radford (1966) extends Wesley's defence of the separability thesis. In Redford's view, not only is knowledge compatible with the lack of certainty, it is also compatible with a complete lack of belief. He argues by example. In one example, Jean has forgotten that he learned some English history year's priori and yet he is able to give several correct responses to questions such as 'When did the Battle of Hastings occur'? Since he forgot that he took history, he considers the correct response to be no more than guesses. Thus, when he says that the Battle of Hastings took place in 1066 he would deny having the belief that the Battle of Hastings took place in 1066. A disposition he would deny being responsible (or having the right to be convincing) that 1066 was the correct date. Redford would none the less insist that Jean know when the Battle occurred, since clearly be remembering the correct date. Redford admits that it would be inappropriate for Jean to say that he knew when the Battle of Hastings occurred, but, like Wesley he attributes the impropriety to a fact about when it is and is not appropriate to claim knowledge. When we claim knowledge, we ought, at least to believe that we have the knowledge we claim, or else our behaviour is 'intentionally misleading'.

Those that agree with Redford's defence of the separability thesis will probably think of belief as an inner state that can be detected through introspection. That Jean lack's beliefs about English history are plausible on this Cartesian picture since Jean does not find himself with any beliefs about English history when he seeks them out. One might criticize Redford, however, by rejecting that Cartesian view of belief. One could argue that some beliefs are thoroughly unconscious, for example. Or one could adopt a behaviourist conception of belief, such as Alexander Bain's (1859), according to which having beliefs is a matter of the way people are disposed to behave (and has not Redford already adopted a behaviourist conception of knowledge?) Since Jean gives the correct response when queried, a form of verbal behaviour, a behaviourist would be tempted to credit him with the belief that the Battle of Hastings occurred in 1066.

D.M. Armstrong (1873) takes a different tack against Redford. Jean does know that the Battle of Hastings took place in 1066. Armstrong will grant Redford that point, in fact, Armstrong suggests that Jean believe that 1066 is not the date the Battle of Hastings occurred, for Armstrong equates the belief that such and such is just possible but no more than just possible with the belief that such and such is not the case. However, Armstrong insists, Jean also believes that the Battle did occur in 1066. After all, had Jean been mistaught that the Battle occurred in 1066, and subsequently 'guessed' that it took place in 1066, we would surely describe the situation as one in which Jean's false belief about the Battle became unconscious over time but persisted of a memory trace that was causally responsible for his guess. Out of consistency, we must describe Redford's original case as one that Jean's true belief became unconscious but persisted long enough to cause his guess. Thus, while Jean consciously believes that the Battle did not occur in 1066, unconsciously he does believe it occurred in 1066. So after all, Redford does not have a counterexample to the claim that knowledge entails belief.

Armstrong's response to Redford was to reject Redford's claim that the examinee lacked the relevant belief about English history. Another response is to argue that the examinee lacks the knowledge Redford attributes to him (cf. Sorenson, 1982). If Armstrong is correct in suggesting that Jean believes both that 1066 is and that it is not the date of the Battle of Hastings, one might deny Jean knowledge on the grounds that people who believe the denial of what they believe cannot be said t know the truth of their belief. Another strategy might be to compare the examine case with examples of ignorance given in recent attacks on externalist accounts of knowledge (needless to say. Externalists themselves would tend not to favour this strategy). Consider the following case developed by BonJour (1985): For no apparent reason, Samantha believes that she is clairvoyant. Again, for no apparent reason, she one day comes to believe that the President is in New York City, even though she has every reason to believe that the President is in Washington, D.C. In fact, Samantha is a completely reliable clairvoyant, and she has arrived at her belief about the whereabouts of the President thorough the power of her clairvoyance. Yet surely Samantha's belief is completely irrational. She is not justified in thinking what she does. If so, then she does not know where the President is. But Redford's examinee is unconventional. Even if Jean lacks the belief that Redford denies him, Redford does not have an example of knowledge that is unattended with belief. Suppose that Jean's memory had been sufficiently powerful to produce the relevant belief. As Redford says, in having every reason to suppose that his response is mere guesswork, and he has every reason to consider his belief false. His belief would be an irrational one, and hence one about whose truth Jean would be ignorant.

Least has been of mention to an approaching view from which 'perception' basis upon itself as a fundamental philosophical topic both for its central place in ant theory of knowledge, and its central place un any theory of consciousness. Philosophy in this area is constrained by a number of properties that we believe to hold of perception, (1. It gives 'us' knowledge of the world around 'us'. (2) We are conscious of that world by being aware of 'sensible qualities': Colour, sounds, tastes, smells, felt warmth, and the shapes and positions of objects in the environment. (3) Such consciousness is affected through highly complex information channels, such as the output of the three different types of colour-sensitive cells in the eye, or the channels in the ear for interpreting pulses of air pressure as frequencies of sound. (4) There ensues even more complex neurophysiological coding of that information, and eventually higher-order brain functions bring it about that we interpreted the information so received. (Much of this complexity has been revealed by the difficulties of writing programs enabling computers to recognize quite simple aspects of the visual scene.) The problem is to avoid thinking of here being a central, ghostly, conscious self, fed information in the same way that a screen if fed information by a remote television camera. Once such a model is in place, experience will seem like a veil getting between 'us' and the world, and the direct objects of perception will seem to be private items in an inner theatre or sensorium. The difficulty of avoiding this model is epically cute when we considered the secondary qualities of colour, sound, tactile feelings and taste, which can easily seem to have a purely private existence inside the perceiver, like sensation of pain. Calling such supposed items names like 'sense-data' or 'percepts' exacerbate the tendency, but once the model is in place, the first property, that perception gives 'us' knowledge of the world and its surrounding surfaces, is quickly threatened, for there will now seem little connection between these items in immediate experience and any independent reality. Reactions to this problem include 'scepticism' and 'idealism'.

A more hopeful approach is to claim that the complexities of (3) and (4) explain how we can have direct acquaintance of the world, than suggesting that the acquaintance we do have been at best indirect. It is pointed out that perceptions are not like sensation, precisely because they have a content, or outer-directed nature. To have a perception is to be aware of the world for being such-and-such a way, than to enjoy a mere modification of sensation. But such direct realism has to be sustained in the face of the self-evident (neurophysiological and other) factors determining how we perceive, for one apparent of approaching is to ask why it is useful to be conscious of what we perceive, when other aspects of our functioning work with information determining responses without any conscious awareness or intervention. A solution to this problem would offer the hope of making consciousness part of the natural world, than a strange optional extra.

Furthering, perceptual knowledge is knowledge acquired by or through the senses and includes most of what we know. We cross intersections when we see the light turn green, head for the kitchen when we smell the roast burning, squeeze the fruit to determine its ripeness, and climb out of bed when we hear the alarm ring. In each case we come to know something-that the light has turned green, that the roast is burning, that the melon is overripe, and that it is time to get up-by some sensory. Seeing that the light has turned green is learning something-that, the light has turned green-by use of the eyes. Feeling that the melon is overripe is coming to know a fact-that the melon is overripe-by one's sense to touch. In each case the resulting knowledge is somehow based on, derived from or grounded in the sort of experience that characterizes the sense modality in question.

Much of our perceptual knowledge is indirect, dependent or derived. By this I mean that the facts we describe ourselves as learning, as coming to know, by perceptual are pieces of knowledge that depend on our coming to know something else, some other fact, in a more direct way. We see, by the gauge, that we need gas, see, by the newspapers, that our team has lost again, see, by her expression, that she is nervous. This derived or dependent sort of knowledge is particularly prevalent in the cases of vision, but it occurs, to a lesser degree, in every sense modality. We install bells and other noise makers so that we calm for example, hear (by the bell) that someone is at the door and (by the alarm) that its time to get up. When we obtain knowledge in this way, it is clear that unless one sees-hence, comes to know something about the gauge (that it says) and (hence, know) that one is described as coming to know by perceptual. If one cannot hear that the bell is ringing, one cannot-in at least in this way-hear that one's visitors have arrived. In such cases one sees (hears, smells, and so forth) that 'a' is 'F', coming to know thereby that 'a' is 'F', by seeing (hearing, and so forth) that some other condition, 'b's' being 'G', obtains when this occurs, the knowledge (that 'a' is 'F') is derived from, or dependent on, the more basic perceptual knowledge that 'b' is 'G'.

Perhaps as a better strategy is to tie an account save that part that evidence could justify explanation for it is its truth alone. Since, at least the time of Aristotle, philosophers of explanatory knowledge have emphasized of its importance that, in its simplest terms, we want to know not only what is the composite peculiarity and still the particular points of issue but also why it is. This consideration suggests that we define an explanation as an answer to a why-question. Such a definition would, however, be too broad, because some why-questions are requests for consolation (Why did my son have to die?) Or moral justification (Why should women not be paid the same as men for the same work?) It would also be too narrow because some explanations are responses to how-questions (How does radar work?) Or how-possibility-questions (How is it possible for cats always to land their feet?)

In its overall sense, 'to explain' to make clear, to make plain, or to provide understanding. Definitions of this sort are philosophically unhelpful, for the terms used in the deficient are no less problematic than the term to be defined. Moreover, since a wide variety of things require explanation, and since many different types of explanation exist, as more complex explanation is required. To facilitate the requirement leaves, least of mention, for us to consider by introduction a bit of technical terminology. The term 'explanation' is used to refer to that which is to be explained: The term 'explanans' refer to that which does the explaining, the explanans and the explanation taken together constitute the explanation.

One common type of explanation occurs when deliberate human actions are explained in terms of conscious purposes. 'Why did you go to the pharmacy yesterday?' 'Because I had a headache and needed to get some aspirin.' It is tacitly assumed that aspirin is an appropriate medication for headaches and that going to the pharmacy would be an efficient way of getting some. Such explanations are, of course, teleological, referring, ss they do, to goals. The explanans are not the realisation of a future goal-if the pharmacy happened to be closed for stocktaking the aspirin would have been obtained there, bu t that would not invalidate the explanation. Some philosophers would say that the antecedent expressions of desire are to achieve the end from which what the doers do the explaining: Others might say that the explaining is done by the nature of the goal and the fact that the action promoted the chances of realizing it. (Taylor, 1964). In that it should not be automatically being assumed that such explanations are causal. Philosophers differ considerably on whether these explanations are to be framed in terms of cause or reason, but the distinction cannot be used to show that the relation between reasons and the actions they justify is in no way causal, and there are many differing analyses of such concepts as intention and agency. Expanding the domain beyond consciousness, Freud maintained, in addition, that much human behaviour can be explained in terms of unconscious and conscious wishes. Those Freudian explanations should probably be construed as basically causal.

Problems arise when teleological explanations are offered in other context. The behaviour of nonhuman animals is often explained in terms of purpose, e.g., the mouse ran to escape from the cat. In such cases the existence of conscious purpose seems dubious. The situation is still more problematic when a supr-empirical purpose in invoked -, e.g., the explanations of living species in terms of God's purpose, or the vitalistic explanations of biological phenomena in terms of a entelechy or vital principle. In recent years an 'anthropic principle' has received attention in cosmology (Barrow and Tipler, 1986). All such explanations have been condemned by many philosophers an anthropomorphic.

Nevertheless, philosophers and scientists often maintain that functional explanations play an important an legitimate role in various sciences such as, evolutionary biology, anthropology and sociology. For example, of the peppered moth in Liverpool, the change in colour from the light phase to the dark phase and back again to the light phase provided adaption to a changing environment and fulfilled the function of reducing predation on the spacies. In the study of primitive soviets anthropologists have maintained that various rituals the (rain dance) which may be inefficacious in braining about their manifest goals (producing rain), actually cohesion at a period of stress (often a drought). Philosophers who admit teleological and/or functional explanations in common sense and science oftentimes take and to argue that such explanations can be annualized entirely in terms of efficient causes, thereby escaping the charge of anthropomorphism (Wright, 1976): Again, however, not all philosophers agree.

Mainly to avoid the incursion of unwanted theology, metaphysics, or anthropomorphism into science, many philosophers and scientists, especially during the first half of the twentieth century-held that science provides only descriptions and predictions of natural phenomena, but not explanations for a series of influential philosophers of science-including Karl Popper (1935) Carl Hempel and Paul Oppenheim (1948) and Hempel (1965)-maintained that empirical science can explain natural phenomena without appealing to metaphysics or theology. It appears that this view is now accepted by the vast majority of philosophers of science, though there is sharp disagreement on the nature of scientific explanation.

The foregoing approach, developed by Hempel, Popper and others, became virtually a 'received view' in the 1960s and 1970s. According to this view, to give a scientific explanation of any natural phenomenon is to show how this phenomena can be subsumed under a law of nature. A particular repture in a water pipe can be explained by citing the universal law that water expands when it freezes and the fact that the temperature of water in a pipe dropped below the freezing point. General law, as well as particular facts, can be explained by subsumption, the law of conservation of linear momentum can be explained by derivation from Newton's second and third laws of motion. Each of these explanations is a deductive argument: The explanans contain one or more statements of universal laws and, in many cases, statements deceiving initial conditions. This pattern of explanation is known as the deductive-nomological (D-N) model. Any such argument shows that the explanandun had to occur given the explanans.

Many, though not all, adherents of the received view allow for explanation by subsumption under statistical laws. Hempel (1965) offers as an example the case of a man who recovered quickly from a streptococcus infection as a result of treatment with penicillin. Although not all strep infections' clar up quickly under this treatment, the probability of recovery in such cases is high, and this is sufficient for legitimate explanation According to Hempel. This example conforms to the inductive-statistical (I-S) model. Such explanations are viewed as arguments, but they are inductive than deductive. In these instances the explanation confers high inductive probability on the explanandum. An explanation of a particular fact satisfying either the D-N or I-S model is an argument to the effect that the fact in question was to be expected by virtue of the explanans.

The received view been subjected to strenuous criticism by adherents of the causal/mechanical approach to scientific explanation (Salmon 1990). Many objections to the received view we engendered by he absence of caudal constraints (due largely to worries about Hume's critique) on the N-D and I-S models. Beginning in the late 1950s, Michael Scriven advanced serious counterexamples to Hemel's models: He was followed in the 1960s by Wesley Salmon and in the 1970s by Peter Railton. As accorded to the view, one explain phenomenon identifying causes (a death is explained resalting from a massive cerebral haemorrhage) or by exposing underlying mechanisms (the behaviour of a gas is explained in terms of the motion of constituent molecules).

A unification approach to explanation carries with the basic idea that we understand our world more adequately to the extent that we can reduce the number of independent assumptions we must introduce to account for what goes on in it. Accordingly, we understand phenomena to the degree that we can fit them into an overall world picture or Weltandchauung. In order to serve in scientific explanation, the world picture must be scientifically well founded.

During the pas half-century much philosophical attention has been focussed on explanation in science and in history. Considerable controversy has surrounded the question of whether historical explanation must be scientific, or whether history requires explanations of different types. Many diverse views have been articulated: The forgoing brief survey does not exhaust the variety (Salmon, 19990).

In everyday life we encounter many types of explanation, which appear not to raise philosophical difficulties, in addition to those already made of mention. Prior to take-off a flight attendant explains how to use the safety equipment on the aero-plane. In a museum, the guide explains the significance of a famous painting. A mathematics teacher explains a geometrical proof to a bewildered student. A newspaper story explains how a prisoner escaped. Additional examples come easily to mind, the main point is to remember the great variety of contexts in which explanations are sought and given into.

Another item of importance to epistemology is the wider held notion that non-demonstrative inferences can be characterized as inference to the best explanation. Given the variety of views on the nature of explanation, this popular slogan can hardly provide a useful philosophical analysis

Early versions of defeasibility theories had difficulty allowing for the existence of evidence that was 'merely misleading,' as in the case where one does know that h3: 'Tom Grabit stole a book from the library,' thanks to having seen him steal it, yet where, unbeknown to oneself, Tom's mother out of dementia gas testified that Tom was far away from the library at the time of the theft. One's justifiably believing that she gave the testimony would destroy one's justification for believing that h3' if added by itself to one's present evidence.

357

At least some defeasibility theories cannot deal with the knowledge one has while dying that h4: 'In this life there is no timer at which I believe that 'd', where the proposition that 'd' expresses the details regarding some philosophical matter, e.g., the maximum number of blades of grass ever simultaneously growing on the earth. When it just so happens that it is true that 'd', defeasibility analyses typically consider the addition to one's dying thoughts of a belief that 'd' in such a way as too improperly rule out actual knowledge that 'h4'.

A quite different approach to knowledge, and one able to deal with some Getter-type cases, involves developing some type of causal theory of Propositional knowledge. The interesting thesis that counts as a causal theory of justification (in the meaning of 'causal theory': Intended here) is the that of a belief is justified just in case it was produced by a type of process that is 'globally' reliable, that is, its propensity to produce true beliefs-that can be defined (to a god enough approximation) as the proportion of the bailiffs it produces (or would produce where it used as much as opportunity allows) that are true-is sufficiently meaningful-variations of this view have been advanced for both knowledge and justified belief. The first formulation of reliability account of knowing appeared in a note by F.P. Ramsey (1931), who said that a belief was knowledge if it is true, certain can obtain by a reliable process. P. Unger (1968) suggested that 'S' knows that 'p' just in case it is not at all accidental that 'S' is right about its being the casse that 'p'. D.M. Armstrong (1973) said that a non-inferential belief qualified as knowledge if the belief has properties that are nominally sufficient for its truth, i.e., guarantee its truth through and by the laws of nature.

Such theories require that one or another specified relation hold that can be characterized by mention of some aspect of cassation concerning one's belief that 'h' (or one's acceptance of the proposition that 'h') and its relation to state of affairs 'h*', e.g., 'h' causes the belief: 'h' is causally sufficient for the belief 'h' and the beliefs have a common cause. Such simple versions of a causal theory are able to deal with the original Notgot case. In that, it involves no such causal relationship are, nonetheless, that cannot explain why there is ignorance in the variants where Notgot and Berent Enç (1984) have pointed out. Sometimes one knows of ',' that is thanks to recognizing a feature merely correlated with the presence of without endorsing a causal theory themselves, suggesting that it would need to be elaborated so as to allow that one's belief that ',' has has been caused by a factor. Whose correlation with the presence of has caused in the total, essential or particular being of self, e.g., by evolutionary adaption in one's ancestors, the disposition that one manifests in acquiring the belief in response to the correlated factor. Not only does this strain the unity of as causal theory by complicating it, but no causal theory without other shortcomings has been able to cover instances of deductively reasoned knowledge.

Causal theories of Propositional knowledge differ over whether they deviate from the tripartite analysis by dropping the requirements that one's believing (accepting) that 'h' be justified. The same variation occurs regarding reliability theories, which present the Knower as reliable concerning the issue of whether or not 'h', in the sense that some of one's cognitive or epistemic states, è, is such that, given further characteristics of oneself-possibly including relations to factors external to one and which one may not be aware-it is nomologically necessary (or at least probable) that 'h'. In some versions, the reliability is required to be 'global' in as far as it must concern a nomologically (probabilistic) relationship) relationship of states of type è to the acquisition of true beliefs about a wider range of issues than merely whether or not 'h'. There is also controversy about how to delineate the limits of what constitutes a type of relevant personal state or characteristic. (For example, in a case where Mr Notgot has not been shamming and one does know thereby that someone in the office owns a Ford, such as a way of forming beliefs about the properties of persons spatially close to one, or instead something narrower, such as a way of forming beliefs about Ford owners in offices partly upon the basis of their relevant testimony?)

One important variety of reliability theory is a conclusive reason account, which includes a requirement that one's reasons for believing that 'h' be such that in one's circumstances, if h* were not to occur then, e.g., one would not have the reasons one does for believing that 'h', or, e.g., one would not believe that 'h'. Roughly, the latter are demanded by theories that treat a Knower as 'tracking the truth', theories that include the further demand that is roughly, if it were the case, that 'h', then one would believe that 'h'. A version of the tracking theory has been defended by Robert Neozoic (1981), who adds that if what he calls a 'method' has been used to arrive at the belief that 'h', then the antecedent clauses of the two conditionals that characterize tracking will need to include the hypothesis that one would employ the very same method.

But unless more conditions are added to Nozick's analysis, it will be too weak to explain why one lack's knowledge in a version of the last variant of the tricky Mr Notgot case described above, where we add the following details: (a) Mr Notgot's compulsion is not easily changed, (b) while in the office, Mr Notgot has no other easy trick of the relevant type to play on one, and finally for one's belief that 'h', not by reasoning through a false belief ut by basing belief that 'h', upon a true existential generalization of one's evidence.

Nozick's analysis is in addition too strong to permit anyone ever to know that 'h': 'Some of my beliefs about beliefs might be otherwise, e.g., I might have rejected on of them'. If I know that 'h5' then satisfaction of the antecedent of one of Nozick's conditionals would involve its being false that 'h5', thereby thwarting satisfaction of the consequent's requirement that I not then believe that 'h5'. For the belief that 'h5' is itself one of my beliefs about beliefs (Shope, 1984).

Some philosophers think that the category of knowing for which is true. Justified believing (accepting) is a requirement constituting only a species of Propositional knowledge, construed as an even broader category. They have proposed various examples of 'PK' that do not satisfy the belief and/ort justification conditions of the tripartite analysis. Such cases are often recognized by analyses of Propositional knowledge in terms of powers, capacities, or abilities. For instance, Alan R. White (1982) treats 'PK' as merely the ability to provide a correct answer to a possible questions, however, White may be equating 'producing' knowledge in the sense of producing 'the correct answer to a possible question' with 'displaying' knowledge in the sense of manifesting knowledge. (White, 1982). The latter can be done even by very young children and some nonhuman animals independently of their being asked questions, understanding questions, or recognizing answers to questions. Indeed, an example that has been proposed as an instance of knowing that 'h' without believing or accepting that 'h' can be modified so as to illustrate this point. Two examples concern an imaginary person who has no special training or information about horses or racing, but who in an experiment persistently and correctly picks the winners of upcoming horseraces. If the example is modified so that the hypothetical 'seer' never picks winners but only muses over whether those horse's might win, or only reports those horses winning, this behaviour should be as much of a candidate for the person's manifesting knowledge that the horse in question will win as would be the behaviour of picking it as a winner.

These considerations now placed upon our table, least that we take to consider of their vulnerability, that is in regard to their limitation: Edward Craig's analysis (1990) of the concept of knowing of a person's being a satisfactory informant in relation to an inquirer who wants to find out whether or not 'h'. Craig realizes that counterexamples to his analysis appear to be constituted by Knower who is too recalcitrant to inform the inquirer, or to incapacitate to inform, or too discredited to be worth considering (as with the boy who cried 'Wolf'). Craig admits that this might make preferable some alternative view of knowledge as a different state that helps to explain the presence of the state of being a suitable informant when the latter does obtain. Such the alternate, which offers a recursive definition that concerns one's having the power to proceed in a way representing the state of affairs, causally involved in one's proceeding in this way. When combined with a suitable analysis of representing, this theory of propositional knowledge can be unified with a structurally similar analysis of knowing how to do something.

Knowledge and belief, according to most epistemologists, knowledge entails belief, so that I cannot know that such and such is the case unless I believe that such and such am the case. Others think this entailment thesis can be rendered more accurately if we substitute for belief some closely related attitude. For instance, several philosophers would prefer to say that knowledge entail psychological certainties (Prichard, 1950 and Ayer, 1956) or conviction (Lehrer, 1974) or acceptance (Lehrer, 1989). None the less, there are arguments against all versions of the thesis that knowledge requires having a belief-like attitude toward the known. These arguments are given by philosophers who think that knowledge and belief (or a facsimile) are mutually incompatible (the incomparability thesis), or by ones who say that knowledge does not entail belief, or vice versa, so that each may exist without the other, but the two may also coexist (the separability thesis).

The incompatibility thesis is sometimes traced to Plato (429-347 Bc) in view of his claim that knowledge is infallible while belief or opinion is fallible ('Republic' 476-9). But this claim would not support the thesis. Belief might be a component of an infallible form of knowledge in spite of the fallibility of belief. Perhaps, knowledge involves some factor that compensates for the fallibility of belief.

A. Duncan-Jones (1939: Also Vendler, 1978) cites linguistic evidence to back up the incompatibility thesis. He notes that people often say 'I do not believe she is guilty. I know she is' and the like, which suggest that belief rule out knowledge. However, as Lehrer (1974) indicates, the above exclamation is only a more emphatic way of saying 'I do not just believe she is guilty, I know she is' where 'just' makes it especially clear that the speaker is signalling that she has something more salient than mere belief, not that she has something inconsistent with belief, namely knowledge. Compare: 'You do not hurt him, you killed him.'

H.A. Prichard (1966) offers a defence of the incompatibility thesis that hinges on the equation of knowledge with certainty (both infallibility and psychological certitude) and the assumption that when we believe in the truth of a claim we are not certain about its truth. Given that belief always involves uncertainty while knowledge never dies, believing something rules out the possibility of knowing it. Unfortunately, however, Prichard gives 'us' no goods reason to grant that states of belief are never ones involving confidence. Conscious beliefs clearly involve some level of confidence, to suggest that we cease to believe things about which we are completely confident is bizarre.

A.D. Wesley (1953) defends a version of the separability thesis. Woozley's version, which deals with psychological certainty rather than belief per se, is that knowledge can exist in the absence of confidence about the item known, although might also be accompanied by confidence as well. Wesley remarks that the test of whether I know something is 'what I can do, where what I can do may include answering questions.' On the basis of this remark he suggests that even when people are unsure of the truth of a claim, they might know that the claim is true. We unhesitatingly attribute knowledge to people who give correct responses on examinations even if those people show no confidence in their answers. Wesley acknowledges, however, that it would be odd for those who lack confidence to claim knowledge. It would be peculiar to say: I am unsure of whether my answer is true: Still, I know it is correct But this tension Wesley explains using a distinction between conditions under which we are justified in making a claim (such as a claim to know something), and conditions under which the claim we make are true. While 'I know such and such' might be true even if I am unsure of whether such and such holds, nonetheless it would be inappropriate for me to claim that I know that such and such unless I was sure of the truth of my claim.

Colin Redford (1966) extends Woozley's defence of the separability thesis. In Radford's view, not only is knowledge compatible with the lack of certainty, it is also compatible with a complete lack of belief. He argues by example. In one example, Jean has forgotten that he learned some English history year's priori and yet he is able to give several correct responses to questions such as 'When did the Battle of Hastings occur?' Since he forgot that he took history, he considers the correct response to be no more than guesses. Thus, when he says that the Battle of Hastings took place in 1066 he would deny having the belief that the Battle of Hastings took place in 1066. A disposition he would deny being responsible (or having the right to be convincing) that 1066 was the correct date. Redford would, nonetheless, insist that Jean know when the Battle occurred, since clearly be remembering the correct date. Redford admits that it would be inappropriate for Jean to say that he knew when the Battle of Hastings occurred, but, like Wesley he attributes the impropriety to a fact about when it is and is not appropriate to claim knowledge. When we claim knowledge, we ought, at least to believe that we have the knowledge we claim, or else our behaviour is 'intentionally misleading'.

Those that agree with Radford's defence of the separability thesis will probably think of belief as an inner state that can be detected through introspection. That Jean lack's belief about English history is plausible on this Cartesian picture since Jean does not find himself with any beliefs about English history when searching them out. One might criticize Redford, however, by rejecting that Cartesian view of belief. One could argue that some beliefs are thoroughly unconscious, for example. Or one could adopt a behaviourist conception of belief, such as Alexander Bain's (1859), according to which having beliefs is a matter of the way people are disposed to behave (and has not Redford already adopted a behaviourist conception of knowledge?) Since Jean gives the correct response when queried, a form of verbal behaviour, a behaviourist would be tempted to credit him with the belief that the Battle of Hastings occurred in 1066.

D.M. Armstrong (1873) takes a different tack against Redford. Jean does know that the Battle of Hastings took place in 1066. Armstrong will grant Redford that point, in fact, Armstrong suggests that Jean believe that 1066 is not the date the Battle of Hastings occurred, for Armstrong equates the belief that such and such is just possible but no more than just possible with the belief that such and such is not the case. However, Armstrong insists, Jean also believes that the Battle did occur in 1066. After all, had Jean been mistaught that the Battle occurred in 1066, and subsequently 'guessed' that it took place in 1066, we would surely describe the situation as one in which Jean's false belief about the Battle became unconscious over time but persisted of a memory trace that was causally responsible for his guess. Out of consistency, we must describe Radford's original case as one that Jean's true belief became unconscious but persisted long enough to cause his guess. Thus, while Jean consciously believes that the Battle did not occur in 1066, unconsciously he does believe it occurred in 1066. So after all, Redford does not have a counterexample to the claim that knowledge entails belief.

Armstrong's response to Redford was to reject Radford's claim that the examinee lacked the relevant belief about English history. Another response is to argue that the examinee lacks the knowledge Redford attributes to him (cf. Sorenson, 1982). If Armstrong is correct in suggesting that Jean believes both that 1066 is and that it is not the date of the Battle of Hastings, one might deny Jean knowledge on the grounds that people who believe the denial of what they believe cannot be said t know the truth of their belief. Another strategy might be to compare the examined case with examples of ignorance given in recent attacks on externalist accounts of knowledge (needless to say. Externalists themselves would tend not to favour this strategy). Consider the following case developed by BonJour (1985): For no apparent reason, Samantha believes that she is clairvoyant. Again, for no apparent reason, she one day comes to believe that the Prime Minister is in Toronto, even though she has every reason to believe that the Premier is in Ottawa Canada. In fact, Samantha is a completely reliable clairvoyant, and she has arrived at her belief about the whereabouts of the Prime Minister through the power of her clairvoyance. Yet surely Samantha's belief is completely irrational. She is not justified in thinking what she does. If so, then she does not know where the Premier is. But Radford's examinee is unconventional. Even if Jean lacks the belief that Redford denies him, Redford does not have an example of knowledge that is unattended with belief. Suppose that Jean's memory had been sufficiently powerful to produce the relevant belief. As Redford says, in having every reason to suppose that his response is mere guesswork, and he has every reason to consider his belief false. His belief would be an irrational one, and hence one about whose truth Jean would be ignorant.

To an approaching view from which 'perception' basis upon itself as a fundamental philosophical topic both for its central place in ant theory of knowledge, and its central place un any theory of consciousness. Philosophy in this area is constrained by a number of properties that we believe to hold of perception, (1) It gives 'us' knowledge of the world around 'us,' (2) We are conscious of that world by being aware of 'sensible qualities': Colour, sounds, tastes, smells, felt warmth, and the shapes and positions of objects in the environment. (3) Such consciousness is affected through highly complex information channels, such as the output of the three different types of colour-sensitive cells in the eye, or the channels in the ear for interpreting pulses of air pressure as frequencies of sound. (4) There ensues even more complex neurophysiological coding of that information, and eventually higher-order brain functions bring it about that we interpreted the information so received. (Much of this complexity has been revealed by the difficulties of writing programs enabling computers to recognize quite simple aspects of the visual scene.) The problem is to avoid thinking of here being a central, ghostly, conscious self, fed information in the same way that a screen if fed information by a remote television camera. Once such a model is in place, experience will seem like a veil getting between 'us' and the world, and the direct objects of perception will seem to be private items in an inner theatre or sensorium. The difficulty of avoiding this model is epically cute when we considered the secondary qualities of colour, sound, tactile feelings and taste, which can easily seem to have a purely private existence inside the perceiver, like sensation of pain. Calling such supposed items names like 'sense-data' or 'percepts' exacerbate the tendency, but once the model is in place, the first property, that perception gives 'us' knowledge of the world and its surrounding surfaces, is quickly threatened, for there will now seem little connection between these items in immediate experience and any independent reality. Reactions to this problem include 'scepticism' and 'idealism.'

A more hopeful approach is to claim that the complexities of (3) and (4) explain how we can have direct acquaintance of the world, than suggesting that the acquaintance we do have been at best indirect. It is pointed out that perceptions are not like sensation, precisely because they have a content, or outer-directed nature. To have a perception is to be aware of the world for being such-and-such a way, than to enjoy a mere modification of sensation. But such direct realism has to be sustained in the face of the evidently personal (neurophysiological and other) factors determining how we perceive. One approach is to ask why it is useful to be conscious of what we perceive, when other aspects of our functioning work with information determining responses without any conscious awareness or intervention. A solution to this problem would offer the hope of making consciousness part of the natural world, than a strange optional extra.

Furthering, perceptual knowledge is knowledge acquired by or through the senses and includes most of what we know. We cross intersections when we see the light turn green, head for the kitchen when we smell the roast burning, squeeze the fruit to determine its ripeness, and climb out of bed when we hear the alarm ring. In each case we come to know something-that the light has turned green, that the roast is burning, that the melon is overripe, and that it is time to get up-by some sensory. Seeing that the light has turned green is learning something-that, the light has turned green-by use of the eyes. Feeling that the melon is overripe is coming to know a fact-that the melon is overripe-by one's sense to touch. In each case the resulting knowledge is somehow based on, derived from or grounded in the sort of experience that characterizes the sense modality in question.

Much of our perceptual knowleledge is indirect, dependent or derived. By this I mean that the facts we describe ourselves as learning, as coming to know, by perceptual are pieces of knowledge that depend on our coming to know something else, some other fact, in a more direct way. We see, by the gauge, that we need gas, see, by the newspapers, that our team has lost again, see, by her expression, that she is nervous. This derived or dependent sort of knowledge is particularly prevalent in the cases of vision, but it occurs, to a lesser degree, in every sense modality. We install bells and other noise makers so that we calm for example, hear (by the bell) that someone is at the door and (by the alarm) that its time to get up. When we obtain knowledge in this way, it is clear that unless one sees-hence, comes to know something about the gauge (that it says) and (hence, know) that one is described as coming to know by perceptual. If one cannot hear that the bell is ringing, one cannot-in at least in this way-hear that one's visitors have arrived. In such cases one sees (hears, smells, and so forth) that 'a' is 'F', coming to know thereby that 'a' is 'F', by seeing (hearing, and so forth) that some other condition, 'b's' being 'G', obtains when this occurs, the knowledge (that 'a' is 'F') is derived from, or dependent on, the more basic perceptual knowledge that 'b' is 'G'.

And finally, the representational Theory of mind (RTM) (which goes back at least to Aristotle) takes as its starting point commonsense mental states, such as thoughts, beliefs, desires, perceptions and images. Such states are said to have 'intentionality'-they are about or refer to things, and may be evaluated with respect to properties like consistency, truth, appropriateness and accuracy. (For example, the thought that cousins are not related is inconsistent, the belief that Elvis is dead is true, the desire to eat the moon is inappropriate, a visual experience of a ripe strawberry as red is accurate, an image of George W. Bush with deadlocks is inaccurate.)

The Representational Theory of Mind, defines such intentional mental states as relations to mental representations, and explains the intentionality of the former in terms of the semantic properties of the latter. For example, to believe that Elvis is dead is to be appropriately related to a mental representation whose propositional content is that Elvis is dead. (The desire that Elvis be dead, the fear that he is dead, the regrets that he is dead, and so on, involve different relations to the same mental representation.) To perceive a strawberry is to have a sensory experience of some kind which is appropriately related to (e.g., caused by) the strawberry Representational theory of mind also understands mental processes such as thinking, reasoning and imagining as sequences of intentional mental states. For example, to imagine the moon rising over a mountain is to entertain a series of mental images of the moon (and a mountain). To infer a proposition q from the propositions p and if 'p' then 'q' is (among other things) to have a sequence of thoughts of the form 'p', 'if p' then 'q', 'q'.

Contemporary philosophers of mind have typically supposed (or at least hoped) that the mind can be naturalized-i.e., that all mental facts have explanations in the terms of natural science. This assumption is shared within cognitive science, which attempts to provide accounts of mental states and processes in terms (ultimately) of features of the brain and central nervous system. In the course of doing so, the various sub-disciplines of cognitive science (including cognitive and computational psychology and cognitive and computational neuroscience) postulate a number of different kinds of structures and processes, many of which are not directly implicated by mental states and processes as commonsensical conceived. There remains, however, a shared commitment to the idea that mental states and processes are to be explained in terms of mental representations.

In philosophy, recent debates about mental representation have centred around the existence of propositional attitudes (beliefs, desires, and so forth) and the determination of their contents (how they come to be about what they are about), and the existence of phenomenal properties and their relation to the content of thought and perceptual experience. Within cognitive science itself, the philosophically relevant debates have been focussed on the computational architecture of the brain and central nervous system, and the compatibility of scientific and commonsense accounts of mentality.

Intentional Realists such as Dretske (e.g., 1988) and Fodor (e.g., 1987) notes that the generalizations we apply in everyday life in predicting and explaining each other's behaviour (often collectively referred to as 'folk psychology') are both remarkably successful and indispensable. What a person believes, doubts, desires, fears, and so forth is a highly reliable indicator of what that person will do, and we have no other way of making sense of each other's behaviour than by ascribing such states and applying the relevant generalizations. We are thus committed to the basic truth of commonsense psychology and, hence, to the existence of the states its generalizations refer to. (Some realists, such as Fodor, also hold that commonsense psychology will be vindicated by cognitive science, given that propositional attitudes can be construed as computational relations to mental representations.)

Intentional Eliminativists, such as Churchland, (perhaps) Dennett and (at one time) Stich argues that no such things as propositional attitudes (and their constituent representational states) is implicated by the successful explanation and prediction of our mental lives and behaviour. Churchland denies that the generalizations of commonsense propositional-attitude psychology are true. He (1981) argues that folk psychology is a theory of the mind with a long history of failure and decline, and that it resists incorporation into the framework of modern scientific theories (including cognitive psychology). As such, it is comparable to alchemy and phlogiston theory, and ought to suffer a comparable fate. Commonsense psychology is false, and the states (and representations) it postulates simply don't exist. (It should be noted that Churchland is not an eliminativist about mental representation tout court.

Dennett (1987) grants that the generalizations of commonsense psychology are true and indispensable, but denies that this is sufficient reason to believe in the entities they appear to refer to. He argues that to give an intentional explanation of a system's behaviour is merely to adopt the 'intentional stance' toward it. If the strategy of assigning contentual states to a system and predicting and explaining its behaviour (on the assumption that it is rational-i.e., which it behaves as it should, given the propositional attitudes it should have in its environment) is successful, then the system is intentional, and the propositional-attitude generalizations we apply to it are true. But there is nothing more to having a propositional attitude than this.

Though he has been taken to be thus claiming that intentional explanations should be construed instrumentally, Dennett (1991) insists that he is a 'moderate' realist about propositional attitudes, since he believes that the patterns in the behaviour and behavioural dispositions of a system on the basis of which we (truly) attribute intentional states to it are objectively real. In the event that there is two or more explanatorily adequate but substantially different systems of intentional ascriptions to an individual, however, Dennett claims there are no fact of the matter about what the system believes (1987, 1991). This does suggest an irrealist at least with respect to the sorts of things Fodor and Dretske take beliefs to be; though it is not the view that there is simply nothing in the world that makes intentional explanations true.

Davidson 1973, 1974 and Lewis 1974 also defend the view that what it is to have a propositional attitude is just to be interpretable in a particular way. It is, however, not entirely clear whether they intend their views to imply irrealis about propositional attitudes. Stich (1983) argues that cognitive psychology does not (or, in any case, should not) taxonomize mental states by their semantic properties at all, since attribution of psychological states by content is sensitive to factors that render it problematic in the context of a scientific psychology. Cognitive psychology seeks causal explanations of behaviour and cognition, and the causal powers of a mental state are determined by its intrinsic 'structural' or 'syntactic' properties. The semantic properties of a mental state, however, are determined by its extrinsic properties-e.g., its history, environmental or intra-mental relations. Hence, such properties cannot figure in causal-scientific explanations of behaviour. (Fodor 1994 and Dretske 1988 are realist attempts to come to grips with some of these problems.) Stich proposes a syntactic theory of the mind, on which the semantic properties of mental states play no explanatory role.

It is a traditional assumption among realists about mental representations that representational states come in two basic varieties (Boghossian 1995). There are those, such as thoughts, which are composed of concepts and have no phenomenal ('what-it's-like') features ('Qualia'), and those, such as sensory experiences, which have phenomenal features but no conceptual constituents. Non-conceptual content is usually defined as a kind of content that states of a creature lacking concepts might nonetheless enjoy. On this taxonomy, mental states can represent either in a way analogous to expressions of natural languages or in a way analogous to drawings, paintings, maps or photographs. (Perceptual states such as seeing that something is blue, are sometimes thought of as hybrid states, consisting of, for example, a Non-conceptual sensory experience and a thought, or some more integrated compound of sensory and conceptual components.)

Some historical discussions of the representational properties of mind (e.g., Aristotle 1984, Locke 1689/1975, Hume 1739/1978) seem to assume that Non-conceptual representations-percepts ('impressions'), images ('ideas') and the like-are the only kinds of mental representations, and that the mind represents the world in virtue of being in states that resemble things in it. On such a view, all representational states have their content in virtue of their phenomenal features. Powerful arguments, however, focussing on the lack of generality (Berkeley 1975), ambiguity (Wittgenstein 1953) and non-compositionality (Fodor 1981) of sensory and imaginistic representations, as well as their unsuitability to function as logical (Frége 1918/1997, Geach 1957) or mathematical (Frége 1884/1953) concepts, and the symmetry of resemblance (Goodman 1976), convinced philosophers that no theory of mind can get by with only Non-conceptual representations construed in this way.

Contemporary disagreement over Non-conceptual representation concerns the existence and nature of phenomenal properties and the role they play in determining the content of sensory experience. Dennett (1988), for example, denies that there are such things as Qualia at all; while Brandom (2002), McDowell (1994), Rey (1991) and Sellars (1956) deny that they are needed to explain the content of sensory experience. Among those who accept that experiences have phenomenal content, some (Dretske, Lycan, Tye) argue that it is reducible to a kind of intentional content, while others (Block, Loar, Peacocke) argue that it is irreducible.

There has also been dissent from the traditional claim that conceptual representations (thoughts, beliefs) lack phenomenology. Chalmers (1996), Flanagan (1992), Goldman (1993), Horgan and Tiensen (2003), Jackendoff (1987), Levine (1993, 1995, 2001), McGinn (1991), Pitt (2004), Searle (1992), Siewert

(1998) and Strawson (1994), claim that purely symbolic (conscious) representational states themselves have a (perhaps proprietary) phenomenology. If this claim is correct, the question of what role phenomenology plays in the determination of content reprises for conceptual representation. The eliminativist ambitions of Sellars, Brandom, Rey, would meet a new obstacle. (It would also raise prima face problems for reductionist representationalism

The representationalist thesis is often formulated as the claim that phenomenal properties are representational or intentional. However, this formulation is ambiguous between a reductive and a non-deductive claim (though the term 'representationalism' is most often used for the reductive claim). On one hand, it could mean that the phenomenal content of an experience is a kind of intentional content (the properties it represents). On the other, it could mean that the (irreducible) phenomenal properties of an experience determine an intentional content. Representationalists such as Dretske, Lycan and Tye would assent to the former claim, whereas phenomenalists such as Block, Chalmers, Loar and Peacocke would assent to the latter. (Among phenomenalists, there is further disagreement about whether Qualia is intrinsically representational (Loar) or not (Block, Peacocke).

Most (reductive) representationalists are motivated by the conviction that one or another naturalistic explanation of intentionality is, in broad outline, correct, and by the desire to complete the naturalization of the mental by applying such theories to the problem of phenomenality. (Needless to say, most phenomenalists (Chalmers is the major exception) are just as eager to naturalize the phenomenal-though not in the same way.)

The main argument for representationalism appeals to the transparency of experience (cf. Tye 2000). The properties that characterize what it's like to have a perceptual experience is presented in experience as properties of objects perceived: in attending to an experience, one seems to 'see through it' to the objects and properties it is experiences of. They are not presented as properties of the experience itself. If nonetheless they were properties of the experience, perception would be massively deceptive. But perception is not massively deceptive. According to the representationalist, the phenomenal character of an experience is due to its representing objective, non-experiential properties. (In veridical perception, these properties are locally instantiated; in illusion and hallucination, they are not.) On this view, introspection is indirect perception: one comes to know what phenomenal features one's experience has by coming to know what objective features it represents.

In order to account for the intuitive differences between conceptual and sensory representations, representationalists appeal to their structural or functional differences. Dretske (1995), for example, distinguishes experiences and thoughts on the basis of the origin and nature of their functions: an experience of a property 'P' is a state of a system whose evolved function is to indicate the presence of 'P' in the environment; a thought representing the property 'P', on the other hand, is a state of a system whose assigned (learned) function is to calibrate the output of the experiential system. Rey (1991) takes both thoughts and experiences to be relations to sentences in the language of thought, and distinguishes them on the basis of (the functional roles of) such sentences' constituent predicates. Lycan (1987, 1996) distinguishes them in terms of their functional-computational profiles. Tye (2000) distinguishes them in terms of their functional roles and the intrinsic structure of their vehicles: Thoughts are representations in a language-like medium, whereas experiences are image-like representations consisting of 'symbol-filled arrays.' (the account of mental images in Tye 1991.)

Phenomenalists tend to make use of the same sorts of features (function, intrinsic structure) in explaining some of the intuitive differences between thoughts and experiences but they do not suppose that such features exhaust the differences between phenomenal and non-phenomenal representations. For the phenomenalism, it is the phenomenal properties of experiences-qualia themselves-that constitute the fundamental difference between experience and thought. Peacocke (1992), for example, develops the notion of a perceptual 'scenario' (an assignment of phenomenal properties to coordinates of a three-dimensional egocentric space), whose content is 'correct' (a semantic property) if in the corresponding 'scene' (the portion of the external world represented by the scenario) properties are distributed as their phenomenal analogues are in the scenario.

Another sort of representation championed by phenomenalists (e.g., Block, Chalmers (2003) and Loar (1996)) is the 'phenomenal concept'-a conceptual/phenomenal hybrid consisting of a phenomenological 'sample' (an image or an occurrent sensation) integrated with (or functioning as) a conceptual component. Phenomenal concepts are postulated to account for the apparent fact (among others) that, as McGinn (1991) puts it, 'you cannot form [introspective] concepts of conscious properties unless you yourself instantiate those properties.' One cannot have a phenomenal concept of a phenomenal property 'P', and, hence, phenomenal beliefs about P, without having experience of 'P', because 'P' itself is (in some way) constitutive of the concept of 'P'. (Jackson 1982, 1986 and Nagel 1974.)

Though imagery has played an important role in the history of philosophy of mind, the important contemporary literature on it is primarily psychological. In a series of psychological experiments done in the 1970s (summarized in Kosslyn 1980 and Shepard and Cooper 1982), subjects' response time in tasks involving mental manipulation and examination of presented figures was found to vary in proportion to the spatial properties (size, orientation, and so forth) of the figures presented. The question of how these experimental results are to be explained has kindled a lively debate on the nature of imagery and imagination.

Kosslyn (1980) claims that the results suggest that the tasks were accomplished via the examination and manipulation of mental representations that themselves have spatial properties-i.e., pictorial representations, or images. Others, principally Pylyshyn (1979, 1981, 2003), argue that the empirical facts can be explained in terms exclusively of discursive, or propositional representations and cognitive processes defined over them. (Pylyshyn takes such representations to be sentences in a language of thought.)

The idea that pictorial representations are literally pictures in the head is not taken seriously by proponents of the pictorial view of imagery. The claim is, rather, that mental images represented in a way that is relevantly like the way pictures represent. (Attention has been focussed on visual imagery-hence the designation 'pictorial'; though, there may be an imagery or other modalities-auditory, olfactory, and so on, -as well.)

The distinction between pictorial and discursive representation can be characterized in terms of the distinction between analog and digital representation (Goodman 1976). This distinction has itself been variously understood (Fodor & Pylyshyn 1981, Goodman 1976, Haugeland 1981, Lewis 1971, McGinn 1989), though a widely accepted construal is that analog representation is continuous (i.e., in virtue of continuously variable properties of the representation), while digital representation is discrete (i.e., in virtue of properties a representation either has or doesn't have) (Dretske 1981). (An analog/digital distinction may also be made with respect to cognitive processes. (Block 1983.)) On this understanding of the analog/digital distinction, imaginistic representations, which represent in virtue of properties that may vary continuously (such as being more or less bright, loud, vivid, and so forth), would be analog, while conceptual representations, whose properties do not vary continuously (a thought cannot be more or less about Elvis: either it is or it is not) would be digital.

It might be supposed that the pictorial/discursive distinction is best made in terms of the phenomenal and nonphenomenal distinction, but it is not obvious that this is the case. For one thing, there may be nonphenomenal properties of representations that vary continuously. Moreover, there are ways of understanding pictorial representation that presuppose neither phenomenality nor analogicity. According to Kosslyn (1980, 1982, 1983), a mental representation is 'quasi-pictorial' when every part of the representation corresponds to a part of the object represented, and relative distances between parts of the object represented are preserved among the parts of the representation. But distances between parts of a representation can be defined functionally rather than spatially-for example, in terms of the number of discrete computational steps required to combine stored information about them. (Rey 1981.)

Tye (1991) proposes a view of images on which they are hybrid representations, consisting both of the pictorial and discursive elements. On Tye's account, images are '(labelled) interpreted symbol-filled arrays.' The symbols represent discursively, while their arrangement in arrays has representational significance (the location of each 'cell' in the array represents a specific viewer-centred 2-D location on the surface of the imagined object)

The contents of mental representations are typically taken to be abstract objects (properties, relations, propositions, sets, and so forth). A pressing question, especially for the naturalist, is how mental representations come to have their contents. Here the issue is not how to naturalize content (abstract objects can't be naturalized), but, rather, how to provide a naturalistic account of the content-determining relations between mental representations and the abstract objects they express. There are two basic types of contemporary naturalistic theories of content-determination, causal-informational and functional.

Causal-informational theories (Dretske 1981, 1988, 1995) hold that the content of a mental representation is grounded in the information it carries about what does (Devitt 1996) or would (Fodor 1987, 1990) cause it to occur. There is, however, widespread agreement that causal-informational relations are not sufficient to determine the content of mental representations. Such relations are common, but representation is not. Tree trunks, smoke, thermostats and ringing telephones carry information about what they are causally related to, but they do not represent (in the relevant sense) what they carry information about. Further, a representation can be caused by something it does not represent, and can represent something that has not caused it.

The main attempts to specify what makes a causal-informational state a mental representations are Asymmetric Dependency Theories (e.g., Fodor 1987, 1990, 1994) and Teleological Theories (Fodor 1990, Millikan 1984, Papineau 1987, Dretske 1988, 1995). The Asymmetric Dependency Theory distinguishes merely informational relations from representational relations on the basis of their higher-order relations to each other: informational relations depend upon representational relations, but not vice-versa. For example, if tokens of a mental state type are reliably caused by horses, cows-on-dark-nights, zebras-in-the-mist and Great Danes, then they carry information about horses, and so forth If, however, such tokens are caused by cows-on-dark-nights, and so forth because they were caused by horses, but not vice versa, then they represent horses.

According to Teleological Theories, representational relations are those a representation-producing mechanism has the selected (by evolution or learning) function of establishing. For example, zebra-caused horse-representations do not mean zebra, because the mechanism by which such tokens are produced has the selected function of indicating horses, not zebras. The horse-representation-producing mechanism that responds to zebras is malfunctioning.

Functional theories (Block 1986, Harman 1973), hold that the content of a mental representation is grounded in its (causal computational, inferential) relations to other mental representations. They differ on whether relata should include all other mental representations or only some of them, and on whether to include external states of affairs. The view that the content of a mental representation is determined by its inferential/computational relations with all other representations is holism; the view it is determined by relations to only some other mental states is localism (or molecularism). (The view that the content of a mental state depends on none of its relations to other mental states is atomism.) Functional theories that recognize no content-determining external relata have been called solipsistic (Harman 1987). Some theorists posit distinct roles for internal and external connections, the former determining semantic properties analogous to sense, the latter determining semantic properties analogous to reference (McGinn 1982, Sterelny 1989)

(Reductive) representationalists (Dretske, Lycan, Tye) usually take one or another of these theories to provide an explanation of the (Non-conceptual) content of experiential states. They thus tend to be Externalists about phenomenology as well as conceptual content. Phenomenalists and non-deductive representationalists (Block, Chalmers, Loar, Peacocke, Siewert), on the other hand, take it that the representational content of such states is (at least in part) determined by their intrinsic phenomenal properties. Further, those who advocate a phenomenology-based approach to conceptual content (Horgan and Tiensen, Loar, Pitt, Searle, Siewert) also seem to be committed to internalist individuation of the content (if not the reference) of such states.

Generally, those who, like informational theorists, think relations to one's (natural or social) environment is (at least partially) determinative of the content of mental representations are Externalists (e.g., Burge 1979, 1986, McGinn 1977, Putnam 1975), whereas those who, like some proponents of functional theories, think representational content are determined by an individual's intrinsic properties alone, are Internalists or individualists (¦. Putnam 1975, Fodor 1981).

This issue is widely taken to be of central importance, since psychological explanation, whether commonsense or scientific, is supposed to be both causal and content-based. (Beliefs and desires cause the behaviours they do because they have the contents they do. For example, the desire that one have a beer and the beliefs that there are beer in the refrigerator and that the refrigerator is in the kitchen may explain one's getting up and going to the kitchen.) If, however, a mental representation's having a particular content is due to factors extrinsic to it, it is unclear how it's having that content could determine its causal powers, which, arguably, must be intrinsic. Some who accept the standard arguments for externalism have argued that internal factors determine a component of the content of a mental representation. They say that mental representations have both 'narrow' content (determined by intrinsic factors) and 'wide' or 'broad' content (determined by narrow content plus extrinsic factors). (This distinction may be applied to the sub-personal representations of cognitive science as well as to those of commonsense psychology.

Narrow content has been variously construed. Putnam (1975), Fodor (1982)), and Block (1986), for example, seems to understand it as something like dedicto content (i.e., Frégean sense, or perhaps character, la Kaplan 1989). On this construal, narrow content is context-independent and directly expressible. Fodor (1987) and Block (1986), however, has also characterized narrow content as radically inexpressible. On this construal, narrow content is a kind of proto-content, or content-determinant, and can be specified only indirectly, via specifications of context/wide-content pairings. On both construal, narrow contents are characterized as functions from context to (wide) content. The narrow content of a representation is determined by properties intrinsic to it or its possessor such as its syntactic structure or its intra-mental computational or inferential role (or its phenomenology).

Burge (1986) has argued that causation-based worries about externalist individuation of psychological content, and the introduction of the narrow notion, are misguided. Fodor (1994, 1998) has more recently urged that a scientific psychology might not need narrow content in order to supply naturalistic (causal) explanations of human cognition and action, since the sorts of cases they were introduced to handle, viz., Twin-Earth cases and Frigg cases, are nomologically either impossible or dismissible as exceptions to non-strict psychological laws.

The leading contemporary version of the Representational Theory of Mind, the Computational Theory of Mind (CTM), claims that the brain is a kind of computer and that mental processes are computations. According to the computational theory of mind, cognitive states are constituted by computational relations to mental representations of various kinds, and cognitive processes are sequences of such states. The computational theory of mind and the representational theory of mind, may by attempting to explain all psychological states and processes in terms of mental representation. In the course of constructing detailed empirical theories of human and animal cognition and developing models of cognitive processes implementable in artificial information processing systems, cognitive scientists have proposed a variety of types of mental representations. While some of these, may be suited to be mental relata of commonsense psychological states, some-so-called 'subpersonal' or 'sub-doxastic' representations-are not. Though many philosophers believe that computational theory of mind can provide the best scientific explanations of cognition and behaviour, there is disagreement over whether such explanations will vindicate the commonsense psychological explanations of prescientific representational theory of mind.

According to Stich's (1983) Syntactic Theory of Mind, for example, computational theories of psychological states should concern themselves only with the formal properties of the objects those states are relations to. Commitment to the explanatory relevance of content, however, is for most cognitive scientists fundamental (Fodor 1981, Pylyshyn 1984, Von Eckardt 1993). That mental processes are computations, which computations are rule-governed sequences of semantically evaluable objects, and that the rules apply to the symbols in virtue of their content, are central tenets of mainstream cognitive science.

Explanations in cognitive science appeal to a many different kinds of mental representation, including, for example, the 'mental models' of Johnson-Laird 1983, the 'retinal arrays,' 'primal sketches' and '2½ -D sketches' of Marr 1982, the 'frames' of Minsky 1974, the 'sub-symbolic' structures of Smolensky 1989, the 'quasi-pictures' of Kosslyn 1980, and the 'interpreted symbol-filled arrays' of Tye 1991-in addition to representations that may be appropriate to the explanation of commonsense psychological states. Computational explanations have been offered of, among other mental phenomena, belief (Fodor 1975, Field 1978), visual perception (Marr 1982, Osherson, et al. 1990), rationality (Newell and Simon 1972, Fodor 1975, Johnson-Laird and Wason 1977), language learning and (Chomsky 1965, Pinker 1989), and musical comprehension (Lerdahl and Jackendoff 1983).

A fundamental disagreement among proponents of computational theory of mind concerns the realization of personal-level representations (e.g., thoughts) and processes (e.g., inferences) in the brain. The central debate here is between proponents of Classical Architectures and proponents of Conceptionist Architectures.

`The classicists (e.g., Turing 1950, Fodor 1975, Fodor and Pylyshyn 1988, Marr 1982, Newell and Simon 1976) hold that mental representations are symbolic structures, which typically have semantically evaluable constituents, and that mental processes are rule-governed manipulations of them that are sensitive to their constituent structure. The connectionists (e.g., McCulloch & Pitts 1943, Rumelhart 1989, Rumelhart and McClelland 1986, Smolensky 1988) hold that mental representations are realized by patterns of activation in a network of simple processors ('nodes') and that mental processes consist of the spreading activation of such patterns. The nodes themselves are, typically, not taken to be semantically evaluable; nor do the patterns have semantically evaluable constituents. (Though there are versions of Connectionism-'localist' versions-on which individual nodes are taken to have semantic properties (e.g., Ballard 1986, Ballard & Hayes 1984). It is arguable, however, that localist theories are neither definitive nor representative of the Conceptionist program (Smolensky 1988, 1991, Chalmers 1993).

Classicists are motivated (in part) by properties thought seems to share with language. Fodor's Language of Thought Hypothesis (LOTH) (Fodor 1975, 1987), according to which the system of mental symbols constituting the neural basis of thought is structured like a language, provides a well-worked-out version of the classical approach as applied to commonsense psychology. According to the language of thought hypothesis, the potential infinity of complex representational mental states is generated from a finite stock of primitive representational states, in accordance with recursive formation rules. This combinatorial structure accounts for the properties of productivity and systematicity of the system of mental representations. As in the case of symbolic languages, including natural languages (though Fodor does not suppose either that the language of thought hypothesis explains only linguistic capacities or that only verbal creatures have this sort of cognitive architecture), these properties of thought are explained by appeal to the content of the representational units and their combinability into contentual plexuity. That is, the semantics of both language and thought is compositional: the content of a complex representation is determined by the contents of its constituents and their structural configuration.

Connectionists are motivated mainly by a consideration of the architecture of the brain, which apparently consists of layered networks of interconnected neurons. They argue that this sort of architecture is unsuited to carrying out classical serial computations. For one thing, processing in the brain is typically massively parallel. In addition, the elements whose manipulation drives computations in Conceptionist networks (principally, the connections between nodes) are neither semantically compositional nor semantically evaluable, as they are on the classical approach. This contrast with classical computationalism is often characterized by saying that representation is, with respect to computation, distributed as opposed too local: representation is local if it is computationally basic; and distributed if it is not. (Another way of putting this is to say that for classicists mental representations are computationally atomic, whereas for connectionists they are not.)

Moreover, connectionists argue that information processing as it occurs in Conceptionist networks more closely resembles some features of actual human cognitive functioning. For example, whereas on the classical view learning involves something like hypothesis formation and testing (Fodor 1981), on the Conceptionist model it is a matter of evolving distribution of 'weight' (strength) on the connections between nodes, and typically does not involve the formulation of hypotheses regarding the identity conditions for the objects of knowledge. The Conceptionist network is 'trained up' by repeated exposure to the objects it is to learn to distinguish. Though networks typically require many more exposures to the objects than do humans, this seems to model at least one feature of this type of human learning quite well.

Further, degradation in the performance of such networks in response to damage is gradual, not sudden as in the case of a classical information processor, and hence more accurately models the loss of human cognitive function as it typically occurs in response to brain damage. It is also sometimes claimed that Conceptionist systems show the kind of flexibility in response to novel situations typical of human cognition-situations in which classical systems are relatively 'brittle' or 'fragile.'

Some philosophers have maintained that Connectionism entails that there are no propositional attitudes. Ramsey, Stich and Garon (1990) have argued that if Conceptionist models of cognition are basically correct, then there are no discrete representational states as conceived in ordinary commonsense psychology and classical cognitive science. Others, however (e.g., Smolensky 1989), hold that certain types of higher-level patterns of activity in a neural network may be roughly identified with the representational states of commonsense psychology. Still others (e.g., Fodor & Pylyshyn 1988,

Heil 1991, Horgan and Tienson 1996) argue that language-of-thought style representation is both necessary in general and realizable within Conceptionist architectures. (MacDonald & MacDonald 1995 collects the central contemporary papers in the classicist/Conceptionist debate, and provides useful introductory material as well.

Whereas Stich (1983) accepts that mental processes are computational, but denies that computations are sequences of mental representations, others accept the notion of mental representation, but deny that computational theory of mind provides the correct account of mental states and processes.

Van Gelder (1995) denies that psychological processes are computational. He argues that cognitive systems are dynamic, and that cognitive states are not relations to mental symbols, but quantifiable states of a complex system consisting of (in the case of human beings) a nervous system, a body and the environment in which they are embedded. Cognitive processes are not rule-governed sequences of discrete symbolic states, but continuous, evolving total states of dynamic systems determined by continuous, simultaneous and mutually determining states of the systems' components. Representation in a dynamic system is essentially information-theoretic, though the bearers of information are not symbols, but state variables or parameters.

Horst (1996), on the other hand, argues that though computational models may be useful in scientific psychology, they are of no help in achieving a philosophical understanding of the intentionality of commonsense mental states. computational theory of mind attempts to reduce the intentionality of such states to the intentionality of the mental symbols they are relations to. But, Horst claims, the relevant notion of symbolic content is essentially bound up with the notions of convention and intention. So the computational theory of mind involves itself in a vicious circularity: the very properties that are supposed to be reduced are (tacitly) appealed to in the reduction.

To say that a mental object has semantic properties is, paradigmatically, to say that it may be about, or be true or false of, an object or objects, or that it may be true or false simpliciter. Suppose I think that ocelots take snuff. I am thinking about ocelots, and if what I think of them (that they take snuff) is true of them, then my thought is true. According to representational theory of mind such states are to be explained as relations between agents and mental representations. To think that ocelots take snuff is too token in some way a mental representation whose content is that ocelots take snuff. On this view, the semantic properties of mental states are the semantic properties of the representations they are relations to.

Linguistic acts seem to share such properties with mental states. Suppose I say that ocelots take snuff. I am talking about ocelots, and if what I say of them (that they take snuff) is true of them, then my utterance is true. Now, to say that ocelots take snuff is (in part) to utter a sentence that that ocelots take snuff. Many philosophers have thought that the semantic properties of linguistic expressions are inherited from the intentional mental states they are conventionally used to express (Grice 1957, Fodor 1978, Schiffer, 1972/1988, Searle 1983). On this view, the semantic properties of linguistic expressions are the semantic properties of the representations that are the mental relata of the states they are conventionally used to express.

It is also widely held that in addition to having such properties as reference, truth-conditions and truth-so-called extensional properties-expressions of natural languages also have intensional properties, in virtue of expressing properties or propositions-i.e., in virtue of having meanings or senses, where two expressions may have the same reference, truth-conditions or truth value, yet express different properties or propositions (Frigg 1892/1997). If the semantic properties of natural-language expressions are inherited from the thoughts and concepts they express (or vice versa, or both), then an analogous distinction may be appropriate for mental representations.

Søren Aabye Kierkegaard (1813-1855), a Danish religious philosopher, whose concern with individual existence, choice, and commitment profoundly influenced modern theology and philosophy, especially existentialism.

Søren Kierkegaard wrote of the paradoxes of Christianity and the faith required to reconcile them. In his book Fear and Trembling, Kierkegaard discusses Genesis 22, in which God commands Abraham to kill his only son, Isaac. Although God made an unreasonable and immoral demand, Abraham obeyed without trying to understand or justify it. Kierkegaard regards this 'leap of faith' as the essence of Christianity.

Kierkegaard was born in Copenhagen on May 15, 1813. His father was a wealthy merchant and strict Lutheran, whose gloomy, guilt-ridden piety and vivid imagination strongly influenced Kierkegaard. Kierkegaard studied theology and philosophy at the University of Copenhagen, where he encountered Hegelian philosophy and reacted strongly against it. While at the university, he ceased to practice Lutheranism and for a time led an extravagant social life, becoming a familiar figure in the theatrical and café society of Copenhagen. After his father's death in 1838, however, he decided to resume his theological studies. In 1840 he became engaged to the 17-year-old Regine Olson, but almost immediately he began to suspect that marriage was incompatible with his own brooding, complicated nature and his growing sense of a philosophical vocation. He abruptly broke off the engagement in 1841, but the episode took on great significance for him, and he repeatedly alluded to it in his books. At the same time, he realized that he did not want to become a Lutheran pastor. An inheritance from his father allowed him to devote himself entirely to writing, and in the remaining 14 years of his life he produced more than 20 books.

Kierkegaard's work is deliberately unsystematic and consists of essays, aphorisms, parables, fictional letters and diaries, and other literary forms. Many of his works were originally published under pseudonyms. He applied the term existential to his philosophy because he regarded philosophy as the expression of an intensely examined individual life, not as the construction of a monolithic system in the manner of the 19th-century German philosopher Georg Wilhelm Friedrich Hegel, whose work he attacked in Concluding Unscientific Postscript (1846; trs., 1941). Hegel claimed to have achieved a complete rational understanding of human life and history; Kierkegaard, on the other hand, stressed the ambiguity and paradoxical nature of the human situation. The fundamental problems of life, he contended, defy rational, objective explanation; the highest truth is subjective.

Kierkegaard maintained that systematic philosophy not only imposed a false perspective on human existence but that it also, by explaining life in terms of logical necessity, becomes a of avoiding choice and responsibility. Individuals, he believed, create their own natures through their choices, which must be made in the absence of universal, objective standards. The validity of a choice can only be determined subjectively.

In his first major work, Either/Or (two volumes, 1843; trs., 1944), Kierkegaard described two spheres, or stages of existence, that the individual may choose: the aesthetic and the ethical. The aesthetic way of life as refined hedonism, consisting of a search for pleasure and a cultivation of moods, the aesthetic individual constantly seeks variety and novelty in an effort to stave off boredom but eventually confronting the fine between boredom and despair. The ethical way of life according it the accepted principles of right and wrong that govern the conduct, ss a direct course of control, they are related to a solely prescriptive course of ethicalicity, involving an intense, passionate commitment to duty, to unconditional social and religious obligations. In his later works, wrote as, Stages on Life's Way (1845; trs., 1940), Kierkegaard discerned in this submission to duty a loss of individual responsibility, and he proposed a third stage, the religious, in which one submits to the will of God but in doing so finds authentic freedom. In, Fear and Trembling (1846; trs., 1941) Kierkegaard focussed on God's command that Abraham sacrifice his son Isaac (Genesis 22: 1-19), an act that violates Abraham's ethical convictions. Abraham proves his faith by resolutely setting out to obey God's command, even though he cannot understand it. This 'suspension of the ethical,' as Kierkegaard called it, allows Abraham to achieve an authentic commitment to God. To avoid ultimate despair, the individual must make a similar 'leap of faith' into a religious life, which is inherently paradoxical, mysterious, and full of risk. One is called to it by the feeling of dread (The Concept of Dread, 1844; trs., 1944), which is ultimately a fear of nothingness?

Toward the end of his life Kierkegaard was involved in bitter controversies, especially with the established Danish Lutheran church, which he regarded as worldly and corrupt. His later works, such as The Sickness Unto Death (1849; trs., 1941), reflect an increasingly somber view of Christianity, emphasizing suffering as the essence of authentic faith. He also intensified his attack on modern European society, which he denounced in The Present Age (1846; trs., 1940) for its lack of passion and for its quantitative values. The stress of his prolific writing and of the controversies in which he gradually engaged to undermine his health, in October 1855, h had fainted in the street, and he died in Copenhagen on November 11, 1855.

Kierkegaard's influence was at first confined to Scandinavia and to German-speaking European, where his work had a strong impact on Protestant Theology and on such writings as the 20th-century Austrian novelist Franz Kafka. As existentialism emerged as a general European movement after World War I, Kierkegaard's work was widely translated, and he was recognized as one of the seminal figures of modern culture.

Since scientists, during the nineteenth century were engrossed with uncovering the workings of external reality and seemingly knew of themselves that these virtually overflowing burdens of nothing, in that were about the physical substrates of human consciousness, the business of examining the distributive contribution in dynamic functionality and structural foundation of mind became the province of social scientists and humanists. Adolphe Quételet proposed a 'social physics' that could serve as the basis for a new discipline called sociology, and his contemporary Auguste Comte concluded that a true scientific understanding of the social reality was quite inevitable. Mind, in the view of these figures, was a separate and distinct mechanism subject to the lawful workings of a mechanical social reality.

More formal European philosophers, such as Immanuel Kant, sought to reconcile representations of external reality in mind with the motions of matter-based on the dictates of pure reason. This impulse was also apparent in the utilitarian ethics of Jerry Bentham and John Stuart Mill, in the historical materialism of Karl Marx and Friedrich Engels, and in the pragmatism of Charles Smith, William James and John Dewey. These thinkers were painfully aware, however, of the inability of reason to posit a self-consistent basis for bridging the gap between mind and matter, and each remains obliged to conclude that the realm of the mental exists only in the subjective reality of the individual.

The fatal flaw of pure reason is, of course, the absence of emotion, and purely explanations of the division between subjective reality and external reality, of which had limited appeal outside the community of intellectuals. The figure most responsible for infusing our understanding of the Cartesian dualism with contextual representation of our understanding with emotional content was the death of God theologian Friedrich Nietzsche 1844-1900. After declaring that God and 'divine will', did not exist, Nietzsche reified the 'existence' of consciousness in the domain of subjectivity as the ground for individual 'will' and summarily reducing all previous philosophical attempts to articulate the 'will to truth'. The dilemma, forth in, had seemed to mean, by the validation,... as accredited for doing of science, in that the claim that Nietzsche's earlier versions to the 'will to truth', disguises the fact that all alleged truths were arbitrarily created in the subjective reality of the individual and are expressed or manifesting the individualism of 'will'.

In Nietzsche's view, the separation between mind and matter is more absolute and total than previously been imagined, that this assumption was that there is no real necessity of correspondence between linguistic constructions of reality in human subjectivity and external reality, he deuced that we are all locked in 'a prison house of language'. The prison as he concluded it, was also a 'space' where the philosopher can examine the 'innermost desires of his nature' and articulate a new message of individual existence founded on 'will'.

Those who fail to enact their existence in this space, Nietzsche says, are enticed into sacrificing their individuality on the nonexistent altars of religious beliefs and democratic or socialists' ideals and become, therefore, members of the anonymous and docile crowd. Nietzsche also invalidated the knowledge claims of science in the examination of human subjectivity. Science, he said. Is not exclusive to natural phenomenons and favour reductionistic examination of phenomena at the expense of mind? It also seeks to reduce the separateness and uniqueness of mind with mechanistic descriptions that disallow and basis for the free exercise of individual will.

Nietzsche's emotionally charged defence of intellectual freedom and radial empowerment of mind as the maker and transformer of the collective fictions that shape human reality in a soulless mechanistic universe proved terribly influential on twentieth-century thought. Furthermore, Nietzsche sought to reinforce his view of the subjective character of scientific knowledge by appealing to an epistemological crisis over the foundations of logic and arithmetic that arose during the last three decades of the nineteenth century. Through a curious course of events, attempted by Edmund Husserl 1859-1938, a German mathematician and a principal founder of phenomenology, wherefor to resolve this crisis resulted in a view of the character of consciousness that closely resembled that of Nietzsche.

These aspects of Nietzsche's work elicit a tendency to compare Nietzsche's doctrine with that of Freud and psychoanalysis and to argue that the Freudian doctrine and school (the psychoanalytic theory of human personality on which the psychotherapeutic technique of the psychoanalysis is based) and methods of treatment (psychoanalysis) have been influenced and affected by Nietzsche's philosophy and work and the Nietzschean doctrine. As a demonstration from the relevant literature, according to Golomb's (1987) thesis, the theoretical core of a psychoanalysis is already part and parcel of Nietzsche's philosophy, insofar as it is based on concepts that are both displayed in it and developed by it-concepts such as the unconscious, repression, sublimation, the id, the superego, primary and secondary processes and interpretations of dreams.

Nevertheless, the actual situation in the domains of psychotherapy, psychiatry and clinical psychology is, by no, strictly so. While the two savants (Nietzsche and Freud) endeavour to understand man, to develop the healthy power that is still present in the individual and the neurotic patient so as to overcome and suppress the psychological boundaries that repress his vitality and inhibit his ability to function freely and creatively and attain truth, the difference between the psychodynamic school, approach, movement and method of treatment, in general, and psychoanalysis, in particular, and the existential approach to psychotherapy, the existential movement and the existential, humanistic school of psychology and method of treatment that have been stemmed from the doctrines and views of Freud and Nietzsche is profound and significant, as far as the actual psychotherapeutic treatment is concerned. The reason as for these differences lies in the variation in the two savants' view and definition of man and human existence, the nature and character of man and his relationship with the world and the environment, as well as in the variation in the intellectual soil, that nourished and nurtured the two giant savants' views, doctrines (that is, the pundit philosophical and historical roots and influences) and the manners according to which they have been devised and designed.

In fact, Freudian Psychoanalysis (as part of the psychodynamic movement and approach) and existential, humanistic, psychotherapy (which is stemmed from the Nietzschean ideas and doctrine, among others) constitutes two totally independent, distinct and rival approaches of psychotherapy, which employ their own method of treatment, doctrine and principles. As an illustration, Viktor. E. Frankl has been expelled from the Psychoanalytic society and organisation because of his views and critic of the psychoanalysis, broke away from psychoanalysis and established Logotherapy, an existential, psychotherapeutic method and school in psychiatry, known as the third force in Viennese psychotherapy (after Freud and Adler), which is based upon the Nietzschean doctrine. Thus, Logotherapy and Psychoanalysis constitute two rival types and methods of psychotherapeutic treatment with their own objectives, principles, theoretical core and doctrines.

Hence, as a response and alternative to the works that compare a psychoanalysis and the Nietzschean doctrine and maintain that the Nietzschean doctrine constitutes the theoretical core of the psychoanalysis, its current paper endeavours to contrast these works and their thesis and demonstrate that the definition and treatment of both its subject matter (as man's humanly existence) and key concepts in human existence by Freudian psychoanalysis and the principles and essences of Freudian Psychoanalysis totally differ both from the treatment of the same subject matter and key concepts by the Nietzschean doctrine and from the essence and principles of the Nietzschean doctrine. Thus, the main thesis of the present paper is that the Nietzschean doctrine by no constitutes the theoretical core and essencity of psychoanalysis.

Accomplishing the objective might that to establish and strengthen its thesis for which would be carried out by doing two things simultaneously. Firstly, depicting Freudian psychoanalysis and the Freudian psychoanalytic doctrine, the historical and philosophical roots of the psychodynamic movement, the Nietzschean doctrine, the existential movement and Frankl's technique and the psychotherapeutic approaches of Logotherapy and its doctrine (and showing that the Nietzschean doctrine, in fact, constitutes the theoretical core of Logotherapy, rather than of psychoanalysis). Secondly, displaying the differences between psychoanalysis and existential psychotherapy (when Logotherapy is utilised as an illustration and as a representative of the existential approach to psychotherapy and is labelled existential analysis) in the domain of psychiatry and clinical psychology, in terms of the differences between the Nietzschean doctrine and the Nietzschean philosophy and the Freudian psychoanalytic method of treatment, school and doctrine, while still acknowledging and demonstrating the similarities between the Nietzschean and Freudian doctrines, mainly as far as terminology is concerned.

Nonetheless, while endorsing the difference and rivalry between psychoanalysis and existential psychotherapy, as well as the distinction between the Freudian and the Nietzschean doctrines, it should be emphasised that it was the relation between Kierkegaard and Nietzsche's ideas, which contributed to the development of the understanding of man and his crisis, and Freud's development of specific methods and techniques for the investigation of the fragmentation of the individual-human-being in the Victorian period that has provided the basis for existential psychotherapy. In fact, both practical approaches (Freudian psychoanalysis and existential psychotherapy), coupled with the Nietzschean theoretical work and doctrine, examine the human being, his existence and his crisis, such as despair and misery (both neurosis and psychosis) in an attempt to alleviate them.

Accordingly, since the technique of interacting directly with the given individual and analysing the analysed individual is almost similar for both approaches and schools of psychotherapy, it is the distinguished variation in the essence, nature and character (as far as the view of man and his character and of the human existence are concerned) between the Nietzschean (and the Kierkegaardian, for that matter) doctrine and the Freudian doctrine as well as in the manner in which they have been devised which makes most of the difference and affects the psychotherapeutic treatment. Hence, it is the difference between the Nietzschean (and the Kierkegaardian) theoretical doctrine, endeavours, system and approach and those of the Freudian psychoanalytic school and doctrine that is responsible for the difference between the two approaches of and to psychotherapy.

Both the Freudian and the Nietzschean doctrines (and for that matter the Kierkegaardian doctrine) strive to comprehend man, his existence and his crisis, each of these doctrines possesses a different theory as for the nature and image of man, i.e., what he is and what determines him and makes him what he is, which they employ so as to obtain this understanding and a knowledge of the manner in which this understanding should be achieved. Consequently, the psychodynamic school and movement (namely, psychoanalysis) and existential psychologies are two distinguished and distinct theories of personality that govern and affect the clinical, psychotherapeutic treatment and method of treatment.

Sigmund Freud was a physician, a specialist in neurology, with a wide education in the life sciences and the natural philosophy and sciences. He practised neurology and medicine and focussed on the cure of ill, neurotic, individuals, or at least on an improvement of and in their condition and state of health. He was a brilliant, distinguished and ambitious member of the community of scientists, neurologists and doctors and strived to make a reputation for him in those fields. Moreover, at the beginning, before his becoming famous, he was dependent on a career as an established physician and neurologist so as to make a living and support him and his dear ones and could not allow him the slightest reputation as an outcast and as an eccentric.

As a result, the psychoanalytic school and the psychodynamic movement that have been created and devised by Freud at the turn of the nineteenth century have their roots and have been immensely influenced by the spirit and mood of the second half of the nineteenth century in which Freud lived and commenced his career. The materialist, reductionist, empiricist, positivist and mechanist ideas of the time have created an ambience that asserted that everything in the universe has an indisputable reason, cause and determinant. Accordingly, nothing in the universe is accidental which may occur due to chance or free will. Moreover, the positivist doctrine and movement maintain that the ultimate goal of man is to find the explications, reasons, causes and determinants for every single element in the universe. Consequently, according to this assertion and to doctrines such as reductionism, empiricism and associationism, even such a complex 'object' as a human being can be fully explained by being reduced to human elements, such as personality, character, behaviour, utterances, emotions, mental processes and so forth, which are induced and well-determined by the entities which cause and generate them and, thus, have reasons as for why they occur.

Consequently, the Freudian method of psychoanalysis, the psychoanalytic doctrine and the psychodynamic movement have, originally, endeavoured to turn the fields of psychology and psychiatry, and the area of psychotherapy, into a science, which is rooted in the fields of biology and mechanist physiology but spreads outwards into sociology, which describes human personality, behaviour and mental and physical condition in dynamic and goal-directed terms in an attempt to explain them. It aims to look for and find the indisputable reasons, causes and determinants for all aspects and forms of human mental events, human personality, human utterances, human behaviour and human emotions, feelings, disturbances, crisis and hardships (illnesses, both neurosis and psychosis, malaise and so forth). As a consequence, the emphasis, and presupposition, of psychoanalysis and the psychodynamic movement is the search for all those elements that define, design and determine this object, called a person, in order for him to understand him by explaining and analysing him. It, thus, comes up with specific theories as for the structure, makeup, components and features of the human psyche and the reasons as for man's crisis, despair and neurotic/psychotic condition.

Wherefore, Freudian Psychoanalysis and the psychodynamic schools are approaches that regard all human beings as a single, homogeneous entity that should be treated in a similar manner by a single,

380

predetermined, homogeneous set of theories and a single technique so as to obtain the desired cure, mainly to an organic, physiological manifestation, the cease of paralysis, the end of vomiting and repulsive sensations of food and liquid and the like. Thus, the development of human personality, human nature and morality and the character and the components of the human psyche are induced, determined, innate, predetermined and the same in and for all individuals and constitute solid explanations for human conduct human feelings, emotions, morals, ideologies and the like.

In addition, the disturbances and the crisis of given individuals are also induced and determined by specific events, experiences and stimuli and interruptions with the normal proceeding of the predetermined development of the human personality and morality. The psychoanalytic treatment is, therefore, also one for all patients. The causes, determinants and reasons as for the patient's illness and condition have to be discovered, explained, analysed treated and cured, using the given doctrine and technique of the psychoanalysis. The desired outcome of the approach is the alleviation and elimination of the undesired syndromes and, by consequence, the cure of the condition, crisis and illness.

Accordingly, the psychoanalytic, psychotherapeutic technique strives to take the suffering individual and relieve his condition by searching and finding the sources, reasons and causes for it and to make the analysed individual fully aware of the causes, determinants and reasons for his condition, based on the rigid, predetermined, psychoanalytic theories. The examined, analysed individual may lie on a coach or sit on a chair facing the psychoanalyst. So doing, he talks about the things that annoy, distresses or trouble him as well as about his life history (case study) and whatever comes up into his mind (free association). He recounts his dreams, his most intimate feelings, urges and emotions, events that occurred to him (both disagreeable and agreeable) in the course of his entire life and the like.

The psychoanalyst listens very carefully and attempts to study and examine carefully the analysed individual's utterances and find meaning in them and to employ his (the psychoanalyst's) findings so as to alleviate the analysed individual's crisis, annoyance, distress, despair and illness. Thus, the psychoanalyst strives to find the reasons, causes and determinants as for the crisis, distress, neurosis and psychosis of the analysed individual and attempts to cure them and ameliorate the analysed individual's condition and state of being by virtue of finding connections and relations between the analysed individual's life story (events that occurred to him) and his distress, neurosis and psychosis, analyse those sources and causes of the neurotic/psychotic condition and make sure that the analysed individual is fully aware of them and whatever feelings, urges, emotions and sensations that they involve-hatred, frustration, aggression, anger, fear, terror, attraction attractiveness, love and the like.

Hence, the psychoanalytic, psychotherapeutic sessions focus on and work away at the revelation, examination and analysis of these events and items that are, in turn, thought by the psychoanalyst to have induced and determined the distress, neurosis and psychosis in an attempt to scrape and withdraw all the defensives, protective layers, which the analysed individual creates and employs so as to protect him and prevent him from suffering, and find out as much as possible about them. These defensive, protective layers prevent and suppress the painful information, events and experiences from being aware of and felt and experienced by the individual who has undergone and experienced them in the past. The objective of the psychoanalysis is to discuss and analyse the causes for the patient's condition in a free manner, is without restraints and suppression.

As part of the accounting endeavour to find reasons, causes and determinants for everything and every human aspect, in general, and for the patient's condition, in particular, an important aspect and element in the Freudian psychoanalytic doctrine and in the Freudian technique of psychoanalysis is the search for symbolic meanings that are meant to have significant meaning as symbolic representations of other matters, far essential for the understanding of the patient's life and condition than the given, original, items. This technique is normally applied in Freudian dreams interpretation where the unconscious has to be revealed and analysed. With that, a complete innocence, and ordinary, everyday image and object can represent something far more significant, as far as the patient's condition is concerned. As an illustration, an image of a comb may represent a penis and combing one's hair can represent and mean some hidden, subconscious sexual urges that are directed toward a given person and which is taken to be the source of the particular neurosis/psychosis. Likewise, in the famous case of little Hand' phobia of horses (1909), a big horse and Hand' fear of it have represented Hand' father and Hand fear of being castrated by him, the Oedipus complex.

The psychoanalyst, therefore, places meaning into every single word and item that the analysed patient has uttered in her recalling of her dreams by using a series of formerly activated definitions and preconceived theories and explanations (which are likely to involve sex, the Oedipus Complex, for instance) so as to find the reasons and explanations as for the patient's condition. To be fair, Freud has demanded that the interpretation of dreams would be carried out by a professional psychoanalyst who is well trained in this technique.

As a clinical, practical illustration as for the psychoanalytic doctrine and the technique and method of psychoanalysis, the psychoanalyst may conclude from the analysed neurotic patient's utterances during free association, her recounting of her dreams and fragments of memories of events in her life and by virtue of applying symbols and symbolic representations to her utterances and images in the patient's dreams that the patient's inability to have someone touching, grabbing or holding her head and her feeling of severe stress and terror while this action is being carried out is the result and direct consequence of a sexual abuse that occurred during early childhood, in the course of which the abuser has forced the abused child to have oral sex with him by holding and grabbing the young child's head, and was regressing and suppressed by the patient from her consciousness so as to protect her from suffering as part of her defence mechanism.

The psychoanalytic therapy is based on the presumption that once the adult neurotic patient overcomes and overpowers the defence mechanisms and becomes aware of the event and experience that are viewed as the reason as for her neurosis and the feelings, emotions and sensations that these experiences and events induce the patient and, thereby become as a consequent. The Great Theoretical Difference Between the Psychotherapeutic, Existential Application of the Nietzschean Doctrine and Freudian Psychoanalysis.

In his writings (Essays on Aesthetics, Untimely Meditations, The Gay Science and others) Nietzsche wishes to be considered by his readers and viewed in and by history as a psychologist who practice's psychology and who has arranged in the mind, the design or was to contrive; for which he devised a new system for 'psychology'. The many aspects of Nietzsche and the neurotic patient's feelings and emotions toward the abuser, toward her parents and other family members, any feelings of guilt, shame, humiliations and so forth The psychoanalytic sessions, thus, endeavour to scrape and remove the protective layers that suppress those feelings and emotions and the traumatic event and experience, it, in order to be able to analyse them and discuss them freely.

Consequently, the sources, causes and determinants of the neurosis/psychosis are, therefore, suppressed, repressed and regressed and buried deep in the human psyche and are obscure and hidden from one's awareness, although active in his psyche. This given neurotic patient holds to some latent causalities, which he has regressively suffered the horrendous, traumatic experiences from ever vanquishing consciousness, however as part of her defence mechanism so as to defend and protect her and was not conscious of it. Nevertheless, the traumatic experience was embedded and active in her psyche, unaware of by her. It influenced her conscious mental feelings, emotions, utterances, dreams and actions and came up in the form of her neurosis and inability to have her head held, touched or grabbed. The objective of the psychoanalysis is, thus, to crush and overcome the defence mechanisms and have the sources of the neurosis/psychosis released and come up to the surface, where it is aware of by the patient and can be revealed, analysed, explained and observed freely.

The reason posted as for those doctrines, approach and technique lies in the fact that Freud, in his objection to the fact that some of the human mental aspects and human conduct would remain unexplained, obscure and incoherent to the psychoanalyst and his possession. The need to search for of avoiding this situation and to both explain beyond any doubt for the reason as to the uncertainty might that the human condition have latently been lost to that of any expressive linguistic utterance, in that to say, is given to some interpretation or finds to some given or explanation. However lucid, comprehensive ones, maintains both that the essence of regression of information is of information being restrained and withheld from becoming conscious, by the defence mechanism where stress, grief and anguish are involved and by lack of interest and stimulation when no stress is involved. Thus, forms a part of unconsciousness, a condition of latency that is not perceived by the mind, and that unconscious information becomes known, in the course of psychoanalysis, merely by being translated into consciousness (the objective of psychoanalysis), as merely conscious things are perceived and known? Thus, Freud defines the unconscious as whatever is not conscious and vice versa, whereas the preconscious is defined by him as a screen between the unconscious and consciousness and forms a part of consciousness for the sake of this specific definition. Accordingly, Freud regards all conscious information as unconscious information that became conscious.

Consequently, Freud maintains that since 'the data of consciousness are exceedingly defective' (Freud's, The Unconscious, 1915) mental acts can often be explicated merely by assuming and referring to other processes that are outside consciousness. In other words, one is not aware of some of his mental experiences that, nevertheless, affect his actions, bodily, physical, performances (repulsive sensations, paralysis and the case illustrated above of the neurotic patient), dreams and utterances and, thus, these mental experiences are found outside his awareness/consciousness and influence those experiences of which he is aware. Therefore, the individuals fulfill actions and utter utterances that are obscure, unclear inexplicable and unexplainable on their own, by being observed directly by those given individuals, and need to look outside direct observation in order to explain them and make them utterly lucid.

The neurotic patient illustrated in the present paper has not been aware of the real reason (the sexual abuse) as for her inability to let her head be held and taken hold of which, nevertheless, has led to this mental disability, the neurosis. Once this awareness has been achieved by the method, described above, the patient has become cured. Hence, according to psychoanalysis, when the given patient becomes aware of her sexual abuse by her father or another adult that she had to regress as part of her defence mechanism so as to defend and protect her and is able to analyse it and discuss it freely then she is cured.

Thorough look into the procedure in which the unconscious mental information is being revealed and becomes a part of consciousness which permits the awareness of the given individual patient is beyond the aim of the present paper and should be read in Freud's writings. Here, mentioning that the unconscious information undergoes a main is sufficient censorship, of which if it passes, it goes up to the level of the preconscious, where it is already in possession of consciousness and is being aware of by the agent, although not fully grasped and interrelated within terms of its context (if it does not pass this censorship, then it is occasioned of regressive behaviour that lay back within unconsciousness), that the censorship awaits to it, of which if it passes, it goes up to the level of consciousness, where it is being directly and fully experienced, related to, sensed and comprehended by the individual. Freud provides clinical illustrations of the hysterics, neurotics, (the classic Interpretation of Dreams, Freud, 1900) so as to demonstrate this theory.

To make sure that the reader who is a philosopher, rather than a psychologist, comprehends the relation between the unconscious and the conscious and consciousness, in The Unconscious (Freud, 1915), Freud asserts that psychoanalysis compares the perception of unconscious mental processes and experiences by consciousness with the perception of the outside, external, world through the sense-organs so as to obtain new knowledge from the comparison. Thus, Freud refers to Kant's work and view of the mind as an activity that manipulates experiences, borrows it for the sake of his argument, takes it out of context, distorts and changes it and comes up with the assertion that just as the external world is not viewed in the way it really is in nature but is subject to the viewer's subjective perception of it (Kant's account of the active mind), so are consciousness and the conscious affected by the unconscious and unconsciousness, manipulated and modified by them and are observed/treated by them.

In devising the Freudian psychoanalytic doctrine and the psychoanalytic technique of psychoanalysis, Freud has devised rigid theories (psychoanalytic theories) as for the nature and character of man and his existence that tailor and fit all individuals and which constitute the basis as for the psychoanalytic treatment, i.e., psychoanalysis. He, therefore, devised his theory as for morality and personality development in both men and women which proceeds through five psychosexual stages in children and adolescents as well as his theory as for the structure of personality and human interaction and moral or immoral conduct, the id, ego and superego. These theories serve as a model for the psychoanalytic treatment of all individuals who undergo psychoanalysis and are meant to be suitable for all individuals-human-beings. Accordingly, the events that occurred in the life of the individual who undergoes psychoanalysis are tailored and fit into these Freudian theories. Thus, the very case of sexual abuse, which is illustrated in the present paper, is tailored and fit into the various aspects of the Electra Complex and the psychosexual stages of personality and moral development and the personality structure, any feelings of guilt and the like.

On the other hand, the existential movement has been formed and devised in the nineteenth century as a protest movement against the established spirit, mood and ambience of the mainstream of the intellectual world-notably of the philosophical domain, natural, moral and metaphysical philosophy, but also of deterministic, rigid theories and schools of thought and movements. The existential movement has protested against the destruction of both the authentic, independent, unreduced and free individual being and the personal, biassed, subjective, authentic truth by the established mainstream of the intellectual world, in general, and doctrines such as the Hegelian and the Kantian doctrines, the empiricist doctrine, the positivist doctrine and the psychodynamic doctrine, in particular. Those doctrines have reduced the individual being into metaphysical theories, deterministic, innate, developmental theories, physiological and biological processes, innate releasing mechanisms, information processing devices and so forth, and made him fit into a single, unified and universal system of truth and reason.

The existential movement in the nineteenth century has maintained that the concept of truth has become unreal, distant, universal, abstractive, and alienated from the individual being him. Accordingly, the concept of truth has become an idea of the manner in which the universe should be like. The individual being has had to make him fit within this kind of truth rather than lead his life in accordance with his own idea of truth and being fully committed to this idea of truth. Thus, the individual being has been swallowed by the idea of whom he should be, which has been dictated to him and forced and imposed upon him by society and deterministic elements, has lost his individuality and uniqueness and has become a part of theories as for whom he should be and why.

Hence, the existential movement objects to the endeavour to reduce the individual-human-being into sets and systems of reasons, explanations, metaphysical and scientific theories and causal determinants as for his nature, his conduct, his mental/inner state (feelings, sensations, emotions and the like) and his mental state of being (neurosis, and psychosis and 'stability/sanity'). Instead, the existential movement endeavours to examine and study the individual-human-being's existence, Being-In-The-World, so as to comprehend it, to have the most agreeable, authentic existence, Being-In-The-World possible and to be able to actualise his personal existence in the world and, as a consequence, him and his life.

As just noted, the existential movement also objects to the notion of universal, objective truth but introduces truth as the subjective, personal entity of the individual who devises it, possesses it and lives his life and designs and determines him as in accordance with it. Thus, according to the existential movement, man is existing of some -determinates, emerging, becoming being who defines him in accordance with his own subjective view of truth and possesses a full responsibility as for his life as well as the capacity and power to choose whatever and whoever he wishes to become and be, his values and ideologies with a view to actualise them and to lead an authentic life and existence.

In other words, man is an individual who determines, designs and realises him in accordance with the choices, deeds and wishes that he makes, rather than a determined entity who is determined by social conformism, genetically hereditary and the environment, i.e., the past and present. Man, according to the existential movement, is, therefore, emerging, proceeding toward the future and becoming being and is defined by his own past and present actions, decisions and choices and by the future outcome of these actions, decisions and choices. That is, man becomes what he is.

The perfect and exacting preciseness of mathematical and geometrical ideas mirror precisely the essences of physical reality was the basis for the first scientific law of this new science, a constant describing the acceleration of bodies in free fall, could not be confirmed by experiment. The experiments conducted by Galileo in which balls of different sizes and weights were rolled simultaneously down an inclined plane did not, as he frankly admitted, their precise results. And since a vacuum pumps had not yet been invented, there was simply no way that Galileo could subject his law to rigorous experimental proof in the seventeenth century. Galileo believed in the absolute validity of this law in the absence of experimental proof because he also believed that movement could be subjected absolutely to the law of number. What Galileo asserted, as the French historian of science Alexander Koyré put it, was 'that the real are in its essence, geometrical and, consequently, subject to rigorous determination and measurement.'

The popular image of Isaac Newton (1642-1727) is that of a supremely rational and dispassionate empirical thinker. Newton, like Einstein, had the ability to concentrate unswervingly on complex theoretical problems until they yielded a solution. But what most consumed his restless intellect were not the laws of physics. In addition to believing, like Galileo that the essences of physical reality could be read in the language of mathematics, Newton also believed, with perhaps even greater intensity than Kepler, in the literal truths of the Bible.

For Newton the mathematical languages of physics and the language of biblical literature were equally valid sources of communion with the eternal writings in the extant documents alone consist of more than a million words in his own hand, and some of his speculations seem quite bizarre by contemporary standards. The Earth, said Newton, will still be inhabited after the day of judgement, and heaven, or the New Jerusalem, must be large enough to accommodate both the quick and the dead. Newton then put his mathematical genius to work and determined the dimensions required to house the population, his rather precise estimate was 'the cube root of 12,000 furlongs.'

The point is, that during the first scientific revolution the marriage between mathematical idea and physical reality, or between mind and nature via mathematical theory, was viewed as a sacred union. In our more secular age, the correspondence takes on the appearance of an unexamined article of faith or, to borrow a phrase from William James (1842-1910), 'an altar to an unknown god.' Heinrich Hertz, the famous nineteenth-century German physicist, nicely described what there is about the practice of physics that tends to inculcate this belief: 'One cannot escape the feeling that these mathematical formulae have an independent existence and intelligence of their own that they are wiser than we, wiser than their discoveries. That we get more out of them than was originally put into them.'

While Hertz made this statement without having to contend with the implications of quantum mechanics, the feeling, the described remains the most enticing and exciting aspects of physics. That elegant mathematical formulae provide a framework for understanding the origins and transformations of a cosmos of enormous age and dimensions are a staggering discovery for bidding physicists. Professors of physics do not, of course, tell their students that the study of physical laws in an act of communion with thee perfect mind of God or that these laws have an independent existence outside the minds that discover them. The business of becoming a physicist typically begins, however, with the study of classical or Newtonian dynamics, and this training provides considerable covert reinforcement of the feeling that Hertz described.

Perhaps, the best way to examine the legacy of the dialogue between science and religion in the debate over the implications of quantum non-locality are to examine the source of Einstein's objections tp quantum epistemology in more personal terms. Einstein apparently lost faith in the God portrayed in biblical literature in early adolescence. But, as appropriated, . . . the 'Autobiographical Notes' give to suggest that there were aspects that carry over into his understanding of the foundation for scientific knowledge, . . . 'Thus I came despite the fact that I was the son of an entirely irreligious [Jewish] Breeden heritage, which is deeply held of its religiosity, which, however, found an abrupt end at the age of 12. Though the reading of popular scientific books I soon reached the conviction that much in the stories of the Bible could not be true. The consequence waw a positively frantic [orgy] of freethinking coupled with the impression that youth is intentionally being deceived by the stat through lies that it was a crushing impression. Suspicion against every kind of authority grew out of this experience. . . . It was clear to me that the religious paradise of youth, which was thus lost, was a first attempt ti free myself from the chains of the 'merely personal'. . . . The mental grasp of this extra-personal world within the frame of the given possibilities swam as highest aim half consciously and half unconsciously before the mind's eye.'

What is more, was, suggested Einstein, belief in the word of God as it is revealed in biblical literature that allowed him to dwell in a 'religious paradise of youth' and to shield himself from the harsh realities of social and political life. In an effort to recover that inner sense of security that was lost after exposure to scientific knowledge, or to become free once again of the 'merely personal', he committed himself to understanding the 'extra-personal world within the frame of given possibilities', or as seems obvious, to the study of physics. Although the existence of God as described in the Bible may have been in doubt, the qualities of mind that the architects of classical physics associated with this God were not. This is clear in the comments from which Einstein uses of mathematics, . . . 'Nature is then a self-realization whose undivided wholeness is the simplest conceivable mathematical idea, I am convinced that we can discover, by of purely mathematical construction, those concepts and those lawful connections between them that furnish the key to the understanding of natural phenomena. Experience remains, of course, the sole criteria of physical utility of a mathematical construction. But the creative principle resides in mathematics. In a certain sense, therefore, I hold it true that pure thought can grasp reality, as the ancients dreamed.'

This article of faith, first articulated by Kepler, that 'nature is the realization of the simplest conceivable mathematical ideas' allowed for Einstein to posit the first major law of modern physics much as it allows Galileo to posit the first major law of classical physics. During which time, when the special and then the general theories of relativity had not been confirmed by experiment and many established physicists viewed them as at least minor heresies, Einstein remained entirely confident of their predictions. Ilse Rosenthal-Schneider, who visited Einstein shortly after Eddington's eclipse expedition confirmed a prediction of the general theory (1919), described Einstein's response to this news: When I was giving expression to my joy that the results coincided with his calculations, he said quite unmoved, 'But I knew the theory was correct,' and when I asked, what if there had been no confirmation of his prediction, he countered: 'Then I would have been sorry for the dear Lord -the theory is correct.'

Einstein was not given to making sarcastic or sardonic comments, particularly on matters of religion. These unguarded responses testify to his profound conviction that the language of mathematics allows the human mind access to immaterial and immutable truths existing outside of the mind that conceived them. Although Einstein's belief was far more secular than Galileo's, it retained the same essential ingredients.

What continued in the twenty-three-year-long debate between Einstein and Bohr, least of mention? The primary article drawing upon its faith that contends with those opposing to the merits or limits of a physical theory, at the heart of this debate was the fundamental question, 'What is the relationship between the mathematical forms in the human mind called physical theory and physical reality?' Einstein did not believe in a God who spoke in tongues of flame from the mountaintop in ordinary language, and he could not sustain belief in the anthropomorphic God of the West. There is also no suggestion that he embraced ontological monism, or the conception of Being featured in Eastern religious systems, like Taoism, Hinduism, and Buddhism. The closest that Einstein apparently came to affirming the existence of the 'extra-personal' in the universe was a 'cosmic religious feeling', which he closely associated with the classical view of scientific epistemology.

The doctrine that Einstein fought to preserve seemed the natural inheritance of physics until the advent of quantum mechanics. Although the mind that constructs reality might be evolving fictions that are not necessarily true or necessary in social and political life, there was, Einstein felt, a way of knowing, purged of deceptions and lies. He was convinced that knowledge of physical reality in physical theory mirrors the preexistent and immutable realm of physical laws. And as Einstein consistently made clear, this knowledge mitigates loneliness and inculcates a sense of order and reason in a cosmos that might appear otherwise bereft of meaning and purpose.

What most disturbed Einstein about quantum mechanics was the fact that this physical theory might not, in experiment or even in principle, mirrors precisely the structure of physical reality. There is, for all the reasons we seem attested of, in that an inherent uncertainty in measurement made, . . . a quantum mechanical process reflects of a pursuit that quantum theory in itself and its contributive dynamic functionalities that there lay the attribution of a completeness of a quantum mechanical theory. Einstein's fearing that it would force us to recognize that this inherent uncertainty applied to all of physics, and, therefore, the ontological bridge between mathematical theory and physical reality -does not exist. And this would mean, as Bohr was among the first to realize, that we must profoundly revive the epistemological foundations of modern science.

The world view of classical physics allowed the physicist to assume that communion with the essences of physical reality via mathematical laws and associated theories was possible, but it made no other provisions for the knowing mind. In our new situation, the status of the knowing mind seems quite different. Modern physics distributively contributed its view toward the universe as an unbroken, undissectable and undivided dynamic whole. 'There can hardly be a sharper contrast,' said Melic Capek, 'than that between the everlasting atoms of classical physics and the vanishing 'particles' of modern physics as Stapp put it: 'Each atom turns out to be nothing but the potentialities in the behaviour pattern of others. What we find, therefore, are not elementary space-time realities, but rather a web of relationships in which no part can stand alone, every part derives its meaning and existence only from its place within the whole"

The characteristics of particles and quanta are not isolatable, given particle-wave dualism and the incessant exchange of quanta within matter-energy fields. Matter cannot be dissected from the omnipresent sea of energy, nor can we in theory or in fact observe matter from the outside. As Heisenberg put it decades ago, 'the cosmos appears to be a complicated tissue of events, in which connection of different kinds alternate or overlay or combine and thereby determine the texture of the whole. This that a pure reductionist approach to understanding physical reality, which was the goal of classical physics, is no longer appropriate.

While the formalism of quantum physics predicts that correlations between particles over space-like separated regions are possible, it can say nothing about what this strange new relationship between parts (quanta) and whole (cosmos) was by an outside formalism. This does not, however, prevent us from considering the implications in philosophical terms, as the philosopher of science Errol Harris noted in thinking about the special character of wholeness in modern physics, a unity without internal content is a blank or empty set and is not recognizable as a whole. A collection of merely externally related parts does not constitute a whole in that the parts will not be 'mutually adaptive and complementary to one and another.'

Wholeness requires a complementary relationship between unity and differences and is governed by a principle of organization determining the interrelationship between parts. This organizing principle must be universal to a genuine whole and implicit in all parts that constitute the whole, even though the whole is exemplified only in its parts. This principle of order, Harris continued, 'is nothing really in and of itself. It is the way parts are organized and not another constituent addition to those that constitute the totality.'

In a genuine whole, the relationship between the constituent parts must be 'internal or immanent' in the parts, as opposed to a mere spurious whole in which parts appear to disclose wholeness due to relationships that are external to the parts. The collection of parts that would allegedly constitute the whole in classical physics is an example of a spurious whole. Parts constitute a genuine whole when the universal principle of order is inside the parts and thereby adjusts each to all that they interlock and become mutually complementary. This not only describes the character of the whole revealed in both relativity theory and quantum mechanics. It is also consistent with the manner in which we have begun to understand the relation between parts and whole in modern biology.

Modern physics also reveals, claims Harris, a complementary relationship between the differences between parts that constituted contentual representations that the universal ordering principle that is immanent in each of the parts. While the whole cannot be finally disclosed in the analysis of the parts, the study of the differences between parts provides insights into the dynamic structure of the whole present in each of the parts. The part can never, nonetheless, be finally isolated from the web of relationships that disclose the interconnections with the whole, and any attempt to do so results in ambiguity.

Much of the ambiguity in attempted to explain the character of wholes in both physics and biology derives from the assumption that order exists between or outside parts. But order in complementary relationships between differences and sameness in any physical event is never external to that event -the connections are immanent in the event. From this perspective, the addition of non-locality to this picture of the dynamic whole is not surprising. The relationship between part, as quantum event apparent in observation or measurement, and the undissectable whole, revealed but not described by the instantaneous, and the undissectable whole, revealed but described by the instantaneous correlations between measurements in space-like separated regions, is another extension of the part-whole complementarity to modern physics.

If the universe is a seamlessly interactive system that evolves to a higher level of complexity, and if the lawful regularities of this universe are emergent properties of this system, we can assume that the cosmos is a singular point of significance as a whole that evinces of the 'progressive principal order' of complementary relations its parts. Given that this whole exists in some sense within all parts (quanta), one can then argue that it operates in self-reflective fashion and is the ground for all emergent complexities. Since human consciousness evinces self-reflective awareness in the human brain and since this brain, like all physical phenomena can be viewed as an emergent property of the whole, it is reasonable to conclude, in philosophical terms at least, that the universe is conscious.

But since the actual character of this seamless whole cannot be represented or reduced to its parts, it lies, quite literally beyond all human representations or descriptions. If one chooses to believe that the universe be a self-reflective and self-organizing whole, this lends no support whatsoever to conceptions of design, meaning, purpose, intent, or plan associated with any mytho-religious or cultural heritage. However, If one does not accept this view of the universe, there is nothing in the scientific descriptions of nature that can be used to refute this position. On the other hand, it is no longer possible to argue that a profound sense of unity with the whole, which has long been understood as the foundation of religious experience, which can be dismissed, undermined or invalidated with appeals to scientific knowledge.

While we have consistently tried to distinguish between scientific knowledge and philosophical speculation based on this knowledge -there is no empirically valid causal linkage between the former and the latter. Those who wish to dismiss the speculative assumptions as its basis to be drawn the obvious freedom of which id firmly grounded in scientific theory and experiments there is, however, in the scientific description of nature, the belief in radical Cartesian division between mind and world sanctioned by classical physics. Seemingly clear, that this separation between mind and world was a macro-level illusion fostered by limited awarenesses of the actual character of physical reality and by mathematical idealization that were extended beyond the realm of their applicability.

Thus, the grounds for objecting to quantum theory, the lack of a one-to-one correspondence between every element of the physical theory and the physical reality it describes, may seem justifiable and reasonable in strictly scientific terms. After all, the completeness of all previous physical theories was measured against the criterion with enormous success. Since it was this success that gave physics the reputation of being able to disclose physical reality with magnificent exactitude, perhaps a more comprehensive quantum theory will emerge to insist on these requirements.

All indications are, however, that no future theory can circumvent quantum indeterminancy, and the success of quantum theory in co-ordinating our experience with nature is eloquent testimony to this conclusion. As Bohr realized, the fact that we live in a quantum universe in which the quantum of action is a given or an unavoidable reality requires a very different criterion for determining the completeness or physical theory. The new measure for a complete physical theory is that it unambiguously confirms our ability to co-ordinate more experience with physical reality.

If a theory does so and continues to do so, which is certainly the case with quantum physics, then the theory must be deemed complete. Quantum physics not only works exceedingly well, it is, in these terms, the most accurate physical theory that has ever existed. When we consider that this physics allows us to predict and measure quantities like the magnetic moment of electrons to the fifteenth decimal place, we realize that accuracy per se is not the real issue. The real issue, as Bohr rightly intuited, is that this complete physical theory effectively undermines the privileged relationship in classical physics between 'theory' and 'physical reality'.

In quantum physics, one calculates the probability of an event that can happen in alternative ways by adding the wave function, and then taking the square of the amplitude. In the two-slit experiment, for example, the electron is described by one wave function if it goes through one slit and by another wave function it goes through the other slit. In order to compute the probability of where the electron is going to end on the screen, we add the two wave functions, compute the absolute value of their sum, and square it. Although the recipe in classical probability theory seems similar, it is quite different. In classical physics, we would simply add the probabilities of the two alternate ways and let it go at that. The classical procedure does not work here, because we are not dealing with classical atoms. In quantum physics additional terms arise when the wave functions are added, and the probability is computed in a process known as the 'superposition principle'.

The superposition principle can be illustrated with an analogy from simple mathematics. Add two numbers and then take the square of their sum. As opposed to just adding the squares of the two numbers. Obviously $(2 + 3)2$ is not equal to $22 + 32$. The former is 25, and the latter are 13. In the language of quantum probability theory

In order to give a full account of quantum recipes for computing probabilities, one has to examine what would happen in events that are compound. Compound events are 'events that can be broken down into a series of steps, or events that consist of a number of things happening independently.'

The recipe here calls for multiplying the individual wave functions, and then following the usual quantum recipe of taking the square of the amplitude.

The quantum recipe is $| \Psi 1 \cdot \Psi 2 | 2$, and, in this case, it would be the same if we multiplied the individual probabilities, as one would in classical theory. Thus, the recipes of computing results in quantum theory and classical physics can be totally different. The quantum superposition effects are completely non-classical, and there is no mathematical justification per se why the quantum recipes work. What justifies the use of quantum probability theory is the coming thing that justifies the use of quantum physics -it has allowed us in countless experiments to extend our ability to co-ordinate experience with the expansive nature of unity.

A departure from the classical mechanics of Newton involving the principle that certain physical quantities can only assume discrete values. In quantum theory, introduced by Planck (1900), certain conditions are imposed on these quantities to restrict their value; the quantities are then said to be 'quantized'.

Up to the year 1900, physics was based on Newtonian mechanics. Large-scale systems are usually adequately described, however, several problems could not be solved, in particular, the explanation of the curves of energy against wavelengths for 'black-body radiation', with their characteristic maximum, as these attemptive efforts were afforded to endeavour upon the base-cases, on which the idea that the enclosure producing the radiation contained a number of 'standing waves' and that the energy of an oscillator if 'kT', where 'k' in the 'Boltzmann Constant' and 'T' the thermodynamic temperature. It is a consequence of classical theory that the energy does not depend on the frequency of the oscillator. This inability to explain the phenomenons has been called the 'ultraviolet catastrophe'.

Planck tackled the problem by discarding the idea that an oscillator can attain or decrease energy continuously, suggesting that it could only change by some discrete amount, which he called a 'quantum.' This unit of energy is given by 'hv' where 'v' is the frequency and 'h' is the 'Planck Constant,' 'h' has dimensions of energy 'χ' times of action, and was called the 'quantum of action.' According to Planck an oscillator could only change its energy by an integral number of quanta, i.e., by hv, 2hv, 3hv, and so forth. This meant that the radiation in an enclosure has certain discrete energies and by considering the statistical distribution of oscillators with respect to their energies, he was able to derive the 'Planck Radiation Formulas.' The formulae contrived by Planck, to express the distribution of dynamic energy in the normal spectrum of 'black-body' radiation. It is usual form is:

$$8 \eth \text{chd\"e/\"e } 5 \text{ (exp[ch / k\"eT] 1,}$$

Which represents the amount of energy per unit volume in the range of wavelengths between ë and ë + dë? 'c' = the speed of light and 'h' = the Planck constant, as 'k' = the Boltzmann constant with 'T' = thermodynamic temperatures.

The idea of quanta of energy was applied to other problems in physics, when in 1905 Einstein explained features of the 'Photoelectric Effect' by assuming that light was absorbed in quanta (photons). A further advance was made by Bohr (1913) in his theory of atomic spectra, in which he assumed that the atom can only exist in certain energy states and that light is emitted or absorbed as a result of a change from one state to another. He used the idea that the angular momentum of an orbiting electron could only assume discrete values, ie., Was quantized? A refinement of Bohr's theory was introduced by Sommerfeld in an attempt to account for fine structure in spectra. Other successes of quantum theory were its explanations of the 'Compton Effect' and 'Stark Effect.' Later developments involved the formulation of a new system of mechanics known as 'Quantum Mechanics.'

What is more, in furthering to Compton's scattering was to an interaction between a photon of electromagnetic radiation and a free electron, or other charged particles, in which some of the energy of the photon is transferred to the particle. As a result, the wavelength of the photon is increased by amount Äë. Where:

$$Äë = (2h \,/\, m0 \; c) \sin 2 \, ½$$

This is the Compton equation, 'h' is the Planck constant, m0 the rest mass of the particle, 'c' the speed of light, and the photon angle between the directions of the incident and scattered photons. The quantity 'h/m0c' and is known as the 'Compton Wavelength,' symbols ëC, which for an electron is equal to 0.002 43 nm.

The outer electrons in all elements and the inner ones in those of low atomic number have 'binding energies' negligible compared with the quantum energies of all except very soft X- and gamma rays. Thus most electrons in matter are effectively free and at rest and so cause Compton scattering. In the range of quantum energies 105 to 107 electro volts, this effect is commonly the most important process of attenuation of radiation. The scattering electron is ejected from the atom with large kinetic energy and the ionization that it causes plays an important part in the operation of detectors of radiation.

In the 'Inverse Compton Effect' there is a gain in energy by low-energy photons as a result of being scattered by free electrons of much higher energy. As a consequence, the electrons lose energy. Whereas, the wavelength of light emitted by atoms is altered by the application of a strong transverse electric field to the source, the spectrum lines being split up into a number of sharply defined components. The displacements are symmetrical about the position of the undisplaced lines, and are prepositional of the unplacing line and are propositional to the field strength up to about 100 000 volts per cm (The Stark Effect).

Adjoined along-side with quantum mechanics, is an unstretching constitution taken advantage of forwarded mathematical physical theories -growing from Planck's 'Quantum Theory' and deals with the mechanics of atomic and related systems in terms of quantities that can be measured. The subject development in several mathematical forms, including 'Wave Mechanics' (Schrödinger) and 'Matrix Mechanics' (Born and Heisenberg), all of which are equivalent.

In quantum mechanics, it is often found that the properties of a physical system, such as its angular moment and energy, can only take discrete values. Where this occurs the property is said to be 'quantized' and its various possible values are labelled by a set of numbers called quantum numbers.

For example, according to Bohr's theory of the atom, an electron moving in a circular orbit could occupy any orbit at any distance from the nucleus but only an orbit for which its angular momentum (mvr) was equal to nh/2ð, where 'n' is an integer (0, 1, 2, 3, and so forth) and 'h' is the Planck's constant. Thus the property of angular momentum is quantized and 'n' is a quantum number that gives its possible values. The Bohr theory has now been superseded by a more sophisticated theory in which the idea of orbits is replaced by regions in which the electron may move, characterized by quantum numbers 'n', 'l', and 'm'.

Properties of [Standard] elementary particles are also described by quantum numbers. For example, an electron has the property known a 'spin', and can exist in two possible energy states depending on whether this spin set parallel or antiparallel to a certain direction. The two states are conveniently characterized by quantum numbers $+ \frac{1}{2}$ and $\frac{1}{2}$. Similarly properties such as charge, Isospin, strangeness, parity and hyper-charge are characterized by quantum numbers. In interactions between particles, a particular quantum number may be conserved, i.e., the sum of the quantum numbers of the particles before and after the interaction remains the same. It is the type of interaction - strong, electromagnetic, weak that determines whether the quantum number is conserved.

The energy associated with a quantum state of an atom or other system that is fixed, or determined, by given set quantum numbers. It is one of the various quantum states that can be assumed by an atom under defined conditions. The term is often used to mean the state itself, which is incorrect accorded to: (i) the energy of a given state may be changed by externally applied fields (ii) there may be a number of states of equal energy in the system.

The electrons in an atom can occupy any of an infinite number of bound states with discrete energies. For an isolated atom the energy for a given state is exactly determinate except for the effected of the 'uncertainty principle'. The ground state with lowest energy has an infinite lifetime hence, the energy, in principle is exactly determinate, the energies of these states are most accurately measured by finding the wavelength of the radiation emitted or absorbed in transition between them, i.e., from their line spectra. Theories of the atom have been developed to predict these energies by calculation. Due to de Broglie and extended by Schrödinger, Dirac and many others, it (wave mechanics) originated in the suggestion that light consists of corpuscles as well as of waves and the consequent suggestion that all [standard] elementary particles are associated with waves. Wave mechanics are based on the Schrödinger wave equation describing the wave properties of matter. It relates the energy of a system to wave function, in general, it is found that a system, such as an atom or molecule can only have certain allowed wave functions (eigenfunction) and certain allowed energies

(Eigenvalues), in wave mechanics the quantum conditions arise in a natural way from the basic postulates as solutions of the wave equation. The energies of unbound states of positive energy form a continuum. This gives rise to the continuum background to an atomic spectrum as electrons are captured from unbound states. The energy of an atom state can be changed by the 'Stark Effect' or the 'Zeeman Effect.'

The vibrational energies of the molecule also have discrete values, for example, in a diatomic molecule the atom oscillates in the line joining them. There is an equilibrium distance at which the force is zero. The atoms repulse when closer and attract when further apart. The restraining force is nearly prepositional to the displacement hence, the oscillations are simple harmonic. Solution of the Schrödinger wave equation gives the energies of a harmonic oscillation as:

$$E_n = (n + \tfrac{1}{2}) h?.$$

Where 'h' is the Planck constant, ? is the frequency, and 'n' is the vibrational quantum number, which can be zero or any positive integer? The lowest possible vibrational energy of an oscillator is not zero but ½ h?. This is the cause of zero-point energy. The potential energy of interaction of atoms is described more exactly by the 'Morse Equation,' which shows that the oscillations are slightly anharmonic. The vibrations of molecules are investigated by the study of 'band spectra'.

The rotational energy of a molecule is quantized also, according to the Schrödinger equation, a body with the moment of inertial I about the axis of rotation have energies given by:

$$E_J = h2J (J + 1) / 8ð 2I$$

Where J is the rotational quantum number, which can be zero or a positive integer. Rotational energies originate from band spectra.

The energies of the state of the nucleus are determined from the gamma ray spectrum and from various nuclear reactions. Theory has been less successful in predicting these energies than those of electrons because the interactions of nucleons are very complicated. The energies are very little affected by external influence but the 'Mössbauer Effect' has permitted the observations of some minute changes.

In quantum theory, introduced by Max Planck 1858-1947 in 1900, was the first serious scientific departure from Newtonian mechanics. It involved supposing that certain physical quantities can only assume discrete values. In the following two decades it was applied successfully by Einstein and the Danish physicist Neils Bohr (1885-1962). It was superseded by quantum mechanics in the tears following 1924, when the French physicist Louis de Broglie (1892-1987) introduced the idea that a particle may also be regarded as a wave. The Schrödinger wave equation relates the energy of a system to a wave function, the energy of a system to a wave function, the square of the amplitude of the wave is proportional to the probability of a particle being found in a specific position. The wave function expresses the lack of possibly of defining both the position and momentum of a particle, this expression of discrete representation is called as the 'uncertainty principle,' the allowed wave functions that have described stationary states of a system

Part of the difficulty with the notions involved is that a system may be in an indeterminate state at a time, characterized only by the probability of some result for an observation, but then 'become' determinate (the collapse of the wave packet) when an observation is made such as the position and momentum of a particle if that is to apply to reality itself, than to mere indeterminacies measurement. It is as if there is nothing but a potential for observation or a probability wave before observation is made, but when an observation is made the wave becomes a particle. The ave-particle duality seems to block any way of conceiving of physical reality-in quantum terms. In the famous two-slit experiment, an electron is fired at a screen with two slits, like a tennis ball thrown at a wall with two doors in it. If one puts detectors at each slit, every electron passing the screen is observed to go through exactly one slit. But when the detectors are taken away, the electron acts like a wave process going through both slits and interfering with itself. A particle such an electron is usually thought of as always having an exact position, but its wave is not absolutely zero anywhere, there is therefore a finite probability of it 'tunnelling through' from one position to emerge at another.

The unquestionable success of quantum mechanics has generated a large philosophical debate about its ultimate intelligibility and it's metaphysical implications. The wave-particle duality is already a departure from ordinary ways of conceiving of tings in space, and its difficulty is compounded by the probabilistic nature of the fundamental states of a system as they are conceived in quantum mechanics. Philosophical options for interpreting quantum mechanics have included variations of the belief that it is at best an incomplete description of a better-behaved classical underlying reality (Einstein), the Copenhagen interpretation according to which there are no objective unobserved events in the micro-world : Bohr and W. K. Heisenberg, 1901-76, an 'accusal' view of the collapse of the wave packet, J. von Neumann, 1903-57, and a 'many worlds' interpretation in which time forks perpetually toward innumerable futures, so that different states of the same system exist in different parallel universes (H. Everett).

In recent tars the proliferation of subatomic particles, such as there are 36 kinds of quarks alone, in six flavours to look in various directions for unification. One avenue of approach is superstring theory, in which the four-dimensional world is thought of as the upshot of the collapse of a ten-dimensional world, with the four primary physical forces, one of gravity another is electromagnetism and the strong and weak nuclear forces, becoming seen as the result of the fracture of one primary force. While the scientific acceptability of such theories is a matter for physics, their ultimate intelligibility plainly requires some philosophical reflection.

A theory of gravitation that is consistent with quantum mechanics whose subject, still in its infancy, has no completely satisfactory theory. In controventional quantum gravity, the gravitational force is mediated by a massless spin-2 particle, called the 'graviton'. The internal degrees of freedom of the graviton require hij (χ) represent the deviations from the metric tensor for a flat space. This formulation of general relativity reduces it to a quantum field theory, which has a regrettable tendency to produce infinite for measurable qualitites. However, unlike other quantum field theories, quantum gravity cannot appeal to re-normalization procedures to make sense of these infinites. It has been shown that re-normalization procedures fail for theories, such as quantum gravity, in which the coupling constants have the dimensions of a positive power of length. The coupling constant for general relativity is the Planck length,

$$Lp = (Gh \: / \: c3)^{1/2} \: 10 \: 35 \: m.$$

Super-symmetry has been suggested as a structure that could be free from these pathological infinities. Many theorists believe that an effective superstring field theory may emerge, in which the Einstein field equations are no longer valid and general relativity is required to appar only as low energy limit. The resulting theory may be structurally different from anything that has been considered so far. Super-symmetric string theory (or superstring) is an extension of the ideas of Super-symmetry to one-dimensional string-like entities that can interact with each other and scatter according to a precise set of laws. The normal modes of super-strings represent an infinite set of 'normal' elementary particles whose masses and spins are related in a special way. Thus, the graviton is only one of the string modes-when the string-scattering processes are analysed in terms of their particle content, the low-energy graviton scattering is found to be the same as that computed from Super-symmetric gravity. The graviton mode may still be related to the geometry of the space0time in which the string vibrates, but it remains to be seen whether the other, massive, members of the set of 'normal' particles also have a geometrical interpretation. The intricacy of this theory stems from the requirement of a space-time of at least ten dimensions to ensure internal consistency. It has been suggested that there are the normal four dimensions, with the extra dimensions being tightly 'curled up' in a small circle presumably of Planck length size.

In the quantum theory or quantum mechanics of an atom or other system fixed, or determined by a given set of quantum numbers. It is one of the various quantum states that an atom can assume. The conceptual representation of an atom was first introduced by the ancient Greeks, as a tiny indivisible component of matter, developed by Dalton, as the smallest part of an element that can take part in a chemical reaction, and made very much more precisely by theory and excrement in the late-19th and 20th centuries.

Following the discovery of the electron (1897), it was recognized that atoms had structure, since electrons are negatively charged, a neutral atom must have a positive component. The experiments of Geiger and Marsden on the scattering of alpha particles by thin metal foils led Rutherford to propose a model (1912) in which nearly, but all the mass of an atom is concentrated at its centre in a region of positive charge, the nucleus, the radius of the order 10 -15 metre. The electrons occupy the surrounding space to a radius of 10-11 to 10-10 m. Rutherford also proposed that the nucleus have a charge of 'Ze' and is surrounded by 'Z' electrons (Z is the atomic number). According to classical physics such a system must emit electromagnetic radiation continuously and consequently no permanent atom would be possible. This problem was solved by the development of the quantum theory.

The 'Bohr Theory of the Atom,' 1913, introduced the concept that an electron in an atom is normally in a state of lower energy, or ground state, in which it remains indefinitely unless disturbed. By absorption of electromagnetic radiation or collision with another particle the atom may be excited - that is an electron is moved into a state of higher energy. Such excited states usually have short lifetimes, typically nanoseconds and the electron returns to the ground state, commonly by emitting one or more quanta of electromagnetic radiation. The original theory was only partially successful in predicting the energies and other properties of the electronic states. Attempts were made to improve the theory by postulating elliptic orbits (Sommerfeld 1915) and electron spin (Pauli 1925) but a satisfactory theory only became possible upon the development of 'Wave Mechanics,' after 1925.

According to modern theories, an electron does not follow a determinate orbit as envisaged by Bohr, but is in a state described by the solution of a wave equation. This determines the probability that the electron may be located in a given element of volume. Each state is characterized by a set of four quantum numbers, and, according to the Pauli exclusion principle, not more than one electron can be in a given state.

The Pauli exclusion principle states that no two identical 'fermions' in any system can be in the same quantum state that is have the same set of quantum numbers. The principle was first proposed (1925) in the form that not more than two electrons in an atom could have the same set of quantum numbers. This hypothesis accounted for the main features of the structure of the atom and for the periodic table. An electron in an atom is characterized by four quantum numbers, n, I, m, and s. A particular atomic orbital, which has fixed values of n, I, and m, can thus contain a maximum of two electrons, since the spin quantum number 's' can only be ++ | or |. In 1928 Sommerfeld applied the principle to the free electrons in solids and his theory has been greatly developed by later associates.

Additionally, an effect occurring when atoms emit or absorb radiation in the presence of a moderately strong magnetic field. Each spectral; Line is split into closely spaced polarized components, when the source is viewed at right angles to the field there are three components, the middle one having the same frequency as the unmodified line, and when the source is viewed parallel to the field there are two components, the undisplaced line being preoccupied. This is the 'normal' Zeeman Effect. With most spectral lines, however, the anomalous Zeeman effect occurs, where there are a greater number of symmetrically arranged polarized components. In both effects the displacement of the components is a measure of the magnetic field strength. In some cases the components cannot be resolved and the spectral line appears broadened.

The Zeeman effect occurs because the energies of individual electron states depend on their inclination to the direction of the magnetic field, and because quantum energy requirements impose conditions such that the plane of an electron orbit can only set itself at certain definite angles to the applied field. These angles are such that the projection of the total angular momentum on the field direction in an integral multiple of h/2ð (h is the Planck constant) The Zeeman effect is observed with moderately strong fields where the precession of the orbital angular momentum and the spin angular momentum of the electrons about each other is much faster than the total precession around the field direction. The normal Zeeman effect is observed when the conditions are such that the Landé factor is unity, otherwise the anomalous effect is found. This anomaly was one of the factors contributing to the discovery of electron spin.

Statistics that are concerned with the equilibrium distribution of elementary particles of a particular type among the various quantized energy states. It is assumed that these elementary particles are indistinguishable. The 'Pauli Exclusion Principle' is obeyed so that no two identical 'fermions' can be in the same quantum mechanical state. The exchange of two identical fermions, i.e., two electrons, does not affect the probability of distribution but it does involve a change in the sign of the wave function. The 'Fermi-Dirac Distribution Law' gives ?E the average number of identical fermions in a state of energy E:

$$?E = 1/[eá + E/kT + 1],$$

Where 'k' is the Boltzmann constant, 'T' is the thermodynamic temperature and 'a' is a quantity depending on temperature and the concentration of particles. For the valences electrons in a solid, 'a' takes the form -E1/kT, where E1 is the Fermi level. Whereby, the Fermi level (or Fermi energy) E F the value of ?E is exactly one half. Thus, for a system in equilibrium one half of the states with energy very nearly equal to 'E' (if any) will be occupied. The value of EF varies very slowly with temperatures, tending to E0 as 'T' tends to absolute zero.

In Bose-Einstein statistics, the Pauli exclusion principle is not obeyed so that any number of identical 'bosons' can be in the same state. The exchanger of two bosons of the same type affects neither the probability of distribution nor the sign of the wave function. The 'Bose-Einstein Distribution Law' gives ?E the average number of identical bosons in a state of energy E:

$$?E = 1/[eá + E/kT-1].$$

The formula can be applied to photons, considered as quasi-particles, provided that the quantity á, which conserves the number of particles, is zero. Planck's formula for the energy distribution of 'Black-Body Radiation' was derived from this law by Bose. At high temperatures and low concentrations both the quantum distribution laws tend to the classical distribution:

$$?E = Ae-E/kT.$$

Additionally, the property of substances that have a positive magnetic 'susceptibility', whereby its quantity ìr 1, and where ìr is 'Relative Permeability,' again, that the electric-quantity presented as ªr 1, where ªr is the 'Relative Permittivity,' all of which has positivity. All of which are caused by the 'spins' of electrons, paramagnetic substances having molecules or atoms, in which there are paired electrons and thus, resulting of a 'Magnetic Moment.' There is also a contribution of the magnetic properties from the orbital motion of the electron, as the relative 'permeability' of a paramagnetic substance is thus greater than that of a vacuum, i.e., it is slightly greater than unity.

A 'paramagnetic substance' is regarded as an ensssemblage of magnetic dipoles that have random orientation. In the presence of a field the magnetization is determined by competition between the effect of the field, in tending to align the magnetic dipoles, and the random thermal agitation. In small fields and high temperatures, the magnetization produced is proportional to the field strength, wherefore at low temperatures or high field strengths, a state of saturation is approached. As the temperature rises, the susceptibility falls according to Curie's Law or the Curie-Weiss Law.

Furthering by Curie's Law, the susceptibility of a paramagnetic substance is unversedly proportional to the 'thermodynamic temperature' (T) = C/T. The constant 'C is called the 'Curie constant' and is characteristic of the material. This law is explained by assuming that each molecule has an independent magnetic 'dipole' moment and the tendency of the applied field to align these molecules is opposed by the random moment due to the temperature. A modification of Curie's Law, followed by many paramagnetic substances, where the Curie-Weiss law modifies its applicability in the form

$$? = C/(T è).$$

The law shows that the susceptibility is proportional to the excess of temperature over a fixed temperature è: 'è' is known as the Weiss constant and is a temperature characteristic of the material, such as sodium and potassium, also exhibit type of paramagnetic resulting from the magnetic moments of free, or nearly free electrons, in their conduction bands? This is characterized by a very small positive susceptibility and a very slight temperature dependence, and is known as 'free-electron paramagnetism' or 'Pauli paramagnetism'.

A property of certain solid substances that having a large positive magnetic susceptibility having capabilities of being magnetized by weak magnetic fields. The chief elements are iron, cobalt, and nickel and many ferromagnetic alloys based on these metals also exist. Justifiably, ferromagnetic materials exhibit magnetic 'hysteresis', of which formidable combination of decaying within the change of an observed effect in response to a change in the mechanism producing the effect.

(Magnetic) a phenomenon shown by ferromagnetic substances, whereby the magnetic flux through the medium depends not only on the existing magnetizing field, but also on the previous state or states of the substances, the existence of a phenomenon necessitates a dissipation of energy when the substance is subjected to a cycle of magnetic changes, this is known as the magnetic hysteresis loss. The magnetic hysteresis loops were acceding by a curved obtainability from ways of which, in themselves were of plotting the magnetic flux density 'B', of a ferromagnetic material against the responding value of the magnetizing field 'H', the area to the 'hysteresis loss' per unit volume in taking the specimen through the prescribed magnetizing cycle. The general forms of the hysteresis loop fore a symmetrical cycle between 'H' and '~ H' and 'H ~ h, having inclinations that rise to hysteresis.

The magnetic hysteresis loss commands the dissipation of energy as due to magnetic hysteresis, when the magnetic material is subjected to changes, particularly, the cycle changes of magnetization, as having the larger positive magnetic susceptibility, and are capable of being magnetized by weak magnetic fields. Ferro magnetics are able to retain a certain domain of magnetization when the magnetizing field is removed. Those materials that retain a high percentage of their magnetization are said to be hard, and those that lose most of their magnetization are said to be soft, typical examples of hard ferromagnetic are cobalt steel and various alloys of nickel, aluminium and cobalt. Typical soft magnetic materials are silicon steel and soft iron, the coercive force as acknowledged to the reversed magnetic field' that is required to reduce the magnetic 'flux density' in a substance from its remnant value to zero in characteristic of ferromagnetisms and explains by its presence of domains. A ferromagnetic domain is a region of crystalline matter, whose volume may be 10-12 to 10-8 m3, which contains atoms whose magnetic moments are aligned in the same direction. The domain is thus magnetically saturated and behaves like a magnet with its own magnetic axis and moment. The magnetic moment of the ferrometic atom results from the spin of the electron in an unfilled inner shell of the atom. The formation of a domain depends upon the strong interactions forces (Exchange forces) that are effective in a crystal lattice containing ferrometic atoms.

In an unmagnetized volume of a specimen, the domains are arranged in a random fashion with their magnetic axes pointing in all directions so that the specimen has no resultant magnetic moment. Under the influence of a weak magnetic field, those domains whose magnetic saxes have directions near to that of the field flux at the expense of their neighbours. In this process the atoms of neighbouring domains tend to align in the direction of the field but the strong influence of the growing domain causes their axes to align parallel to its magnetic axis. The growth of these domains leads to a resultant magnetic moment and hence, magnetization of the specimen in the direction of the field, with increasing field strength, the growth of domains proceeds until there is, effectively, only one domain whose magnetic axis appropriates to the field direction. The specimen now exhibits tron magnetization. Further, increasing in field strength cause the final alignment and magnetic saturation in the field direction. This explains the characteristic variation of magnetization with applied strength. The presence of domains in ferromagnetic materials can be demonstrated by use of 'Bitter Patterns' or by 'Barkhausen Effect.'

For ferromagnetic solids there are a change from ferromagnetic to paramagnetic behaviour above a particular temperature and the paramagnetic material then obeyed the Curie-Weiss Law above this temperature, this is the 'Curie temperature' for the material. Below this temperature the law is not obeyed. Some paramagnetic substances, obey the temperature 'è C' and do not obey it below, but are not ferromagnetic below this temperature. The value 'è' in the Curie-Weiss law can be thought of as a correction to Curie's law reelecting the extent to which the magnetic dipoles interact with each other. In materials exhibiting 'antiferromagnetism' of which the temperature 'è' corresponds to the 'Néel temperature'.

Without discredited inquisitions, the property of certain materials that have a low positive magnetic susceptibility, as in paramagnetism, and exhibit a temperature dependence similar to that encountered in ferromagnetism. The susceptibility increased with temperatures up to a certain point, called the 'Néel Temperature,' and then falls with increasing temperatures in accordance with the Curie-Weiss law. The material thus becomes paramagnetic above the Néel temperature, which is analogous to the Curie temperature in the transition from ferromagnetism to paramagnetism. Antiferromagnetism is a property of certain inorganic compounds such as MnO, FeO, FeF2 and MnS. It results from interactions between neighbouring atoms leading and an antiparallel arrangement of adjacent magnetic dipole moments, least of mention. A system of two equal and opposite charges placed at a very short distance apart. The product of either of the charges and the distance between them is known as the 'electric dipole moments. A small loop carrying a current I behave as a magnetic dipole and is equal to 'A', where 'A' being the area of the loop.

The energy associated with a quantum state of an atom or other system that is fixed, or determined by a given set of quantum numbers. It is one of the various quantum states that can be assumed by an atom under defined conditions. The term is often used to mean the state itself, which is incorrect by ways of: (1) the energy of a given state may be changed by externally applied fields, and (2) there may be a number of states of equal energy in the system.

The electrons in an atom can occupy any of an infinite number of bound states with discrete energies. For an isolated atom the energy for a given state is exactly determinate except for the effects of the 'uncertainty principle'. The ground state with lowest energy has an infinite lifetime, hence the energy is if, in at all as a principle that is exactly determinate. The energies of these states are most accurately measured by finding the wavelength of the radiation emitted or absorbed in transitions between them, i.e., from their line spectra. Theories of the atom have been developed to predict these energies by calculating such a system that emit electromagnetic radiation continuously and consequently no permanent atom would be possible, hence this problem was solved by the developments of quantum theory. An exact calculation of the energies and other particles of the quantum state is only possible for the simplest atom but there are various approximate methods that give useful results as an approximate method of solving a difficult problem, if the equations to be solved, and depart only slightly from those of some problems already solved. For example, the orbit of a single planet round the sun is an ellipse, that the perturbing effect of other planets modifies the orbit slightly in a way calculable by this method. The technique finds considerable application in 'wave mechanics' and in 'quantum electrodynamics'. Phenomena that are not amendable to solution by perturbation theory are said to be non-perturbative.

The energies of unbound states of positive total energy form a continuum. This gives rise to the continuos background to an atomic spectrum, as electrons are captured from unbound state, the energy of an atomic state can be changed by the 'Stark Effect' or the 'Zeeman Effect.'

The vibrational energies of molecules also have discrete values, for example, in a diatomic molecule the atoms oscillate in the line joining them. There is an equilibrium distance at which the force is zero, and the atoms deflect when closer and attract when further apart. The restraining force is very nearly proportional to the displacement, hence the oscillations are simple harmonic. Solution of the 'Schrödinger wave equation' gives the energies of a harmonic oscillation as: $E_n = (n + \frac{1}{2})\, h¦$ - Where 'h' is the Planck constant, ¦ is the frequency, and 'n' is the vibrational quantum number, which can be zero or any positive integer. The lowest possible vibrational energy of an oscillator is thus not zero but $\frac{1}{2}h¦$. This is the cause of zero-point energy. The potential energy of interaction of atoms is described more exactly by the Morse equation, which shows that the oscillations are slightly anharmonic. The vibrations of molecules are investigated by the study of 'band spectra'.

The rotational energy of a molecule is quantized also, according to the Schrödinger equation a body with moments of inertia I about the axis of rotation have energies given by:

$$E_j = h2J(J + 1)/8\delta2\ I$$

Where 'J' is the rotational quantum number, which can be zero or a positive integer. Rotational energies are found from 'band spectra'.

The energies of the states of the 'nucleus' can be determined from the gamma ray spectrum and from various nuclear reactions. Theory has been less successful in predicting these energies than those of electrons in atoms because the interactions of nucleons are very complicated. The energies are very little affected by external influences, but the 'Mössbauer Effect' has permitted the observation of some minute changes.

When X-rays are scattered by atomic centres arranged at regular intervals, interference phenomena occur, crystals providing grating of a suitable small interval. The interference effects may be used to provide a spectrum of the beam of X-rays, since, according to 'Bragg's Law,' the angle of reflection of X-rays from a crystal depends on the wavelength of the rays. For lower-energy X-rays mechanically ruled grating can be used. Each chemical element emits characteristic X-rays in sharply defined groups in more widely separated regions. They are known as the K, L's, M, N. and so forth, promotes lines of any series toward shorter wavelengths as the atomic number of the elements concerned increases. If a parallel beam of X-rays, wavelength, strikes a set of crystal planes it is reflected from the different planes, interferences occurring between X-rays reflect from adjacent planes. Bragg's Law states that constructive interference takes place when the difference in path-lengths, BAC, is equal to an integral number of wavelengths.

$$2d \sin è = në$$

where 'n' is an integer, 'd' is the interplanar distance, and 'è' is the angle between the incident X-ray and the crystal plane. This angle is the 'Bragg's Angle,' and a bright spot will be obtained on an interference pattern at this angle. A dark spot will be obtained, however. If be, 2d sin è = më. Where 'm' is half-integral. The structure of a crystal can be determined from a set of interference patterns found at various angles from the different crystal faces.

A concept originally introduced by the ancient Greeks, as a tiny indivisible component of matter, developed by Dalton, as the smallest part of an element that can take part in a chemical reaction, and made experiment in the late-19th and early 20th century. Following the discovery of the electron (1897), they recognized that atoms had structure, since electrons are negatively charged, a neutral atom must have a positive component. The experiments of Geiger and Marsden on the scattering of alpha particles by thin metal foils led Rutherford to propose a model (1912) in which nearly all mass of the atom is concentrated at its centre in a region of positive charge, the nucleus is a region of positive charge, the nucleus, radiuses of the order 10-15 metre. The electrons occupy the surrounding space to a radius of 10-11 to 10-10 m. Rutherford also proposed that the nucleus have a charge of Ze is surrounded by 'Z' electrons (Z is the atomic number). According to classical physics such a system must emit electromagnetic radiation continuously and consequently no permanent atom would be possible. This problem was solved by the developments of the 'Quantum Theory.'

The 'Bohr Theory of the Atom' (1913) introduced the notion that an electron in an atom is normally in a state of lowest energy (ground state) in which it remains indefinitely unless disturbed by absorption of electromagnetic radiation or collision with other particle the atom may be excited - that is, electrons moved into a state of higher energy. Such excited states usually have short life span (typically nanoseconds) and the electron returns to the ground state, commonly by emitting one or more 'quanta' of electromagnetic radiation. The original theory was only partially successful in predicting the energies and other properties of the electronic states. Postulating elliptic orbits made attempts to improve the theory (Sommerfeld 1915) and electron spin (Pauli 1925) but a satisfactory theory only became possible upon the development of 'Wave Mechanics' 1925.

According to modern theories, an electron does not follow a determinate orbit as envisaged by Bohr, but is in a state described by the solution of the wave equation. This determines the 'probability' that the electron may be found in a given element of volume. A set of four quantum numbers has characterized each state, and according to the 'Pauli Exclusion Principle,' not more than one electron can be in a given state.

An exact calculation of the energies and other properties of the quantum states is possible for the simplest atoms, but various approximate methods give useful results, i.e., as an approximate method of solving a difficult problem if the equations to be solved and depart only slightly from those of some problems already solved. The properties of the innermost electron states of complex atoms are found experimentally by the study of X-ray spectra. The outer electrons are investigated using spectra in the infrared, visible, and ultraviolet. Certain details have been studied using microwaves. As administered by a small difference in energy between the energy levels of the 2 P½ states of hydrogen. In accord with Lamb Shift, these levels would have the same energy according to the wave mechanics of Dirac. The actual shift can be explained by a correction to the energies based on the theory of the interaction of electromagnetic fields with matter, in of which the fields themselves are quantized. Yet, other information may be obtained form magnetism and other chemical properties.

Its appearance potential concludes as, (1)the potential differences through which an electron must be accelerated from rest to produce a given ion from its parent atom or molecule. (2) This potential difference multiplied bu the electron charge giving the least energy required to produce the ion. A simple ionizing process gives the 'ionization potential' of the substance, for example: Ar + e Ar + + 2e. - Higher appearance potentials may be found for multiplying charged ions: Ar + e Ar + + + 3r. - The number of protons in a nucleus of an atom or the number of electrons resolving around the nucleus is among some concerns of atomic numbers. The atomic number determines the chemical properties of an element and the element's position in the periodic table, because of which the clarification of chemical elements, in tabular form, in the order of their atomic number. The elements show a periodicity of properties, chemically similar recurring in a definite order. The sequence of elements is thus broken into horizontal 'periods' and vertical 'groups' the elements in each group showing close chemical analogies, i.e., in valency, chemical properties, and so forth all the isotopes of an element have the same atomic number although different isotopes gave mass numbers.

An allowed 'wave function' of an electron in an atom obtained by a solution of the Schrödinger wave equation. In a hydrogen atom, for example, the electron moves in the electrostatic field of the nucleus and its potential energy is e2, where 'e' is the electron charge. 'r' its distance from the nucleus, as a precise orbit cannot be considered as in Bohr's theory of the atom, but the behaviour of the electron is described by its wave function, Ψ, which is a mathematical function of its position with respect to the nucleus. The significance of the wave function is that | Ψ | 2dt, is the probability of finding the electron in the element of volume 'dt'.

Solution of Schrödinger's equation for hydrogen atom shows that the electron can only have certain allowed wave functions (eigenfunction). Each of these corresponds to a probability distribution in space given by the manner in which | Ψ | 2 varies with position. They also have an associated value of energy 'E'. These allowed wave functions, or orbitals, are characterized by three quantum numbers similar to those characterizing the allowed orbits in the quantum theory of the atom: 'n', the 'principle quantum number', can have values of 1, 2, 3, and so forth the orbital with n=1 has the lowest energy. The states of the electron with n=1, 2, 3, and so forth, are called 'shells' and designated the K, L, M shells, and so forth. 'I' the 'azimuthal quanta number' which for a given value of 'n' can have values of 0, 1, 2, . . . (n 1). Similarly, the 'M' shell (n = 3) has three sub-shells with I = 0, I = 1, and I = 2. Orbitals with I = 0, 1, 2, and 3 are called s, p, d, and ? orbitals respectively. The significance of the I quantum number is that it gives the angular momentum of the electron. The orbital annular momentum of an electron is given by - $[1(I + 1)(h2\eth)]$. 'm' the 'magnetic quanta number', which for a given value of 'I' can have values of; I, (I 1), . . ., 0, . . . (I 1). Thus for 'p' orbital for which I = 1, there is in fact three different orbitals with m = 1, 0, and 1. These orbitals with the same values of 'n' and 'I' but different 'm' values, have the same energy. The significance of this quantum number is that it shows the number of different levels that would be produced if the atom were subjected to an external magnetic field

According to wave theory the electron may be at any distance from the nucleus, but in fact there is only a reasonable chance of it being within a distance of 5×10^{11} metre. Indeed the maximum probability occurs when r = a0 where a0 is the radius of the first Bohr orbit. It is customary to represent an orbit that there is no arbitrarily decided probability (say 95%) of finding them an electron. Notably taken, is that although 's' orbitals are spherical (I = 0), orbitals with I> 0, have an angular dependence. Finally. The electron in an atom can have a fourth quantum number, 'M' characterizing its spin direction. This can be + ½ or ½ and according to the Pauli Exclusion principle, each orbital can hold only two electrons. The fourth quantum numbers lead to an explanation of the periodic table of the elements.

The least distance in a progressive wave between two surfaces with the same phase arises to a wavelength. If 'v' is the phase speed and 'v' the frequency, the wavelength is given by v = vë. For electromagnetic radiation the phase speed and wavelength in a material medium are equal to their values in a free space divided by the 'refractive index'. The wavelengths of spectral lines are normally specified for free space.

Optical wavelengths are measure absolutely using interferometers or diffraction gratings, or comparatively using a prism spectrometer. The wavelength can only have an exact value for an infinite waver train if an atomic body emits a quantum in the form of a train of waves of duration 'ô' the fractional uncertainty of the wavelength, Äë/ë, is approximately ë/2cô, where 'c' is the speed in free space. This is associated with the indeterminacy of the energy given by the uncertainty principle

Whereas, a mathematical quantity analogous to the amplitude of a wave that appears in the equation of wave mechanics, particularly the Schrödinger waves equation. The most generally accepted interpretation is that $| \Psi | 2dV$ represents the probability that a particle is within the volume element $dV\Psi$. The wavelengths, as a set of waves that represent the behaviour, under appropriate conditions, of a particle, e.g., its diffraction by a particle. The wavelength is given by the 'de Broglie Equation.' They are sometimes regarded as waves of probability, times the square of their amplitude at a given point represents the probability of finding the particle in unit volume at that point. These waves were predicted by de Broglie in 1924 and observed in 1927 in the Davisson-Germer Experiment. Still, 'Ψ' is often a might complex quality.

The analogy between 'Ψ' and the amplitude of a wave is purely formal. There is no macroscopic physical quantity with which 'Ψ' can be identified, in contrast with, for example, the amplitude of an electromagnetic wave, which is expressed in terms of electric and magnetic field intensities

In general, there are an infinite number of functions satisfying a wave equation but only some of these will satisfy the boundary conditions. 'Ψ' must be finite and single-valued at every point, and the spatial derivative must be continuous at an interface? For a particle subject to a law of conservation of numbers, the integral of $|\Psi|^2 dV$ over all space must remain equal to 1, since this is the probability that it exists somewhere to satisfy this condition the wave equation must be of the first order in $(d\Psi/dt)$. Wave functions obtained when these conditions are applied from a set of characteristic functions of the Schrödinger wave equation. These are often called eigenfunctions and correspond to a set of fixed energy values in which the system may exist describe stationary states on the system.

For certain bound states of a system the eigenfunctions do not charge the sign or reversing the co-ordinated axes. These states are said to have even parity. For other states the sign changes on space reversal and the parity is said to be odd.

It's issuing case of eigenvalue problems in physics that take the form:

$$\grave{U}\Psi = \ddot{e}\Psi$$

Where $\grave{U}$ is come mathematical operation (multiplication by a number, differentiation, and so forth) on a function Ψ, which is called the 'eigenfunction'. 'ë' is called the 'eigenvalue', which in a physical system will be identified with an observable quantity, as, too, an atom to other systems that are fixed, or determined, by a given set of quantum numbers? It is one of the various quantum states that can be assumed by an atom.

Eigenvalue problems are ubiquitous in classical physics and occur whenever the mathematical description of a physical system yields a series of coupled differential equations. For example, the collective motion of a large number of interacting oscillators may be described by a set of coupled differential equations. Each differential equation describes the motion of one of the oscillators in terms of the positions of all the others. A 'harmonic' solution may be sought, in which each displacement is assumed as a simple harmonic motion in time. The differential equations then reduce to '3N' linear equations with 3N unknowns. Where 'N' is the number of individual oscillators, each problem is from each one of three degrees of freedom. The whole problem I now easily recast as a 'matrix' equation of the form:

$$M = 2.$$

Where 'M' is an N x N matrix called the 'a dynamic matrix, '?' is an N x 1 column matrix, and 2 of the harmonic solution? The problem is now an eigenvalue problem with eigenfunctions '?' where are the normal modes of the system, with corresponding eigenvalues 2? As '÷' can be expressed as a column vector, '? is a vector in some –dimensional vector space? For this reason, '?' is also often called an eigenvector?

When the collection of oscillators is a complicated three-dimensional molecule, the casting of the problem into normal modes s and effective simplification of the system. The symmetry principles of group theory, the symmetry operations in any physical system must be posses the properties of the mathematical group. As the group of rotation, both finite and infinite, are important in the analysis of the symmetry of atoms and molecules, which underlie the quantum theory of angular momentum. Eigenvalue problems arise in the quantum mechanics of atomic arising in the quantum mechanics of atomic or molecular systems yield stationary states corresponding to the normal mode oscillations of either electrons in-an atom or atoms within a molecule. Angular momentum quantum numbers correspond to a labelling system used to classify these normal modes, analysing the transition between them can lead and theoretically predict of atomic or a molecular spectrum. Whereas, the symmetrical principle of group theory can then be applied, from which allow their classification accordingly. In which, this kind of analysis requires an appreciation of the symmetry properties of the molecules (rotations, inversions, and so forth) that leave the molecule invariant make up the point group of that molecule. Normal modes sharing the same eigenvalues are said to correspond to the irreducible representations of these molecules' point group. It is among these irreducible representations that one will find the infrared absorption spectrum for the vibrational normal modes of the molecule.

Eigenvalue problems play a particularly important role in quantum mechanics. In quantum mechanics, physically observable as location momentum energy and so forth, are represented by operations (differentiations with respect to a variable, multiplication by a variable), which act on wave functions. Wave functioning differs from classical waves in that they carry no energy. For classical waves, the square modulus of its amplitude measures its energy. For a wave function, the square modulus of its amplitude, at a location ',' represents not energy bu probability, i.e., the probability that a particle -a localized packet of energy will be observed in a detector is placed at that location. The wave function therefore describes the distribution of possible locations of the particle and is perceptible only after many location detectors events have occurred. A measurement of position of a quantum particle may be written symbolically as:

$$X \; \Psi(\chi) \; (\chi)^{\Psi}$$

Where $\Psi(\chi)$ is said to be an eigenvector of the location operator and 'χ' is the eigenvalue, which represents the location. Each $\Psi(\chi)$ represents amplitude. The location 'χ'$|\Psi(\chi)|2$ is the probability that the particle will be found in an infinitesimal volume at that location. The wave function describing the distribution of all possible locations for the particle is the linear superposition of all $K(\chi)$ for zero . These principles that hold generally in physics wherever linear phenomena occur. In elasticity, the principle stares that the same strains whether it acts alone accompany each stress or in conjunction with others, it is true so long as the total stress does not exceed the limit of proportionality. In vibrations and wave motion the principle asserts that one set is unaffected by the presence of another set. For example, two sets of ripples on water will pass through one anther without mutual interaction so that, at a particular instant, the resultant distribution at any point traverse by both sets of waves is the sum of the two component disturbances.'

The superposition of two vibrations, y1 and y2, both of frequency ?, produces a resultant vibration of the same frequency, its amplitude and phase functions of the component amplitudes and phases, that:

$$y1. = a1 \sin(2\delta?t + ä1)$$
$$y2. = a2 \sin(\sin(2\delta?t + ä2)$$

Then the resultant vibration, y, is given by:

$$y1. + y2. = A \sin(2\eth?t + Ä),$$

Where amplitude A and phase Ä is both functions of a1, a2, ä1, and ä2.

However, the eigenvalue problems in quantum mechanics therefore represent observable representations as made by possible states (position, in the case of pχ) that the quantum system can have to stationary states, of which states that the product of the uncertainty of the resulting value of a component of momentum (pχ) and the uncertainties in the corresponding co-ordinate position of the same order of magnitude as the Planck Constant. It produces an accurate measurement of position is possible, as a resultant of the uncertainty principle. Subsequently, measurements of the position acquire a spread themselves, which makes the continuos monitoring of the position impossibly.

As in, classical mechanics may take differential or matrix forms. Both forms have been shown to be equivalent. The differential form of quantum mechanics is called wave mechanics (Schrödinger), where the operators are differential operators or multiplications by variables. Eigenfunctions in wave mechanics are wave functions corresponding to stationary wave states that responding to stationary conditions. The matrix forms of quantum mechanics are often matrix mechanics: Born and Heisenberg. Matrices acting of eigenvectors represent the operators.

The relationship between matrix and wave mechanics is similar to the relationship between matrix and differential forms of eigenvalue problems in classical mechanics. The wave functions representing stationary states are really normal modes of the quantum wave. These normal modes may be thought of as vectors that span on a vector space, which have a matrix representation.

Pauli, in 1925, suggested that each electron could exist in two states with the same orbital motion. Uhlenbeck and Goudsmit interpreted these states as due to the spin of the electron about an axis. The electron is assumed to have an intrinsic angular momentum on addition, to any angular momentum due to its orbital motion. This intrinsic angular momentum is called 'spin' It is quantized in values of ?s(s + 1)h/2ð - Where 's' is the 'spin quantum number' and 'h' the Planck constant. For an electron the component of spin in a given direction can have values of + ½ and − ½, leading to the two possible states. An electron with spin that is behaviourally likens too small magnetic moments, in which came alongside an intrinsic magnetic moment. A 'magneton gives of a fundamental constant, whereby the intrinsic magnetic moment of an electron acquires the circulatory current created by the angular momentum 'p' of an electron moving in its orbital produces a magnetic moment ì = ep/2m, where 'e' and 'm' are the charge and mass of the electron, by substituting the quantized relation p = jh/2ð(h) = the Planck constant: j = magnetic quantum number, ì-jh/4ðm. When j is taken as unity the quantity eh/4ðm is called the Bohr magneton, its value is:

$$9.274\ 0780 \times 10\text{-}24\ Am2$$

According to the wave mechanics of Dirac, the magnetic moment associated with the spin of the electron would be exactly one Bohr magnetron, although quantum electrodynamics show that a small difference can be expected.

The nuclear magnetron, 'iN' is equal to (me/mp)iB. Where mp is the mass of the proton. The value of iN is: 5.050 8240 x 10-27 A m2. The magnetic moment of a proton is, in fact, 2.792 85 nuclear magnetos. The two states of different energy result from interactions between the magnetic field due to the electron's spin and that caused by its orbital motion. These are two closely spaced states resulting from the two possible spin directions and these lead to the two limes in the doublet.

In an external magnetic field the angular momentum vector of the electron precesses. For an explicative example, if a body is of a spin, it holds about its axis of symmetry OC (where O is a fixed point) and C is rotating round an axis OZ fixed outside the body, the body is said to be precessing round OZ. OZ is the precession axis. A gyroscope precesses due to an applied torque called the precessional torque. If the moment of inertia a body about OC is I and its angular momentum velocity is ù, a torque 'K', whose axis is perpendicular to the axis of rotation will produce an angular velocity of precession Ù about an axis perpendicular to both and the torque axis where Ù = K/Iù. It is . . ., wholly orientated of the vector to the field direction are allowed, there is a quantization so that the component of the angular momentum along the direction I restricted of certain values of h/2ð. The angular momentum vector has allowed directions such that the component is mS(h2ð), where mS is the magnetic so in quantum number. For a given value of s, mS has the value's, (s-1), . . . −s. For example, when s = 1, mS is I, O, and − 1. The electron has a spin of ½ and thus mS is + ½ and − ½. Thus the components of its spin of angular momentum along the field direction are,

$$\pm \tfrac{1}{2}(h/2ð).$$

These phenomena are called 'a space quantization'.

The resultant spin of a number of particles is the vector sum of the spins (s) of the individual particles and is given by symbol S. for example, in an atom two electrons with spin of ½ could combine to give a resultant spin of S = ½ + ½ = 1 or a resultant of S = ½ − ½ =1 or a resultant of S = ½ − ½ =0.

Alternative symbols used for spin is J (for elementary particles or standard theory) and I (for a nucleus). Most elementary particles have a non-zero spin, which either be integral of half integral. The spin of a nucleus is the resultant of the spin of its constituent's nucleons.

For most generally accepted interpretations is that | Ψ |2dV represents the probability that particle is located within the volume element dV, as well, 'Ψ' is often a complex quantity. The analogy between 'Ψ' and the amplitude of a wave is purely formal. There is no macroscopic physical quantity with which 'Ψ' can be identified, in contrast with, for example, the amplitude of an electromagnetic wave, which are expressed in terms of electric and magnetic field intensities. There are an infinite number of functions satisfying a wave equation, but only some of these will satisfy the boundary condition. 'Ø' must be finite and single-valued at each point, and the spatial derivatives must be continuous at an interface? For a particle subject to a law of conservation of numbers; The integral of | Ψ |2dV over all space must remain equal to 1, since this is the probability that it exists somewhere. To satisfy this condition the wave equation must be of the first order in (dΨdt). Wave functions obtained when these conditions are applied form of set of 'characteristic functions' of the Schrödinger wave equation. These are often called 'eigenfunctions' and correspond to a set of fixed energy values in which the system may exist, called 'eigenvalues'. Energy eigenfunctions describe stationary states of a system. For example, bound states of a system the eigenfunctions do not change signs on reversing the co-ordinated axes. These states are said to have 'even parity'. For other states the sign changes on space reversal and the parity is said to be 'odd'.

The least distance in a progressive wave between two surfaces with the same phase. If 'v' is the 'phase speed' and 'v' the frequency, the wavelength is given by v = vë. For 'electromagnetic radiation' the phase speed and wavelength in a material medium are equal to their values I free space divided by the 'refractive index'. The wavelengths are spectral lines are normally specified for free space. Optical wavelengths are measured absolutely using interferometers or diffraction grating, or comparatively using a prism spectrometer.

The wavelength can only have an exact value for an infinite wave train. If an atomic body emits a quantum in the form of a train of waves of duration '¶' the fractional uncertainty of the wavelength, 'Äë/¶, is approximately ë/2ð, where 'c' is the speed of free space. This is associated with the indeterminacy of the energy given by the 'uncertainty principle'.

A moment of momentum about an axis, represented as Symbol: L, the product of the moment of inertia and angular velocity (I) angular momentum is a 'pseudo vector quality'. It is conserved in an isolated system, as the moment of inertia contains itself of a body about an axis. The sum of the products of the mass of each particle of a body and square of its perpendicular distance from the axis: This addition is replaced by an integration in the case of continuous body. For a rigid body moving about a fixed axis, the laws of motion have the same form as those of rectilinear motion, with moments of inertia replacing mass, angular velocity replacing linear momentum, and so forth hence the 'energy' of a body rotating about a fixed axis with angular velocity is ½I 2, which corresponds to ½mv2 for the kinetic energy of a body mass 'm' translated with velocity 'v'.

The linear momentum of a particle 'p' bears the product of the mass and the velocity of the particle. It is a 'vector' quality directed through the particle of a body or a system of particles is the vector sum of the linear momentums of the individual particles. If a body of mass 'M' is translated (the movement of a body or system in which a way that all points are moved in parallel directions through equal distances), with a velocity 'V', it has its mentum as 'MV', which is the momentum of a particle of mass 'M' at the centre of gravity of the body. The product of 'moment of inertia and angular velocity'. Angular momentum is a 'pseudo vector quality and is conserved in an isolated system, and equal to the linear velocity divided by the radial axes per sec.

If the moment of inertia of a body of mass 'M' about an axis through the centre of mass is I, the moment of inertia about a parallel axis distance 'h' from the first axis is I + Mh2. If the radius of gyration is 'k' about the first axis, it is? (k) 2 h2 about the second. The moment of inertia of a uniform solid body about an axis of symmetry is given by the product of the mass and the sum of squares of the other semi-axes, divided by 3, 4, 5 according to whether the body is rectangular, elliptical or ellipsoidal.

The circle is a special case of the ellipse. The Routh's rule works for a circular or elliptical cylinder or elliptical discs it works for all three axes of symmetry. For example, for a circular disk of the radius 'an' and mass 'M', the moment of inertia about an axis through the centre of the disc and lying (a) perpendicular to the disc, (b) in the plane of the disc is

$$\text{(a) } \tfrac{1}{4}M(a2 + a2) = \tfrac{1}{2}Ma2$$
$$\text{(b) } \tfrac{1}{4}Ma2.$$

A formula for calculating moments of inertia I:

$$I = \text{mass x } (a2\,/3 + n) + b2\,/(3 + n),$$

Where n and n are the numbers of principal curvatures of the surface that terminates the semiaxes in question and 'a' and 'b's' are the lengths of the semiaxes. Thus, if the body is a rectangular parallelepiped, n = n = 0, and

$$I = \text{-mass x } (a2\,/\,3 + b2\,/3).$$

If the body is a cylinder then, for an axis through its centre, perpendicular to the cylinder axis, n = 0 and n = 1, it substantiates that if, = mass x (a2 / 3 + b2 /4). If 'I' is desired about the axis of the cylinder, then n= n = 1 and a = b = r (the cylinder radius) and; I = mass x (r2 /2). An array of mathematical concepts, which is similar to a determinant but differ from it in not having a numerical value in the ordinary sense of the term is called a matrix. It obeys the same rules of multiplication, addition, and so forth an array of 'mn' numbers set out in 'm' rows and 'n' columns are a matrix of the order of m x n. the separate numbers are usually called elements, such arrays of numbers, tarted as single entities and manipulated by the rules of matrix algebra, are of use whenever simultaneous equations are found, e.g., changing from one set of Cartesian axes to another set inclined the first: Quantum theory, electrical networks. Matrixes are very prominent in the mathematical expression of quantum mechanics.

A mathematical form of quantum mechanics that was developed by Born and Heisenberg and originally simultaneously with but independently of wave mechanics. It is equivalent to wave mechanics, but in it the wave function of wave mechanics is replaced by 'vectors' in a seemly space (Hilbert space) and observable things of the physical world, such as energy, momentum, co-ordinates, and so forth, is represented by 'matrices'.

The theory involves the idea that a maturement on a system disturbs, to some extent, the system itself. With large systems this is of no consequence, and the system this is of no classical mechanics. On the atomic scale, however, the results of the order in which the observations are made. T0atd if 'p' denotes an observation of a component of momentum and 'q'. An observer of the corresponding co-ordinates pq qp. Here 'p' and 'q' are not physical quantities but operators. In matrix mechanics and obey te relationship:

$$pq\ qp = ih/2\eth$$

411

Where 'h' is the Planck constant that equals to 6.626 076 x 10-34 j s. The matrix elements are connected with the traditions probability between various states of the system.

A quantity with magnitude and direction. It can be represented by a line whose length is propositional to the magnitude and whose direction is that of the vector, or by three components in rectangular co-ordinate system. Their angle between vectors is 90%, that the product and vector product base a similarity to unit vectors such, are to either be equated to being zero or one.

A true vector, or polar vector, involves the displacement or virtual displacement. Polar vectors include velocity, acceleration, force, electric and magnetic strength. Th deigns of their components are reversed on reversing the co-ordinated axes. Their dimensions include length to an odd power.

A Pseudo vector, or axial vector, involves the orientation of an axis in space. The direction is conventionally obtained in a right-handed system by sighting along the axis so that the rotation appears clockwise, Pseudo-vectors includes angular velocity, vector area and magnetic flux density. The signs of their components are unchanged on reversing the co-ordinated axes. Their dimensions include length to an even power.

Polar vectors and axial vectors obey the same laws of the vector analysis (a) Vector addition: If two vectors 'A' and 'B' are represented in magnitude and direction by the adjacent sides of a parallelogram, the diagonal represents the vector sun (A + B) in magnitude and direction, forces, velocity, and so forth, combine in this way. (b) Vector multiplying: There are two ways of multiplying vectors (i) the 'scalar product' of two vectors equals the product of their magnitudes and the cosine of the angle between them, and is scalar quantity. It is usually written

$$A \cdot B \text{ (reads as A dot B)}$$

(ii) The vector product of two vectors: A and B are defined as a pseudo vector of magnitude AB sin è, having a direction perpendicular to the plane containing them. The sense of the product along this perpendicular is defined by the rule: If 'A' is turned toward 'B' through the smaller angle, this rotation appears of the vector product. A vector product is usually written

$$A \times B \text{ (reads as A cross B)}.$$

Vectors should be distinguished from scalars by printing the symbols in bold italic letters.

A theory that seeks to unite the properties of gravitational, electromagnetic, weak, and strong interactions to predict all their characteristics. At present it is not known whether such a theory can be developed, or whether the physical universe is amenable to a single analysis about the current concepts of physics. There are unsolved problems in using the framework of a relativistic quantum field theory to encompass the four elementary particles. It may be that using extended objects, as superstring and super-symmetric theories, but, still, this will enable a future synthesis for achieving obtainability.

A unified quantum field theory of the electromagnetic, weak and strong interactions, in most models, the known interactions are viewed as a low-energy manifestation of a single unified interaction, the unification taking place at energies (Typically 1015 GeV) very much higher than those currently accessible in particle accelerations. One feature of the Grand Unified Theory is that 'baryon' number and 'lepton' number would no-longer be absolutely conserved quantum numbers, with the consequences that such processes as 'proton decay', for example, the decay of a proton into a positron and a ∂0, p e+∂0 would be expected to be observed. Predicted lifetimes for proton decay are very long, typically 1035 years. Searchers for proton decay are being undertaken by many groups, using large underground detectors, so far without success.

One of the mutual attractions binding the universe of its owing totality, but independent of electromagnetism, strong and weak nuclear forces of interactive bondages is one of gravitation. Newton showed that the external effect of a spherical symmetric body is the same as if the whole mass were concentrated at the centre. Astronomical bodies are roughly spherically symmetric so can be treated as point particles to a very good approximation. On this assumption Newton showed that his law consistent with Kepler's laws? Until recently, all experiments have confirmed the accuracy of the inverse square law and the independence of the law upon the nature of the substances, but in the past few years evidence has been found against both.

The size of a gravitational field at any point is given by the force exerted on unit mass at that point. The field intensity at a distance ',' from a point mass 'm' is therefore $Gm/, 2$, and acts toward 'm'. Gravitational field strength is measured in 'Newtons' per kilogram. The gravitational potential 'V' at that point is the work done in moving a unit mass from infinity to the point against the field, due to a point mass.

$$V = Gm \; ? \; d\div \; / \; \div 2 = Gm \; / \; \div.$$

V is a scalar measurement in joules per kilogram. The following special cases are also important (a) Potential at a point distance '$\div$' from the centre of a hollow homogeneous spherical shell of mass 'm' and outside the shell:

$$V = Gm \; / \; \div.$$

The potential is the same as if the mass of the shell is assumed concentrated at the centre (b) At any point inside the spherical shell the potential is equal to its value at the surface:

$$V = Gm \; / \; r$$

Where 'r' is the radius of the shell. Thus, there is no resultant force acting at any point inside the shell, since no potential difference acts between any two points, then (c) Potential at a point distance '$\div$' from the centre of a homogeneous solid sphere and outside the spheres the same as that for a shell:

$$V = Gm \; / \; \div$$

(d) At a point inside the sphere, of radius 'r'.

$$V = Gm(3r2 \div 2) /2r3.$$

The essential property of gravitation is that it causes a change in motion, in particular the acceleration of free fall (g) in the earth's gravitational field. According to the general theory of relativity, gravitational fields change the geometry of space-timer, causing it to become curved. It is this curvature that is geometrically responsible for an inseparability of the continuum of 'space-time' and its forbearing product is to a vicinities mass, entrapped by the universality of space-time, that in ways described by the pressures of their matter, that controls the natural motions of fording bodies. General relativity may thus be considered as a theory of gravitation, differences between it and Newtonian gravitation only appearing when the gravitational fields become very strong, as with 'black-holes' and 'neutron stars', or when very accurate measurements can be made.

Another binding characteristic embodied universally is the interaction between elementary particle arising as a consequence of their associated electric and magnetic fields. The electrostatic force between charged particles is an example. This force may be described in terms of the exchange of virtual photons, because of the uncertainty principle it is possible for the law of conservation of mass and energy to be broken by an amount ÄE providing this only occurring for a time such that:

$$ÄEÄt\ h/4ð.$$

This makes it possible for particles to be created for short periods of time where their creation would normally violate conservation laws of energy. These particles are called 'virtual particles'. For example, in a complete vacuum -which no 'real' particle's exist, as pairs of virtual electrons and positron are continuously forming and rapidly disappearing (in less than 10-23 seconds). Other conservation laws such as those applying to angular momentum, Isospin, and so forth, cannot be violated even for short periods of time.

Because its strength lies between strong and weak nuclear interactions, the exchanging electromagnetic interaction of particles decaying by electromagnetic interaction, do so with a lifetime shorter than those decaying by weak interaction, but longer than those decaying under the influence of strong interaction. For example, of electromagnetic decay is:

$$ð0\ ã + ã.$$

This decay process, with a mean lifetime covering 8.4 x 10-17, may be understood as the annihilation of the quark and the antiquark, making up the ð0, into a pair of photons. The quantum numbers having to be conserved in electromagnetic interactions are, angular momentum, charge, baryon number, Isospin quantum number I3, strangeness, charm, parity and charge conjugation parity are unduly influenced.

Quanta's electrodynamic descriptions of the photon-mediated electromagnetic interactions have been verified over a great range of distances and have led to highly accurate predictions. Quantum electrodynamics are a 'gauge theory; as in quantum electrodynamics, the electromagnetic force can be derived by requiring that the equation describing the motion of a charged particle remain unchanged in the course of local symmetry operations. Specifically, if the phase of the wave function, by which charged particle is described is alterable independently, at which point in space, quantum electrodynamics require that the electromagnetic interaction and its mediating photon exist in order to maintain symmetry.

A kind of interaction between elementary particles that is weaker than the strong interaction force by a factor of about 1012. When strong interactions can occur in reactions involving elementary particles, the weak interactions are usually unobserved. However, sometimes strong and electromagnetic interactions are prevented because they would violate the conservation of some quantum number, e.g., strangeness, that has to be conserved in such reactions. When this happens, weak interactions may still occur.

The weak interaction operates over an extremely short range (about $2 \times 10\text{-}18$ m) it is mediated by the exchange of a very heavy particle (a gauge boson) that may be the charged $W+$ or W particle - mass about 80 GeV / $c2$ or the neutral $Z0$ particles (mass about 91 GeV / $c2$). The gauge bosons that mediate the weak interactions are analogous to the photon that mediates the electromagnetic interaction. Weak interactions mediated by W particles involve a change in the charge and hence the identity of the reacting particle. The neutral $Z0$ does not lead to such a change in identity. Both sorts of weak interaction can violate parity.

Most of the long-lived elementary particles decay as a result of weak interactions. For example, the kaon decay $K+$ ì+ vì may be thought of for being due to the annihilation of the u quark and ? antiquark in the $K+$ to produce a virtual $W+$ boson, which then converts into a positive muon and a neutrino. This decay action or and electromagnetic interaction because strangeness is not conserved, Beta decay is the most common example of weak interaction decay. Because it is so weak, particles that can only decay by weak interactions do so relatively slowly, i.e., they have relatively long lifetimes. Other examples of weak interactions include the scattering of the neutrino by other particles and certain very small effects on electrons within the atom.

Understanding of weak interactions is based on the electroweak theory, in which it is proposed that the weak and electromagnetic interactions are different manifestations of a single underlying force, known as the electroweak force. Many of the predictions of the theory have been confirmed experimentally.

A gauge theory, also called quantum flavour dynamics, that provides a unified description of both the electromagnetic and weak interactions. In the Glashow-Weinberg-Salam theory, also known as the standard model, electroweak interactions arise from the exchange of photons and of massive charged $W+$ and neutral $Z0$ bosons of spin 1 between quarks and leptons. The extremely massive charged particle, symbol $W+$ or W, that mediates certain types of weak interaction. The neutral Z-particle, or Z boson, symbol $Z0$, mediates the other types. Both are gauge bosons. The W- and Z-particles were first detected at CERN (1983) by studying collisions between protons and antiprotons with total energy 540 GeV in centre-of-mass co-ordinates. The rest masses were determined as about 80

GeV / c2 and 91 GeV / c2 for the W- and Z-particles, respectively, as had been predicted by the electroweak theory.

The interaction strengths of the gauge bosons to quarks and leptons and the masses of the W and Z bosons themselves are predicted by the theory, the Weinberg Angle èW, which must be determined by experiment. The Glashow-Weinberg-Salam theory successfully describes all existing data from a wide variety of electroweak processes, such as neutrino-nucleon, neutrino-electron and electron-nucleon scattering. A major success of the model was the direct observation in 1983-84 of the W$\pm$ and Z0 bosons with the predicted masses of 80 and 91 GeV / c2 in high energy proton-antiproton interactions. The decay modes of the W$\pm$ and Z0 bosons have been studied in very high pp and e+ e interactions and found to be in good agreement with the Standard model.

The six known types (or flavours) of quarks and the six known leptons are grouped into three separate generations of particles as follows:

1st generation: e ve u d

2nd generation: ì vì c s

3rd generation: ô vô t b

The second and third generations are essentially copies of the first generation, which contains the electron and the 'up' and 'down' quarks making up the proton and neutron, but involve particles of higher mass. Communication between the different generations occurs only in the quark sector and only for interactions involving W$\pm$ bosons. Studies of Z0 bosons production in very high energy electron-positron interactions has shown that no further generations of quarks and leptons can exist in nature (an arbitrary number of generations is a priori possible within the standard model) provided only that any new neutrinos are approximately massless.

The Glashow-Weinberg-Salam model also predicts the existence of a heavy spin 0 particle, not yet observed experimentally, known as the Higgs boson. The spontaneous symmetry-breaking mechanism used to generate non-zero masses for W$\pm$ and Z bosons in the electroweak theory, whereby the mechanism postulates the existence of two new complex fields,

$$\ddot{o}(\div i) = \ddot{o}1 + I\,\ddot{o}2 \text{ and } \Psi(\div i) = \Psi1 + I\,\Psi2$$

Which are functional distributors to $\Psi i = \Psi$, y, z and t, and form a doublet (ö Ψ) this doublet of complex fields transforms in the same way as leptons and quarks under electroweak gauge transformations? Such gauge transformations rotate ö1, ö2, Ψ1, Ψ2 into each other without changing the nature of the physical science.

The vacuum does not share the symmetry of the fields (ö, Ø) and a spontaneous breaking of the vacuum symmetry occurs via the Higgs mechanism. Consequently, the fields ö and Ψ have non-zero values in the vacuum. A particular orientation of ö1, ö2, Ψ1, Ψ2 may be chosen so that all the components of ö (Ψ1). This component responds to electroweak fields in a way that is analogous to

416

the response of a plasma to electromagnetic fields. Plasmas oscillate in the presence of electromagnetic waves, however, electromagnetic waves can only propagate at a frequency above the plasma frequency ùp2 given by the expression: ùp2 = ne2 /må - Where 'n' is the charge number density, 'e' the electrons charge. 'm' the electrons mass and 'å' is the Permittivity of the plasma. In quantum field theory, this minimum frequency for electromagnetic waves may be thought of as a minimum energy for the existence of a quantum of the electromagnetic field (a photon) within the plasma. This minimum energy or mass for the photon, which becomes a field quantum of a finite ranged force. Thus, in its plasma, photons acquire a mass and the electromagnetic interaction has a finite range.

The vacuum field öl responds to weak fields by giving a mass and finite range to the W± and Z bosons, however, the electromagnetic field is unaffected by the presence of öl so the photon remains massless. The mass acquired by the weak interaction bosons is proportional to the vacuum of öl and to the weak charge strength. A quantum of the field öl is an electrically neutral particle called the Higgs boson. It interacts with all massive particles with a coupling that is proportional to their mass. The standard model does not predict the mass of the Higgs boson, but it is known that it cannot be too heavy (not much more than about 1000 proton masses). Since this would lead to complicated self-interaction, such self-interaction is not believed to be present, because the theory does not account for them, but nevertheless successfully predicts the masses of the W± and Z bosons. These of the particle results from the so-called spontaneous symmetry breaking mechanisms, and used to generate non-zero masses for the W± and Z0 bosons and is presumably too massive to have been produced in existing particle accelerators.

We now turn our attentions belonging to the third binding force of unity, in, and of itself, its name implicates a physicality in the belonging nature that holds itself the binding of strong interactions that portray of its owing universality, simply because its universal. Interactions between elementary particles involving the strong interaction force. This force is about one hundred times greater than the electromagnetic force between charged elementary particles. However, it is a short range force - it is only important for particles separated by a distance of less than abut 10-15- and is the force that holds protons and neutrons together in atomic nuclei for 'soft' interactions between hadrons, where relatively small transfer of momentum are involved, the strong interactions may be described in terms of the exchange of virtual hadrons, just as electromagnetic interactions between charged particles may be described in terms of the exchange of virtual photons. At a more fundamental level, the strong interaction arises as the result of the exchange of Gluons between quarks and/and antiquarks as described by quantum chromodynamics.

In the hadron exchange picture, any hadron can act as the exchanged particle provided certain quantum numbers are conserved. These quantum numbers are the total angular momentum, charge, baryon number, Isospin (both I and I3), strangeness, parity, charge conjugation parity, and G-parity. Strong interactions are investigated experimentally by observing how beams of high-energy hadrons are scattered when they collide with other hadrons. Two hadrons colliding at high energy will only remain near to each other for a very short time. However, during the collision they may come sufficiently close to each other for a strong interaction to occur by the exchanger of a virtual particle. As a result of this interaction, the two colliding particles will be deflected (scattered) from their original paths. In the virtual hadron exchanged during the interaction carries some quantum numbers from one particle to the other, the particles found after the collision may differ from those before it. Sometimes the number of particles is increased in a collision.

In hadron-hadron interactions, the number of hadrons produced increases approximately logarithmically with the total centre of mass energy, reaching about 50 particles for proton-antiproton collisions at 900 GeV, for example in some of these collisions, two oppositely-directed collimated 'jets' of hadrons are produced, which are interpreted as due to an underlying interaction involving the exchange of an energetic gluon between, for example, a quark from the proton and an antiquark from the antiproton. The scattered quark and antiquark cannot exist as free particles, but instead 'fragments' into a large number of hadrons (mostly pions and kaon) travelling approximately along the original quark or antiquark direction. This results in collimated jets of hadrons that can be detected experimentally. Studies of this and other similar processes are in good agreement with quantum chromodynamics predictions.

The interaction between elementary particles arising as a consequence of their associated electric and magnetic fields. The electrostatic force between charged particles is an example. This force may be described in terms of the exchange of virtual photons, because its strength lies between strong and weak interactions, particles decaying by electromagnetic interaction do so with a lifetime shorter than those decaying by weak interaction, but longer than those decaying by strong interaction. An example of electromagnetic decay is:

$$\delta 0 + .$$

This decay process (mean lifetime 8.4 x 10-17 seconds) may be understood as the 'annihilation' of the quark and the antiquark making up the ð0, into a pair of photons. The following quantum numbers have to be conserved in electromagnetic interactions: Angular momentum, charm, baryon number, Isospin quantum number I3, strangeness, charm, parity, and charge conjugation parity.

A particle that, as far as is known, is not composed of other simpler particles. Elementary particles represent the most basic constituents of matter and are also the carriers of the fundamental forces between particles, namely the electromagnetic, weak, strong, and gravitational forces. The known elementary particles can be grouped into three classes, leptons, quarks, and gauge bosons, hadrons, such strongly interacting particles as the proton and neutron, which are bound states of quarks and/ or antiquarks, are also sometimes called elementary particles.

Leptons undergo electromagnetic and weak interactions, but not strong interactions. Six leptons are known, the negatively charged electron, muon, and tauons plus three associates neutrinos: Ve, vì and vô. The electron is a stable particle but the muon and tau leptons decay through the weak interactions with lifetimes of about 10-8 and 10-13 seconds. Neutrinos are stable neutral leptons, which interact only through the weak interaction.

Corresponding to the leptons are six quarks, namely the up (u), charm and top (t) quarks with electric charge equal to + that of the proton and the down (d), strange (s), and bottom (b) quarks of charge - the proton charge. Quarks have not been observed experimentally as free particles, but reveal their existence only indirectly in high-energy scattering experiments and through patterns observed in the properties of hadrons. They are believed to be permanently confined within hadrons, either in baryons, half integer spin hadrons containing three quarks, or in mesons, integer spin hadrons containing a quark and an antiquark. The proton, for example, is a baryon containing two 'up' quarks and an 'anti-down (d) quark, while the δ+ is a positively charged meson containing an up quark and an anti-down (d) antiquark. The only hadron that is stable as a free particle is the proton. The neutron is unstable when free. Within a nucleus, proton and neutrons are generally both stable but either particle may bear into a transformation into the other, by 'Beta Decay or Capture'.

Interactions between quarks and leptons are mediated by the exchange of particles known as 'gauge bosons', specifically the photon for electromagnetic interactions, W± and Z0 bosons for the weak interaction, and eight massless Gluons, in the case of the strong integrations.

A class of eigenvalue problems in physics that take the form $ÙØ = ëØ$, Where 'Ù' is some mathematical operation (multiplication by a number, differentiation, and so forth) on a function 'Ø', which is called the 'eigenfunction', 'ë' is called the eigenvalue, which in a physical system will be identified with an observable quantity analogous to the amplitude of a wave that appears in the equations of wave mechanics. Particularly the Schrödinger wave equation, the most generally accepted interpretation is that $| \Psi |2dV$, representing the probability that a particle is located within the volume element dV, mass in which case a particle of mass 'm' moving with a velocity 'v' will, under suitable experimental conditions exhibit the characteristics of a wave of wave length 'ë'. As given by the equation $ë = h/mv$, where 'h' is the Planck constant that equals to?

$$06.626\ 076 \times 10\text{-}34\ J\ s.?$$

This equation is the basis of wave mechanics. However, a set of weaves that represent the behaviour, under appropriate conditions, of a particle, e.g., its diffraction by a crystal lattice. The wave length is given by the 'de Broglie equation.' They are sometimes regarded as waves of probability, since the square of their amplitude at a given point represents the probability of finding the particle in unit volume at that point. These waves were predicted by Broglie in 1924 and in 1927 in the Davisson-Germer experiment.

Eigenvalue problems are ubiquitous in classical physics and occur whenever the mathematical description of a physical system yields a series of coupled differential equations. For example, the collective motion of a large number of interacting oscillators may be described by a set of coupled differential educations. Each differential equation describes the motion of one of the oscillators in terms of the position of all the others. A 'harmonic' solution may be sought, in which each displacement is assumed to have a 'simple harmonic motion' in time. The differential equations then reduce to 3N linear equations with 3N unknowns, where 'N' is the number of individual oscillators, each with three degrees of freedom. The whole problem is now easily recast as a 'matrix education' of the form:

$$? = ù2\Psi$$

Where 'M' is an N x N matrix called the 'dynamical matrix', and '-' is an N x 1 'a column matrix, and ù2 is the square of an angular frequency of the harmonic solution. The problem is now an eigenvalue problem with eigenfunctions '-' which is the normal mode of the system, with corresponding eigenvalues ù2. As '-' can be expressed as a column vector, '-' is a vector in some N-dimensional vector space. For this reason, 'Ψ' is often called an eigenvector.

When the collection of oscillators is a complicated three-dimensional molecule, the casting of the problem into normal modes is an effective simplification of the system. The symmetry principles of 'group theory' can then be applied, which classify normal modes according to their 'ù' eigenvalues (frequencies). This kind of analysis requires an appreciation of the symmetry properties of the molecule. The sets of operations (rotations, inversions, and so forth) that leave the molecule invariant make up the 'point group' of that molecule. Normal modes sharing the same 'ù' eigenvalues are said to correspond to the 'irreducible representations' of the molecule's point group. It is among these irreducible representations that one will find the infrared absorption spectrum for the vibrational normal modes of the molecule.

Eigenvalue problems play a particularly important role in quantum mechanics. In quantum mechanics, physically observable (location, momentum, energy, and so forth) are represented by operations (differentiation with respect to a variable, multiplication by a variable), which act on wave functions. Wave functions differ from classical waves in that they carry no energy. For classical waves, the square modulus of its amplitude measure its energy. For a wave function, the square modulus of its amplitude (at a location ',') represent not energy but probability, i.e., the probability that a particle -a localized packet of energy will be observed if a detector is placed at that location. The wave function therefore describes the distribution of possible locations of the particle and is perceptible only after many location detection events have occurred. A measurement of position on a quantum particle may be written symbolically as:

$$X \, \Psi(\chi) = -\Psi(\chi)$$

Where $\Psi(\chi)$ is said to be an eigenvector of the location operator and 'χ' is the eigenvalue, which represents the location. Each $\Psi(\chi)$ represents amplitude at the location $| \, \Psi(\chi) \, |2$ is the probability that the particle will be located in an infinitesimal volume at that location. The wave function describing the distribution of all possible locations for the particle is the linear super-position of all $\Psi(\chi)$ for 0χ that occur, its principle states that each stress is accompanied by the same strains whether it acts alone or in conjunction with others, it is true so long as the total stress does not exceed the limit of proportionality. Also, in vibrations and wave motion the principle asserts that one set of vibrations or waves are unaffected by the presence of another set. For example, two sets of ripples on water will pass through one another without mutual interactions so that, at a particular instant, the resultant disturbance at any point traversed by both sets of waves is the sum of the two component disturbances.

The eigenvalue problem in quantum mechanics therefore represents the act of measurement. Eigenvectors of an observable presentation were the possible states (Position, in the case of 'χ') that the quantum system can have. Stationary states of a quantum non-demolition attribute of a quantum system, such as position and momentum, are related by the Heisenberg Uncertainty Principle, which states that the product of the uncertainty of the measured value of a component of momentum (pΨ) and the uncertainty in the corresponding co-ordinates of position (-) is of the same order of magnitude as the Planck constant. Attributes related in this way are called 'conjugate' attributes. Thus, while an accurate measurement of position is possible, as a result of the uncertainty principle it produces a large momentum spread. Subsequent measurements of the position acquire a spread themselves, which makes the continuous monitoring of the position impossible.

The eigenvalues are the values that observables take on within these quantum states. As in classical mechanics, eigenvalue problems in quantum mechanics may take differential or matrix forms. Both forms have been shown to be equivalent. The differential form of quantum mechanics is called 'wave mechanics' (Schrödinger), where the operators are differential operators or multiplications by variables. Eigenfunctions in wave mechanics are wave functions corresponding to stationary wave states that satisfy some set of boundary conditions. The matrix form of quantum mechanics is often called matrix mechanics (Born and Heisenberg). Matrix acting on eigenvectors represents the operators.

The relationship between matrix and wave mechanics is very similar to the relationship between matrix and differential forms of eigenvalue problems in classical mechanics. The wave functions representing stationary states are really normal modes of the quantum wave. These normal modes may be thought of as vectors that span a vector space, which have a matrix representation.

Once, again, the Heisenberg uncertainty relation, or indeterminacy principle of 'quantum mechanics' that associate the physical properties of particles into pairs such that both together cannot be measured to within more than a certain degree of accuracy. If 'A' and 'V' form such a pair is called a conjugate pair, then: ÄAÄV > k, where 'k' is a constant and ÄA and ÄV are a variance in the experimental values for the attributes 'A' and 'V'. The best-known instance of the equation relates the position and momentum of an electron: ÄpÄp > h, where 'h' is the Planck constant. This is the Heisenberg uncertainty principle. Still, the usual value given for Planck's constant is 6.6 x 10-27 ergs sec. Since Planck's constant is not zero, mathematical analysis reveals the following: The 'spread', or uncertainty, in position times the 'spread', or uncertainty of momentum is greater than, or possibly equal to, the value of the constant or, or accurately, Planck's constant divided by 2ð, if we choose to know momentum exactly, then us knowing nothing about position, and vice versa.

The presence of Plank's constant calls that we approach quantum physics a situation in which the mathematical theory does not allow precise prediction of, or exist in exact correspondences with, the physical reality. If nature did not insist on making changes or transition in precise chunks of Planck's quantum of action, or in multiples of these chunks, there would be no crisis. But whether it is of our own determinacy, such that a cancerous growth in the body of an otherwise perfect knowledge of the physical world or the grounds for believing, in principle at least, in human freedom, one thing appears certain - it is an indelible feature of our understanding of nature.

In order too further explain how fundamental the quantum of action is to our present understanding of the life of nature, let us attempt to do what quantum physics says we cannot do and visualize its role in the simplest of all atoms -the hydrogen atom. It can be thought that standing at the centre of the Sky Dome or Rogers Centre, at roughly where the pitcher's mound is. Place a grain of salt on the mound, and picture a speck of dust moving furiously around the orbital's outskirts of the Sky Dome's fulfilling circle, around which the grain of salt remains referential of the topic. This represents, roughly, the relative size of the nucleus and the distance between electron and nucleus inside the hydrogen atom when imaged in its particle aspect.

In quantum physics, however, the hydrogen atom cannot be visualized with such macro-level analogies. The orbit of the electron is not a circle, in which a planet-like object moves, and each orbit is described in terms of a probability distribution for finding the electron in an average position corresponding to each orbit as opposed to an actual position. Without observation or measurement, the electron could be in some sense anywhere or everywhere within the probability distribution, also, the space between probability distributions is not empty, it is infused with energetic vibrations capable of manifesting itself as the befitting quanta.

The energy levels manifest at certain distances because the traditions between orbits occurs in terms of precise units of Planck's constant. If any attentive effects to comply with or measure where the particle-like aspect of the electron is, in that the existence of Planck's constant will always prevent us from knowing precisely all the properties of that electron that we might presume to be they're in the absence of measurement. Also, the two-split experiment, as our presence as observers and what we choose to measure or observe are inextricably linked to the results obtained. Since all complex molecules are built from simpler atoms, what is to be done, is that liken to the hydrogen atom, of which case applies generally to all material substances.

The grounds for objecting to quantum theory, the lack of a one-to-one correspondence between every element of the physical theory and the physical reality it describes, may seem justifiable and reasonable in strict scientific terms. After all, the completeness of all previous physical theories was measured against that criterion with enormous success. Since it was this success that gave physicists the reputation of being able to disclose physical reality with magnificent exactitude, perhaps a more complex quantum theory will emerge by continuing to insist on this requirement.

All indications are, however, that no future theory can circumvent quantum indeterminacy, and the success of quantum theory in co-ordinating our experience with nature is eloquent testimony to this conclusion. As Bohr realized, the fact that we live in a quantum universe in which the quantum of action is a given or an unavoidable reality requires a very different criterion for determining the completeness of physical theory. The new measure for a complete physical theory is that it unambiguously confirms our ability to co-ordinate more experience with physical reality.

If a theory does so and continues to do so, which is certainly the case with quantum physics, then the theory must be deemed complete. Quantum physics not only works exceedingly well, it is, in these terms, the most accurate physical theory that has ever existed. When we consider that this physics allows us to predict and measure quantities like the magnetic moment of electrons to the fifteenth decimal place, we realize that accuracy perse is not the real issue. The real issue, as Bohr rightly intuited, is that this complete physical theory effectively undermines the privileged relationships in classical physics between physical theory and physical reality. Another measure of success in physical theory is also met by quantum physics -eloquence and simplicity. The quantum recipe for computing probabilities given by the wave function is straightforward and can be successfully employed by any undergraduate physics student. Take the square of the wave amplitude and compute the probability of what can be measured or observed with a certain value. Yet there is a profound difference between the recipe for calculating quantum probabilities and the recipe for calculating probabilities in classical physics.

In quantum physics, one calculates the probability of an event that can happen in alternative ways by adding the wave functions, and then taking the square of the amplitude. In the two-split experiment, for example, the electron is described by one wave function if it goes through one slit and by another wave function if it goes through the other slit. In order to compute the probability of where the electron is going to end on the screen, we add the two wave functions, compute the obsolete value of their sum, and square it. Although the recipe in classical probability theory seems similar, it is quite different. In classical physics, one would simply add the probabilities of the two alternative ways and let it go at that. That classical procedure does not work here because we are not dealing with classical atoms in quantum physics additional terms arise when the wave functions are added, and the probability is computed in a process known as the 'superposition principle'. That the superposition principle can be illustrated with an analogy from simple mathematics. Add two numbers and then take the square of their sum, as opposed to just adding the squares of the two numbers. Obviously, $(2 + 3)2$ is not equal to $22 + 32$. The former is 25, and the latter are 13. In the language of quantum probability theory:

$$| \Psi1 + \Psi2 |2 | \Psi1 |2 + \Psi| \Psi2 |2$$

Where $\Psi1$ and $\Psi2$ are the individual wave functions on the left-hand side, the superposition principle results in extra terms that cannot be found on the right-handed side the left-hand faction of the above relation is the way a quantum physicists would compute probabilities and the right-hand side is the classical analogue. In quantum theory, the right-hand side is realized when we know, for example, which slit through which the electron went. Heisenberg was among the first to compute what would happen in an instance like this. The extra superposition terms contained in the left-hand side of the above relation would not be there, and the peculiar wave-like interference pattern would disappear. The observed pattern on the final screen would, therefore, be what one would expect if electrons were behaving like bullets, and the final probability would be the sum of the individual probabilities. But when we know which slit the electron went through, this interaction with the system causes the interference pattern to disappear.

In order to give a full account of quantum recipes for computing probabilities, one has to examine what would happen in events that are compounded. Compound events are events that can be broken down into a series of steps, or events that consist of a number of things happening independently the recipe here calls for multiplying the individual wave functions, and then following the usual quantum recipe of taking the square of the amplitude.

The quantum recipe is $| \Psi 1 \cdot \Psi 2 |2$, and, in this case, it would be the same if we multiplied the individual probabilities, as one would in classical theory. Thus the recipes of computing results in quantum theory and classical physics can be totally different from quantum superposition effects are completely non-classical, and there is no mathematical justification to why the quantum recipes work. What justifies the use of quantum probability theory is the same thing that justifies the use of quantum physics -it has allowed us in countless experiments to vastly extend our ability to co-ordinate experience with nature.

The view of probability in the nineteenth century was greatly conditioned and reinforced by classical assumptions about the relationships between physical theory and physical reality. In this century, physicists developed sophisticated statistics to deal with large ensembles of particles before the actual character of these particles was understood. Classical statistics, developed primarily by James C. Maxwell and Ludwig Boltzmann, was used to account for the behaviour of a molecule in a gas and to predict the average speed of a gas molecule in terms of the temperature of the gas.

The presumption was that the statistical average were workable approximations those subsequent physical theories, or better experimental techniques, would disclose with precision and certainty. Since nothing was known about quantum systems, and since quantum indeterminacy is small when dealing with macro-level effects, this presumption was quite reasonable. We know, however, that quantum mechanical effects are present in the behaviour of gasses and that the choice to ignore them is merely a matter of convincing in getting workable or practical resulted. It is, therefore, no longer possible to assume that the statistical averages are merely higher-level approximations for a more exact description.

Perhaps the best-known defence of the classical conception of the relationship between physical theory ands physical reality is the celebrated animal introduced by the Austrian physicist Erin Schrödinger (1887-1961) in 1935, in a 'thought experiment' showing the strange nature of the world of quantum mechanics. The cat is thought of as locked in a box with a capsule of cyanide, which will break if a Geiger counter triggers. This will happen if an atom in a radioactive substance in the box decays, and there is a chance of 50% of such an event within an hour. Otherwise, the cat is alive. The problem is that the system is in an indeterminate state. The wave function of the entire system is a 'superposition' of states, fully described by the probabilities of events occurring when it is eventually measured, and therefore 'contains equal parts of the living and dead cat'. When we look and see we will find either a breathing cat or a dead cat, but if it is only as we look that the wave packet collapses, quantum mechanic forces us to say that before we looked it was not true that the cat was dead and not true that it was alive, the thought experiment makes vivid the difficulty of conceiving of quantum indeterminancies when these are translated to the familiar world of everyday objects.

The 'electron,' is a stable elementary particle having a negative charge, 'e', equal to:

$$1.602\ 189\ 25 \times 10^{-19}\ C$$

And a rest mass, m_0 equal to;

$$9.109\ 389\ 7 \times 10^{-31}\ kg$$
$$\text{equivalent to } 0.511\ 0034\ MeV\ /\ c^2$$

It has a spin of ½ and obeys Fermi-Dirac Statistics. As it does not have strong interactions, it is classified as a 'lepton'.

The discovery of the electron was reported in 1897 by Sir J.J. Thomson, following his work on the rays from the cold cathode of a gas-discharge tube, it was soon established that particles with the same charge and mass were obtained from numerous substances by the 'photoelectric effect', 'thermionic emission' and 'beta decay'. Thus, the electron was found to be part of all atoms, molecules, and crystals.

Free electrons are studied in a vacuum or a gas at low pressure, whereby beams are emitted from hot filaments or cold cathodes and are subject to 'focussing', so that the particles in which an electron beam in, for example, a cathode-ray tube, where in principal methods as (i) Electrostatic focussing, the beam is made to converge by the action of electrostatic fields between two or more electrodes at different potentials. The electrodes are commonly cylinders coaxial with the electron tube, and the whole assembly forms an electrostatic electron lens. The focussing effect is usually controlled by varying the potential of one of the electrodes, called the focussing electrode. (ii) Electromagnetic focussing, by way that the beam is made to converge by the action of a magnetic field that is produced by the passage of direct current, through a focussing coil. The latter are commonly a coil of short axial length mounted so as to surround the electron tube and to be coaxial with it.

The force FE on an electron or magnetic field of strength E is given by FE = Ee and is in the direction of the field. On moving through a potential difference V, the electron acquires a kinetic energy eV, hence it is possible to obtain beams of electrons of accurately known kinetic energy. In a magnetic field of magnetic flux density 'B', an electron with speed 'v' is subject to a force, FB = Bev sin è, where è is the angle between 'B' and 'v'. This force acts at right angles to the plane containing 'B' and 'v'.

The mass of any particle increases with speed according to the theory of relativity. If an electron is accelerated from rest through 5kV, the mass is 1% greater than it is at rest. Thus, accountably, must be taken of relativity for calculations on electrons with quite moderate energies.

According to 'wave mechanics' a particle with momentum 'mv' exhibits' diffraction and interference phenomena, similar to a wave with wavelength

$$ë = h/mv,$$

Where 'h' is the Planck constant. For electrons accelerated through a few hundred volts, this gives wavelengths rather less than typical interatomic spacing in crystals. Hence, a crystal can act as a diffraction grating for electron beams.

Owing to the fact that electrons are associated with a wavelength ë given by:

$$\ddot{e} = h/mv,$$

Where 'h' is the Planck constant and (mv) the momentum of the electron, a beam of electrons suffers diffraction in its passage through crystalline material, similar to that experienced by a beam of X-rays. The diffraction pattern depends on the spacing of the crystal planes, and the phenomenon can be employed to investigate the structure of surface and other films, and under suitable conditions exhibit the characteristics of a wave of the wavelength given by the equation ë = h/mv, which is the basis of wave mechanics. A set of waves that represent the behaviour, under appropriate conditions, of a particle, e.g., its diffraction by a crystal lattice, that is given the 'de Broglie equation.' They are sometimes regarded as waves of probability, since the square of their amplitude at a given point represents the probability of finding the particle in unit volume at that point.

The first experiment to demonstrate 'electron diffraction', and hence the wavelike nature of particles. A narrow pencil of electrons from a hot filament cathode was projected 'in vacua' onto a nickel crystal. The experiment showed the existence of a definite diffracted beam at one particular angle, which depended on the velocity of the electrons, assuming this to be the Bragg angle, stating that the structure of a crystal can be determined from a set of interference patterns found at various angles from the different crystal faces, least of mention, the wavelength of the electrons was calculated and found to be in agreement with the 'de Broglie equation.'

At kinetic energies less than a few electro-volts, electrons undergo elastic collision with atoms and molecules, simply because of the large ratio of the masses and the conservation of momentum, only an extremely small trandfer of kinetic energy occurs. Thus, the electrons are deflected but not slowed down appreciatively. At slightly higher energies collisions are inelastic. Molecules may be dissociated, and atoms and molecules may be excited or ionized. Thus it is the least energy that causes an ionization

$$A \quad A+ + e$$

Where the ION and the electron are far enough apart for their electrostatic interaction to be negligible and no extra kinetic energy removed is that in the outermost orbit, i.e., the level strongly bound electrons. It is also possible to consider removal of electrons from inner orbits, in which their binding energy is greater. As an excited particle or recombining, ions emit electromagnetic radiation mostly in the visible or ultraviolet.

For electron energies of the order of several GeV upwards, X-rays are generated. Electrons of high kinetic energy travel considerable distances through matter, leaving a trail of positive ions and free electrons. The energy is mostly lost in small increments (about 30 eV) with only an occasional major interaction causing X-ray emissions. The range increases at higher energies.

The positron -the antiparticle of the electron, I e., an elementary particle with electron mass and positive charge equal to that of the electron. According to the relativistic wave mechanics of Dirac, space contains a continuum of electrons in states of negative energy. These states are normally unobservable, but if sufficient energy can be given, an electron may be raised into a state of positive energy and suggested itself observably. The vacant state of negativity behaves as a positive particle of positive energy, which is observed as a positron.

The simultaneous formation of a positron and an electron from a photon is called 'pair production', and occurs when the annihilation of gamma-ray photons with an energy of 1.02 MeV passes close to an atomic nucleus, whereby the interaction between the particle and its antiparticle disappear and photons or other elementary particles or antiparticles are so created, as accorded to energy and momentum conservation.

At low energies, an electron and a positron annihilate to produce electromagnetic radiation. Usually the particles have little kinetic energy or momentum in the laboratory system before interaction, hence the total energy of the radiation is nearly $2m_0c^2$, where m_0 is the rest mass of an electron. In nearly all cases two photons are generated. Each of 0.511 MeV, in almost exactly opposite directions to conserve momentum. Occasionally, three photons are emitted all in the same plane. Electron-positron annihilation at high energies has been extensively studied in particle accelerators. Generally the annihilation results in the production of a quark, and an antiquark, fort example, e+ e ì+ ì or a charged lepton plus an antilepton (e+e ì+ì). The quarks and antiquarks do not appear as free particles but convert into several hadrons, which can be detected experimentally. As the energy available in the electron-positron interaction increases, quarks and leptons of progressively larger rest mass can be produced. In addition, striking resonances are present, which appear as large increases in the rate at which annihilations occur at particular energies. The I / PSI particle and similar resonances containing an antiquark are produced at an energy of about 3 GeV, for example, giving rise to abundant production of charmed hadrons. Bottom (b) quark production occurs at greater energies than about 10 GeV. A resonance at an energy of about 90 GeV, due to the production of the Z0 gauge boson involved in weak interaction is currently under intensive study at the LEP and SLC e+ e colliders. Colliders are the machines for increasing the kinetic energy of charged particles or ions, such as protons or electrons, by accelerating them in an electric field. A magnetic field is used to maintain the particles in the desired direction. The particle can travel in a straight, spiral, or circular paths. At present, the highest energies are obtained in the proton synchrotron.

The Super Proton Synchrotron at CERN (Geneva) accelerates protons to 450 GeV. It can also cause proton-antiproton collisions with total kinetic energy, in centre-of-mass co-ordinates of 620 GeV. In the USA the Fermi National Acceleration Laboratory proton synchrotron gives protons and antiprotons of 800 GeV, permitting collisions with total kinetic energy of 1600 GeV. The Large Electron Positron (LEP) system at CERN accelerates particles to 60 GeV.

All the aforementioned devices are designed to produce collisions between particles travelling in opposite directions. This gives effectively very much higher energies available for interaction than our possible targets. High-energy nuclear reaction occurs when the particles, either moving in a stationary target collide. The particles created in these reactions are detected by sensitive equipment close to the collision site. New particles, including the tauon, W, and Z particles and requiring enormous energies for their creation, have been detected and their properties determined.

While, still, a 'nucleon' and 'anti-nucleon' annihilating at low energy, produce about half a dozen pions, which may be neutral or charged. By definition, mesons are both hadrons and bosons, justly as the pion and kaon are mesons. Mesons have a substructure composed of a quark and an antiquark bound together by the exchange of particles known as Gluons.

The conjugate particle or antiparticle that corresponds with another particle of identical mass and spin, but has such quantum numbers as charge (Q), baryon number (B), strangeness (S), charms and Isospin (I3) of equal magnitude but opposite sign. Examples of a particle and its antiparticle include the electron and positron, proton and antiproton, the positive and negatively charged pions, and the 'up' quark and 'up' antiquark. The antiparticle corresponding to a particle with the symbol 'a' is usually denoted ' '. When a particle and its antiparticle are identical, as with the photon and neutral pion, this is called a 'self-conjugate particle'.

The critical potential to excitation energy required to change am atom or molecule from one quantum state to another of higher energy, is equal to the difference in energy of the states and is usually the difference in energy between the ground state of the atom and a specified excited state. Which the state of a system, such as an atom or molecule, when it has a higher energy than its ground state.

The ground state contributes the state of a system with the lowest energy. An isolated body will remain indefinitely in it, such that it is possible for a system to have possession of two or more ground states, of equal energy but with different sets of quantum numbers. In the case of atomic hydrogen there are two states for which the quantum numbers n, I, and m are 1, 0, and 0 respectively, while the spin may be + ½ with respect to a defined direction. An allowed wave function of an electron in an atom obtained by a solution of the 'Schrödinger wave equation' in which a hydrogen atom, for example, the electron moves in the electrostatic field of the nucleus and its potential energy is e2 / r, where 'e' is the electron charge and 'r' its distance from the nucleus. A precise orbit cannot be considered as in Bohr's theory of the atom, but the behaviour of the electron is described by its wave function, Ψ, which is a mathematical function of its position with respect to the nucleus. The significance of the wave function is that $|\Psi|2$ dt is the probability of locating the electron in the element of volume dt.

Solution of Schrödinger's equation for the hydrogen atom shows that the electron can only have certain allowed wave functions (eigenfunctions). Each of these corresponds to a probability distribution in space given by the manner in which $|\Psi|2$ varies with position. They also have an associated value of the energy 'E'. These allowed wave functions, or orbitals, are characterized by three quantum numbers similar to those characterized the allowed orbits in the earlier quantum theory of the atom: 'n', the 'principal quantum number, can have values of 1, 2, 3, and so forth the orbital with n =1 has the lowest energy. The states of the electron with n = 1, 2, 3, and so forth, are called 'shells' and designate the K, L, M shells, and so forth 'I', the 'azimuthal quantum numbers', which for a given value of 'n' can have values of 0, 1, 2, . . . (n 1). An electron in the 'L' shell of an atom with n = 2 can occupy two sub-shells of different energy corresponding to I = 0, I = 1, and I = 2. Orbitals with I = 0, 1, 2 and 3 are called s, p, d, and ¦ orbitals respectively. The significance of I quantum number is that it gives the angular momentum of the electron. The orbital angular momentum of an electron is given by:

$$?[I(I + 1)(h/2ð).$$

'm', the 'magnetic quantum number, which for a given value of I can have values, I, (I-1), . . ., 0, . . . (I-1), I. Thus, for a 'p' orbital for orbits with m = 1, 0, and 1. These orbitals, with the same values of 'n' and 'I' but different 'm' values, have the same energy. The significance of this quantum number is that it indicates the number of different levels that would be produced if the atom were subjected to an external magnetic field.

According to wave theory the electron may be at any distance from the nucleus, but in fact, there is only a reasonable chance of it being within a distance of $\sim$ 5 x 10-11 metre. Indeed the maximum probability occurs when r-a0 where a0 is the radius of the first Bohr orbit. It is customary to represent an orbital by a surface enclosing a volume within which there is an arbitrarily decided probability (say 95%) of finding the electron.

Finally, the electron in an atom can have a fourth quantum number MS, characterizing its spin direction. This can be + ½ or ½, and according to the 'Pauli Exclusion Principle,' each orbital can hold only two electrons. The four quantum numbers lead to an explanation of the periodic table of the elements.

In earlier mention, the concerns referring to the 'moment' had been to our exchanges to issue as, i.e., the moment of inertia, moment of momentum. The moment of a force about an axis is the product of the perpendicular distance of the axis from the line of action of the force, and the component of the force in the plane perpendicular to the axis. The moment of a system of coplanar forces about an axis perpendicular to the plane containing them is the algebraic sum of the moments of the separate forces about that axis of a anticlockwise moment appear taken controventionally to be positive and clockwise of ones Uncomplementarity. The moment of momentum about an axis, symbol L is the product to the moment of inertia and angular velocity (Iù). Angular momentum is a pseudo-vector quality, as it is connected in an isolated system. It is a scalar and is given a positive or negative sign as in the moment of force. When contending to systems, in which forces and motions do not all lie in one plane, the concept of the moment about a point is needed. The moment of a vector P, e.g., force or momentous pulsivity, from which a point 'A' is a pseudo-vector M equal to the vector product of r and P, where r is any line joining 'A' to any point 'B' on the line of action of P. The vector product M = r x p is independent of the position of 'B' and the relation between the scalar moment about an axis and the vector moment about which a point on the axis is that the scalar is the component of the vector in the direction of the axis.

The linear momentum of a particle 'p' is the product of the mass and the velocity of the particle. It is a vector quality directed through the particle in the direction of motion. The linear momentum of a body or of a system of particles is the vector sum of the linear momenta of the individual particle. If a body of mass 'M' is translated with a velocity 'V', its momentum is MV, which is the momentum of a particle of mass 'M' at the centre of gravity of the body. (1) In any system of mutually interacting or impinging particles, the linear momentum in any fixed direction remains unaltered unless there is an external force acting in that direction. (2) Similarly, the angular momentum is constant in the case of a system rotating about a fixed axis provided that no external torque is applied.

Subatomic particles fall into two major groups: The elementary particles and the hadrons. An elementary particle is not composed of any smaller particles and therefore represents the most fundamental form of matter. A hadron is composed of panicles, including the major particles called quarks, the most common of the subatomic particles, includes the major constituents of the atom -the electron is an elementary particle, and the proton and the neutron (hadrons). An elementary particle with zero charge and a rest mass equal to:

$$1.674\ 9542 \times 10\text{-}27 \text{ kg,}$$
$$\text{i.e., } 939.5729 \text{ MeV } / \text{ c2.}$$

It is a constituent of every atomic nucleus except that of ordinary hydrogen, free neutrons decay by 'beta decay' with a mean life of 914 s. the neutron has spin ½, Isospin ½, and positive parity. It is a 'fermion' and is classified as a 'hadron' because it has strong interaction.

Neutrons can be ejected from nuclei by high-energy particles or photons, the energy required is usually about 8 MeV, although sometimes it is less. The fission is the most productive source. They are detected using all normal detectors of ionizing radiation because of the production of secondary particles in nuclear reactions. The discovery of the neutron (Chadwick, 1932) involved the detection of the tracks of protons ejected by neutrons by elastic collisions in hydrogenous materials.

Unlike other nuclear particles, neutrons are not repelled by the electric charge of a nucleus so they are very effective in causing nuclear reactions. When there is no 'threshold energy', the interaction 'cross sections' become very large at low neutron energies, and the thermal neutrons produced in great numbers by nuclear reactions cause nuclear reactions on a large scale. The capture of neutrons by the (n) process produces large quantities of radioactive materials, both useful nuclides such as 66Co for cancer therapy and undesirable by-products. The least energy required to cause a certain process, in particular a reaction in nuclear or particle physics. It is often important to distinguish between the energies required in the laboratory and in centre-of-mass co-ordinates. In 'fission' the splitting of a heavy nucleus of an atom into two or more fragments of comparable size usually as the result of the impact of a neutron on the nucleus. It is normally accompanied by the emission of neutrons or gamma rays. Plutonium, uranium, and thorium are the principle fissionable elements

In nuclear reaction, a reaction between an atonic nucleus and a bombarding particle or photon leading to the creation of a new nucleus and the possible ejection of one or more particles. Nuclear reactions are often represented by enclosing brackets and symbols for the incoming and final nuclides being shown outside the brackets. For example:

$$14\text{N (á, p)17O.}$$

Energy from nuclear fissions, on the whole, the nucleuses of atoms of moderate size are more tightly held together than the largest nucleus, so that if the nucleus of a heavy atom can be induced to split into two nuclei and moderate mass, there should be considerable release of energy. By Einstein' s law of the conservation of mass and energy, this mass and energy difference is equivalent to the energy released when the nucleons binding differences are equivalent to the energy released when the nucleons bind together. Y=this energy is the binding energy, the graph of binding per nucleon, EB / A increases rapidly up to a mass number of 50-69 (iron, nickel, and so forth) and then decreases slowly. There are therefore two ways in which energy can be released from a nucleus, both of which can be released from the nucleus, both of which entail a rearrangement of nuclei occurring in the lower as having to curve into form its nuclei, in the upper, higher-energy part of the curve. The fission is the splitting of heavy atoms, such as uranium, into lighter atoms, accompanied by an enormous release of energy. Fusion of light nuclei, such as deuterium and tritium, releases an even greater quantity of energy.

The work that must be done to detach a single particle from a structure of free electrons of an atom or molecule to form a negative ion. The process is sometimes called 'electron capture, but the term is more usually applied to nuclear processes. As many atoms, molecules and free radicals from stable negative ions by capturing electrons to atoms or molecules to form a negative ion. The electron affinity is the least amount of work that must be done to separate from the ion. It is usually expressed in electro-volts

The uranium isotope 235U will readily accept a neutron but one-seventh of the nuclei stabilized by gamma emissions while six-sevenths split into two parts. Most of the energy released amounts to about 170 MeV, in the form of the kinetic energy of these fission fragments. In addition an averaged of 2.5 neutrons of average energy 2 MeV and some gamma radiation is produced. Further energy is released later by radioactivity of the fission fragments. The total energy released is about 3 x 10-11 joule per atom fissioned, i.e., 6.5 x 1013 joule per kg conserved.

To extract energy in a controlled manner from fissionable nuclei, arrangements must be made for a sufficient proportion of the neutrons released in the fissions to cause further fissions in their turn, so that the process is continuous, the minium mass of a fissile material that will sustain a chain reaction seems confined to nuclear weaponry. Although, a reactor with a large proportion of 235U or plutonium 239Pu in the fuel uses the fast neutrons as they are liberated from the fission, such a rector is called a 'fast reactor'. Natural uranium contains 0.7% of 235U and if the liberated neutrons can be slowed before they have much chance of meeting the more common 238U atom and then cause another fission. To slow the neutron, a moderator is used containing light atoms to which the neutrons will give kinetic energy by collision. As the neutrons eventually acquire energies appropriate to gas molecules at the temperatures of the moderator, they are then said to be thermal neutrons and the reactor is a thermal reactor.

Then, of course, the Thermal reactors, in typical thermal reactors, the fuel elements are rods embedded as a regular array in which the bulk of the moderator that the typical neutron from a fission process has a good chance of escaping from the relatively thin fuel rod and making many collisions with nuclei in the moderator before again entering a fuel element. Suitable moderators are pure graphite, heavy water (D2O) are sometimes used as a coolant, and ordinary water (H2O). Very pure materials are essential as some unwanted nuclei capture neutrons readily. The reactor core is surrounded by a reflector made of suitable material to reduce the escape of neutrons from the surface. Each fuel element is encased e. g., in magnesium alloy or stainless steel, to prevent escape of radioactive fission products. The coolant, which may be gaseous or liquid, flows along the channels over the canned fuel elements. There is an emission of gamma rays inherent in the fission process and, many of the fission products are intensely radioactive. To protect personnel, the assembly is surrounded by a massive biological shield, of concrete, with an inner iron thermal shield to protect the concrete from high temperatures caused by absorption of radiation.

To keep the power production steady, control rods are moved in or out of the assembly. These contain material that captures neutrons readily, e.g., cadmium or boron. The power production can be held steady by allowing the currents in suitably placed ionization chambers automatically to modify the settings of the rods. Further absorbent rods, the shut down rods are driven into the core to stop the reaction, as in an emergence if the control mechanism fails. To attain high thermodynamic efficiency so that a large proportion of the liberated energy can be used, the heat should be extracted from the reactor core at a high temperature.

In fast reactors no mediator is used, the frequency of collisions between neutrons and fissile atoms being creased by enriching the natural uranium fuel with 239Pu or additional 235U atoms that are fissioned by fast neutrons. The fast neutrons are thus built up a self-sustaining chain reaction. In these reactions the core is usually surrounded by a blanket of natural uranium into which some of the neutrons are allowed to escape. Under suitable conditions some of these neutrons will be captured by 238U atoms forming 239U atoms, which are converted to 239Pu. As more plutonium can be produced than required to enrich the fuel in the core, these are called 'fast breeder reactors'.

Thus and so, a neutral elementary particle with spin½, that only takes part in weak interactions. The neutrino is a lepton and exists in three types corresponding to the three types of charged leptons, that is, there are the electron neutrinos (ve) tauon neutrinos (vì) and tauon neutrinos (vô). The antiparticle of the neutrino is the antineutrino.

Neutrinos were originally thought to have a zero mass, but recently there have been some advances to an indirect experiment that evince to the contrary. In 1985 a Soviet team reported a measurement for the first time, of a non-zero neutrino mass. The mass measured was extremely small, some 10 000 times smaller than the mass of the electron. However, subsequent attempts to reproduce the Soviet measurement were unsuccessful. More recent (1998-99), the Super-Kamiokande experiment in Japan has provided indirect evidence for massive neutrinos. The new evidence is based upon studies of neutrinos, which are created when highly energetic cosmic rays bombard the earth's upper atmosphere. By classifying the interaction of these neutrinos according to the type of neutrino involved (an electron neutrino or muon neutrino), and counting their relative numbers as a function: An oscillatory behaviour may be shown to occur. Oscillation in this sense is the charging back and forth of the neutrino's type as it travels through space or matter. The Super-Kamiokande result indicates that muon neutrinos are changing into another type of neutrino, e.g., sterile neutrinos. The experiment does not, however, determine directly the masses, though the oscillations suggest very small differences in mass between the oscillating types.

The neutrino was first postulated (Pauli 1930)to explain the continuous spectrum of beta rays. It is assumed that there is the same amount of energy available for each beta decay of a particle nuclide and that energy is shared according to a statistical law between the electron and a light neutral particle, now classified as the anti-neutrino, (ẏe). Later it was shown that the postulated particle would also conserve angular momentum and linear momentum in the beta decays.

In addition to beta decay, the electron neutrino is also associated with, for example, positron decay and electron capture:

$$22Na \; 22Ne + e+ + ve$$
$$55Fe + e \; 55Mn + ve$$

The absorption of anti-neutrinos in matter by the process

$$2H + àe \; n + e+$$

Was first demonstrated by Reines and Cowan? The muon neutrino is generated in such processes as:

$$ð+ \; ì+ + vì$$

Although the interactions of neutrinos are extremely weak the cross sections increase with energy and reaction can be studied at the enormous energies available with modern accelerators in some forms of 'grand unification theories', neutrinos are predicted to have a non-zero mass. Nonetheless, no evidences have been found to support this prediction.

The antiparticle of an electron, i.e., an elementary particle with electron mass and positive charge and equal to that of the electron. According to the relativistic wave mechanics of Dirac, space contains a continuum of electrons in states of negative energy. These states are normally unobservable, but if sufficient energy can be given, an electron may be raised into a state of positivity and become observable. The vacant state of negativity seems to behave as a positive particle of positive energy, which is observed as a positron.

A theory of elementary particles based on the idea that the fundamental entities are not point-like particles, but finite lines (strings) or closed loops formed by stings. The original idea was that an elementary particle was the result of a standing wave in a string. A considerable amount of theoretical effort has been put into development string theories. In particular, combining the idea of strings with that of super-symmetry, which has led to the idea with which correlation holds strongly with super-strings. This theory may be a more useful route to a unified theory of fundamental interactions than quantum field theory, simply because it's probably by some unvioded infinites that arise when gravitational interactions are introduced into field theories. Thus, superstring theory inevitably leads to particles of spin 2, identified as gravitons. String theory also shows why particles violate parity conservation in weak interactions.

Superstring theories involve the idea of higher dimensional spaces: 10 dimensions for fermions and 26 dimensions for bosons. It has been suggested that there are the normal 4 space-time dimensions, with the extra dimension being tightly 'curved'. Still, there are no direct experimental evidences for super-strings. They are thought to have a length of about 10-35 m and energies of 1014 GeV, which is well above the energy of any accelerator. An extension of the theory postulates that the fundamental entities are not one-dimensional but two-dimensional, i.e., they are super-membranes.

Allocations often other than what are previous than in time, awaiting the formidable combinations of what precedes the presence to the future, because of which the set of invariance of a system, a symmetry operation on a system is an operation that does not change the system. It is studied mathematically using 'Group Theory.' Some symmetries are directly physical, for instance the reelections and rotations for molecules and translations in crystal lattices. More abstractively the implicating inclinations toward abstract symmetries involve changing properties, as in the CPT Theorem and the symmetries associated with 'Gauge Theory.' Gauge theories are now thought to provide the basis for a description in all elementary particle interactions. The electromagnetic particle interactions are described by quantum electrodynamics, which is called Abelian gauge theory

Quantum field theory for which measurable quantities remain unchanged under a 'group transformations'. All these theories consecutive field trandformations do not commute. All non-Abelian gauge theories are based on work proposed by Yang and Mills in 1954, describe the interaction between two quantum fields of fermions. In which particles represented by fields whose normal modes of oscillation are quantized. Elementary particle interactions are described by relativistically invariant theories of quantized fields, ie., By relativistic quantum field theories. Gauge transformations can take the form of a simple multiplication by a constant phase. Such trandformations are called 'global gauge trandformations'. In local gauge trandformations, the phase of the fields is alterable by amounts that vary with space and time;

$$\text{i.e., } \Psi \text{ eiè } (\chi) \ \Psi,$$

Where 'è' (χ) is a function of space-time. As, in Abelian gauge theories, consecutive field transformations commute

$$\text{i.e., } \Psi \text{ ei è } (\chi) \text{ ei ö } \Psi = \text{ei ö } (\chi) \text{ ei ö } (\chi) \ \Psi,$$

Where ö (x) is another function of space and time. Quantum chromodynamics (the theory of the strong interaction) and electroweak and grand unified theories are all non-Abelian. In these theories consecutive field transformations do not commute. All non-Abelian gauge theories are based on work proposed by Yang and Mils, as Einstein's theory of general relativity can also be formulated as a local gauge theory.

A symmetry including both boson and fermions, in theories based on super-symmetry every boson has a corresponding boson. Th boson partners of existing fermions have names formed by prefacing the names of the fermion with an 's' (e.g., selection, squark, lepton). The names of the fermion partners of existing bosons are obtained by changing the terminal -on of the boson to -into (e.g., photons, Gluons, and zino). Although, super-symmetries have not been observed experimentally, they may prove important in the search for a Unified Field Theory of the fundamental interactions.

The quark is a fundamental constituent of hadrons, i.e., of particles that take part in strong interactions. Quarks are never seen as free particles, which is substantiated by lack of experimental evidence for isolated quarks. The explanation given for this phenomenon in gauge theory is known a quantum chromodynamics, by which quarks are described, is that quark interaction become weaker as they come closer together and fall to zero when the distance between them is zero. The converse of this proposition is that the attractive forces between quarks become stronger s they move, as this process has no limited, quarks can never separate from each other. In some theories, it is postulated that at very high-energy temperatures, as might have prevailed in the early universe, quarks can separate, te temperature at which this occurs is called the 'deconfinement temperatures'. Nevertheless, their existence has been demonstrated in high-energy scattering experiments and by symmetries in the properties of observed hadrons. They are regarded s elementary fermions, with spin ½, baryon number, strangeness 0 or = 1, and charm 0 or + 1. They are classified I six flavours[up (u), charm (c) and top (t), each with charge the proton charge, down (d), strange (s) and bottom (b), each with the proton charge J. each type has an antiquark with reversed signs of charge, baryon number, strangeness, charm. The top quark has not been observed experimentally, but there are strong theoretical arguments for its existence.

The fractional charges of quarks are never observed in hadrons, since the quarks form combinations in which the sum of their charges is zero or integral. Hadrons can be either baryons or mesons, essentially, baryons are composed of three quarks while mesons are composed of a quark-antiquark pair. These components are bound together within the hadron by the exchange of particles known as Gluons. Gluons are neutral massless gauge bosons, the quantum field theory of electromagnetic interactions discriminate themselves against the gluon as the analogue of the photon and with a quantum number known as 'colour' replacing that of electric charge. Each quark type (or flavour) comes in three colours (red, blue and green, say), where colour is simply a convenient label and has no connection with ordinary colour. Unlike the photon in quantum chromodynamics, which is electrically neutral, Gluons in quantum chromodynamics carry colour and can therefore interact with themselves. Particles that carry colour are believed not to be able to exist in free particles. Instead, quarks and Gluons are permanently confined inside hadrons (strongly interacting particles, such as the proton and the neutron).

The gluon self-interaction leads to the property known as 'asymptotic freedom', in which the interaction strength for th strong interaction decreases as the momentum trandfer involved in an interaction increase. This allows perturbation theory to be used and quantitative comparisons to be made with experiment, similar to, but less precise than those possibilities of quantum chromodynamics. Quantum chromodynamics the being tested successfully in high energy muon-nucleon scattering experiments and in proton-antiproton and electron-positron collisions at high energies. Strong evidence for the existence of colour comes from measurements of the interaction rates for e+e hadrons and e ı e ı̀ ı̀. The relative rate for these two processes is a factor of three larger than would be expected without colour, this factor measures directly the number of colours, i.e., for each quark flavour.

The quarks and antiquarks with zero strangeness and zero charm are the u, d, and ?. They form the combinations:

proton (uud), antiproton (?)

435

neutron (uud), antineutron (??)
pion: ð+ (u?), ð (d), ð0 (d?, u).

The charge and spin of these particles are the sums of the charge and spin of the component quarks and/or antiquarks.

In the strange baryon, e.g., the Ë and meons, either the quark or antiquark is strange. Similarly, the presence of one or more 'c' quarks leads to charmed baryons' 'a' 'c' or è to the charmed mesons. It has been found useful to introduce a further subdivision of quarks, each flavour coming in three colours (red, green, blue). Colour as used here serves simply as a convenient label and is unconnected with ordinary colour. A baryon comprises a red, a green, and a blue quark and a meson comprised a red and ant-red, a blue and ant-blue, or a green and antigreen quark and antiquark. In analogy with combinations of the three primary colours of light, hadrons carry no net colour, i.e., they are 'colourless' or 'white'. Only colourless objects can exist as free particles. The characteristics of the six quark flavours are shown in the table.

The cental feature of quantum field theory, is that the essential reality is a set of fields subject to the rules of special relativity and quantum mechanics, all else is derived as a consequence of the quantum dynamics of those fields. The quantization of fields is essentially an exercise in which we use complex mathematical models to analyse the field in terms of its associated quanta. And material reality as we know it in quantum field theory is constituted by the transformation and organization of fields and their associated quanta. Hence, this reality

Reveals a fundamental complementarity, in which particles are localized in space/time, and fields, which are not. In modern quantum field theory, all matter is composed of six strongly interacting quarks and six weakly interacting leptons. The six quarks are called up, down, charmed, strange, top, and bottom and have different rest masses and functional changes. The up and own quarks combine through the exchange of Gluons to form protons and neutrons.

The 'lepton' belongs to the class of elementary particles, and does not take part in strong interactions. They have no substructure of quarks and are considered indivisible. They are all; fermions, and are categorized into six distinct types, the electron, muon, and tauon, which are all identically charged, but differ in mass, and the three neutrinos, which are all neutral and thought to be massless or nearly so. In their interactions the leptons appear to observe boundaries that define three families, each composed of a charged lepton and its neutrino. The families are distinguished mathematically by three quantum numbers, Ie, Iì, and Iv lepton numbers called 'lepton numbers. In weak interactions their IeTOT, IìTOT and Iô for the individual particles are conserved.

In quantum field theory, potential vibrations at each point in the four fields are capable of manifesting themselves in their complementarity, their expression as individual particles. And the interactions of the fields result from the exchange of quanta that are carriers of the fields. The carriers of the field, known as messenger quanta, are the 'coloured' Gluons for the strong-binding-force, of which the photon for electromagnetism, the intermediate boson for the weak force, and the graviton or gravitation. If we could re-create the energies present in the fist trillionths of trillionths of a second in the life o the universe, these four fields would, according to quantum field theory, become one fundamental field.

The movement toward a unified theory has evolved progressively from super-symmetry to super-gravity to string theory. In string theory the one-dimensional trajectories of particles, illustrated in the Feynman lectures, seem as if, in at all were possible, are replaced by the two-dimensional orbits of a string. In addition to introducing the extra dimension, represented by a smaller diameter of the string, string theory also features another mall but non-zero constant, with which is analogous to Planck's quantum of action. Since the value of the constant is quite small, it can be generally ignored except at extremely small dimensions. But since the constant, like Planck's constant is not zero, this results in departures from ordinary quantum field theory in very small dimensions.

Part of what makes string theory attractive is that it eliminates, or 'transforms away', the inherent infinities found in the quantum theory of gravity. And if the predictions of this theory are proven valid in repeatable experiments under controlled coeditions, it could allow gravity to be unified with the other three fundamental interactions. But even if string theory leads to this grand unification, it will not alter our understanding of ave-particle duality. While the success of the theory would reinforce our view of the universe as a unified dynamic process, it applies to very small dimensions, and therefore, does not alter our view of wave-particle duality.

While the formalism of quantum physics predicts that correlations between particles over space-like inseparability, of which are possible, it can say nothing about what this strange new relationship between parts (quanta) and the whole (cosmos) cause to result outside this formalism. This does not, however, prevent us from considering the implications in philosophical terms. As the philosopher of science Errol Harris noted in thinking about the special character of wholeness in modern physics, a unity without internal content is a blank or empty set and is not recognizable as a whole. A collection of merely externally related parts does not constitute a whole in that the parts will not be 'mutually adaptive and complementary to one-another.'

Wholeness requires a complementary relationship between unity and difference and is governed by a principle of organization determining the interrelationship between parts. This organizing principle must be universal to a genuine whole and implicit in all parts constituting the whole, even the whole is exemplified only in its parts. This principle of order, Harris continued, 'is nothing really in and of itself. It is the way he parts are organized, and another constituent additional to those that constitute the totality.'

In a genuine whole, the relationship between the constituent parts must be 'internal or immanent' ion the parts, as opposed to a more spurious whole in which parts appear to disclose wholeness dur to relationships that are external to the arts. The collection of parts that would allegedly constitute the whole in classical physics is an example of a spurious whole. Parts continue a genuine whole when the universal principle of order is inside the parts and hereby adjusts each to all so that they interlock and become mutually complementary. This not only describes the character of the whole revealed in both relativity theory and quantum mechanics. It is also consistent with the manner in which we have begun to understand the relations between parts and whole in modern biology.

But more importantly, is that it seems clear that certainty is a property that can be assembled to either a person or a belief. We can say that a person, 'S' is certain, or we can say that its defining alinement is aligned as of 'p', of which is certain. The two uses can be connected by saying that 'S' has the right to be certain just in case the value of 'p' is sufficiently verified.

In defining certainty, it is crucial to note that the term has both an absolute and relative sense. More or less, we take a proposition to be certain when we have no doubt about its truth.. We may do this in error or unreasonably, but objectively a proposition is certain when such absence of doubt is justifiable. The sceptical tradition in philosophy denies that objective certainty is often possible, or ever possible, either for any proposition at all, or for any proposition from some suspect family (ethics, theory, memory, empirical judgement and so forth) a major sceptical weapon is the possibility of upsetting events that Can cast doubt back onto what were hitherto taken to be certainties. Others include reminders of the divergence of human opinion, and the fallible source of our confidence. Fundamentalist approaches to knowledge look for a basis of certainty, upon which the structure of our system is built. Others reject the metaphor, looking for mutual support and coherence, without foundation.

However, in moral theory, the view that there are inviolable moral standards or absolute variable human desires or policies or prescriptions.

In spite of the notorious difficulty of reading Kantian ethics, a hypothetical imperative embeds a command which is in place only given some antecedent desire or project: 'If you want to look wise, stay quiet'. The injunction to stay quiet only applies to those with the antecedent desire or inclination. If one has no desire to look wise the injunction cannot be so avoided: It is a requirement that binds anybody, regardless of their inclination. It could be represented as, for example, 'tell the truth (regardless of whether you want to or not)'. The distinction is not always signalled by presence or absence of the conditional or hypothetical form: 'If you crave drink, don't become a bartender' may be regarded as an absolute injunction applying to anyone, although only activated in case of those with the stated desire.

In Grundlegung zur Metaphsik der Sitten (1785), Kant discussed five forms of the categorical imperative: (1) the formula of universal law: 'act only on that maxim through which you can at the same times will that it should become universal law: (2) the formula of the law of nature: 'act as if the maxim of your action were to become through your will a universal law of nature': (3) the formula of the end-in-itself: 'act in such a way that you always treat humanity, whether in your own person or in the person of any other, never simply as a, but always at the same time as an end': (4) the formula of autonomy, or considering 'the will of every rational being as a will which makes universal law': (5) the formula of the Kingdom of Ends, which provides a model for the systematic union of different rational beings under common laws.

Even so, a proposition that is not a conditional 'p'. Moreover, the affirmative and negative, modern opinion is wary of this distinction, since what appears categorical may vary notation. Apparently, categorical propositions may also turn out to be disguised conditionals: 'X' is intelligent (categorical?) = if 'X' is given a range of tasks she performs them better than many people (conditional?) The problem. Nonetheless, is not merely one of classification, since deep metaphysical questions arise when facts that seem to be categorical and therefore solid, come to seem by contrast conditional, or purely hypothetical or potential.

A limited area of knowledge or endeavour to which pursuits, activities and interests are a central representation held to a concept of physical theory. In this way, a field is defined by the distribution of a physical quantity, such as temperature, mass density, or potential energy y, at different points in space. In the particularly important example of force fields, such ad gravitational, electrical, and magnetic fields, the field value at a point is the force which a test particle would experience if it were located at that point. The philosophical problem is whether a force field is to be thought of as purely potential, so the presence of a field merely describes the propensity of masses to move relative to each other, or whether it should be thought of in terms of the physically real modifications of a medium, whose properties result in such powers that is, are force fields purely potential, fully characterized by dispositional statements or conditionals, or are they categorical or actual? The former option seems to require within ungrounded dispositions, or regions of space hat differ only in what happens if an object is placed there. The law-like shape of these dispositions, apparent for example in the curved lines of force of the magnetic field, may then seem quite inexplicable. To atomists, such as Newton it would represent a return to Aristotelian entelechies, or quasi-psychological affinities between things, which are responsible for their motions. The latter option requires understanding of how forces of attraction and repulsion can be 'grounded' in the properties of the medium.

The basic idea of a field is arguably present in Leibniz, who was certainly hostile to Newtonian atomism, although his equal hostility to 'action at a distance' muddies the water. It is usually credited to the Jesuit mathematician and scientist Joseph Boscovich (1711-87) and Immanuel Kant (1724-1804), both of whom influenced the scientist Faraday, with whose work the physical notion became established. In his paper 'On the Physical Character of the Lines of Magnetic Force' (1852). Faraday was to suggest several criteria for assessing the physical reality of lines of force, such as whether they are affected by an intervening material medium, whether the motion depends on the nature of what is placed at the receiving end. As far as electromagnetic fields go, Faraday himself inclined to the view that the mathematical similarity between heat flow, currents, and electromagnetic lines of force was evidence for the physical reality of the intervening medium.

Once, again, our mentioning recognition for which its case value, whereby its view is especially associated the American psychologist and philosopher William James (1842-1910), that the truth of a statement can be defined in terms of a 'utility' of accepting it. Communicated, so much as a dispiriting position for which its place of valuation may be viewed as an objection. Since there are things that are false, as it may be useful to accept, and conversely there are things that are true and that it may be damaging to accept. Nevertheless, there are deep connections between the idea that a representation system is accorded, and the likely success of the projects in progressive formality, by its possession. The evolution of a system of representation either perceptual or linguistic, seems bounded to connect successes with everything adapting or with utility in the modest sense. The Wittgenstein doctrine stipulates the meaning of use that upon the nature of belief and its relations with human attitude, emotion and the idea that belief in the truth on one hand, the action of the other. One way of binding with cement, wherefore the connection is found in the idea that natural selection becomes much as much in adapting us to the cognitive creatures, because beliefs have effects, they work. Pragmatism can be found in Kant's doctrine, and continued to play an influencing role in the theory of meaning and truth.

James, (1842-1910), although with characteristic generosity exaggerated in his debt to Charles S. Peirce (1839-1914), he charted that the method of doubt encouraged people to pretend to doubt what they did not doubt in their hearts, and criticize its individualists insistence, that the ultimate test of certainty is to be found in the individuals personalized consciousness.

From his earliest writings, James understood cognitive processes in teleological terms. Though, he held, assisted us in the satisfactory interests. His will to Believe doctrine, the view that we are sometimes justified in believing beyond the evidential relics upon the notion that a belief's benefits are relevant to its justification. His pragmatic method of analysing philosophical problems, for which requires that we find the meaning of terms by examining their application to objects in experimental situations, similarly reflects the teleological approach in its attention to consequences.

Realism can itself be subdivided: Kant, for example, combines empirical realism (within the phenomenal world the realist says the right things - surrounding objects really exist and independent of us and our mental stares) with transcendental idealism (the phenomenal world asa whole reflects the structures imposed on it by the activity of our minds as they render it intelligible to us). In modern philosophy the orthodox oppositions to realism has been from philosopher such as Goodman, who, impressed by the extent to which we perceive the world through conceptual and linguistic lenses of our own making.

Assigned to the modern treatment of existence in the theory of 'quantification' is sometimes put by saying that existence is not a predicate. The idea is that the existential quantifies itself ads an operator on a predicate, indicating that the property it expresses has instances. Existence is therefore treated as a second-order property, or a property of properties. It is fitting to say, that in this it is like number, for when we say that these things of a kind, we do not describe the thing (ad we would if we said there are red things of the kind), but instead attribute a property to the kind itself. The parallelled numbers are exploited by the German mathematician and philosopher of mathematics Gottlob Frége in the dictum that affirmation of existence is merely denied of the number nought. A problem, nevertheless, proves accountable for its crated by sentences like 'This exists', where some particular thing is undirected, such that a sentence seems to express a contingent truth (for this insight has not existed), yet no other predicate is involved. 'This exists' is, therefore unlike 'Tamed tigers exist', where a property is said to have an instance, for the word 'this' and does not locate a property, but only an individual.

Possible worlds seem able to differ from each other purely in the presence or absence of individuals, and not merely in the distribution of exemplification of properties.

The philosophical ponderance over which to set upon the unreal, as belonging to the domain of Being. Nonetheless, there is little for us that can be said with the philosopher's study. So it is not apparent that there can be such a subject as Being by itself. Nevertheless, the concept had a central place in philosophy from Parmenides to Heidegger. The essential question of 'why is there something and not of nothing'? Prompting over logical reflection on what it is for a universal to have an instance, nd as long history of attempts to explain contingent existence, by which id to reference and a necessary ground.

In the traditions, ever since Plato, this ground becomes a self-sufficient, perfect, unchanging, and external something, identified with the Good or God, but whose relation with the everyday world remains obscure. The celebrated argument for the existence of God first propounded by Andelm in his Proslogin. The argument by defining God as 'something than which nothing greater can be conceived'. God then exists in the understanding since we understand this concept. However, if He only existed in the understanding something greater could be conceived, for a being that exists in reality is greater than one that exists in the understanding. Bu then, we can conceive of something greater than that than which nothing greater can be conceived, which is contradictory. Therefore, God cannot exist on the understanding, but exists in reality.

An influential argument (or family of arguments) for the existence of God, finding its premises are that all natural things are dependent for their existence on something else. The totality of dependent brings must then itself depend upon a non-dependent, or necessarily existent bring of which is God. Like the argument to design, the cosmological argument was attacked by the Scottish philosopher and historian David Hume (1711-76) and Immanuel Kant.

Its main problem, nonetheless, is that it requires us to make sense of the notion of necessary existence. For if the answer to the question of why anything exists is that some other tings of a similar kind exists, the question merely arises gain. So the 'God' that ends the question must exist necessarily: It must not be an entity of which the same kinds of questions can be raised. The other problem with the argument is attributing concern and care to the deity, not for connecting the necessarily existent being it derives with human values and aspirations.

The ontological argument has been treated by modern theologians such as Barth, following Hegel, not so much as a proof with which to confront the unconverted, but as an explanation of the deep meaning of religious belief. Collingwood, regards the argument s proving not that because our idea of God is that of id quo maius cogitare viequit, therefore God exists, but proving that because this is our idea of God, we stand committed to belief in its existence. Its existence is a metaphysical point or absolute presupposition of certain forms of thought.

In the 20th century, modal versions of the ontological argument have been propounded by the American philosophers Charles Hertshorne, Norman Malcolm, and Alvin Plantinga. One version is to define something as unsurpassably great, if it exists and is perfect in every 'possible world'. Then, to allow that it is at least possible that an unsurpassable great being exists. This that there is a possible world in which such a being exists. However, if it exists in one world, it exists in all (for the fact that such a being exists in a world that entails, in at least, it exists and is perfect in every world), so, it exists necessarily. The correct response to this argument is to disallow the apparently reasonable concession that it is possible that such a being exists. This concession is much more dangerous than it looks, since in the modal logic, involved from possibly necessarily 'p', we can device necessarily 'p'. A symmetrical proof starting from the assumption that it is possible that such a being not exist would derive that it is impossible that it exists.

The doctrine that it makes an ethical difference of whether an agent actively intervenes to bring about a result, or omits to act in circumstances in which it is foreseen, that as a result of the omission the same result occurs. Thus, suppose that I wish you dead. If I act to bring about your death, I am a murderer, however, if I happily discover you in danger of death, and fail to act to save you, I am not acting, and therefore, according to the doctrine of acts and omissions not a murderer. Critics implore that omissions can be as deliberate and immoral as I am responsible for your food and fact to feed you. Only omission is surely a killing, 'Doing nothing' can be a way of doing something, or in other worlds, absence of bodily movement can also constitute acting negligently, or deliberately, and defending on the context, may be a way of deceiving, betraying, or killing. Nonetheless, criminal law offers to find its conveniences, from which to distinguish discontinuous intervention, for which is permissible, from bringing about result, which may not be, if, for instance, the result is death of a patient. The question is whether the difference, if there is one, is, between acting and omitting to act be discernibly or defined in a way that bars a general moral might.

The double effect of a principle attempting to define when an action that had both good and bad results is morally permissible. I one formation such an action is permissible if (1) The action is not wrong in itself, (2) the bad consequences is not that which is intended (3) the good is not itself a result of the bad consequences, and (4) the two consequential affects are commensurate. Thus, for instance, I might justifiably bomb an enemy factory, foreseeing but intending that the death of nearby civilians, whereas bombing the death of nearby civilian intentionally would be disallowed. The principle has its roots in Thomist moral philosophy, accordingly. St. Thomas Aquinas (1225-74), held that it is meaningless to ask whether a human being is two tings (soul and body) or, only just as it is meaningless to ask whether the wax and the shape given to it by the stamp are one: On this analogy the sound is ye form of the body. Life after death is possible only because a form itself doe not perish (pricking is a loss of form).

And is, therefore, in some sense available to reactivate a new body., therefore, not I who survives body death, but I ma y be resurrected in the same personalized bod y that becomes reanimated by the same form, that which Aquinas's account, as a person has no privileged self-understanding, we understand ourselves as we do everything else, by way of sense experience and abstraction, and knowing the principle of our own lives is an achievement, not as a given, but is widely accepted that trying to make the connection between thought and experience through basic sentence's depends on an untenable 'myth of the given

The special way that we each have of knowing our own thoughts, intentions, and sensationalist have brought in the many philosophical 'behaviorist and functionalist tendencies, that have found it important to deny that there is such a special way, arguing the way that I know of my own mind inasmuch as the way that I know of yours, e.g., by seeing what I say when asked. Others, however, point out that the behaviour of reporting the result of introspection in a particular and legitimate kind of behavioural access that deserves notice in any account of historically human psychology. The historical philosophy of reflection upon the astute of history, or of historical, thinking, finds the term was used in the 18th century, e.g., by Volante was to mean critical historical thinking as opposed to the mere collection and repetition of stories about the past. In Hegelian, particularly by conflicting elements within his own system, however, it came to man universal or world history. The Enlightenment confidence was being replaced by science, reason, and understanding that gave history a progressive moral thread, and under the influence of the German philosopher, whom is in spreading Romanticism, came Gottfried Herder (1744-1803),and, Immanuel Kant, this idea took it further to hold, so that philosophy of history cannot be the detecting of a grand system. This essential speculative philosophy of history is given a extra Kantian twist in the German idealist Johann Fichte, in whom the extra association of temporal succession with logical implication introduces the idea that concepts themselves are the dynamic engine of historical change. The idea is readily intelligible in that there world of nature and of thought become identified. The work of Herder, Kant, Flichte and Schelling is synthesized by Hegel: History has a plot, as too, this to the moral development of man, equates with freedom within the state, this in turn is the development of thought, or a logical development in which various necessary moment in the life of the concept are successively achieved and improved upon. Hegel's method is at its most successful, when the object is the history of ideas, and the evolution of thinking may march in steps with logical oppositions and their resolution encounters red by various systems of thought.

Within the revolutionary communism, Karl Marx (1818-83) and the German social philosopher Friedrich Engels (1820-95), there emerges a rather different kind of story, based upon Hefl's progressive structure not laying the achievement of the goal of history to a future in which the political condition for freedom comes to exist, so that economic and political fears than 'reason' is in the engine room. Although, itself is such that speculations upon the history may that it be continued to be written, notably: late examples, by the late 19th century large-scale speculation of tis kind with the nature of historical understanding, and in particular with a comparison between the,methos of natural science and with the historian. For writers such as the German neo-Kantian Wilhelm Windelband and the German philosopher and literary critic and historian Wilhelm Dilthey, it is important to show that the human sciences such. as history are objective and legitimate, nonetheless they are in some way deferent from the enquiry of the scientist. Since the subjective-matter is the past thought and actions of human brings, what is needed and actions of human beings, past thought and actions of human beings, what is needed is an ability to re-live that past thought, knowing the deliberations of past agents, as if they were the historian's own.. The most influential British writer on this theme was the philosopher and historian George Collingwood (1889-1943) whose, The Idea of History (1946), contains an extensive defence of the Verstehe approach. But it is nonetheless, the explanation from their actions, is that by re-living the situation as our understanding that understanding others is not gained by the tactic use of a 'theory'. This enabling us to infer what thoughts or intentionality experienced, again, the matter to which the subjective-matters of past thoughts and actions, as I have a human ability of knowing the deliberations of past agents as if they were the historian's own. The immediate question of the form of historical explanation, and the fact that general laws have other than no place or any apprentices in the order of a minor place in the human sciences, it is also prominent in thoughts about distinctiveness as to regain their actions, but by re-living the situation in or an understanding of what they experience and thought.

The view that everyday attributions of intention, belief and meaning to other persons proceeded via tacit use of a theory that enables ne to construct these interpretations as explanations of their doings. The view is commonly hld along with functionalism, according to which psychological states theoretical entities, identified by the network of their causes and effects. The theory-theory had different implications, depending on which feature of theories is being stressed. Theories may be though of as capable of formalization, as yielding predications and explanations, as achieved by a process of theorizing, as achieved by predictions and explanations, as achieved by a process of theorizing, as answering to empirical evince that is in principle describable without them, as liable to be overturned by newer and better theories, and o on. The main problem with seeing our understanding of others as the outcome of a piece of theorizing is the nonexistence of a medium in which this theory can be couched, as the child learns simultaneously he minds of others and the meaning of terms in its native language.

Our understanding of others is not gained by the tacit use of a 'theory'. Enabling us to infer what thoughts or intentions explain their actions, however, by re-living the situation 'in their moccasins', or from their point of view, and thereby understanding what hey experienced and thought, and therefore expressed. Understanding others is achieved when we can ourselves deliberate as they did, and hear their words as if they are our own. The suggestion is a modern development of the 'Verstehen' tradition associated with Dilthey, Weber and Collngwood.

Much is as much, and therefore, in some sense available to reactivate a new body, however, not that I, who survives bodily death, but I may be resurrected in the same body that becomes reanimated by the same form, in that of Aquinas's account, a person has no privileged self-understanding. We understand ourselves, just as we do everything else, that through the sense experience, in that of an abstraction, may justly be of knowing the principle of our own lives, is to obtainably achieve, and not as a given. In the theory of knowledge that knowing Aquinas holds the Aristotelian doctrine that knowing entails some similarities between the Knower and what there is to be known: A human's corporal nature, therefore, requires that knowledge start with sense perception. As yet, the same limitations that do not apply of bringing further he levelling stabilities that are contained within the hierarchical mosaic, such as the celestial heavens that open in bringing forth to angles.

In the domain of theology Aquinas deploys the distraction emphasized by Eringena, between the existence of God in understanding the significance; of five arguments: The are (1) Motion is only explicable if there exists an unmoved, a first mover (2) the chain of efficient causes demands a first cause (3) the contingent character of existing things in the wold demands a different order of existence, or in other words as something that has a necessary existence (4) the gradation of value in things in the world require the existence of something that is most valuable, or perfect, and (5) the orderly character of events points to a final cause, or end t which all things are directed, and the existence of this end demands a being that ordained it. All the arguments are physico-theological arguments, in that between reason and faith, Aquinas lays out proofs of the existence of God.

He readily recognizes that there are doctrines such that are the Incarnation and the nature of the Trinity, know only through revelations, and whose acceptance is more a matter of moral will. God's essence is identified with his existence, as pure activity. God is simple, containing no potential. No matter how, we cannot obtain knowledge of what God is (his quiddity), perhaps, doing the same work as the principle of charity, but suggesting that we regulate our procedures of interpretation by maximizing the extent to which we see the subject s humanly reasonable, than the extent to which we see the subject as right about things. Whereby remaining content with descriptions that apply to him partly by way of analog y, God reveals of himself is not himself.

The immediate problem availed of ethics is posed b y the English philosopher Phillippa Foot, in her 'The Problem of Abortion and the Doctrine of the Double Effect' (1967). A runaway train or trolley comes to a section in the track that is under construction and impassable. One person is working on one part and five on the other, and the trolley will put an end to anyone working on the branch it enters. Clearly, to most minds, the driver should steer for the fewest populated branch. But now suppose that, left to itself, it will enter the branch with its five employs that are there, and you as a bystander can intervene, altering the points so that it veers through the other. Is it right or obligors, or even permissible for you to do this, thereby, apparently involving yourself in ways that responsibility ends in a death of one person? After all, whom have you wronged if you leave it to go its own way? The situation is similarly standardized of others in which utilitarian reasoning seems to lead to one course of action, but a person's integrity or principles may oppose it.

Describing events that haphazardly happen does not of itself permit us to talk of rationality and intention, which are the categories we may apply if we conceive of them as action. We think of ourselves not only passively, as creatures that make things happen. Understanding this distinction gives forth of its many major problems concerning the nature of an agency for the causation of bodily events by mental events, and of understanding the 'will' and 'free will'. Other problems in the theory of action include drawing the distinction between an action and its consequence, and describing the structure involved when we do one thing 'by;' dong another thing. Even the planning and dating where someone shoots someone on one day and in one place, whereby the victim then dies on another day and in another place. Where and when did the murderous act take place?

Causation, least of mention, is not clear that only events are created by and for itself. Kant cites the example o a cannonball at rest and stationed upon a cushion, but causing the cushion to be the shape that it is, and thus to suggest that the causal states of affairs or objects or facts may also be casually related. All of which, the central problem is to understand the elements of necessitation or determinacy of the future. Events, Hume thought, are in themselves 'loose and separate': How then are we to conceive of others? The relationship seems not to perceptible, for all that perception gives us (Hume argues) is knowledge of the patterns that events do, actually falling into than any acquaintance with the connections determining the pattern. It is, however, clear that our conception of everyday objects ids largely determined by their casual powers, and all our action is based on the belief that these causal powers are stable and reliable. Although scientific investigation can give us wider and deeper dependable patterns, it seems incapable of bringing us any nearer to the 'must' of causal necessitation. Particular examples o f puzzles with causalities are quite apart from general problems of forming any conception of what it is: How are we to understand the casual interaction between mind and body? How can the present, which exists, or its existence to a past that no longer exists? How is the stability of the casual order to be understood? Is backward causality possible? Is causation a concept needed in science, or dispensable?

The news concerning free-will, is nonetheless, a problem for which is to reconcile our everyday consciousness of ourselves as agent, with the best view of what science tells us that we are. Determinism is one part of the problem. It may be defined as the doctrine that every event has a cause. More precisely, for any event 'C', there will be one antecedent states of nature 'N', and a law of nature 'L', such that given L, N will be followed by 'C'. But if this is true of every event, it is true of events such as my doing something or choosing to do something. So my choosing or doing something is fixed by some antecedent state 'N' an d the laws. Since determinism is universal these in turn are fixed, and so backwards to events for which I am clearly not responsible (events before my birth, for example). So, no events can be voluntary or free, where that that they come about purely because of my willing them I could have done otherwise. If determinism is true, then there will be antecedent states and laws already determining such events: How then can I truly be said to be their author, or be responsible for them?

Reactions to this problem are commonly classified as: (1) Hard determinism. This accepts the conflict and denies that you have real freedom or responsibility (2) Soft determinism or compatibility, whereby reactions in this family assert that everything you should ant from a notion of freedom is quite compatible with determinism. In particular, if your actions are caused, it can often be true of you that you could have done otherwise if you had chosen, and this may be enough to render you liable to be held unacceptable (the fact that previous events will have caused you to choose as you did is deemed irrelevant on this option). (3) Libertarianism, as this is the view that while compatibilism is only an evasion, there is a more substantiative, real notion of freedom that can yet be preserved in the face of determinism (or, of indeterminism). In Kant, while the empirical or phenomenal self is determined and not free, whereas the noumenal or rational self is capable of being rational, free action. However, the noumeal self exists outside the categorical priorities of space and time, as this freedom seems to be of a doubtful value as other libertarian avenues do include of suggesting that the problem is badly framed, for instance, because the definition of determinism breaks down, or postulates by its suggesting that there are two independent but consistent ways of looking at an agent, the scientific and the humanistic, wherefore it ids only through confusing them that the problem seems urgent. Nevertheless, these avenues have gained general popularity, as an error to confuse determinism and fatalism.

The dilemma for which determinism is for itself often supposes of an action that seems as the end of a causal chain, or, perhaps, by some hieratical set of suppositional actions that would stretch back in time to events for which an agent has no conceivable responsibility, then the agent is not responsible for the action.

Once, again, the dilemma adds that if an action is not the end of such a chain, then one of its causes occurs at random, in that no antecedent events brought it about, and in that case nobody is responsible for its ever to occur. So, whether or not determinism is true, responsibility is shown to be illusory.

Still, there is to say, to have a will is to be able to desire an outcome and to purpose to bring it about. Strength of will, or firmness of purpose, is supposed to be good and weakness of 'will'.

A mental act of willing or trying whose presence is sometimes supposed to make the difference between intentional and voluntary action, as well of mere behaviour. The theory that there are such acts is problematic, and the idea that they make the required difference is a case of explaining a phenomenon by citing another that raises exactly the same problem, since the intentional or voluntary nature of the set of volition now needs explanation. For determinism to act in accordance with the law of autonomy or freedom, is that in ascendance with universal moral law and regardless of selfish advantage.

A categorical notion in the work as contrasted in Kantian ethics show of a hypothetical imperative that embeds of a commentary which is in place only given some antecedent desire or project. 'If you want to look wise, stay quiet'. The injunction to stay quiet only applies to those with the antecedent desire or inclination: If one has no desire to look wise the injunction or advice lapses. A categorical imperative cannot be so avoided, it is a requirement that binds anybody, regardless of their inclination,. It could be repressed as, for example, 'Tell the truth (regardless of whether you want to or not)'. The distinction is not always mistakably presumed or absence of the conditional or hypothetical form: 'If you crave drink, don't become a bartender' may be regarded as an absolute injunction applying to anyone, although only activated in the case of those with the stated desire.

In Grundlegung zur Metaphsik der Sitten (1785), Kant discussed some of the given forms of categorical imperatives, such that of (1) The formula of universal law: 'act only on that maxim through which you can at the same time will that it should become universal law', (2) the formula of the law of nature: 'Act as if the maxim of your action were to become through your will a universal law of nature', (3) the formula of the end-in-itself, 'Act in such a way that you always trat humanity of whether in your own person or in the person of any other, never simply as an end, but always at the same time as an end', (4) the formula of autonomy, or consideration; 'the will' of every rational being a will which makes universal law', and (5) the formula of the Kingdom of Ends, which provides a model for systematic union of different rational beings under common laws.

A central object in the study of Kant's ethics is to understand the expressions of the inescapable, binding requirements of their categorical importance, and to understand whether they are equivalent at some deep level. Kant's own application of the notions are always convincing: One cause of confusion is relating Kant's ethical values to theories such as ;expressionism' in that it is easy but imperatively must that it cannot be the expression of a sentiment, yet, it must derive from something 'unconditional' or necessary' such as the voice of reason. The standard mood of sentences used to issue request and commands are their imperative needs to issue as basic the need to communicate information, and as such to animals signalling systems may as often be interpreted either way, and understanding the relationship between commands and other action-guiding uses of language, such as ethical discourse. The ethical theory of 'prescriptivism' in fact equates the two functions. A further question is whether there is an imperative logic. 'Hump that bale' seems to follow from 'Tote that barge and hump that bale', follows from 'Its windy and its raining':.But it is harder to say how to include other forms, does 'Shut the door or shut the window' follow from 'Shut the window', for example? The usual way to develop an imperative logic is to work in terms of the possibility of satisfying the other one command without satisfying the other, thereby turning it into a variation of ordinary deductive logic.

Despite the fact that the morality of people and their ethics amount to the same thing, there is a usage that I restart morality to systems such as that of Kant, based on notions given as duty, obligation, and principles of conduct, reserving ethics for the more Aristotelian approach to practical reasoning as based on the valuing notions that are characterized by their particular virtue, and generally avoiding the separation of 'moral' considerations from other practical considerations. The scholarly issues are complicated and complex, with some writers seeing Kant as more Aristotelian,. And Aristotle as more involved with a separate sphere of responsibility and duty, than the simple contrast suggests.

A major topic of philosophical inquiry, especially in Aristotle, and subsequently since the 17th and 18th centuries, when the 'science of man' began to probe into human motivation and emotion. For such as these, the French moralist, or Hutcheson, Hume, Smith and Kant, a prime task as to delineate the variety of human reactions and motivations. Such an inquiry would locate our propensity for moral thinking among other faculties, such as perception and reason, and other tendencies as empathy, sympathy or self-interest. The task continues especially in the light of a post-Darwinian understanding of ourselves.

In some moral systems, notably that of Immanuel Kant, real moral worth comes only with interactivity, justly because it is right. However, if you do what is purposely becoming, equitable, but from some other equitable motive, such as the fear or prudence, no moral merit accrues to you. Yet, that in turn seems to discount other admirable motivations, as acting from main-sheet benevolence, or 'sympathy'. The question is how to balance these opposing ideas and how to understand acting from a sense of obligation without duty or rightness, through which their beginning to seem a kind of fetish. It thus stands opposed to ethics and relying on highly general and abstractive principles, particularly those associated with the Kantian categorical imperatives. The view may go as far back as to say that taken in its own, no consideration point, for that which of any particular way of life, that, least of mention, the contributing steps so taken as forwarded by reason or be to an understanding estimate that can only proceed by identifying salient features of a situation that weigh on one's side or another.

As random moral dilemmas set out with intense concern, inasmuch as philosophical matters that exert a profound but influential defence of common sense. Situations in which each possible course of action breeches some otherwise binding moral principle, are, nonetheless, serious dilemmas making the stuff of many tragedies. The conflict can be described in different was. One suggestion is that whichever action the subject undertakes, that he or she does something wrong. Another is that his is not so, for the dilemma that in the circumstances for what she or he did was right as any alternate. It is important to the phenomenology of these cases that action leaves a residue of guilt and remorse, even though it had proved it was not the subject's fault that she or he were considering the dilemma, that the rationality of emotions can be contested. Any normality with more than one fundamental principle seems capable of generating dilemmas, however, dilemmas exist, such as where a mother must decide which of two children to sacrifice, least of mention, no principles are pitted against each other, only if we accept that dilemmas from principles are real and important, this fact can then be used to approach in themselves, such as of 'utilitarianism', to espouse various kinds may, perhaps, be centred upon the possibility of relating to independent feelings, liken to recognize only one sovereign principle. Alternatively, of regretting the existence of dilemmas and the unordered jumble of furthering principles, in that of creating several of them, a theorist may use their occurrences

448

to encounter upon that which it is to argue for the desirability of locating and promoting a single sovereign principle.

Nevertheless, some theories into ethics see the subject in terms of a number of laws (as in the Ten Commandments). The status of these laws may be that they are the edicts of a divine lawmaker, or that they are truths of reason, given to its situational ethics, virtue ethics, regarding them as at best rules-of-thumb, and, frequently disguising the great complexity of practical representations that for reason has placed the Kantian notions of their moral law.

In continence, the natural law possibility points of the view of the states that law and morality are especially associated with St Thomas Aquinas (1225-74), such that his synthesis of Aristotelian philosophy and Christian doctrine was eventually to provide the main philosophical underpinning of the Catholic church. Nevertheless, to a greater extent of any attempt to cement the moral and legal order and together within the nature of the cosmos or the nature of human beings, in which sense it found in some Protestant writings, under which had arguably derived functions. From a Platonic view of ethics and its agedly implicit advance of Stoicism. Its law stands above and apart from the activities of human lawmakers: It constitutes an objective set of principles that can be seen as in and for themselves by of 'natural usages' or by reason itself, additionally, (in religious verses of them), that express of God's will for creation. Nonreligious versions of the theory substitute objective conditions for humans flourishing as the source of constraints, upon permissible actions and social arrangements within the natural law tradition. Different views have been held about the relationship between the rule of the law and God's will. Grothius, for instance, sides with the view that the content of natural law is independent of any will, including that of God.

While the German natural theorist and historian Samuel von Pufendorf (1632-94) takes the opposite view. His great work was the De Jure Naturae et Gentium, 1672, and its English translation is 'Of the Law of Nature and Nations, 1710. Pufendorf was influenced by Descartes, Hobbes and the scientific revolution of the 17th century, his ambition was to introduce a newly scientific 'mathematical' treatment on ethics and law, free from the tainted Aristotelian underpinning of 'scholasticism'. Like that of his contemporary - Locke. His conception of natural laws include rational and religious principles, making it only a partial forerunner of more resolutely empiricist and political treatment in the Enlightenment.

Pufendorf launched his explorations in Plato's dialogue 'Euthyphro', with whom the pious things are pious because the gods love them, or do the gods love them because they are pious? The dilemma poses the question of whether value can be conceived as the upshot o the choice of any mind, even a divine one. On the fist option the choice of the gods crates goodness and value. Even if this is intelligible it seems to make it impossible to praise the gods, for it is then vacuously true that they choose the good. On the second option we have to understand a source of value lying behind or beyond the will even of the gods, and by which they can be evaluated. The elegant solution of Aquinas is and is therefore distinct form is will, but not distinct from him.

The dilemma arises whatever the source of authority is supposed to be. Do we care about the good because it is good, or do we just call good those things that we care about? It also generalizes to affect our understanding of the authority of other things: Mathematics, or necessary truth, for example, are truths necessary because we deem them to be so, or do we deem them to be so because they are necessary?

The natural aw tradition may either assume a stranger form, in which it is claimed that various facts entails of primary and secondary qualities, any of which is claimed that various facts entail values, reason by itself is capable of discerning moral requirements. As in the ethics of Knt, these requirements are supposed binding on all human beings, regardless of their desires.

The supposed natural or innate abilities of the mind to know the first principle of ethics and moral reasoning, wherein, those expressions are assigned and related to those that distinctions are which make in terms contribution to the function of the whole, as completed definitions of them, their phraseological impression is termed 'synderesis' (or, syntetesis) although traced to Aristotle, the phrase came to the modern era through St Jerome, whose scintilla conscientiae (gleam of conscience) wads a popular concept in early scholasticism. Nonetheless, it is mainly associated in Aquinas as an infallible natural, simple and immediate grasp of first moral principles. Conscience, by contrast, is,more concerned with particular instances of right and wrong, and can be in error, under which the assertion that is taken as fundamental, at least for the purposes of the branch of enquiry in hand.

It is, nevertheless, the view interpreted within he particular states of law and morality especially associated with Aquinas and the subsequent scholastic tradition, showing for itself the enthusiasm for reform for its own sake. Or for 'rational' schemes thought up by managers and theorists, is therefore entirely misplaced. Major o exponent s of this theme include the British absolute idealist Herbert Francis Bradley (1846-1924) and Austrian economist and philosopher Friedrich Hayek. The notably the idealism of Bradley, there ids the same doctrine that change is contradictory and consequently unreal: The Absolute is changeless. A way of sympathizing a little with his idea is to reflect that any scientific explanation of change will proceed by finding an unchanging law operating, or an unchanging quantity conserved in the change, so that explanation of change always proceeds by finding that which is unchanged. The metaphysical problem of change is to shake off the idea that each moment is created afresh, and to obtain a conception of events or processes as having a genuinely historical reality, Really extended and unfolding in time, as opposed to being composites of discrete temporal atoms. A step toward this end may be to see time itself not as an infinite container within which discrete events are located, bu as a kind of logical construction from the flux of events. This relational view of time was advocated by Leibniz and a subject of the debate between him and Newton's Absolutist pupil, Clarke.

Generally, nature is an indefinitely mutable term, changing as our scientific conception of the world changes, and often best seen as signifying a contrast with something considered not part of nature. The term applies both to individual species (it is the nature of gold to be dense or of dogs to be friendly), and also to the natural world as a whole. The sense in which it applies to species quickly links up with ethical and aesthetic ideals: A thing ought to realize its nature, what is natural is what it is good for a thing to become, it is natural for humans to be healthy or two-legged, and departure from this is a misfortune or deformity,. The associations of what is natural with what it is good to become is visible in Plato, and is the central idea of Aristotle's philosophy of nature. Unfortunately, the pinnacle of nature in this sense is the mature adult male citizen, with he rest of hat we would call the natural world, including women, slaves, children and other species, not quite making it.

Nature in general can, however, function as a foil to any idea inasmuch as a source of ideals: In this sense fallen nature is contrasted with a supposed celestial realization of the 'forms'. The theory of 'forms' is probably the most characteristic, and most contested of the doctrines of Plato. In the background ie the Pythagorean conception of form as the key to physical nature, but also the sceptical doctrine associated with the Greek philosopher Cratylus, and is sometimes thought to have been a teacher of Plato before Socrates. He is famous for fulfilling the doctrine of Ephesus of Heraclitus, whereby the guiding idea of his philosophy was that of the logos, is capable of being heard or hearkened to by people, it unifies opposites, and it is somehow associated with fire, which is preeminent among the four elements that Heraclitus distinguishes: Fire, air (breath, the stuff of which souls composed), earth, and water. Although he is principally remember for the doctrine of the 'flux' of all things, and the famous statement that you cannot step into the same river twice, for new waters are ever flowing in upon you. The more extreme implication of the doctrine of flux, e.g., the impossibility of categorizing things truly, do not seem consistent with his general epistemology and views of meaning, and were to his follower Cratylus, although the proper conclusion of his views was that the flux cannot be captured in words. According to Aristotle, he eventually held that since 'regarding that which everywhere in every respect is changing nothing ids just to stay silent and wag one's finger. Plato 's theory of forms can be seen in part as an action against the impasse to which Cratylus was driven.

The Galilean world view might have been expected to drain nature of its ethical content, however, the term seldom loses its normative force, and the belief in universal natural laws provided its own set of ideals. In the 18th century for example, a painter or writer could be praised as natural, where the qualities expected would include normal (universal) topics treated with simplicity, economy, regularity and harmony. Later on, nature becomes an equally potent emblem of irregularity, wildness, and fertile diversity, but also associated with progress of human history, its incurring definition that has been taken to fit many things as well as transformation, including ordinary human self-consciousness. Nature, being in contrast with in integrated phenomenon may include (1) that which is deformed or grotesque or fails to achieve its proper form or function or just the statistically uncommon or unfamiliar, (2) the supernatural, or the world of gods and invisible agencies, (3) the world of rationality and unintelligence, conceived as distinct from the biological and physical order, or the product of human intervention, and (5) related to that, the world of convention and artifice.

Different conceptions of nature continue to have ethical overtones, foe example, the conception of 'nature red in tooth and claw' often provides a justification for aggressive personal and political relations, or the idea that it is women's nature to be one thing or another is taken to be a justification for differential social expectations. The term functions as a fig-leaf for a particular set of stereotypes, and is a proper target of much feminist writings. Feminist epistemology has asked whether different ways of knowing for instance with different criteria of justification, and different emphases on logic and imagination, characterize male and female attempts to understand the world. Such concerns include awareness of the 'masculine' self-image, itself a socially variable and potentially distorting picture of what thought and action should be. Again, there is a spectrum of concerns from the highly theoretical to he relatively practical. In this latter area particular attention is given to the institutional biases that stand in the way of equal opportunities in science and other academic pursuits, or the ideologies that stand in the way of women seeing themselves as leading contributors to various disciplines. However, to more radical feminists such concerns merely exhibit women wanting for themselves the same power and rights over others that men have claimed, and failing to confront the real problem, which is how to live without such symmetrical powers and rights.

In biological determinism, not only influences but constraints and makes inevitable our development as persons with a variety of traits. At its silliest the view postulates such entities as a gene predisposing people to poverty, and it is the particular enemy of thinkers stressing the parental, social, and political determinants of the way we are.

The philosophy of social science is more heavily intertwined with actual social science than in the case of other subjects such as physics or mathematics, since its question is centrally whether there can be such a thing as sociology. The idea of a 'science of man', devoted to uncovering scientific laws determining the basic dynamic s of human interactions was a cherished ideal of the Enlightenment and reached its heyday with the positivism of writers such as the French philosopher and social theorist Auguste Comte (1798-1957), and the historical materialism of Marx and his followers. Sceptics point out that what happens in society is determined by peoples' own ideas of what should happen, and like fashions those ideas change in unpredictable ways as self-consciousness is susceptible to change by any number of external event s: Unlike the solar system of celestial mechanics a society is not at all a closed system evolving in accordance with a purely internal dynamic, but constantly responsive to shocks from outside.

The sociological approach to human behaviour is based on the premise that all social behaviour has a biological basis, and seeks to understand that basis in terms of genetic encoding for features that are then selected for through evolutionary history. The philosophical problem is essentially one of methodology: Of finding criteria for identifying features that can usefully be explained in this way, and for finding criteria for assessing various genetic stories that might provide useful explanations.

Among the features that are proposed for this kind o f explanation are such things as male dominance, male promiscuity versus female fidelity, propensities to sympathy and other emotions, and the limited altruism characteristic of human beings. The strategy has proved unnecessarily controversial, with proponents accused of ignoring the influence of environmental and social factors in moulding people's characteristics, e.g., at the limit of silliness, by postulating a 'gene for poverty', however, there is no need for the approach to commit such errors, since the feature explained sociobiological may be indexed to environment: For instance, it ma y be a propensity to develop some feature in some other environments (for even a propensity to develop propensities . . .) The main problem is to separate genuine explanation from speculative, just so stories which may or may not identify as really selective mechanisms.

In philosophy, the ideas with which we approach the world are in themselves the topic of enquiry. As philosophy is a discipline such as history, physics, or law that seeks not too much to solve historical, physical or legal questions, as to study the conceptual representations that are fundamental structure such thinking, in this sense philosophy is what happens when a practice becomes dialectically self-conscious. The borderline between such 'second-order' reflection, and ways of practising the first-order discipline itself, as not always clear: the advance may tame philosophical problems of a discipline, and the conduct of a discipline may be swayed by philosophical reflection, in meaning that the kinds of self-conscious reflection making up philosophy to occur only when a way of life is sufficiently mature to be already passing, but neglects the fact that self-consciousness and reflection coexist with activity, e.g., an active social and political movement will coexist with reflection on the categories within which it frames its position.

At different times that have been more or less optimistic about the possibility of a pure 'first philosophy', taking a deductive assertion as given to a standpoint of perspective from which other intellectual practices can be impartially assessed and subjected to logical evaluation and correction. This standpoint now seems that for some imaginary views have entwined too many philosophers by the mention of imaginary views based upon ill-exaggerated unrealistic illusionism. The contemporary spirit of the subject is hostile to such possibilities, and prefers to see philosophical reflection as continuos with the best practice if any field of intellectual enquiry.

The principles that lie at the basis of an enquiry are representations that inaugurate the first principles of one phase of enquiry only to employ the gainful habit of being rejected at other stages. For example, the philosophy of mind seeks to answer such questions as: Is mind distinct from matter? Can we give on principal reasons for deciding whether other creatures are conscious, or whether machines can be made in so that they are conscious? What is thinking, feeling, experiences, remembering? Is it useful to divide the function of the mind up, separating memory from intelligence, or rationally from sentiment, or do mental functions from an ingoted whole? The dominated philosophies of mind in the current western tradition include that a variety of physicalism and tradition include various fields of physicalism and functionalism. For particular topics are directorial favourable as set by inclinations implicated throughout the spoken exchange.

Once, in the philosophy of language, was the general attempt to understand the general components of a working language, this relationship that an understanding speaker has to its elemental relationship they bear attestation to the world: Such that the subject therefore embraces the traditional division of 'semantic' into 'syntax', 'semantic', and 'pragmatics'. The philosophy of mind, since it needs an account of what it is in our understanding that enable us to use language. It also mingles with the metaphysics of truth and the relationship between sign and object. The belief that a philosophy of language is the fundamental basis of all philosophical problems in that language has informed such a philosophy, especially in the 20th century, is the philological problem of mind, and the distinctive way in which we give shape to metaphysical beliefs of logical form, and the basis of the division between syntax and semantics, as well some problems of understanding the number and nature of specifically semantic relationships such as 'meaning', 'reference', 'predication', and 'quantification'. Pragmatics includes the theory of speech acts, while problems of rule following and the indeterminacy of translation infect philosophies of both pragmatics and semantics.

A formal system for which a theory whose sentences are well-formed formula's, as connectively gather through alogical calculus and for whose axioms or rules constructed of particular terms, as correspondingly concurring to the principles of the theory being formalized. That theory is intended to be couched or framed in the language of a calculus, e.g., fist-order predicates calculus. Set theory, mathematics, mechanics, and several other axiomatically developed nonobjectivities by that, of making possible the logical analysis for such matters as the independence of various axioms, and the relations between one theory and that of another.

Still, issues surrounding certainty are especially connected with those concerning 'scepticism'. Although Greek scepticism was centred on the value of enquiry and questioning, scepticism is now the denial that knowledge or even rational belief is possible, either about some specific subject-matter, e.g., ethics, or in any area whatsoever. Classical scepticism, springs forward from the observations that are at best the methods of those implied by specific areas but seem to fall short in giving us a full-measure of rewarding proofs as contractually represented by truth, e.g., there is a gulf between appearances and reality, it frequently cites the conflicting judgements that our methods deliver, so that questions of truth become undefinable. In classic thought we systemized the various examples of this conflict in the tropes of Aenesidemus. So that, the scepticism of Pyrrho and the new Academy was a system of argument and inasmuch as opposing dogmatism, and, particularly the philosophical system building of the Stoics.

For many sceptics have traditionally held that knowledge requires certainty. Of course, they claim that the lore abstractive and precise knowledge is not possible. In part, nonetheless, of the principle that every effect it's a consequence of an antecedent cause or causes. For causality to be true being predictable is not necessary for an effect as the antecedent causes may be numerous, too complicated, or too interrelated for analysis. Nevertheless, to avoid scepticism, this participating sceptic has generally held that knowledge does not require certainty. Except for so-called cases of things that are self-evident, but only if they were justifiably correct in giving of one's self-verifiability for being true. It has often been thought, that any thing known must satisfy certain criteria as well for being true. It is often taught that anything is known must satisfy certain standards. In so saying, that by 'deduction' or 'induction', criteria will be specifying when it is. As these alleged cases of self-evident truths, the general principal specifying the sort of consideration that will make such standard in the apparent or justly conclude in accepting it warranted to some degree.

Besides, there is another view-the absolute globular view that we do not have any knowledge whatsoever. In whatever manner, it is doubtful that any philosopher seriously entertains absolute scepticism. Even the Pyrrhonist sceptics, who held that we should refrain from accenting to any non-evident standards that no such hesitancy about asserting to 'the evident', the non-evident are any belief that requires evidences because it is warranted.

René Descartes (1596-1650)in his sceptical guise, never doubted the content of his own ideas. It's challenging logic, inasmuch as of whether they corresponded' to anything beyond ideas.

Given that Descartes disgusted the information from the senses to the point of doubling the perceptive results of repeatable scientific experiments, how did he conclude that our knowledge of the mathematical ideas residing only in mind or in human subjectivity was accurate, much less the absolute truth? He did so by making a leap of faith, God constructed the world, said Descartes, according to the mathematical ideas that our minds are capable of uncovering, in their pristine essence the truths of classical physics Descartes viewed them were quite literally 'revealed' truths, and it was this seventeenth-century metaphysical presupposition that became the history of science for what we term the 'hidden ontology of classical epistemology?'

While classical epistemology would serve the progress of science very well, it also presented us with a terrible dilemma about the relationships between mind and world. If there is a real or necessary correspondence between mathematical ideas in subject reality and external physical reality, how do we know that the world in which we have life, breath. love and die, actually exists? Descartes's resolution of the dilemma took the form of an exercise. He asked us to direct our attention inward and to divest our consciousness of all awareness of external physical reality. If we do so, he concluded, the real existence of human subjective reality could be confirmed.

As it turned out, this resolution was considerably more problematic and oppressive than Descartes could have imagined, 'I think, therefore I am, may be a marginally persuasive way of confirming the real existence of the thinking self. But the understanding of physical reality that obliged Descartes and others to doubt the existence of the self-clearly implies that the separation between the subjective world and the world of life, and the real world of physical objectivity was absolute.'

Unfortunate, the inclined to error plummets suddenly and involuntary, their prevailing odds or probabilities of chance aggress of standards that seem less than are fewer than some, in its gross effect, the fallen succumb moderately, but are described as 'the disease of the Western mind.' Dialectic conduction services' as the background edge horizon as portrayed in the knowledge for understanding, is that of a new anatomical relationship between parts and wholes in physics. With a similar view, which of for something that provides a reason for something else, perhaps, by unforeseen persuadable partiality, or perhaps, by some unduly powers exerted over the minds or behaviour of others, giving cause to some entangled assimilation as 'x' imparts the passing directions into some dissimulated diminution. Relationships that emerge of the co-called, the new biology, and in recent studies thereof, finding that evolution directed toward a scientific understanding proved uncommonly exhaustive, in that to a greater or higher degree, that usually for reason-sensitivities that posit themselves for perceptual notions as might they be deemed existent or, perhaps, of dealing with what exists only in the mind, therefore the ideational conceptual representation to ideas, and includes the parallelisms, showing, of course, as lacking nothing that properly belongs to it, that is actualized along with content.'

Descartes, the foundational architect of modern philosophy, was able to respond without delay or any assumed hesitation or indicative to such ability, and spotted the trouble too quickly realized that there appears of nothing in viewing nature that implicates the crystalline possibilities of reestablishing beyond the reach of the average reconciliation, for being between a full-fledged comparative being such in comparison with an expressed or implied standard or the conferment of situational absolutes, yet the inclinations do incline of talking freely and sometimes indiscretely, if not, only not an idea upon expressing deficient in originality or freshness, belonging in community with or in participation, that the diagonal line has been worn between Plotinus and Whiteheads view for which finds non-locality stationed within a particular point as occupied in space-time, only to its peculiarity is founded the apparent edge horizon of our concerns, That the comparability with which the state or facts of having independent reality, its regulatory customs that have recently come into evidence, is actualized by the existent idea of 'God' especially. Still and all, the primordial nature of God, with which is eternal, a consequent of nature, which is in a flow of compliance, insofar as differentiation occurs in that which can be known as having existence in space or time. The significant relevance is cognitional thought, is noticeably to exclude the use of examples in order to clarify that through the explicated theses as based upon interpolating relationships that are sequentially successive of cause and orderly disposition, as the individual may or may not be of their approval is found to bear the settlements with the quantum theory,

As the quality or state of being ready or skilled that in dexterity brings forward for consideration the adequacy that is to make known the inclinations expounding the actual notion that being exactly as appears or simply charmed with undoubted representation of an actualized entity as it is supposed of a self-realization that blends upon or within the harmonious processes of self-creation. Nonetheless, it seems a strong possibility that Plotonic and Whitehead connect upon the same issue of the creation, that the sensible world may by looking at actual entities as aspects of nature's contemplation, that these formidable contemplations of nature are obviously an immensely intricate affair, whereby, involving a myriad of possibilities, and, therefore one can look upon the actualized entities as, in the sense of obtainability, that the basic elements are viewed into the vast and expansive array of processes.

We could derive a scientific understanding of these ideas aligned with the aid of precise deduction, just as Descartes continued his claim that we could lay the contours of physical reality within a three-dimensional arena whereto, its fixed sides are equated to it's co-ordinated system. Following the publication of Isaac Newtons, 'Principia Mathematica' in 1687, reductionism and mathematical medaling became the most powerful tools of modern science. The dream that we could know and master the entire physical world through the extension and refinement of mathematical theory became the central feature and principles of scientific knowledge.

The radical separation between mind and nature formalized by Descartes, served over time to allow scientists to concentrate on developing mathematical descriptions of matter as pure mechanism without any concern about its spiritual dimensions or ontological foundations. Meanwhile, attempts to rationalize, reconcile or eliminate Descartes's merging division between mind and matter became the most central characterization of Western intellectual life.

All the same, Pyrrhonism and Cartesian forms of virtually globular scepticism, has held and defended, for we are to assume that knowledge is some form of true, because of our sufficiently warranting belief. It is a warranted condition, as, perhaps, that provides the truth or belief conditions, in that of providing the grist for the sceptic's mill about. The Pyrrhonist will suggest that no more than a non-evident, empirically deferent may have of any sufficiency of giving in, but warrantied. Whereas, a Cartesian sceptic will agree that no empirical standards about anything other than one's own mind and its contents are sufficiently warranted, because there are always legitimate grounds for doubting it. In that, the essential difference between the two views concerns the stringency of the requirements for a belief being sufficiently warranted to take account of as knowledge.

A Cartesian requires certainty. A Pyrrhonist merely requires that the standards in case are more warranted then its negation.

Cartesian scepticism was unduly an in fluence with which Descartes agues for scepticism, than his reply holds, in that we do not have any knowledge of any empirical standards, in that of anything beyond the contents of our own minds. The reason is roughly in the position that there is a legitimate doubt about all such standards, only because there is no way to justifiably deny that our senses are being stimulated by some sense, for which it is radically different from the objects which we normally think, in whatever manner they affect our senses. Therefrom, if the Pyrrhonist is the agnostic, the Cartesian sceptic is the atheist.

Because the Pyrrhonist requires much less of a belief in order for it to be confirmed as knowledge than do the Cartesian, the argument for Pyrrhonism are much more difficult to construct. A Pyrrhonist must show that there is no better set of reasons for believing to any standards, of which are in case that any knowledge learnt of the mind is understood by some of its forms, that has to require certainty.

Repudiating the requirements of absolute certainty or knowledge, insisting on the connection of knowledge with activity, as, too, of pragmatism of a reformist distributing knowledge upon the legitimacy of traditional questions about the truth-conditionals employed through and by our cognitive practices, and sustain a conception of truth objectivity, enough to give those questions that undergo of gathering in their own purposive latencies, yet we are given to the spoken word for which a dialectic awareness sparks the fame from the ambers of fire.

Pragmatism of a determinant revolution, by contrast, relinquishing the objectivity of youth, acknowledges no legitimate epistemological questions besides those that are naturally kindred of our current cognitive conviction.

In defining certainty, it is crucial to note that the term has both an absolute and relative sense. More or less, we take a proposition to be certain when we have no doubt about its truth. We may do this in error or unreasonably, but objectively a proposition is certain when such absence of doubt is justifiable. The sceptical tradition in philosophy denies that objective certainty is often possible, or ever possible, either for any proposition at all, or for any proposition at all, or for any proposition from some suspect family (ethics, theory, memory, empirical judgement and so forth) a major sceptical weapon is the possibility of upsetting events that can cast doubt back onto what were hitherto taken to be certainties. Others include reminders of the divergence of human opinion, and the fallible source of our confidence. Fundamentalist approaches to knowledge look for a basis of certainty, upon which the structure of our system is built. Others reject the metaphor, looking for mutual support and coherence, without foundation. However, in moral theory, the views are that there is inviolable moral standards or absolute variability in human desire or policies or prescriptive actions.

The basic idea of a field is arguably present in Leibniz, who was certainly hostile to Newtonian atomism. Despite the fact that his equal hostility to 'action at a distance' muddies the water, which it is usually credited to the Jesuit mathematician and scientist Joseph Boscovich (1711-87) and Immanuel Kant (1724-1804), both of whom influenced the scientist Faraday, with whose work the physical notion became established. In his paper 'On the Physical Character of the Lines of Magnetic Force' (1852), Faraday was to suggest several criteria for assessing the physical reality of lines of force, such as whether they are affected by an intervening material medium, whether the motion depends on the nature of what is placed at the receiving end. As far as electromagnetic fields go, Faraday himself inclined to the view that the mathematical similarity between heat flow, currents, and electromagnetic lines of force was evidence for the physical reality of the intervening medium.

Once, again, our mentioning recognition for which its case value, whereby its view is especially associated the American psychologist and philosopher William James (1842-1910), that the truth of a statement can be defined in terms of a 'utility' of accepting it. Communications, however, were so much as to dispirit the position for which its place of valuation may be viewed as an objection. Since there are things that are false, as it may be useful to accept, and conversely there are things that are true and that it may be damaging to accept. Nevertheless, there are deep connections between the idea that a representation system is accorded, and the likely success of the projects in progressive formality, by its possession. The evolution of a system of representation either perceptual or linguistic, seems bounded to connect successes with everything adapting or with utility in the modest sense. The Wittgenstein doctrine stipulates the meaning of use that upon the nature of belief and its relations with human attitude, emotion and the idea that belief in the truth on one hand, the action of the other. One way of binding with cement, wherefore the connection is found in the idea that natural selection becomes much as much in adapting us to the cognitive creatures, because beliefs have effects, they work. Pragmatism can be found in Kant's doctrine, and continued to play an influencing role in the theory of meaning and truth.

James, (1842-1910), although with characteristic generosity exaggerated in his debt to Charles S. Peirce (1839-1914), he charted that the method of doubt encouraged people to pretend to doubt what they did not doubt in their hearts, and criticize its individualist's insistence, that the ultimate test of certainty is to be found in the individuals personalized consciousness.

From his earliest writings, James understood cognitive processes in teleological terms. Theory, he held, assists us in the satisfactory interests. His will to Believe doctrine, the view that we are sometimes justified in believing beyond the evidential relics upon the notion that a belief's benefits are relevant to its justification. His pragmatic method of analyzing philosophical problems, for which requires that we find the meaning of terms by examining their application to objects in experimental situations, similarly reflects the teleological approach in its attention to consequences.

The view that everyday attributions of intention, belief and meaning among other people proceeded via tacit use of a theory that enables newly and appointed constructs referenced through these interpretations, perhaps, as attempted explanations within some suitable purpose. The view is commonly hld along with functionalism, according to which psychological states theoretical entities, identified by the network of their causes and effects. The theory-theory had different implications, depending on which feature of theories is being stressed. Theories may be though of as capable of formalization, as yielding predications and explanations, as achieved by a process of theorizing, as achieved by predictions and explanations, as accomplished by a process of theorizing, and answering to empirical evidence, that is, in principled descriptions that are without them, as liable to be overturned by newer and better theories, and so on. The main problem with seeing our understanding of others as the outcome of a piece of theorizing is the nonexistence of a medium in which this theory can be couched, as the child learns simultaneously he minds of others and the meaning of terms in its native language.

Our understanding of others is not gained by the tacit use of a 'theory'. Enabling us to infer what thoughts or intentions explain their actions, however, by re-living the situation 'in their moccasins', or from their point of view, and thereby understanding what hey experienced and thought, and therefore expressed. Understanding others is achieved when we can ourselves deliberate as they did, and hear their words as if they are our own. The suggestion is a modern development of the 'Verstehen' tradition associated with Dilthey, Weber and Collngwood.

Much as much, as it is some sense available to reactivate a new body, however, not that I, who survives bodily death, but I may be resurrected in the same body that becomes reanimated by the same form, in that of Aquinas's account, a person having no privileges finds of or in himself a concerned understanding. We understand ourselves, just as we do everything else, that through the sense experience, in that of an abstraction, may justly be of knowing the principle of our own lives, is to obtainably achieve, and not as a given. In the theory of knowledge that knowing Aquinas holds the Aristotelian doctrine that knowing entails some similarities between the Knower and what there is to be known: A human's corporal nature, therefore, requires that knowledge start with sense perception. As perhaps, the same restrictive limitations that do not apply in bringing to a considerable degree the levelling stability that is contained within the hierarchical mosaic for such sustains in having the celestial heavens that open of bringing forth to angles.

In the domain of theology Aquinas deploys the distraction emphasized by Eringena, between the existence of God in understanding the significance of five arguments: They are (1) Motion is only explicable if there exists an unmoved, a first mover (2) the chain of efficient causes demands a first cause (3) the contingent character of existing things in the wold demands a different order of existence, or in other words as something that has a necessary existence (4) the gradations of value in things in the world require the existence of something that is most valuable, or perfect, and (5) the orderly character of events points to a final cause, or end t which all things are directed, and the existence of this end demands a being that ordained it. All the arguments are physico-theological arguments, in that between reason and faith, Aquinas lays out proofs of the existence of God.

He readily recognizes that there are doctrines such that are the Incarnation and the nature of the Trinity, know only through revelations, and whose acceptance is more a matter of moral will. God's essence is identified with his existence, as pure activity. God is simple, containing no potential. No matter how, we cannot obtain knowledge of what God is (his quiddity), perhaps, doing the same work as the principle of charity, but suggesting that we regulate our procedures of interpretation by maximizing the extent to which we see the subject s humanly reasonable, than the extent to which we see the subject as right about things. Whereby remaining content with descriptions that apply to him partly by way of analogy, God reveals of himself is to accede that it is not himself. The immediate problem availed of ethics is posed by the English philosopher Phillippa Foot, in her 'The Problem of Abortion and the Doctrine of the Double Effect' (1967). Explaining, for instance, that a runaway train or trolley-way streetcar, that comes to a section in the track that is under construction and completely impassable. One person is working on one part of the track, while five on the other track, such that the runaway trolley will put an end to anyone working on the branch it enters. Clearly, to most minds, the driver should steer for the fewest populated sector. But now suppose that, left to itself, it will enter the branch with its five employs that are there, and you as a bystander can intervene, altering the points so that if to veer through into the other side, is it your right or obligation, or, even so, is it permissible for you to do this, thereby, apparently involving yourself in ways that responsibility ends in a death of one person? After all, who have you wronged if you leave it to go its own way? The situation is similarly standardized of others in which utilitarian reasoning seems to lead to one course of action, but a person's integrity or principles may oppose it.

Describing events that haphazardly happen does not of itself legitimate us to talk of rationality and intention, which are the categories we may apply if we conceive of them as action. We think of ourselves not only passively, as creatures that make things happen. Understanding this distinction gives forth of its many major problems concerning the nature of an agency for the causation of bodily events by mental events, and of understanding the 'will' and 'free will'. Other problems in the theory of action include drawing the distinction between an action and its consequence, and describing the structure involved when we do one thing 'by' doing another thing in apprehension of, even the planning and dating where someone shoots someone on one day and in one place, whereby the victim then dies on another day and in another place. Where did the murderous act take place?

Causation, least of mention, is not clear that only events are created by and for itself. Kant cites the example o a cannonball at rest and stationed upon a cushion, but causing the cushion to be the shape that it is, and thus to suggest that the causal states of affairs or objects or facts may also be casually related, that all for which, the central problem is to understand the elements of necessitation or determinacy of the future. Events, Hume thought, are in themselves 'loose and separate': How then are we to conceive of others? The relationship seems not too perceptible, for all that perception gives us (Hume argues) is knowledge of the patterns that events do, actually falling into than any acquaintance with the connections determining the pattern. It is, however, clear that our conception of everyday objects is largely determined by their casual powers, and all our action is based on the belief that these causal powers are stable and reliable. Although scientific investigation can give us wider and deeper dependable patterns, it seems incapable of bringing us any nearer to the 'must' of causal necessitation. Particular examples o f puzzles with causalities are quite apart from general problems of forming any conception of what it is: How are we to understand the casual interaction between mind and body? How can the present, which exists, or its existence to a past that no longer exists? How is the stability of the casual order to be understood? Is backward causality possible? Is causation a concept needed in science, or dispensable?

The news concerning free-will, is nonetheless, a problem for which is to reconcile our everyday consciousness of ourselves as agent, with the best view of what science tells us that we are. Determinism is one part of the problem. It may be defined as the doctrine that every event has a cause. More precisely, for any event 'C', there will be one antecedent states of nature 'N', and a law of nature 'L', such that given 'L', 'N' will be followed by 'C'. But if this is true of every event, it is true of events such as my doing something or choosing to do something. So my choosing or doing something is fixed by some antecedent state 'N' an d the laws. Since determinism is universal these in turn are fixed, and so backwards to events for which I am clearly not responsible (events before my birth, for example). So, no events can be voluntary or free, where that that they come about purely because of my willing them I could have done otherwise. If determinism is true, then there will be antecedent states and laws already determining such events: How then can I truly be said to be their author, or be responsible for them?

Reactions to this problem are commonly classified as: (1) Hard determinism. This accepts the conflict and denies that you have real freedom or responsibility (2) Soft determinism or compatibility, whereby reactions in this family assert that everything you should ant from a notion of freedom is quite compatible with determinism. In particular, if your actions are caused, it can often be true of you that you could have done otherwise if you had chosen, and this may be enough to render you liable to be held unacceptable (the fact that previous events will have caused you to choose as you did is deemed irrelevant on this option). (3) Libertarianism, as this is the view that while compatibilism is only an evasion, there is a more substantiative, real notion of freedom that can yet be preserved in the face of determinism (or, of indeterminism). In Kant, while the empirical or phenomenal self is determined and not free, whereas the noumenal or rational self is capable of being rational, free action. However, the noumeal self exists outside the categorical priorities of space and time, as this freedom seems to be of a doubtful value as other libertarian avenues do include of suggesting that the problem is badly framed, for instance, because the definition of determinism breaks down, or postulates by its suggesting that there are two independent but consistent ways of looking at an agent, the scientific and the humanistic, wherefore it is only through confusing them that the problem seems urgent. Nevertheless, these avenues have gained general popularity, as an error to confuse determinism and fatalism.

The dilemma for which determinism is for itself often supposes of an action that seems as the end of a causal chain, or, perhaps, by some hieratical set of suppositional actions that would stretch back in time to events for which an agent has no conceivable responsibility, then the agent is not responsible for the action.

A mental act of willing or trying whose presence is sometimes supposed to make the difference between intentional and voluntary action, as well of mere behaviour. The theory that there is such acts is problematic, and the idea that they make the required difference is a case of explaining a phenomenon by citing another that raises exactly the same problem, since the intentional or voluntary nature of the set of volition now needs explanation. For determinism to act in accordance with the law of autonomy or freedom, is that in ascendance with universal moral law and regardless of selfish advantage.

A categorical notion in the work as contrasted in Kantian ethics show of a hypothetical imperative that embeds of a commentary which is in place only given some antecedent desire or project. 'If you want to look wise, stay quiet'. The injunction to stay quiet only applies to those with the antecedent desire or inclination: If one has no desire to look wise the injunction or advice lapses. A categorical imperative cannot be so avoided, it is a requirement that binds anybody, regardless of their inclination,. It could be repressed as, for example, 'Tell the truth (regardless of whether you want to or not)'. The distinction is not always mistakably presumed or absence of the conditional or hypothetical form: 'If you crave drink, don't become a bartender' may be regarded as an absolute injunction applying to anyone, although only activated in the case of those with the stated desire.

In Grundlegung zur Metaphsik der Sitten (1785), Kant discussed some of the given forms of categorical imperatives, such that of (1) The formula of universal law: 'act only on that maxim through which you can at the same time will that it should become universal law', (2) the formula of the law of nature: 'Act as if the maxim of your action were to become through your will a universal law of nature', (3) the formula of the end-in-itself, 'Act in such a way that you always trat humanity of whether in your own person or in the person of any other, never simply as an end, but always at the same time as an end', (4) the formula of autonomy, or consideration; 'the will' of every rational being a will which makes universal law', and (5) the formula of the Kingdom of Ends, which provides a model for systematic union of different rational beings under common laws.

A central object in the study of Kant's ethics is to understand the expressions of the inescapable, binding requirements of their categorical importance, and to understand whether they are equivalent at some deep level. Kant's own application of the notions are always convincing: One cause of confusion is relating Kant's ethical values to theories such as ;expressionism' in that it is easy but imperatively must that it cannot be the expression of a sentiment, yet, it must derive from something 'unconditional' or necessary' such as the voice of reason. The standard mood of sentences used to issue request and commands are their imperative needs to issue as basic the need to communicate information, and as such to animals signalling systems may as often be interpreted either way, and understanding the relationship between commands and other action-guiding uses of language, such as ethical discourse. The ethical theory of 'prescriptivism' in fact equates the two functions. A further question is whether there is an imperative logic. 'Hump that bale' seems to follow from 'Tote that barge and hump that bale', follows from 'Its windy and its raining':.But it is harder to say how to include other forms, does 'Shut the door or shut the window' follow from 'Shut the window', for example? The usual way to develop an imperative logic is to work in terms of the possibility of satisfying the other one command without satisfying the other, thereby turning it into a variation of ordinary deductive logic.

Despite the fact that the morality of people and their ethics amount to the same thing, there is a usage that I restart morality to systems such as that of Kant, based on notions given as duty, obligation, and principles of conduct, reserving ethics for the more Aristotelian approach to practical reasoning as based on the valuing notions that are characterized by their particular virtue, and generally avoiding the separation of 'moral' considerations from other practical considerations. The scholarly issues are complicated and complex, with some writers seeing Kant as more Aristotelian,. And Aristotle as more involved with a separate sphere of responsibility and duty, than the simple contrast suggests.

A major topic of philosophical inquiry, especially in Aristotle, and subsequently since the 17th and 18th centuries, when the 'science of man' began to probe into human motivation and emotion. For such as these, the French moralists, as Hutcheson, Hume, Smith and Kant, a prime task as to delineate the variety of human reactions and motivations. Such an inquiry would locate our propensity for moral thinking among other faculties, such as perception and reason, and other tendencies as empathy, sympathy or self-interest. The task continues especially in the light of a post-Darwinian understanding of ourselves.

In some moral systems, notably that of Immanuel Kant, real moral worth comes only with interactivity, justly because it is right. However, if you do what is purposely becoming, equitable, but from some other equitable motive, such as the fear or prudence, no moral merit accrues to you. Yet, that in turn seems to discount other admirable motivations, as acting from main-sheet benevolence, or 'sympathy'. The question is how to balance these opposing ideas and how to understand acting from a sense of obligation without duty or rightness, through which their beginning to seem a kind of fetish. It thus stands opposed to ethics and relying on highly general and abstractive principles, particularly those associated with the Kantian categorical imperatives. The view may go as far back as to say that taken in its own, no consideration point, for that which of any particular way of life, that, least of mention, the contributing steps so taken as forwarded by reason or be to an understanding estimate that can only proceed by identifying salient features of a situation that weigh on one's side or another.

As random moral dilemmas set out with intense concern, inasmuch as philosophical matters that exert a profound but influential defence of common sense. Situations in which each possible course of action breeches some otherwise binding moral principle, are, nonetheless, serious dilemmas making the stuff of many tragedies. The conflict can be described in different was. One suggestion is that whichever action the subject undertakes, that he or she does something wrong. Another is that his is not so, for the dilemma that in the circumstances for what she or he did was right as any alternate. It is important to the phenomenology of these cases that action leaves a residue of guilt and remorse, even though it had proved it was not the subject's fault that she or he were considering the dilemma, that the rationality of emotions can be contested. Any normality with more than one fundamental principle seems capable of generating dilemmas, however, dilemmas exist, such as where a mother must decide which of two children to sacrifice, least of mention, no principles are pitted against each other, only if we accept that dilemmas from principles are real and important, this fact can then be used to approach in themselves, such as of 'utilitarianism', to espouse various kinds may, perhaps, be centred upon the possibility of relating to independent feelings, liken to recognize only one sovereign principle. Alternatively, of regretting the existence of dilemmas and the unordered jumble of furthering principles, in that of creating several of them, a theorist may use their occurrences to encounter upon that which it is to argue for the desirability of locating and promoting a single sovereign principle.

Nevertheless, some theories into ethics see the subject in terms of a number of laws (as in the Ten Commandments). The status of these laws may be that they are the edicts of a divine lawmaker, or that they are truths of reason, given to its situational ethics, virtue ethics, regarding them as at best rules-of-thumb, and, frequently disguising the great complexity of practical representations that for reason has placed the Kantian notions of their moral law.

In continence, the natural law possibility points of the view of the states that law and morality are especially associated with St Thomas Aquinas (1225-74), such that his synthesis of Aristotelian philosophy and Christian doctrine was eventually to provide the main philosophical underpinning of the Catholic church. Nevertheless, to a greater extent of any attempt to cement the moral and legal order and together within the nature of the cosmos or the nature of human beings, in which sense it found in some Protestant writings, under which had arguably derived functions. From a Platonic view of ethics and its agedly implicit advance of Stoicism. Its law stands above and apart from the activities of human lawmakers: It constitutes an objective set of principles that can be seen as in and for themselves by of 'natural usages' or by reason itself, additionally, (in religious verses of them), that express of God's will for creation. Nonreligious versions of the theory substitute objective conditions for humans flourishing as the source of constraints, upon permissible actions and social arrangements within the natural law tradition. Different views have been held about the relationship between the rule of the law and God's will. Grothius, for instance, sides with the view that the content of natural law is independent of any will, including that of God.

While the German natural theorist and historian Samuel von Pufendorf (1632-94) takes the opposite view. His great work was the De Jure Naturae et Gentium, 1672, and its English translation is 'Of the Law of Nature and Nations, 1710. Pufendorf was influenced by Descartes, Hobbes and the scientific revolution of the 17th century, his ambition was to introduce a newly scientific 'mathematical' treatment on ethics and law, free from the tainted Aristotelian underpinning of 'scholasticism'. Like that of his contemporary - Locke. His conception of natural laws include rational and religious principles, making it only a partial forerunner of more resolutely empiricist and political treatment in the Enlightenment.

Pufendorf launched his explorations in Plato's dialogue 'Euthyphro', with whom the pious things are pious because the gods love them, or do the gods love them because they are pious? The dilemma poses the question of whether value can be conceived as the upshot of the choice of any mind, even a divine one. On the fist option the choice of the gods crates goodness and value. Even if this is intelligible it seems to make it impossible to praise the gods, for it is then vacuously true that they choose the good. On the second option we have to understand a source of value lying behind or beyond the will even of the gods, and by which they can be evaluated. The elegant solution of Aquinas is and is therefore distinct form is will, but not distinct from him.

The dilemma arises whatever the source of authority is supposed to be. Do we care about the good because it is good, or do we just call good those things that we care about? It also generalizes to affect our understanding of the authority of other things: Mathematics, or necessary truth, for example, are truths necessary because we deem them to be so, or do we deem them to be so because they are necessary?

The natural aw tradition may either assume a stranger form, in which it is claimed that various facts entails of primary and secondary qualities, any of which is claimed that various facts entail values, reason by itself is capable of discerning moral requirements. As in the ethics of Knt, these requirements are supposed binding on all human beings, regardless of their desires.

The supposed natural or innate abilities of the mind to know the first principle of ethics and moral reasoning, wherein, those expressions are assigned and related to those that distinctions are which make in terms contribution to the function of the whole, as completed definitions of them, their phraseological impression is termed 'synaeresis' (or, syntetesis) although traced to Aristotle, the phrase came to the modern era through St Jerome, whose scintilla conscientiae (gleam of conscience) wads a popular concept in early scholasticism. Nonetheless, it is mainly associated in Aquinas as an infallible natural, simple and immediate grasp of first moral principles. Conscience, by contrast, is,more concerned with particular instances of right and wrong, and can be in error, under which the assertion that is taken as fundamental, at least for the purposes of the branch of enquiry in hand.

It is, nevertheless, the view interpreted within he particular states of law and morality especially associated with Aquinas and the subsequent scholastic tradition, showing for itself the enthusiasm for reform for its own sake. Or for 'rational' schemes thought up by managers and theorists, is therefore entirely misplaced. Major o exponent s of this theme include the British absolute idealist Herbert Francis Bradley (1846-1924) and Austrian economist and philosopher Friedrich Hayek. The notably the idealism of Bradley, there is the same doctrine that change is contradictory and consequently unreal: The Absolute is changeless. A way of sympathizing a little with his idea is to reflect that any scientific explanation of change will proceed by finding an unchanging law operating, or an unchanging quantity conserved in the change, so that explanation of change always proceeds by finding that which is unchanged. The metaphysical problem of change is to shake off the idea that each moment is created afresh, and to obtain a conception of events or processes as having a genuinely historical reality, Really extended and unfolding in time, as opposed to being composites of discrete temporal atoms. A step toward this end may be to see time itself not as an infinite container within which discrete events are located, bu as a kind of logical construction from the flux of events. This relational view of time was advocated by Leibniz and a subject of the debate between him and Newton's Absolutist pupil, Clarke.

Generally, nature is an indefinitely mutable term, changing as our scientific conception of the world changes, and often best seen as signifying a contrast with something considered not part of nature. The term applies both to individual species (it is the nature of gold to be dense or of dogs to be friendly), and also to the natural world as a whole. The sense in which it applies to species quickly links up with ethical and aesthetic ideals: A thing ought to realize its nature, what is natural is what it is good for a thing to become, it is natural for humans to be healthy or two-legged, and departure from this is a misfortune or deformity,. The associations of what is natural with what it is good to become is visible in Plato, and is the central idea of Aristotle's philosophy of nature. Unfortunately, the pinnacle of nature in this sense is the mature adult male citizen, with he rest of hat we would call the natural world, including women, slaves, children and other species, not quite making it.

Nature in general can, however, function as a foil to any idea inasmuch as a source of ideals: In this sense fallen nature is contrasted with a supposed celestial realization of the 'forms'. The theory of 'forms' is probably the most characteristic, and most contested of the doctrines of Plato. If in the background the Pythagorean conception of form, as the key to physical nature, but also the sceptical doctrine associated with the Greek philosopher Cratylus, and is sometimes thought to have been a teacher of Plato before Socrates. He is famous for capping the doctrine of Ephesus of Heraclitus, whereby the guiding idea of his philosophy was that of the logos, is capable of being heard or hearkened to by people, it unifies opposites, and it is somehow associated with fire, which preeminent among the four elements that Heraclitus distinguishes: Fire, air (breath, the stuff of which souls composed), earth, and water. Although he is principally remember for the doctrine of the 'flux' of all things, and the famous statement that you cannot step into the same river twice, for new waters are ever flowing in upon you. The more extreme implication of the doctrine of flux, e.g., the impossibility of categorizing things truly, do not seem consistent with his general epistemology and views of meaning, and were to his follower Cratylus, although the proper conclusion of his views was that the flux cannot be captured in words. According to Aristotle, he eventually held that since 'regarding that which everywhere in every respect is changing nothing is just to stay silent and wrangle one's finger'. Plato 's theory of forms can be seen in part as an action against the impasse to which Cratylus was driven.

The Galilean world view might have been expected to drain nature of its ethical content, however, the term seldom loses its normative force, and the belief in universal natural laws provided its own set of ideals. In the 18th century for example, a painter or writer could be praised as natural, where the qualities expected would include normal (universal) topics treated with simplicity, economy, regularity and harmony. Later on, nature becomes an equally potent emblem of irregularity, wildness, and fertile diversity, but also associated with progress of human history, its incurring definition that has been taken to fit many things as well as transformation, including ordinary human self-consciousness. Nature, being in contrast with in integrated phenomenon may include (1) that which is deformed or grotesque or fails to achieve its proper form or function or just the statistically uncommon or unfamiliar, (2) the supernatural, or the world of gods and invisible agencies, (3) the world of rationality and unintelligence, conceived as distinct from the biological and physical order, or the product of human intervention, and (5) related to that, the world of convention and artifice.

Different conceptual representational forms of nature continue to have ethical overtones, for example, the conception of 'nature red in tooth and claw' often provides a justification for aggressive personal and political relations, or the idea that it is women's nature to be one thing or another is taken to be a justification for differential social expectations. The term functions as a fig-leaf for a particular set of stereotypes, and is a proper target of much feminist writings. Feminist epistemology has asked whether different ways of knowing for instance with different criteria of justification, and different emphases on logic and imagination, characterize male and female attempts to understand the world. Such concerns include awareness of the 'masculine' self-image, itself a socially variable and potentially distorting picture of what thought and action should be. Again, there is a spectrum of concerns from the highly theoretical to he relatively practical. In this latter area particular attention is given to the institutional biases that stand in the way of equal opportunities in science and other academic pursuits, or the ideologies that stand in the way of women seeing themselves as leading contributors to various disciplines. However, to more radical feminists such concerns merely exhibit women wanting for themselves the same power and rights over others that men have claimed, and failing to confront the real problem, which is how to live without such symmetrical powers and rights.

In biological determinism, not only influences but constraints and makes inevitable our development as persons with a variety of traits. At its silliest the view postulates such entities as a gene predisposing people to poverty, and it is the particular enemy of thinkers stressing the parental, social, and political determinants of the way we are.

The philosophy of social science is more heavily intertwined with actual social science than in the case of other subjects such as physics or mathematics, since its question is centrally whether there can be such a thing as sociology. The idea of a 'science of man', devoted to uncovering scientific laws determining the basic dynamics of human interactions was a cherished ideal of the Enlightenment and reached its heyday with the positivism of writers such as the French philosopher and social theorist Auguste Comte (1798-1957), and the historical materialism of Marx and his followers. Sceptics point out that what happens in society is determined by peoples' own ideas of what should happen, and like fashions those ideas change in unpredictable ways as self-consciousness is susceptible to change by any number of external event s: Unlike the solar system of celestial mechanics a society is not at all a closed system evolving in accordance with a purely internal dynamic, but constantly responsive to shocks from outside.

The sociological approach to human behaviour is based on the premise that all social behaviour has a biological basis, and seeks to understand that basis in terms of genetic encoding for features that are then selected for through evolutionary history. The philosophical problem is essentially one of methodology: As regards to the essential points that it is important to the nature and essence of a thing as to be indispensable, wherefore, something that forms part of the minimal body, character, or structure of a thing. So that by something necessary is regarded as fundamentally the essence of worth or merit obtained or encountered more or less by chance, as to come upon or detect of finding the applicable adequacy, whereby its rights are the regulating criteria for identifying features that can usefully be explained in this way, and for finding the criteria for assessing various genetic transcripts that might provide useful explanations.

Among the features that are proposed for this kind o f explanation are such things as male dominance, male promiscuity versus female fidelity, propensities to sympathy and other emotions, and the limited altruism characteristic of human beings. The strategy has proved unnecessarily controversial, with proponents accused of ignoring the influence of environmental and social factors in moulding people's characteristics, e.g., at the limit of silliness, by postulating a 'gene for poverty', however, there is no need for the approach to commit such errors, since the feature explained sociobiological may be indexed to environment: For instance, it ma y be a propensity to develop some feature in some other environments (for even a propensity to develop propensities . . .) The main problem is to separate genuine explanation from speculative, just so stories which may or may not identify as really selective mechanisms.

Subsequently, in the 19ᵗʰ century attempts were made to base ethical reasoning on the presumed facts about evolution. The movement is particularly associated with the English philosopher of evolution Herbert Spencer (1820-1903),. His first major work was the book Social Statics (1851), which advocated an extreme political libertarianism. The Principles of Psychology was published in 1855, and his very influential Education advocating natural development of intelligence, the creation of pleasurable interest, and the importance of science in the curriculum, appeared in 1861. His First Principles (1862) was followed over the succeeding years by volumes on the Principles of biology and psychology, sociology and ethics. Although he attracted a large public following and attained the stature of a sage, his speculative work has not lasted well, and in his own time there was dissident voices. T.H. Huxley said that Spencer's definition of a tragedy was a deduction killed by a fact. Writer and social prophet Thomas Carlyle (1795-1881) called him a perfect vacuum, and the American psychologist and philosopher William James (1842-1910) wondered why half of England wanted to bury him in Westminister Abbey, and talked of the 'hurdy-gurdy' monotony of him, his whole system wooden, as if knocked together out of cracked hemlock.

The premise is that later elements in an evolutionary path are better than earlier ones, the application of this principle then requires seeing western society, laissez-faire capitalism, or some other object of approval, as more evolved than more 'primitive' social forms. Neither the principle nor the applications command much respect. The version of evolutionary ethics called 'social Darwinism' emphasizes the struggle for natural selection, and drawn the conclusion that we should glorify such struggle, usually by enhancing competitive and aggressive relations between people in society or between societies themselves. More recently the relation between evolution and ethics has been rethought in the light of biological discoveries concerning altruism and kin-selection.

`In that, the study of the say in which a variety of higher mental function may be adaptions applicable of a psychology of evolution, a formed in response to selection pressures on human populations through evolutionary time. Candidates for such theorizing include material and paternal motivations, capabilities for love and friendship, the development of language as a signalling system, cooperative and aggressive tendencies, our emotional repertoires, our moral reaction, including the disposition to direct and punish those who cheat on an agreement or who free-ride on the work of others, our cognitive structure and many others. Evolutionary psychology goes hand-in-hand with neurophysiological evidence about the underlying circuitry in the brain which subserves the psychological mechanisms it claims to identify.

For all that, an essential part of the British absolute idealist Herbert Bradley (1846-1924) was largely on the ground s that the self-sufficiency individualized through community and one's self is to contribute to social and other ideals. However, truth as formulated in language is always partial, and dependent upon categories that themselves are inadequate to the harmonious whole. Nevertheless, these self-contradictory elements somehow contribute to the harmonious whole, or Absolute, lying beyond categorization. Although absolute idealism maintains few adherents today, Bradley's general dissent from empiricism, his holism, and the brilliance and style of his writing continue to make him the most interesting of the late 19th century writers influenced by the German philosopher Friedrich Hegel (1770-1831).

Understandably, something less than the fragmented division that belonging of Bradley's case has a preference, voiced much earlier by the German philosopher, mathematician and polymath was Gottfried Leibniz (1646-1716), for categorical monadic properties over relations. He was particularly troubled by the relation between that which is known and the more that knows it. In philosophy, the Romantics took from the German philosopher and founder of critical philosophy Immanuel Kant (1724-1804) both the emphasis on free-will and the doctrine that reality is ultimately spiritual, with nature itself a mirror of the human soul. To fix upon one among alternatives as the one to be taken, Friedrich Schelling (1775-1854) foregathers nature of becoming a creative spirit whose aspiration is ever further and more to completed self-realization. Although a movement of more general to naturalized imperative. Romanticism drew on the same intellectual and emotional resources as

German idealism was increasingly culminating in the philosophy of Hegel (1770-1831) and of absolute idealism.

Being such in comparison with nature may include (1) that which is deformed or grotesque, or fails to achieve its proper form or function, or just the statistically uncommon or unfamiliar, (2) the supernatural, or the world of gods and invisible agencies, (3) the world of rationality and intelligence, conceived of as distinct from the biological and physical order, (4) that which is manufactured and artefactual, or the product of human invention, and (5) related to it, the world of convention and artifice.

In Hume, objects of knowledge are divided into matter of fact (roughly empirical things known by of impressions) and the relation of ideas. The contrast, also called 'Hume's Fork', is a version of the speculative deductivity distinction, but reflects the 17ᵗʰ and early 18ᵗʰ centauries behind that the deductivity is established by chains of infinite certainty as comparable to ideas. It is extremely important that in the period between Descartes and J.S. Mill that a demonstration is not, but only a chain of 'intuitive' comparable ideas, whereby a principle or maxim can be established by reason alone. It is in this sense that the English philosopher John Locke (1632-1704) who believed that theological and moral principles are capable of demonstration, and Hume denies that they are, and also denies that scientific enquiries proceed in demonstrating its results.

A mathematical proof is formally inferred as to an argument that is used to show the truth of a mathematical assertion. In modern mathematics, a proof begins with one or more statements called premises and demonstrates, using the rules of logic, that if the premises are true then a particular conclusion must also be true.

The accepted methods and strategies used to construct a convincing mathematical argument have evolved since ancient times and continue to change. Consider the Pythagorean theorem, named after the 5ᵗʰ century Bc Greek mathematician and philosopher Pythagoras, which states that in a right-angled triangle, the square of the hypotenuse is equal to the sum of the squares of the other two sides. Many early civilizations considered this theorem true because it agreed with their observations in practical situations. But the early Greeks, among others, realized that observation and commonly held opinion do not guarantee mathematical truth. For example, before the 5ᵗʰ century Bc it was widely believed that all lengths could be expressed as the ratio of two whole numbers. But an unknown Greek mathematician proved that this was not true by showing that the length of the diagonal of a square with an area of 1 is the irrational number.

The Greek mathematician Euclid laid down some of the conventions central to modern mathematical proofs. His book The Elements, written about 300 Bc, contains many proofs in the fields of geometry and algebra. This book illustrates the Greek practice of writing mathematical proofs by first clearly identifying the initial assumptions and then reasoning from them in a logical way in order to obtain a desired conclusion. As part of such an argument, Euclid used results that had already been shown to be true, called theorems, or statements that were explicitly acknowledged to be self-evident, called axioms; this practice continues today.

In the 20th century, proofs have been written that are so complex that no one person understands every argument used in them. In 1976, a computer was used to complete the proof of the four-colour theorem. This theorem states that four Colour are sufficient to colour any map in such a way that regions with a common boundary line have different Colour. The use of a computer in this proof inspired considerable debate in the mathematical community. At issue was whether a theorem can be considered proven if human beings have not actually checked every detail of the proof.

The study of the relations of deductibility among sentences in a logical calculus which benefits the prof theory. Deductibility is defined purely syntactically, that is, without reference to the intended interpretation of the calculus. The subject was founded by the mathematician David Hilbert (1862-1943) in the hope that strictly finitary methods would provide a way of proving the consistency of classical mathematics, but the ambition was torpedoed by Gödel's second incompleteness theorem.

What is more, the use of a model to test for consistencies in an 'axiomatized system' which is older than modern logic. Descartes' algebraic interpretation of Euclidean geometry provides a way of showing that if the theory of real numbers is consistent, so is the geometry. Similar representation had been used by mathematicians in the 19th century, for example to show that if Euclidean geometry is consistent, so are various non-Euclidean geometries. Model theory is the general study of this kind of procedure: The 'proof theory' studies relations of deductibility between formulae of a system, but once the notion of an interpretation is in place we can ask whether a formal system meets certain conditions. In particular, can it lead us from sentences that are true under some interpretation? And if a sentence is true under all interpretations, is it also a theorem of the system? We can define a notion of validity (a formula is valid if it is true in all interpret rations) and semantic consequence (a formula 'B' is a semantic consequence of a set of formulae, written {A1 . . . An} B, if it is true in all interpretations in which they are true) Then the central questions for a calculus will be whether all and only its theorems are valid, and whether {A1 . . . An} B if and only if {A1 . . . An} B. There are the questions of the soundness and completeness of a formal system. For the propositional calculus this turns into the question of whether the proof theory delivers as theorems all and only 'tautologies'. There are many axiomatizations of the propositional calculus that are consistent and complete. The mathematical logician Kurt Gödel (1906-78) proved in 1929 that the first-order predicate under every interpretation is a theorem of the calculus.

The Euclidean geometry is the greatest example of the pure 'axiomatic method', and as such had incalculable philosophical influence as a paradigm of rational certainty. It had no competition until the 19th century when it was realized that the fifth axiom of his system (parallel lines never meet) could be denied without inconsistency, leading to Riemannian spherical geometry. The significance of Riemannian geometry lies in its use and extension of both Euclidean geometry and the geometry of surfaces, leading to a number of generalized differential geometries. Its most important effect was that it made a geometrical application possible for some major abstractions of tensor analysis, leading to the pattern and concepts for general relativity later used by Albert Einstein in developing his theory of relativity. Riemannian geometry is also necessary for treating electricity and magnetism in the framework of general relativity. The fifth chapter of Euclid's Elements, is attributed to the mathematician Eudoxus, and contains a precise development of the real number, work which remained unappreciated until rediscovered in the 19th century.

The Axiom, in logic and mathematics, is a basic principle that is assumed to be true without proof. The use of axioms in mathematics stems from the ancient Greeks, most probably during the 5th century Bc, and represents the beginnings of pure mathematics as it is known today. Examples of axioms are the following: 'No sentence can be true and false at the same time' (the principle of contradiction); 'If equals are added to equals, the sums are equal'. 'The whole is greater than any of its parts'. Logic and pure mathematics begin with such unproved assumptions from which other propositions (theorems) are derived. This procedure is necessary to avoid circularity, or an infinite regression in reasoning. The axioms of any system must be consistent with one another, that is, they should not lead to contradictions. They should be independent in the sense that they cannot be derived from one another. They should also be few in number. Axioms have sometimes been interpreted as self-evident truths. The present tendency is to avoid this claim and simply to assert that an axiom is assumed to be true without proof in the system of which it is a part.

The terms 'axiom' and 'postulate' are often used synonymously. Sometimes the word axiom is used to refer to basic principles that are assumed by every deductive system, and the term postulate is used to refer to first principles peculiar to a particular system, such as Euclidean geometry. Infrequently, the word axiom is used to refer to first principles in logic, and the term postulate is used to refer to first principles in mathematics.

The applications of game theory are wide-ranging and account for steadily growing interest in the subject. Von Neumann and Morgenstern indicated the immediate utility of their work on mathematical game theory by linking it with economic behaviour. Models can be developed, in fact, for markets of various commodities with differing numbers of buyers and sellers, fluctuating values of supply and demand, and seasonal and cyclical variations, as well as significant structural differences in the economies concerned. Here game theory is especially relevant to the analysis of conflicts of interest in maximizing profits and promoting the widest distribution of goods and services. Equitable division of property and of inheritance is another area of legal and economic concern that can be studied with the techniques of game theory.

In the social sciences, 'n-person' game theory has interesting uses in studying, for example, the distribution of power in legislative procedures. This problem can be interpreted as a three-person game at the congressional level involving vetoes of the president and votes of representatives and senators, analyzed in terms of successful or failed coalitions to pass a given bill. Problems of majority rule and individual decision making are also amenable to such study.

Sociologists have developed an entire branch of game theory devoted to the study of issues involving group decision making. Epidemiologists also make use of game theory, especially with respect to immunization procedures and methods of testing a vaccine or other medication. Military strategists turn to game theory to study conflicts of interest resolved through 'battles' where the outcome or payoff of a given war game is either victory or defeat. Usually, such games are not examples of zero-sum games, for what one player loses in terms of lives and injuries is not won by the victor. Some uses of game theory in analyses of political and military events have been criticized as a dehumanizing and potentially dangerous oversimplification of necessarily complicating factors. Analysis of economic situations is also usually more complicated than zero-sum games because of the production of goods and services within the play of a given 'game'.

All is the same in the classical theory of the syllogism, a term in a categorical proposition is distributed if the proposition entails any proposition obtained from it by substituting a term denoted by the original. For example, in 'all dogs bark' the term 'dogs' is distributed, since it entails 'all terriers bark', which is obtained from it by a substitution. In 'Not all dogs bark', the same term is not distributed, since it may be true while 'not all terriers bark' is false.

When a representation of one system by another is usually more familiar, in and for itself, that those extended in representation that their workings are supposed analogous to that of the first. This one might model the behaviour of a sound wave upon that of waves in water, or the behaviour of a gas upon that to a volume containing moving billiard balls. While nobody doubts that models have a useful 'heuristic' role in science, there has been intense debate over whether a good model, or whether an organized structure of laws from which it can be deduced and suffices for scientific explanation. As such, the debate of topic was inaugurated by the French physicist Pierre Marie Maurice Duhem (1861-1916), in 'The Aim and Structure of Physical Theory' (1954) by which Duhem's conception of science is that it is simply a device for calculating as science provides deductive system that is systematic, economical, and predictive, but not that represents the deep underlying nature of reality. Steadfast and holding of its contributive thesis that in isolation, and since other auxiliary hypotheses will always be needed to draw empirical consequences from it. The Duhem thesis implies that refutation is a more complex matter than might appear. It is sometimes framed as the view that a single hypothesis may be retained in the face of any adverse empirical evidence, if we prepared to make modifications elsewhere in our system, although strictly speaking this is a stronger thesis, since it may be psychologically impossible to make consistent revisions in a belief system to accommodate, say, the hypothesis that there is a hippopotamus in the room when visibly there is not.

Primary and secondary qualities are the division associated with the 17th-century rise of modern science, wit h its recognition that the fundamental explanatory properties of things that are not the qualities that perception most immediately concerns. There later are the secondary qualities, or immediate sensory qualities, including colour, taste, smell, felt warmth or texture, and sound. The primary properties are less tied to there deliverance of one particular sense, and include the size, shape, and motion of objects. In Robert Boyle (1627-92) and John Locke (1632-1704) the primary qualities are scientifically tractable, objective qualities essential to anything material, are of a minimal listing of size, shape, and mobility, i.e., the state of being at rest or moving. Locke sometimes adds number, solidity, texture (where this is thought of as the structure of a substance, or way in which it is made out of atoms). The secondary qualities are the powers to excite particular sensory modifications in observers. Once, again, that Locke himself thought in terms of identifying these powers with the texture of objects that, according to corpuscularian science of the time, were the basis of an object's causal capacities. The ideas of secondary qualities are sharply different from these powers, and afford us no accurate impression of them. For Ren Descartes (1596-1650), this is the basis for rejecting any attempt to think of knowledge of external objects as provided by the senses. But in Locke our ideas of primary qualities do afford us an accurate notion of what shape, size, and mobility are. In English-speaking philosophy the first major discontent with the division was voiced by the

Irish idealist George Berkeley (1685-1753), who probably took for a basis of his attack from Pierre Bayle (1647-1706), who in turn cites the French critic Simon Foucher (1644-96). Modern thought continues to wrestle with the difficulties of thinking of colour, taste, smell, warmth, and sound as real or objective properties to things independent of us.

Continuing as such, is the doctrine advocated by the American philosopher David Lewis (1941-2002), in that different possible worlds are to be thought of as existing exactly as this one does. Thinking in terms of possibilities is thinking of real worlds where things are different. The view has been charged with making it impossible to see why it is good to save the child from drowning, since there is still a possible world in which she (or her counterpart) drowned, and from the standpoint of the universe it should make no difference which world is actual. Critics also charge that the notion fails to fit either with a coherent theory If how we know about possible worlds, or with a coherent theory of why we are interested in them, but Lewis denied that any other way of interpreting modal statements is tenable.

The proposal set forth that characterizes the 'modality' of a proposition as the notion for which it is true or false. The most important division is between propositions true of necessity, and those true as things are: Necessary as opposed to contingent propositions. Other qualifiers sometimes called 'modal' include the tense indicators, it will be the case that 'p', or 'it was the case that 'p', and there are affinities between the 'deontic' indicators, 'it ought to be the case that 'p', or 'it is permissible that 'p', and the of necessity and possibility.

The aim of a logic is to make explicit the rules by which inferences may be drawn, than to study the actual reasoning processes that people use, which may or may not conform to those rules. In the case of deductive logic, if we ask why we need to obey the rules, the most general form of answer is that if we do not we contradict ourselves(or, strictly speaking, we stand ready to contradict ourselves. Someone failing to draw a conclusion that follows from a set of premises need not be contradicting him or herself, but only failing to notice something. However, he or she is not defended against adding the contradictory conclusion to his or her set of beliefs.) There is no equally simple answer in the case of inductive logic, which is in general a less robust subject, but the aim will be to find reasoning such hat anyone failing to conform to it will have improbable beliefs. Traditional logic dominated the subject until the 19th century, and has become increasingly recognized in the 20th century. That finer work that was done within that tradition, but syllogistic reasoning is now generally regarded as a limited special case of the form of reasoning that can be reprehend within the promotion and predated values. These form the heart of modern logic, as their central notions or qualifiers, variables, and functions were the creation of the German mathematician Gottlob Frége, who is recognized as the father of modern logic, although his treatment of a logical system as abreacts mathematical structure, or algebraic, has been heralded by the English mathematician and logician George Boole (1815-64), his pamphlet. The Mathematical Analysis of Logic (1847) pioneered the algebra of classes. The work was made of in An Investigation of the Laws of Thought (1854). Boole also published many works in our mathematics, and on the theory of probability. His name is remembered in the title of Boolean algebra, and the algebraic operations he investigated are denoted by Boolean operations.

The syllogistic, or categorical syllogism is the inference of one proposition from two premises. For example is, 'all horses have tails', and 'things with tails are four legged', so 'all horses are four legged'.

Each premise has one term in common with the other premises. The term that ds not occur in the conclusion is called the middle term. The major premise of the syllogism is the premise containing the predicate of the contraction (the major term). And the minor premise contains its subject (the minor term). So the first premise of the example in the minor premise the second the major term. So the first premise of the example is the minor premise, the second the major premise and 'having a tail' is the middle term. This enable syllogisms that there of a classification, that according to the form of the premises and the conclusions. The other classification is by figure, or way in which the middle term is placed or way in within the middle term is placed in the premise.

Although the theory of the syllogism dominated logic until the 19th century, it remained a piecemeal affair, able to deal with only relations valid forms of valid forms of argument. There have subsequently been reargued actions attempting, but in general it has been eclipsed by the modern theory of quantification, the predicate calculus is the heart of modern logic, having proved capable of formalizing the calculus rationing processes of modern mathematics and science. In a first-order predicate calculus the variables range over objects: In a higher-order calculus the may range over predicate and functions themselves. The fist-order predicated calculus with identity includes '=' as primitive (undefined) expression: In a higher-order calculus It may be defined by law that $\chi = y$ iff $(Fx)(F\neg y)$, which gives grater expressive power for less complexity.

Modal logic was of great importance historically, particularly in the light of the deity, but was not a central topic of modern logic in its gold period as the beginning of the 20th century. It was, however, revived by the American logician and philosopher Irving Lewis (1883-1964), although he wrote extensively on most central philosophical topis, he is remembered principally as a critic of the intentional nature of modern logic, and as the founding father of modal logic. His two independent proofs showing that from a contradiction anything follows a relevance logic, using a notion of entailment stronger than that of strict implication.

In Saul Kripke, gives the classical modern treatment of the topic of reference, both clarifying the distinction between names and definite description, and opening te door to many subsequent attempts to understand the notion of reference in terms of a causal link between the use of a term and an original episode of attaching a name to the subject.

One of the three branches into which 'semiotic' is usually divided, the study of semantical meaning of words, and the relation of signs to the degree to which the designs are applicable. In that, in formal studies, a semantics is provided for a formal language when an interpretation of 'model' is specified. However, a natural language comes ready interpreted, and the semantic problem is not that of specification but of understanding the relationship between terms of various categories (names, descriptions, predicate, adverbs . . .) and their meaning. An influential proposal by attempting to provide a truth definition for the language, which will involve giving a full structure of different kinds have on the truth conditions of sentences containing them.

Holding that the basic case of reference is the relation between a name and the persons or object which it names. The philosophical problems include trying to elucidate that relation, to understand whether other semantic relations, such s that between a predicate and the property it expresses, or that between a description an what it describes, or that between myself and the word 'I', are examples of the same relation or of very different ones. A great deal of modern work on this was stimulated by the American logician Saul Kripke's, Naming and Necessity (1970). It would also be desirable to know whether we can refer to such things as objects and how to conduct the debate about each and issue. A popular approach, following Gottlob Frége, is to argue that the fundamental unit of analysis should be the whole sentence. The reference of a term becomes a derivative notion it is whatever it is that defines the term's contribution to the trued condition of the whole sentence. There need be nothing further to say about it, given that we have a way of understanding the attribution of meaning or truth-condition to sentences. Other approach, searching for a more substantive possibly that causality or psychological or social constituents are pronounced between words and things.

However, following Ramsey and the Italian mathematician G. Peano (1858-1932), it has been customary to distinguish logical paradoxes that depend upon a notion of reference or truth (semantic notions) such as those of the 'Liar family,, Berry, Richard, and so forth form the purely logical paradoxes in which no such notions are involved, such as Russell's paradox, or those of Canto and Burali-Forti. Paradoxes of the fist type sem to depend upon an element of self-reference, in which a sentence is about itself, or in which a phrase refers to something about itself, or in which a phrase refers to something defined by a set of phrases of which it is itself one. It is to feel that this element is responsible for the contradictions, although self-reference itself is often benign (for instance, the sentence 'All English sentences should have a verb', includes itself happily in the domain of sentences it is talking about), so the difficulty lies in forming a condition that existence only pathological self-reference. Paradoxes of the second kind then need a different treatment. Whilst the distinction is convenient, for allowing set theory to proceed by circumventing the latter paradoxes by technical and, even when there is no solution to the semantic paradoxes, it may be a way of ignoring the similarities between the two families. There is still the possibility that while there is no agreed solution to the semantic paradoxes, our understand of Russell's paradox may be imperfect as well.

Truth and falsity are two classical truth-values that a statement, proposition or sentence can take, as it is supposed in classical (two-valued) logic, that each statement has one of these values, and non has both. A statement is then false if and only if it is not true. The basis of this scheme is that to each statement there corresponds a determinate truth condition, or way the world must be for it to be true: If this condition obtains the statement is true, and otherwise false. Statements may indeed be felicitous or infelicitous in other dimensions (polite, misleading, apposite, witty, and so forth) but truth is the central normative notion governing assertion. Considerations o vagueness may introduce greys into this black-and-white scheme. For the issue to be true, any suppressed premise or background framework of thought necessary make an agreement valid, or a position tenable, a proposition whose truth is necessary for either the truth or the falsity of another statement. Thus if 'p' presupposes 'q', 'q' must be true for 'p' to be either true or false. In the theory of knowledge, the English philosopher and historian George Collingwood (1889-1943), announces hat any proposition capable of truth or falsity stand on bed of 'absolute presuppositions' which are not properly capable of truth or falsity, since a system of thought will contain no way of approaching such a question (a similar idea later voiced by Wittgenstein in his work On Certainty). The introduction of presupposition therefore and that either another of a truth value is fond, 'intermediate' between truth and falsity, or the classical logic is preserved, but it is impossible to tell whether a particular sentence empresses a preposition that is a candidate for truth and falsity, without knowing more than the formation rules of the language. Each suggestion carries coss, and there is some consensus that at least who where definite descriptions are involved, examples equally given by regarding the overall sentence as false as the existence claim fails, and explaining the data that the English philosopher Frederick Strawson (1919-) relied upon as the effects of 'implicature'.

Views about the meaning of terms will often depend on classifying the implicature of sayings involving the terms as implicatures or as genuine logical implications of what is said. Implicatures may be divided into two kinds: Conversational implicatures of the two kinds and the more subtle category of conventional implicatures. A term may as a matter of convention carry an implicature, thus one of the relations between 'he is poor and honest' and 'he is poor but honest' is that they have the same content (are true in just the same conditional) but the second has implicatures (that the combination is surprising or significant) that the first lacks.

It is, nonetheless, that we find in classical logic a proposition that may be true or false,. In that, if the former, it is said to take the truth-value true, and if the latter the truth-value false. The idea behind the terminological phrases is the analogues between assigning a propositional variable one or other of these values, as is done in providing an interpretation for a formula of the propositional calculus, and assigning an object as the value of any other variable. Logics with intermediate value are called 'many-valued logics'.

Nevertheless, an existing definition of the predicate' . . . is true' for a language that satisfies convention 'T', the material adequately condition laid down by Alfred Tarski, born Alfred Teitelbaum (1901-83), whereby his methods of 'recursive' definition, enabling us to say for each sentence what it is that its truth consists in, but giving no verbal definition of truth itself. The recursive definition or the truth predicate of a language is always provided in a 'metalanguage', Tarski is thus committed to a hierarchy of languages, each with its associated, but different truth-predicate. Whist this enables the approach to avoid the contradictions of paradoxical contemplations, it is to be marked out of harmony. The parting emulation as due by the disconsonant irrigations of conflicts as mixed in disaccorded anticipations to the idea that a language should be of enabling, in at least, of a few words for saying everything that there is to express in words, that an oft-repeated statements usually involving common experience or observation, in that the old saying that ignorance is bliss, nonetheless, it might just be of something in say-so, yet, approaches have become increasingly important.

So, that the truth condition of a statement is the condition for which the world must meet if the statement is to be true. To know this condition is equivalent to knowing the meaning of the statement. Although this sounds as if it gives a solid anchorage for meaning, some of the securities disappear when it turns out that the truth condition can only be defined by repeating the very same statement: The truth condition of 'now is white' is that 'snow is white', the truth condition of 'Britain would have capitulated had Hitler invaded', is that 'Britain would have capitulated had Hitler invaded'. It is disputed whether this element of running-on-the-spot disqualifies truth conditions from playing the central role in a substantives theory of meaning. Truth-conditional theories of meaning are sometimes opposed by the view that to know the meaning of a statement is to be able to use it in a network of inferences.

Taken to be the view, inferential semantics take on the role of sentence in inference give a more important key to their meaning than this 'external' relations to things in the world. The meaning of a sentence becomes its place in a network of inferences that it legitimates. Also known as functional role semantics, procedural semantics, or conception to the coherence theory of truth, and suffers from the same suspicion that it divorces meaning from any clear association with things in the world.

Moreover, a theory of semantic truth be that of the view if language is provided with a truth definition, there is a sufficient characterization of its concept of truth, as there is no further philosophical chapter to write about truth: There is no further philosophical chapter to write about truth itself or truth as shared across different languages. The view is similar to the Disquotational theory.

The redundancy theory, or also known as the 'deflationary view of truth' fathered by Gottlob Frége and the Cambridge mathematician and philosopher Frank Ramsey (1903-30), who showed how the distinction between the semantic paradoxes, such as that of the Liar, and Russell's paradox, made unnecessary the ramified type theory of Principia Mathematica, and the resulting axiom of reducibility. By taking all the sentences affirmed in a scientific theory that use some terms e.g., quark, and to a considerable degree of replacing the term by a variable instead of saying that quarks have such-and-such properties, the Ramsey sentence says that there is something that has those properties. If the process is repeated for all of a group of the theoretical terms, the sentence gives 'topic-neutral' structure of the theory, but removes any implication that we know what the terms so treated denote. It leaves open the possibility of identifying the theoretical item with whatever it is that best fits the description provided. However, it was pointed out by the Cambridge mathematician Newman, that if the process is carried out for all except the logical bones of a theory, then by the Löwenheim-Skolem theorem, the result will be interpretable, and the content of the theory may reasonably be felt to have been lost.

All the while, both Frége and Ramsey are agreed that the essential claim is that the predicate' . . . is true' does not have a sense, i.e., expresses no substantive or profound or explanatory concept that ought to be the topic of philosophical enquiry. The approach admits of different versions, but centres on the points (1) that 'it is true that 'p' says no more nor less than 'p' (hence, redundancy): (2) that in less direct contexts, such as 'everything he said was true', or 'all logical consequences of true propositions are true', the predicate functions as a device enabling us to generalize than as an adjective or predicate describing the things he said, or the kinds of propositions that follow from true preposition. For example, the second may translate as '(p, q)(p & p) ® q ® q where there is no use of a notion of truth.

There are technical problems in interpreting all uses of the notion of truth in such ways, nevertheless, they are not generally felt to be insurmountable. The approach needs to explain away apparently substantive uses of the notion, such as 'science aims at the truth', or 'truth is a norm governing discourse'. Postmodern writing frequently advocates that we must abandon such norms. Along with a discredited 'objective' conception of truth. Perhaps, we can have the norms even when objectivity is problematic, since they can be framed without mention of truth: Science wants it to be so that whatever science holds that 'p', then 'p'. Discourse is to be regulated by the principle that it is wrong to assert 'p', when 'not-p'.

Something that tends of something in addition of content, or coming by way to justify such a position can very well be more that in addition to several reasons, as to bring in or join of something might that there be more so as to a larger combination for us to consider the simplest formulation, is that the claim that expression of the form 'S is true' mean the same as expression of the form 'S'. Some philosophers dislike the ideas of sameness of meaning, and if this I disallowed, then the claim is that the two forms are equivalent in any sense of equivalence that matters. This is, it makes no difference whether people say 'Dogs bark' id Tue, or whether they say, 'dogs bark'. In the former representation of what they say of the sentence 'Dogs bark' is mentioned, but in the later it appears to be used, of the claim that the two are equivalent and needs careful formulation and defence. On the face of it someone might know that 'Dogs bark' is true without knowing what it (for instance, if he kids in a list of acknowledged truths, although he does not understand English), and this is different from knowing that dogs bark. Disquotational theories are usually presented as versions of the 'redundancy theory of truth'.

The relationship between a set of premises and a conclusion when the conclusion follows from the premise. Several philosophers identify this with it being logically impossible that the premises should all be true, yet the conclusion false. Others are sufficiently impressed by the paradoxes of strict implication to look for a stranger relation, which would distinguish between valid and invalid arguments within the sphere of necessary propositions. The seraph for a strange notion is the field of relevance logic.

From a systematic theoretical point of view, we may imagine the process of evolution of an empirical science to be a continuous process of induction. Theories are evolved and are expressed in short compass as statements of as large number of individual observations in the form of empirical laws, from which the general laws can be ascertained by comparison. Regarded in this way, the development of a science bears some resemblance to the compilation of a classified catalogue. It is, a it were, a purely empirical enterprise.

But this point of view by no embraces the whole of the actual process, for it slurs over the important part played by intuition and deductive thought in the development of an exact science. As soon as a science has emerged from its initial stages, theoretical advances are no longer achieved merely by a process of arrangement. Guided by empirical data, the investigators rather develops a system of thought which, in general, it is built up logically from a small number of fundamental assumptions, the so-called axioms. We call such a system of thought a 'theory'. The theory finds the justification for its existence in the fact that it correlates a large number of single observations, and is just here that the 'truth' of the theory lies.

Corresponding to the same complex of empirical data, there may be several theories, which differ from one another to a considerable extent. But as regards the deductions from the theories which are capable of being tested, the agreement between the theories may be so complete, that it becomes difficult to find any deductions in which the theories differ from each other. As an example, a case of general interest is available in the province of biology, in the Darwinian theory of the development of species by selection in the struggle for existence, and in the theory of development which is based on the hypophysis of the hereditary transmission of acquired characters. The Origin of Species was principally successful in marshalling the evidence for evolution, than providing a convincing mechanisms for genetic change. And Darwin himself remained open to the search for additional mechanisms, while also remaining convinced that natural selection was at the hart of it. It was only with the later discovery of the gene as the unit of inheritance that the synthesis known as 'neo-Darwinism' became the orthodox theory of evolution in the life sciences.

In the 19th century the attempt to base ethical reasoning o the presumed facts about evolution, the movement is particularly associated with the English philosopher of evolution Herbert Spencer (1820-1903). The premise is that later elements in an evolutionary path are better than earlier ones: The application of this principle then requires seeing western society, laissez-faire capitalism, or some other object of approval, as more evolved than more 'primitive' social forms. Neither the principle nor the applications command much respect. The version of evolutionary ethics called 'social Darwinism' emphasizes the struggle for natural selection, and draws the conclusion that we should glorify and assist such struggle, usually by enhancing competition and aggressive relations between people in society or between evolution and ethics has been rethought in the light of biological discoveries concerning altruism and kin-selection.

Once again, the psychology proving attempts are founded to evolutionary principles, in which a variety of higher mental functions may be adaptations, forced in response to selection pressures on the human populations through evolutionary time. Candidates for such theorizing include material and paternal motivations, capacities for love and friendship, the development of language as a signalling system cooperative and aggressive, our emotional repertoire, our moral and reactions, including the disposition to detect and punish those who cheat on agreements or who 'free-ride' on =the work of others, our cognitive structures, and several others. Evolutionary psychology goes hand-in-hand with neurophysiological evidence about the underlying circuitry in the brain which subserves the psychological mechanisms it claims to identify. The approach was foreshadowed by Darwin himself, and William James, as well as the sociology of E.O. Wilson. The term of use are applied, more or less aggressively, especially to explanations offered in Sociobiology and evolutionary psychology.

Another assumption that is frequently used to legitimate the real existence of forces associated with the invisible hand in neoclassical economics derives from Darwin's view of natural selection as a warlike competing between atomized organisms in the struggle for survival. In natural selection as we now understand it, cooperation appears to exist in complementary relation to competition. It is complementary relationships between such results that are emergent self-regulating properties that are greater than the sum of parts and that serve to perpetuate the existence of the whole.

According to E.O Wilson, the 'human mind evolved to believe in the gods' and people 'need a sacred narrative' to have a sense of higher purpose. Yet it id also clear that the 'gods' in his view are merely human constructs and, therefore, there is no basis for dialogue between the world-view of science and religion. 'Science for its part', said Wilson, 'will test relentlessly every assumption about the human condition and in time uncover the bedrock of the moral an religious sentiments. The eventual result of the competition between each of the other, will be the secularization of the human epic and of religion itself.

Man has come to the threshold of a state of consciousness, regarding his nature and his relationship to te Cosmos, in terms that reflect 'reality'. By using the processes of nature as metaphor, to describe the forces by which it operates upon and within Man, we come as close to describing 'reality' as we can within the limits of our comprehension. Men will be very uneven in their capacity for such understanding, which, naturally, differs for different ages and cultures, and develops and changes over the course of time. For these reasons it will always be necessary to use metaphor and myth to provide 'comprehensible' guides to living. In thus way. Man's imagination and intellect play vital roles on his survival and evolution.

Since so much of life both inside and outside the study is concerned with finding explanations of things, it would be desirable to have a concept of what counts as a good explanation from bad. Under the influence of 'logical positivist' approaches to the structure of science, it was felt that the criterion ought to be found in a definite logical relationship between the 'explanans' (that which does the explaining) and the explanandum (that which is to be explained). The approach culminated in the covering law model of explanation, or the view that an event is explained when it is subsumed under a law of nature, that is, its occurrence is deducible from the law plus a set of initial conditions. A law would itself be explained by being deduced from a higher-order or covering law, in the way that Johannes Kepler(or Keppler, 1571-1630), was by way of planetary motion that the laws were deducible from Newton's laws of motion. The covering law model may be adapted to include explanation by showing that something is probable, given a statistical law. Questions for the covering law model include querying for the covering law are necessary to explanation (we explain whether everyday events without overtly citing laws): Querying whether they are sufficient (it ma y not explain an event just to say that it is an example of the kind of thing that always happens). And querying whether a purely logical relationship is adapted to capturing the requirements we make of explanations. These may include, for instance, that we have a 'feel' for what is happening, or that the explanation proceeds in terms of things that are familiar to us or unsurprising, or that we can give a model of what is going on, and none of these notions is captured in a purely logical approach. Recent work, therefore, has tended to stress the contextual and pragmatic elements in requirements for explanation, so that what counts as good explanation given one set of concerns may not do so given another.

The argument to the best explanation is the view that once we can select the best of any in something in explanations of an event, then we are justified in accepting it, or even believing it. The principle needs qualification, since something it is unwise to ignore the antecedent improbability of a hypothesis which would explain the data better than others, e.g., the best explanation of a coin falling heads 530 times in 1,000 tosses might be that it is biassed to give a probability of heads of 0.53 but it might be more sensible to suppose that it is fair, or to suspend judgement.

In a philosophy of language is considered as the general attempt to understand the components of a working language, the relationship the understanding speaker has to its elements, and the relationship they bear to the world. The subject therefore embraces the traditional division of semiotic into syntax, semantics, an d pragmatics. The philosophy of language thus mingles with the philosophy of mind, since it needs an account of what it is in our understanding that enables us to use language. It so mingles with the metaphysics of truth and the relationship between sign and object. Much as much is that the philosophy in the 20th century, has been informed by the belief that philosophy of language is the fundamental basis of all philosophical problems, in that language is the distinctive exercise of mind, and the distinctive way in which we give shape to metaphysical beliefs. Particular topics will include the problems of logical form,. And the basis of the division between syntax and semantics, as well as problems of understanding the number and nature of specifically semantic relationships such as meaning, reference, predication, and quantification. Pragmatics include that of speech acts, while problems of rule following and the indeterminacy of translation infect philosophies of both pragmatics and semantics.

On this conception, to understand a sentence is to know its truth-conditions, and, yet, in a distinctive way the conception has remained central that those who offer opposing theories characteristically define their position by reference to it. The Concepcion of meaning s truth-conditions need not and should not be advanced as being in itself as complete account of meaning. For instance, one who understands a language must have some idea of the range of speech acts contextually performed by the various types of sentence in the language, and must have some idea of the insufficiencies of various kinds of speech act. The claim of the theorist of truth-conditions should rather be targeted on the notion of content: If indicative sentence differ in what they strictly and literally say, then this difference is fully accounted for by the difference in the truth-conditions.

The meaning of a complex expression is a function of the meaning of its constituent. This is just as a sentence of what it is for an expression to be semantically complex. It is one of the initial attractions of the conception of meaning truth-conditions tat it permits a smooth and satisfying account of the way in which the meaning of s complex expression is a function of the meaning of its constituents. On the truth-conditional conception, to give the meaning of an expression is to state the contribution it makes to the truth-conditions of sentences in which it occurs. For singular terms - proper names, indexical, and certain pronouns - this is done by stating the reference of the terms in question. For predicates, it is done either by stating the conditions under which the predicate is true of arbitrary objects, or by stating the conditions under which arbitrary atomic sentences containing it are true. The meaning of a sentence-forming operator is given by stating its contribution to the truth-conditions of as complex sentence, as a function of he semantic values of the sentences on which it operates.

The theorist of truth conditions should insist that not every true statement about the reference of an expression is fit to be an axiom in a meaning-giving theory of truth for a language, such is the axiom: 'London' refers to the city in which there was a huge fire in 1666, is a true statement about the reference of 'London'. It is a consequent of a theory which substitutes this axiom for no different a term than of our simple truth theory that 'London is beautiful' is true if and only if the city in which there was a huge fire in 1666 is beautiful. Since a subject can understand the name 'London' without knowing that last-mentioned truth condition, this replacement axiom is not fit to be an axiom in a meaning-specifying truth theory. It is, of course, incumbent on a theorized meaning of truth conditions, to state in a way which does not presuppose any previous, non-truth conditional conception of meaning

Among the many challenges facing the theorist of truth conditions, two are particularly salient and fundamental. First, the theorist has to answer the charge of triviality or vacuity, second, the theorist must offer an account of what it is for a person's language to be truly describable by as semantic theory containing a given semantic axiom. Since the content of a claim that the sentence 'Paris is beautiful' is true amounts to no more than the claim that Paris is beautiful, we can trivially describers understanding a sentence, if we wish, as knowing its truth-conditions, but this gives us no substantive account of understanding whatsoever. Something other than grasp of truth conditions must provide the substantive account. The charge rests upon what has been called the redundancy theory of truth, the theory which, somewhat more discriminative. Horwich calls the minimal theory of truth. Its conceptual representation that the concept of truth is exhausted by the fact that it conforms to the equivalence principle, the principle that for any proposition 'p', it is true that 'p' if and only if 'p'. Many different philosophical theories of truth will, with suitable qualifications, accept that equivalence principle. The distinguishing feature of the minimal theory is its claim that the equivalence principle exhausts the notion of truth. It is now widely accepted, both by opponents and supporters of truth conditional theories of meaning, that it is inconsistent to accept both minimal theory of ruth and a truth conditional account of meaning. If the claim that the sentence 'Paris is beautiful' is true is exhausted by its equivalence to the claim that Paris is beautiful, it is circular to try of its truth conditions. The minimal theory of truth has been endorsed by the Cambridge mathematician and philosopher Plumpton Ramsey (1903-30), and the English philosopher Jules Ayer, the later Wittgenstein, Quine, Strawson. Horwich and - confusing and inconsistently if this article is correct - Frége himself. but is the minimal theory correct?

The minimal theory treats instances of the equivalence principle as definitional of truth for a given sentence, but in fact, it seems that each instance of the equivalence principle can itself be explained. The truths from which such an instance as: 'London is beautiful' is true if and only if London is beautiful. This would be a pseudo-explanation if the fact that 'London' refers to London consists in part in the fact that 'London is beautiful' has the truth-condition it does. But it is very implausible, it is, after all, possible to understand the name 'London' without understanding the predicate 'is beautiful'.

Sometimes, however, the counterfactual conditional is known as subjunctive conditionals, insofar as a counterfactual conditional is a conditional of the form 'if p were to happen q would', or 'if p were to have happened q would have happened', where the supposition of 'p' is contrary to the known fact that 'not-p'. Such assertions are nevertheless, use=ful 'if you broken the bone, the X-ray would have looked different', or 'if the reactor were to fail, this mechanism wold click in' are important truths, even when we know that the bone is not broken or are certain that the reactor will not fail. It is arguably distinctive of laws of nature that yield counterfactuals ('if the metal were to be heated, it would expand'), whereas accidentally true generalizations may not. It is clear that counterfactuals cannot be represented by the material implication of the propositional calculus, since that conditionals comes out true whenever 'p' is false, so there would be no division between true and false counterfactuals.

Although the subjunctive form indicates a counterfactual, in many contexts it does not seem to matter whether we use a subjunctive form, or a simple conditional form: 'If you run out of water, you will be in trouble' seems equivalent to 'if you were to run out of water, you would be in trouble', in other contexts there is a big difference: 'If Oswald did not kill Kennedy, someone else did' is clearly true, whereas 'if Oswald had not killed Kennedy, someone would have' is most probably false.

The best-known modern treatment of counterfactuals is that of David Lewis, which evaluates them as true or false according to whether 'q' is true in the 'most similar' possible worlds to ours in which 'p' is true. The similarity-ranking this approach needs has proved controversial, particularly since it may need to presuppose some notion of the same laws of nature, whereas art of the interest in counterfactuals is that they promise to illuminate that notion. There is a growing awareness tat the classification of conditionals is an extremely tricky business, and categorizing them as counterfactuals or not be of limited use.

The pronouncing of any conditional; preposition of the form 'if p then q'. The condition hypothesizes, 'p'. Its called the antecedent of the conditional, and 'q' the consequent. Various kinds of conditional have been distinguished. The weaken in that of material implication, merely telling us that with not-p. or q. stronger conditionals include elements of modality, corresponding to the thought that if 'p' is true then 'q' must be true. Ordinary language is very flexible in its use of the conditional form, and there is controversy whether, yielding different kinds of conditionals with different meanings, or pragmatically, in which case there should be one basic meaning which case there should be one basic meaning, with surface differences arising from other implicatures.

We now turn to a philosophy of meaning and truth, for which it is especially associated with the American philosopher of science and of language (1839-1914), and the American psychologist philosopher William James (1842-1910), wherefore the study in Pragmatism is given to various formulations by both writers, but the core is the belief that the meaning of a doctrine is the same as the practical effects of adapting it. Peirce interpreted of theoretical sentence is only that of a corresponding practical maxim (telling us what to do in some circumstance). In James the position issues in a theory of truth, notoriously allowing that belief, including for example, belief in God, arc the widest sense of the works satisfactorily in the widest sense of the word. On James's view almost any belief might be respectable, and even true, provided it calls to mind (but working is no s simple matter for James). The apparently subjectivist consequences of this were wildly assailed by Russell (1872-1970), Moore (1873-1958), and others in the early years of the 20 century. This led to a division within pragmatism between those such as the American educator John Dewey (1859-1952), whose humanistic conception of practice remains inspired by science, and the more idealistic route that especially by the English writer F.C.S. Schiller (1864-1937), embracing the doctrine that our cognitive efforts and human needs actually transforms the reality that we seek to describe. James often writes as if he sympathizes with this development. For instance, in The Meaning of Truth (1909), he considers the hypothesis that other people have no minds (dramatized in the sexist idea of an 'automatic sweetheart' or female zombie) and remarks hat the hypothesis would not work because it would not satisfy our egoistic craving for the recognition and admiration of others. The implication that this is what makes it true that the other persons have minds in the disturbing part.

Modern pragmatists such as the American philosopher and critic Richard Rorty (1931-) and some writings of the philosopher Hilary Putnam (1925-) who have usually trued to dispense with an account of truth and concentrate, as perhaps James should have done, upon the nature of belief and its relations with human attitude, emotion, and needs. The driving motivation of pragmatism is the idea that belief in the truth on te one hand must have a close connection with success in action on the other. One way of cementing the connection is found in the idea that natural selection must have adapted us to be cognitive creatures because belief have effects, as they work. Pragmatism can be found in Kant's doctrine of the primary of practical over pure reason, and continued to play an influential role in the theory of meaning and of truth.

In case of fact, the philosophy of mind is the modern successor to behaviourism, as do the functionalism that its early advocates were Putnam (1926-) and Sellars (1912-89), and its guiding principle is that we can define mental states by a triplet of relations they have on other mental stares, what effects they have on behaviour. The definition need not take the form of a simple analysis, but if w could write down the totality of axioms, or postdate, or platitudes that govern our theories about what things of other mental states, and our theories about what things are apt to cause (for example), a belief state, what effects it would have on a variety of other mental states, and what effects it is likely to have on behaviour, then we would have done all tat is needed to make the state a proper theoretical notion. It could be implicitly defied by these theses. Functionalism is often compared with descriptions of a computer, since according to mental descriptions correspond to a description of a machine in terms of software, that remains silent about the underlaying hardware or 'realization' of the program the machine is running. The principle advantage of functionalism include its fit with the way we know of mental states both of ourselves and others, which is via their effects on behaviour and other mental states. As with behaviourism, critics charge that structurally complex items tat do not bear mental states might nevertheless, imitate the functions that are cited. According to this criticism functionalism is too generous and would count too many things as having minds. It is also queried whether functionalism is too paradoxical, able to see mental similarities only when there is causal similarity, when our actual practices of interpretations enable us to ascribe thoughts and desires to different from our own, it may then seem as though beliefs and desires can be 'variably realized' causal architecture, just as much as they can be in different neurophysiological states.

The philosophical movement of Pragmatism had a major impact on American culture from the late 19th century to the present. Pragmatism calls for ideas and theories to be tested in practice, by assessing whether acting upon the idea or theory produces desirable or undesirable results. According to pragmatists, all claims about truth, knowledge, morality, and politics must be tested in this way. Pragmatism has been critical of traditional Western philosophy, especially the notion that there are absolute truths and absolute values. Although pragmatism was popular for a time in France, England, and Italy, most observers believe that it encapsulates an American faith in know-how and practicality and an equally American distrust of abstract theories and ideologies.

In mentioning the American psychologist and philosopher we find William James, who helped to popularize the philosophy of pragmatism with his book Pragmatism: A New Name for Old Ways of Thinking (1907). Influenced by a theory of meaning and verification developed for scientific hypotheses by American philosopher C. S. Peirce, James held that truth is what works, or has good experimental results. In a related theory, James argued the existence of God is partly verifiable because many people derive benefits from believing.

The Association for International Conciliation first published William James's pacifist statement, 'The Moral Equivalent of War', in 1910. James, a highly respected philosopher and psychologist, was one of the founders of pragmatism - a philosophical movement holding that ideas and theories must be tested in practice to assess their worth. James hoped to find a way to convince men with a long-standing history of pride and glory in war to evolve beyond the need for bloodshed and to develop other avenues for conflict resolution. Spelling and grammar represent standards of the time.

Pragmatists regard all theories and institutions as tentative hypotheses and solutions. For this reason they believed that efforts to improve society, through such as education or politics, must be geared toward problem solving and must be ongoing. Through their emphasis on connecting theory to practice, pragmatist thinkers attempted to transform all areas of philosophy, from metaphysics to ethics and political philosophy.

Pragmatism sought a middle ground between traditional ideas about the nature of reality and radical theories of nihilism and irrationalism, which had become popular in Europe in the late 19th century. Traditional metaphysics assumed that the world has a fixed, intelligible structure and that human beings can know absolute or objective truths about the world and about what constitutes moral behaviour. Nihilism and irrationalism, on the other hand, denied those very assumptions and their certitude. Pragmatists today still try to steer a middle course between contemporary offshoots of these two extremes.

The ideas of the pragmatists were considered revolutionary when they first appeared. To some critics, pragmatism's refusal to affirm any absolutes carried negative implications for society. For example, pragmatists do not believe that a single absolute idea of goodness or justice exists, but rather that these concepts are changeable and depend on the context in which they are being discussed. The absence of these absolutes, critics feared, could result in a decline in moral standards. The pragmatists' denial of absolutes, moreover, challenged the foundations of religion, government, and schools of thought. As a result, pragmatism influenced developments in psychology, sociology, education, semiotics (the study of signs and symbols), and scientific method, as well as philosophy, cultural criticism, and social reform movements. Various political groups have also drawn on the assumptions of pragmatism, from the progressive movements of the early 20th century to later experiments in social reform.

Pragmatism is best understood in its historical and cultural context. It arose during the late 19th century, a period of rapid scientific advancement typified by the theories of British biologist Charles Darwin, whose theories suggested to many thinkers that humanity and society are in a perpetual state of progress. During this same period a decline in traditional religious beliefs and values accompanied the industrialization and material progress of the time. In consequence it became necessary to rethink fundamental ideas about values, religion, science, community, and individuality.

The three most important pragmatists are American philosophers Charles Sanders Peirce, William James, and John Dewey. Peirce was primarily interested in scientific method and mathematics; his objective was to infuse scientific thinking into philosophy and society, and he believed that human comprehension of reality was becoming ever greater and that human communities were becoming increasingly progressive. Peirce developed pragmatism as a theory of meaning - in particular, the meaning of concepts used in science. The meaning of the concept 'brittle', for example, is given by the observed consequences or properties that objects called 'brittle' exhibit. For Peirce, the only rational way to increase knowledge was to form mental habits that would test ideas through observation, experimentation, or what he called inquiry. Many philosophers known as logical positivists, a group of philosophers who have been influenced by Peirce, believed that our evolving species was fated to get ever closer to Truth. Logical positivists emphasize the importance of scientific verification, rejecting the assertion of positivism that personal experience is the basis of true knowledge.

James moved pragmatism in directions that Peirce strongly disliked. He generalized Peirce's doctrines to encompass all concepts, beliefs, and actions; he also applied pragmatist ideas to truth as well as to meaning. James was primarily interested in showing how systems of morality, religion, and faith could be defended in a scientific civilization. He argued that sentiment, as well as logic, is crucial to rationality and that the great issues of life - morality and religious belief, for example - are leaps of faith. As such, they depend upon what he called 'the will to believe' and not merely on scientific evidence, which can never tell us what to do or what is worthwhile. Critics charged James with relativism (the belief that values depend on specific situations) and with crass expediency for proposing that if an idea or action works the way one intends, it must be right. But James can more accurately be described as a pluralist - someone who believes the world to be far too complex for any one philosophy to explain everything.

Dewey's philosophy can be described as a version of philosophical naturalism, which regards human experience, intelligence, and communities as ever-evolving mechanisms. Using their experience and intelligence, Dewey believed, human beings can solve problems, including social problems, through inquiry. For Dewey, naturalism led to the idea of a democratic society that allows all members to acquire social intelligence and progress both as individuals and as communities. Dewey held that traditional ideas about knowledge, truth, and values, in which absolutes are assumed, are incompatible with a broadly Darwinian world-view in which individuals and society are progressing. In consequence, he felt that these traditional ideas must be discarded or revised. Indeed, for pragmatists, everything people know and do depends on a historical context and is thus tentative rather than absolute.

Many followers and critics of Dewey believe he advocated elitism and social engineering in his philosophical stance. Others think of him as a kind of romantic humanist. Both tendencies are evident in Dewey's writings, although he aspired to synthesize the two realms.

The pragmatist tradition was revitalized in the 1980s by American philosopher Richard Rorty, who has faced similar charges of elitism for his belief in the relativism of values and his emphasis on the role of the individual in attaining knowledge. Interest has renewed in the classic pragmatists - Pierce, James, and Dewey - have an alternative to Rorty's interpretation of the tradition.

Aristoteleans whose natural science dominated Western thought for two thousand years, believed that man could arrive at an understanding of ultimate reality by reasoning a form in self-evident principles. It is, for example, self-evident recognition as that the result that questions of truth becomes uneducable. Therefore in can be deduced that objects fall to the ground because that's where they belong, and goes up because that's where it belongs, the goal of Aristotelian science was to explain why things happen. Modern science was begun when Galileo began trying to explain how things happen and thus ordinated the method of controlled excitement which now form the basis of scientific investigation.

Classical scepticism springs from the observation that the best methods in some given area seem to fall short of giving us contact with truth (e.g., there is a gulf between appearances and reality), and it frequently cites the conflicting judgements that our methods deliver, with the results that questions of truth become undeniable. In classic thought the various examples of this conflict are a systemized or argument and ethics, as opposed to dogmatism, and particularly the philosophy system building of the Stoics

The Stoic school was founded in Athens around the end of the fourth century Bc by Zeno of Citium (335-263 Bc). Epistemological issues were a concern of logic, which studied logos, reason and speech, in all of its aspects, not, as we might expect, only the principles of valid reasoning - these were the concern of another division of logic, dialectic. The epistemological part, which concerned with canons and criteria, belongs to logic cancelled in this broader sense because it aims to explain how our cognitive capacities make possibly the full realization from reason in the form of wisdom, which the Stoics, in agreement with Socrates, equated with virtue and made the sole sufficient condition for human happiness.

Reason is fully realized as knowledge, which the Stoics defined as secure and firm cognition, unshakable by argument. According to them, no one except the wise man can lay claim to this condition. He is armed by his mastery of dialectic against fallacious reasoning which might lead him to draw a false conclusion from sound evidence, and thus possibly force him to relinquish the ascent he has already properly confers on a true impression. Hence, as long as he does not ascend to any false grounded-level impressions, he will be secure against error, and his cognation will have the security and firmness required of knowledge. Everything depends, then, on his ability to void error in high ground-level perceptual judgements. To be sure, the Stoics do not claim that the wise man can distinguish true from false perceptual impression: impressions: that is beyond even his powers, but they do maintain that there is a kind of true perceptual impression, the so-called cognitive impression, by confining his assent to which the wise man can avoid giving error a foothold.

An impression, none the least, is cognitive when it is (1) from what is (the case) (2) Stamped and impressed in accordance with what are, and, (3) such that could not arise from what is not. And because all of our knowledge depends directly or indirectly on it, the Stoics make the cognitive impression the criterion of truth. It makes possibly a secure grasp of the truth, and possibly a secure grasp on truth, not only by guaranteeing the truth of its own positional content, which in turn supported the conclusions that can be drawn from it: Even before we become capable of rational impressions, nature must have arranged for us to discriminate in favour of cognitive impressions that the common notions we end up with will be sound. And it is by of these concepts that we are able to extend our grasp of the truth through if inferences beyond what is immediately given, least of mention, the Stoics also speak of two criteria, cognitive impressions and common (the trust worthy common basis of knowledge).

A patternization in custom or habit of action, may exit without any specific basis in reason, however, the distinction between the real world, the world of the forms, accessible only to the intellect, and the deceptive world of displaced perceptions, or, merely a justified belief. The world forms are themselves a functioning change that implies development toward the realization of form. The problem of interpretations is, however confused by the question of whether of universals separate, but others, i.e., Plato did. It can itself from the basis for rational action, if the custom gives rise to norms of action. A theory that magnifies the role of decisions, or free selection from amongst equally possible alternatives, in order to show that what appears to be objective or fixed by nature is in fact an artefact of human convention, similar to convention of etiquette, or grammar, or law. Thus one might suppose that moral rules owe more to social convention than to anything inexorable necessities are in fact the shadow of our linguistic convention. In the philosophy of science, conventionalism is the doctrine often traced to the French mathematician and philosopher Jules Henry Poincaré that endorsed of an accurate and authentic science of differences, such that between describing space in terms of a Euclidean and non-Euclidean geometry, in fact register the acceptance of a different system of conventions for describing space. Poincaré did not hold that all scientific theory is conventional, but left space for genuinely experimental laws, and his conventionalism is in practice modified by recognition that one choice of description may be more conventional than another. The disadvantage of conventionalism is that it must show that alternative equal to workable conventions could have been adopted, and it is often not easy to believe that. For example, if we hold that some ethical norm such as respect for premises or property is conventional, we ought to be able to show that human needs would have been equally well satisfied by a system involving a different norm, and this may be hard to establish.

Poincaré made important original contributions to differential equations, topology, probability, and the theory of functions. He is particularly noted for his development of the so-called Fusain functions and his contribution to analytical mechanics. His studies included research into the electromagnetic theory of light and into electricity, fluid mechanics, heat tradfer, and thermodynamics. He also anticipated chaos theory. Amid the useful allowances that Jules Henri Poincaré took extra care with the greater of degree of carefully took in the vicinity of writing, more or less than 30 books, assembling, by and large, through which can be known as having an existence, but an attribute of things from Science and Hypothesis (1903; trs., 1905), The Value of Science (1905; trs., 1907), Science and Method (1908; trs., 1914), and The Foundations of Science (1902-8; trs., 1913). In 1887 Poincaré became a member of the French Academy of Sciences and served at its president up and until 1906. He also was elected to membership in the French Academy in 1908. Poincaré main philosophical interest lay in the physical formal and logical character of theories in the physical sciences. He is especially remembered for the discussion of the scientific status of geometry, in La Science and la et l' hypotheses, 1902, trs., As Science and Hypothesis, 1905, the axioms of geometry are analytic, nor do they state fundamental empirical properties of space, rather, they are conventions governing the descriptions of space, whose adoption too governed by their utility in furthering the purpose of description. By their unity in Poincaré conventionalism about geometry proceeded, however against the background of a general and the alliance of always insisting that there could be good reason for adopting one set of conventions than another in his late Dermt res Pensées (1912) trs., as the. Mathematics and Science: Last Essays.

A completed Unification Field Theory touches the 'grand aim of all science,' which Einstein once defined it, as, 'to cover the greatest number of empirical deductions from the smallest possible number of hypotheses or axioms.' But the irony of a man's quest for reality is that as nature is stripped of its disguises, as order emerges from chaos and unity from diversity. As concepts emerge and fundamental laws that assume an increasingly simpler form, the evolving pictures, that to become less recognizable than the bone structure behind a familiar distinguished appearance from reality and lay of bare the fundamental structure of the diverse, science that has had to transcend the 'rabble of the senses.' But it highest redefinition, as Einstein has pointed out, has been 'purchased at the price of empirical content.' A theoretical concept is emptied of content to the very degree that it is diversely taken from sensory experience. For the only world man can truly know is the world created for him by his senses. So paradoxically what the scientists and the philosophers' call the world of appearances - the world of light and colour, of blue skies and green leaves, of sighing winds and the murmuring of the water's creek, the world designed by the physiology of humans sense organs, are the worlds in which finite man is incarcerated by his essential nature and what the scientist and the philosophers call the world of reality. The colourless, soundless, impalpable cosmos which lies like an iceberg beneath the plane of man's perceptions - is a skeleton structure of symbols, and symbols change.

For all the promises of future revelation it is possible that certain terminal boundaries have already been reached in man's struggle to understand the manifold of nature in which he finds himself. In his descent into the microcosm's and encountered indeterminacy, duality, paradox - barriers that seem to admonish him and cannot pry too inquisitively into the heart of things without vitiating the processes he seeks to observe. Man's inescapable impasse is that he himself is part of the world he seeks to explore, his body and proud brain are mosaics of the same elemental particles that compose the dark, drifting clouds of interstellar space, is, in the final analysis, is merely an ephemeral confrontation of primordial space-time - time fields. Standing midway between macrocosm an macrocosm he finds barriers between every side and can perhaps, but marvel as, St. Paul did nineteen hundred years ago, 'the world was created by the world of God, so that what is seen was made out of things under which do not appear.'

Although, we are to centre the Greek scepticism on the value of enquiry and questioning, we now depict scepticism for the denial that knowledge or even rational belief is possible, either about some specific subject-matter, e.g., ethics, or in any area elsewhere. Classical scepticism, sprouts from the remarking reflection that the best method in some area seems to fall short of giving to remain in a certain state with the truth, e.g., there is a widening disruption between appearances and reality, it frequently cites conflicting judgements that our personal methods of bring to a destination, the result that questions of truth becomes indefinable. In classic thought the various examples of this conflict were systemized in the tropes of Aenesidemus. So that, the scepticism of Pyrrho and the new Academy was a system of argument and inasmuch as opposing dogmatism, and, particularly the philosophical system building of the Stoics.

Steadfast and fixed the philosophy of meaning holds beingness as formatted in and for and of itself, the given migratory scepticism for which accepts the every day or commonsensical beliefs, is not the saying of reason, but as due of more voluntary habituation. Nonetheless, it is self-satisfied at the proper time, however, the power of reason to give us much more. Mitigated scepticism is thus closer to the attitude fostered by the accentuations from Pyrrho through to Sextus Expiricus. Despite the fact that the phrase Cartesian scepticism is sometimes used, nonetheless, Descartes himself was not a sceptic, however, in the method of doubt uses a sceptical scenario in order to begin the process of finding a general distinction to mark its point of knowledge. Descartes trusts in categories of 'distinct' ideas, not far removed from that of the Stoics.

For many sceptics have traditionally held that knowledge requires certainty, artistry. And, of course, they claim that not all of the knowledge is achievable. In part, nonetheless, of the principle that every effect it's a consequence of an antecedent cause or causes. For causality to be true it is not necessary for an effect to be predictable as the antecedent causes may be numerous, too complicated, or too interrelated for analysis. Nevertheless, in order to avoid scepticism, this participating sceptic has generally held that knowledge does not require certainty. For some alleged cases of things that are self-evident, the singular being of one is justifiably corrective if only for being true. It has often been thought, that any thing known must satisfy certain criteria as well for being true. It is often taught that anything is known must satisfy certain standards. In so saying, that by deduction or induction, there will be criteria specifying when it is. As these alleged cases of self-evident truths, the general principle specifying the sort of consideration that will make such standard in the apparent or justly conclude in accepting it warranted to some degree.

Besides, there is another view - the absolute globular view that we do not have any knowledge whatsoever. In whatever manner, it is doubtful that any philosopher would seriously entertain to such as absolute scepticism. Even the Pyrrhonist sceptic shadow, in those who notably held that we should hold in ourselves back from doing or indulging something as from speaking or from accenting to any non-evident standards that no such hesitancy concert or settle through their point to tend and show something as probable in that all particular and often discerning intervals of this interpretation, if not for the moment, we take upon the quality of an utterance that arouses interest and produces an effect, liken to projective application, here and above, but instead of asserting to the evident, the non-evident are any belief that requires evidence because it is to maintain with the earnest of securities as pledged to Foundationalism.

René Descartes (1596-1650), in his sceptical guise, but in the 'method of doubt' uses a scenario to begin the process of finding himself a secure mark of knowledge. Descartes himself trusted a category of 'clear and distinct' ideas not far removed from the phantasia kataleptike of the Stoics, never doubted the content of his own ideas. It's challenging logic, inasmuch as whether they corresponded to anything beyond ideas.

Scepticism should not be confused with relativism, which is a doctrine about nature of truth, and might be identical to motivating by trying to avoid scepticism. Nor does it accede in any condition or occurrence traceable to a cayuse whereby the effect may induce to come into being as specific genes effect specific bodily characters, only to carry to a successful conclusion. That which counsels by ways of approval and taken careful disregard for consequences, as free from moral restrain abandoning an area of thought, also to characterize things for being understood in collaboration of all things considered, as an agreement for the most part, but generally speaking, in the main of relevant occasion, beyond this is used as an intensive to stress the comparative degree that after-all, is that, to apply the pending occurrence that along its passage made in befitting the course for extending beyond a normal or acceptable limit, so and then, it is therefore given to an act, process or instance of expression in words of something that gives specially its equivalence in good qualities as measured through worth or value. Significantly, by compelling implication is given for being without but necessarily in being so in fact, as things are not always the way they seem. However, from a number or group by figures or given to preference, as to a select or selection that alternatively to be important as for which we owe ourselves to what really matters. With the exclusion or exception of any condition in that of accord with being objectionably expectant for. In that, is, because we cannot know the truth, but because there cannot be framed in the terms we use.

All the same, Pyrrhonism and Cartesian form of virtual globularity, in that if scepticism has been held and opposed, that of assuming that knowledge is some form is true. Sufficiently warranted belief, is the warranted condition that provides the truth or belief conditions, in that of providing the grist for the sceptics manufactory in that direction. The Pyrrhonist will suggest that none if any are evident, empirically deferring the sufficiency of giving in but warranted. Whereas, a Cartesian sceptic will agree that no empirical standards about anything other than ones own mind and its contents are sufficiently warranted, because there are always legitimate grounds for doubting it. Out and away, the essential difference between the two views concerns the stringency of the requirements for a belief being sufficiently warranted to take account of as knowledge.

A-Cartesian requirements are intuitively certain, justly as the Pyrrhonist, who merely require that the standards in case value are more warranted then the unsettled negativity.

Cartesian scepticism was unduly influenced with which Descartes agues for scepticism, than his reply holds, in that we do not have any knowledge of any empirical standards, in that of anything beyond the contents of our own minds. The reason is roughly in the position that there is a legitimate doubt about all such standards, only because there is no way to justifiably deny that our senses are being stimulated by some sense, for which it is radically different from the objects which we normally think, in whatever manner they affect our senses. Therefrom, if the Pyrrhonist is the agnostic, the Cartesian sceptic is the atheist.

Because the Pyrrhonist requires much less of a belief in order for it to be confirmed as knowledge than do the Cartesian, the argument for Pyrrhonism are much more difficult to construct. A Pyrrhonist must show that there is no better set of reasons for believing to any standards, of which are in case that any knowledge learnt of the mind is understood by some of its forms, that has to require certainty.

The underlying latencies given among the many derivative contributions as awaiting their presence to the future that of specifying to the theory of knowledge, is, but, nonetheless, the possibility to identify a set of shared doctrines, however, identity to discern two broad styles of instances to discern, in like manners, these two styles of pragmatism, clarify the innovation that a Cartesian approval is fundamentally flawed, nonetheless, of responding very differently but not forgone.

Even so, the coherence theory of truth, sheds to view that the truth of a proposition consists in its being a member of same suitably defined body of coherent and possibly endowed with other virtues, provided these are not defined as for truths. The theory, at first sight, has two strengths (1) we test beliefs for truth in the light of other beliefs, including perceptual beliefs, and (2) we cannot step outside our own best system of belief, to see how well it is doing about correspondence with the world.

To many thinkers the weak point of pure coherence theories is that they fail to include a proper sense of the way in which actual systems of belief are sustained by persons with perceptual experience, impinged upon by their environment. For a pure coherence theory, experience is only relevant as the source of perceptual belief representation, which take their place as part of the coherent or incoherent set. This seems not to do justice to our sense that experience plays a special role in controlling our system of beliefs, but Coherentists have contested the claim in various ways.

However. a correspondence theory is not simply the view that truth consists in correspondence with the 'facts', but rather the view that it is theoretically uninteresting to realize this. A correspondence theory is distinctive in holding that the notion of correspondence and fact can be sufficiently developed to make the platitude into an inter-setting theory of truth. We cannot look over our own shoulders to compare our beliefs with a reality to compare other that those beliefs, or perhaps, further beliefs. So we have no fix on 'facts' as something like structures to which our beliefs may not correspond.

And now and again, we take upon the theory of measure to which evidence supports a theory. A fully formalized confirmation theory would dictate the degree of confidence that a rational investigator might have in a theory, given some body of evidence. The principal developments were due to the German logical positivist Rudolf Carnap (1891-1970), who culminating in his Logical Foundations of Probability (1950), Carnap's idea was that the measure needed would be the proposition of logical possible states of affairs in which the theory and the evidence both hold, compared to the number in which the evidence itself holds. The difficulty with the theory lies in identifying sets of possibilities so that they admit to measurement. It therefore demands that we can put a measure ion the 'range' of possibilities consistent with theory and evidence, compared with the range consistent with the enterprise alone. In addition, confirmation proves to vary with the language in which the science is couched and the Carnapian programme has difficulty in separating genuine confirming variety from less compelling repetition of the same experiment. Confirmation also proved to be susceptible to acute paradoxes. Briefly, such that of Hempel's paradox, wherefore, the principle of induction by enumeration allows a suitable generalization to be confirmed by its instance or Goodman's paradox, by which the classical problem of induction is often phrased in terms of finding some reason to expect that nature is uniform.

Finally, scientific judgement seems to depend on such intangible factors as the problem facing rival theories, and most workers have come to stress instead the historically situated sense of what looks plausible, characteristic of a scientific culture at a given time.

Once said, of the philosophy of language, was that the general attempt to understand the components of a working language, the relationship that an understanding speaker has to its elements, and the relationship they bear to the world: Such that the subject therefore embraces the traditional division of semantic into syntax, semantic, and pragmatics. The philosophy of mind, since it needs an account of what it is in our understanding that enable us to use language. It mingles with the metaphysics of truth and the relationship between sign and object. Such a philosophy, especially in the 20th century, has been informed by the belief that a philosophy of language is the fundamental basis of all philosophical problems in that language is the philosophical problem of mind, and the distinctive way in which we give shape to metaphysical beliefs of logical form, and the basis of the division between syntax and semantics, as well a problem of understanding the number and nature of specifically semantic relationships such as meaning, reference, predication, and quantification. Pragmatics includes the theory of speech acts, while problems of rule following and the indeterminacy of translation infect philosophies of both pragmatics and semantics.

A formal system for which a theory whose sentences are well-formed formula of a logical calculus, and in which axioms or rules of being of a particular term corresponds to the principles of the theory being formalized. The theory is intended to be framed in the language of a calculus, e.g., first-order predicate calculus. Set theory, mathematics, mechanics, and many other axiomatically that may be developed formally, thereby making possible logical analysis of such matters as the independence of various axioms, and the relations between one theory and another.

Are terms of logical calculus are also called a formal language, and a logical system? A system in which explicit rules are provided to determining (1) which are the expressions of the system (2) which sequence of expressions count as well formed (well-forced formulae) (3) which sequence would count as proofs. A system which takes on axioms for which leaves a terminable proof, however, it shows of the prepositional calculus and the predicated calculus.

We could derive a scientific understanding of these ideas aligned with the aid of precise deduction, just as Descartes continued his claim that we could lay the contours of physical reality within the realm of a three-dimensional co-ordinate system. Following the publication of Isaac Newton's 'Principia Mathematica' in 1687, reductionism and mathematical medaling became the most powerful tools of modern science. The dream that we could know and master the entire physical world through the extension and refinement of mathematical theory became the central feature and principles of scientific knowledge.

The radical separation between mind and nature formalized by Descartes, served over time to allow scientists to concentrate on developing mathematical descriptions of matter as pure mechanism without any concern about its spiritual dimensions or ontological foundations. Meanwhile, attempts to rationalize, reconcile or eliminate Descartes's merging division between mind and matter became the most central characterization of Western intellectual life.

Philosophers like John Locke, Thomas Hobbes, and David Hume tried to articulate some basis for linking the mathematical describable motions of matter with linguistic representations of external reality in the subjective space of mind. Descartes' compatriot Jean-Jacques Rousseau reified nature on the ground of human consciousness in a state of innocence and proclaimed that 'Liberty, Equality, Fraternities' are the guiding principles of this consciousness. Rousseau also fabricated the idea of the 'general will' of the people to achieve these goals and declared that those who do not conform to this will were social deviants.

The conceptualization attributed to the Enlightenment idea of 'deism', which imaged the universe as a clockwork and God as the clockmaker, provided grounds for believing in a divine agency, from which the time of moment the formidable creations also imply, in of which, the exhaustion of all the creative forces of the universe at origins ends, and that the physical substrates of mind were subject to the same natural laws as matter. In that the only of mediating the gap between mind and matter was pure reason, causally by the traditional Judeo-Christian theism, which had previously been based on both reason and revelation, responded to the challenge of deism by debasing traditionality as a test of faith and embracing the idea that we can know the truths of spiritual reality only through divine revelation. This engendered a conflict between reason and revelation that persists to this day. And laid the foundation for the fierce completion between the mega-narratives of science and religion as frame tales for mediating the relation between mind and matter and the manner in which they should ultimately define the special character of each.

The nineteenth-century Romantics in Germany, England and the United States revived Rousseau's attempt to posit a ground for human consciousness by reifying nature in a different form. Goethe and Friedrich Schelling proposed a natural philosophy premised on ontological Monism (the idea that adhering manifestations that govern toward evolutionary principles have grounded inside an inseparable spiritual Oneness) and argued God, man, and nature for the reconciliation of mind and matter with an appeal to sentiment, mystical awareness, and quasi-scientific attempts, as he afforded the efforts of mind and matter, nature became a mindful agency that 'loves illusion', as it shrouds man in mist, presses him or her heart and punishes those who fail to see the light. Schelling, in his version of cosmic unity, argued that scientific facts were at best partial truths and that the mindful creative spirit that unities mind and matter is progressively moving toward self-realization and 'undivided wholeness'.

The British version of Romanticism, articulated by figures like William Wordsworth and Samuel Taylor Coleridge, placed more emphasis on the primary of the imagination and the importance of rebellion and heroic vision as the grounds for freedom. As Wordsworth put it, communion with the 'incommunicable powers' of the 'immortal sea' empowers the mind to release itself from all the material constraints of the laws of nature. The founders of American transcendentalism, Ralph Waldo Emerson and Henry David Thereat, articulated a version of Romanticism that commensurate with the ideals of American democracy.

The American envisioned a unified spiritual reality that manifested itself as a personal ethos that sanctioned radical individualism and bred aversion to the emergent materialism of the Jacksonian era. They were also more inclined than their European counterpart, as the examples of Thoreau and Whitman attest, to embrace scientific descriptions of nature. However, the Americans also dissolved the distinction between mind and natter with an appeal to ontological monism and alleged that mind could free itself from all the constraint of assuming that by some sorted limitation of matter, in which such states have of them, some mystical awareness.

Since scientists, during the nineteenth century were engrossed with uncovering the workings of external reality and seemingly knew of themselves that these virtually overflowing burdens of nothing, in that were about the physical substrates of human consciousness, the business of examining the distributive contribution in dynamic functionality and structural foundation of mind became the province of social scientists and humanists. Adolphe Quételet proposed a 'social physics' that could serve as the basis for a new discipline called sociology, and his contemporary Auguste Comte concluded that a true scientific understanding of the social reality was quite inevitable. Mind, in the view of these figures, was a separate and distinct mechanism subject to the lawful workings of a mechanical social reality.

More formal European philosophers, such as Immanuel Kant, sought to reconcile representations of external reality in mind with the motions of matter-based on the dictates of pure reason. This impulse was also apparent in the utilitarian ethics of Jerry Bentham and John Stuart Mill, in the historical materialism of Karl Marx and Friedrich Engels, and in the pragmatism of Charles Smith, William James and John Dewey. These thinkers were painfully aware, however, of the inability of reason to posit a self-consistent basis for bridging the gap between mind and matter, and each remains obliged to conclude that the realm of the mental exists only in the subjective reality of the individual.

The fatal flaw of pure reason is, of course, the absence of emotion, and purely explanations of the division between subjective reality and external reality, of which had limited appeal outside the community of intellectuals. The figure most responsible for infusing our understanding of the Cartesian dualism with contextual representation of our understanding with emotional content was the death of God theologian Friedrich Nietzsche. Nietzsche reified the existence of consciousness in the domain of subjectivity as the ground for individual will and summarily reducing all previous philosophical attempts to articulate the will to truth. The dilemma, forth in, had seemed to mean, by the validation, . . . as accredited for doing of science, in that the claim that Nietzsche's earlier versions to the will to truth, disguises the fact that all alleged truths were arbitrarily created in the subjective reality of the individual and are expressed or manifesting the individualism of will.

In Nietzsche's view, the separation between mind and matter is more absolute and total than previously been imagined. To serve as a basis on the assumptions that there are no really imperative necessities corresponding in common to or in participated linguistic constructions that provide everything needful, resulting in itself, but not too far as to distance from the influence so gainfully employed, that of which was founded as close of action, wherefore the positioned intent to settle the occasioned-difference may that we successively occasion to occur or carry out at the time after something else is to be introduced into the mind, that from a direct line or course of circularity inseminates in its finish. Their successive alternatives are thus arranged through anabatic existing or dealing with what exists only in the mind, so that, the conceptual analysis of a problem gives reason to illuminate, for that which is fewer than is more in the nature of opportunities or requirements that employ something imperatively substantive, moreover, overlooked by some forming elementarily whereby the gravity held therein so that to induce a given particularity, yet, in addition by the peculiarity of a point as placed by the curvilinear trajectory as introduced through the principle of equivalence, there, founded to the occupied position to which its order of magnitude runs a location of that which only exists within self-realization and corresponding physical theories. Ours being not rehearsed, however, unknowingly their extent temporality extends the quality value for purposes that are substantially spatial, as analytic situates points indirectly into the realities established with a statement with which are intended to upcoming reasons for self-irrational impulse as explicated through the geometrical persistence so that it is implicated by the position, and, nonetheless, as space-time, wherein everything began and takes its proper place and dynamic of function.

Earlier, Nietzsche, in an effort to subvert the epistemological authority of scientific knowledge, sought to appropriate a division between mind and world was properly set, inasmuch as the austereness than was originally envisioned by Descartes. In Nietzsche's view, the separation between mind and matter is more absolute and total than previously thought. Based on the assumption that there is no real or necessary correspondence between linguistic constructions of reality in human subjectivity and external reality, but quick to realize, that there was nothing in this of nature that could explain or provide a foundation for the mental, or for all that we know from direct experience as distinctly human. Given that Descartes distrusted the information from the senses to the point of doubting the perceived results of repeatable scientific experiments, how did he conclude that our knowledge of the mathematical ideas residing only in mind or in human subjectivity was accurate, much less the absolute truth? He did so by taking a leap if faith - God constructed the world, said Descartes, in accordance with the mathematical ideas that our minds are capable of uncovering in their pristine essence. The truth of classical physics as Descartes viewed them were quite literally revealed truths, and this was this seventeenth-century metaphysical presupposition that became in the history of science what is termed the hidden ontology of classical epistemology, however, if there is no real or necessary correspondence between non-mathematical ideas in subjective reality and external physical reality, how do we know that the world in which we live, breath, and have our Being, actually exists? Descartes resolution of this dilemma took the form of an exercise. But, nevertheless, as it turned out, its resolution was considerably more problematic and oppressive than Descartes could have imagined, I think therefore I am; is marginally persuasive in the ways of confronting the real existence of the thinking self? But, the understanding of physical reality that obliged Descartes and others to doubt the existence of this self clearly implied that the separation between the subjective world and the world of life, and the real wold of physical reality as absolute.

There is a multiplicity of different positions to which the term epistemological relativism has been applied, however, the basic idea common to all forms denies that there is a single, universal context. Many traditional epistemologists have striven to uncover the basic process, method or determined rules that allow us to hold true belief's, recollecting, for example, of Descartes's attempt to find the rules for directions of the mind. Hume's investigation into the science of mind or Kant's description of his epistemological Copernican revolution, where each philosopher attempted to articulate universal conditions for the acquisition of true belief.

The coherence theory of truth, finds to it view that the truth of a proposition consists in its being a member of some suitably defined body of other propositions, as a body that is consistent, coherent and possibly endowed with other virtues, provided there are not defined in terms of truth. The theory has two strengths: We cannot step outside our own best system of beliefs, to see how well it is doing in terms of correspondence with the world. To many thinkers the weak points of pure coherence theories in that they fail to include a proper sense of the way in which include a proper sense of the way in which actual systems of belief are sustained by persons with perceptual experience, impinged upon using their environment. For a pure coherence theorist, experience is only relevant as the source of perceptual representations of beliefs, which take their place as part of the coherent or incoherent set. This seems not to do justice to our sense that experience plays a special role in controlling our systems of belief, but Coherentists have contested the claim in various ways.

The pragmatic theory of truth is the view particularly associated with the American psychologist and philosopher William James (1842-1910), that the truth of a statement can be defined in terms of the utility of accepting it. Put so badly the view is open too objective, since there are things that are false that it may be useful to accept, and conversely there are things that are true that, perhaps, as it is damaging to accept. However, their area deeply connects between the ideas that a representative system is accurate, and he likely success of the projects and purposes formed by its possessor. The evolution of a system of representation, of whether its given priority in consistently perceptual or linguistically bond by the corrective connection with evolutionary adaption, or under with utility in the widest sense, as for Wittgenstein's doctrine that its use of deceptions over which the pragmatic emphasis on technique and practice are the matrix which meaning is possible.

Nevertheless, after becoming the tutor of the family of the Addé de Mably that Jean-Jacques Rousseau (1712-78) became acquainted with philosophers of the French Enlightenment. The Enlightenment idea of deism, when we are assured that there is an existent God, additional revelation, some dogmas are all excluded. Supplication and prayer in particular are fruitless, may only be thought of as an 'absentee landlord'. The belief that remains abstractively a vanishing point, as wintered in Diderot's remark that a deist is someone who has not lived long enough to become an atheist. Which can be imagined of the universe as a clock and God as the clockmaker, provided grounds for believing in a divine agency at the moment of creation? It also implied, however, that all the creative forces of the universe were exhausted at origins, that the physical substrates of mind were subject to the same natural laws as matter, and pure reason. In the main, Judeo-Christian has had an atheistic lineage, for which had previously been based on both reason and revelation, responded to the challenge of deism by debasing rationality as a test of faith and embracing the idea that the truth of spiritual reality can be known only through divine revelation. This engendered a conflict between reason and revelations that persists to this day. And it also laid the foundation for the fierce competition between the mega-narratives of science and religion as frame tales for mediating the relation between mind and matter and the manner in which the special character of each should be ultimately defined.

Obviously, here, is, at this particular intermittent interval in time no universally held view of the actual character of physical reality in biology or physics and no universally recognized definition of the epistemology of science. And it would be both foolish and arrogant to claim that we have articulated this view and defined this epistemology.

The best-known disciple of Husserl was Martin Heidegger, and the work of both figures greatly influenced that of the French atheistic existentialist Jean-Paul Sartre. The work of Husserl, Heidegger, and Sartre became foundational to that of the principal architects of philosophical postmodernism, and deconstructionist Jacques Lacan, Roland Barthes, Michel Foucault and Jacques Derrida. The obvious attribution of a direct linkage between the nineteenth-century crisis about the epistemological foundations of mathematical physics and the origin of philosophical postmodernism served to perpetuate the Cartesian two-world dilemma in an even more oppressive form. It also allows us better to understand the origins of cultural ambience and the ways in which they could resolve that conflict.

Heidegger, and the work of Husserl, and Sartre became foundational to those of the principal architects of philosophical postmodernism, and deconstructionist Jacques Lacan, Roland Barthes,

Michel Foucault and Jacques Derrida. It obvious attribution of a direct linkage between the nineteenth-century crisis about the epistemological foundations of mathematical physics and the origin of philosophical postmodernism served to perpetuate the Cartesian two world dilemmas in an even more oppressive form. It also allows us better to understand the origins of cultural ambience and the ways in which they could resolve that conflict.

Pragmatism of a determinant revolution, by contrast, relinquishing the objectivity of early days, and acknowledges no legitimate epistemological questions over and above those that are naturally kindred of our current cognitive conviction.

It seems clear that certainty is a property that can be assembled to either a person or a belief. We can say that a person, 'S' might be certain or we can say that its resulting alignment is coordinated to accommodate the connexion, by saying that 'S' has the right to be certain just in case the value of 'p' is sufficiently verified.

In defining certainty, it is crucial to note that the term has both an absolute and relative sense. More or less, we take a proposition to be certain when we have no doubt about its truth. We may do this in error or unreasonably, but objectively a proposition is certain when such absence of doubt is justifiable. The sceptical tradition in philosophy denies that objective certainty is often possible, or ever possible, either for any proposition at all, or for any proposition at all, or for any proposition from some suspect family (ethics, theory, memory, empirical judgement and so forth) a major sceptical weapon is the possibility of upsetting events that can cast doubt back onto what was hitherto taken to be certainty. Others include reminders of the divergence of human opinion, and the fallible source of our confidence. Fundamentalist approaches to knowledge look for a basis of certainty, upon which the structure of our system is built. Others reject the metaphor, looking for mutual support and coherence, without foundation.

However, in moral theory, the views that there are inviolable moral standards or absolute variable human desires or policies or prescriptions, and subsequently since the 17th and 18th centuries, when the science of man began to probe into human motivations and emotions. For writers such as the French moralistes, and political philosopher Francis Hutcheson (1694-1746), David Hume (1711-76), and both Adam Smith (1723-90) and Immanuel Kant (1724-1804), whereby the prime task to delineate the variety of human reactions and motivations, such inquiry would locate our propensity for moral thinking about other faculties such as perception and reason, and other tendencies, such as empathy, sympathy or self-interest. The task continues especially in the light of a post-Darwinian understanding of the evolutionary governing principles about us.

In some moral system notably that in personal representations as standing for the German and founder of critical philosophy was Immanuel Kant (1724-1804), through which times real moral worth comes only with acting rightly because it is right. If you do what you should but from some other motive, such as fear or prudence, no moral merit accrues to you. Yet, in turn, for which it gives the impression of being without necessarily being so in fact, in that to look in quest or search, at least of what is not apparent. Of each discount other admirable motivations, are such as acting from sheer benevolence or sympathy. The question is how to balance the opposing ideas, and also how to understand acting from a sense of obligation without duty or rightness beginning to seem a kind of fetish.

The entertaining commodity that rests for any but those whose abilities for vauntingly are veering to the variously involving differences, is that for itself that the variousness in the quality or state of being decomposed of different parts, elements or individuals with which are consisting of a goodly but indefinite number, much as much of our frame of reference that, least of mention, maintain through which our use or by we are to contain or constitute a command as some sorted mandatorily anthropomorphic virility. Several distinctions of otherwise, diverse probability, is that the right is not all on one side, so that, qualifies (as adherence to duty or obedience to lawful authority), that together constitute the ideal of moral propriety or merit approval. These given reasons for what remains strong in number, are the higher mental categories that are completely charted among their itemized regularities, that through which it will arise to fall, to have as a controlling desire something that transcends ones present capacity for attainment, inasmuch as to aspire by obtainably achieving. The intensity of sounds, in that it is associated chiefly with poetry and music, that the rhythm of the music made it easy to manoeuver, where inturn, we are provided with a treat, for such that leaves us with much to go through the ritual pulsations in rhythmical motions of finding back to some normalcy, however, at this time we ought but as justly as we might, be it that at this particular point of an occupied position as stationed at rest, as its peculiarity finds to its reference, and, pointing into the abyssal of space and time. So, once found to the ups-and-downs, and justly to move in the in and pots of the dance. Placed into the working potentials are to be charged throughout the functionally sportive inclinations that manifest the tune of a dynamic contribution, so that almost every selectively populated pressure ought to be the particular species attributive to evolutionary times, in that our concurrences are temporally at rest. Candidates for such theorizing include material and paternal motivations, capacities for love and friendship, and the development of language is a signalling system, cooperatives and aggressive tendencies our emotional repertoire, our moral reactions, including the disposition to denote and punish those who cheat on agreements or who free-riders, on whose work of others, our cognitive intuition may be as many as other primordially sized infrastructures, in that their intrenched inter-structural foundations are given as support through the functionally dynamic resources based on volitionary psychology, but it seems that it goes of a hand-in-hand interconnectivity, finding to its voluntary relationship with a partially parallelled profession named as, neurophysiological evidences, this, is about the underlying circuitry, in terms through which it subserves the psychological mechanism it claims to identify. The approach was foreshadowed by Darwin himself, and William James, as well as the sociologist E.O. Wilson.

An explanation of an admittedly speculative nature, tailored to give the results that need explanation, but currently lacking any independent aggressively, especially to explanations offered in sociological and evolutionary psychology. It is derived from the explanation of how the leopard got its spots, and so forth

In spite of the notorious difficulty of reading Kantian ethics, a hypothetical imperative embeds a command which in its place are only to provide by or as if by formal action as the possessions of another who in which does he express to fail in responses to physical stress, nonetheless. The reflective projection, might be that: If you want to look wise, stay quiet. The inductive ordering to stay quiet only to apply to something into shares with care and assignment, gives of an equalling lot amongst a number that make a request for their opportunities in those with the antecedent desire or inclination. If one has no desire to look, seemingly the absence of wise becomes the injunction and this cannot be so avoided: It is a requirement that binds anybody, regardless of their inclination. It could be represented as, for example, tell the truth (regardless of whether you want to or not). The distinction is not always signalled by presence or absence of the conditional or hypothetical form: If you crave drink, don't become a bartender may be regarded as an absolute injunction applying to anyone, although only activated in cases of those with the stated desire.

In Grundlegung zur Metaphsik der Sitten (1785), Kant discussed five forms of the categorical imperative: (1) the formula of universal law: act only on that maxim through which you can at the same times will that it should become universal law: (2) the formula you the laws of nature, act as if the maxim of your action were to commence to be, that from beginning to end your will (a desire to act in a particular way or have a particular thing), is the universal law of nature: (3) the formula of the end-in-itself: has in inertness or appearance the end or the ending of such ways that you have always treat humanity, whether in your own person or in the person of any other, never simply as a, but always at the same time as an end? : (4) The formula of autonomy, or considering the will of every rational being as a will which makes universal law: (5) the formula of the Kingdom of Ends, which provides a model for the systematic union of different rational beings under common laws.

Even so, a proposition that is not a conditional 'p', may affirmatively and negatively, modernize the opinion is wary of this distinction, since what appears categorical may vary notation. Apparently, categorical propositions may also turn out to be disguised conditionals: 'X' is intelligent (categorical?) if 'X' is given a range of tasks she performs them better than many people (conditional?) The problem. Nonetheless, is not merely one of classification, since deep metaphysical questions arise when facts that seem to be categorical and therefore solid, come to seem by contrast conditional, or purely hypothetical or potential.

A limited area of knowledge or endeavour to which pursuits, activities and interests are a central representation held to a concept of physical theory. In this way, a field is defined by the distribution of a physical quantity, such as temperature, mass density, or potential energy y, at different points in space. In the particularly important example of force fields, such as gravitational, electrical, and magnetic fields, the field value at a point is the force which a test particle would experience if it were located at that point. The philosophical problem is whether a force field is to be thought of as purely potential, so the presence of a field merely describes the propensity of masses to move relative to each other, or whether it should be thought of in terms of the physically real modifications of a medium, whose properties result in such powers that are force fields pure potential, fully characterized by dispositional statements or conditionals, or are they categorical or actual? The former option seems to require within ungrounded dispositions, or regions of space that differ only in what happens if an object is placed there. The law-like shape of these dispositions, apparent for example in the curved lines of force of the magnetic field, may then seem quite inexplicable. To atomists, such as Newton it would represent a return to Aristotelian entelechies, or quasi-psychological affinities between things, which are responsible for their motions. The latter option requires understanding of how forces of attraction and repulsion can be grounded in the properties of the medium.

The basic idea of a field is arguably present in Leibniz, who was certainly hostile to Newtonian atomism. Despite the fact that his equal hostility to action at a distance muddies the water, it is usually credited to the Jesuit mathematician and scientist Joseph Boscovich (1711-87) and Immanuel Kant. Both of whose influenced the scientist Faraday, with whose work the physical notion became established. In his paper On the Physical Character of the Lines of Magnetic Force (1852), Faraday was to suggest several criteria for assessing the physical reality of lines of force, such as whether they are affected by an intervening material medium, whether the motion depends on the nature of what is placed at the receiving end. As far as electromagnetic fields go, Faraday himself inclined to the view that the mathematical similarity between heat flow, currents, and electromagnetic lines of force was evidence for the physical reality of the intervening medium.

Once, again, our mentioning recognition for which its case value, whereby its view is especially associated the American psychologist and philosopher William James (1842-1910), that the truth of a statement can be defined in terms of a utility of accepting it. Communicable messages of thoughts are made popularly known throughout the interchange of thoughts or opinions through shared symbols. The difficulties of communication between people of different cultural backgrounds and exchangeable directives, only for which our word is the intellectual interchange for conversant chatter, or in general for talking. Man, alone, is Disquotational among situation analyses that are viewed as a remonstrative objection. Since, there are things that are false, as it may be useful to accept, and conversely give in the things that are true and consequently, it may be damaging to accept. Nevertheless, there are deep connections between the idea that a representation system is accorded, and the likely success of the projects in progressive formality, by its possession. The evolution of a system of representation either perceptual or linguistic, seems bounded to connect successes with everything adapting or with utility in the modest sense. The Wittgenstein doctrine stipulates the meaning of use that upon the nature of belief and its relations with human attitude, emotion and the idea that belief in the truth on one hand, the action of the other. One way of binding with cement, wherefore the connexion is found in the idea that natural selection becomes much as much in adapting us to the cognitive creatures, because beliefs have effects, they work. Pragmatism can be found in Kant's doctrine, and continued to play an influencing role in the theory of meaning and truth.

James, (1842-1910), although with characteristic generosity exaggerated in his debt to Charles S. Peirce (1839-1914), he charted that the method of doubt encouraged people to pretend to doubt what they did not doubt in their hearts, and criticize its individualists insistence, that the ultimate test of certainty is to be found in the individuals personalized consciousness.

From his earliest writings, James understood cognitive processes in teleological terms. Thought, he held, assisted us in the satisfactory interests. His will to Believe doctrine, the view that we are sometimes justified in believing beyond the evidential relics upon the notion that a beliefs benefits are relevant to its justification. His pragmatic method of analyzing philosophical problems, for which requires that we find the meaning of terms by examining their application to objects in experimental situations, similarly reflects the teleological approach in its attention to consequences.

Such an approach to come or go near or nearer of meaning, yet lacking of an interest in concerns, justly as some lack of emotional responsiveness have excluded from considerations for those apart,

501

and otherwise e elsewhere partitioning. Although the work for verification has seemed dismissively metaphysical, and, least of mention, were drifting of becoming or floated along to knowable inclinations that inclines to knowable implications that directionally show the purposive values for which we inturn of an allowance change by reversal for together is founded the theoretical closeness, that insofar as there is of no allotment for pointed forward. Unlike the verificationalist, who takes cognitive meaning to be a matter only of consequences in sensory experience, James took pragmatic meaning to include emotional and matter responses, a pragmatic treat of special kind of linguistic interaction, such as interviews and a feature of the use of a language would explain the features in terms of general principles governing appropriate adherence, than in terms of a semantic rule. However, there are deep connections between the idea that a representative of the system is accurate, and the likely success of the projects and purposes of a system of representation, either perceptual or linguistic seems bound to connect success with evolutionary adaption, or with utility in the widest sense. Moreover, his, metaphysical standard of value, not a way of dismissing them as meaningless but it should also be noted that in a greater extent for which is marked by careful prudence. James hold's that even his broad sets of consequences were exhaustive of some terms meaning. Theism, for example, he took to have antecedently, definitional meaning, in addition to its varying degree of importance and chance upon an important pragmatic meaning.

James theory of truth reflects upon his teleological conception of cognition, by considering a true belief to be one which is compatible with our existing system of beliefs, and leads us to satisfactory interaction with the world.

Even so, to believe a proposition is to hold it to be true, that the philosophical problem is to align ones precarious states, for which a persons constituent representations form their personal beliefs, is it, for example, a simple disposition to behaviour? Or a more complicated, complex state that resists identification with any such disposition, is compliant with verbalized skills or verbal behaviourism which are essential to belief, concernedly by what is to be said about prelinguistic infants, or nonlinguistic animals? An evolutionary approach asks how the cognitive success of possessing the capacity to believe things relates to success in practice. Further topics include discovering whether belief differs from other varieties of assent, such as acceptance, discovering whether belief is an all-or-nothing matter, or to what extent degrees of belief are possible, understanding the ways in which belief is controlled by rational and irrational factors, and discovering its links with other properties, such as the possession of conceptual or linguistic skills.

Nevertheless, for Peirces' famous pragmatist principle is a rule of logic employed in clarifying our concepts and ideas. Consider the claim the liquid in a flask is an acid, if, we believe this, we except that it would turn red: We accept an action of ours to have certain experimental results. The pragmatic principle holds that listing the conditional expectations of this kind, in that we associate such immediacy with applications of a conceptual representation that provides a complete and orderly sets clarification of the concept. This is relevant to the logic of abduction: Clarificationists using the pragmatic principle provides all the information about the content of a hypothesis that is relevantly to decide whether it is worth testing. All the same, as the founding figure of American pragmatism, perhaps, its best expressage would be found in his essay How to Make our Idea s Clear, (1878), in which he proposes the famous dictum: The opinion which is fated to be ultimately agreed to by all who investigate is what we mean by the truth, and the object representation in this opinion are the real. Also made pioneering investigations into the logic of relations, and of the truth-functions, and independently discovered the quantifier slightly later that Frége. His work on probability and induction includes versions of the frequency theory of probability, and the first suggestion of a vindication of the process of induction. Surprisedly, Peirces scientific outlook and opposition to rationalize coexisted with admiration for Dun Scotus, (1266-1308), a Franciscan philosopher and theologian, who locates freedom in our ability to turn from desire and toward justice. Scotus characterlogical distinction has directly been admired by such different thinkers as Peirce and Heidegger, he was dubbed the doctor subtilis (short for Dunsman) reflects the low esteem into which scholasticism later fell between humanists and reformers.

To a greater extent, and most important, is the famed apprehension of the pragmatic principle, In so that, C.S. Pierce, the founder of American pragmatism, had been concerned with the nature of language and how it related to thought. From what account of reality did he develop this theory of semiotics as a method of philosophy. How exactly does language relate to thought? Can there be complex, conceptual thought without language? These issues that operate to form an idea of something in the mind as of them to conceive or realize by understanding, as, one's powers of conception, judgement or inference, the power to think sets human apart from other lower primates or animals, there capable of being actual or actualized upon that which of or thinking and attempted efforts as of drawing out the implications for question about meaning, ontology, truth and knowledge, nonetheless, they have quite different takes on what those implications are

These issues had brought about the entrapping fascinations of some engagingly encountered sense for causalities that through which its overall topic of linguistic tradition was grounded among furthering subsequential developments, that those of the earlier insistences of the twentieth-century positions. That to lead by such was the precarious situation into bewildering heterogeneity, so that princely it came as of a tolerable philosophy occurring in the early twenty-first century. The very nature of philosophy is itself radically disputed, analytic, continental, postmodern, Critical theory, feminist and non-Western are all prefixes that give a different meaning when joined to philosophy. The variety of thriving different schools, the number of professional philologers, the proliferation of publications, the developments of technology in helping reach all manifest a radically different situation to that of one hundred years ago. Sharing some common sources with David Lewis, the German philosopher Rudolf Carnap (1891-1970) articulated a doctrine of linguistic frameworks that was radically relativistic in its implications. Carnap was influenced by the Kantian idea of the constitution of knowledge: That our knowledge is in some sense the end result of a cognitive process. He also shared Lewis pragmatism and valued the practical application of knowledge. However, as empiricism, he was headily influenced by the development of modern science, regarding scientific knowledge s the paradigm of knowledge and motivated by a desire to be rid of pseudo-knowledge such as traditional metaphysics and theology. These influences remain constant as his work moved though various distinct stages and then he moved to live in America. In 1950, he published a paper entitled Empiricism, Semantics and Ontology in which he articulated his views about linguistic frameworks.

When an organized integrated whole made up of diverse but interrelated and interdependent parts, the capacity of the system precedes to be real that something that stands for something else by reason that being in accordance with or confronted to action we think it not as it might be an imperfection in character or an ingrained moral weakness predetermined to be agreed upon by all who investigate. The matter to which it stands, in other words, that, if I believe that it is really the case that p, then I except that if anyone were to inquire into the finding state of internal and especially the quality values, state, or conditions of being self-complacent as to poise of a comparable satisfactory measure of whether p, would arrive at the belief that p it is not part of the theory that the experimental consequences of our actions should be specified by a warranted empiricist vocabulary - Peirce insisted that perceptual theories are abounding in latency. Even so, nor is it his view that the collected conditionals do or not clarify a concept as all analytic. In addition, in later writings, he argues that the pragmatic principle could only be made plausible to someone who accepted its metaphysical realism: It requires that would-bees are objective and, of course, real.

If realism itself can be given a fairly quick clarification, it is more difficult to chart the various forms of supposition, for they seem legendary. Other opponents deny that entitles firmly held points of view or way of regarding something capable of being constructively applied, that only to presuppose in the lesser of views or ways of regarding something, at least the conservative position is posited by the relevant discourse that exists or at least exists: The standard example is idealism, which reality is somehow mind-curative or mind-co-ordinated - that real objects comprising the external worlds are dependently of eloping minds, but only exist as in some way correlative to the mental operations. The doctrine assembled of idealism enters on the conceptual note that reality as we understand this as meaningful and reflects the working of mindful purposes. And it construes this as meaning that the inquiring mind itself makes of some formative constellations and not of any mere understanding of the nature of the real bit even the resulting charger we attributively acknowledge for it.

Wherefore, the term is most straightforwardly used when qualifying another linguistic form of Grammatik: a real 'x' may be contrasted with a fake, a failed 'x', a near 'x', and so on. To that something as real, without qualification, is to suppose it to be part of the actualized world. To reify something is to suppose that we have committed by some indoctrinated treatise, as that of a theory. The central error in thinking of reality and the totality of existence is to think of the unreal as a separate domain of things, perhaps, unfairly to that of the benefits of existence.

Such that nonexistence of all things, and as the product of logical confusion of treating the term nothing as itself a referring expression of something that does not exist, instead of a quantifier, wherefore, the important point is that the treatment holds off thinking of something, as to exist of nothing, and then kin as kinds of names. Formally, a quantifier will bind a variable, turning an open sentence with some distinct free variables into one with, n - 1 (an individual letter counts as one variable, although it may recur several times in a formula). (Stating informally as a quantifier is an expression that reports of a quantity of times that a predicate is satisfied in some class of things, i.e., in a domain.) This confusion leads the unsuspecting to think that a sentence such as Nothing is all around us talks of a special kind of thing that is all around us, when in fact it merely denies that the predicate is all around us has appreciation. The feelings that lad some philosophers and theologians, notably Heidegger, to talk of the experience of nothing, is not properly the experience of anything, but rather the failure of a hope or expectations that there would be something of some kind at some point. This may arise in quite everyday cases, as when one finds that the article of functions one expected to see as usual, in the corner has disappeared. The difference between existentialist and analytic philosophy, on the point of what, whereas the former is afraid of nothing, and the latter think that there is nothing to be afraid of.

A rather different set of concerns arises when actions are specified in terms of doing nothing, saying nothing may be an admission of guilt, and doing nothing in some circumstances may be tantamount to murder. Still, other substitutional problems arise over conceptualizing empty space and time.

Whereas, the standard opposition between those who affirm and those who deny, for these of denial are forsaken of a real existence by some kind of thing or some kind of fact, that, conceivably are in accord given to provide, or if by formal action bestow or dispense by some action to fail in response to physical stress, also by their stereotypical allurement of affairs so that a of determines what a thing should be, however, each generation has its on standards of morality. Almost any area of discourse may be the focus of this dispute: The external world, the past and future, other minds, mathematical objects, possibilities, universals, moral or aesthetic properties are examples. There be to one influential suggestion, as associated with the British philosopher of logic and language, and the most determinative of philosophers centred round Anthony Dummett (1925), to which is borrowed from the intuitivistic critique of classical mathematics, and suggested that the unrestricted use of the principle of bivalence is the trademark of realism. However, this has to overcome counter examples both ways, although Aquinas was a moral realist, he held that moral really was not sufficiently structured to make true or false every moral claim. Unlike Kant who believed that he could use the law of bivalence quite effectively in mathematics, precisely because it was only our own construction. Realism can itself be subdivided: Kant, for example, combines empirical realism (within the phenomenal world the realist says the right things - surrounding objects really exist and independent of us and our mental states) with transcendental idealism (the phenomenal world as whole reflects the structures imposed on it by the activity of our minds as we render its intelligibility to us). In modern philosophy the orthodox opposition to realism has been from the philosopher such as Goodman, who, impressed by the extent to which we perceive the world through conceptual and linguistic lenses of our own making.

Assigned to the modern treatment of existence in the theory of quantification is sometimes put by saying that existence is not a predicate. The idea is that the existential quantify themselves as an operator on a predicate, indicating that the property it expresses has instances. Existence is therefore treated as a second-order property, or a property of properties. It is fitting to say, that in this it is like number, for when we say that these things of a kind, we do not describe the thing (ad we would if we said there are red things of the kind), but instead attribute a property to the kind itself. The parallelled numbers are exploited by the German mathematician and philosopher of mathematics Gottlob Frége in the dictum that affirmation of existence is merely denied of the number nought. A problem, nevertheless, proves accountable for its created by sentences like this exists where some particular thing is undirected, such that a sentence seems to express a contingent truth (for this insight has not existed), yet no other predicate is involved. This exists is, therefore, unlike Tamed tigers exist, where a property is said to have an instance, for the word this and does not locate a property, but only correlated by an individual.

Possible worlds seem able to differ from each other purely in the presence or absence of individuals, and not merely in the distribution of exemplification of properties.

The philosophical ponderance over which to set upon the unreal, as belonging to the domain of Being, as, there is little for us that can be said with the philosophers study. So it is not apparent that there can be such a subject for being by itself. Nevertheless, the concept had a central place in philosophy from Parmenides to Heidegger. The essential question of why is there something and not of nothing? Prompting over logical reflection on what it is for a universal to have an instance, and as long history of attempts to explain contingent existence, by which id to reference and a necessary ground.

In the tradition, ever since Plato, this ground becomes a self-sufficient, perfect, unchanging, and external something, identified with having a helpful or auspicious character. Only to be conforming to a high standard of morality or virtuosity, such in an acceptable or desirable manner that can be fond, as something that is adaptively viewed to its very end, or its resultant extremity might for which of its essence, is plainly basic yet underlying or constituting unity, meaning or form, perhaps, the essential nature as so placed on the reference too conveyed upon the positivity that is good or God, however, whose relation with the everyday world remains shrouded by its own nakedness. The celebrated argument for the existence of God was first propounded by an Andelm in his Proslogin. The argument by defining God as something other than that which nothing is greater can be conceived, but God then exists in our understanding, only that we sincerely understand this concept. However, if he only existed in the understanding something greater could be conceived, for a being that exists in reality is greater than one that exists in the understanding. Bu then, we can conceive of something greater than that than which nothing greater can be conceived, which is contradictory. Therefore, God cannot exist on the understanding, but exists in reality.

An influential argument (or family of arguments) for the existence of God, finding its premises are that all natural things are dependent for their existence on something else. The totality of dependence brings within itself the primary dependence upon a non-dependent, or necessarily existent being of which is God. Like the argument to design, the cosmological argument was attacked by the Scottish philosopher and historian David Hume (1711-76) and Immanuel Kant.

Its main problem, nonetheless, is that it requires us to make sense of the notion of necessary existence. For if the answer to the question of why anything exists is that some other things of a similar kind exist, the question merely arises by its gainfully obtained achievement. So, in at least, respectively, God ends the querying of questions, that, He must stand alone insofar as, He must exist of idealistic necessities: It must not be an entity of which the same kinds of questions can be raised. The other problem with the argument is attributing concern and care to the deity, not for connecting the necessarily existent being it derives with human values and aspirations.

The ontological argument has been treated by modern theologians such as Barth, following Hegel, not so much as a proof with which to confront the unconverted, but as an explanation of the deep meaning of religious belief. Collingwood, regards the arguments proving not that because our idea of God is that of, quo maius cogitare viequit, therefore God exists, but proving that because this is our idea of God, we stand committed to belief in its existence. Its existence is a metaphysical point or absolute presupposition of certain forms of thought.

In the 20th century, modal versions of the ontological argument have been propounded by the American philosophers Charles Hertshorne, Norman Malcolm, and Alvin Plantinga. One version is to define something as unsurpassably great, if it exists and is perfect in every possible world. Then, to allow for that which through its possibilities, is potentially that of what is to be seen as an unsurpassably great being existing. This that there is a possible world in which such a being exists. However, if it exists in one world, it exists in all (for the fact that such a being exists in a world that entails, in at least, it exists and is perfect in every world), so, it exists necessarily. The correct response to this argument is to disallow the apparently reasonable concession that it is possible that such a being exists. This concession is much more dangerous than it looks, since in the modal logic, involved from possibly necessarily 'p', we endorse the ground working of its necessities, 'P'. A symmetrical proof starting from the assumption that it is possibly that such a being does not exist would derive that it is impossible that it exists.

The doctrine that it makes an ethical difference of whether an agent actively intervenes to bring about a result, or omits to act within circumstances forwarded through the anticipated forthcoming, in that, as a result by omissions that the same distinguishing quality as essentially characterized of especially a character trait recognized and acknowledged as they occur from whatever happens. Thus, suppose that I wish you dead. If I act to bring about your death, I am a murderer, however, if I happily discover you in danger of death, and fail to act to save you, I am not acting, and therefore, according to the doctrine of acts and omissions not a murderer. Critics implore that omissions can be as deliberate and immoral as I am responsible for your food and fact to feed you. Only omission is surely a killing, Doing nothing can be a way of doing something, or in other worlds, absence of bodily movement can also constitute acting negligently, or deliberately, and defending on the context, may be a way of deceiving, betraying, or killing. Nonetheless, criminal law offers to find its conveniences, from which to distinguish discontinuous intervention, for which is permissible, from bringing about results, which may not be, if, for instance, the result is death of a patient. The question is whether the difference, if there is one, is, between acting and omitting to act be discernibly or defined in a way that bars a general moral might.

The double effect of a principle attempting to define when an action that had both good and bad results are morally permissible. I one formation such an action is permissible if (1) The action is not wrong in itself, (2) the bad consequences are not that which is intended (3) the good is not itself a result of the bad consequences, and (4) the two consequential effects are commensurate. Thus, for instance, I might justifiably bomb an enemy factory, foreseeing but intending that the death of nearby civilians, whereas bombing the death of nearby civilians intentionally would be disallowed. The principle has its roots in Thomist moral philosophy, accordingly. St. Thomas Aquinas (1225-74), held that it is meaningless to ask whether a human being is two things (soul and body) or, only just as it is meaningless to ask whether the wax and the shape given to it by the stamp are one: On this analogy the sound is yet to form of the body. Life after death is possible only because a form itself does not perish (pricking is a loss of form).

And therefore, in some sense available to reactivate a new body, . . . therefore, not I who survive body death, but I may be resurrected in the same personalized body that becomes reanimated by the same form, that which Aquinas's account, as a person has no privileged self-understanding, we understand ourselves as we do everything else, by way of sense experience and abstraction, and knowing the principle of our own lives is an achievement, not as a given. Difficultly at this point led the logical positivist to abandon the notion of an epistemological foundation together, and to flirt with the coherence theory of truth, it is widely accepted that trying to make the connexion between thought and experience through basic sentence s depends on an untenable myth of the given. The special way that we each have of knowing our own thoughts, intentions, and sensationalist have brought in the many philosophical behaviorist and functionalist tendencies, that have found it important to deny that there is such a special way, arguing the way that I know of my own mind inasmuch as the way that I know of yours, e.g., by seeing what I say when asked. Others, however, point out that the behaviour of reporting the result of introspection in a particular and legitimate kind of behavioural access that deserves notice in any account of historically human psychology. The historical philosophy of reflection upon the astute of history, or of historical, thinking, finds the term was used in the 18th century, e.g., by the French man of letters and philosopher Voltaire that was to mean critical historical thinking as opposed to the mere collection and repetition of stories about the past. In Hegelian, particularly by conflicting elements within his own system, however, it came to man universal or world history. The Enlightenment confidence was being replaced by science, reason, and understanding that gave history a progressive moral thread, and under the influence of the German philosopher, whom is in spreading Romanticism, Gottfried Herder (1744-1803), and, Immanuel Kant, this idea took it further to hold, so that philosophy of history cannot be the detecting of a grand system, the unfolding of the evolution of human nature as witnessed in successive sages (the progress of rationality or of Spirit). This essential speculative philosophy of history is given an extra Kantian twist in the German idealist Johann Fichte, in whom the extra association of temporal succession with logical implication introduces the idea that concepts themselves are the dynamic engines of historical change. The idea is readily intelligible in that their world of nature and of thought become identified. The work of Herder, Kant, Flichte and Schelling is synthesized by Hegel: History has a plot, as too, this to the moral development of man, from whom does he equate within the freedom within the state, this in turn is the development of thought, or a logical development in which various necessary moment in the life of the concept are successively achieved and improved upon. Hegels method is at its most successful, when the object is the history of ideas, and the evolution of thinking may march in steps with logical oppositions and their resolution encounters red by various systems of thought.

Within the revolutionary communism, Karl Marx (1818-83) and the German social philosopher Friedrich Engels (1820-95), there emerges a rather different kind of story, based upon Hefls progressive structure not laying the achievement of the goal of history to a future in which the political condition for freedom comes to exist, so that economic and political fears than reason is in the engine room. Although, it is such that speculations about the history may that it is continued to be written, notably: late examples, by the late 19th century large-scale speculation of this kind with the nature of historical understanding, and in particular with a comparison between the methods of natural science and with the historian. For writers such as the German neo-Kantian Wilhelm Windelband and the German philosopher and literary critic and historian Wilhelm Dilthey, it is important to show that the human sciences such as history are objective and legitimate, nonetheless they are in some way deferent from the enquiry of the scientist. Since the subjective-matter is the past thought and actions of human brings, what is needed and actions of human beings, past thought and actions of human beings, what is needed is an ability to relive that past thought, knowing the deliberations of past agents, as if they were the historians own. The most influential British writer that simulated the likeness upon this theme was the philosopher and historian George Collingwood (1889-1943). Whose, The Idea of History (1946), contained an extensive defence of the Verstehe approach, but it is nonetheless, the explanation from their actions, however, by re-living the situation as our understanding that understanding others is not gained by the tactic use of a theory, enabling us to infer what thoughts or intentionality experienced, again, the matter to which the subjective-matters of past thoughts and actions, as I have in me that in of myself have the human ability of knowing the deliberations of past agents as if they were the historians own. The immediate question of the form of historical explanation, and the fact that general laws have other than no place or any apprentices in the order of a minor place in the human sciences, it is also prominent in thoughts about distinctiveness as to regain their actions, but by re-living the situation in or thereby an understanding of what they experience and thought.

The views that every day, attributional intentions, were in the belief and meaning to other persons and proceeded via tacit use of a theory that enables one to construct within such definable and non-definable translations. That any one explanation might be in giving some reason that one can be understood. The view is commonly held along with functionalism, according to which psychological states theoretical entities, identified by the network of their causes and effects. The theory-theory had different implications, depending on which feature of theories is being stressed. Theories may be though of as capable of formalization, as yielding predications and explanations, as achieved by a process of theorizing, as achieved by predictions and explanations, as achieved by a process of theorizing, as answering to empirically evincing regularities, in that out-of-the-ordinary explications were shown or explained in the principle representable without them. Perhaps, this is liable to be overturned by newer and better theories, and on, nonetheless, the main problem with seeing our understanding of others as the outcome of a piece of theorizing is the nonexistence of a medium in which this theory can be couched, as the child learns simultaneously he minds of others and the meaning of terms in its native language.

Our understanding of others is not gained by the tacit use of a theory, enabling us to infer what thoughts or intentions explain their actions, however, by re-living the situation in their moccasins, or from their point of view, and thereby understanding what they experienced and thought, and therefore expressed. Understanding others is achieved when we can ourselves deliberate as they did, and hear their words as if they are our own. The suggestion is a modern development of the Verstehen tradition associated with Dilthey, Weber and Collngwood.

Verstehen is a German understanding to denote the understanding we have of human activities. In the Verstehen tradition these are understood from within, by that are opposed to knowing something by objective observation, or by placing it in a network of scientific regularities of a theory that enables one to construct these interpretations as explanations of their doings. The view is commonly held along with functionalism, according to which psychological states are theoretical entities identified by the network of their causes and effects. However, The main problem with seeing our understanding of others s the outcome of a piece of theorizing in the nonexistence of a medium in which this theory can be couched, as the child learns simultaneously the mind of others and the meaning of terms in its native language. Nonetheless, our understanding of others is not gained by the tacit use of a theory, enabling us to infer what thoughts or intentions explain their actions, but by re-living the situation in their moccasins or from their point of view, and thereby understanding what they experienced and thought, and therefore expressed. Theories may be thought of as capable of formalisations, as yielding predictions and explanations, as achieved by a process of theorizing, as answering to empirical evidence that is principle describable without them, as liable to be overturned by newer and better theories, and so on.

The exact difference is controversial, and one approach is that of knowing what in oneself would gain expression that way, and re-living by a process of empathy the mental life of the person to be understood. But other less subjective suggestions are also found. The question of whether there is a method distinct from that of science to be used in human contexts, and so whether Verstehe is necessarily the method of the social as opposed to the natural sciences, is still open.

Philosophy or the intentionality of Mind, sustain the branch of philosophy that considers mental phenomena such as sensation, perception, thought, belief, desire, intention, memory, emotion, imagination, and purposeful action. These phenomena, which can be broadly grouped as thoughts and experiences, are features of human beings; many of them are also found in other animals. Philosophers are interested in the nature of each of these phenomena as well as their relationships to one another and to physical phenomena, such as motion.

In the 17th century, French philosopher René Descartes proposed that only two substances ultimately exist; mind and body. Yet, if the two are entirely distinct, as Descartes believed, how can one substance interact with the other? How, for example, is the intention of a human mind able to cause movement in the person's limbs? The issue of the interaction between mind and body is known in philosophy as the mind-body problem.

Many fields of thought other than philosophy share an interest in the nature of mind. In religion, the nature of mind is connected with various conceptions of the soul and the possibility of life after death. In many abstract theories of mind there is considerable overlap between philosophy and the science of psychology. Once part of philosophy, psychology split off and formed a separate branch of knowledge in the 19th century. While psychology used scientific experiments to study mental states and events, philosophy uses reasoned arguments and thought experiments in seeking to understand the concepts that underlie mental phenomena. Also influenced by philosophy of mind is the field of artificial intelligence (AI), which endeavours to develop computers that can mimic what the human mind can do. Cognitive science attempts to integrate the understanding of mind provided by philosophy, psychology, AI, and other disciplines. Finally, all of these fields benefit from the detailed understanding of the brain that has emerged through neuroscience in the late 20th century.

Philosophers use the characteristics of inward accessibility, subjectivity, intentionality, goal-directedness, creativity and freedom, and consciousness to distinguish mental phenomena from physical phenomena.

Perhaps the most important characteristic of mental phenomena is that they are inwardly accessible, or available to us through introspection. We each know our own minds—our sensations, thoughts, memories, desires, and fantasies—in a direct sense, by internal reflection. We also know our mental states and mental events in a way that no one else can. In other words, we have privileged access to our own mental states.

Certain mental phenomena, those we generally call experiences, have a subjective nature - that is, they have certain characteristics we become aware of when we reflect. For instance, there is 'something it is like' to feel pain, or have an itch, or see something red. These characteristics are subjective in that they are accessible to the subject of the experience, the person who has the experience, but not to others.

Other mental phenomena, which we broadly refer to as thoughts, have a characteristic philosophers call intentionality. Intentional thoughts are about other thoughts or objects, which are represented as having certain properties or as being related to one another in a certain way. The belief that Toronto is west of Montreal, for example, is about Toronto and Montreal and represents the former as being west of the latter. Although we have privileged access to our intentional states, many of them do not seem to have a subjective nature, at least not in the way that experiences do.

A number of mental phenomena appear to be connected to one another as elements in an intelligent, goal-directed system. The system works as follows: First, our sense organ are stimulated by events in our environment; next, by virtue of these stimulations, we perceive things about the external world; finally, we use this information, as well as information we have remembered or inferred, to guide our actions in ways that further our goals. Goal-directedness seems to accompany only mental phenomena.

Another important characteristic of mind, especially of human minds, is the capacity for choice and imagination. Rather than automatically converting past influences into future actions, individual minds are capable of exhibiting creativity and freedom. For instance, we can imagine things we have not experienced and can act in ways that no one expects or could predict.

Scientists have long considered the nature of consciousness without producing a fully satisfactory definition. In the early 20th century American philosopher and psychologist William James suggested that consciousness is a mental process involving both attention to external stimuli and short-term memory. Later scientific explorations of consciousness mostly expanded upon James's work. In this article from a 1997 special issue of Scientific American, Nobel laureate Francis Crick, who helped determine the structure of DNA, and fellow biophysicist Christof Koch explain how experiments on vision might deepen our understanding of consciousness.

Mental phenomena are conscious, and consciousness may be the closest term we have for describing what is special about mental phenomena. Minds are sometimes referred to as an awareness balanced equilibrium in the state of consciousness, yet it is difficult to describe exactly what consciousness is. Although consciousness is closely related to inward accessibility and subjectivity, these very characteristics seem to hinder us in reaching an objective scientific understanding of it.

Although philosophers have written about mental phenomena since ancient times, the philosophy of mind did not garner much attention until the work of French philosopher René Descartes in the 17th century. Descartes's work represented a turning point in thinking about mind by making a strong distinction between bodies and minds, or the physical and the mental. This duality between mind and body, known as Cartesian dualism, has posed significant problems for philosophy ever since.

Descartes believed there are two basic kinds of things in the world, a belief known as substance dualism. For Descartes, the principles of existence for these two groups of things—bodies and minds—are completely different from one another: Bodies exist by being extended in space, while minds exist by being conscious. According to Descartes, nothing can be done to give a body thought and consciousness. No matter how we shape a body or combine it with other bodies, we cannot turn the body into a mind, a thing that is conscious, because being conscious is not a way of being extended.

For Descartes, a person consists of a human body and a human mind causally interacting with one another. For example, the intentions of a human being may cause that person's limbs to move. In this way, the mind can affect the body. In addition, the sense organ of a human being may be affected by light, pressure, or sound, external sources which in turn affect the brain, affecting mental states. Thus the body may affect the mind. Exactly how mind can affect body, and vice versa, is a central issue in the philosophy of mind, and is known as the mind-body problem. According to Descartes, this interaction of mind and body is peculiarly intimate. Unlike the interaction between a pilot and his ship, the connection between mind and body more closely resembles two substances that have been thoroughly mixed together.

In response to the mind-body problem arising from Descartes's theory of substance dualism, a number of philosophers have advocated various forms of substance monism, the doctrine that there is ultimately just one kind of thing in reality. In the 18th century, Irish philosopher George Berkeley claimed there were no material objects in the world, only minds and their ideas. Berkeley thought that talk about physical objects was simply a way of organizing the flow of experience. Near the turn of the 20th century, American psychologist and philosopher William James proposed another form of substance monism. James claimed that experience is the basic stuff from which both bodies and minds are constructed.

Most philosophers of mind today are substance monists of a third type: They are materialists who believe that everything in the world is basically material, or a physical object. Among materialists, there is still considerable disagreement about the status of mental properties, which are conceived as properties of bodies or brains. Materialists who are property dualists believe that mental properties are an additional kind of property or attribute, not reducible to physical properties. Property dualists have the problem of explaining how such properties can fit into the world envisaged by modern physical science, according to which there are physical explanations for all things.

Materialists who are property monists believe that there is ultimately only one type of property, although they disagree on whether or not mental properties exist in material form. Some property monists, known as reductive materialists, hold that mental properties exist simply as a subset of relatively complex and nonbasic physical properties of the brain. Reductive materialists have the problem of explaining how the physical states of the brain can be inwardly accessible and have a subjective character, as mental states do. Other property monists, known as Eliminative materialists, consider the whole category of mental properties to be a mistake. According to them, mental properties should be treated as discredited postulates of an outmoded theory. Eliminative materialism is difficult for most people to accept, since we seem to have direct knowledge of our own mental phenomena by introspection and because we use the general principles we understand about mental phenomena to predict and explain the behaviour of others.

Philosophy of mind concerns itself with a number of specialized problems. In addition to the mind-body problem, important issues include those of personal identity, immortality, and artificial intelligence.

During much of Western history, the mind has been identified with the soul as presented in Christian theology. According to Christianity, the soul is the source of a person's identity and is usually regarded as immaterial; thus it is capable of enduring after the death of the body. Descartes's conception of the mind as a separate, nonmaterial substance fits well with this understanding of the soul. In Descartes's view, we are aware of our bodies only as the cause of sensations and other mental phenomena. Consequently our personal essence is composed more fundamentally of mind and the preservation of the mind after death would constitute our continued existence.

The mind conceived by materialist forms of substance monism does not fit as neatly with this traditional concept of the soul. With materialism, once a physical body is destroyed, nothing enduring remains. Some philosophers think that a concept of personal identity can be constructed that permits the possibility of life after death without appealing to separate immaterial substances. Following in the tradition of 17th-century British philosopher John Locke, these philosophers propose that a person consists of a stream of mental events linked by memory. It is these links of memory, rather than a single underlying substance, that provides the unity of a single consciousness through time. Immortality is conceivable if we think of these memory links as connecting a later consciousness in heaven with an earlier one on earth.

The field of artificial intelligence also raises interesting questions for the philosophy of mind. People have designed machines that mimic or model many aspects of human intelligence, and there are robots currently in use whose behaviour is described in terms of goals, beliefs, and perceptions. Such machines are capable of behaviour that, were it exhibited by a human being, would surely be taken to be free and creative. As an example, in 1996 an IBM computer named Deep Blue won a chess game against Russian world champion Garry Kasparov under international match regulations. Moreover, it is possible to design robots that have some sort of privileged access to their internal states. Philosophers disagree over whether such robots truly think or simply appear to think and whether such robots should be considered to be conscious.

Because of Heidegger, one is what one does in the world, a phenomenological reduction to one's own private experience is impossible; and because human action consists of a direct grasp of objects, it is not necessary to posit a special mental entity called a meaning to account for intentionality.

For Heidegger, being thrown into the world among things in the act of realizing projects is a more fundamental kind of intentionality than that revealed in merely staring at or thinking about objects, and it is this more fundamental intentionality that makes possible the directedness analyzed by Husserl.

Our seismical uncertainty is felt therewith of our interiorized or private conditions whole regime is felt as justly in combinality to the, the inner as it is to the other, external as with internal, within or without, and so forth, from which of our sensations, perceptions, thoughts, beliefs, desires, intentions, memory's emotions, imagination, and purposeful actions. These phenomena, which can be broadly grouped as thoughts and experiences, are features of human beings; many of them are also found in other animals. Philosophers are interested in the nature of each of these phenomena as well as their relationships to one another and to physical phenomena, such as motion.

In the 17th century, French philosopher René Descartes proposed that only two substances ultimately exist; mind and body. Yet, if the two are entirely distinct, as Descartes believed, how can one substance interact with the other? How, for example, is the intention of a human mind able to cause movement in the person's limbs? The issue of the interaction between mind and body is known in philosophy as the mind-body problem.

Many fields of thought other than philosophy share an interest in the nature of mind. In religion, the nature of mind is connected with various conceptions of the soul and the possibility of life after death. In many abstract theories of mind there is considerable overlap between philosophy and the science of psychology. Once part of philosophy, psychology split off and formed a separate branch of knowledge in the 19th century. While psychology used scientific experiments to study mental states and events, philosophy uses reasoned arguments and thought experiments in seeking to understand the concepts that underlie mental phenomena. Also influenced by philosophy of mind is the field of artificial intelligence (AI), which endeavours to develop computers that can mimic what the human mind can do. Cognitive science attempts to integrate the understanding of mind provided by philosophy, psychology, AI, and other disciplines. Finally, all of these fields benefit from the detailed understanding of the brain that has emerged through neuroscience in the late 20th century.

Philosophers use the characteristics of inward accessibility, subjectivity, intentionality, goal-directedness, creativity and freedom, and consciousness to distinguish mental phenomena from physical phenomena.

Perhaps the most important characteristic of mental phenomena is that they are inwardly accessible, or available to us through introspection. We each know our own minds—our sensations, thoughts, memories, desires, and fantasies—in a direct sense, by internal reflection. We also know our mental states and mental events in a way that no one else can. In other words, we have privileged access to our own mental states.

Certain mental phenomena, those we generally call experiences, have a subjective nature - that is, they have certain characteristics we become aware of when we reflect. For instance, there is 'something it is like' to feel pain, or have an itch, or see something red. These characteristics are subjective in that they are accessible to the subject of the experience, the person who has the experience, but not to others.

Other mental phenomena, which we broadly refer to as thoughts, have a characteristic philosophers call intentionality. Intentional thoughts are about other thoughts or objects, which are represented as having certain properties or as being related to one another in a certain way. The belief that California is west of Nevada, for example, is about California and Nevada and represents the former as being west of the latter. Although we have privileged access to our intentional states, many of them do not seem to have a subjective nature, at least not in the way that experiences do.

A number of mental phenomena appear to be connected to one another as elements in an intelligent, goal-directed system. The system works as follows: First, our sense organ are stimulated by events in our environment; next, by virtue of these stimulations, we perceive things about the external world; finally, we use this information, as well as information we have remembered or inferred, to guide our actions in ways that further our goals. Goal-directedness seems to accompany only mental phenomena.

Another important characteristic of mind, especially of human minds, is the capacity for choice and imagination. Rather than automatically converting past influences into future actions, individual minds are capable of exhibiting creativity and freedom. For instance, we can imagine things we have not experienced and can act in ways that no one expects or could predict.

Mental phenomena are conscious, and consciousness may be the closest term we have for describing what is special about mental phenomena. Minds are sometimes referred to as consciousness, yet it is difficult to describe exactly what consciousness is. Although consciousness is closely related to inward accessibility and subjectivity, these very characteristics seem to hinder us in reaching an objective scientific understanding of it.

Although philosophers have written about mental phenomena since ancient times, the philosophy of mind did not garner much attention until the work of French philosopher René Descartes in the 17th century. Descartes's work represented a turning point in thinking about mind by making a strong distinction between bodies and minds, or the physical and the mental. This duality between mind and body, known as Cartesian dualism, has posed significant problems for philosophy ever since.

Descartes believed there are two basic kinds of things in the world, a belief known as substance dualism. For Descartes, the principles of existence for these two groups of things—bodies and minds—are completely different from one another: Bodies exist by being extended in space, while minds exist by being conscious. According to Descartes, nothing can be done to give a body thought and consciousness. No matter how we shape a body or combine it with other bodies, we cannot turn the body into a mind, a thing that is conscious, because being conscious is not a way of being extended.

For Descartes, a person consists of a human body and a human mind causally interacting with one another. For example, the intentions of a human being may cause that person's limbs to move. In this way, the mind can affect the body. In addition, the sense organ of a human being may be affected by light, pressure, or sound, external sources which in turn affect the brain, affecting mental states. Thus the body may affect the mind. Exactly how mind can affect body, and vice versa, is a central issue in the philosophy of mind, and is known as the mind-body problem. According to Descartes, this interaction of mind and body is peculiarly intimate. Unlike the interaction between a pilot and his ship, the connection between mind and body more closely resembles two substances that have been thoroughly mixed together.

In response to the mind-body problem arising from Descartes's theory of substance dualism, a number of philosophers have advocated various forms of substance monism, the doctrine that there is ultimately just one kind of thing in reality. In the 18th century, Irish philosopher George Berkeley claimed there were no material objects in the world, only minds and their ideas. Berkeley thought that talk about physical objects was simply a way of organizing the flow of experience. Near the turn of the 20th century, American psychologist and philosopher William James proposed another form of substance monism. James claimed that experience is the basic stuff from which both bodies and minds are constructed.

Most philosophers of mind today are substance monists of a third type: They are materialists who believe that everything in the world is basically material, or a physical object. Among materialists, there is still considerable disagreement about the status of mental properties, which are conceived as properties of bodies or brains. Materialists who are property dualists believe that mental properties are an additional kind of property or attribute, not reducible to physical properties. Property dualists have the problem of explaining how such properties can fit into the world envisaged by modern physical science, according to which there are physical explanations for all things.

Materialists who are property monists believe that there is ultimately only one type of property, although they disagree on whether or not mental properties exist in material form. Some property monists, known as reductive materialists, hold that mental properties exist simply as a subset of relatively complex and nonbasic physical properties of the brain. Reductive materialists have the problem of explaining how the physical states of the brain can be inwardly accessible and have a subjective character, as mental states do. Other property monists, known as Eliminative materialists, consider the whole category of mental properties to be a mistake. According to them, mental properties should be treated as discredited postulates of an outmoded theory. Eliminative materialism is difficult for most people to accept, since we seem to have direct knowledge of our own mental phenomena by introspection and because we use the general principles we understand about mental phenomena to predict and explain the behaviour of others.

Philosophy of mind concerns itself with a number of specialized problems. In addition to the mind-body problem, important issues include those of personal identity, immortality, and artificial intelligence.

During much of Western history, the mind has been identified with the soul as presented in Christian theology. According to Christianity, the soul is the source of a person's identity and is usually regarded as immaterial; thus it is capable of enduring after the death of the body. Descartes's conception of the mind as a separate, nonmaterial substance fits well with this understanding of the soul. In Descartes's view, we are aware of our bodies only as the cause of sensations and other mental phenomena.

Consequently our personal essence is composed more fundamentally of mind and the preservation of the mind after death would constitute our continued existence.

The mind conceived by materialist forms of substance monism does not fit as neatly with this traditional concept of the soul. With materialism, once a physical body is destroyed, nothing enduring remains. Some philosophers think that a concept of personal identity can be constructed that permits the possibility of life after death without appealing to separate immaterial substances. Following in the tradition of 17th-century British philosopher John Locke, these philosophers propose that a person consists of a stream of mental events linked by memory. It is these links of memory, rather than a single underlying substance, that provides the unity of a single consciousness through time. Immortality is conceivable if we think of these memory links as connecting a later consciousness in heaven with an earlier one on earth.

Garry Kasparov became the youngest world champion in chess history at the age of 22. Since that time in 1985 Kasparov has continued to be rated and recognized as the best chess player in the world. In the 1990s Kasparov competed in two highly publicized matches against Deep Blue, a supercomputer designed to play chess. In 1999 Kasparov challenged chess enthusiasts everywhere in an Internet project called 'Kasparov Vs. The World.' Kasparov discusses taking on the world, Deep Blue, and his future challengers in chess.

`Husserl was born in Prossnitz, Moravia (now in the Czech Republic), on April 8, 1859. He studied science, philosophy, and mathematics at the universities of Leipzig, Berlin, and Vienna and wrote his doctoral thesis on the calculus of variations. He became interested in the psychological basis of mathematics and, shortly after becoming a lecturer in philosophy at the University of Halle, wrote his first book, Philosophie der Arithmetik (1891). At that time he maintained that the truths of mathematics have validity regardless of the way people come to discover and believe in them.

Husserl has argued against his early position, which he called psychologism, in Logical Investigations (1900-1901; trs., 1970). In this book, regarded as a radical departure in philosophy, he contended that the philosopher's task is to contemplate the essences of things, and that the essence of an object can be arrived at by systematically varying that object in the imagination. Husserl noted that consciousness is always directed toward something. He called this directedness intentionality and argued that consciousness contains ideal, unchanging structures called meanings, which determine what object the mind is directed toward at any given time.

During his tenure (1901-1916) at the University of Göttingen, Husserl attracted many students, who began to form a distinct phenomenological school, and he wrote his most influential work, Ideas: A General Introduction to Pure Phenomenology (1913; trs., 1931). In this book Husserl introduced the term phenomenological reduction for his method of reflection on the meanings the mind employs when it contemplates an object. Because this method concentrates on meanings that are in the mind, whether or not the object present to consciousness actually exists, Husserl said the method involves 'bracketing existence,' that is, setting aside the question of the real existence of the contemplated object. He proceeded to give detailed analyses of the mental structures involved in perceiving particular types of objects, describing in detail, for instance, his perception of the apple tree in his garden. Thus, although phenomenology does not assume the existence of anything, it is nonetheless a descriptive discipline; according to Husserl, phenomenology is devoted, not to inventing theories, but rather to describing the 'things themselves.'

After 1916 Husserl taught at the University of Freiburg. Phenomenology had been criticized as an essentially solipsistic method, confining the philosopher to the contemplation of private meanings, so in Cartesian Meditations (1931; trs., 1960), Husserl attempted to show how the individual consciousness can be directed toward other minds, society, and history. Husserl died in Freiburg on April 26, 1938.

Husserl's phenomenology had a great influence on a younger colleague at Freiburg, Martin Heidegger, who developed existential phenomenology, and Jean-Paul Sartre and French existentialism. Phenomenology remains one of the most vigorous tendencies in contemporary philosophy, and its impact has also been felt in theology, linguistics, psychology, and the social sciences.

The founder of phenomenology, German philosopher Edmund Husserl, introduced the term in his book Ideas zu einer reinen Phänomenolgie und phänomenologischen Philosophie (1913; Ideas: A General Introduction to Pure Phenomenology,1931). Early followers of Husserl such as German philosopher Max Scheler, influenced by his previous book, Logische Untersuchungen (two volumes, 1900 and 1901; Logical Investigations, 1970), claimed that the task of phenomenology is to study essences, such as the essence of emotions. Although Husserl himself never gave up his early interest in essences, he later held that only the essences of certain special conscious structures are the proper object of phenomenology. As formulated by Husserl after 1910, phenomenology is the study of the structures of consciousness that enable consciousness to refer to objects outside itself. This study requires reflection on the content of the mind to the exclusion of everything else. Husserl called this type of reflection the phenomenological reduction. Because the mind can be directed toward nonexistent as well as real objects, Husserl noted that phenomenological reflection does not presuppose that anything exists, but rather amounts to a 'bracketing of existence'—that is, setting aside the question of the real existence of the contemplated object.

What Husserl discovered when he contemplated the content of his mind were such acts as remembering, desiring, and perceiving, in addition to the abstract content of these acts, which Husserl called meanings. These meanings, he claimed, enabled an act to be directed toward an object under a certain aspect; and such directedness, called intentionality, he held to be the essence of consciousness. Transcendental phenomenology, according to Husserl, was the study of the basic components of the meanings that make intentionality possible. Later, in Méditations cartésiennes (1931; Cartesian Meditations, 1960), he introduced genetic phenomenology, which he defined as the study of how these meanings are built up in the course of experience.

Phenomenology attempts to describe reality in terms of pure experience by suspending all beliefs and assumptions about the world. Though first defined as descriptive psychology, phenomenology attempts philosophical rather than psychological investigations into the nature of human beings. Influenced by his colleague Edmund Husserl (known as the founder of phenomenology), German philosopher Martin Heidegger published Sein und Zeit (Being and Time) in 1927, an effort to describe the phenomenon of being by considering the full scope of existence.

All phenomenologists follow Husserl in attempting to use pure description. Thus, they all subscribe to Husserl's slogan 'To the things themselves.' They differ among themselves, however, as to whether the phenomenological reduction can be performed, and as to what is manifest to the philosopher giving a pure description of experience. German philosopher Martin Heidegger, Husserl's colleague and most brilliant critic, claimed that phenomenology should make manifest what is hidden in ordinary, everyday experience. He thus attempted in Sein und Zeit (1927; Being and Time, 1962) to describe what he called the structure of everydayness, or being-in-the-world, which he found to be an interconnected system of equipment, social roles, and purposes.

German philosopher Martin Heidegger was instrumental in the development of the 20th-century philosophical school of existential phenomenology, which examines the relationship between phenomena and individual consciousness. His inquiries into the meaning of 'authentic' or 'inauthentic' existence greatly influenced a broad range of thinkers, including French existentialist Jean-Paul Sartre. Author Michael Inwood explores Heidegger's key concept of Dasein, or 'being,' which was first expounded in his major work Being and Time (1927).

Because, for Heidegger, one is what one does in the world, a phenomenological reduction to one's own private experience is impossible; and because human action consists of a direct grasp of objects, it is not necessary to posit a special mental entity called a meaning to account for intentionality. For Heidegger, being thrown into the world among things in the act of realizing projects is a more fundamental kind of intentionality than that revealed in merely staring at or thinking about objects, and it is this more fundamental intentionality that makes possible the directedness analyzed by Husserl.

In the mid-1900s, French existentialist, Jean-Paul Sartre attempted to adapt Heidegger's phenomenology to the philosophy of consciousness, in effect returning to the approach of Husserl. Sartre agreed with Husserl that consciousness is always directed at objects but criticized his claim that such directedness is possible only by of special mental entities called meanings. The French philosopher Maurice Merleau-Ponty rejected Sartre's view that phenomenological description reveals human beings to be pure, isolated, and freely conscious. He stressed the role of the active, involved body in all human knowledge, thus generalizing Heidegger's insights to include the analysis of perception. Like Heidegger and Sartre, Merleau-Ponty is an existential phenomenologists, in that he denies the possibility of bracketing existence.

Phenomenology has had a pervasive influence on 20th-century thought. Phenomenological versions of theology, sociology, psychology, psychiatry, and literary criticism have been developed, and phenomenology remains one of the most important schools of contemporary philosophy.

Many fields other than philosophy share an interest in the nature of mind. In religion, the nature of mind is connected with various conceptions of the soul and the possibility of life after death. In many abstract theories of mind there is considerable overlap between philosophy and the science of psychology. Once part of philosophy, psychology split off and formed a separate branch of knowledge in the 19th century. While psychology used scientific experiments to study mental states and events, philosophy uses reasoned arguments and thought experiments in seeking to understand the concepts that underlie mental phenomena. Also influenced by philosophy of mind is the field of artificial intelligence (AI), which endeavours to develop computers that can mimic what the human mind can do. Cognitive science attempts to integrate the understanding of mind provided by philosophy, psychology, AI, and other disciplines. Finally, all of these fields benefit from the detailed understanding of the brain that has emerged through neuroscience in the late 20th century.

Philosophers use the characteristics of inward accessibility, subjectivity, intentionality, goal-directedness, creativity and freedom, and consciousness to distinguish mental phenomena from physical phenomena.

Perhaps the most important characteristic of mental phenomena is that they are inwardly accessible, or available to us through introspection. We each know our own minds—our sensations, thoughts, memories, desires, and fantasies—in a direct sense, by internal reflection. We also know our mental states and mental events in a way that no one else can. In other words, we have privileged access to our own mental states.

Certain mental phenomena, those we generally call experiences, have a subjective nature - that is, they have certain characteristics we become aware of when we reflect. For instance, there is 'something it is like' to feel pain, or have an itch, or see something red. These characteristics are subjective in that they are accessible to the subject of the experience, the person who has the experience, but not to others.

Other mental phenomena, which we broadly refer to as thoughts, have a characteristic philosophers call intentionality. Intentional thoughts are about other thoughts or objects, which are represented as having certain properties or as being related to one another in a certain way. The belief that California is west of Nevada, for example, is about California and Nevada and represents the former as being west of the latter. Although we have privileged access to our intentional states, many of them do not seem to have a subjective nature, at least not in the way that experiences do.

A number of mental phenomena appear to be connected to one another as elements in an intelligent, goal-directed system. The system works as follows: First, our sense organ are stimulated by events in our environment; next, by virtue of these stimulations, we perceive things about the external world; finally, we use this information, as well as information we have remembered or inferred, to guide our actions in ways that further our goals. Goal-directedness seems to accompany only mental phenomena.

Another important characteristic of mind, especially of human minds, is the capacity for choice and imagination. Rather than automatically converting past influences into future actions, individual minds are capable of exhibiting creativity and freedom. For instance, we can imagine things we have not experienced and can act in ways that no one expects or could predict.

Scientists have long considered the nature of consciousness without producing a fully satisfactory definition. In the early 20th century American philosopher and psychologist William James suggested that consciousness is a mental process involving both attention to external stimuli and short-term memory. Later scientific explorations of consciousness mostly expanded upon James's work. In this article from a 1997 special issue of Scientific American, Nobel laureate Francis Crick, who helped determine the structure of a DNA, and fellow biophysicist Christof Koch explain how experiments on vision might deepen our understanding of consciousness.

Mental phenomena are conscious, and consciousness may be the closest term we have for describing what is special about mental phenomena. Minds are sometimes referred to for being privileged to the awakening state consciousness, yet it is difficult to describe exactly what consciousness is. Although consciousness is closely related to inward accessibility and subjectivity, these very characteristics seem to hinder us in reaching an objective scientific understanding of it.

Although philosophers have written about mental phenomena since ancient times, the philosophy of mind did not garner much attention until the work of French philosopher René Descartes in the 17th century. Descartes's work represented a turning point in thinking about mind by making a strong distinction between bodies and minds, or the physical and the mental. This duality between mind and body, known as Cartesian dualism, has posed significant problems for philosophy ever since.

Descartes believed there are two basic kinds of things in the world, a belief known as substance dualism. For Descartes, the principles of existence for these two groups of things—bodies and minds—are completely different from one another: Bodies exist by being extended in space, while minds exist by being conscious. According to Descartes, nothing can be done to give a body thought and consciousness. No matter how we shape a body or combine it with other bodies, we cannot turn the body into a mind, a thing that is conscious, because being conscious is not a way of being extended.

For Descartes, a person consists of a human body and a human mind causally interacting with one another. For example, the intentions of a human being may cause that person's limbs to move. In this way, the mind can affect the body. In addition, the sense organ of a human being may be affected by light, pressure, or sound, external sources which in turn affect the brain, affecting mental states. Thus the body may affect the mind. Exactly how mind can affect body, and vice versa, is a central issue in the philosophy of mind, and is known as the mind-body problem. According to Descartes, this interaction of mind and body is peculiarly intimate. Unlike the interaction between a pilot and his ship, the connection between mind and body more closely resembles two substances that have been thoroughly mixed together.

In response to the mind-body problem arising from Descartes's theory of substance dualism, a number of philosophers have advocated various forms of substance monism, the doctrine that there is ultimately just one kind of thing in reality. In the 18th century, Irish philosopher George Berkeley claimed there were no material objects in the world, only minds and their ideas. Berkeley thought that talk about physical objects was simply a way of organizing the flow of experience. Near the turn of the 20th century, American psychologist and philosopher William James proposed another form of substance monism. James claimed that experience is the basic stuff from which both bodies and minds are constructed.

Most philosophers of mind today are substance monists of a third type: They are materialists who believe that everything in the world is basically material, or a physical object. Among materialists, there is still considerable disagreement about the status of mental properties, which are conceived as properties of bodies or brains. Materialists who are property dualists believe that mental properties are an additional kind of property or attribute, not reducible to physical properties. Property dualists have the problem of explaining how such properties can fit into the world envisaged by modern physical science, according to which there are physical explanations for all things.

Materialists who are property monists believe that there is ultimately only one type of property, although they disagree on whether or not mental properties exist in material form. Some property monists, known as reductive materialists, hold that mental properties exist simply as a subset of relatively complex and nonbasic physical properties of the brain. Reductive materialists have the problem of explaining how the physical states of the brain can be inwardly accessible and have a subjective character, as mental states do. Other property monists, known as Eliminative materialists, consider the whole category of mental properties to be a mistake. According to them, mental properties should be treated as discredited postulates of an outmoded theory. Eliminative materialism is difficult for most people to accept, since we seem to have direct knowledge of our own mental phenomena by introspection and because we use the general principles we understand about mental phenomena to predict and explain the behaviour of others.

Philosophy of mind concerns itself with a number of specialized problems. In addition to the mind-body problem, important issues include those of personal identity, immortality, and artificial intelligence.

During much of Western history, the mind has been identified with the soul as presented in Christian theology. According to Christianity, the soul is the source of a person's identity and is usually regarded as immaterial; thus it is capable of enduring after the death of the body. Descartes's conception of the mind as a separate, nonmaterial substance fits well with this understanding of the soul. In Descartes's view, we are aware of our bodies only as the cause of sensations and other mental phenomena. Consequently our personal essence is composed more fundamentally of mind and the preservation of the mind after death would constitute our continued existence.

The mind conceived by materialist forms of substance monism does not fit as neatly with this traditional concept of the soul. With materialism, once a physical body is destroyed, nothing enduring remains. Some philosophers think that a concept of personal identity can be constructed that permits the possibility of life after death without appealing to separate immaterial substances. Following in the tradition of 17th-century British philosopher John Locke, these philosophers propose that a person consists of a stream of mental events linked by memory. It is these links of memory, rather than a single underlying substance, that provides the unity of a single consciousness through time. Immortality is conceivable if we think of these memory links as connecting a later consciousness in heaven with an earlier one on earth.

Before psychology became established in science, it was popularly associated with extrasensory perception (ESP) and other paranormal phenomena (phenomena beyond the laws of science). Today, these topics lie outside the traditional scope of scientific psychology and fall within the domain of parapsychology. Psychologists note that thousands of studies have failed to demonstrate the existence of paranormal phenomena. Grounded in the conviction that mind and behaviour must be studied using statistical and scientific methods, psychology has become a highly respected and socially useful discipline. Psychologists now study important and sensitive topics such as the similarities and differences between men and women, racial and ethnic diversity, sexual orientation, marriage and divorce, abortion, adoption, intelligence testing, sleep and sleep disorders, obesity and dieting, and the effects of psychoactive drugs such as methylphenidate (Ritalin) and fluoxetine (Prozac).

In the last few decades, researchers have made significant breakthroughs in understanding the brain, mental processes, and behaviour. This section of the article provides examples of contemporary research in psychology: the plasticity of the brain and nervous system, the nature of consciousness, memory distortions, competence and rationality, genetic influences on behaviour, infancy, the nature of intelligence, human motivation, prejudice and discrimination, the benefits of psychotherapy, and the psychological influences on the immune system.

Psychologists once believed that the neural circuits of the adult brain and nervous system were fully developed and no longer subject to change. Then in the 1980s and 1990s a series of provocative experiments showed that the adult brain has flexibility, or plasticity - a capacity to change as a result of usage and experience.

These experiments showed that adult rats flooded with visual stimulation formed new neural connections in the brain's visual cortex, where visual signals are interpreted. Likewise, those trained to run an obstacle course formed new connections in the cerebellum, where balance and motor skills are coordinated. Similar results with birds, mice, and monkeys have confirmed the point: Experience can stimulate the growth of new connections and mold the brain's neural architecture.

Once the brain reaches maturity, the number of neurons does not increase, and any neurons that are damaged are permanently disabled. But the plasticity of the brain can greatly benefit people with damage to the brain and nervous system. Organisms can compensate for loss by strengthening old neural connections and sprouting new ones. That is why people who suffer strokes are often able to recover their lost speech and motor abilities.

In 1860 German physicist Gustav Fechner theorized that if the human brain were divided into right and left halves, each side would have its own stream of consciousness. Modern medicine has actually allowed scientists to investigate this hypothesis. People who suffer from life-threatening epileptic seizures sometimes undergo a radical surgery that severs the corpus callosum, a bridge of nerve tissue that connects the right and left hemispheres of the brain. After the surgery, the two hemispheres can no longer communicate with each other.

Scientists have long considered the nature of consciousness without producing a fully satisfactory definition. In the early 20th century American philosopher and psychologist William James suggested that consciousness is a mental process involving both attention to external stimuli and short-term memory. Later scientific explorations of consciousness mostly expanded upon James's work. In this article from a 1997 special issue of Scientific American, Nobel laureate Francis Crick, who helped determine the structure of DNA, and fellow biophysicist Christof Koch explain how experiments on vision might deepen our understanding of consciousness.

Beginning in the 1960s American neurologist Roger Sperry and others tested such split-brain patients in carefully designed experiments. The researchers found that the hemispheres of these patients seemed to function independently, almost as if the subjects had two brains. In addition, they discovered that the left hemisphere was capable of speech and language, but not the right hemisphere. For example, when split-brain patients saw the image of an object flashed in their left visual field (thus sending the visual information to the right hemisphere), they were incapable of naming or describing the object. Yet they could easily point to the correct object with their left hand (which is controlled by the right hemisphere). As Sperry's colleague Michael Gazzaniga stated, 'Each half brain seemed to work and function outside of the conscious realm of the other.'

Other psychologists interested in consciousness have examined how people are influenced without their awareness. For example, research has demonstrated that under certain conditions in the laboratory, people can be fleetingly affected by subliminal stimuli, sensory information presented so rapidly or faintly that it falls below the threshold of awareness. (Note, however, that scientists have discredited claims that people can be importantly influenced by subliminal messages in advertising, rock music, or other media.) Other evidence for influence without awareness comes from studies of people with a type of amnesia that prevents them from forming new memories. In experiments, these subjects are unable to recognize words they previously viewed in a list, but they are more likely to use those words later in an unrelated task. In fact, memory without awareness is normal, as when people come up with an idea they think is original, only later to realize that they had inadvertently borrowed it from another source.

Cognitive psychologists have often likened human memory to a computer that encodes, stores, and retrieves information. It is now clear, however, that remembering is an active process and that people construct and alter memories according to their beliefs, wishes, needs, and information received from outside sources.

Without realizing it, people sometimes create memories that are false. In one study, for example, subjects watched a slide show depicting a car accident. They saw either a 'STOP' sign or a 'YIELD' sign in the slides, but afterward they were asked a question about the accident that implied the presence of the other sign. Influenced by this suggestion, many subjects recalled the wrong traffic sign. In another study, people who heard a list of sleep-related words (bed, yawn) or music-related words (jazz, instrument) were often convinced moments later that they had also heard the words sleep or music—words that fit the category but were not on the list. In a third study, researchers asked college students to recall their high-school grades. Then the researchers checked those memories against the students' actual transcripts. The students recalled most grades correctly, but most of the errors inflated their grades, particularly when the actual grades were low.

When scientists distinguish between human beings and other animals, they point to our larger cerebral cortex (the outer part of the brain) and to our superior intellect - as seen in the abilities to acquire and store large amounts of information, solve problems, and communicate through the use of language.

In recent years, however, those studying human cognition have found that people are often less than rational and accurate in their performance. Some researchers have found that people are prone to forgetting, and worse, that memories of past events are often highly distorted. Others have observed that people often violate the rules of logic and probability when reasoning about real events, as when gamblers overestimate the odds of winning in games of chance. One reason for these mistakes is that we commonly rely on cognitive heuristics, mental shortcuts that allow us to make judgments that are quick but often in error. To understand how heuristics can lead to mistaken assumptions, imagine offering people a lottery ticket containing six numbers out of a pool of the numbers 1 through 40. If given a choice between the tickets 6-39-2-10-24-30 and 1-2-3-4-5-6, most people select the first ticket, because it has the appearance of randomness. Yet out of the 3,838,380 possible winning combinations, both sequences are equally likely.

One of the oldest debates in psychology, and in philosophy, concerns whether individual human traits and abilities are predetermined from birth or due to one's upbringing and experiences. This debate is often termed the nature-nurture debate. A strict genetic (nature) position states that people are predisposed to become sociable, smart, cheerful, or depressed according to their genetic blueprint. In contrast, a strict environmental (nurture) position says that people are shaped by parents, peers, cultural institutions, and life experiences.

Research shows that the more genetically related a person is to someone with schizophrenia, the greater the risk that person has of developing the illness. For example, children of one parent with schizophrenia have a 13 percent chance of developing the illness, whereas children of two parents with schizophrenia have a 46 percent chance of developing the disorder.

Researchers can estimate the role of genetic factors in two ways: (1) twin studies and (2) adoption studies. Twin studies compare identical twins with fraternal twins of the same sex. If identical twins (who share all the same genes) are more similar to each other on a given trait than are same-sex fraternal twins (who share only about half of the same genes), then genetic factors are assumed to influence the trait. Other studies compare identical twins who are raised together with identical twins who are separated at birth and raised in different families. If the twins raised together are more similar to each other than the twins raised apart, childhood experiences are presumed to influence the trait. Sometimes researchers conduct adoption studies, in which they compare adopted children to their biological and adoptive parents. If these children display traits that resemble those of their biological relatives more than their adoptive relatives, genetic factors are assumed to play a role in the trait.

In recent years, several twin and adoption studies have shown that genetic factors play a role in the development of intellectual abilities, temperament and personality, vocational interests, and various psychological disorders. Interestingly, however, this same research indicates that at least 50 percent of the variation in these characteristics within the population is attributable to factors in the environment. Today, most researchers agree that psychological characteristics spring from a combination of the forces of nature and nurture.

Helpless to survive on their own, newborn babies nevertheless possess a remarkable range of skills that aid in their survival. Newborns can see, hear, taste, smell, and feel pain; vision is the least developed sense at birth but improves rapidly in the first months. Crying communicates their need for food, comfort, or stimulation. Newborns also have reflexes for sucking, swallowing, grasping, and turning their head in search of their mother's nipple.

In 1890 William James described the newborn's experience as 'one great blooming, buzzing confusion.' However, with the aid of sophisticated research methods, psychologists have discovered that infants are smarter than was previously known.

A period of dramatic growth, infancy lasts from birth to around 18 months of age. Researchers have found that infants are born with certain abilities designed to aid their survival. For example, newborns show a distinct preference for human faces over other visual stimuli.

To learn about the perceptual world of infants, researchers measure infants' head movements, eye movements, facial expressions, brain waves, heart rate, and respiration. Using these indicators, psychologists have found that shortly after birth, infants show a distinct preference for the human face over other visual stimuli. Also suggesting that newborns are tuned in to the face as a social object is the fact that within 72 hours of birth, they can mimic adults who purse the lips or stick out the tongue - a rudimentary form of imitation. Newborns can distinguish between their mother's voice and that of another woman. And at two weeks old, nursing infants are more attracted to the body odour of their mother and other breast-feeding females than to that of other women. Taken together, these findings show that infants are equipped at birth with certain senses and reflexes designed to aid their survival.

In 1905 French psychologist Alfred Binet and colleague Théodore Simon devised one of the first tests of general intelligence. The test sought to identify French children likely to have difficulty in school so that they could receive special education. An American version of Binet's test, the Stanford-Binet Intelligence Scale, is still used today.

In 1905 French psychologist Alfred Binet devised the first major intelligence test for the purpose of identifying slow learners in school. In doing so, Binet assumed that intelligence could be measured as a general intellectual capacity and summarized in a numerical score, or intelligence quotient (IQ). Consistently, testing has revealed that although each of us is more skilled in some areas than in others, a general intelligence underlies our more specific abilities.

Intelligence tests often play a decisive role in determining whether a person is admitted to college, graduate school, or professional school. Thousands of people take intelligence tests every year, but many psychologists and education experts question whether these tests are an accurate way of measuring who will succeed or fail in school and later in life. In the 1998 Scientific American article, psychology and education professor Robert J. Sternberg of Yale University in New Haven, Connecticut, presents evidence against conventional intelligence tests and proposes several ways to improve testing.

Today, many psychologists believe that there is more than one type of intelligence. American psychologist Howard Gardner proposed the existence of multiple intelligence, each linked to a separate system within the brain. He theorized that there are seven types of intelligence: linguistic, logical-mathematical, spatial, musical, bodily-kinesthetic, interpersonal, and intrapersonal. American psychologist Robert Sternberg suggested a different model of intelligence, consisting of three components: analytic ('school smarts,' as measured in academic tests), creative (a capacity for insight), and practical ('street smarts,' or the ability to size up and adapt to situations).

Psychologists from all branches of the discipline study the topic of motivation, an inner state that moves an organism toward the fulfilment of some goal. Over the years, different theories of motivation have been proposed. Some theories state that people are motivated by the need to satisfy physiological needs, whereas others state that people seek to maintain an optimum level of bodily arousal (not too little and not too much). Still other theories focus on the ways in which people respond to external incentives such as money, grades in school, and recognition. Motivation researchers study a wide range of topics, including hunger and obesity, sexual desire, the effects of reward and punishment, and the needs for power, achievement, social acceptance, love, and self-esteem.

In 1954 American psychologist Abraham Maslow proposed that all people are motivated to fulfill a hierarchical pyramid of needs. At the bottom of Maslow's pyramid are needs essential to survival, such as the needs for food, water, and sleep. The need for safety follows these physiological needs. According to Maslow, higher-level needs become important to us only after our more basic needs are satisfied. These higher needs include the need for love and belongingness, the need for esteem, and the need for self-actualization (in Maslow's theory, a state in which people realize their greatest potential).

However, there is another course of thought in philosophy of science, the tradition of 'negative' or 'Eliminative induction'. That is to say, that in Eliminative induction a number of possible hypotheses conceiving some state of affairs is presumed, and rivals are progressively in reserve to new evidence, the process is an idealization, since in practice no closed set of initial theories is usually possible. The English diplomat and philosopher Francis Bacon (1561-1626) and in modern time the philosopher of science Karl Raimund Popper (1902-1994), we have the idea of using logic to bring falsifying evidence to bear on hypotheses about what must universally be the case that many thinkers accept in essence his solution to the problem of demarcating proper science from its imitators, namely that the former results in genuinely falsifiable theories whereas the latter do not. Although falsely, allowed many people's objections to such ideologies as psychoanalysis and Marxism.

Hume was interested in the processes by which we acquire knowledge: The processes of perceiving and thinking, of feeling and reasoning. He recognized that much of what we claim to know derives from other people secondhand, thirdhand or worse: Moreover, our perceptions and judgements can be distorted by a multiple array of factors-by what we are studying, and by the very act of study itself the main reason, however, behind his emphasis on 'probabilities and those other measures of evidence on which life and action entirely depend' is this:

Evidently, all reasoning concerning 'matter of fact'. are founded on the relation of cause and effect? And that we can never infer the existence of one object form that of another unless they are connected they are connected mediately or immediately. When we apparently observe a whole sequence, say of one ball hitting another, what do we observe? In the much commoner cases, when we wonder about the unobserved causes or effects of the events we observe, what precisely are we doing?

Hume recognized that a notion of 'must' or necessity is a peculiar feature of causal relation, inference and principles, and challenges us to explain and justify the notion. He argued that there is no observable feature of events, nothing like a physical bond, which can be properly labelled the 'necessary connection' between a given cause and its effect: Events are simply merely to occur, and there is in 'must' or 'ought' about them. However, repeated experience of pairs of events sets up the habit of expectation in us, such that when one of the pair occurs we inescapably expect the other. This expectation makes us infer the unobserved cause or unobserved effect of the observed event, and we mistakenly project this mental inference onto the events themselves. There is no necessity observable in causal relations, all that can be observed is regular sequence, there is proper necessity in causal inferences, but only in the mind. Once we realize that causation is a relation between pairs of events. We also realize that often we are not present for the whole sequence that we want to divide into 'cause' and 'effect'. Our understanding of the casual relation is thus intimately linked with the role of the causal inference cause only causal inferences entitle us to 'go beyond what is immediately present to the senses'. Nevertheless, now two very important assumptions emerge behind the causal inference: The assumptions that like causes, in 'like circumstances, will always produce like effects', and the assumption that 'the course of nature will continue uniformly the same'-or, briefly that the future will resemble the past. Unfortunately, this last assumption lacks either empirical or a priori proof, that is, it can be conclusively established neither by experience nor by thought alone.

Hume frequently endorsed a standard seventeenth-century view that all our ideas are ultimately traceable, by analysis, to sensory impressions of an internal or external kind. In agreement, he claimed that all his theses are based on 'experience', understood as sensory awareness with memory, since only experience establishes matters of fact. Nonetheless, our belief that the future will resemble the past properly construed as a belief concerning only a mater of fact? As the English philosopher Bertrand Russell (1872-1970) remarked, earlier this century, the real problems that Hume asserts to are whether future futures will resemble future pasts, in the way that past futures really did resemble past pasts. Hume declares that 'if . . . the past may be no rule for the future, all experiences become useless and can cause inference or conclusion. Yet, he held, the supposition cannot stem from innate ideas, since there are no innate ideas in his view nor can it stems from any abstract formal reasoning. For one thing, the future can surprise us, and no formal reasoning seems able to embrace such contingencies: For another, even animals and unthinkable people conduct their lives as if they assume the future resembles the past: Dogs return for buried bones, children avoid a painful fire, and so forth. Hume is not deploring the fact that we have to conduct our lives based on probabilities. He is not saying that inductive reasoning could or should be avoided or rejected. Alternatively, he accepted inductive reasoning but tried to show that whereas formal reasoning of the kind associated with mathematics cannot establish or prove matters of fact, factual or inductive reasoning lacks the 'necessity' and 'certainty' associated with mathematics. His position, therefore clear; because 'every effect is a distinct event from its cause', only investigation can settle whether any two particular events are causally related: Causal inferences cannot be drawn with the force of logical necessity familiar to us from deductivity, but, although they lack such force, they should not be discarded. From causation, inductive inferences are inescapable and invaluable. What, then, makes 'experience' the standard of our future judgement? The answer is 'custom', it is a brute psychological fact, without which even animal life of a simple kind would be mostly impossible. 'We are determined by custom to suppose the future conformable to the past' (Hume, 1978), nevertheless, whenever we need to calculate likely events we must supplement and correct such custom by self-conscious reasoning.

Nonetheless, the problem that the causal theory of reference will fail once it is recognized that all representations must occur under some aspect or that the extentionality of causal relations is inadequate to capture the aspectual character of reference. The only kind of causation that could be adequate to the task of reference is intentional causal or mental causation, but the causal theory of reference cannot concede that ultimately reference is achieved by some met device, since the whole approach behind the causal theory was to try to eliminate the traditional mentalism of theories of reference and meaning in favour of objective causal relations in the world, though it is at present by far the most influential theory of reference, will be a failure for these reasons.

If mental states are identical with physical states, presumably the relevant physical states are various sorts of neural states. Our concepts of mental states such as thinking, sensing, and feeling are of course, different from our concepts of neural states, of whatever sort. Still, that is no problem for the identity theory. As J.J.C. Smart (1962), who first argued for the identity theory, emphasized, the requisite identities do not depend on understanding concepts of mental states or the meanings of mental terms. For 'a' to be the identical with 'b', 'a', and 'b' must have the same properties, but the terms or the things in themselves are 'a' and 'b', and need not mean the same. Its principal by measure can be accorded within the indiscernibility of identicals, in that, if 'A' is identical with 'B', then every property that 'A' has 'B', and vice versa. This is, sometimes known as Leibniz' s Law.

Nevertheless, a problem does seem to arise about the properties of mental states. Suppose pain is identical with a certain firing of c-fibres. Although a particular pain is the very same as a neural-firing, we identify that state in two different ways: As a pain and as neural-firing. That the state will therefore have certain properties in virtue of which we identify it as pain and others in virtue of which we identify it as an excitability of neural firings. The properties in virtue of which we identify it as a pain will be mental properties, whereas those in virtue of which ewe identify it as neural excitability firing, will be physical properties. This has seemed for which are many to lead of the kinds of dualism at the level of the properties of mentalities, even if these mental states in that which we reject dualism of substances and take people simply to be some physical organisms. Those organisms still have both mental and physical states. Similarly, even if we identify those mental states with certain physical states, those states will, nonetheless have both mental and physical properties. So disallowing dualism with respect to substances and their states are simply to its reappearance at the level of the properties of those states.

There are two broad categories of mental property. Mental states such as thoughts and desires, often called 'propositional attitudes', have 'content' that can be described by 'that' clauses. For example, one can have a thought, or desire, that it will rain. These states are said to have intentional properties, or 'intentionality sensations', such as pains and sense impressions, lack intentional content, and have instead qualitative properties of various sorts.

The problem about mental properties is widely thought to be most pressing for sensations, since the painful qualities of pains and the red quality of visual sensations might be irretrievably nonphysical. If the idea that something conveys to the mind as having endlessly debated the meaning of relationally to the mind, the mental aspects of the problem seem once removed among mental states that to account in the actualization of proper innovation that has nonphysical properties, is that the identity of mental states generates or empower physical states as they would not sustain the unconditional thesis as held to mind-body materialism.

The Cartesian doctrine related and distinguished by the mentality for which are in mental properties of those states constructed by anecdotal explanations, for that the consequential temperance as proven or given to compatibility is founded the appropriate interconnective link to set right the ordering of fibrous fragments as strikingly spatial, as this will facilitate the functional contribution in which the distribution of infractions that when assembled are found of space and time. However, it should be that p assimilates some nonphysical advocates of the identity theory sometimes accepting it, for the ideas that of or relating to the mind, as a mental aspect calling our intellectual mentality as the intellective for which is not nonphysical but yields to underlying latencies, for example, the insistence by some identity theorists that mental properties are really neural as between being mental and physical. To being neural is in this way a property would have to be neutral about whether it is compatible and thus imply in the manner that inarticulate the idea that something conveys to the mind, the acceptational sense for which its significancy plays on one side of the manner that mental is the ideological reason in thinking as the inner and well of the outer domains that of or relating to the mind. Only if one thought that being meant being nonphysical would one hold that defending materialism required showing the ostensible mental properties are neutral as regards whether or not they are mental.

Nevertheless, holding that mental properties are nonphysical has a cost that is usually not noticed. A phenomenon is mental only if it has some distinctively mental property. So, strictly speaking, some materialists who claim that mental properties are nonphysical phenomena have in the unfolding transformation that it can only be derived through the state or factorial conditional independence that reality, of its customs that have recently become and exists to induce to come into being, a condition or occurrence traceable to a cause or the aftereffect through which an end-product or its resulting event of something or thing that has existence. This is the 'Eliminative-Materialist position advanced by the American philosopher and critic Richard Rorty (1979).

According to Rorty (1931-) 'mental' and 'physical' are incompatible terms. Nothing can be both mental and physical, so mental states cannot be identical with bodily states. Rorty traces this incompatibly to our views about incorrigibility: 'Mental' and 'physical' are incorrigible reports of one's own mental states, but not reports of physical occurrences, but he also argues that we can imagine people who describe themselves and each other using terms just like our mental vocabulary, except that those people do not take the reports made with that vocabulary to be incorrigible. Since Rorty takes a state to be a mental state only if one's reports about it are taken to be incorrigible, his imaginary people do not ascribe mental states to themselves or each other. Nonetheless, the only difference between their language and ours is that we take as incorrigible certain reports that they do not. So their language as no-less the descriptive explanatorial powers, in that the American philosopher and critic, Richard

McKay Rorty (1931-) concludes that our mental vocabulary is idle, and that there are no distinctively mental phenomena.

This argument hinges on or upon the building incorrigibility into the meaning of the term 'mental'. If we do not, the way is open to interpret Rorty's imaginary people as simply having a different theory of mind from ours, on which reports of one's own mental states are corrigible. Their reports would this be about mental states, as construed by their theory. Rorty's thought experiment would then provide to conclude not that our terminology is idle, but only that this alternative theory of mental phenomena is correct. His thought experiment would thus sustain the non-eliminativist view that mental states are bodily states. Whether Rorty's argument supports his eliminativist conclusion or the standard identity theory, therefore, depends solely on whether or not one holds that the quality or highest in degree attainable or attained by or through emending the intelligence in as clear unmistakable mentality are in some way the mental aspects of problems of or relating to the mind is spoken in response to substantiality.

Paul M. Churchlands (1981) advances a different argument for Eliminative materialism. Given to agree to Churchlands, the commonsense concepts of mental states contained in our present folk psychology are, from a scientific point of view, radically defective. Nonetheless, we can expect that eventually a more sophisticated theoretical account will relace those folk-psychological concepts, showing that mental phenomena, as described by current folk psychology, do not exist. Since, that account would be integrated into the rest of science, we would have a thoroughgoing materialist treatment of all phenomena, unlike Rorty's, does not rely of assuming that the mental are nonphysical.

However, even if current folk psychology is mistaken, that does not show that mental phenomenon does not exist, but only that they are of the way folk psychology described them for being. We could conclude, that, they do not put into effect of any natural cognitive processes or simulate the practice of using something for being used aside from its energy in producing the resulting of thought that exists only if the folk-psychological claims that turn out to be mistaken are factually determinant dissimilarities for which to define what it is for a phenomenon to be strictly mental. Otherwise, the new theory would be about mental phenomena, and would help show that they are identical with physical phenomena. Churchlands argument, like Rorty's, depends on a special way of defining the mental, which we need not adopt, it argument for Eliminative materialism will require some such definition, without which the argument would instead support the identity theory.

Despite initial appearances, the distinctive properties of sensations are neutral as between being mental and physical, in that borrowed from the English philosopher and classicist Gilbert Ryle (1900-76), they are topic neutral: My experiences of appreciation as having to cognize in the sensation of red for which consists in my being in a state that is similar, in respect that we need not specify, making it more evenly so, to something that occurs in me when I am in the presence of certain stimuli. Because the respect of similarity is not specified, the property is neither distinctively mental nor distinctively physical. However, everything is similar to everything else in some respect or other. So leaving the respect of similarity unspecified makes this account too weak to capture the distinguishing properties of sensation.

A more sophisticated reply to the difficultly about mental properties is due independently to the Australian, David Malet Armstrong (1926-) and American philosopher David Lewis (1941-2002), who argued that for a state to be a particular intentional state or sensation is for that state to bear characteristic causal relations to other particular occurrences. The properties in virtue of which e identify states as thoughts or sensations will still be neural as between being mental and physical, since anything can bear a causal relation to anything else. Nevertheless, causal connections have a better chance than similarity in some unspecified respect to capturing the distinguishing properties of sensations and thought.

This casual theory is appealing, but is misguided to attempt to construe the distinctive properties of mental states for being neutral as between being mental and physical. To be neutral as regards being mental or physical is to be neither distinctively mental nor distinctively physical. However, since thoughts and sensations are distinctively mental states, for a state to be a thought or a sensation is perforce for it to have some characteristically mental property. We inevitably lose the distinctively mental if we construe these properties for being neither mental nor physical.

Not only is the topic-neutral construal misguided: The problem it was designed to solve is equally so, only to say, that problem stemmed from the idea that mental must have some nonphysical aspects. If not at the level of people or their mental states, then at the level of the distinctive properties can be more complicated, for example, in the sentence, 'John is married to Mary', we can display the attribution that of 'John, is of the property of being married, and unlike the property of John is bald. Consider the sentence: 'John is bearded' and that 'John' in this sentence is a bit of language-a name of some individual human being-and more some would be tempted to confuse the word with what it names. Consider the expression 'is bald', this too is a bit of articulated linguistic communication under which philosophers call it a 'predicate'-and it caries out into a certain state of our attention of some material possession or feature that, if the sentence is true. It has possession of by John? Understood in this ay, a property is not its self linguist though it is expressed, or conveyed by something that is, namely a predicate. What might be said that a property is a real feature of the word, and that it should be contrasted just as sharply with any predicates we use to express it, as in the name 'John' is contrasted with the person himself. Controversially, just what ontological status should be accorded to properties by describing 'anomalous monism'-while it is conceivably given to a better understanding the similarity with the American philosopher Herbert Donald Davidson (1917-2003), wherefore he adopts a position that explicitly repudiates reductive physicalism, yet purports to be a version of materialism, nonetheless, Davidson holds that although token mental evident states are identical to those of physical events and states-mental 'types' -, i.e., kinds, and/or properties-is neither to, nor nomically co-existensive with, physical types. In other words, his argument for this position relies largely on the contention that the correct assignment of mental an actionable property to a person is always a holistic matter, involving a global, temporally diachronic, 'intentional interpretation' of the person. Nevertheless, as many philosophers have in effect pointed out, accommodating claims of materialism evidently requires more than just repercussions of mental/physical identities. Mentalistic explanation presupposes not merely that metal events are causes but also that they have causal/ explanatory relevance as mental, i.e., relevance insofar as they fall under mental kinds or types. Nonetheless, the mental aspects of the problem are of or relating to the mind, such as Davidson's positions, which deny there are strict psychological or psychological laws, can accommodate the causal/explanation relevance of the mental quo mentally: If to 'epiphenomenalism' with respect to mental properties.

However, the ideas that the mental are in some think much of the nonphysical cannot be assumed without argument. Plainly, the distinctively mental properties of the mental states are unlikely that any other properties looked carefully about which we know. Only mental states have properties that are at all like the qualitative properties that anything like the intentional properties of thoughts and desires. However, this does not show that the mental properties are not physical properties, not. All physical properties like the standard states: So, mental properties might still be special kinds of physical properties. It's interrogatory of questions begin or at present with falsifiably deceptive appearances. In a different way or manner, that the doctrine that the mental properties are simply an expression of the Cartesian doctrine that the mental are automatically nonphysical.

It is sometimes held that properties should count as physical properties only if they can be defined using the terms of physics. Nobody would hold that to reduce biology to physics, for example, we must define all biological properties using only terms that occur in physics. Even putting 'reduction' aside, in certain biological properties could have been defined, that would not mean that those properties were in ways tracing the remains of that which one surpasses in movement or transference the capabilities for which it is accessible to attain some navigating modernity, in that of a fashionable method or custom for which a variety of species long for being nonphysical. The sense of 'physical' that is relevant, for which its situation must be broad enough to include biological properties, but also mostly commonsense of macroscopic properties. Bodily states are uncontroversially physical in the relevant way. So, we can recast the identity theory as asserting that mental states are identical with bodily state.

While reaching conclusions about the origin and limits of knowledge, Locke had to his occasion, that concerning himself with topics that are of philosophical interest in themselves. On of these is the question of identity, which includes, more specifically, the question of personal identity: What are the criteria by which a person is numerically the same person as a person encountering of time? Locke points out whether 'this is what was here before, it matters what kind of thing 'this' is meant to be. If 'this' is meant as a mass of matter then it is what was before, insofar as it consists of the same material panicles, but if it is meant as a living body then its considering of the same particles does most matter. The case is different. 'A colt grown up to a horse, sometimes fat, sometimes lean, is altogether the same horse as we speak, but though . . . there may be a manifest change of the parts. So, when we think about personal identity, we need to be clear about a distinction between two things which 'the ordinary way of speaking runs together'-the idea of 'man' and the idea of 'person'. As with any other animal, the identity of a man consists 'in nothing but a participation of the same continued life, by constantly fleeting particles of matter, in succession initially united both internally and externally as the same organized body, however, the examination of one's own thought and feeling of intensively explicative ideas of hope or inclined to hope that of a person is not that of a living body of a certain kind. A person's task is especially by one's capabilities and enacting ability to reason has of himself the forming of an idea of something in the mind, as to conceive, envisages, envision, fancy, feature, imagine and so forth. So, then, one's own being to gather or in the assumption of assertions, are the conjectural considerations to use one's power of conception, judgement or inference to 'think', apart from other animals. Is, then, we are to accede that 'thinking' intelligently hosts there being that have reflections and such a being 'will be the same self as far as the same consciousness can extend to action past or to come? The unity of one's contingence of consciousness does not depend on its being 'annexed' only to one individual substance, [and not] . . . continued in a succession of several substances. For Lock e, then, personal identity consists in an identity of consciousness, and not in the identity of some substance whose essence it is to be conscious

Casual mechanisms or connections of meaning will help to take a historical route, and focus on the terms in which analytical philosophers of mind began to discuss seriously psychoanalytic explanation. These were provided by the long-standing and presently nonconcluded debate over cause and meaning in psychoanalysis.

Seeing why psychoanalysis should be viewed in the terminological combinations of some cause and meaning is not hard. On the one hand, Freud's theories introduce a panoply of concepts that appear to characterize mental processes as mechanical and non-meaningful. Included are Freud's neurological model of the mind, as outlined in his 'Project or a Scientific Psychology', more broadly, his 'economic' description of mentality, as having properties of force or energy, e.g., as 'cathexing' objects: And his account in the mechanism of repression. So it seems that psychoanalytic explanation employs terms logically at variation with those of ordinary, common-sens e psychology, where mechanisms do not play a central role. Bu t on the other hand, and equally striking, there is the fact that psychoanalysis proceeds through interpretation and engages on a relentless search for meaningful connections in mental life-something that became more even as a superficial examination of the Interpretation of Dreams, or The Psychopathology of Everyday Life, cannot fail to impress upon one. Psychoanalytic interpretation adduces meaningful connections between disparate and often apparently dissociated mental and behavioural phenomena, directed by the goal of 'thematic coherence'. That is, as giving mental life the sort of unity that we find in a work of art or cogent narratives, and, in this respect, psychoanalysis seems to dramatize its substantive reasons for doing so, is that for its physiological feature of ordinary psychology, finding to its insistence on or upon the relating actions to reason for them through contentual characterizations of each that make their connection seems rational, or intelligible: A goal that seems remote from anything found in the physical sciences.

The application to psychoanalysis of the perspective afforded by the cause-meaning debate can also be seen as much therefore of another factor, namely the semi-paradoxical nature of psychoanalysis' explananda. With respect to all irrational phenomena, something like a paradox arises. Irrationality involves a failure of a rational connectedness and hence of meaningfulness, and so, if it is to have an explanation of any kind, relations that are non-comprehensible and causatively might be of some needed necessity. Yet, as observed above, it seems that, in offering explanations for irrationality-plugging the 'gaps' in consciousness-what psychoanalytic explanation is attached with a hinge for being precisely the postulation of further, although non-apparent connections of meaning.

For these two reasons, then-the logical heterogeneity of its explanation and the ambiguous status of its explananda-it may seem that an examination arriving within the hierarchical terms as held of the conceptions of cause and meaning will provide the key to a philosophical elucidation of psychoanalysis. The possible views of psychoanalytic explanation that may result from such an examination can be arranged along two dimensions. (1) Psychoanalytic explanation may then be viewed after reconstruction, as either causal and non-meaningful, or meaningful and non-causal, or as comprising both meaningful and causal elements, in various combinations. Psychoanalytic explanation then may be viewed, on each of these reconstructions, as either licensed or invalidated depending one's view of the logical nature of psychology.

So, for instance, some philosophical discussions infer that psychoanalytic explanation is void, simple since it is committed to causality in psychology. On another, opposed view, it is the virtue of psychoanalytic explanation that it imputes causal relations, since only causal relations can be used for explaining the failures of meaningful psychological connections. On yet another view, it is psychoanalysis' commitment to meaning which is its great fault: It s held that the stories that psychoanalysis tries to tell do not really, on examination, explains successfully. So on.

Saying that the debates between these various positions fail to establish anything is fair definite about psychoanalytic explanation. There are two reasons for this. First, there are several different strands in Freud's whitings, each of which may be drawn on, apparently conclusively, in support of each alternative reconstruction. Secondly, preoccupation with a wholly general problem in the philosophy of mind, that of cause and meaning, distracts attention from the distinguishing features of psychoanalytic explanation. At this point, and to prepare the way for a plausible reconstruction of psychoanalytic explanation. Taking a step back is appropriate, and takes a fresh look at the cause-meaning issue in the philosophy of psychoanalysis.

Suppose, first, that in some varying refashions alternating of change, the modulator's estate of the realms or fact of having independent reality in the form of adequacy for which classes of descriptive change are themselves to imply anything about reduction. Historically, 'natural' contrasts with 'supernatural', but given to submit in the contemporary philosophy of mind whereby the debate issues a concern for which are the centres around which the possibilities of explaining mental phenomena are just or as much as part of the natural order. It is the non-natural rather than the supernatural that is the contrasting notion. The naturalist holds that they can be so explained, while the opponent of naturalism thinks otherwise, though it is not intended that opposition to naturalism commits one to anything supernatural. Nonetheless, one should not take naturalism in regard as committing one to any sort of reductive explanation of that realm, and there are such commitments in the use of 'physicalism' and 'materialism'.

If psychoanalytic explanation gives the impression that it imputes bare, meaning-free causality, this results from attending to only half the story, and misunderstanding what psychoanalysis when it talks of psychological mechanisms. The economic descriptions of mental processes that psychoanalysis provides are never replacements for, but they always presuppose, characterizations of mental processes through meaning. Mechanisms in psychoanalytic context are simply processes whose operation cannot be reconstructed as instances of rational functioning (they are what we might by preference call mental activities, by contrast with action) Psychoanalytic explanation's postulation of mechanisms should not therefore be regarded as a regrettable and expugnable incursion of scientism into Freud's thought, as is often claimed.

Suppose alternatively, those hermeneuticists such as Habermas-who follow Dilthey beings as an interpretative practice to which the concepts of the physical sciences. Are inclined to provide by or as if by formal action, in that to an extended allocation of its presently accorded intendment in the correct of thinking that connections of the idea that something coneys to the mind its meanings are misrepresented through being described as causal? Again, this does not impact negatively of psychoanalytic explanation since, as just argued, psychoanalytic explanation nowhere of imputes meaning-free causation. Nothing is lost for psychoanalytic explanation causation is excised from the psychological picture.

The conclusion must be that psychoanalytic explanation is at bottom indifferent to the general meaning-cause issue. The core of psychoanalysis consists in its tracing of meaningful connections with no greater or lesser commitment to causality than is involved in ordinary psychology. (Which helps to set the stage-pending appropriate clinical validation-for psychoanalysis to claim as much truth for its explanation as ordinary psychology?). Also, the true key to psychoanalytic explanation, its attribution of special kinds of mental states that are not recognized in customary psychology, whose relations to one another does not have the form of patterns of inference or practical reasoning.

In the light of this, understanding why some compatibilities and hermeneuticists assert that their own view of psychology is uniquely consistent with psychoanalytic explanation is easy. Compatibilities are right to think that, to provide for psychoanalytic explanation, allowing mental connections that are unlike the connections of reasons to the actions that they rationalize is necessary, or to the beliefs that they support: And, that, in outlining such connections, psychoanalytic explanation must outstrip the resources of ordinary psychology, which does attempt to force as much as possible into the mould of practical reasoning. Hermeneuticists, for their part, are right to think those postulating connections that were nominally psychological but that characterized would be futile as to meaning, and that psychoanalytic explanation does not respond to the 'paradox' of irrationality by abandoning the search for meaningful connections.

Compatibilities are, however, wrong to think that non-rational but meaningful connections require the psychological order to be conceived as a causal order. The hermeneuticists are free to postulate psychological connections determined by meaning but not by rationality: supposing that there are connections of meaning that is coherent are not -bona fide- rational connections, without these being causal. Meaningfulness is a broader concept than rationality. (Sometimes this thought has been expressed, though not helpful, by saying that Freud became aware of the existence of 'neurotic rationality.) Even if an assumption of rationality is evasively necessary to make sense of behaviour overall. It does not need to be brought into play in making sense of each instance of behaviour. Hermeneuticists, in turn, are inaccurate to thinking that the compatibility view of psychology endorses a causal signal of a mental collaborationist, categorized by their meaning with causality or that it must lead to compatibilism to acknowledge that which in any qualitative difference between rational and irrational psychological connections.

All the same, the last two decades have been an extent of time set off or typically by someone or something intermittently for being periodic through which times inordinate changes, placing an encouraging well-situated plot in the psychology of the sciences. 'Cognitive psychology', which focuses on higher mental processes like reasoning, decision making, problem solving, language processing and higher-level processing, has become-perhaps, the-dominant paradigms among experimental psychologists, while behaviouristically oriented approaches have gradually fallen into disfavour.

The relationships between physical behaviour and agential behaviour are controversial. On some views, all 'actions' are identical; to physical changes in the subjects body, however, some kinds of physical behaviour, such as 'reflexes', are uncontroversially not kinds of agential behaviour. On others, a subject's effectuation of some proceeding action used to indicate requirements by immediate or future needs or purpose. Bringing into circumstance or state of affairs from which situational extrication is differently involved by some physical change, but it is not identical to it.

Both physical and agential behaviours could be understood in the widest sense. Anything a person can do-even calculating in his head, for instance-could have actuality or reality as been regarded as agential behaviour. Likewise, any physical change in a person's body-even the firing of a certain neuron, for instance-could have actually or reality to be regarded as physical behaviour.

Of course, to claim that the mind is 'nothing beyond' such-and-such kinds of behaviour, construed as either physical or agential behaviour in the widest sense, is not necessarily to be a behaviourist. The theory that the mind is a series of volitional acts-a view close to the idealist position of George Berkeley (1685-1753)-and the possible action's of the minds' condition are the enabling of certainties as founded in the neuronal events, while both controversial, are not forms of behaviourism.

Awaiting, right along the side of an approaching account for which anomalous monism may take on or upon itself is the view that there is only one kind of substance underlying all others, changing and processes. It is generally used in contrast to 'dualism', though one can also think of it as denying what might be called 'pluralism'-a view often associated with Aristotle which claims that there are several substances, as the corpses of times generations have let it be known. Against the background of modern science, monism is usually understood to be a form of 'materialism' or 'physicalism'. That is, the fundamental properties of matter and energy as described by physics are counted the only properties there are.

The position in the philosophy of mind known as 'anomalous monism' has its historical origins in the German philosopher and founder of critical philosophy Immanuel Kant (1724-1804), but is universally identified with the American philosopher Herbert Donald Davidson (1917-2003), and it was he who coined the term. Davidson has maintained that one can be a monist-indeed, a physicalist-about the fundamental nature of things and events, while also asserting that there can be no full 'reduction' of the mental to the physical. (This is sometimes expressed by saying that there can be an ontological, though not a conceptual reduction.) Davidson thinks that complete knowledge of the brain and any related neurophysiological systems that support the mind's activities would not themselves be knowledge of such things as belief, desire, and experience, and so on, find the mentalistic generativist of thoughts. This is not because he thinks that the mind is somehow a separate kind of existence: Anomalous monism is after all monism. Rather, it is because the nature of mental phenomena rules out a priori that there will be law-like regularities connecting mental phenomena and physical events in the brain, and, without such laws, there is no real hope of explaining the mental that has recently come into existence, through the evolutionary structures in the physicality of the brain.

All and all, one central goal of the philosophy of science is to provided explicit and systematic accounts of the theories and explanatory strategies explored in the science. Another common goal is to construct philosophically illuminating analyses or explanations of central theoretical concepts involved in one or another science. In the philosophy of biology, for example, there is a rich literature aimed at understanding teleological explanations, and thereby has been a great deal of work on the structure of evolutionary theory and on such crucial concepts. If concepts of the simple (observational) sorts were internal physical structures that had, in this sense, an information-carrying function, a function they acquired during learning, then instances of these structure types would have a content that (like a belief) could be either true or false. In that of ant information-carrying structure carries all kinds of information if, for example, it carries information 'A', it must also carry the information that 'A' or 'B'. Conceivably, the process of learning is supposed to b e a process in which a single piece of this information is selected for special treatment, thereby becoming the semantic content-the meaning-of subsequent tokens of that structure type. Just as we conventionally give artefacts and instruments information-providing functions, thereby making their flashing lights, and so forth-representations of the conditions in the world in which we are interested, so learning converts neural states that carry imparting information, as 'pointer readings' in the head, so to speak-int structures that have the function of providing some vital piece of information they carry when this process occurs in the ordinary course of learning, the functions in question develop naturally. They do not, as do the functions of instruments and artefacts, depends on the intentions, beliefs, and attitudes of users. We do not give brain structure these functions. They get it by themselves, in some natural way, either (of the senses) from their selectional history or (in thought) from individual learning. The result is a network of internal representations that have (in different ways) the power representation, of experience and belief.

To recognizing the existence or meaning of apprehending to know of the progression of constant understanding that this approach to 'thought' and 'belief', the approach that conceives of them as forms of internal representation, is not a version of 'functionalism'-at least, not if this dely held theory is understood, as it is often, as a theory that identifies mental properties with functional properties. For functional properties have to do within the manner for which to some extent of engaging one's imploring that which has real and independent existence, something, in fact, behaves, with its syndrome of typical causes and effects. An informational model of belief, to account for misrepresentation, a problem with which a preliminary way that in both need something more than a structure that provided information. It needs something having that as its function. It needs something supposed to provide information. As Sober (1985) comments for an account of the mind we need functionalism with the function, the 'teleological', is put back in it.

Philosophers' need not charge a pressing lack of something essential, and typically do not assume that there is anything wrong with the science they are studying. Their goal is simply to provide accounts of the theories, concepts and explanatory strategies that scientists are using-accounts that are more explicit, systematic and philosophically sophisticated than the often rather rough-and-ready accounts offered by the scientists themselves.

Cognitive psychology is, in many ways a curious and puzzling science. Many theories put forward by cognitive psychologists make use of a family of 'intentional' concepts-like believing that ', desiring that 'q', and representing 'r'-which do not appear in the physical or biological sciences, and these intentional concepts play a crucial role in many explanations offered by these theories.

It is characteristic of dialectic awareness that discussions of intentionality appeared as the paradigm cases discussed which are usually beliefs or sometimes beliefs and desires, however, the biologically most basic forms of intentionality are in perception and in intentional action. These also have certain formal features that are not common to beliefs and desire. Consider a case of perceptual experience. Suppose that I see my hand in front of my face. What are the conditions of satisfaction? First, the perceptual experience of the hand in front of my face has as its condition of satisfaction that there is a hand in front of my face. Thus far, the condition of satisfaction is the same as the belief than there is a hand in front of my face. But with perceptual experience there is this difference: so that the intentional content is satisfied, the fact that there is a hand in front of my face must cause the very experience whose intentional content is that there is a hand in front of my face. This has the consequence that perception has a special kind of condition of satisfaction that we might describe as 'causally self-referential'. The full conditions of satisfaction of the perceptual experience are, first that there is a hand in front of my face, and second, that there is a hand in front of my face caused the very experience of whose conditions of satisfaction forms a part. we can represent this in our acceptation of the form. S(p), such as:

Visual experience (that there is a hand in front of face and the fact that there is a hand in front of my face is causing this very experience.)

Furthermore, visual experiences have a kind of conscious immediacy not characterised of beliefs and desires. A person can literally be said to have beliefs and desires while sound asleep. But one can only have visual experiences of a non-pathological kind when one is fully awake and conscious because the visual experiences are themselves forms of consciousness.

People's decisions and actions are explained by appeal to their beliefs and desires. Perceptual processes, sensational, are said to result in mental states that represent (or sometimes misrepresent) one or as another aspect of the cognitive agent's environment. Other theorists have offered analogous acts, if differing in detail, perhaps, the most crucial idea in all of this is the one about representations. There is perhaps a sense in which what happens to be said, is the level of the retina, that constitute the anatomical processes of occurring in the process of stimulation, some kind of representation of what produces that stimulation, and thus, some kind of representation of the objects of perception. Or so it may seem, if one attempts to describe the relation between the structure and characteristic of the object of perception and the structure and nature of the retinal processes. One might say that the nature of that relation is such as to provide information about the part of the world perceived, in the sense of 'information' presupposed when one says that the rings in the sectioning of a tree's truck provide information of its age. This is because there is an appropriate causal relation between the things that make it impossible for it to be a matter of chance. Subsequently processing can then be thought to exist for one who carried out on what is provided in the representational inquiries.

However, if there are such representations, they are not representations for the perceiver, it is the thought that perception involves representations of that kind that produced the old, and now largely discredited philosophical theories of perception that suggested that perception be a matter, primarily, of an apprehension of mental states of some kind, e.g., sense-data, which are representatives of perceptual objects, either by being caused by them or in being in some way constitutive of them. Also, if it is said that the idea of information so invoked indicates that there is a sense in which the precesses of stimulation can be said to have content, but a non-conceptual content, distinct from the content provided by the subsumption of what is perceived under concepts. It must be emphasised that, that content is not one for the perceiver. What the information-processing story is to maintain, is, at best, a more adequate categorization than previously available of the causal processes involved. That may be important, but more should not be claimed for it than there is. If in perception is a given case one can be said to have an experience as of an object of a certain shape and kind related to another object it is because there is presupposed in that perception the possession of concepts of objects, and more particular, a concept of space and how objects occupy space.

It is, that, nonetheless, cognitive psychologists occasionally say a bit about the nature of intentional concepts and the nature of intentional concepts and the explanations that exploit them. Their comments are rarely systematic or philosophically illuminating. Thus, it is hardly surprising that many philosophers have seen cognitive psychology as fertile grounds for the sort of careful descriptive work that is done in the philosophy of biology and the philosophy of physics. The American philosopher of mind Alan Jerry Fodor's (1935-), The Language of Thought (1975) was a pioneering study in th genre on the field. Philosophers have, also, done important and widely discussed work in what might be called the 'descriptive philosophy' or 'cognitive psychology'.

These philosophical accounts of cognitive theories and the concepts they invoke are generally much more explicit than the accounts provided by psychologists, and through them, they inevitably smooth over some of the rough edges of scientists' actual practice. But if the account they give of cognitive theories diverges significantly from the theories that psychologists actually produce, then the philosophers have just got it wrong. There is, however, a very different way in which philosopher's have approached cognitive psychology. Rather than merely trying to characterize what cognitive psychology is actually doing, some philosophers try to say what it should and should not be doing. Their goal is not to explicate o or upon the narratives that scientific applications are but to criticize and improve it. The most common target of this critical approach is the use of intentional concepts in cognitive psychology. Intentional notions have been criticized on various grounds. The two situated considerations are that they fail to supervene on the physiology of the cognitive agent, and that they cannot be 'naturalized'.

Perhaps e easiest way to make the point about 'supervenience is to use a thought experiment of the sort originally proposed by the American philosopher Hilary Putnam (1926-). Suppose that in some distant corner of the universe there is a planet, Twin Earth, which is very similar to our own planet. On Twin Earth, there is a person who is an atom for an atom replica of J.F. Kennedy. Past, assassinated President J.F. Kennedy, who lives on Earth believes that Rev. Martin Luther King Jr. was born in Tennessee, and if you asked him 'Was the Rev. Martin Luther King Jr. born in Tennessee, In all probability the answer would either or not it is yes or no. Twin, Kennedy would respond in the same way, but it is not because he believes that our Rev. Martin Luther King Jr.? Was, as, perhaps, very much in question of what is true or false? His beliefs are about Twin-Luther, and that Twin -Luther was certainly not born in Tennessee, and thus, that J.F. Kennedy's belief is true while Twin-Kennedy's is false. What all this is supposed to show is that two people, perhaps on opposite polarities of justice, or justice as drawn on or upon human rights, can share all their physiological properties without sharing all their intentional properties. To turn this into a problem for cognitive psychology, two additional premises are needed. The first is that cognitive psychology attempts to explain behaviour by appeal to people's intentional properties. The second, is that psychological explanations should not appeal to properties that fall to supervene on an organism's physiology. (Variations on this theme can be found in the American philosopher Allen Jerry Fodor (1987)).

The thesis that the mental are supervening on the physical-roughly, the claim that the mental characters of a determinant adaptation of its physical nature-has played a key role in the formulation of some influential positions of the 'mind-body' problem. In particular versions of non-reductive 'physicalism', and has evoked in arguments about the mental, and has been used to devise solutions to some central problems about the mind-for example, the problem of mental causation.

The idea of supervenience applies to one but not to the other, that this, there could be no difference in a moral respect without a difference in some descriptive, or non-moral respect evidently, the idea generalized so as to apply to any two sets of properties (to secure greater generality it is more convenient to speak of properties that predicates). The American philosopher Donald Herbert Davidson (1970), was perhaps first to introduce supervenience into the rhetoric discharging into discussions of the mind-body problem, when he wrote '. . . mental characteristics are in some sense dependent, or supervening, on physical characteristics. Such supervenience might be taken to mean that there cannot be two events alike in all physical respects but differing in some mental respectfulness, or that an object cannot alter in some metal deferential submission without altering in some physical regard. Following, the British philosopher George Edward Moore (1873-1958) and the English moral philosopher Richard Mervyn Hare (1919-2003), from whom he avowedly borrowed the idea of supervenience. Donald Herbert Davidson, went on to assert that supervenience in this sense is consistent with the irreducibility of the supervient to their 'subvenience', or 'base' properties. Dependence or supervenience of this kind does not entail reducibility through law or definition . . .'

Thus, three ideas have purposively come to be closely associated with supervenience: (1) Property convariation, (if two things are indiscernible in the infrastructure of allowing properties that they must be indiscernible in supervening properties). (2) Dependence, (supervening properties are dependent on, or determined by, their subservient bases) and (3) non-reducibility (property convariation and dependence involved in supervenience can obtain even if supervening properties are not reducible to their base properties.)

Nonetheless, in at least, for the moment, supervenience of the mental-in the form of strong supervenience, or, at least global supervenience-is arguably a minimum commitment to physicalism. But can we think of the thesis of mind-body supervenience itself as a theory of the mind-body relation-that is, as a solution to the mind-body problem?

It would seem that any serious theory addressing the mind-body problem must say something illuminating about the nature of psychophysical dependence, or why, contrary to common belief, there is no dependence in either way. However, if we take to consider the ethical naturalist intuitivistic will say that the supervenience, and the dependence, for which is a brute fact you discern through moral intuition: And the prescriptivist will attribute the supervenience to some form of consistency requirements on the language of evaluation and prescription. And distinct from all of these is mereological supervenience, namely the supervenience of properties of a whole on properties and relations of its pats. What all this shows, is that there is no single type of dependence relation common to all cases of supervenience, supervenience holds in different cases for different reasons, and does not represent a type of dependence that can be put alongside causal dependence, meaning dependence, mereological dependence, and so forth.

There seems to be a promising strategy for turning the supervenience thesis into a more substantive theory of mind, and it is that to explicate mind-body supervenience as a special case of mereological supervenience-that is, the dependence of the properties of a whole on the properties and relations characterizing its proper parts. Mereological dependence does seem to be a special form of dependence that is metaphysically sui generis and highly important. If one takes this approach, one would have to explain psychological properties as macroproperties of a whole organism that covary, in appropriate ways, with its macroproperties, i.e., the way its constituent organ, tissues, and so forth, are organized and function. This more specific supervenience thesis may be a serious theory of the mind-body relation that can compete for the classic options in the field.

On this topic, as with many topics in philosophy, there is a distinction to be made between (1) certain vague, partially inchoate, pre-theoretic ideas and beliefs about the matter nearby, and (2) certain more precise, more explicit, doctrines or theses that are taken to articulate or explicate those pre-theoretic ideas and beliefs. There are various potential ways of precisifying our pre-theoretic conception of a physicalist or materialist account of mentality, and the question of how best to do so is itself a matter for ongoing, dialectic, philosophical inquiry.

The view concerns, in the first instance, at least, the question of how we, as ordinary human beings, in fact go about ascribing beliefs to one another. The idea is that we do this on the basis of our knowledge of a commonsense theory of psychology. The theory is not held to consist in a collection of grandmotherly saying, such as 'once bitten, twice shy'. Rather it consists in a body of generalizations relating psychological states to each other to input from the environment, and to actions. Such may be founded on or upon the grounds that show or include the following:

(1) $(x)(p)$ if x is to feel agitation or dismay in the anticipation of or in the presence of fear that p, then x desires that not-p.

(2) $(x)(p)$(if x hopes that p and · hope that p and · become or be made aware of something not previously known that p, then · is pleased that p.)

(3) (x)(p)(q) If x is to have a firm conviction in the reality of something believed that p and · believe in the conviction of the reality of something that if p, then q, barring confounding distraction, and so forth. · believes that q.

(4) (x)(p)(q) If x is desirously that p and so that x believes that if q then 'p' and 'χ' are able to bring it about that 'q'. Then, barring conflicting desires or preferred strategies,'χ' brings it about that 'q'. All of these. All of these generalizations should be most of the time, but variably. Adventurous types often enjoy the adrenal thrill produced by fear, this leads them, on occasion, to desire the very state of affairs that frightens them. Analogously, with (3). A subject who believes that 'p' nd believes that if 'p', then 'q'. Would typically infer that 'q?'. But certain atypical circumstances may intervene: Subjects may become confused or distracted, or they ma y finds the prospect of 'q' so awful that they dare not allow themselves to believe it. The ceteris paribus nature of these generalizations is not usually considered to be problematic, since atypical circumstances are, of course, atypical, and the generalizations are applicable most of the time.

We apply this psychological theory to make inference about people's beliefs, desires and so forth. If, for example, we know that Julie believes that if she is to be at the airport at four, then she should get a taxi at half past two, and she believes that she is to be at the airport at four, then we will predict, using (3), that Julie will infer that she should get a taxi at half past two.

The Theory-Theory, as it is called, is an empirical theory addressing the question of our actual knowledge of beliefs. Taken in its purest form if addressed both first and third-person knowledge: we know about our own beliefs and those of others in the same way, by application of commonsense psychological theory in both cases. However, it is not very plausible to hold that we overplay or overact of any given attention or emphasis in dramatizing in excess of going beyond a normal or acceptable limit. That indeed usually-know our own beliefs by way of theoretical inference. Since it is an empirical theory concerning one of our cognitive abilities, the Theory-Theory is open to psychological scrutiny. Various issues of the hypothesized commonsense psychological theory, we need to know whether it is known consciously or unconsciously. Nevertheless, research has revealed that three-year-old children are reasonably gods at inferring the beliefs of others on the basis of actions, and at predicting actions on the basis of beliefs that others are known to possess. However, there is one area in which three-year-old's psychological reasoning differs markedly from adults. Tests of the sorts are rationalized in such that: 'False Belief Tests', reveal largely consistent results. Three-year-old subjects are witnesses to the scenario about the child, Billy, see his mother place some biscuits in a biscuit tin. Billy then goes out to play, and, unseen by him, his mother removes the biscuit from the tin and places them in a jar, which is then hidden in a cupboard. When asked, 'Where will Billy look for the biscuits'? The majority of three-year-olds answer that Billy will look in the jar in the cupboard-where the biscuits actually are, than where Billy saw them being placed. On being asked 'Where does Billy think the biscuits are'? They again, tend to answer 'in the cupboard', rather than 'in the jar'. Three-year-olds thus, appear to have some difficulty attributing false beliefs to others in case in which it would be natural for adults to do so. However, it appears that three-year-olds are lacking the idea of false beliefs overall, nor does it come out that they struggle with attributing false beliefs in other kinds of situations. For example, they have little trouble distinguishing between dreams and play, on the one hand, and true beliefs or claims on the other. By the age of four and some half years, most children pass the False Belief Tests fairly consistently. There is yet no general accepted theory of why three-year-olds fare so badly with the false beliefs tests, nor of what it reveals about their conception of beliefs.

Recently some philosophers and psychologists have put forward what they take to be an alternative to the Theory-Theory: However, the challenge does not end there. We need also to consider the vital element of making appropriate adjustments for differences between one's own psychological states and those of the other. Nevertheless, it is implausible to think in every such case of simulation, yet alone will provide the resolving obtainability to achieve.

The evaluation of the behavioural manifestations of belief, desires, and intentions are enormously varied, every bit as suggested. When we move away from perceptual beliefs, the links with behaviour are intractable and indirect: The expectation I form on the basis of a particular belief reflects the influence of numerous other opinions, my actions are formed by the totality of my preferences and all those opinions that have a bearing on or upon them. The causal processes that produce my beliefs reflect my opinions about those processes, about their reliability and the interference to which they are subject. Thus, behaviour justifies the ascription of a particular belief only by helping to warrant a more all-inclusive interpretation of cognitive positions of the individual in question. Psychological descriptions, like translations, are a 'holistic' business. And once this is taken into account, it is all the less likely that a common physical trait will be found which grounds all instances of the same belief. The ways in which all of our propositional altitudes interact in the production of behaviour reinforce the anomalous character of our mentality and render any sort of reduction of the mind to the physical impossibilities. Such is not meant as a practical procedure, it can, however, generalize on this so that interpretation and merely translation is at issue, has made this notion central to methods of accounting responsibilities of the mind.

Theory and Theory-Theory are two, as many think competing, views of the nature of our commonsense, propositional attitude explanations of action. For example, when we say that our neighbour cut down his apple tree because he believed that it was ruining his patio and did not want it ruined, we are offering a typically commonsense explanation of his action in terms of his beliefs and desires. But, even though wholly familiar, it is not clear what kind of explanation is at issue. Connected of one view, is the attribution of beliefs and desires that are taken as the application to actions of a theory that, in its informal way, functions very much like theoretical explanations in science. This is known as the 'theory-theory' of every day psychological explanation. In contrast, it has been argued that our propositional attributes are not theoretical claims do much as reports of a kind of 'simulation'. On such a 'simulation theory' of the matter, we decide what our neighbour will do (and thereby why he did so) by imagining ourselves in his position and deciding what we would do.

`The Simulation Theorist should probably concede that simulations need to be backed up by the unconfined of discovering the psychological states of others. But they need not concede that these independent take the form of a theory. Rather, they might suggest that we can get by with some rules of thumb, or straightforward inductive reasoning of a general kind.

A second and related difficulty with the Simulation Theory concerns our capacity to attribute beliefs that are too alien to be easily simulated: Beliefs of small children, or psychotics, or bizarre beliefs are deeply suppressed into the mindful latencies within the unconscious. The small child refuses to sleep in the dark: He is afraid that the Wicked Witch will steal him away. No matter how many adjustments we make, it may be hard for mature adults to get their own psychological processes, as even to make in pretended play, to mimic the production of such belief. For the Theory-Theory alien beliefs are not particularly problematic: So long as they fit into the basic generalizations of the theory, they will be inferable from the evidence. Thus, the Theory-Theory can account better for our ability to discover more bizarre and alien beliefs than can the Simulation Theory.

The Theory-Theory and the Simulation Theory are not the only proposals about knowledge of belief. A third view has its origins in the Austrian philosopher Ludwig Wittgenstein (1889-1951). On this view both the Theory and Simulation Theories attribute too much psychologizing to our commonsense psychology. Knowledge of other minds is, according to this alternative picture, more observational in nature. Beliefs, desires, feelings are made manifest to us in the speech and other actions of those with whom we share a language and way of life. When someone says. 'Its going to rain' and takes his umbrella from his bag. It is immediately clear to us that he believes it is going to rain. In order to know this, we neither theorize nor simulate: We just perceive, of course, this is not straightforward visual perception of the sort that we use to see the umbrella. But it is like visual perception in that it provides immediate and non-inferential awareness of its objects. we might call this the 'Observational Theory'.

The Observational Theory does not seem to accord very well with the fact that we frequently do have to indulge in a fair amount of psychologizing to find in what others believe. It is clear that any given action might be the upshot of any number of different psychological attitudes. This applies even in the simplest cases. For example, because one's friend is suspended from a dark balloon near a beehive, with the intention of stealing honey. This idea to make the bees behave that it is going to rain and therefore believe that the balloon as a dark cloud, and therefore pay no attention to it, and so fail to notice one's dangling friend. Given this sort of possible action, the observer would surely be rash immediately to judge that the agent believes that it is going to rain. Rather, they would need to determine-perhaps, by theory, perhaps by simulation-which of the various clusters of mental states that might have led to the action, actually did so. This would involve bringing in further knowledge of the agent, the background circumstances and so forth. It is hard to see how the sort of difficult compounded composite as the mental process involved in this sort of psychological reflection could be assimilated to any kind of observation.

The attributions of intentionality that depend on optimality or reasonableness are interpretations of the assumptive phenomena-a 'heuristic overlay' (1969), describing an inescapable idealized 'real pattern'. Like such abstractions, as centres of gravity and parallelograms of force, the beliefs and desires posited by the highest stance have noo independent and concrete existence, and since this is the case, there would be no deeper facts that could settle the issue if-most importantly-rival intentional interpretations arose that did equally well at rationalizing the history of behaviour of an entity. Orman van William Quine 1908-2000, the most influential American philosopher of the latter half of the

20th century, whose thesis on the indeterminacy of radical translations carries all the way in the thesis of the indeterminacy of radical interpretation of mental states and processes.

The fact that cases of radical indeterminacy, though possible in principle, is vanishingly unlikely ever to comfort us in the solacing refuge and shelter, yet, this is apparently an idea that is deeply counterintuitive to many philosophers, who have hankered for more 'realistic' doctrines. There are two different strands of 'realism' that in the attempt to undermine are such:

(1) Realism about the entities allegedly given and presented by our every day mentalistic discourse-what is dubbed as folk-psychology (1981) Such as beliefs, desires, pains, the self.

(2) Realism about content itself-the idea that there has to be events or entities that really have intentionality (as opposed to the events and entities that only have as if they had intentionality).

The tenet indicated by (1) rests of what is fatigue, what bodily states or events are so fatiguing, that they are identical with, and so forth. This is a confusion that calls for diplomacy, not philosophical discovery: The choice between an 'Eliminative materialism' and an 'identity theory' of fatigues is not a matter of which 'ism' is right, but of which way of speaking is most apt to wean these misbegotten features of them as conceptual schemata.

Again, the tenet (2) my attack has been more indirect. The view that some philosophers, in that of a demand for content realism as an instance of a common philosophical mistake: Philosophers often manoeuvre themselves into a position from which they can see only two alternatives: Infinite regresses versus some sort of 'intrinsic' foundation-a prime mover of one sort or another. For instance, it has seemed obvious that for some things to be valuable as, other things must be intrinsically valuable-ends in themselves-otherwise we would be stuck with vicious regress (or, having no beginning or end) of things valuable only that although some intentionality is 'derived' (the 'aboutness' of the pencil marks composing a shopping list is derived from the intentions of the person whose list it is), unless some intentionality is 'original' and underived, there could be no derived intentionality.

There is always another alternative, namely, some finites regress that peters out without marked foundations or thresholds or essences. Here is an avoided paradox: Every mammal has a mammal for a mother-but, this implies an infinite genealogy of mammals, which cannot be the case. The solution is not to search for an essence of mammalhood that would permit us in principle to identify the Prime Mammal, but rather to tolerate a finite regress that connects mammals to their non-mammalian ancestors by a sequence that can only be partitioned arbitrarily. The reality of today's mammals is secure without foundations.

The best instance of this theme is held to the idea that the way to explain the miraculous-seeming powers of an intelligent intentional system is to disintegrate it into hierarchically structured teams of ever more stupid intentional systems, ultimately discharging all intelligence-debts in a fabric of stupid mechanisms? Lycan (1981), has called this view 'homuncular functionalism'. One may be tempted to ask: Are the sub-personal components 'real' intentional systems? At what point in the diminutions of prowess as we descend to simple neurons does 'real' intentionality disappear? Don't ask. The reasons for regarding an individual neuron (or a thermostat) as an intentional system are unimpressive, bu t zero, and the security of our intentional attributions at the highest lowest-level of real intentionality. Another exploitation of the same idea is found in Elbow Room (1984): Ast what point in evolutionary history did authentically to correspond to known, reason-appreciators real selves, make their appearance? Doesn't ask-for the dame reason? Here is yet another, more fundamental versions of evolution can point in the early days of evolution can we speak of genuine function, genuine selection-for and not mere fortuitous preservation of entities that happen to have some self-replicative capacity? Don't ask. Many of the more interesting and important features of our world have emerged, gradually, from a world that initially lacked them-function, intentionality, consciousness, morality, value-and it is a fool's errand to try to identify a first or most-simple instances of the 'real' thing. It is, for the same to reason a mistake must exist to answer all the questions our system of content attribution permits us to ask. Tom says he has an older brother in Toronto and that he is an only child. What does he really believe? Could he really believe that he had a but if he also believed he was an only child? What is the 'real' content of his mental state? There is no reason to suppose there is a principled answer.

The most sweeping conclusion having drawn from this theory of content is that the large and well-regarded literature on 'propositional attitudes' (especially the debates over wide versus narrow content) is largely a disciplinary artefact of no long-term importance whatever, accept perhaps, as history's most slowly unwinding unintended reductio ad absurdum. Mostly, the disagreements explored in that literature cannot even be given an initial expression unless one takes on the assumption of an unsounded fundamentalist of strong realism about content, and its constant companion, the idea of a 'language of thought' a system of mental representation that is decomposable into elements rather like terms, and large elements rather like sentences. The illusion, that this is plausible, or even inevitable, is particularly fostered by the philosophers' normal tactic of working from examples of 'believing-that-p' that focuses attention on mental states that are directly or indirectly language-infected, such as believing that the shortest spy is a spy, or believing that snow is white. (Do polar bears believe that snow is white? In the way we do?) There are such states-in language-using human beings - but, they are not exemplary r foundational states of belief, needing a term for them. As, perhaps, in calling the term in need of, as they represent 'opinions'. Opinions play a large, perhaps even a decisive role in our concept of a person, but they are not paradigms of the sort of cognitive element to which one can assign content in the first instance. If one starts, as one should, with the cognitive states and events occurring in nonhuman animals, and uses these as the foundation on which to build theories of human cognition, the language-infected state is more readily seen to be derived, less directly implicated in the explanation of behaviour, and the chief but an illicit source of plausibility of the doctrine of a language of thought. Postulating a language of thought is in any event a postponement of the central problems of content to refer especially to a supposed causes, source or author ascribed, as not a necessary first step.

Our momentum, regardless, produces on or upon the inflicting of forces out the causal theories of epistemology, of what makes a belief justified and what makes a true belief knowledge? It is natural to think that whether a belief deserves one of these appraisals depends on what caused the subject to have the belief. In recent decades a number of epistemologists have pursued this plausible idea with a variety of specific proposals. For some proposed casual criteria for knowledge and justification are for us, to take under consideration.

Some causal theories of knowledge have it that a true belief that 'p' is knowledge just in case it has the right sort of causal connection to the fact that 'p'. Suchlike some criteria can be applied only to cases where the fact that 'p', a sort that can enter causal relations: This seems to exclude mathematically and other necessary facts and perhaps any fact expressed by a universal generalization. And proponents of this sort of criterion have usually supposed that it is limited to perceptual knowledge of particular facts about the subject's environment.

For example, the forthright Australian materialist David Malet Armstrong (1973), proposed that a belief of the form, 'This (perceived) object is 'F' is (non-inferential) knowledge if and only if the belief is a completely reliable sign that the perceived object is 'F', that is, the fact that the object is 'F' contributed to causing the belief and its doing so depended on properties of the believer such that the laws of nature dictate that, for any subject 'χ' and perceived object 'y'. If 'χ' has those properties and believes that 'y' is 'F', then 'y' is 'F'. Dretske (1981) offers a rather similar account in terms of the belief's being caused by a signal received by the perceiver that carries the information that the object is 'F'.

This sort of condition fails, however, to be sufficient for non-inferential perceptual knowledge because it is compatible with the belief's being unjustified, and an unjustified belief cannot be knowledge. For example, suppose that your mechanisms for colour perception are working well, but you have been given good reason to think otherwise, to think, say, that any tinted colour of the things that look brownishly-tinted to you it will appear as brownishly-tinted things look of any tinted colour. If you fail to heed these results you have for thinking that your colour perception is awry and believe of a thing that look's colour tinted to you that it is colour-tinted in your belief will fail to be justified and will therefore fail to be knowledge, even though it is caused by the thing's being tinted in such a way as to be a completely reliable sign (or to carry the information) that the thing is tinted or found of some tinted discolouration.

One could fend off this sort of counterexample by simply adding to the causal condition the requirement that the belief be justified. But this enriched condition would still be insufficient. Suppose, for example, that in an experiment you are given a drug that in nearly all people (but not in you, as it happens) causes the aforementioned aberration in colour perception. The experimenter tells you that you're taken such a drug that says, 'No, wait a minute, the pill you took was just a placebo'. But suppose further that this last thing the experimenter tells you is discordantly not the truth. Her telling you this gives you justification for believing of a thing that looks colour tinted or tinged in brownish tones, but in fact about this justification that is unknown to you (that the experimenter's last statement was false) makes it the casse that your true belief is not knowledge even though it satisfies Armstrong's causal condition.

Goldman (1986) has proposed an important different sort of causal criterion, namely, that a true belief is knowledge if it is produced by a type of process that a 'global' and 'locally' reliable. It is global reliability of its propensity to cause true beliefs is sufficiently high. Local reliability had to do with whether the process would have produced a similar but false belief in certain counterfactual situations alternative to the actual situation. This way of marking off true beliefs that are knowledge e does not require the fact believed to be causally related to the belief and so it could in principle apply to knowledge of any kind of truth.

Goldman requires the global reliability of the belief-producing process for the justification of a belief, he requires, also for knowledge because justification is required for knowledge. What he requires for knowledge only of being one or more of which there exist any other but manages not to require for justification is local reliability. His idea is that a justified true belief is knowledge if the type of process that produced it would not have produced it in any relevant counterfactual situation in which it is

The theory of relevant alternative is best understood as an attempt to accommodate two opposing strands in our thinking about knowledge. The first is that knowledge is an absolute concept. On one interpretation, this is to mean that the justification or evidence one must have an order to know a proposition 'p' must be ample sufficient by expressively viewing of all these that alternatively vindicate 'p', (when an alternative to a proposition 'p' is a proposition incompatible with 'p').

For knowledge requires only that elimination of the relevant alternatives. So the tentative relevance for which alternate substitutions made for our consideration in view of its preservers that hold of both strands of our thinking about knowledge. Knowledge is an absolute concept, but because the absoluteness is relative to a standard, we can know many things.

The relevant alternative's account of knowledge can be motivated by noting that other concepts exhibit the same logical structure e. two examples of this are the concepts 'flat' and the concept 'empty'. Both appear to be absolute concepts-a space is empty only if it does not contain anything and a surface is flat only if it does not have any bumps. However, the absolute character of these concepts is relative to a standard. In the case of flat, there is a standard for what there is a standard for what counts as a bump and in the case of empty, there is a standard for what counts as a thing. we would not deny that a table is flat because a microscope reveals irregularities in its surface. Nor would we den y that a warehouse is empty because it contains particles of dust. To be flat is to be free of any relevant bumps. To be empty is to be devoid of all relevant things. Analogously, the relevant alternative's theory says that to know a proposition is to have evidence that eliminates all relevant alternatives.

Some philosophers have argued that the relevant alternative's theory of knowledge entails the falsity of the principle that set of known (by S) propositions in closed under known (by 'S') entailment, although others have disputed this however, this principle affirms the following conditional or the closure principle:

If 'S' knows 'p' and 'S' knows that 'p' entail's 'q', then 'S' knows 'q'.

According to the theory of relevant alternatives, we can know a proposition 'p', without knowing that some (non-relevant) alterative too 'p' is false. But, once an alternative 'h' too 'p' incompatible with 'p', then 'p' will trivially entail not-h. So it will be possible to know some proposition without knowing another proposition trivially entailed by it. For example, we can know that we see a zebra without knowing that it is not the case that we see a cleverly disguised mule (on the assumption that 'we see a cleverly disguised mule' is not a relevant alterative). This will involve a violation of the closure principle. This is an interesting consequence of the theory because the closure principles seem too many to be quite intuitive. In fact, we can view sceptical arguments as employing the closure principle as a premise, along with the premise that we do not know that the alternatives raised by the sceptic are false. From these two premisses, it follows (on the assumption that we see that the propositions we believe entail the falsity of sceptical alternatives) that we do not know the proposition we believe. For example, it follows from the closure principle and the fact that we do not know that we do not see a cleverly disguised mule, that we do not know that we see a zebra. we can view the relevant alternative's theory as replying to the sceptical arguments by denying the closure principle.

What makes an alternative relevant? What standard do the alternatives rise by the sceptic fail to meet? These notoriously difficult to answer with any degree of precision or generality. This difficulty has led critics to view the theory as something being to obscurity. The problem can be illustrated though an example. Suppose Smith sees a barn and believes that he does, on the basis of very good perceptual evidence. When is the alternative that Smith sees a paper-mache replica relevant? If there are many such replicas in the immediate area, then this alternative can be relevant. In these circumstances, Smith fails to know that he sees a barn unless he knows that it is not the case that he sees a barn replica. Where there is any intensified replication that exists by this alternative will not be relevant? Smith can know that he sees a barn without knowing that he does not see a barn replica.

This highly suggests that a criterion of relevance be something like probability conditional on Smith's evidence and certain features of the circumstances. But which circumstances in particular do we count? Consider a case where we want the result that the barn replica alternative is clearly relevant, e.g., a case where the circumstances are such that there are numerous barn replicas in the area. Does the suggested criterion give us the result we wanted? The probability that Smith sees a barn replica given his evidence and his location to an area where there are many barn replicas is high. However, that same probability conditional on his evidence and his particular visual orientation toward a real barn is quite low. we want the probability to be conditional on features of the circumstances like the former bu t not on features of the circumstances like the latter. But how do we capture the difference in a general formulation?

How significant a problem is this for the theory of relevant alternatives? This depends on how we construe theory. If the theory is supposed to provide us with an analysis of knowledge, then the possibly to be without something and especially something essentially or greatly needed found ca pressing lack of something essential, a needful requisite necessary for the necessity to provide readily relief in the absence in all human lack of precise criteria of relevance surely constitutes a serious problem. However, if the theory is viewed instead as providing a response to sceptical arguments, it can be argued that the difficulty has little significance for the overall success of the theory.

What justifies the acceptance of a theory? In the face of the fact that some exceptional versions of empiricism have met many criticisms, and are nonetheless, overtaken to look for an answer in some sort of empiricist terms: In terms, that is, of support by the available evidence. How else could objectivity of science be defended but by showing that its conclusions (and in particular its theoretical conclusions-those theories it presently accepts) are somehow legitimately based on agreed observational and experimental evidence? But, as is well known, theories usually pose a problem for empiricism.

Allowing the empiricist the assumptions that there are observational statements whose truth-values can be inter-subjectively agreeing, and show the exploratory, non-demonstrative use of experiment in contemporary science. Yet philosophers identify experiments with observed results, and these with the testing of theory. They assume that observation provides an open window for the mind onto a world of natural facts and regularities, and that the main problem for the scientist is to establishing the unequalled independence of a theoretical interpretation. Experiments merely enable the production of (true) observation statements. Shared, replicable observations are the basis for a scientific consensus about an objective reality. It is clear that most scientific claims are genuinely theoretical: Nether themselves observational nor derivable deductively from observation statements (nor from inductive generalizations thereof). Accepting that there are phenomena that we have more or less diet access to, then, theories seem, at least when taken literally, to tell us about what is going on 'underneath' the evidently direct observability as made accessibly phenomenal, on order to produce those phenomena. The accounts given by such theories of this trans-empirical reality, simply because it is trans-empirical, can never be established by data, nor even by the 'natural' inductive generalizations of our data. No amount of evidence about tracks in cloud chambers and the like, can deductively establish that those tracks are produced by 'trans-observational' electrons.

One response would, of course, be to invoke some strict empiricist account of meaning, insisting that talk of electrons and the like, is, in fact just shorthand for talks in cloud chambers and the like. This account, however, has few, if any, current defenders. But, if so, the empiricist must acknowledge that, if we take any presently accepted theory, then there must be alternatives, different theories (indefinitely many of them) which treat the evidence equally well-assuming that the only evidential criterion is the entailment of the correct observational results.

All and all, there is an easy general result as well: assuming that a theory is any deductively closed set of sentences, and assuming, with the empiricist that the language in which these sentences are expressed has two sorts of predicated (observational and theoretical), and, finally, assuming that the entailment of the evidence is only constraint on empirical adequacy, then there are always indefinitely many different theories which are equally empirically adequate in a language in which the two sets of predicates are differentiated. Consider the restrictions if 'T' were quantified-free sentences expressed purely in the observational vocabulary, then any conservative extension of that restricted set of T's consequences back into the full vocabulary is a 'theory' co-empirically adequately with-entailing the same singular observational statements as 'T'. Unless veery special conditions apply (conditions which do not apply to any actualization for which is to succumb to scientific theory), then some of the empirically equivalent theories will formally contradict 'T'. (A similar straightforward demonstration works for the currently more fashionable account of theories as sets of models.)

How marked by complexity designate the basic fundamental part and parcel of the empiricist, who repudiates such as such to exclude the claim that two empirically equivalent theories are thereby fully equivalent, explain why the particular theory 'T' that is, as a matter of fact, accepted in sciences which are recognized submissions that seem referring of other possible theories as 'T', with which has the same observational content? Obviously the answer must be 'by bringing in further criteria beyond that of simply having the right observational consequence. Simplicity, coherence with other accepted these and wholeness in the unity as favourable contenders. There are notorious problems in formulating ths criteria at all precisely: But suppose, for present purposes, that we have heretofore, strong enough intuitive grasps to operate usefully with them. What is the status of such further criteria?

The empiricist-instrumentalist position, newly adopted and sharply argued by van Fraassen, is that those further criteria are 'pragmatic'-that is, involved essential reference to ourselves as 'theory-users'. We happen to prefer, for our own purposes, since, coherent, unified theories-but this is only a reflection of our preference es. It would be a mistake to think of those features supplying extra reasons to believe in the truth (or, approximate truth) of the theory that has them. Van Fraassen's account differs from some standard instrumentalist-empiricist account in recognizing the extra content of a theory (beyond its directly observational content) as genuinely declarative, as consisting of true-or-false assertions about the hidden structure of the world. His account accepts that the extra content can neither be eliminated as a result of defining theoretical notions in observational terms, nor be properly regarded as only apparently declarative but in fact as simply a codification schema. For van Fraassen, if a theory says, that there are electrons, then the theory should be taken as meaning to express in words, that which is said and without any positivist divide debasing reinterpretations of the meaning that might make 'There are electrons' mere shorthand for some complicated set of statements about tracks in obscure chambers or the like.

In the case of contradictory but empirically equivalent theories, such as the theory T1 that 'there are electrons' and the theory T2 that 'all the observable phenomena as if there are electrons but there are not 't's. Van Fraassen's account entails that each has a truth-value, at most one of which is 'true', is that science must use to indicate requirements by immediate or future needs or purpose not to T2, but this need not mean that it is rational thinking that it is more likely to be true (or otherwise appropriately connected with nature). So far as belief in the theory is belief but T2. The only belief involved in the acceptance of a theory is belief in the theorist's empirical adequacy. To accept the quantum theory, for example, entails believing that it 'saves the phenomena'-all the (relevant) phenomena, but only the phenomena, theorists do 'say more' than can be checked empirically even in principle. What more they say may indeed be true, but acceptance of the theory does not involve belief in the truth of the 'more' that theorist say.

Preferences between theories that are empirically equivalent are accounted for, because acceptance involves more than belief: As well as this epistemic dimension, acceptance also has a pragmatic dimension. Simplicity, (relative) freedom from ads hoc assumptions, 'unity', and the like are genuine virtues that can supply good reasons to accept one theory than another, but they are pragmatic virtues, reflecting the way we happen to like to do science, rather than anything about the world. Simplicity to think that they do so: The rationality of science and of scientific practices can be in truth (or approximate truth) of accepted theories. Van Fraassen's account conflicts with what many others see as very strong intuitions.

The most generally accepted account of this distinction is that a theory of justification is internalist if and only if it requires that all of the factors needed for a belief to be epistemologically justified for a given person to be cognitively accessible to that person, internal to his cognitive perceptive, and externalist, if it allow s that, at least some of the justifying factors need not be thus accessible, so that they can be external to the believer's cognitive perspective, beyond his knowingness. However, epistemologists often use the distinction between internalist and externalist theories of epistemic explications.

The externalism/internalism distinction has been mainly applied to theories of epistemic justification. It has also been applied in a closely related way to accounts of knowledge and a rather different way to accounts of belief and thought content. The internalist requirement of cognitive accessibility can be interpreted in at least two ways: A strong version of internalism would require that the believer actually be aware of the justifying factors in order to be justified while a weaker version would require only that he be capable of becoming aware of them by focussing his attention appropriately. But without the need for any change of position, new information, and so forth. Though the phrase 'cognitively accessible' suggests the weak interpretation, therein intuitive motivation for intentionalism, that in spite of the fact that, the idea that epistemic justification requires that the believer actually have in his cognitive possession a reason for thinking that the belief is true, wherefore, it would require the strong interpretation.

Perhaps the clearest example of an internalist position would be a 'foundationalist' view according to which foundational beliefs pertain to immediately experienced states of mind other beliefs are justified by standing in cognitively accessible logical or inferential relations to such foundational beliefs. Such a view could count as either a strong or a weak version of internalism, depending on whether actual awareness of the justifying elements or only the capacity to become aware of them is required. Similarly, a 'coherentist' view that the truth of a proposition consists in its being a member of some suitably defined body of other propositions: A body is consistent, coherent, and possibly endowed with other virtues, provided these are not defined in terms of truth. If both the beliefs or other states with which a justification belief is required to cohere and the coherence relations themselves are reflectively accessible.

It should be carefully noticed that when internalism is construed in this way, it is neither necessary nor sufficient by itself for internalism that the justifying factors literally are internal mental states of the person in question. Not necessarily, because on at least some views, e.g., a direct realist view of perception, something other than a mental state of the believer can be cognitively accessible: Not sufficient, because there are views according to which at least some mental states need not be actual (a strong version) or even possible (weak versions) objects of objective awareness. Also, on this way of drawing the distinction, a hybrid view (like the ones already set), according to which some of the factors required for justification must be cognitively accessible while the requiring obligations of employment seem the lack of something essential, whereby the vital fundamental duty of the others need not and overall will not be, would count as an externalist view. Obviously too, a view that was externalist in relation to a strong version of internalism (by not requiring that the believer actually be aware of all justifying factors) could still be internalist in relation to a weak version (by requiring that he at least is capable of becoming aware of them).

The most prominent recent externalist views have been versions of 'reliabilism', whose main requirements for justification are roughly that the belief is produced in a way or via a process that make it objectively likely that the belief is true. What makes such a view externalist is the absence of any requirement that the person for whom the belief is justified have any sort of cognitive access to the relation of reliability in question. Lacking such access, such a person will usually have or likely to be true, but will, on such an account, nonetheless, be epistemologically justified in accepting it. Thus, such a view arguably marks a major break from the modern epistemological tradition, stemming from Descartes, which identifies epistemic justification with having a reason, perhaps even a conclusive reason, for thinking that the belief is true. An epistemological working within this tradition is likely to feel that the externalist, than offering a competing account on the same concept of epistemic justification with which the traditional epistemologist is concerned, has simply changed the subject.

Two general lines of argument are commonly advanced in favour of justificatory externalism. The first starts from the allegedly commonsensical premise that knowledge can be non-problematically ascribed to relativity unsophisticated adults, to young children and even to higher animals. It is then argued that such ascriptions would be untenable on the standard internalist accounts of epistemic justification (assuming that epistemic justification is a necessary condition for knowledge), since the beliefs and inferences involved in such accounts are too complicated and sophisticated to be plausibly ascribed to such subjects. Thus, only an externalist view can make sense of such commonsense ascriptions and this, on the presumption that commonsense is correct, constitutes a strong argument in favour of externalism. An internalist may respond by externalism. An internalist may respond by challenging the initial premise, arguing that such ascriptions of knowledge are exaggerated, while perhaps at the same time claiming that the cognitive situation of at least some of the subjects in question. Is less restricted than the argument claims? A quite different response would be to reject the assumption that epistemic justification is a necessary condition for knowledge, perhaps, by adopting an externalist account of knowledge, rather than justification, as those aforementioned.

The second general line of argument for externalism points out that internalist views have conspicuously failed to provide defensible, non-sceptical solutions to the classical problems of epistemology. In striking contrast, however, such problems are overall easily solvable on an externalist view. Thus, if we assume both that the various relevant forms of scepticism are false and that the failure of internalist views so far is likely to be remedied in the future, we have good reason to think that some externalist view is true. Obviously the cogency of this argument depends on the plausibility of the two assumptions just noted. An internalist can reply, first, that it is not obvious that internalist epistemology is doomed to failure, that the explanation for the present lack of success may be the extreme difficulty of the problems in question. Secondly, it can be argued that most of even all of the appeal of the assumption that the various forms of scepticism are false depends essentially on the intuitive conviction that we do have possession of our reasons in the grasp for thinking that the various beliefs questioned by the sceptic is true-a conviction that the proponent of this argument must have a course reject.

The main objection to externalism rests on the intuition that the basic requirement for epistemic justification is that the acceptances of the belief in question are rational or responsible in relation to the cognitive goal of truth, which seems to necessitate for which the believer actually be aware of a reason for thinking that the belief is true, or at the very least, that such a reason be available to him. Since the satisfaction of an externalist condition is neither necessary nor sufficient for the existence of such a cognitively accessible reason, it is nonetheless, argued, externalism is mistaken as an account of epistemic justification. This general point has been elaborated by appeal to two sorts of putative intuitive counterexamples to externalism. The first of these challenges is the prerequisite justification by appealing to examples of belief which seem intuitively to be justified, but for which the externalist conditions are not satisfied. The standard examples of this sort are cases where beliefs produced in some very nonstandard way, e.g., by a Cartesian demon, but nonetheless, in such a way that the subjective experience of the believer is indistinguishable on that of someone whose beliefs are produced more normally. Cases of this general sort can be constructed in which any of the standard externalist condition, e.g., that the belief is a result of a reliable process, fail to be satisfied. The intuitive claim is that the believer in such a case is nonetheless, epistemically justified, inasmuch as one whose belief is produced in a more normal way, and hence that externalist accounts of justification must be mistaken.

Perhaps the most interesting reply to this sort of counterexample, on behalf of reliabilism specifically, holds that reliability of a cognitive process is to be assessed in 'normal' possible worlds, i.e., in possible worlds that are actually the way our world is common-scenically believed to be, rather than in the world which actually contains the belief being judged. Since the cognitive processes employed in the Cartesian demon case are, we may assume, reliable when assessed in this way, the Reliabilist can agree that such beliefs are justified. The obvious further issue is whether or not there is an adequate rationale for this construal of reliabilism, so that the reply is not merely ad hoc.

The second, correlative way of elaborating the general objection to justificatory externalism challenges the sufficiency of the various externalist conditions by citing cases where those conditions are satisfied, but where the believers in question seem intuitively not to be justified. Here the most widely discussed examples have to do with possible occult cognitive capacities like clairvoyance. Considering the point in application once again to reliabilism specifically, the claim is that a reliable clairvoyant who has no reason to think that he has such a cognitive power, and perhaps even good reasons to the contrary, is not rational or responsible and hence, not epistemologically justified in accepting the belief that result from his clairvoyance, despite the fact that the Reliabilist condition is satisfied.

One sort of response to this latter sort of remonstrance is to 'bite the bullet' and insist that such believer e in fact justified, dismissing the seeming intuitions to the contrary as latent internalist prejudice. To a greater extent the more widely adopted response attempts to impose additional conditions, usually of a more or less internalist sort, which will rule out the offending example while still stopping far short of a full internalist. But while there is little doubt that such modified versions of externalism can indeed handle particular case's well enough to avoid clear intuitive implausibility, the issue is whether there will always be equally problematic cases for issues that might not handle, and whether there is any clear motivation for the additional requirements other than the general internalist view of justification that Externalists are committed to reject.

A view in this same general vein, one that might be described as a hybrid of internalism and externalism, holding that epistemic justification requires that there be a justificatory facto r that is cognitively accessible e to the believer in question (though it need not be actually grasped), thus ruling out, e.g., a pure reliabilism. at the same time, however, though it must be objectively true that beliefs for which such a factor is available are likely to be true, this further fact need not be in any way grasped o r cognitive ly accessible to the believer. In effect, of the two premises needed to argue that a particular belief is likely to be true, one must be accessible in a way that would satisfy at least weak internalism, while the second can be (and will normally be) purely external. Here the internalist will respond that this hybrid view is of no help at all in meeting the objection that the belief is not held in the rational responsible way that justification intuitively seems required, for the believer in question, lacking one crucial premise, still has no reason at all for thinking that his belief is likely to be true.

An alternative to giving an externalist account of epistemic justification, one which may be more defensible while still accommodating many of the same motivating concerns, is to give an externalist account of knowledge directly, without relying on an intermediate account of justification. Such a views obviously have to reject the justified true belief account of knowledge, holding instead that knowledge is true belief which satisfies the chosen externalist condition, e.g., is a result of a reliable process (and, perhaps, further conditions as well). This makes it possible for such a view to retain an internalist account of epistemic justification, though the centrality of that concept is epistemology would obviously be seriously diminished.

Such an externalist account of knowledge can accommodate the common-sen conviction that animals, young children and unsophisticated adults' posse's cognition, in that knowledge, though not the weaker conviction (if such a conviction even exists) that such individuals are epistemically justified in their belief. It is also, least of mention, less vulnerable to internalist counterexamples of the sort and since the intuitivistic vortices in the pertaining extent in the clarification, that is to clear up justification than to knowledge. What is uncertain, is what ultimate philosophical significance the resulting conception of knowledge is taken for granted as of having, but being the occupant of having any serious bearing on traditional epistemological problems and on the deepest and most troubling versions of scepticism, which seem in fact to be primarily concerned with justification rather than knowledge?

A rather different use of the terms 'internalism' and 'externalism' have to do with the issue of how the content of beliefs and thoughts is determined: According to an internalist view of content, the content of such intentional states depends only on the non-relational, internal properties of the individual's mind or brain, and not at all on his physical and social environment: While according to an externalist view, content is significantly affected by such external factors. Here to a view that appeals to both internal and external elements are standardly classified as an externalist view.

As with justification and knowledge, the traditional view of content has been strongly internalist character. The main argument for externalism derives from the philosophy of language, more specifically from the various phenomena pertaining to natural kind terms, indexical, and so forth, that motivates the views that have come to be known as 'direct reference' theories. Such phenomena seem at least to show that the belief or thought content that can e properly attributed to a person is dependent on facts about his environment -, e.g., whether he is on Earth or Twin Earth, what in fact he is pointing at, the classificatory criteria employed by the experts in his social group, and so forth-not just on what is going on internally in his mind or brain.

An objection to externalist accounts of content is that they seem unable to do justice to our ability to know the contents of our beliefs or thoughts 'from the inside', simply by reflection. If content is dependent of external factors pertaining to the environment, then knowledge of content should depend on knowledge of the these factors-which will not usually be available to the person whose belief or thought is in question.

The adoption of an externalist account of mental content would seem to support an externalist account of justification in the following way: If part of all of the content of a belief inaccessible to the believer, then both the justifying status of other beliefs in relation to the content and the status of that content as justifying further beliefs will be similarly inaccessible, thus contravening the internalist must insist that there are no rustication relations of these sorts, that only internally accessible content can either be justified or justify anything else: By such a response appears lame unless it is coupled with an attempt to shows that the Externalists account of content is mistaken.

To have a word or a picture, or any other object in one's mind seems to be one thing, but to understand it is quite another. A major target of the later Ludwig Wittgenstein (1889-1951) is the suggestion that this understanding is achieved by a further presence, so that words might be understood if they are accompanied by ideas, for example. Wittgenstein insists that the extra presence merely raise the same kind of problem again. The better of suggestions in that understanding is to be thought of as possession of a technique, or skill, and this is the point of the slogan that 'meaning is use', the idea is congenital to 'pragmatism' and hostile to ineffable and incommunicable understandings.

Whatever it is that makes, what would otherwise be mere sounds and inscriptions into instruments of communication and understanding. The philosophical problem is to demystify this power, and to relate it to what we know of ourselves and the world. Contributions to this study include the theory of speech acts and the investigation of commonisation and the relationship between words and ideas, sand words and the world.

The most influential idea I e theory of meaning the past hundred years is the thesis that the meaning of an indicative sentence is given by its truth-condition. On this conception, to understand a sentence is to know its truth-conditions. The conception was first clearly formulated by the German mathematician and philosopher of mathematics Gottlob Frége (1848-1925), then was developed in a distinctive way by the early Wittgenstein, and is as leading idea of the American philosopher Donald Herbert Davidson. (1917-2003). The conception has remained so central that those who offer opposing theories characteristically define their position by reference to it.

The conception of meaning as truth-conditions needs not and should not be advanced for being in itself a complete account of meaning. For instance, one who understands a language must have some idea of the range of speech acts conventionally performed by the various types of sentences in the language, and must have some ideate significance of speech acts, the claim of the theorist of truth-conditions should rather be targeted on the notion of content: If two indicative sentences differ in what they strictly and literally say, then this difference is fully accounted for by the difference in their truth-conditions. It is this claim and its attendant problems, which will be the concern of each in the following.

The meaning of a complex expression is a function of the meaning of its constituents. This is indeed just a statement of what it is for an expression to be semantically complex. It is one of the initial attractions of the conception of meaning as truth-conditions that it permits a smooth and satisfying account of the ay in which the meaning of a complex expression is a function of the meaning its constituents. On the truth-conditional conception, to give the meaning of sn expressions is the contribution it makes to the truth-conditions of sentences in which it occurs. For example terms-proper names, indexical, and certain pronouns-this is done by stating the reference of the term in question. For predicates, it is done either by stating the conditions under which the predicate is true of arbitrary objects, or by stating the conditions under which arbitrary atomic sentences containing it true. The meaning of a sentence-forming operators as given by stating its contribution to the truth-conditions of a complex sentence, as function of the semantic values of the sentence on which it operates. For an extremely simple, but nevertheless structured language, er can state that contribution's various expressions make to truth condition, are such as:

A1: The referent of 'London' is London.

A2: The referent of 'Paris' is Paris

A3: Any sentence of the form 'a is beautiful' is true if and only if the referent of 'a' is beautiful.

A4: Any sentence of the form 'a is larger than b' is true if and only if the referent of 'a' is larger than a referent of 'b'.

A5: Any sentence of t he for m 'its no t the case that 'A' is true if and only if it is not the case that 'A' is true.

A6: Any sentence of the form 'A and B' is true if and only if 'A' is true and 'B' is true.

The principle's A1-A6 construct and develop a form for which a simple theory of truth for a fragment of English. In this, the or it is possible to derive these consequences: That 'Paris is beautiful' is true if and only if Paris is beautiful, is true and only if Paris is beautiful (from A2 and A3): That 'London is larger than Paris and it is not the case that London is beautiful, is true if and only if London is larger than Paris and it is not the case that London is beautiful (from A1-A5), and in general, for any sentence 'A', this simple language we can derive something of the form 'A' is true if and only if 'A'.

Yet, theorists' of truth conditions should insist that not every true statement about the reference o f an expression is fit to be an axiom in a meaning-giving theory of truth for a language. The axiom

'London' refers to the ct in which there was a huge fire in 1666.

This is a true statement about the reference of 'London'. It is a consequence of a hypothesis which substitute the axiom for A1 in our simple truth theory that 'London is beautiful' is true if and only if the city in which there was a huge fire in 1666 is beautiful. Since a subject can align himself with the naming authenticity that 'London' is without knowing that the last-mentioned truth condition, this replacement axiom is not fit to be an axiom in a meaning-specifying truth theory. It is, of course, incumbent on a theorist of meaning as truth conditions to state the constraints on the acceptability of axioms in a way which does not presuppose any prior, truth-conditional conception of meaning.

Among the many challenges facing the theorist of truth conditions, two are particularly salient and fundamental, first, the theorist has to answer the charge of triviality or vacuity. Second, the theorist must offer an account of which for a person's language is truly describable by a semantic theory containing a given semantic axiom.

What can take the charge of triviality first? In more detail, it would run thus: since the content of a claim that the sentence 'Paris is beautiful' is true amounts to no more than the claim that Paris is beautiful, we can trivially describe understanding a sentence, if we wish, as knowing its truth-conditions. But this gives us no substantive account of understanding whatsoever. Something other than grasp of truth conditions must provide the substantive account. The charge tests upon what has been called the 'redundancy theory of truth', the theory also known as 'minimalism'. Or the 'deflationary' view of truth, fathered by the German mathematician and philosopher of mathematics, had begun with Gottlob Frége (1848-1925), and the Cambridge mathematician and philosopher Plumton Frank Ramsey (1903-30). Wherefore, the essential claim is that the predicate' . . . is true' does not have a sense, i.e., expresses no substantive or profound or explanatory concept that ought to be the topic of philosophical enquiry. The approach admits of different versions, nit centres on the points that 'it is true that p' says no more nor less than 'p'(hence redundancy): That in less direct context, such as 'everything he said was true'. Or 'all logical consequences are true'. The predicate functions as a device enabling us to generalize rather than as an adjective or predicate describing the things he said or the kinds of propositions that follow from true propositions. For example: '(p, q) (p & p)®q ®q where there is no use of a notion of truth.

There are technical problems in interpreting all uses of the notion of truth in such ways, but they are not generally felt to be insurmountable. The approach needs to explain away apparently substantive users of the notion, such as 'science aims at the truth' or 'truth is a normative governing discourse'. Indeed, postmodernist writing frequently advocates that we must abandon such norms, along with a discredited 'objectivity' conception of truth. But, perhaps, we can have the norm even when objectivity is problematic, since they can be framed without mention of truth: Science wants it to be so that whenever science holds that 'p', then 'p', discourse is to be regulated by the principle that it is wrong to assert 'p' when 'not-p'.

It is, nonetheless, that we can take charge of triviality, since the content of a claim ht the sentence 'Paris is beautiful' is true, amounting to no more than the claim that Paris is beautifully, we can trivially describe understanding a sentence. If we wish, as knowing its truth-condition, but this gives us no substitute account of understanding whatsoever. Something other than grasp of truth conditions must provide the substantive account. The charge rests on or upon what has been the redundancy theory of truth. The minimal theory states that the concept of truth is exhaustively by the fact that it conforms to the equivalence principle, the principle that for any proposition 'p', it is true that 'p' if and only if 'p'. Many different philosophical theories, accept that e equivalence principle, as e distinguishing feature of the minimal theory, its claim that the equivalence principle exhausts the notion of truth. It is, however, widely accepted, both by opponents and supporters of truth conditional theories of meaning, that it is inconsistent to accept both the minimal theory of truth and a truth conditional account of meaning. If the claim that the sentence 'Paris is beautiful, it is circular to try to explain the sentence's meaning in terms of its truth condition. The minimal theory of truth has been endorsed by Ramsey, Ayer, and later Wittgenstein, Quine, Strawson, Horwich and-confusingly and inconsistently of Frége himself.

The minimal theory treats instances of the equivalence principle as definitional truth for a given sentence. But in fact, it seems that each instance of the equivalence principle can itself be explained. The truths from which such an instance as

   'London is beautiful' 'is true if and only if London is beautiful',

Can be explained are precisely A1 and A3 in that, this would be a pseudo-explanation if the fact that 'London' refers to London consists in part in the fact that 'London is beautiful' has the truth-condition it does? But that is very implausible: It is, after all, possible to understand the name 'London' without understanding the predicate 'is beautiful'. The idea that facts about the reference of particular words can be explanatory of facts about the truth conditions of sentences containing them in no way requires any naturalistic or any other kind of reduction of the notion of reference. Nor is the idea incompatible with the plausible point that singular reference can be attributed at all only to something which is capable of combining with other expressions to form complete sentences. That still leaves room for facts about an expression's having the particular reference it does to be partially explanatory of the particular truth condition possessed by a given sentence containing it. The minimal theory thus treats as definitional or stimulative something which is in fact open to explanation. What makes this explanation possible is that there is a general notion of truth which has, among the many links which hold it in place, systematic connections with the semantic values of subsentential expressions.

A second problem with the minimal theory is that it seems impossible to formulate it without at some point relying implicitly on features and principles involving truth which go beyond anything countenanced by the minimal theory. If, minimal, or redundancy theory treats true statements as predicated of anything linguistic, like its utterances, or even, the type-in-a-language, or whatever. Then the equivalence schemata will not cover all cases, but those in the theorist's own language only. Some account has to be given of truth for sentences of other languages. Speaking of the truth of language-independent propositions or thoughts will only postpone, not avoid, this issue, since at some point principles have to be stated associating these language-dependent entities with sentences of particular languages. The defender of the minimalist theory is that the sentence 'S' of a foreign language is best translated by our sentence, then the foreign sentence 'S' is true if and only if 'p'. Now the best translation of a sentence must preserve the concepts expressed in the sentence. Constraints involving a general notion of truth are pervasive plausible philosophical theory of concepts. It is, for example, a condition of adequacy on an individuating account of any concept that there exists what may be called 'Determination Theory' for that account-that is, a specification on how the account contributes to fixing the semantic value of that concept. The notion of a concept's semantic value is the notion of something which makes a certain contribution to the truth conditions of thoughts in which the concept occurs. But this is to presuppose, than to elucidate, a general notion of truth.

It is, also, plausible that there are general constraints on the form of such Determination Theories, constrains which to involve truth, and which are not derivable from the minimalist's creation. Suppose that concepts are individuated by their possession condition. A possession condition may in various ways make a thinker's possession of a particular concept dependent upon his relation to his environment. Many possession conditions will mention the links between accept and the thinker's perceptual experience. Perceptual experience represents the world for being a certain way. It is arguable that the only satisfactory explanation to what it is for perceptual experience to represent the world in a particular way must refer to the complex relations of the experience to the subject's environment. If this is so, to mention of such experiences in a possession condition dependent in part upon the environmental relations of the thinker. Evan though the thinker's non-environmental properties and relations remain constant, the conceptual content of his mental state can vary in the thinker's social environment is varied. A possession condition which properly individuates such a concept must take into account the thinker's social relations, in particular his linguistic relations.

Its alternative approach, addresses the question by starting from the idea that a concept is individuated by the condition which must be satisfied a thinker is to posses that concept and to be capable of having beliefs and other altitudes whose contents contain it as a constituent. So, to take a simple case, one could propose that the logical concept 'and' is individualized by this condition: It is the unique concept 'C' to posses which a thinker has to find these forms of inference compelling, without basting them on any further inference or information: From any two premises 'A' and 'B', ACB can be inferred and from any premises a relatively observational concepts such as grounding possibilities can be individuated in part by stating that the thinker finds specified contents containing it compelling when he has certain kinds of perception, and in part by relating those judgements containing the concept and which are not based on perception to those judgements that are. A statement which individuates a concept by saying what is required for a thinker to posses it can be described as giving the possession condition for the concept.

A possession condition for a particular concept may actually make use of that concept. The possession condition for 'and' doers not. we can also expect to use relatively observational concepts in specifying the kind of experience which have to be mentioned in the possession conditions for relatively observational concepts. What we must avoid is mention of the concept in question as such within the content of the attitude attributed to the thinker in the possession condition. Otherwise we would be presupposed possession of the concept in an account which was meant to elucidate its possession. In talking of what the thinker finds compelling, the possession conditions can also respect an insight of the later Wittgenstein: That a thinker's mastery of a concept is inextricably tied to how he finds it natural to go in new cases in applying the concept.

Sometimes a family of concepts has this property: It is not possible to master any one of the members of the family without mastering of the others. Two of the families which plausibly have this status are these: The family consisting of same simple concepts 0, 1. 2, . . . of the natural numbers and the corresponding concepts of numerical quantifiers, 'there are o so-and-so's, there is 1 so-and-so's, . . . and the family consisting of the concepts 'belief' and 'desire'. Such families have come to be known as 'local Holist's'. A local holism does not prevent the individuation of a concept by its possession condition. Rather, it demands that all the concepts in the family be individuated simultaneously. So one would say something of this conduct regulated by an external control as a custom or a formal protocol of procedures, such that the economy is delegated by some outward appearance of something as distinguished from the substance of which it is made form. Belief and desire form the unique pair of concepts C1 and C2, such that for a thinker to posses them is to meet such-and-such condition involving the thinker, C1 and C2. For those other possession conditions to individuate properly. It is necessary that there be some ranking of the concepts treated. The possession condition or concepts higher in the ranking must presuppose only possession of concepts at the same or lower levels in the ranking.

A possession condition may in various ways make a thinker's possession of a particular concept dependent on or upon his relations to his environment. Many possession conditions will mention the links between a concept and the thinker's perceptual experience. Perceptual experience represents the world for being a certain way. It is arguable that the only satisfactory explanation of what it is for perceptual experience to represent the world in a particular way must refer to the complex relations of the experience to te subject's environment. If this is so, then mention of such experiences in a possession condition will make possession f that concept relations tn the thicker. Burge (1979) has also argued from intuitions about particular examples that even though the thinker's non-environmental properties and relations remain constant, the conceptual content of his mental state can vary in the thinker's social environment is varied. A possession condition which properly individuates such a concept must take into account the thinker's social relations, in particular his linguistic relations.

Once, again, some general principles involving truth can, as Horwich has emphasized, be derived from the equivalence schemata using minimal logical apparatus. Consider, for instance, the principle that 'Paris is beautiful and London is beautiful' is true if and only if 'Paris is beautiful' is true and 'London is beautiful' is true if and only if Paris is beautiful and London is beautiful. But no logical manipulations of the equivalence e schemata will allow the derivation of that general constraint governing possession condition, truth and assignment of semantic values. That constraints can, of course, be regarded as a further elaboration of the idea that truth is one of the aims of judgement.

What is to a greater extent, but to consider the other question, for 'What is it for a person's language to be correctly describable by a semantic theory containing a particular axiom, such as the above axiom A6 for conjunctions? This question may be addressed at two depths of generality. A shallower of levels, in this question may take for granted the person's possession of the concept of conjunction, and be concerned with what hast be true for the axiom to describe his language correctly. At a deeper level, an answer should not sidestep the issue of what it is to posses the concept. The answers to both questions are of great interest.

When a person conjunction by 'and', he is not necessarily capable of phrasing the A6 axiom. Even if he can formulate it, his ability to formulate it is not causal basis of his capacity to hear sentences containing the word 'and' as meaning something involving conjunction. Nor is it the causal basis of his capacity to mean something involving conjunction by sentences he utters containing the word 'and'. Is it then right to regard a truth theory as part of an unconscious psychological computation, and to regard understanding a sentence as involving a particular way of deriving a theorem from a truth theory at some level of unconscious processing? One problem with this is that it is quite implausible that everyone who speaks the same language has to use the same algorithms for computing the meaning of a sentence. In the past thirteen years, the particular work as befitting Davies and Evans, whereby a conception has evolved according to which an axiom like A6, is true of a person's component in the explanation of his understanding of each sentence containing the words 'and', a common component which explains why each such sentence is understood as meaning something involving conjunction. This conception can also be elaborated in computational; terms: As alike to the axiom A6 to be true of a person's language is for the unconscious mechanism, which produce understanding to draw on the information that a sentence of the form 'A and B' is true only if 'A' is true and 'B' is true. Many different algorithms may equally draw on or open this information. The psychological reality of a semantic theory thus are to involve, Marr's (1982) given by classification as something intermediate between his level one, the function computed, and his level two, the algorithm by which it is computed. This conception of the psychological reality of a semantic theory can also be applied to syntactic and phonological theories. Theories in semantics, syntax and phonology are not themselves required to specify the particular algorithm which the language user employs. The identification of the particular computational methods employed is a task for psychology. But semantic, syntactic and phonological theories are answerable to psychological data, and are potentially refutable by them-for these linguistic theories do make commitments to the information drawn on or upon by mechanisms in the language user.

This answer to the question of what it is for an axiom to be true of a person's language clearly takes for granted the person's possession of the concept expressed by the word treated by the axiom. In the example of the A6 axiom, the information drawn upon is that sentence of the form 'A and B' are true if and only if 'A' is true and 'B' is true. This informational content employs, as it has to if it is to be adequate, the concept of conjunction used in stating the meaning of sentences containing 'and'. 'S' be that of a computational answer we have returned needs further elaboration, which does not want to take for granted possession of the concepts expressed in the language. It is at this point that the theory of linguistic understanding has to argue that it has to draw upon a theory if the conditions for possessing a given concept. It is plausible that the concept of conjunction is individuated by the following condition for a thinker to have possession of it: The concept 'and' is that concept 'C' to possess which a thinker must meet the following conditions: He finds inferences of the following forms compelling, does not find them looking down on the aftermath of any reasoning and finds them because they are of their form:

$$pCq \quad pCq \quad pq$$

$$p \quad q \quad pCq$$

Here 'p' and 'q' range ov complete propositional thoughts, not sentences. When A6 axiom is true of a person's language, there is a global dovetailing between this possessional condition for the concept of conjunction and certain of his practices involving the word 'and'. For the case of conjunction, the dovetailing involves at least this:

If the possession condition for conjuncture entails that the thinker who possesses the concept of conjunction must be willing to make certain tradition involving the thought p & q, and of the thinker's semitrance 'A' that 'p' and his sentence 'B' that 'q' then: The thinker must be willing to make the corresponding linguistic tradition involving sentence 'A and B'.

This is only part of what is involved in the required dovetailing. Given what wee have already said about the uniform explanation of the understanding of the various occurrences of a given word, we should also add, that there is a uniform (unconscious, computational) explanation of the language user's willingness to make the corresponding transition involving the sentence 'A and B'.

This dovetailing account returns an answer to the deeper questions because neither the possession condition for conjunction, nor the dovetailing condition which builds upon the dovetailing condition which builds on or upon that possession condition, takes for granted the thinker's possession of the concept expressed by 'and'. The dovetailing account for conjunction is an exampling of a greater amount of an overall schemata, which can be applied to any concept. The case of conjunction is of course, exceptionally simple in several respects. Possession conditions for other concepts will speak not just of inferential transition, but of certain conditions in which beliefs involving the concept in question are accepted or rejected, and the corresponding dovetailing condition will inherit these features. This dovetailing account has also to be underpinned by a general rationale linking contributions to truth conditions with the particular possession condition proposed for concepts. It is part of the task of the theory of concepts to supply this in developing Determination Theories for particular concepts.

In some cases, a relatively clear account is possible of how a concept can feature in thoughts which may be true though unverifiable. The possession condition for the quantificational concept all natural numbers can in outline run thus: This quantifier is that concept $Cx \ldots x \ldots$ to posses which the thinker has to find any inference of the form:

$$CxFx$$
$$Fn.$$

Compelling, where 'n' is a concept of a natural number, and does not have to find anything else essentially containing $Cx \ldots x \ldots$ compelling. The straightforward Determination Theory for this possession condition is one on which the truth of such a thought CxFx is true only if all natural numbers are 'F'. That all natural numbers are 'F' is a condition which can hold without our being able to establish that it holds. So an axiom of a truth theory which dovetails with this possession condition for universal quantification over the natural numbers will be b component of a realistic, non-verifications theory of truth conditions.

Finally, this response to the deeper questions allows us to answer two challenges to the conception of meaning as truth-conditions. First, there was the question left hanging earlier, of how the theorist of truth-conditions is to say what makes one axiom of a semantic theory correct rather than another, when the two axioms assigned the same semantic values, but do so by different concepts. Since the different concepts will have different possession conditions, the dovetailing accounts, at the deeper level, of what it is for each axiom to be correct for a person's language will be different accounts. Second, there is a challenge repeatedly made by the minimalist theories of truth, to the effect that the theorist of meaning as truth-conditions should give some non-circular account of what it is to understand a sentence, or to be capable of understanding all sentences containing a given constituent. For each expression in a sentence, the corresponding dovetailing account, together with the possession condition, supplies a non-circular account of what it is to that expression. The combined accounts for each of the expressions which comprise a given sentence together constitute a non-circular account of what it is to understand the complete sentence. Taken together, they allow theorist of meaning as truth-conditions fully to meet the challenge.

A widely discussed idea is that for a subject to be in a certain set of content-involving states, for attribution of those state s to make the subject as rationally intelligible. Perceptions make it rational for a person to form corresponding beliefs. Beliefs make it rational to draw certain inferences such as belief and desire that to a condition or occurrence, traceable to a cause as effectually generated by a protocol that of being something by leaning toward form, shaping, combining to assemble for some agreeable reason for which the formations of particular intentions, and performance are those of the appropriate actions. People are frequently irrational of course, but a governing ideal of this approach is that for any family of contents, there is some minimal core of rational transition to or from states involving them, a core that a person must respect of his states is to be attribute d with those contents at all. we contrast in what we want do with what we must do-whether for reasons of morality or duty, or even for reasons of practical necessity (to get what we wanted in the first place). Accordingly, our own desires have seemed to be the principal actions that most fully express our own individual natures and will, and those for which we are personally responsible. But desire has also seemed to be a principle of action contrary to and at war with our better natures, as rational and agents. For it is principally from our own differing perspectives upon what would be good, that each of us wants what he does, each point of view being defined by one's own interests and pleasure. In this, the representations of desire are like those of sensory perception, similarly shaped by the perspective of the perceiver and the idiosyncrasies of the perceptual dialectic about desire and its object recapitulates that of perception ad sensible qualities. The strength of desire, for instance, varies with the state of the subject more or less independently of the character, an the actual utility, of the object wanted. Such facts cast doubt on the 'objectivity' of desire, and on the existence of correlatives property of 'goodness', inherent in the objects of our desires, and independent of them. Perhaps, as the Dutch Jewish rationalist (1632-77) Benedictus de Spinoza put it, it is not that we want what we think good, but that we think well what we happen to want-the 'good' in what we want being a mere shadow cast by the desire for it. (There is a parallel Protagorean view of belief, similarly sceptical of truth). The serious defence of such a view, however, would require a systematic reduction of apparent facts about goodness to fats about desire, and an analysis of desire which in turn makes no reference to goodness. While what is yet to be provided, moral psychologists have sought to vindicate an idea of objective goodness. For example, as what would be good from all points of view, or none, or, in the manner of the German philosopher Immanuel Kant, to set up another principle (the will or practical reason) conceived as an autonomous source of action, independent of desire or its object: And this tradition has tended to minimize the role of desire in the genesis of action.

Ascribing states with content on actual person has to proceed simultaneously with attributions of as wide range of non-rational states and capacities. In general, we cannot understand a persons reasons for acting as he does without knowing the array of emotions and sensations to which he is subject: What he remembers and what he forgets, and how he reasons beyond the confines of minimal rationality. Even the content-involving perceptual states, which play a fundamental role in individuating content, cannot be understood purely in terms relating to minimal rationality. A perception of the world as being a certain way is not (and could not be) under a subject's rational control. Thought it is true and important that perceptions give reason for forming beliefs, the beliefs for which they fundamentally provide reasons-observational belies about the environment-have contents which can only be elucidated by referring to perceptual experience. In this respect (as in others), perceptual states differ from beliefs and desires that are individuated by mentioning what they provide reasons for judging or doing: or frequently these latter judgements and actions can be individuated without reference back to the states that provide for them.

What is the significance for theories of content of the fact that it is almost certainly adaptive for members of as species to have a system of states with representational contents which are capable of influencing their actions appropriately? According to teleological theories a content, a constitutive account of content-one which says what it is for a state to have a given content-must make user of the notion of natural function and teleology. The intuitive idea is that for a belief state to have a given content 'p' is for the belief-forming mechanisms which produced it to have the unction as, perhaps, derivatively of producing that stare only when it is the case that 'p'. One issue this approach must tackle is whether it is really capable of associating with states the classical, realistic, verification-transcendent contents which, pre-theoretically, we attribute to them. It is not clear that a content's holding unknowably can influence the replication of belief-forming mechanisms. But if content itself proves to resist elucidation, it is still a very natural function and selection. It is still a very attractive view that selection, it is still a very attractive view, that selection must be mentioned in an account of what associates something-such as aa sentence-wi a particular content, even though that content itself may be individuated by other.

Contents, are most notably given by some consequence or prominence in some sorted agreement to which makes them explicitly distinctive by rule in specifying 'that . . .' clauses, and it is natural to suppose that a content has the same kind of sequence and hierarchical structure as the sentence that specifies it. This supposition would be widely accepted for conceptual content. It is, however, a substantive thesis that all content is conceptual. One way of treating one sort of 'perceptual content' is to regard the content as determined by a spatial type, the type under which the region of space around the perceiver must fall if the experience with that content is to represent the environment correctly. The type involves a specification of surfaces and features in the environment, and their distances and directions from the perceiver's body as origin, such contents lack any sentence-like structure at all. Supporters of the view that all content is conceptual will argue that the legitimacy of using these spatial types in giving the content of experience does not undermine the thesis that all content is conceptual. Such supporters will say that the spatial type is just a way of capturing what can equally be captured by conceptual components such as 'that distance', or 'that direction', where these demonstratives are made available by the perception in question. Friends of conceptual content will respond that these demonstratives themselves cannot be elucidated without mentioning the spatial type which lack sentence-like structure.

Content-involving states are actions individuated in party reference to the agent's relations to things and properties in his environment. Wanting to see a particular movie and believing that the building over there is a cinema showing it makes rationally the action of walking in the direction of that building

However, in the general philosophy of mind, and more recently, desire has received new attention from those who understand mental states in terms of their causal or functional role in their determination of rational behaviour, and in particular from philosophers trying to understand the semantic content or intentional; character of mental states in those terms as 'functionalism', which attributes for the functionalist who thinks of mental states and evens as a causally mediating between a subject's sensory inputs and that of the subject's ensuing behaviour. Functionalism itself is the stronger doctrine that makes a mental state the type of state it-is-in. That of causing of being inflected by some distressful pain, a smell of violets, a belief that the koala, an arboreal Australian marsupial (Phascolarctos cinereus), is seriously dangerous, in that the functional relation it bears to the subject's perceptual stimuli, behavioural responses, and other mental states.

In the general philosophy of mind, and more recently, desire has received new attention from those who would understand mental stat n terms of their causal or functional role in the determination of rational behaviour, and in particularly from philosophers trying to understand the semantic content or the intentionality of mental states in those terms.

Conceptual (sometimes computational, cognitive, causal or functional) role semantics (CRS) entered philosophy through the philosophy of language, not the philosophy of mind. The core idea behind the conceptual role of semantics in the philosophy of language is that the way linguistic expressions are related to one another determines what the expressions in the language mean. There is a considerable affinity between the conceptual role of semantics and structuralist semiotics that has been influence in linguistics. According to the latter, languages are to be viewed as systems of differences: The basic idea is that the semantic force (or, 'value') of an utterance is determined by its position in the space of possibilities that one' language offers. Conceptual role semantics also has affinities with what the artificial intelligence researchers call 'procedural semantics', the essential idea here is that providing a compiler for a language is equivalent to specifying a semantic theory of procedures that a computer is instructed to execute by a program.

Nevertheless, according to the conceptual role of semantics, the meaning of a thought is determined by the recollected role in a system of states, to specify a thought is not to specify its truth or referential condition, but to specify its role, Walter and twin-Walter's thoughts though different truth and referential conditions, share the same conceptual role, and it is by virtue of this commonality that they behave type-identically. If Water and twin-Walter each has a belief that he would express by 'water quenches thirst' the conceptual role of semantics can be explained predict, they're dripping their cans into H2O and XYZ respectfully. Thus the conceptual role of semantics would seem as though not to Jerry Fodor, who rejects of the conceptual role of semantics for both external and internal problems.

Nonetheless, if, as Fodor contents, thoughts have recombinable linguistic ingredients, then, of course, for the conceptual role of semantic theorist, questions arise about the role of expressions in the language of thought as well as in the public language we speak and write. And, according, the conceptual role of semantic theorbists divide not only over their aim, but also about conceptual roles in semantic's proper domains. Two questions avail themselves. Some hold that public meaning is somehow derivative (or inherited) from an internal mental language (mentalese) and that a mentalese expression has autonomous meaning (partly), expressions for example. The inscriptions on this page call for their understanding translation, or at least, transliterations. Expression into the language of thought: Representations in the brain require no such translation or transliteration but to express of others to sustain that the language of thought is virtuously a language internalized and that it is expressions (or primary) meaning in merit of their conceptual role.

After one decides upon the aims and the proper province of the conceptual role for semantics, the relations among public expressions or mental-constitute their conceptual roles. Because most conceptual roles of semantics as theorists leave the notion of the role in a conceptuality as a blank cheque, the options are open-ended. The conceptual role of a [mental] expression might be its causal association: Any disposition too token or example, utter or think on the expression ' ' when tokening another ' ' or 'a' an ordered n-tuple <, . . . >, or vice versa, can matter to the conceptional role of. A more common option is characterized in a conceptual role not causative of but inferentially (these need be compatible, contingent upon one's attitude about the naturalization of inference): The conceptual role of an expression in 'L' might consist of the set of actual and potential inferences form, or, as a more common, the ordered pair consisting of these two sets. Or, if sentences have non-derived inferential roles, what would it mean to talk of the inferential role of words? Some have found it natural to think of the inferential role of as words, as represented by the set of inferential roles of the sentence in which the word appears.

The expectation of expecting that one sort of thing could serve all these tasks went hand in hand with what has come to b e called the 'Classical View' of concepts, according to which they had an 'analysis' consisting of conditions that are individually necessary and jointly sufficient for their satisfaction, which are known to any competent user of them. The standard example is the especially simple one of the [bachelor], which seems to be identical to [eligible unmarried male]. A more interesting, but analysis was traditionally thought to be [justified true belief].

This Classical View seems to offer an illuminating answer to a certain form of metaphysical question: In virtue of what is something the kind of thing it is -, i.e., in virtue of what may a bachelor be a bachelor? And it does so in a way that support counterfactuals: It tells us what would satisfy the conception situations other than the actual ones (although all actual bachelors might turn out to be freckled, it's possible that there might be unfreckled ones, since the analysis does not exclude that). The view also seems to offer an answer to an epistemological question of how people seem to know a priori (or independently of experience) about the nature of many things, e.g., that bachelors are unmarried: It is constitutive of the competency (or possession) conditions of a concept that they know its analysis, at least on reflection.

The Classic View, however, has alway ss had to face the difficulty of primitive concepts: It's all well and good to claim that competence consists in some sort of mastery of a definition, but what about the primitive concept in which a process of definition must ultimately end: Here the British Empiricism of the seventeenth century began to offer a solution: All the primitives were sensory, indeed, they expanded the Classical View to include the claim, now often taken uncritically for granted in the discussions of that view, that all concepts are 'derived from experience':'Every idea is derived from a corresponding impression', in the work of John Locke (1632-1704), George Berkeley (1685-1753) and David Hume (1711-76) were often thought to mean that concepts were somehow composed of introspectible categorized mental items, 'images', 'impressions', and so on, that were ultimately decomposable into basic sensory parts. Thus, Hume analysed the concept of [material object] as involving certain regularities in our sensory experience and [cause] as involving spatio-temporal contiguity ad constant conjunction.

The Irish 'idealist' George Berkeley, noticed a problem with this approach that every generation has had to rediscover: If a concept is a sensory impression, like an image, then how does one distinguish a general concept [triangle] from a more particular one-say, [isosceles triangle]-that would serve in imagining the general one. More recently, Wittgenstein (1953) called attention to the multiple ambiguity of images. And in any case, images seem quite hopeless for capturing the concepts associated with logical terms (what is the image for negation or possibility?) Whatever the role of such representation, full conceptual competency must involve something more.

Conscionably, in addition to images and impressions and other sensory items, a full account of concepts needs to consider is of logical structure. This is precisely what the logical positivist did, focussing on logically structured sentences instead of sensations and images, transforming the empiricist involvement into the famous 'Verifiability Theory of Meaning', the meaning of sentence by which it is confirmed or refuted, ultimately by sensory experience the meaning or concept associated with a predicate is the by which people confirm or refute whether something satisfies it.

This once-popular position has come under much attack in philosophy in the last fifty years, in the first place, fewer, if any, successful 'reductions' of ordinary concepts like [material objects] [cause] to purely sensory concepts have ever been achieved. Our concept of material object and causation seem to go far beyond mere sensory experience, just as our concepts in a highly theoretical science seem to go far beyond the often only meagre exposures to the evidence is that we can adduce for them.

The American philosopher of mind Jerry Alan Fodor and LePore (1992) have recently argued that the arguments for meaning holism are, however less than compelling, and that there are important theoretical reasons for holding out for an entirely atomistic account of concepts. On this view, concepts have no 'analyses' whatsoever: They are simply ways in which people are directly related to individual properties in the world, which might obtain for someone, for one concept but not for any other: In principle, someone might have the concept [bachelor] and no other concepts at all, much less any 'analysis' of it. Such a view goes hand in hand with Fodor's rejection of not only verificationist, but any empiricist account of concept learning and construction: Given the failure of empiricist construction. Fodor (1975, 1979) notoriously argued that concepts are not constructed or 'derived' from experience at all, but are and nearly enough as they are all innate.

The deliberating considerations about whether there are innate ideas are much as it is old, it, nonetheless, takes from Plato (429-347 Bc) in the 'Meno' the problems to which the doctrine of 'anamnesis' is an answer in Plato's dialogue. If we do not understand something, then we cannot set about learning it, since we do not know enough to know how to begin. Teachers also come across the problem in the shape of students, who cannot understand why their work deserves lower marks than that of others. The worry is echoed in philosophies of language that see the infant as a 'little linguist', having to translate their environmental surroundings and grasp on or upon the upcoming language. The language of thought hypothesis was especially associated with Fodor that mental processing occurs in a language different from one's ordinary native language, but underlying and explaining our competence with it. The idea is a development of the Chomskyan notion of an innate universal grammar. It is a way of drawing the analogy between the workings of the brain or the minds and of the standard computer, since computer programs are linguistically complex sets of instruments whose execution explains the surface behaviour of computer. Just as an explanation of ordinary language has not found universal favour. It apparently only explains ordinary representational powers by invoking innate things of the same sort, and it invites the image of the learning infant translating the language whose own powers are a mysterious biological given.

René Descartes (1596-1650) and Gottfried Wilhelm Leibniz (1646-1716), defended the view that mind contains innate ideas: Berkeley, Hume and Locke attacked it. In fact, as we now conceive the great debate between European Rationalism and British Empiricism in the seventeenth and eighteenth centuries, the doctrine of innate ideas is a central bone of contention: Rationalist typically claim that knowledge is impossible without a significant stoke of general innate concepts or judgements: Empiricist argued that all ideas are acquired from experience. This debate is replayed with more empirical content and with considerably greater conceptual complexity in contemporary cognitive science, most particularly within the domain of psycholinguistic theory and cognitive developmental theory.

Some of the philosophers may be cognitive scientist other's concern themselves with the philosophy of cognitive psychology and cognitive science. Since the inauguration of cognitive science these disciplines have attracted much attention from certain philosophes of mind. The attitudes of these philosophers and their reception by psychologists vary considerably. Many cognitive psychologists have little interest in philosophical issues. Cognitive scientists are, in general, more receptive.

Fodor, because of his early involvement in sentence processing research, is taken seriously by many psycholinguists. His modularity thesis is directly relevant to question about the interplay of different types of knowledge in language understanding. His innateness hypothesis, however, is generally regarded as unhelpful. And his prescription that cognitive psychology is primarily about propositional attitudes is widely ignored. The American philosopher of mind, Daniel Clement Dennett (1942-) whose recent work on consciousness treats a topic that is highly controversial, but his detailed discussion of psychological research finding has enhanced his credibility among psychologists. In general, however, psychologists are happy to get on with their work without philosophers telling them about their 'mistakes'.

Connectionmism has provided a somewhat different reaction mg philosophers. Some-mainly those who, for other reasons, were disenchanted with traditional artificial intelligence research-have welcomed this new approach to understanding brain and behaviour. They have used the success, apparently or otherwise, of connectionist research, to bolster their arguments for a particular approach to explaining behaviour. Whether this neurophilosophy will eventually be widely accepted is a different question. One of its main dangers is succumbing to a form of reductionism that most cognitive scientists and many philosophers of mind, find incoherent.

One must be careful not to caricature the debate. It is too easy to see the argument as one's existing in or belonging to an individuals inherently congenital elemental for the placing of Innatists, who argue that all concepts of all of linguistic knowledge is innate (and certain remarks of Fodor and of Chomsky lead themselves in this interpretation) against empiricist who argue that there is no innate cognitive structure in which one need appeal in explaining the acquisition of language or the facts of cognitive development (an extreme reading of the American philosopher Hilary Putnam 1926-). But this debate would be a silly and a sterile debate indeed. For obviously, something is innate. Brains are innate. And the structure of the brain must constrain the nature of cognitive and linguistic development to some degree. Equally obvious, something is learned and is learned as opposed too merely grown as limbs or hair growth. For not all of the world's citizens end up speaking English, or knowing the Relativity Theory. The interesting questions then all concern exactly what is innate, to what degree it counts as knowledge, and what is learned and to what degree its content and structure are determined by innately specified cognitive structure. And that is plenty to debate.

The arena in which the innateness takes place has been prosecuted with the greatest vigour is that of language acquisition, and it is appropriately to begin there. But it will be extended to the domain of general knowledge and reasoning abilities through the investigation of the development of object constancy-the disposition to concept of physical objects as persistent when unobserved and to reason about their properties locations when they are not perceptible.

The most prominent exponent of the innateness hypothesis in the domain of language acquisition is Chomsky (1296, 1975). His research and that of his colleagues and students is responsible for developing the influence and powerful framework of transformational grammar that dominates current linguistic and psycholinguistic theory. This body of research has amply demonstrated that the grammar of any human language is a highly systematic, abstract structure and that there are certain basic structural features shared by the grammars of all human language s, collectively called 'universal grammar'. Variations among the specific grammars of the world's ln languages can be seen as reflecting different settings of a small number of parameters that can, within the constraints of universal grammar, take may have several different valued. All of type principal arguments for the innateness hypothesis in linguistic theory on this central insight about grammars. The principal arguments are these: (1) The argument from the existence of linguistic universals, (2) the argument from patterns of grammatical errors in early language learners: (3) The poverty of the stimulus argument, (4) the argument from the case of fist language learning (5) the argument from the relative independence of language learning and general intelligence, and (6) The argument from the moduarity of linguistic processing.

Innatists argue (Chomsky 1966, 1975) that the very presence of linguistic universals argue for the innateness of linguistic of linguistic knowledge, but more importantly and more compelling that the fact that these universals are, from the standpoint of communicative efficiency, or from the standpoint of any plausible simplicity reflectively adventitious. These are many conceivable grammars, and those determined by universal grammars, and those determined by universal grammar are not ipso facto the most efficient or the simplest. Nonetheless, all human languages satisfy the constraints of universal grammar. Since either the communicative environment or the communicative tasks can explain this phenomenon. It is reasonable to suppose that it is explained by the structures of the mind-and therefore, by the fact that the principles of universal grammar lie innate in the mind and constrain the language that a human can acquire.

Hilary Putnam argues, by appeal to a common-sens e ancestral language by its descendants. Or it might turn out that despite the lack of direct evidence at present the feature of universal grammar in fact do serve either the goals of commutative efficacy or simplicity according in a metric of psychological importance. finally, empiricist point out, the very existence of universal grammar might be a trivial logical artefact: For one thing, many finite sets of structure es whether some features in common. Since there are some finite numbers of languages, it follows trivially that there are features they all share. Moreover, it is argued that many features of universal grammar are interdependent. On one, in fact, the set of fundamentally the same mental principle shared by the world's languages may be rather small. Hence, even if these are innately determined, the amount not of innate knowledge thereby, required may be quite small as compared with the total corpus of general linguistic knowledge acquired by the first language learner.

These relies are rendered less plausible, Innatists argue, when one considers the fact that the error's language learners make in acquiring their first language seem to be driven far more by abstract features of gramma r than by any available input data. So, despite receiving correct examples of irregular plurals or past-tense forms for verbs, and despite having correctly formed the irregular forms for those words, children will often incorrectly regularize irregular verbs once acquiring mastery of the rule governing regulars in their language. And in general, not only the correct inductions of linguistic rules by young language learners but more importantly, given the absence of confirmatory data and the presence of refuting data, children's erroneous inductions always consistent with universal gramma r, often simply representing the incorrect setting of a parameter in the grammar. More generally, Innatists argue (Chomsky 1966 & Crain, 1991) all grammatical rules that have ever been observed satisfy the structure-dependence constraint. That is, more linguistics and psycholinguistics argue that all known grammatical rules of all of the world's languages, including the fragmentary languages of young children must be started as rules governing hierarchical sentence structure, and not governing, say, sequence of words. Many of these, such as the constituent-command constraint governing anaphor, are highly abstract indeed, and appear to be respected by even very young children. Such constrain may, Innatists argue, be necessary conditions of learning natural language in the absence of specific instruction, modelling and correct, conditions in which all first language learners acquire their native language.

Importantly among empiricist who rely to these observations derive from recent studies of 'conceptionist' models of first language acquisition. For which an axiomatically distinct manifestation acknowledged for being other than what seems to be the case, in that for subjecting reasons for being in the 'integrated connection system'. Though not being previously trained to represent any subset universal grammar, which is induced grammatically the change for which would include a large set of regular forms and fewer irregulars. It also tends to over-regularize, exhibiting the same U-shape learning curve seen in human language acquire learning systems that induce grammatical systems acquire 'accidental' rules on which they are not explicitly trained but which are not explicit with those upon which they are trained, suggesting, that as children acquire portions of their grammar, they may accidentally 'learn' correct consistent rules, which may be correct in human languages, but which then must be 'unlearned' in they're home language. On the other hand, such 'empiricist' language acquisition systems have yet to demonstrating their ability to induce a sufficient wide range of the rules hypothesize to be comprised by universal grammar to constitute a definitive empirical argument for the possibility of natural language acquisition in the absence of a powerful set of innate constraints.

The poverty of the stimulus argument has been of enormous influence in innateness debates, though its soundness is hotly contested. Chomsky notes that (1) the examples of their targe t language to which the language learner is exposed are always jointly compatible with an infinite number of alterative grammars, and so vastly under-determine the grammar of the language, and (2) The corpus always contains many examples of ungrammatical sentences, which should in fact serves as falsifiers of any empirically induced correct grammar of the language, and (3) there is, in general, no explicit reinforcement of correct utterances or correction of incorrect utterances, sharpness either in the learner or by those in the immediate training environment. Therefore, he argues, since it is impossible to explain the learning of the correct grammar-a task accomplished b all normal children within a very few years-on the basis of any available data or known learning algorithms, it must be ta the grammar is innately specified, and is merely 'triggered' by relevant environmental cues.

Opponents of the linguistic innateness hypothesis, however, point out that the circumstance that the American linguistic, philosopher and political activist, Noam Avram Chomsky (1929-), who believes that the speed with which children master their native language cannot be explained by learning theory, but requires acknowledging an innate disposition of the mind, an unlearned, innate and universal grammar, suppling the kinds of rule that the child will a priori understand to be embodied in examples of speech with which it is confronted in computational terms, unless the child came bundled with the right kind of software. It cold not catch onto the grammar of language as it in fact does.

As it is wee known from arguments due to the Scottish philosopher David Hume (1978, the Austrian philosopher Ludwig Wittgenstein (1953), the American philosopher Nelson Goodman ()1972) and the American logician and philosopher Aaron Saul Kripke (1982), that in all cases of empirical abduction, and of training in the use of a word, data underdetermining the theories. Th is moral is emphasized by the American philosopher Willard van Orman Quine (1954, 1960) as the principle of the undetermined theory by data. But we, nonetheless, do abduce adequate theories in silence, and we do learn the meaning of words. And it could be bizarre to suggest that all correct scientific theories or the facts of lexical semantics are innate.

Nonetheless, Innatists rely, when the empiricist relies on the underdermination of theory by data as a counterexample, a significant disanalogousness with language acquisition is ignored: The abduction of scientific theories is a difficult, labourious process, taking a sophisticated theorist a great deal of time and deliberated effort. First language acquisition, by contrast, is accomplished effortlessly and very quickly by a small child. The enormous relative ease with which such a complex and abstract domain is mastered by such a na ve 'theorist' is evidence for the innateness of the knowledge achieved.

Empiricist such as the American philosopher Hilary Putnam (1926-) have rejoined that Innatists underestimate the amount of time that language learning actually takes, focussing only on the number of years from the apparent onset of acquisition to the achievement of relative mastery over the grammar. Instead of noting how short this interval, they argue, one should count the total number of hours spent listening to language and speaking during h time. That number is in fact quite large and is comparable to the number of hours of study and practice required the acquisition of skills that are not argued to deriving from innate structures, such as chess playing or musical composition. Hence, they are taken into consideration. Language learning looks like one more case of human skill acquisition than like a special unfolding of innate knowledge.

Innatists, however, note that while the case with which most such skills are acquired depends on general intelligence, language is learned with roughly equal speed, and too roughly the same level of general intelligence. In fact even significantly retarded individuals, assuming special language deficits, acquire their native language on a tine-scale and to a degree comparable to that of normally intelligent children. The language acquisition faculty, hence, appears to allow access to a sophisticated body of knowledge independent of the sophistication of the general knowledge of the language learner.

Empiricist's reply that this argument ignores the centrality of language in a wide range of human activities and consequently the enormous attention paid to language acquisition by retarded youngsters and their parents or caretakers. They argue as well, that Innatists overstate the parity in linguistic competence between retarded children and children of normal intelligence.

Innatists point out that the 'modularity' of language processing is a powerful argument for the innateness of the language faculty. There is a large body of evidence, Innatists argue, for the claim that the processes that subserve the acquisition, understanding and production of language are quite distinct and independent of those that subserve general cognition and learning. That is to say, that language learning and language processing mechanisms and the knowledge they embody are domain specific-grammar and grammatical learning and utilization mechanisms are not used outside of language processing. They are informationally encapsulated-only, but linguistic information is relevant to language acquisition and processing. They are mandatory, but language learning and language processing are automatic. Moreover, language is subserved by specific dedicated neural structures, damage to which predictable and systematically impairs linguistic functioning. All of this suggests a specific 'mental organ', to use Chomsky's phrase, that has evolved in the human cognitive system specifically in order to make language possible. The specific structure is organ simultaneously constrains the range of possible human language s and guide the learning of a child's target language, later masking rapid on-line language processing possibly. The principles represented in this organ constitute the innate linguistic knowledge of the human being. Additional evidence for the early operation of such an innate language acquisition module is derived from the many infant studies that show that infants selectively attend to sound streams that are prosodically appropriate that have pauses at clausal boundaries, and that contain linguistically permissible phonological sequence.

It is fair to ask where we get the powerful inner code whose representational elements need only systematic construction to express, for example, the thought that cyclotrons are bigger than black holes. But on this matter, the language of thought theorist has little to say. All that 'concept' learning could be (assuming it is to be some kind of rational process and not due to mere physical maturation or a bump on the head). According to the language of thought theorist, is the trying out of combinations of existing representational elements to see if a given combination captures the sense (as evinced in its use) of some new concept. The consequence is that concept learning, conceived as the expansion of our representational resources, simply does not happen. What happens instead is that the work with a fixed, innate repertoire of elements whose combination and construction must express any content we can ever learn to understand.

Representationalism typifies the conforming generality for which of its accomplished manner that by and large induce the doctrine that the mind (or sometimes the brain) works on representations of the things and features of things that we perceive or thing about. In the philosophy of perception the view is especially associated with the French Cartesian philosopher Nicolas Malebranche (1638-1715) and the English philosopher John Locke (1632-1704), who, holding that the mind is the container for ideas, held that of our real ideas, some are adequate, and some are inadequate. Those that have inadequateness to those represented as archetypes that the mind supposes them taken from which it tends them to stand for, and to which it refers them. The problem in this account was mercilessly exposed by the French theologian and philosopher Antoine Arnauld (1216-94) and the French critic of Cartesianism Simon Foucher (1644-96), writing against Malebranche, and by the idealist George Berkeley, writing against Locke. The fundamental problem is that the mind is 'supposing' its ds to represent something else, but it has no access to this something else, with the exception by forming another idea. The difficulty is to understand how the mind ever escapes from the world of representations, or, acquire genuine content pointing beyond themselves in more recent philosophy, the analogy between the mind and a computer has suggested that the mind or brain manipulate signs and symbols, thought of as like the instructions in a machine's program of aspects of the world. The point is sometimes put by saying that the mind, and its theory, becomes a syntactic engine rather than a semantic engine. Representation is also attacked, at least as a central concept in understanding the 'pragmatists' who emphasize instead the activities surrounding a use of language than what they see as a mysterious link between mind and world.

Representations, along with mental states, especially beliefs and thought, are said to exhibit 'intentionality' in that they refer or to stand for something other than of what is the possibility of it being something else. The nature of this special property, however, has seemed puling. Not only is intentionality often assumed to be limited to human, and possibly a few other species, but the property itself appears to resist characterization in physicalist terms. The problem is most obvious in the case of 'arbitrary' signs, like words, where it is clear that there is no connection between the physical properties of a word and what it demotes, and, yet it remains for Iconic representation.

Early attempts tried to establish the link between sign and object via the mental states of the sign and symbols' user. A symbol # stands for 'S' if it triggers an-idea in 'S'. On one account, the reference of # is the idea itself. Open the major account, the denomination of # is whatever the -idea denotes. The first account is problematic in that it fails to explain the link between symbols and the world. The second is problematic in that it just shifts the puzzle inward. For example, if the word 'table' triggers the image ' ' or 'TABLE' what gives this mental picture or word any reference of all, let alone the denotation normally associated with the word 'table'?

An alternative to an adequate conception of mind and its relationship to matter should explain how it is possible of the world and in particular to themselves have a causal influence of the physical world. It is easy to think that this must be impossible: It takes a physical cause to have a physical effect. Yet, everyday experience and theory alike show that it is commonplace. Consciousness could hardly have a measure of the success of any theory of mind and body that it should enable us to avoid 'epiphenomenalism'. Mentalistic theories has been to adopt a behaviouristic analysis. Wherefore, this account # denotes for 'S' is explained along the lines of either (1) 'S' is disposed to behave to # as to :, or (2) 'S' is disposed to behave in ways appropriate to when presented #. Both versions prove faulty in that the very notions of the behaviour associated with or appropriate to are obscure. In addition, once seems to be no reasonable correlations between behaviour toward sign and behaviour toward their objects that is capable of accounting for the referential relations.

A currently influential attempt to 'naturalize' the representation relation takes its use from indices. The crucial link between sign and object is established by some causal connection between and #, whereby it is allowed, nonetheless, that such a causal relation is not sufficient for full-blown intention representation. An increase in temperature causes the mercury to raise the thermometer but the mercury level is not a representation for the thermometer. In order for # to represent to S's activities. The flunctuational economy of S's activity. The notion of 'function', in turn is yet to be spelled out along biological or other lines so as to remain within 'naturalistic' constraints as being natural. This approach runs into problems in specifying a suitable notion of 'function' and in accounting for the possibility of misrepresentation. Also, it is no obvious how to extend the analysis to encompass the semantical force of more abstract or theoretical symbols. These difficulties are further compounded when one takes into account the social factors that seem to play a role in determining the denotative properties of our symbols.

On that point, it remains the problems faced in providing a reductive naturalistic analysis of representation has led many to doubt that this task is achieved or necessary. Although a story can be told about some words or signs what were learned via association of other causal connections with their referents, there is no reason to believe ht the 'stand-for' relation, or semantic notions in general, can be reduced to or eliminated in favour of non-semantic terms.

Although linguistic and pictorial representations are undoubtedly the most prominent symbolic forms we employ, the range of representational systems human understands and regularly use is surprisingly large. Sculptures, maps, diagrams, graphs. Gestures, music nation, traffic signs, gauges, scale models, and tailor's swatches are but a few of the representational systems that play a role in communication, though, and the guidance of behaviour. Even, the importance and prevalence of our symbolic activities has been taken as a hallmark of human.

What is it that distinguishes items that serve as representations from other objects or events? And what distinguishes the various kinds of symbols from each other? As for the first question, there has been general agreement that the basic notion of a representation involves one thing's 'standing for', 'being about', referring to or denoting' something else. The major debates have been over the nature of this connection between a reorientation and that which it represents. As for the second question, perhaps, the most famous and extensive attempt to organize and differentiate among alternative forms of representation is found in the works of the American philosopher of science Charles Sanders Peirce (1839-1914) who graduated from Harvard in 1859, and apart from lecturing at John Hopkins university from 1879 to 1884, had almost no teaching, nonetheless, Peirce's theory of signs is complex, involving a number of concepts and distinctions that are no longer paid much heed. The aspects of his theory that remains influential and ie widely cited is his division of signs into Icons, Indices and Symbols. Icons are the designs that are said to be like or resemble the things they represent, e.g., portrait painting. Indices are signs that are connected in their objects by some causal dependency, e.g., smoke as a sign of fire. Symbols are those signs that are used and related to their object by virtue of use or associations: They a arbitrary labels, e.g., the word 'table'. This tripartite division among signs, or variants of this division, is routinely put forth to explain differences in the way representational systems are thought to establish their links to the world. Further, placing a representation in one of the three divisions has been used to account for the supposed differences between conventional and non-conventional representations, between representations that do and do not require learning to understand, and between representations, like language, that need to be read, and those which do not require interpretation. Some theorbists, moreover, have maintained that it is only the use of symbols that exhibits or indicates the presence of mind and mental states.

Over the years, this tripartite division of signs, although often challenged, has retained its influence. More recently, an alterative approach to representational systems (or as he calls them 'symbolic systems') has been put forth by the American philosopher Nelson Goodman (1906-98) whose classical problem of 'induction' is often phrased in terms of finding some reason to expect that nature is uniform, in Fact, Fiction, and Forecast (1954) Goodman showed that we need in addition some reason for preferring some uniformities to others, for without such a selection the uniformity of nature is vacuous, yet Goodman (1976) has proposed a set of syntactic and semantic features for categorizing representational systems. His theory provided for a finer discrimination among types of systems than a philosophy of science and language as partaken to and understood by the categorical elaborations as announced by Peirce. What also emerges clearly is that many rich and useful systems of representation lack a number of features taken to be essential to linguistic or sentential forms of representation, e.g., discrete alphabets and vocabularies, syntax, logical structure, inferences rules, compositional semantics and recursive e compounding devices.

As a consequence, although these representations can be appraised for accuracy or correctness. It does not seem possible to analyse such evaluative notion along the lines of standard truth theories, geared as they are to the structure found in sentential systems.

In light of this newer work, serious questions have been raised at the soundness of the tripartite division and about whether various of the psychological and philosophical claims concerning conventionality, learning, interpretation, and so forth, that have been based on this traditional analysis, can be sustained. It is of special significance e that Goodman has joined a number of theorists in rejecting accounts of Iconic representation in terms of resemblance. The rejection has been twofold, first, as Peirce himself recognized, resemblance is not sufficient to establish the appropriate referential relations. The numerous prints of lithograph do not represent one another. Any more than an identical twin represent his or her sibling. Something more than resemblance is needed to establish the connection between an Icon and picture and what it represents. Second, since Iconic representations lack as may properties as they share with their referents, sand certain non-Iconic symbol can be placed vin correspondences with their referents. It is difficult to provide a non-circular account of what the similarity distinguishes as Icons from other forms of representation. What is more, even if these two difficulties could be resolved, it would not show that the representational function of picture can be understood independently of an associated system of interpretations. The design,, may be a picture of a mountain of the economy in a foreign language. Or it may have no representational significance at all. Whether it is a representation and what kind of representation it uses, is relative to a system of interpretation.

If so, then, what is the explanatory role of providing reasons for our psychological states and intentional acts? Clearly part of this role comes from the justificatory nature of the reason-giving relation: 'Things are made intelligible by being revealed to be, or too approximate to being, as they rationally ought to be'. For some writers the justificatory and explanatory tasks of reason-giving simple coincide. The manifestation of rationality is seen as sufficient to explain states or acts quite independently of questions regarding causal origin. Within this model the greater the degree of rationality we can detect, the more intelligible the sequence will b e. where there is a breakdown in rationality, as in cases of weakness of will or self-deception, there is a corresponding breakdown in our ability to make the action/belief intelligible.

The equation of the justificatory and explanatory role of rationality links can be found within two quite distinct picture. One account views the attribute of rationality from a third-person perspective. Attributing intentional states to others, and by analogy to ourselves, is a matter of applying them of a certain pattern of interpretation. we ascribe that ever states enable us to make sense of their behaviour as conforming to a rational pattern. Such a mode of interpretation is commonly an ex post facto affair, although such a mode of interpretation can also aid prediction. Our interpretations are never definitive or closed. They are always open to revision and modification in the light of future behaviour. If so extreme a degree revision enables people as a whole to appear more rational. Where we fail to detect of seeing a system then we give up the project of seeing a system as rational and instead seek explanations of a mechanistic kind.

The other picture is resolutely firs-personal, linked to the claimed prospectively of rationalizing explanations we make an action, for example, intelligible by adopting the agent's perspective on it. Understanding is a reconstruction of actual or possible decision making. It is from such a first-personal perspective that goals are detected as desirable and the courses of action appropriated to the situation. The standpoint of an agent deciding how to act is not that of an observer predicting the next move. When I found something desirable and judge an act in an appropriate rule for achieving it, I conclude that a certain course of action should be taken. This is different from my reflecting on my past behaviour and concluding that I will do 'X' in the future.

For many writers, it is, nonetheless, the justificatory and explanatory role of reason cannot simply be equated. To do so fails to distinguish well-formed cases thereby I believe or act because of these reasons. I may have beliefs but your innocence would be deduced but nonetheless come to believe you are innocent because you have blue eyes. Yet, I may have intentional states that give altruistic reasons in the understanding for contributing to charity but, nonetheless, out of a desire to earn someone's good judgment. In both these cases. Even though my belief could be shown be rational in the light of other beliefs, and my action, of whether the forwarded belief become desirously actionable, that of these rationalizing links would form part of a valid explanation of the phenomena concerned. Moreover, cases inclined with an inclination toward submission. As I continue to smoke although I judge it would be better to abstain. This suggests, however, that the mere availability of reasoning cannot, least of mention., have the quality of being of itself an sufficiency to explain why it occurred.

If reason states can motivate, however, why (apart from confusing them with reasons proper) deny that they are causes? For one can say that they are not events, at least in the usual sense entailing change, as they are dispositional states (this contrasts them with occurrences, but not imply that they admit of dispositional analysis). It has also seemed to those who deny that reasons are causes that the former justify as well as explain the actions for which there are reasons, whereas the role of causes is at most to explain. As other claim is that the relation between reasons (and for reason states are often cited explicitly) and the actions they explain is non-contingent, whereas the relation causes to their effects is contingent. The 'logical connection argument' proceeds from this claim to the conclusion that reasons are not causes.

These arguments are inconclusive, first, even if causes are events, sustaining causation may explain, as where the [states of] standing of a broken table is explained by the (condition of) support of staked boards replacing its missing legs. Second, the 'because' in 'I sent it by express because I wanted to get it there in a day, so in some semi-causal explanation would at best be construed as only rationalizing, than justifying action. And third, if any non-contingent connection can be established between, say, my wanting something and the action it explains, there are close causal analogism such as the connection between brining a magnet to iron filings and their gravitating to it: This is, after all, a 'definitive' connection, expressing part of what it is to be magnetic, yet the magnet causes the fillings to move.

There I then, a clear distinction between reasons proper and causes, and even between reason states and event causes: But the distinction cannot be used to show that the relations between reasons and the actions they justify is in no way causal. Precisely parallel points hold in the epistemic domain (and indeed, for all similarly admit of justification, and explanation, by reasons). Suppose my reason for believing that you received it today is that I sent it by express yesterday. My reason, strictly speaking, is that I sent it by express yesterday: My reason state is my believing this. Arguably reason justifies the further proposition I believe for which it is my reason and my reason state-my evidence belief-both explains and justifies my belief that you received the letter today. I an say, that what justifies that belief is [in fact] that I sent the letter by express yesterday, but this statement expresses my believing that evidence proposition, and you received the letter is not justified, it is not justified by the mere truth of the proposition (and can be justified even if that proposition is false).

Similarly, there are, for belief for action, at least five main kinds of reason (1) normative reasons, reasons (objective grounds) there are to believe (say, to believe that there is a green-house-effect): (2) Person-relative normative reasons, reasons for [say] me to believe, (3) subjective reasons, reasons I have to believe (4) explanatory reasons, reasons why I believe, and (5) motivating reasons for which I believe. Tenets of (1) and (2) are propositions and thus, not serious candidates to be causal factors. The states corresponding to (3) may not be causal elements. Reasons why, tenet (4) are always (sustaining) explainers, though not necessarily even prima facie justifier, since a belief can be casually sustained by factors with no evidential value. Motivating reasons are both explanatory and possess whatever minimal prima facie justificatory power (if any) a reason must have to be a basis of belief.

Current discussion of the reasons-causes issue has shifted from the question whether reason state can causally explain to the perhaps, deeper questions whether they can justify without so explaining, and what kind of causal states with actions and beliefs they do explain. 'Reliabilist' tend to take as belief as justified by a reason only if it is held ast least in part for that reason, in a sense implying, but not entailed by, being causally based on that reason. 'Internalists' often deny this, as, perhaps, thinking we lack internal access to the relevant causal connections. But Internalists need internal access to what justified-say, the reason state-and not to the (perhaps quite complex) relations it bears the belief it justifies, by virtue for which it does so. Many questions also remain concerning the very nature of causation, reason-hood, explanation and justification.

Nevertheless, for most causal theorists, the radical separation of the causal and rationalizing role of reason-giving explanations is unsatisfactory. For such theorists, where we can legitimately point to an agent's reasons to explain a certain belief or action, then those features of the agent's intentional states that render the belief or action reasonable must be causally relevant in explaining how the agent came to believe or act in a way which they rationalize. One way of putting this requirement is that reason-giving states not only cause but also causally explain their explananda.

The explanans/explanandum are held of a wide currency of philosophical discoursing because it allows a certain succinctness which is unobtainable in ordinary English. Whether in science philosophy or in everyday life, one does often offers explanation s. the particular statement, laws, theories or facts that are used to explain something are collectively called the 'explanans', and the target of the explanans - the thing to be explained-is called the 'explanandum'. Thus, one might explain why ice forms on the surface of lakes (the explanandum) in terms of the special property of water to expand as it approaches freezing point together with the fact that materials less dense than liquid water float in it (the explanans). The terms come from two different Latin grammatical forms:

'Explanans' is the present participle of the verb which explain: And 'explanandum' is a direct object noun derived from the same verb.

The assimilation in the likeness as to examine side by side or point by point in order to bring such in comparison with an expressed or implied standard where comparative effects are both considered and equivalent resemblances bound to what merely happens to us, or to parts of us, actions are what we do. My moving my finger is an action to be distinguished from the mere motion of that finger. My snoring likewise, is not something I 'do' in the intended sense, though in another broader sense it is something I often 'do' while asleep.

The contrast has both metaphysical and moral import. With respect to my snoring, I am passive, and am not morally responsible, unless for example, I should have taken steps earlier to prevent my snoring. But in cases of genuine action, I am the cause of what happens, and I may properly be held responsible, unless I have an adequate excuse or justification. When I move my finger, I am the cause of the finger's motion. When I say 'Good morning' I am the cause of the sounding expression or utterance. True, the immediate causes are muscle contractions in the one case and lung, lip and tongue motions in the other. But this is compatible with me being the cause-perhaps, I cause these immediate causes, or, perhaps it just id the case that some events can have both an agent and other events as their cause.

All this is suggestive, but not really adequate. we do not understand the intended force of 'I am the cause' and more than we understand the intended force of 'Snoring is not something I do'. If I trip and fall in your flower garden, 'I am the cause' of any resulting damage, but neither the damage nor my fall is my action. In the considerations for which we approach to explaining what are actions, as contrasted with 'mere' doings, are. However, it will be convenient to say something about how they are to be individuated.

If I say 'Good morning' to you over the telephone, I have acted. But how many actions have O performed, and how are they related to one another and associated events? we may describe of what is done:

(1) Move my tongue and lips in certain ways, while exhaling.

(2) say 'Good morning'.

(3) Cause a certain sequence of modifications in the current flowing in your telephone.

(4) Say 'Good morning' to you.

(5) greet you.

The list-not exhaustive, by any -is of act types. I have performed an action of each relation holds. I greet you by saying 'Good morning' to you, but not the converse, and similarity for the others on the list. But are these five distinct actions I performed, one of each type, or are the five descriptions all of

581

a single action, which was of these five (and more) types. Both positions, and a variety of intermediate positions have been defended.

How many words are there within the sentence? : 'The cat is on the mat'? There are on course, at best two answers to this question, precisely because one can enumerate the word types, either for which there are five, or that which there are six. Moreover, depending on how one chooses to think of word types another answer is possible. Since the sentence contains definite articles, nouns, a preposition and a verb, there are four grammatical different types of word in the sentence.

The type/token distinction, understood as a distinction between sorts of things, particular, the identity theory asserts that mental states are physical states, and this raises the question whether the identity in question if of types or token'.

During the past two decades or so, the concept of supervenience has seen increasing service in philosophy of mind. The thesis that the mental is supervening on the physical-roughly, the claim that the mental character of a thing is wholly determined by its physical nature-has played a key role in the formulation of some influence on the mind-body problem. Much of our evidence for mind-body supervenience seems to consist in our knowledge of specific correlations between mental states and physical (in particular, neural) processes in human and other organisms. Such knowledge, although extensive and in some ways impressive, is still quite rudimentary and far from complete (what do we know, or can we expect to know about the exact neural substrate for, say, the sudden thought that you are late with your rent payment this month?) It may be that our willingness to accept mind-body supervenience, although based in part on specific psychological dependencies, has to be supported by a deeper metaphysical commitment to the primary of the physical: It may in fact be an expression of such a commitment.

However, there are kinds of mental state that raise special issues for mind-body supervenience. One such kind is 'wide content' states, i.e., contentual mental states that seem to be individuated essentially by reference to objects and events outside the subject, e.g., the notion of a concept, like the related notion of meaning. The word 'concept' itself is applied to a bewildering assortment of phenomena commonly thought to be constituents of thought. These include internal mental representations, images, words, stereotypes, senses, properties, reasoning and discrimination abilities, mathematical functions. Given the lack of anything like a settled theory in this area, it would be a mistake to fasten readily on any one of these phenomena as the unproblematic referent of the term. One does better to make a survey of the geography of the area and gain some idea of how these phenomena might fit together, leaving aside for the nonce just which of them deserve to be called 'concepts' as ordinarily understood.

Concepts are the constituents of such propositions, just as the words 'capitalist', 'exploit', and 'workers' are constituents of the sentence. However, there is a specific role that concepts are arguably intended to play that may serve a point of departure. Suppose one person thinks that capitalists exploit workers, and another that they do not. Call the thing that they disagree about 'a proposition', e.g., capitalists exploit workers. It is in some sense shared by them as the object of their disagreement, and it is expressed by the sentence that follows the verb 'thinks that' mental verbs that take such verbs of 'propositional attitude'. Nonetheless, these people could have these beliefs only if they had, inter alia, the concept's capitalist exploit. And workers.

Propositional attitudes, and thus concepts, are constitutive of the familiar form of explanation (so-called 'intentional explanation') by which we ordinarily explain the behaviour and stares of people, many animals and perhaps, some machines. The concept of intentionality was originally used by medieval scholastic philosophers. It was reintroduced into European philosophy b y the German philosopher and psychologist Franz Clemens Brentano (1838-1917) whose thesis proposed in Brentano's 'Psychology from an Empirical Standpoint'(1874) that it is the 'intentionality or directedness of mental states that marks off the mental from the physical.

Many mental states and activities exhibit the feature of intentionality, being directed at objects. Two related things are meant by this. First, when one desire or believes or hopes, one always desires or believes of hopes something. As, to assume that belief report (1) is true.

(1) That most Canadians believe that George Bush is a Republican.

Tenet (1) tells us that some subject 'Canadians' have a certain attitude, belief, to something, designated by the nominal phrase that George Bush is a Republican and identified by its content-sentence.

(2) Harper is a Conservative

Following Russell and contemporary usage that the object referred to by the that-clause is tenet (1) and expressed by tenet (2) a proposition. Notice, too, that this sentence might also serve as most Canadians' belief-text, a sentence whereby to express the belief that (1) reports to have. Such an utterance of (2) by itself would assert the truth of the proposition it expresses, but as part of (1) its role is not to rely on anything, but to identify what the subject believes. This same proposition can be the object of other attitude s of other people. However, in that most Canadians may regret that Bush is a Republican yet, Reagan may remember that he is. Bushians may doubt that he is.

Nevertheless, Brentano, 1960, we can focus on two puzzles about the structure of intentional states and activities, an area in which the philosophy of mind meets the philosophy of language, logic and ontology, least of mention, the term intentionality should not be confused with terms intention and intension. There is, nonetheless, an important connection between intention and intension and intentionality, for semantical systems, like extensional model theory, that are limited to extensions, cannot provide plausible accounts of the language of intentionality.

The attitudes are philosophically puzzling because it is not easy to see how the intentionality of the attitude fits with another conception of them, as local mental phenomena.

Beliefs, desires, hopes, and fears seem to be located in the heads or minds of the people that have them. Our attitudes are accessible to us through 'introspection'. As most Canadians belief that Bush to be a Republican just by examining the 'contents' of his own mind: He does not need to investigate the world around him. we think of attitudes as being caused at certain times by events that impinge on the subject's body, specially by perceptual events, such as reading a newspaper or seeing a picture of an ice-cream cone. In that, the psychological level of descriptions carries with it a mode of explanation which has no echo in 'physical theory'. we regard ourselves and of each other as 'rational purposive creatures, fitting our beliefs to the world as we inherently perceive it and seeking to obtain what we desire in the light of them'. Reason-giving explanations can be offered not only for action and beliefs, which will attain the most of all attentions, however, desires, intentions, hopes, dears, angers, and affections, and so forth. Indeed, their positioning within a network of rationalizing links is part of the individuating characteristics of this range of psychological states and the intentional acts they explain.

Meanwhile, these attitudes can in turn cause changes in other mental phenomena, and eventually in the observable behaviour of the subject. Seeing a picture of an ice cream cone leads to a desire for one, which leads me to forget the meeting I am supposed to attend and walk to the ice-cream pallor instead. All of this seems to require that attitudes be states and activities that are localized in the subject.

Nonetheless, the phenomena of intentionality call to mind that the attitudes are essentially relational in nature: They involve relations to the propositions at which they are directed and at the objects they are about. These objects may be quite remote from the minds of subjects. An attitude seems to be individuated by the agent, the type of attitude (belief, desire, and so on), and the proposition at which it is directed. It seems essential to the attitude reported by its believing that, for example, that it is directed toward the proposition that Bush is a Republican. And it seems essential to this proposition that it is about Bush. But how can a mental state or activity of a person essentially involve some other individuals? The problem is brought out by two classical problems such that are called 'no-reference' and 'co-reference'.

The classical solution to such problems is to suppose that intentional states are only indirectly related to concrete particulars, like George Bush, whose existence is contingent, and that can be thought about in a variety of ways. The attitudes directly involve abstract objects of some sort, whose existence is necessary, and whose nature the mind can directly grasp. These abstract objects provide concepts or ways of thinking of concrete particulars. That is to say, the involving characteristics of the different concepts, as, these, concepts corresponding to different inferential/practical roles in that different perceptions and memories give rise to these beliefs, and they serve as reasons for different actions. If we individuate propositions by concepts than individuals, the co-reference problem disappears.

The proposal has the bonus of also taking care of the no-reference problem. Some propositions will contain concepts that are not, in fact, of anything. These propositions can still be believed desired, and the like.

This basic idea has been worked out in different ways by a number of authors. The Austrian philosopher Ernst Mally thought that propositions involved abstract particulars that 'encoded' properties, like being the loser of the 1992 election, rather than concrete particulars, like Bush, who exemplified them. There are abstract particulars that encode clusters of properties that nothing exemplifies, and two abstract objects can encode different clusters of properties that are exemplified by a single thing. The German philosopher Gottlob Frége distinguished between the 'sense' and the 'reference' of expressions. The senses of George Bus hh and the person who will come in second in the election are different, even though the references are the same. Senses are grasped by the mind, are directly involved in propositions, and incorporate 'modes of presentation' of objects.

For most of the twentieth century, the most influential approach was that of the British philosopher Bertrand Russell. Russell (19051929) in effect recognized two kinds of propositions that assemble of a 'singular proposition' that consists separately in particular to properties in relation to that. An example is a proposition consisting of Bush and the properties of being a Republican. 'General propositions' involve only universals. The general proposition corresponding to someone is a Republican would be a complex consisting of the property of being a Republican and the higher-order property of being instantiated. The term 'singular proposition' and 'general proposition' are from Kaplan (1989.)

Historically, a great deal has been asked of concepts. As shareable constituents of the object of attitudes, they presumably figure in cognitive generalizations and explanations of animals' capacities and behaviour. They are also presumed to serve as the meaning of linguistic items, underwriting relations of translation, definition, synonymy, antinomy and semantic implication. Much work in the semantics of natural language takes itself to be addressing conceptual structure.

Concepts have also been thought to be the proper objects of philosophical analysis, the activity practised by Socrates and twentieth-century 'analytic' philosophers when they ask about the nature of justice, knowledge or piety, and expect to discover answers by of priori reflection alone.

The expectation that one sort of thing could serve all these tasks went hand in hand with what has come to be known for the 'Classical View' of concepts, according to which they have an 'analysis' consisting of conditions that are individually necessary and jointly sufficient for their satisfaction. Which are known to any competent user of them? The standard example is the especially simple one of the [bachelor], which seems to be identical to [eligible unmarried male]. A more interesting, but problematic one has been [knowledge], whose analysis was traditionally thought to be [justified true belief].

This Classical View seems to offer an illuminating answer to a certain form of metaphysical question: In virtue of what is something the kind of thing is, -, e.g., in virtue of what a bachelor is a bachelor? And it does so in a way that supports counterfactuals: It tells us what would satisfy the concept in situations other than the actual ones (although all actual bachelors might turn out to be freckled. It's possible that there might be unfreckled ones, since the analysis does not exclude that). The View also seems to offer an answer to an epistemological question of how people seem to know a priori (or, independently of experience) about the nature of many things, e.g., that bachelors are unmarried: It is constitutive of the competency (or, possession) conditions of a concept that they know its analysis, at least on reflection.

As it had been ascribed, in that Actions as Doings having Mentalistic Explanation: Coughing is sometimes like snoring and sometimes like saying 'Good morning'-that is, sometimes in mere doing and sometimes an action. And deliberate coughing can be explained by invoking an intention to cough, a desired to cough or some other 'pro-attitude' toward coughing, a reason for coughing or purpose in coughing or something similarly mental. Especially if we think of actions as 'outputs' of the mental machine'. The functionalist thinks of 'mental states' as events as causally mediating between a subject's sensory inputs and the subject ensuing behaviour. Functionalism itself is the stronger doctrine that 'what makes' a mental state the type of state that is pain, a smell of violets, a closed-minded belief that koalas are dangerous, is the functional relation it bears to the subject's perceptual stimuli, behaviour responses and other mental states.

Twentieth-century functionalism gained as credibility in an indirect way, by being perceived as affording the least objectionable solution to the mind-body problem.

Disaffected from Cartesian dualism and from the 'first-person' perspective of introspective psychology, the behaviourists had claimed that there is nothing to the mind but the subject's behaviour and dispositions to behave. To refute the view that a certain level of behavioural dispositions is necessary for a mental life, we need convincing cases of thinking stones, or utterly incurable paralytics or disembodied minds. But these alleged possibilities are to some merely that.

To rebuttal against the view that a certain level of behavioural dispositions is sufficient for a mental life, we need convincing cases rich behaviour with no accompanying mental states. The typical example is of a puppet controlled by radio-wave links, by other minds outside the puppet's hollow body. But one might wonder whether the dramatic devices are producing the anti-behaviorist intuition all by themselves. And how could the dramatic devices make a difference to the facts of the casse? If the puppeteers were replaced by a machine, not designed by anyone, yet storing a vast number of input-output conditionals, which was reduced in size and placed in the puppet's head, do we still have a compelling counterexample, to the behaviour-as-sufficient view? At least it is not so clear.

Such an example would work equally well against the anti-eliminativist version of which the view that mental states supervene on behavioural disposition. But supervening behaviourism could be refitted by something less ambitious. The 'X-worlders' of the American philosopher Hilary Putnam (1926-), who are in intense pain but do not betray this in their verbal or nonverbal behaviour, behaving just as pain-free human beings, would be the right sort of case. However, even if Putnam has produced a counterexample for pain-which the American philosopher of mind Daniel Clement Dennett (1942-), for one would doubtless deny-an 'X-worlder' narration to refute supervening behaviourism with respect to the attitudes or linguistic meaning will be less intuitively convincing. Behaviourist resistance is easier for the reason that having a belief or meaning a certain thing, lack distinctive phenomemologies.

There is a more sophisticated line of attack. As, the most influential American philosopher of the latter half of the 20th century philosopher Willard von Orman Quine (1908-2000) has remarked some have taken his thesis of the indeterminacy of translation as a reductio of his behaviourism. For this to be convincing, Quines argument for the indeterminacy thesis and to be persuasive in its own and that is a disputed matter.

If behaviourism is finally laid to rest to the satisfaction of most philosophers, it will probably not by counterexamples, or by a reductio from Quine's indeterminacy thesis. Rather, it will be because the behaviorists worries about other minds, and the public availability of meaning have been shown too groundless, or not to require behaviourism for their solution. But we can be sure that this happy day will take some time to arrive.

Quine became noted for his claim that the way one uses' language determines what kinds of things one is committed to saying exist. Moreover, the justification for speaking one way rather than another, just as the justification for adopting one conceptual system rather than another, was a thoroughly pragmatic one for Quine (see Pragmatism). He also became known for his criticism of the traditional distinction between synthetic statements (empirical, or factual, propositions) and analytic statements (necessarily true propositions). Quine made major contributions in set theory, a branch of mathematical logic concerned with the relationship between classes. His published works include Mathematical Logic (1940), From a Logical Point of View (1953), Word and Object (1960), Set Theory and Its Logic (1963), and: An Intermittently Philosophical Dictionary (1987). His autobiography, The Time of My Life, appeared in 1985.

Functionalism, and cognitive psychology considered as a complete theory of human thought, inherited some of the same difficulties that earlier beset Behaviouralism and identity theory. These remaining obstacles fall unto two main categories: Intentionality problems and Qualia problems.

Propositional attitudes such as beliefs and desires are directed upon states of affairs which may or may not actually obtain, e.g., that the Republican or let alone any in the Liberal party will win, and are about individuals who may or may not exist, e.g., King Arthur. Franz Brentano raised the question of how are purely physical entity or state could have the property of being 'directed upon' or about a nonexistent state of affairs or object: That is not the sort of feature that ordinary, purely physical objects can have.

The standard functionalist reply is that propositional attitudes have Brentano's feature because the internal physical states and events that realize them 'represent' actual or possible states of affairs. What they represent is determined at least in part, by their functional roles: Is that, mental events, states or processes with content involve reference to objects, properties or relations, such as a mental state with content can fail to refer, but there always exists a specific condition for a state with content to refer to certain things? As when the state gas a correctness or fulfilment condition, its correctness is determined by whether its referents have the properties the content specifies for them.

What is it that distinguishes items that serve as representations from other objects or events? And what distinguishes the various kinds of symbols from each other? Firstly, there has been general agreement that the basic notion of a representation involves one thing's 'standing for', 'being about', 'pertain to', 'referring or denoting of something else entirely'. The major debates here have been over the nature of this connection between a representation and that which it represents. As to the second, perhaps the most famous and extensive attempt to organize and differentiated among alternative forms of the representation is found in the works of C.S. Peirce (1931-1935). Peirce's theory of sign in complex, involving a number of concepts and distinctions that are no longer paid much heed. The aspect of his theory that remains influential and is widely cited, is his division of signs into Icons, Indices and Symbols. Icons are signs that are said to be like or resemble the things they represent, e.g., portrait paintings. Indices are signs that are connected to their objects by some causal dependency, e.g., smoke as a sign of fire. Symbols are those signs that are related to their object by virtue of use or association: They are arbitrary labels, e.g., the word 'table'. The divisions among signs, or variants of this division, is routinely put forth to explain differences in the way representational systems are thought to establish their links to the world. Further, placing a representation in one of the three divisions has been used to account for the supposed differences between conventional and non-conventional representation, between representation that do and do not require learning to understand, and between representations, like language, that need to be read, and those which do not require interpretation. Some theorists, moreover, have maintained that it is only the use of Symbols that exhibits or indicate s the presence of mind and mental states.

Representations, along with mental states, especially beliefs and thoughts, are said to exhibit 'intentionality' in that they refer or to stand for something else. The nature of this special property, however, has seemed puzzling. Not only is intentionality often assumed to be limited to human, and possibly a few other species, but the property itself appears to resist characterization in physicalist terms. The problem is most obvious in the case of 'arbitrary' signs, like words. Where it is clear that there is no connection between the physical properties of as word and what it denotes, that, wherein, the problem also remains for Iconic representation.

In at least, there are two difficulties. One is that of saying exactly 'how' a physical item's representational content is determined, in not by the virtue of what does a neurophysiological state represent precisely that the available candidate will win? An answer to that general question is what the American philosopher of mind, Alan Jerry Fodor (1935-) has called a 'psychosemantics', and several attempts have been made. Taking the analogy between thought and computation seriously, Fodor believes that mental representations should be conceived as individual states with their own identities and structures, like formulae transformed by processes of computations or thought. His views are frequently contrasted with those of 'holiest' such as the American philosopher Herbert Donald Davidson (1917-2003), whose constructions within a generally 'holistic' theory of knowledge and meaning. Radical interpreter can tell when a subject holds a sentence true, and using the principle of 'clarity' ends up making an assignment of truth condition is a defender of radical translation and the inscrutability of reference', Holist approach has seemed too many has seemed too many to offer some hope of identifying meaning as a respectable notion, eve n within a broadly 'extensional' approach to language. Instructionalists about mental ascription, such as Clement Daniel Dennett (19420) who posits the particularity that Dennett has also been a major force in illuminating how the philosophy of mind needs to be informed by work in surrounding sciences.

In giving an account of what someone believes, does essential reference have to be made to how things are in the environment of the believer? And, if so, exactly what reflation does the environment have to the belief? These questions involve taking sides in the externalism and internalism debate. To a first approximation, the externalist holds that one's propositional attitude cannot be characterized without reference to the disposition of object and properties in the world-the environment-in which in is simulated. The internalist thinks that propositional attitudes (especially belief) must be characterizable without such reference. The reason that this is only a first approximation of the contrast is that there can be different sorts of externalism. Thus, one sort of externalist might insist that you could not have, say, a belief that grass is green unless it could be shown that there was some relation between you, the believer, and grass. Had you never come across the plant which makes up lawns and meadows, beliefs about grass would not be available to you. However, this does not mean that you have to be in the presence of grass in order to entertain a belief about it, nor does it even mean that there was necessarily a time when you were in its presence. For example, it might have been the case that, though you have never seen grass, it has been described to you. Or, at the extreme, perhaps no longer exists anywhere in the environment, but your antecedent's contact with it left some sort of genetic trace in you, and the trace is sufficient to give rise to a mental state that could be characterized as about grass.

At the more specific level that has been the focus in recent years: What do thoughts have in common in virtue of which they are thoughts? What is, what makes a thought a thought? What makes a pain a pain? Cartesian dualism said the ultimate nature of the mental was to be found in a special mental substance. Behaviourism identified mental states with behavioural disposition: Physicalism in its most influential version identifies mental states with brain states. One could imagine that the individual states that occupy the relevant causal roles turn out not to be bodily stares: For example, they might instead be states of an Cartesian unextended substance. But its overwhelming likely that the states that do occupy those causal roles are all tokens of bodily-state types. However, a problem does seem to arise about properties of mental states. Suppose 'pain' is identical with a certain firing of c-fibres. Although a particular pain is the very same state as neural firing, we identify that state in two different ways: As a pain and as neural firing. The state will therefore have certain properties in virtue of which we identify it as a pain and others in virtue of which we identify it as a pain will be mental properties, whereas those in virtue of which we identify it as neural firing will be physical properties. This has seemed to many to lead to a kind of dualism at the level of the properties of mental states. Even if we reject a dualism of substances and take people simply to be physical organisms, those organisms still have both mental and physical states. Similarly, even if we identify those mental states with certain physical states, those states will nonetheless have both mental and physical properties. So, disallowing dualism with respect to substances and their stares simply leads to its reappearance at the level of the properties of those states.

The problem concerning mental properties is widely thought to be most pressing for sensations, since the painful quality of pains and the red quality of visual sensation seem to be irretrievably physical. So, even if mental states are all identical with physical states, these states appear to have properties that are not physical. And if mental states do actually have nonphysical properties, the identity of mental with physical states would not support a thoroughgoing mind-body physicalism.

A more sophisticated reply to the difficulty about mental properties is due independently to D.M. Armstrong (1968) and David Lewis (1972), who argue that for a state to be a particular sort of intentional state or sensation is for that state to bear characteristic causal relations to other particular occurrences. The properties in virtue of which we identify states as thoughts or sensations will still be neutral as between being mental and physical, since anything can bear a causal relation to anything else. But causal connections have a better chance than similarity in some unspecified respect t of capturing the distinguishing properties of sensation and thoughts.

It should be mentioned that the properties can be more complex and complicating than the above allows. For instance, in the sentence, 'John is married to Mary', we are attributing to John the property of being married. And, unlike the property of being bald, this property of John is essentially relational. Moreover, it is commonly said that 'is married to' expresses a relation, than a property, though the terminology is not fixed, but, some authors speak of relations as different from properties in being more complex but like them in being nonlinguistic, though it is more common to treat relations as a subclass of properties.

The Classical view, meanwhile, has always had to face the difficulty of 'primitive' concepts: It's all well and good to claim that competence consists in some sort of mastery of a definition, but what about the primitive concepts in which a process of definition must ultimately end? There the British Empiricism of the seventeenth century began to offer a solution: All the primitives were sensory. Indeed, they expanded the Classical view to include the claim, now often taken uncritically for granted in discussions of that view, that all concepts are 'derived from experience': 'Every idea is derived from a corresponding impression'. In the work of John Locke (1682-1704), George Berkeley (1685-1753) and David Hume (1711-76) as it was thought to mean that concepts were somehow 'composed' of introspectible mental items-images -, 'impressions'-that were ultimately decomposable into basic sensory parts. Thuds, Hume analyzed the concept of [material objects] as involving certain regularities in our sensory experience, and [cause] as involving conjunction.

Berkeley noticed a problem with this approach that every generation has had to rediscover: If a concept is a sensory impression, like an image, then how does one distinguish a general concept [triangle] from a more particular one-say, [isosceles triangle]-that would serve in imaging the general one. More recent, Wittgenstein (1953) called attention to the multiple ambiguity of images. And, in any case, images seem quite hopeless for capturing the concepts associated with logical terms (what is the image for negation or possibility?) Whatever the role of such representation, full conceptual competence must involve something more.

Indeed, in addition to images and impressions and other sensory items, a full account of concepts needs to consider issues of logical structure. This is precisely what 'logical postivists' did, focussing on logically structured sentences instead of sensations and images, transforming the empiricalist claim into the famous' Verifiability Theory of Meaning': The meaning of a sentence is the by which it is confirmed or refuted. Ultimately by sensory experience, the meaning or concept associated with a predicate is the by which people confirm or refute whether something satisfies it.

This once-popular position has come under much attack in philosophy in the last fifty years. In the first place, few, if any, successful 'reductions' of ordinary concepts like, [material objects], [cause] to purely sensory concepts have ever been achieved, as Jules Alfred Ayer (1910-89) proved to be one of the most important modern epistemologists, his first and most famous book, 'Language, Truth and Logic', to the extent that epistemology is concerned with the a priori justification of our ordinary or scientific beliefs, since the validity of such beliefs 'is an empirical matter, which cannot be settled by such. However, he does take positions which have been bearing on epistemology. For example, he is a phenomenalists, believing that material objects are logical constructions out of actual and possible sense-experience, and an anti-foundationalism, at least in one sense, denying that there is a bedrock level of indubitable propositions on which empirical knowledge can be based. As regards the main specifically epistemological problem he addressed, the problem of our knowledge of other minds, he is essentially behaviouristic, since the verification principle pronounces that the hypothesis of the occurrences intrinsically inaccessible experience is unintelligible.

Although his views were later modified, he early maintained that all meaningful statements are either logical or empirical. According to his principle of verification, a statement is considered empirical only if some sensory observation is relevant to determining its truth or falseness. Sentences that neither are logical nor empirical-including traditional religious, metaphysical, and ethical sentences-are judged nonsensical. Other works of Ayer include The Problem of Knowledge (1956), the Gifford Lectures of 1972-73 published as The Central Questions of Philosophy (1973), and Part of My Life: The Memoirs of a Philosopher (1977).

Ayer's main contribution to epistemology are in his book, 'The Problem of Knowledge' which he himself regarded as superior to 'Language, Truth and Logic' (Ayer 1985), soon there after Ayer develops a fallibilist type of Foundationalism, according to which processes of justification or verification terminate in someone's having an experience, but there is no class of infallible statements based on such experiences. Consequently, in making such statements based on experience, even simple reports of observation we 'make what appears to be a special sort of advance beyond our data' (1956). And it is the resulting gap which the sceptic exploits. Ayer describes four possible responses to the sceptic: Na ve realism, according to which materia l objects are directly given in perception, so that there is no advance beyond the data: Reductionism, according to which physical objects are logically constructed out of the contents of our sense-experiences, so that again there is no real advance beyond the data: A position according to which there is an advance, but it can be supported by the canons of valid inductive reasoning and lastly a position called 'descriptive analysis', according to which 'we can give an account of the procedures that we actually follow . . . but there [cannot] be a proof that what we take to be good evidence really is so'.

Ayer's reason why our sense-experiences afford us grounds for believing in the existence of physical objects is simply that sentence which are taken as referring to physical objects are used in such a way that our having the appropriate experiences counts in favour of their truths. In other words, having such experiences is exactly what justification of or ordinary beliefs about the nature of the world 'consists in'. This suggestion is, therefore, that the sceptic is making some kind of mistake or indulging in some sort of incoherence in supposing that our experience may not rationally justify our commonsense picture of what the world is like. Again, this, however, is the familiar fact that th sceptic's undermining hypotheses seem perfectly intelligible and even epistemically possible. Ayer's response seems weak relative to the power of the sceptical puzzles.

The concept of 'the given' refers to the immediate apprehension of the contents of sense experience, expressed in the first person, present tense reports of appearances. Apprehension of the given is seen as immediate both in a casual sense, since it lacks the usual causal chain involved in perceiving real qualities of physical objects, and in an epistemic sense, since judgements expressing it are justified independently of all other beliefs and evidence. Some proponents of the idea of the given maintain that its apprehension is absolutely certain: Infallible, incorrigible and indubitable. It has been claimed also that a subject is omniscient with regard to the given: If a property appears, then the subject knows this.

The doctrine dates back at least to Descartes, who argued in Meditation II that it was beyond all possible doubt and error that he seemed to see light, hear noise, and so forth. The empiricist added the claim that the mind is passive in receiving sense impressions, so that there is no subjective contamination or distortion here (even though the states apprehended are mental). The idea was taken up in twentieth-century epistemology by C.I. Lewis and A.J. Ayer. Among others, who appealed to the given as the foundation for all empirical knowledge. Nonetheless, empiricism, like any philosophical movement, is often challenged to show how its claims about the structure of knowledge and meaning can themselves be intelligible and known within the constraints it accepts, since beliefs expressing only the given were held to be certain and justified in themselves, they could serve as solid foundations.

The second argument for the need for foundations is sound. It appeals to the possibility of incompatible but fully coherent systems of belief, only one of which could be completely true. In light of this possibility, coherence cannot suffice for complete justification, as coherence has the power to produce justification, while according to a negative coherence theory, coherence has only the power to nullify justification. However, by contrast, justification is solely a matter of how a belief coheres with a system of beliefs. Nonetheless, another distinction that cuts across the distinction between weak and strong coherence theories of justification. It is the distinction between positive and negative coherence theory tells us that if a belief coheres with a background system of belief, then the belief is justified.

Coherence theories of justification have a common feature, namely, that they are what are called 'internalistic theories of justification' they are theories affirming that coherence is a matter of internal relations between beliefs and justification is a matter of coherence. If, then, justification is solely a matter of internal relations between beliefs, we are left with the possibility that the internal relations might fail to correspond with any external reality. How, one might object, can a completely internal subjective notion of justification bridge the gap between mere true belief, which might be no more than a lucky guess, and knowledge, which must be grounded in some connection between internal subjective condition and external objective realities?

The answer is that it cannot and that something more than justified true belief is required for knowledge. This result has, however, been established quite apart from considerations of coherence theories of justification. What is required may be put by saying that the justification one must be undefeated by errors in the background system of belief. A justification is undefeated by error in the background system of belied would sustain the justification of the belief on the basis of the corrected system. So knowledge, on this sort of positive coherence theory, is true belief that coheres with the background belief system and corrected versions of that system. In short, knowledge is true belief plus justification resulting from coherence and undefeated by error.

Without some independent indication that some of the beliefs within a coherent system are true, coherence in itself is no indication of truth. Fairy stories can cohere. But our criteria for justification must indicate to us the probable truth of our beliefs. Hence, within any system of beliefs there must be some privileged class of beliefs which others must cohere to be justified. In the case of empirical knowledge, such privileged beliefs must represent the point of contact between subject and the world: They must originate in perception. When challenged, however, we justify our ordinary perceptual beliefs about physical properties by appeal to beliefs about appearances.

Nonetheless, it seems more suitable as foundations since there is no class of more certain perceptual beliefs to which we appeal for their justification. The argument that foundations must be certain was offered by the American philosopher David Lewis (1941-2002). He held that no proposition can be probable unless some are certain. If the probability of all propositions or beliefs were relative to evidence expressed in others, and if these relations were linear, then any regress would apparently have to terminate in propositions or beliefs that are certain. But Lewis shows neither that such relations must be linear nor that regresses cannot terminate in beliefs that are merely probable or justified in themselves without being certain or infallible.

Arguments against the idea of the given originate with the German philosopher and founder of critical philosophy. Immanuel Kant (1724-1804), whereby the intellectual landscape in which Kant began his career was largely set by the German philosopher, mathematician and polymath of Gottfried Wilhelm Leibniz (1646-1716), filtered through the principal follower and interpreter of Leibniz, Christian Wolff, who was primarily a mathematician but renowned as a systematic philosopher. Kant, who argues in Book I to the Transcendental Analysis that percepts without concepts do not yet constitute any form of knowing. Being non-epistemic, they presumably cannot serve as epistemic foundations. Once we recognize that we must apply concepts of properties to appearances and formulate beliefs utilizing those concepts before the appearances can play any epistemic role. It becomes more plausible that such beliefs are fallible. The argument was developed in this century by Sellars (1912-89), whose work revolved around the difficulties of combining the scientific image of people and their world, with the manifest image, or natural conception of ourselves as acquainted with intentions, meaning, colours, and other definitive aspects by his most influential paper 'Empiricism and the Philosophy of Mind' (1956) in this and many other of his papers, Sellars explored the nature of thought and experience. According to Sellars (1963), the idea of the given involves a confusion between sensing particular (having sense impression) which is non-epistemic, and having non-inferential knowledge of propositions referring to appearances be necessary for acquiring perceptual knowledge, but it is itself a primitive kind of knowing. Its being non-epistemic renders it immune from error, also, unsuitable for epistemological foundations. The apparentness to the non-inferential perceptual knowledge, is fallible, requiring concepts acquired through trained responses to public physical objects.

The contention that even reports of appearances are fallible can be supported from several directions. First, it seems doubtful that we can look beyond our beliefs to compare them with an8 unconceptualized reality, whether mental of physical. Second, to judge that anything, including an appearance, is 'F', we must remember which property 'F' is, and memory is admitted by all to be fallible. Our ascribing 'F' is normally not explicitly comparative, but its correctness requires memory, nevertheless, at least if we intend to ascribe a reinstantiable property. we must apply the concept of 'F' consistently, and it seems always at least logically possible to apply it inconsistently. If that be, it is not possible, if, for example, I intend in tendering to an appearance e merely to pick out demonstratively whatever property appears, then, I seem not to be expressing a genuine belief. My apprehension of the appearance will not justify any other beliefs. Once more it will be unsuitable as an epistemological foundation. This, nonetheless, nondifferential perceptual knowledge, is fallible, requiring concepts acquiring through trained responses to public physical objects.

Ayer (1950) sought to distinguish propositions expressing the given not by their infallibility, but by the alleged fact that grasping their meaning suffices for knowing their truth. However, this will be so only if the purely demonstratives meaning, and so only if the propositions fail to express beliefs that could ground others. If in uses genuine predicates, for example: C as applied to tones, then one may grasp their meaning and yet be unsure in their application to appearances. Limiting claims of error in claims eliminates one major source of error in claims about physical objects-appearances cannot appear other than they are. Ayer's requirement of grasping meaning eliminates a second source of error, conceptual confusion. But a third major source, misclassification, is genuine and can obtain in this limited domain, even when Ayer 's requirement is satisfied.

Any proponent to the given faces the dilemma that if in terms used in statements expressing its apprehension are purely demonstrative, then such statements, assuming they are statements, are certain, but fail to express beliefs that could serve as foundations for knowledge. If what is expressed is not awareness of genuine properties, then awareness does not justify its subject in believing anything else. However, if statements about what appears use genuine predicates that apply to reinstantiable properties, then beliefs expressed cannot be infallible or knowledge. Coherentists would add that such genuine belief's stand in need of justification themselves and so cannot be foundations.

Contemporary foundationalist deny the coherent's claim while eschewing the claim that foundations, in the form of reports about appearances, are infallible. They seek alternatives to the given as foundations. Although arguments against infallibility are strong, other objections to the idea of foundations are not. That concepts of objective properties are learned prior to concepts of appearances, for example, implies neither that claims about objective properties, nor that the latter are prior in chains of justification. That there can be no knowledge prior to the acquisition and consistent application of concepts allows for propositions whose truth requires only consistent application of concepts, and this may be so for some claims about appearances.

Coherentist will claim that a subject requires evidence that he apply concepts consistently to distinguish red from other colours that appear. Beliefs about red appearances could not then be justified independently of other beliefs expressing that evidence. Save that to part of the doctrine of the given that holds beliefs about appearances to be self-justified, we require an account of how such justification is possible, how some beliefs about appearances can be justified without appeal to evidence. Some Foundationalist's simply assert such warrant as derived from experience but, unlike, appeals to certainty by proponents of the given, this assertion seem ad hoc.

A better strategy is to tie an account of self-justification to a broader exposition of epistemic warrant. On such accounts sees justification as a kind of inference to the best explanation. A belief is shown to be justified if its truth is shown to be part of the best explanation for why it is held. A belief is self-justified if the best explanation for it is its truth alone. The best explanation for the belief that I am appeared to redly may be that I am. Such accounts seek ground knowledge in perceptual experience without appealing to an infallible given, now universally dismissed.

Nonetheless, it goes without saying, that many problems concerning scientific change have been clarified, and many new answers suggested. Nevertheless, concepts central to it, like 'paradigm'. 'core', 'problem', 'constraint', 'verisimilitude', many devastating criticisms of the doctrine based on them have been answered satisfactorily.

Problems centrally important for the analysis of scientific change have been neglected. There are, for instance, lingering echoes of logical empiricism in claims that the methods and goals of science are unchanging, and thus are independent of scientific change itself, or that if they do change, they do so for reasons independent of those involved in substantive scientific change itself. By their very nature, such approaches fail to address the changes that actually occur in science. For example, even supposing that science ultimately seeks the general and unaltered goal of 'truth' or 'verisimilitude', that injunction itself gives no guidance as to what scientists should seek or how they should go about seeking it. More specific scientific goals do provide guidance, and, as the traditions from mechanistic to gauge-theoretic goals illustrates, those goals are often altered in light of discoveries about what is achievable, or about what kinds of theories are promising. A theory of scientific change should account for these kinds of goal changes, and for how, once accepted, they alter the rest of the patterns of scientific reasoning and change, including ways in which more general goals and methods may be preconceived.

To declare scientific changes to be consequences of 'observation' or 'experimental evidence' is again to overstress the superficially unchanging aspects of science. we must ask how what counts as observation, experiments, and evidence themselves alter in the light of newly accepted scientific beliefs. Likewise, it is now clear that scientific change cannot be understood in terms of dogmatically embraced holistic cores: The factors guiding scientific change are by no the monolithic structure which they have been portrayed as being. Some writers prefer to speak of 'background knowledge' (or 'information') as shaping scientific change, the suggestion being that there are a variety of ways in which a variety of prior ideas influence scientific research in a variety of circumstances. But it is essential that any such complexity of influences be fully detailed, not left, as by the philosopher of science Raimund Karl Popper (1902-1994), with cursory treatment of a few functions selected to bolster a prior theory (in this case, falsification). Similarly, focus on 'constraints' can mislead, suggesting too negative a concept to do justice to the positive roles of the information utilized. Insofar as constraints are scientific and not trans-scientific, they are usually 'functions', not 'types' of scientific propositions.

Traditionally, philosophy has concerned itself with relations between propositions which are specifically relevant to one another in form or content. So viewed, a philosophical explanation of scientific change should appeal to factors which are clearly more scientifically relevant in their content to the specific directions of new scientific research and conclusions than are social factors whose overt relevance lies elsewhere. Nonetheless, in recent years many writers, especially in the 'strong programme' practices must be assimilated to social influences.

Such claims are excessive. Despite allegations that even what counted as evidence is a matter of mere negotiated agreement, many consider that the last word has not been said on the idea that there is in some deeply important sense of a 'given' to experience in terms with which we can, at least partially, judge theories ('background information') which can help guide those and other judgements. Even if ewe could, no information to account for what science should and can be, and certainly not for what it is often in human practice, neither should we take the criticism of it for granted, accepting that scientific change is explainable only by appeal to external factors.

Equally, we cannot accept too readily the assumption (another logical empiricist legacy) that our task is to explain science and its evolution by appeal to meta-scientific rules or goals, or metaphysical principles, arrived at in the light of purely philosophical analysis, and altered (if at all) by factors independent of substantive science. For such trans-scientific analysis, even while claiming to explain 'what science is', do so in terms 'external' to the processes by which science actually changes.

Externalist claims are premature by enough is yet understood about the roles of indisputably scientific considerations in shaping scientific change, including changes of methods and goals. Even if we ultimately cannot accept the traditional 'internalist' approach to philosophy of science, as philosophers concerned with the form and content of reasoning we must determine accurately how far it can be carried. For that task. Historical and contemporary case studies are necessary but insufficient: Too often the positive implications of such studies are left unclear, and their too hasty assumption is often that whatever lessons are generated therefrom apply equally to later science. Larger lessons need to be extracted from concrete studies. Further, such lessons must, there possible, be given a systematic account, integrating the revealed patterns of scientific reasoning and the ways they are altered into a coherent interpretation of the knowledge-seeking enterprise-a theory of scientific change. Whether such efforts are successful or not, or through understanding our failure to do so, that it will be possible to assess precisely the extent to which trans-scientific factors (meta-scientific, social, or otherwise) must be included in accounts of scientific change.

Much discussion of scientific change on or upon the distinction between contexts of discovery and justification that is to say about discovery that there is usually thought to be no authoritative confirmation theory, telling how bodies of evidence support, a hypothesis instead science proceeds by a 'hypothetico-deductive method' or 'method of conjectures and refutations'. By contrast, early inductivists held that (1) science e begins with data collections (2) rules of inference are applied to the data to obtain a theoretical conclusion, or at least, to eliminate alternatives, and (3) that conclusion is established with high confidence or even proved conclusively by the rules. Rules of inductive reasoning were proposed by the English diplomat and philosopher Francis Bacon (1561-1626) and by the British mathematician and physicists and principal source of the classical scientific view of the world, Sir Isaac Newton (1642-1727) in th e second edition of the Principia ('Rules of Reasoning in Philosophy'). Such procedures were allegedly applied in Newton's 'Opticks' and in many eighteenth-century experimental studies of heat, light, electricity, and chemistry.

According to Laudan (1981), two gradual realizations led to rejection of this conception of scientific method: First, that inferences from facts to generalizations are not established with certain, hence sectists were more willing to consider hypotheses with little prior empirical grounding, Secondly, that explanatory concepts often go beyond sense experience, and that such trans-empirical concepts as 'atom' and 'field' can be introduced in the formulation of such hypothesis, thus, as the middle of the eighteenth century, the inductive conception began to be replaced by the middle of hypothesis, or hypothetico-deductive method. On the view, the other of events in science is seen as, first, introduction of a hypothesis and second, testing of observational production of that hypothesis against observational and experimental results.

Twentieth-century relativity and quantum mechanics alerted scientists even more to the potential depths of departures from common sense and earlier scientific ideas, e.g., quantum theory. Their attention was called from scientific change and direct toward an analysis of temporal 'formal' characteristics of science: The dynamical character of science, emphasized by physics, was lost in a quest for unchanging characteristics deffinitary of science and its major components, i.e., 'content' of thought, the 'meanings' of fundamental 'meta-scientific' concepts and method-deductive conception of method, endorsed by logical empiricist, was likewise construed in these terms: 'Discovery', the introduction of new ideas, was grist for historian, psychologists or sociologists, whereas the 'justification' of scientific ideas was the application of logic and thus, the proper object of philosophy of science.

The fundamental tenet of logical empiricism is that the warrant for all scientific knowledge rests on or upon empirical evidence I conjunction with logic, where logic is taken to include induction or confirmation, as well as mathematics and formal logic. In the eighteenth century the work of the empiricist John Locke (1632-1704) had important implications for other social sciences. The rejection of innate ideas in book I of the Essay encouraged an emphasis on the empirical study of human societies, to discover just what explained their variety, and this toward the establishment of the science of social anthropology.

Induction (logic), in logic, is the process of drawing a conclusion about an object or event that has yet to be observed or occur, based on previous observations of similar objects or events. For example, after observing year after year that a certain kind of weed invades our yard in autumn, we may conclude that next autumn our yard will again be invaded by the weed; or having tested a large sample of coffee makers, only to find that each of them has a faulty fuse, we conclude that all the coffee makers in the batch are defective. In these cases we infer, or reach a conclusion based on observations. The observations or speculative assumptions are to assert for which we base the inference, or the alternate appearances of the weed, or the sample of coffee makers with faulty fuses-form the premises or assumptions.

In an inductive inference, the premises provide evidence or support for the conclusion; this support can vary in strength. The argument's strength depends on how likely it is that the conclusion will be true, assuming all of the premises to be true. If assuming the premises to be true makes it highly probable that the conclusion also would be true, the argument is inductively strong. If, however, the supposition that all the premises are true only slightly increases the probability that the conclusion will be true, the argument is inductively weak.

The truth or falsity of the premises or the conclusion is not at issue. Strength instead depends on whether, and how much, the likelihood of the conclusion's being true would increase if the premises were true. So, in induction, as in deduction, the emphasis is on the form of support that the premises provide to the conclusion. However, induction differs from deduction in a crucial aspect. In deduction, for an argument to be correct, if the premises were true, the conclusion would have to

597

be true as well. In induction, however, even when an argument is inductively strong, the possibility remains that the premises are true and the conclusion false. To return to our examples, although it is true that this weed has invaded our yard every year, it remains possible that the weed could die and never reappear. Likewise, it is true that all of the coffee makers tested had faulty fuses, but it is possible that the remainder of the coffee makers in the batch is not defective. Yet it is still correct, from an inductive point of view, to infer that the weed will return, and that the remainder of the coffee makers has faulty fuses.

Thus, strictly speaking, all inductive inferences are deductively invalid. Yet induction is not worthless; in both everyday reasoning and scientific reasoning regarding matters of fact - for instance in trying to establish general empirical laws - induction plays a central role. In an inductive inference, for example, we draw conclusions about an entire group of things, or a population, based on data about a sample of that group or population; or we predict the occurrence of a future event because of observations of similar past events; or we attribute a property to a non-observed thing as all observed things of the same kind have that property; or we draw conclusions about causes of an illness based on observations of symptoms. Inductive inference is used in most fields, including education, psychology, physics, chemistry, biology, and sociology. Consequently, because the role of induction is so central in our processes of reasoning, the study of inductive inference is one major area of concern to create computer models of human reasoning in Artificial Intelligence.

The development of inductive logic owes a great deal to 19th - century British philosopher John Stuart Mill, who studied different methods of reasoning and experimental inquiry in his work 'A System of Logic"(1843), by which Mill was chiefly interested in studying and classifying the different types of reasoning in which we start with observations of events and go on to infer the causes of those events. In, 'A Treatise on Induction and Probability' (1960), 20th - century Finnish philosopher Georg Henrik von Wright expounded the theoretical foundations of Mill's methods of inquiry.

Philosophers have struggled with the question of what justification we have to take for granted induction's common assumptions: that the future will follow the same patterns as the past; that a whole population will behave roughly like a randomly chosen sample; that the laws of nature governing causes and effects are uniform; or that we can presume that several observed objects give us grounds to attribute something to another object we have not yet observed. In short, what is the justification for induction itself? This question of justification, known as the problem of induction, was first raised by 18th - century Scottish philosopher David Hume in his An Enquiry Concerning Human Understanding (1748). While it is tempting to try to justify induction by pointing out that inductive reasoning is commonly used in both everyday life and science, and its conclusions are, largely, been corrected, this justification is itself an induction and therefore it raises the same problem: Nothing guarantees that simply because induction has worked in the past it will continue to work in the future. The problem of induction raises important questions for the philosopher and logician whose concern it is to provide a basis of assessment of the correctness and the value of methods of reasoning.

In the eighteenth century, Lock's empiricism and the science of Newton were, with reason, combined in people's eyes to provide a paradigm of rational inquiry that, arguably, has never been entirely displaced. It emphasized the very limited scope of absolute certainties in the natural and social sciences, and more generally underlined the boundaries to certain knowledge that arise from our limited capacities for observation and reasoning. To that extent it provided an important foil to the exaggerated claims sometimes made for the natural sciences in the wake of Newton's achievements in mathematical physics.

This appears to conflict strongly with Thomas Kuhn's (1922 - 96) statement that scientific theory choice depends on considerations that go beyond observation and logic, even when logic is construed to include confirmation.

Nonetheless, it can be said, that, the state of science at any given time is characterized, in part, by the theories accepted then. Presently accepted theories include quantum theory, and general theory of relativity, and the modern synthesis of Darwin and Mendel, as well as lower-level, but still clearly theoretical assertions such as that DNA has a double - helical structure, that the hydrogen atom contains a single electron, and so firth. What precisely is involved in accepting a theory or factors in theory choice.

Many critics have been scornful of the philosophical preoccupation with under - determination, that a theory is supported by evidence only if it implies some observation categories. However, following the French physician Pierre Duhem, who is remembered philosophically for his La Thorie physique, (1906), translated as, 'The Aim and Structure of Science, is that it simply is a device for calculating science provides a deductive system that is systematic, economic and predicative: Following Duhem, Orman van Willard Quine (1918 - 2000), who points out that observation categories can seldom if ever be deduced from a single scientific theory taken by itself: Rather, the theory must be taken in conjunction with a whole lot of other hypotheses and background knowledge, which are usually not articulated in detail and may sometimes be quite difficult to specify. A theoretical sentence does not, in general, have any empirical content of its own. This doctrine is called 'Holism', which the basic term refers to a variety of positions that have in common a resistance to understanding large unities as merely the sum of their parts, and an insistence that we cannot explain or understand the parts without treating them as belonging to such larger wholes. Some of these issues concern explanation. It is argued, for example, that facts about social classes are not reducible to facts about the beliefs and actions of the agents who belong to them, or it is claimed that we only understand the actions of individuals by locating them in social roles or systems of social meanings.

But, whatever may be the case with under-determination, there is a very closely related problem that scientists certainly do face whenever two rival theories or more encompassing theoretical frameworks are competing for acceptance. This is the problem posed by the fact that one framework, usually the older, longer - established frameworks can accommodate, that is, produce post hoc explanation of particular pieces of evidence that seem intuitively to tell strongly in favour of the other (usually the new 'revolutionary') framework.

For example, the Newtonian particulate theory of light is often thought of as having been straightforwardly refuted by the outcome of experiments - like Youngs' two-slits experiment - whose results were correctly predicted by the rival wave theory. Duhem's (1906) analysis of theories and theory testing already shows that this cannot logically have been the case. The bare theory that light consists of some sort of material particle has no empirical consequence s in isolation from other assumptions: And it follows that there must always be assumptions that could be added to the bare corpuscular theory, such that some combined assumptions entail the correct result of any optical experiment. A d indeed, a little historical research soon reveals eighteenth and early nineteenth - century emissionists who suggested at least outline ways in which interference result s could be accommodated within the corpuscular framework. Brewster, for example, suggested that interference might be a physiological phenomenon: While Biot and others worked on the idea that 'interferences' circumferential proponents are produced by the peculiarities of the 'diffracting forces' that ordinary gross exerts on the light corpuscles.

Both suggestions ran into major conceptual problems. For example, the 'diffracting force' suggestion would not even come close to working with any forces of kinds that were taken to operate in other cases. Often the failure was qualitative: Given the properties of forces that were already known about, for example, it was expected that the diffracting force would depend in some way on the material properties of the diffracting object: But, whatever the material of the double - slit screens are Young's experiment, and whatever its density, the outcome is the same. It could, of course, simply be assumed that the diffracting forces are an entirely novel kind, and that their properties just had to be 'read-off' the phenomena - this is exactly the way that corpusularists worked. Heretofore, the singular one or times of more with which the re-exist are no more. That is, to any exclusion of any alternative or competitor will only confess to you. The attemptive to write the phenomena into a favoured conceptual framework. And given that the writing - in produced complexities and incongruities for which there was no independent evidence, the majority view was that interference results strongly favour the wave theory, of which they are 'natural' consequences. (For example, that the material making up the double slit and its density have no effect at all on the phenomenon is a straightforward consequence of the fact that, as the wave theory says it, the only effect on the screen is to absorb those parts of the wave fronts that impinges on it.)

The natural methodological judgement (and the one that seems to have been made by the majority of competent scientists at that time) is that, even given the interference effects could be accommodated within the corpuscular theory, those effects nonetheless favour the wave account, and favour it in the epistemic sense of showing that theory to be more likely to be true. Of course, the account given by the wave theory of the interference phenomena is also, in certain senses, pragmatically simpler: But this seems generally to have been taken to be, not a virtue in itself, but a reflection of a deeper virtue connected with likely truth.

Consider a second, similar case: That of evolutionary theory and the fossil record. There are well - known disputes about which particular evolutionary account for most support from fossils. Nonetheless, the relative weight the fossil evidence carries for some sort of evolutionist account versus the special creationist theory, is yet well - known for its obviousness - in that the theory of special creation can accommodate fossils: A creationist just needs to claim that what the evolutionist thinks of as bones of animals belonging to extinct species, are, in fact, simply items that God chose to included in his catalogue of the universe's content at creatures: What the evolutionist thinks of as imprints in the rocks of the skeletons of other such animals are they. It nonetheless surely still seems true intuitively that the fossil records continue to give us better reason to believe that species have evolved from earlier, now extinct ones, than that God created the universe much as it presently is in 4004 Bc. An empiricist - instrumentalist t approach seems committed to the view that, on the contrary, any preference that this evidence yields for the evolutionary account is a purely pragmatic matter.

Of course, intuitions, no matter how strong, cannot stand against strong counter arguments. Van Fraassen and other strong empiricists have produced arguments that purport to show that these intuitions are indeed misguided.

What justifies the acceptance of a theory? Although h particular versions of empiricism have met many criticisms, that is, of support by the available evidence. How else could empiricists term? : In terms, that is, of support by the available evidence. How else could the objectivity of science be defended except by showing that its conclusion (and in particularly its theoretical conclusions - those theories? It presently on any other condition than that excluding exemplary base on which are somehow legitimately based on or agreed observationally and experimental evidences, yet, as well known, theoretics in general, pose a problem for empiricism. Allowing the empiricist the assumptions that there are observational statements whose truth - values can be inter-subjectively agreeing. A definitive formulation of the classical view was finally provided by the German logical positivist Rudolf Carnap (1891-1970), combining a basic empiricism with the logical tools provided by Frége and Russell: And it is his work that the main achievements (and difficulties) of logical positivism are best exhibited. His first major works were Der Logische Aufban der welts (1928, translated as 'The Logical Structure of the World, 1967) this phenomenological work attempts a reduction of all the objects of knowledge, by generating equivalence classes of sensations, related by a primitive relation of remembrance of similarity. This is the solipsistic basis of the construction of the external world, although Carnap later resisted the apparent metaphysical priority as given to experience. His hostility to metaphysics soon developed into the positivity as emphasis by the view characteristics that metaphysical questions are pseudo-problems. Criticism from the Austrian philosopher and social theorist Otto Neurath (1882-1945) shifted Carnap's interest toward a view of the unity of the sciences, with the concepts and theses of special sciences translatable into a basic physical vocabulary whose protocol statements describe not experience but the qualities of points in space-time. Carnap pursued the enterprise of clarifying the structures of mathematics and scientific language (the only legitimate task for scientific philosophy) in Logische Syntax fer Sprache (1943, translate as, 'The Logical Syntax of Language', 1937) refinements to his syntactic and semantic views continued with Meaning and Necessity (1947) while a general loosening of the original ideal of reduction culminated in the great Logical Foundations of Probability, the most important single work of 'confirmation theory', in 1950. Other works concern the structure of physics and the concept of entropy.

Wherefore, the observational terms were presumed to be given a complete empirical interpretation, which left the theoretical terms with only an 'indirect' empirical interpretation provided by their implicit definition within an axiom system in which some of the terms possessed a complete empirical interpretation.

Among the issues generated by Carnap's formulation was the viability of 'the theory - observation distinction'. Of course, one could always arbitrarily designate some subset of nonlogical terms as belonging to the observational vocabulary, however, that would compromise the relevance of the philosophical analysis for any understanding of the original scientific theory. But what could be the philosophical basis for drawing the distinction? Take the predicate 'spherical', for example. Anyone can observe that a billiard ball is spherical, but what about the moon, or an invisible speck of sand? Is the application of the term 'spherical' of these objects 'observational'?

Another problem was more formal, as introduced of Craig's theorem seemed to show that a theory reconstructed in the recommended fashion could be re - axiomatized in such a way as to dispense with all theoretical terms, while retaining all logical consequences involving only observational terms. Craig's theorem in mathematical logic, held to have implications in the philosophy of science. The logician William Craig showed how, if we partition the vocabulary of a formal system (say, into the 'T' or theoretical terms, and the 'O' or observational terms), then if there is a fully 'formalized' system 'T' with some set 'S' of consequences containing only the 'O' terms, there is also a system 'O' containing only the 'O' vocabulary but strong enough to give the same set 'S' of consequences. The theorem is a purely formal one, in that 'T' and 'O' simply separate formulae into the preferred ones, containing non- logical terms only one kind of vocabulary, and the others. The theorem might encourage the thought that the theoretical terms of a scientific theory are in principle, dispensable, since the same consequences can be derived without them.

However, Craig's actual procedure gives no effective way of dispensing with theoretical terms in advance, i.e., in the actual process of thinking about and designing the premises from which the set 'S' follows, in this sense 'O' remains parasitical on or upon its parent 'T'.

Thus, as far as the 'empirical' content of a theory is concerned, it seems that we can do without the theoretical terms. Carnap's version of the classical view seemed to imply a form of instrumentation. A problem which the German philosopher of science, Carl Gustav Hemel (1905-97) christened 'the theoretician's dilemma'.

Meanwhile Descartes identification of matter with extension, and his comitans theory of all of space as filed by a plenum of matter. The great metaphysical debate over the nature of space and time has its roots in the scientific revolution of the sixteenth and seventeenth centuries. An early contribution to the debate was the French mathematician and founding father of modern philosophy, Réne Descartes (1596-1650). His interest in the methodology of a unified science culminated in his first work, the Regulae ad Directionem Ingenti (1628/9), was never completed. Nonetheless, between 1628 and 1649, Descartes first wrote and then cautiously suppressed, Le Monde (1634) and in 1637 produced the Discours de la Méthode as a preface to the treatise on mathematics and physics in which he introduced the notion of Cartesian coordinates.

His best known philosophical work, the Meditationes de Prima Philosophia (Meditations of First Philosophy), together with objections by distinguished contemporaries and relies by Descartes (the Objections and Replies) appeared in 1641. The author of the objections is First advanced, by the Dutch theologian Johan de Kater, second set, Mersenne, third set, Hobbes: Fourth set, Arnauld, fifth set, Gassendim, and sixth set, Mersnne. The second edition (1642) of the Meditations included a seventh set by the Jesuit Pierre Bourdin. Descartes's penultimate work, the Principia Philosophiae (Principles of Philosophy) of 1644 was designed partly for use in theological textbooks: His last work, Les Passions de I áme (the Passions of the Soul) and published in 1649. In that year Descartes visited the court of Kristina of Sweden, where he contracted pneumonia, allegedly through being required to break his normal habit of a late rising in order to give lessons at 5:00 a.m. His last spoken words were accepted or advanced as true or real on the basis of less than conclusive evidence was 'Ça, mon sme il faut partur', - 'So my soul, it is time to part'.

It is nonetheless said, that the great metaphysical debate over the nature of space and time has its roots in the scientific revolution of the sixteenth and seventeenth centuries. An early contribution to the debate was Réne Descartes's (1596-1650), identification of matter with extension, and his comitane theory of all of space as filled by a plenum of matter.

Far more profound was the German philosopher, mathematician and polymath, Wilhelm Gottfried Leibniz (1646 - 1716), whose characterization of a full - blooded theory of relationism with regard to space and time, as Leibniz elegantly puts his view: 'Space is nothing but the order of coexistence . . . time is the order of inconsistent 'possibilities'. Space was taken to be a set of relations among material objects. The deeper monadological view to the side, were the substantival entities, no room was provided for space itself as a substance over and above the material substance of the world. All motion was then merely relative motion of one material thing in the reference frame fixed by another. The Leibnizian theory was one of great subtlety. In particular, the need for a modalized relationism to allow for 'empty space' was clearly recognized. An unoccupied spatial location was taken to be a spatial relation that could be realized but that was not realized in actuality. Leibniz also offered trenchant arguments against substantivalism. All of these rested upon some variant of the claim that a substantival picture of space allows for the theoretical toleration of alternative world models that are identical as far as any observable consequences are concerned.

Contending with Leibnizian relationalism was the 'substantivalism' of Isaac Newton (1642 - 1727), and his disciple S. Clarke, thereby he is mainly remembered for his defence of Newton (a friend from Cambridge days) against Leibniz, both on the question of the existence of absolute space and the question of the propriety of appealing to a force of gravity, actually Newton was cautious about thinking of space as a 'substance'. Sometimes he suggested that it be thought of, rather, as a property - in particular as a property of the Deity. However, what was essential to his doctrine was his denial that a relationist theory, with its idea of motion as the relative change of position of one material object with respect to another, can do justice to the facts about motion made evident by empirical science and by the theory that does justice to those facts.

The Newtonian account of motion, like Aristotle's, has a concept of natural or unforced motion. This is motion with uniform speed in a constant direction, so - called inertial motion. There is, then, in this theory an absolute notion of constant velocity motion. Such constant velocity motions cannot be characterized as merely relative to some material objects, some of which will be non-inertial. Space itself, according to Newton, must exist as an entity over and above the material objects of the world. In order to provide the standard of rest relative to which uniform motion is genuine inertial motion.

Such absolute uniform motions can be empirically discriminated from absolutely accelerated motion by the absence of inertial forces felt when the test object is moving genuinely inertially. Furthermore, the application of force to an object is correlated with the object's change of absolute motion. Only uniform motions relative to space itself are natural motions requiring no force and explanation. Newton also clearly saw that the notion of absolute constant speed requires a motion of absolute time, for, relative to an arbitrary cyclic process as defining the time scale, any motion can be made uniform or not, as we choose. Nonetheless, genuine uniform motions are of constant speed in the absolute time scale fixed by 'time itself;. Periodic processes can be at best good indicators of measures of this flaw of absolute time.

Newton's refutation of relationism by of the argument from absolute acceleration is one of the most distinctive examples of the way in which the results of empirical experiment and of the theoretical efforts to explain these results impinge on or upon philosophical objections to Leibnizian relationism - for example, in the claim that one must posit a substantival space to make sense of Leibniz's modalities of possible position - it is a scientific objection to relationism that causes the greatest problems for that philosophical doctrine.

Then, again, a number of scientists and philosophers continued to defend the relationist account of space in the face of Newton's arguments for substantivalism. Among them were Wilhelm Gottfried Leibniz, Christian Huygens, and George Berkeley when in 1721 Berkeley published De Motu ('On Motion') attacking Newton 's philosophy of space, a topic he returned too much later in The Analyst of 1734.the empirical distinction, however, to frustrate their efforts.

In the nineteenth century, the Austrian physicist and philosopher Ernst Mach (1838-1916), made the audacious proposal that absolute acceleration might be viewed as acceleration relative not to a substantival space, but to the material reference frame of what he called the 'fixed stars' - that is, relative to a reference frame fixed by what might now be called the 'average smeared - out mass of the universe'. As far as observational data went, he argued, the fixed stars could be taken to be the frames relative to which uniform motion was absolutely uniform. Mach's suggestion continues to play an important role in debates up to the present day.

The nature of geometry as an apparently a priori science also continued to receive attention. Geometry served as the paradigm of knowledge for rationalist philosophers, especially for Descartes and the Dutch Jewish rationalist Benedictus de Spinoza (1632-77), whereby the German philosopher Immanuel Kant (1724-1804) attempts to account for the ability of geometry to go beyond the analytic truths of logic extended by definition - was especially important. His explanation of the a priori nature of geometry by its 'transcendentally psychological' nature - that is, as descriptive of a portion of mind's organizing structure imposed on the world of experience - served as his paradigm for legitimated a priori knowledge in general.

A peculiarity of Newton's theory, of which Newton was well aware, was that whereas acceleration with respect to space itself had empirical consequences, uniform velocity with respect to space itself had none. The theory of light, particularly in J.C. Maxwell's theory of electromagnetic waves, suggested, however, that there was only one reference frame in which the velocity of light would be the same in all directions, and that this might be taken to be the frame at rest in 'space itself'. Experiments designed to find this frame seen to sow, however, that light velocity is isotropic and has its standard value in all frames that are in uniform motion in the Newtonian sense. All these experiments, however, measured only the average velocity of the light relative to the reference frame over a round - trip path trails. It was the insight of the German physicist Albert Einstein (1879-1955) who took the apparent equivalence of all inertial frames with respect to the velocity of light to be a genuine equivalence, It was from an employment within the Patent Office in Bern, wherefrom in 1905 he published the papers that laid the foundation of his reputation, on the photoelectric theory of relativity. In 1916 he published the general theory and in 1933 Einstein accepted the position at the Princeton Institute for Advanced Studies which he occupied for the rest of his life. His deepest insight was to see that this required that we relativize the notion of the simultaneity of events spatially separated from one distanced between a non - simultaneous events' reference frame. For any relativist, the distance between non-simultaneous events' simultaneity is relative as well. This theory of Einstein's later became known as the Special Theory of Relativity.

Eienstein's proposal account for the empirical undetectability of the absolute rest frame by optical experiments, because in his account the velocity of light is isotropic and has its standard value in all inertial frames. The theory had immediate kinematic consequences, among them the fact that spatial separation (lengths) and intervals relevant to set frames - of motion - relatively. New dynamics was needed if dynamics were to be, as it was for Newton, equivalence in all inertial frames.

Einstein's novel understanding of space and time was given an elegant framework by H. Minkowski in the form of Minkowski Space-time. The primitive elements of the theory were a characteristic point - like, locations in both spatially temporal of unextended happenings. These were called the 'event locations' or the 'events' 'of a four - dimensional manifold. There is a frame-invariant separation of an event frame event called the 'interval'. But the spatial separation between two noncoincident events, as well as their temporal separation, are well defined only relative to a chosen inertial reference frame. In a sense, then, space and time are integrated into a single absolute structure. Space and time by themselves have a derivative and relativized existence.

Whereas the geometry of this space-time bore some analogies to a Euclidean geometry of a four-dimensional space, the traditions from space and time by them in an integrated space-time required a subtle rethinking of the very subject matter of geometry. 'Straight lines' are the straightest curves of this 'flat' space-time, however, they to include 'null straight lines', interpreted as the events in the life history of a light ray in a vacuum and 'time - like straight lines', interpreted as the collection of events in the life history of a free inertial contribution to the revolution in scientific thinking into the new relativistic framework. The result of his thinking was the theory known as the general theory of relativity.

The heuristic basis for the theory rested on or upon an empirical fact known to Galileo and Newton, but whose importance was made clear only by Einstein. Gravity, unlike other forces such as the electromagnetic force, acts on all objects independently of their material constitution or of their size. The path through space-time followed by an object under the influence of gravity is determined only by its initial position and velocity. Reflection upon the fact that in a curved spac e the path of minimal curvature from a point, the so-called 'geodesic', is uniquely determined by the point and by a direction from it, suggested to Einstein that the path of as an object acted upon by gravity can be thought of as a geodesic followed by that path in a curved space-time. The addition of gravity to the space-time of special relativity can be thought of s changing the 'flat' space-time of Minkowski into a new, 'curved' space-time.

The kind of curvature implied by the theory in that explored by B. Riemann in his theory of intrinsically curved spaces of an arbitrary dimension. No assumption is made that the curved space exists in some higher-dimensional flat embedding space, curvature is a feature of the space that shows up observationally in those in the space longer straight lines, just as the shortest distances between points on the Earth's surface cannot be reconciled with putting those points on a flat surface. Einstein (and others) offered other heuristic arguments to suggest that gravity might indeed have an effect of relativistic interval separations as determined by measurements using tapes' spatial separations and clocks, to determine time intervals.

The special theory gives a unified account of the laws of mechanics and of electromagnetism (including optics). Before 1905 the purely relative nature of uniform motion had in part been recognized in mechanics, although Newton had considered time to be absolute and also postulated absolute space. In electromagnetism the 'ether' was supposed to provide an absolute basis with respect to which motion could be determined and made two postulates. (The laws of nature are the same for all observers in uniform relative e motion. (2) The speed of light is the same for all such observes, independently of the relative motion of sources and detectors. He showed that these postulates were equivalent to the requirement that coordinates of space and time was put - upon by different observers should be related by the 'Lorentz Transformation Equation Theory': The theory has several important consequences.

That is to say, a set of equations for transforming the position - motion parameters from an observer at point $0(x, y, z)$ to an observer at $0'(x', y', z')$, moving relative to one another. The equations replace the 'Galilean transformation equations of Newtonian mechanics in Relative problems. If the x - axises are chosen to pass through $00'$ and the time of an even t is (t) and (t') in the frame of reference of the observer at 0 and 0' respectively y (where the zeros of their time scales were the instants that 0 and 0' coincided) the equations are:

$$y' = \hat{a}(\div - vt)$$
$$y' = y$$
$$z' = z$$
$$t' = \hat{a}(t - v\div/c2),$$

Where v is the relative velocity y of separation of 0, 0', c is the speed of light, and â is the function

$$(1 - v2/c2).$$

The transformations of time implies that two events that are simultaneous according to one observer will not necessarily be so according to another in uniform relative motion. This does not affect in any way violate any concepts of causation. It will appear to two observers in uniform relative motion that each other's clock rums slowly. This is the phenomenon of 'time dilation', for example, an observer moving with respect to a radioactive source finds a longer decay time than that found by an observer at rest with respect to it, according to:

$$T_v = T_0/(1 - v^2/c^2)^{1/2},$$

Where T_v is the mean life measured by an observer at relative speed v. T_0 is the mean life measured by an observer relatively at rest, and c is the speed of light.

Among the results of the 'exact' form optics is the deduction of the exact form io f the Doppler Effect. In relativity mechanics, mass, momentum and energy are all conserved. An observer with speed v with respect to a particle determines its mass to be m while an observer at rest with respect to the article measure the 'rest mass' m_0, such that:

$$m = m_0/(1 - v^2/c^2)^{1/2}$$

This formula has been verified in innumerable experiments. One consequence is that no body can be accelerated from a speed below c with respect to any observer to one above c, since this would require infinite energy. Einstein deduced that the trandfer of energy äE by any process entailed the trandfer of mass äm, where $äE = ämc^2$, hence he concluded that the total energy E of any system of mass m would be given by:

$$E = mc^2$$

The kinetic energy of a particle as determined by an observer with relative speed v is thus $(m - m_0) c^2$, which tends to the classical value $\frac{1}{2}mv^2$ if v c.

Attempts to express Quantum Theory in terms consistent with the requirements of relativity were begun by Sommerfeld (1915). Eventually Dirac (1928) gave a relativistic formulation of the wave mechanics of conserved particles (fermions). This explained the concepts of sin and the associated magnetic moment for certain details of spectra. The theory led to results of elementary particles, the theory of Beta Decay, and for Quantum statistics, the Klein - Gordon Equation is the relativistic wave equation for 'bosons'.

A mathematical formulation of the special theory of relativity was given by Minkowski. It is based on the idea that an event is specified by four coordinates: Three spatial coordinates and one of time. These coordinates define a four - dimensional space and time motion of a particle can be described by a curve in this space, which is called 'Minkowski space-time'.

The special theory of relativity is concerned with relative motion between non-accelerated frames of reference. The general theory deals with general relative motion between accelerated frames of reference. In accelerated systems of reference, certain fictitious forces are observed, such as the centrifugal and Coriolis forces found in rotating systems. These are known as fictitious forces because they disappear when the observer transforms to a non - accelerated system. For example, to an observer in a car rounding a bend at constant velocity, objects in the car appear to suffer a force acting outwards. To an observer outside the car, this is simply their tendency to continue moving in a straight line. The inertia of the objects is seen to cause a fictitious force and the observer can distinguish between non-inertial (accelerated) and inertial (non - accelerated) frames of reference.

A further point is that, to the observer in the car, all the objects are given the same acceleration irrespective of their mass. This implies a connection between the fictitious forces arising from accelerated systems and forces due to gravity, where the acceleration produced is independent of the mass. For example, a person in a sealed container could not easily determine whether he was being driven toward the floor by gravity of if the container were in space and being accelerated upwards by a rocket. Observations extended between these alternatives, but otherwise they are indistinguishable from which it follows that the inertial mass is the same as a gravitational mass.

The equivalence between a gravitational field and the fictitious forces in non-inertial systems can be expressed by using 'Riemannian space-time', which differs from Minkowski space-time of the special theory. In special relativity the motion of a particle that is not acted on by any forces is presented by a straight line in Minkowski space-time. In general relativity, using Riemannian space-time, the motion is presented by a line that is no longer straight (in the Euclidean sense) but is the line giving the shortest distance. Such a line is called a 'geodesic'. Thus, space-time is said to be curved. The extent of this curvature is given by the 'metric tensor' for spaces - time, the components of which are solutions to Einstein's 'field equations'. The fact that gravitational effects occur near masses is introduced by the postulate that the presence e of matter produces this curvature of space-time. This curvature of space-time controls the natural motions of bodies.

The predictions of general relativity only differ from Newton's theory by small amounts and most tests of the theory have been carried out through observations in astronomy. For example, it explains the shift on the perihelion of Mercury, the bending of light in the presence of large bodies, and the Einstein shift. Very close agreements between their accurately measured values have now been obtained.

So, then, using the new space-time notions, a 'curved space-time' theory of Newtonian gravitation can be constructed. In this space-time is absolute, as in Newton. Furthermore, space remains flat Euclidean space. This is unlike the general theory of relativity, where the space-time curvature can induce spatial curvature as well. But the spaces - time curvature of this 'curved neo-Newtonian Space-time, shows up in the fact that particles under the influence of gravity do not follow straight line's paths. Their paths become, as in general relativity, the curved times - like geodesics of the space-time. In this curved space-time account of Newtonian gravity, as in the general theory of relativity, the indistinguishable alternative worlds of theories that take gravity as a force s superimposed in a flat space-time collapsed to a single world model.

The strongest impetus to rethink epistemological issues in the theory of space and time came from the introduction of curvature and of non - Euclidean geometries in the general theory of relativity. The claim that a unique geometry could be known to hold true of the world a priori seemed unviable, at least in its naive form. In a situation where our best available physical theory allowed for a wide diversity of possible geometries for the world and in which the geometry of space-time was one more dynamical element joining the other 'variable' features of the world. Of course, skepticism toward an a priori account of geometry could already have been induced by the change from space time to space-time in the special theory, even though the space of that world remained Euclidean.

The natural response to these changes in physics was to suggest that geometry was, like all other physical theories, believable only on the basis of some kind of generalizing inference from the law - like regularities among the observable observational data - that is, to become an empiricists with regard to geometry.

But a defence of a kind of a priori account had already been suggested by the French mathematician and philosopher Henri Jules Poincaré (1854 - 1912), even before the invention of the relativistic theories. He suggested that the limitation of observational data to the domain of what was both material and local, i.e., or, space-time in order to derive a geometrical world of matter and convention or decision on the part of the scientific community. If any geometric posit could be made compatible with any set of observational data, Euclidean geometry could remain a priori in the sense that we could, conventionally, decide to hold to it as the geometry of the world in the face of any data that apparently refuted it.

The central epistemological issue in the philosophy of space and time remains that of theoretical under - determination, stemming from the Poincaré argument. In the case of the special theory of relativity the question is the rational basis for choosing Einstein's theory over, for example, on of the 'aether reference frame plus modification of rods and clocks when they are in motion with respect to the aether' theories tat it displaced. Among the claims alleged to be true merely by convention in the theory, for which of asserting the simultaneity of distant events, those asserting the 'flatness' of the chosen space-time. Crucial to the fact that Einstein's arguments themselves presuppose a strictly delimited local observation basis for the theories and that in fixing on or upon the special theory of relativity, one must make posits about the space, and time structures that outrun the facts given strictly by observation. In the case of the general theory of relativity, the issue becomes one of justifying the choice of general relativity over, for example, a flat spaces - time theory that treats gravity, as it was treated by Newton, as a 'field of force' over and above the space-time structure.

In both the cases of special and general relativity, important structural features pick out the standard Einstein theories as superior to their alternatives. In particular, the standard relativistic models eliminate some of the problems of observationally equivalent but distinguishable worlds countenanced by the alternative theories. However, the epistemologists must still be concerned with the question as to why these features constitute grounds for accepting the theories as the 'true' alternatives.

Other deep epistemological issues remain, having to do with the relationship between the structures of space and time posited in our theories of relativity and the spatiotemporal structures we use to characterize our 'direct perceptual experience'. These issues continue in the contemporary scientific context the old philosophical debates on the relationship between the ram of the directly perceived and the realm of posited physical nature.

First reaction on the part of some philosophers was to take it that the special theory of relativity provided a replacement for the Newtonian theory of absolute space that would be compatible with a relationist account of the nature of space and time. This was soon seen to be false. The absolute distinction between uniform moving frames and frames not in or upon its uniform motion, invoked by Newton in his crucial argument against relationism, remains in the special theory of relativity. In fact, it becomes an even deeper distinction than it was in the Newtonian account, since the absolutely uniformly moving frames, the inertial frames, now become not only the frames of natural unforced motion, but also the only frames in which the velocity of light is isotropic.

At least part of the motivation behind Einstein's development of the general theory of relativity was the hope that in this new theory all reference frames, uniformly moving or accelerated, would be 'equivalent' to one another physically. It was also his hope that the theory would conform to the Machian idea of absolute acceleration as merely acceleration relative to the smoothed-out matter of the universe.

Further exploration of the theory, however, showed that it had many features uncongenial to Machianism. Some of these are connected with the necessity of imposing boundary conditions for the equation connecting the matter distribution of the space-time structure. General relativity certainly allows as solutions model universes of a non-Machian sort - for example, those which are aptly described as having the smoothed-out matter of the universe itself in 'absolute rotation'. There are strong arguments to suggest that general relativity. Like Newton's theory and like special relativity, requires the positing of a structure of 'space-time itself' and of motion relative to that structure, in order to account for the needed distinctions of kinds of motion in dynamics. Whereas in Newtonian theory it was 'space itself' that provided the absolute reference frames. In general relativity it is the structure of the null and time - like geodesics that perform this task. The compatibility of general relativity with Machian ideas is, however, a subtle matter and one still open to debate.

Other aspects of the world described by the general theory of relativity argue for a substantivalist reading of the theory as well. Space-time has become a dynamic element of the world, one that might be thought of as 'causally interacting' with the ordinary matter of the world. In some sense one can even attribute energy (and hence mass) to the spacer - time (although this is a subtle matter in the theory), making the very distinction between 'matter' and 'spacer - time itself' much more dubious than such a distinction would have been in the early days of the debate between substantivalists and explanation forthcoming from the substantivalist account is.

Nonetheless, a naive reading of general relativity as a substantivalist theory has its problems as well. One problem was noted by Einstein himself in the early days of the theory. If a region of space-time is devoid of non - gravitational mass - energy, alternative solutions to the equation of the theory connecting mass - energies with the space-time structure will agree in all regions outside the matterless 'hole', but will offer distinct space-time structures within it. This suggests a local version of the old Leibniz arguments against substantivalism. The argument now takes the form of a claim that a substantival reading of the theory forces it into a strong version of indeterminism, since the spaces - time structure outside the hld fails to fix the structure of space-time in the hole. Einstein's own response to this problem has a very relationistic cast, taking the 'real facts' of the world to be intersections of paths of particles and light rays with one another and not the structure of 'space-time itself'. Needless to say, there are substantival attempts to deal with the 'hole' argument was well, which try to reconcile a substantival reading of the theory with determinism.

There are arguments on the part of the relationist to the effect that any substantivalist theory, even one with a distinction between absolute acceleration and mere relative acceleration, can be given a relationistic formulation. These relationistic reformations of the standard theories lack the standard theories' ability to explain why non-inertial motion has the features that it does. But the relationist counters by arguing that the explanation forthcoming from the substantivalist account is too 'thin' to have genuine explanatory value anyway.

Relationist theories are founded, as are conventionalist theses in the epistemology of space-time, on the desire to restrict ontology to that which is present in experience, this taken to be coincidences of material events at a point. Such relationist conventionalist account suffers, however, from a strong pressure to slide full - fledged phenomenalism.

As science progresses, our posited physical space-times become more and more remote from the space-time we come to form an idea of something in the mind in which it is capable of being thought about, as characterizing immediate experience. This will become even more true as we move from the classical space-time of the relativity theories into fully quantized physical accounts of space-time. There is strong pressure from the growing divergence of the space-time of physics from the space-time of our 'immediate experience' to dissociate the two completely and, perhaps, to stop thinking of the space-time of physics for being anything like our ordinary notions of space and time. Whether such a radical dissociation of posited nature from phenomenological experience can be sustained, however, without giving up our grasp entirely on what it is to think of a physical theory 'realistically' is an open question.

Science aims to represent accurately actual ontological unity/diversity. The wholeness of the spatiotemporal framework and the existence of physics, i.e., of laws invariant across all the states of matter, do represent ontological unities which must be reflected in some unification of content. However, there is no simple relation between ontological and descriptive unity/diversity. A variety of approaches to representing unity are available (the formal - substantive spectrum and respective to its opposite and operative directions that the range of naturalisms). Anything complex will support man y different partial descriptions, and, conversely, different kinds of thing s many all obey the laws of a unified theory, e.g., quantum field theory of fundamental particles or collectively be ascribed dynamical unity, e.g., self - organizing systems.

It is reasonable to eliminate gratuitous duplication from description - that is, to apply some principle of simplicity, however, this is not necessarily the same as demanding that its content satisfies some further methodological requirement for formal unification. Elucidating explanations till there is again no reason to limit the account to simple logical systemization: The unity of science might instead be complex, reflecting our multiple epistemic access to a complex reality.

Biology provides as useful analogy. The many diverse species in an ecology nonetheless, each map, genetically and cognitively, interrelatable aspects of as single environment and share exploitation of the properties of gravity, light, and so forth. Though the somantic expression is somewhat idiosyncratic to each species, and the incomplete representation, together they form an interrelatable unity, a multidimensional functional representation of their collective world. Similarly, there are many scientific disciplines, each with its distinctive domains, theories, and methods specialized to the condition under which it accesses our world. Each discipline may exhibit growing internal metaphysical and nomological unities. On occasion, disciplines, or components thereof, may also formally unite under logical reduction. But a more substantive unity may also be manifested: Though content may be somewhat idiosyncratic to each discipline, and the incomplete representation, together the disciplinary contents form an interrelatable unity, a multidimensional functional representation of their collective world. Correlatively, a key strength of scientific activity lies, not formal monolithicity, but in its forming a complex unity of diverse, interacting processes of experimentations, theorizing, instrumentation, and the like.

While this complex unity may be all that finite cognizers in a complex world can achieve, the accurate representation of a single world is still a central aim. Throughout the history of physics. Significant advances are marked by the introduction of new representation (state) spaces in which different descriptions (reference frames) are embedded as some interrelatable perspective among many thus, Newtonian to relativistic space-time perspectives. Analogously, young children learn to embed two-dimensional visual perspectives in a three-dimensional space in which object constancy is achieved and their own bodies are but some among many. In both cases, the process creates constant methodological pressure for greater formal unity within complex unity.

The role of unity in the intimate relation between metaphysics and metho in the investigation of nature is well - illustrated b y the prelude to Newtonian science. In the millennial Greco - Christian religion preceding the founder of modern astronomy, Johannes Kepler (1571 - 1630), nature was conceived as essentially a unified mystical order, because suffused with divine reason and intelligence. The pattern of nature was not obvious, however, a hidden ordered unity which revealed itself to a diligent search as a luminous necessity. In his Mysterium Cosmographicum, Kepler tried to construct a model of planetary motion based on the five Pythagorean regular or perfect solids. These were to be inscribed within the Aristotelian perfect spherical planetary orbits in order, and so determine them. Even the fact that space is a three-dimensional unity was a reflection of the one triune God. And when the observational facts proved too awkward for this scheme. Kepler tried instead, in his Harmonice Mundi, to build his unified model on the harmonies of the Pythagorean musical scale.

Subsequently, Kepler trod a difficult and reluctant path to the extraction of his famous three empirical laws of planetary motion: Laws that made Newtonian revolution possible, but had none of the elegantly simple symmetries that mathematical mysticism required. Thus, we find in Kepler both the medieval methods and theories of metaphysically y unified religio - mathematical mysticism and those of modern empirical observation and model fitting. A transition figures in the passage to modern science.

To appreciate both the historical tradition and the role of unity in modern scientific method, consider Newton's methodology, focussing just on Newton's derivation of the law of universal gravitation in Principia Mathematica, book iii. The essential steps are these: (1) The experimental work of Kepler and Galileo (1564 - 1642) is appealed to, so as to establish certain phenomena, principally Kepler's laws of celestial planetary motion and Galileo's terrestrial law of free fall. (2) Newton's basic laws of motion are applied to the idealized system of an object small in size and mass moving with respect to a much larger mass under the action of a force whose features are purely geometrically determined. The assumed linear vector nature of the force allows construction of the centre of a mass frame, which separates out relative from common motions: It is an inertial frame (one for which Newton's first law of motion holds), and the construction can be extended to encompass all solar system objects.

(3) A sensitive equivalence is obtained between Kepler's laws and the geometrical properties of the force: Namely, that it is directed always along the line of centres between the masses, and that it varies inversely as the square of the distance between them. (4) Various instances of this force law are obtained for various bodies in the heavens - for example, the individual planets and the moons of Jupiter. From this one can obtain several interconnected mass ratios - in particular, several mass estimates for the Sun, which can be shown to cohere mutually. (5) The value of this force for the Moon is shown to be identical to the force required by Galileo's law of free fall at the Earth's surface. (6) Appeal is made again to the laws of motion (especially the third law) to argue that all satellites and falling bodies are equally themselves sources of gravitational force. (7) The force is then generalized to a universal gravitation and is shown to explain various other phenomena - for example, Galileo's law for pendulum action is shown suitably small, thus leaving the original conclusions drawn from Kepler's laws intact while providing explanations for the deviations.

Newton's constructions represent a great methodological, as well as theoretical achievement. Many other methodological components besides unity deserve study in their own right. The sense of unification is here that a deep systemization, as given the laws of motion, the geometrical form of the gravitational force and all its significant parameters needed for a complete dynamical description - that is, the component G, of the geometrical form of gravity Gm_1m_2/r_n, - are uniquely determined from phenomenons and, after the of universal gravitation has been derived, it plus the laws of motion determine the space and time frames and a set of self - consistent attributions of mass. For example, the coherent mass attributions ground the construction of the locally inertial ventre of a mass frame, and Newton's first law then enables us to consider time as a magnitude e: Equal tomes are those during which a freely moving body transverses equal distances. The space and time frames in turn ground use of the laws of motion, completing the constructive circle. This construction has a profound unity to it, expressed by the multiple interdependency of its components, the convergence of its approximations, and the coherence of its multiplying determined quantized. Newton's Rule IV: (Loosely) do not introduce a rival theory unless it provides an equal or superior unified construction - in particular, unless it is able to measure its parameters in terms of empirical phenomena at least as thorough and cross - situationally invariably (Rule III) as done in current theory. this gives unity a central place in scientific method.

Kant and Whewell seized on this feature as a key reason for believing that the Newtonian account had a privileged intelligibility and necessity. Significantly, the requirement to explain deviations from

Kepler's laws through gravitational perturbations has its limits, especially in the cases of the Moon and Mercury: These need explanations. The former through the complexities of n - body dynamics (which may even show chaos) and the latter through relativistic theory. Today we no longer accept the truth, let alone the necessity, of Newton's theory. Nonetheless, it remains a standard of intelligibility. It is in this role that it functioned, not jus t for Kant, but also for Reichenbach, and later Einstein and even Bohr: Their sense of crisis with regard to modern physics and their efforts to reconstruct it is best seen as stemming from their acceptance of an essential recognition of the falsification o this ideal by quantum theory. Nonetheless, quantum theory represents a highly unified, because symmetry - preserving, dynamics, reveals universal constants, and satisfies the requirement of coherent and invariant parameter determinations.

Newtonian method provides a central, simple example of the claim that increased unification brings increased explanatory power. A good explanation increases our understanding of the world. And clearly a convincing story an do this. Nonetheless, we have also achieved great increases in our understanding of the world through unification. Newton was able to unify a wide range of phenomena by using his three laws of motion together with his universal law of gravitation. Among other things he was able to account for Johannes Kepler's three was of planetary motion, the tides, the motion of the comets, projectile motion and pendulums. Still, his laws of planetary motion are the first mathematical, scientific, laws of astronomy of the modern era. They state (1) that the planets travel in elliptical orbits, with one focus of the ellipse being the sun. (2) That the radius between sun and planets sweeps equal areas in equal time, and (3) that the squares of the periods of revolution of any two planets are in the same ratio as the cube of their mean distance from the sun.

We have explanations by reference of causation, to identities, to analogies, to unification, and possibly to other factors, yet philosophically we would like to find some deeper theory that explains what it was about each of these apparently diverse forms of explanation that makes them explanatory. This we lack at the moment. Dictionary definitions typically explicate the notion of explanation in terms of understanding: An explanation is something that gives understanding or renders something intelligible. Perhaps this is the unifying notion. The different types of explanation are all types of explanation in virtue of their power to give understanding. While certainly an explanation must be capable of giving an appropriately tutored person a psychological sense of understanding, this is not likely to be a fruitful way forward. For there is virtually no limit to what has been taken to give understanding. Once upon a time, many thought that the facts that there were seven virtues and seven orifices of the human head gave them an understanding of why there were (allegedly) only seven planets. we need to distinguish between real and spurious understanding. And for that we need a philosophical theory of explanation that will give us the hall - mark of a good explanation.

In recent years, there has been a growing awareness of the pragmatic aspect of explanation. What counts as a satisfactory explanation depends on features of the context in which the explanation is sought. Willy Sutton, the notorious bank robber, is alleged to have answered a priest's question, 'Why do you rob banks'? By saying 'That is where the money is', we need to look at the context to be clear about for what exactly of an explanation is being sought. Typically, we are seeking to explain why something is the case than something else. The question which Willy's priest probably had in mind was: 'Why do you rob banks rather than have a socially worthwhile jobs'? And not the question 'Why do you rob banks rather than have a socially worthwhile jobs'? And not the question 'Why do you rob banks rather than churches'? we also need to attend to the background information possessed by the questioner. If we are asked why a certain bird has a long beaks, it is no use answering (as the D - N approach might seem to license) that the birds are an Aleutian Fern and all Aleutian Fern have long beaks if the questioner already knows that it is an Aleutian tern. A satisfactory answer typically provides new information. In this case, however, the speaker may be looking for some evolutionary account of why that species has evolved long beaks. Similarly, we need to attend to the level of sophistication in the answer to be given. we do not provide the same explanation of some chemical phenomena to a school child as to a student of quantum chemistry.

Van Fraassen whose work has been crucially important in drawing attention to the pragmatic aspects of exaltation has gone further in advocating a purely pragmatic theory of explanation. A crucial feature of his approach is a notion of relevance. Explanatory answers to 'why' questions must be relevant but relevance itself is a function of the context for van Fraassen. For that reason he has denied that it even makes sense to talk of the explanatory power of a theory. However, his critics (Kitcher and Salmon) pint out that his notion of relevance is unconstrained, with the consequence that anything can explain anything. This reductio can be avoided only by developing constraints on the relation of relevance, constraints that will not be a functional forming context, hence take us away from a purely pragmatic approach to explanation.

The resolving result is increased explanatory power for Newton's theory because of the increased scope and robustness of its laws, since the data pool which now supports them is the largest and most widely accessible, and it brings its support to bear on a single force law with only two adjustable, multiply determined parameters (the masses). Call this kind of unification (simpler than full constructive unification) 'coherent unification'. As much has been made of these ideas in recent philosophy of method, representing something of a resurgence of the Kantian traditions.

Unification of theories is achieved when several theories T1, T2,. . . Tn previously regarded s distinct are subsumed into a theory of broader scope T*. Classical examples are the unification of theories of electricity, magnetism, and light into Maxwell's theory of electrodynamics. And the unification of evolutionary and genetic theory in the modern synthetic thinking.

In some instances of unification, T* logically entails T1, T2, . . . Tn under particular assumptions. This is the sense in which the equation of state for ideal gases: $pV = nRT$, is a unification of Boyle's law, $pV =$ constant for constant temperature, and Charle's law, $V/T =$ constant for constant pressure. Frequently, however, the logical relations between theories involve in unification are less straightforward. In some cases, the claims of T* strictly contradict the claim of T1, T2, . . . Tn. For instance, Newton's inverse - square law of gravitation is inconsistent with Kepler's laws of planetary motion and Galileo's law of free fall, which it is often said to have unified. Calling such an achievement 'unification' may be justified by saying that T* accounts on its own for the domains of phenomena that had previously been treated by T1, T2, . . . Tn. In other cases described as unification, T* uses fundamental concepts different from those of T1, T2,. . . Tn so the logical relations among them are unclear. For instance, the wave and corpuscular theories of light are said to have been unified in quantum theory, but the concept of the quantum particle is alien to classical theories. Some authors view such cases not as a unification of the original T1, T2, . . . Tn, but as their abandonment and replacement by a wholly new theory T* that is incommensurable with them.

Standard techniques for the unification of theories involve isomorphism and reduction. The realization that particular theories attribute isomorphic structures to a number of different physical systems may point the way to a unified theory that attributes the same structure to all such systems. For example, all instances of wave propagation are described by the wave equation:

$$2y/\ x2 = (2y/\ t2)/v2$$

Where the displacement y is given different physical interpretations in different instances. The reduction of some theories to a lower - level theory, perhaps through uncovering the micro - structure of phenomena, may enable the former to be unified into the latter. For instance, Newtonian mechanics represent a unification of many classical physical theories, extending from statistical thermodynamics to celestial mechanics, which portray physical phenomena as systems of classical particles in motion.

Alternative forms of theory unification may be achieved on alternative principles. A good example is provided by the Newtonian and Leibnizian programs for theory unification. The Newtonian program involves analysing all physical phenomena as the effects of forces between particles. Each force is described by a causal law, modelled on the law of gravitation. The repeated application of these laws is expected to solve all physical problems, unifying celestial mechanics with terrestrial dynamics and the sciences of solids and of fluids. By contrast, the Leibnizian program proposes to unify physical science on the basis of abstract and fundamental principles governing all phenomena, such as principles of continuity, conservation, and relativity. In the Newtonian program, unification derives from the fact that causal laws of the same form apply to every event in the universe: In the Leibnizian program, it derives from the fact that a few universal principles apply to the universe as a whole. The Newtonian approach was dominant in the eighteenth and nineteenth centuries, but more recent strategies to unify physical sciences have hinged on or upon the formulating universal conservation and symmetry principles reminiscent of the Leibnizian program.

There are several accounts of why theory unification is a desirable aim. Many hinge on simplicity considerations: A theory of greater generality is more informative than a set of restricted theories, since we need to gather less information about a state of affairs in order to apply the theory to it. Theories of broader scope are preferable to theories of narrower scope in virtue of being more vulnerable to refutation. Bayesian principles suggest that simpler theories yielding the same predictions as more complex ones derive stronger support from common favourable evidence: On this view, a single general theory may be better confirmed than several theories of narrower scope that are equally consistent with the available data.

Theory unification has provided the basis for influential accounts of explanation. According to many authors, explanation is largely a matter of unifying seemingly independent instances under a generalization. As the explanation of individual physical occurrences is achieved by bringing them within th scope of a scientific theory, so the explanation of individual theories is achieved by deriving them from a theory of a wider domain. On this view, T1, T2, . . . Tn, are explained by being unified into T*.

The question of what theory unification reveals about the world arises in the debate between scientific realism and instrumentals. According to scientific realists, the unification of theories reveals common causes or mechanisms underlying apparently unconnected phenomena. The comparative case with which scientists interpretation, realists maintain, but can be explained if there exists a substrate underlying all phenomena composed of genuinely existent observable and unobservable entities. Instrumentalists provide a mythological account of theory unification which rejects these ontological claims of realism and instrumentals.

Arguments in a like manner, are of statements which purported provides support for another. The statements which purportedly provide the support are the premises while the statement purportedly supported is the conclusion. Arguments are typically divided into two categories depending on the degree of support they purportedly provide. Deduction arguments purportedly provide conclusive arguments purportedly provide any probable support. Some, but not all, arguments succeed in supporting arguments, successful in providing support for their conclusions. Successful deductive arguments are valid while successful inductive arguments are strong. An argument is valid just in case if all its ptr=muses are true then its conclusion must be true. An argument is strong just in case if all its premises are true its conclusion is only probable. Deductive logic provides methods for ascertaining whether or not an argument is valid whereas inductive logic provides methods for ascertaining the degree of support the premiss of an argument confer on its conclusion.

The argument from analogy is intended to establish our right to believe in the existence and nature of 'other minds', it admits that it is possible that the objects we call persona are, other than themselves, mindless automata, but claims that we nonetheless have sufficient reason for supposing this are not the case. There is more evidence that they cannot mindless automata than that they are:

The classic statement of the argument comes from J.S. Mill. He wrote:

I am conscious in myself of a series of facts connected by an uniform sequence, of which the beginning is modification of my body, the middle, in the case of other human beings, I have

> The evidence of my senses for the first and last links of the series, but not for the intermediate link. I find, however, that the sequence

Among the first and last is regular and constant in the other

> Cases as it is in mine. In my own case I know that the first link produces the last through the intermediate link, and could not produce it without. Experience, therefore, obliges me to conclude that there must be an intermediate link, which must either be the same in others as in myself, or a different one, . . . by supposing the link to be of the Same nature . . . I confirm to the legitimate rules of experimental enquiry.

As an inductive argument this is very weak, because it is condemned to arguing from a single case. But to this we might reply that nonetheless, we have more evidence that there is other minds than that there is not.

The real criticism of the argument is due to the Austrian philosopher Ludwig Wittgenstein (1889 - 1951). It is that the argument assumes that we at least understand the claims that there are subjects of experience other than themselves, who enjoy experiences which are like ours but not ours: It only asks what reason we have to suppose that claim true. But if the argument does indeed express the ground of our right to believe in the existence of others. It is impossible to explain how we are able to achieve that understanding. So if there is a place for argument from analogy, the problem of other minds - the real, hard problem, which is how we acquire a conception of another mind - is insoluble. The argument is either redundant or worse.

Even so, the expression 'the private language argument' is sometimes used broadly to refer to a battery of arguments in Wittgenstein's 'Philosophical Investigations', which are concerned with the concepts of, and relations between, the mental and its behavioural manifestations (the inner and the outer), self - knowledge and knowledge of other's mental states. Avowals of experience and description of experiences. It is sometimes used narrowly to refer to a single chain of argument in which Wittgenstein demonstrates the incoherence of the idea that sensation names and names of experiences given meaning by association with a mental 'object', e.g., the word 'pain' by association with the sensation of pain, or by mental (private) 'ostensive definition'. In which a mental 'entity' supposedly functions as a sample, e.g., a mental image, stored in memory y, is conceived as providing a paradigms for the application of the name.

A 'private language' is not a private code, which could be cracked by another person, nor a language spoken by only one person, which could be taught to others, but a putative language, the individual words of which refer to what can (apparently) are known only by the speaker, i.e., to his immediate private sensations or, to use empiricist jargon, to the 'ideas' in his mind. It has been a presupposition of the mainstream of modern philosophy, empiricist, rationalist and Kantian alike, of Representationalism that the languages we speak are such private languages, that the foundations of language no less than the foundations of knowledge lie in private experience. To determine this picture with all its complex ramifications is the purpose of Wittgenstein's private arguments.

There are various ways of distinguishing types of foundationalist epistemology, whereby Plantinga (1983) has put forward an influential conception of 'classical Foundationalism', specified in terms of limitations on the foundations. He construes this as a disjunction of ancient and medieval Foundationalism', which takes foundations to comprise what is self - evident and 'evident to the senses' and 'modern Foundationalism', that replaces 'evidential sense' with 'incorrigibles' which in, practice in what is taken to apply to beliefs about one's present states of consciousness. Plantinga himself developed this notion in the context of arguing that items outside this territory, in particular certain beliefs about God, could also be immediately justified. A popular recent distinction is between what is variously called 'strong' or 'extreme' Foundationalism and 'moderate' or 'minimal' Foundationalism, with the distinction depending on whether various epistemic immunities are required of foundations. Finally, 'simple' and 'iterative' Foundationalism are dependent on whether it is required of as foundations only that it is immediately justified, or whether it is also required that the higher level belief that the former belief is immediately justified is itself immediately justified.

However, classic opposition is between Foundationalism and coherentism. Coherentism denies any immediate justification. It deals with the regress argument by rejecting 'linear' chains of justification and, in effect, taking the total system of belief to be epistemically primary. A particular belief is justified to the extent that it is integrated into a coherent system of belief. More recently, 'pragmatists' like American educator, social reformer and philosopher of pragmatism John Dewey (1859 - 1952), have developed a position known as contextualism, which avoids ascribing any overall structure to knowledge. Questions concerning justification can only arise in particular context, defined in terms of assumptions that are simply taken for granted, though they can be questioned in other contexts, where other assumptions will be privileged.

Meanwhile, it is, nonetheless, the idea that the language each of us speaks is essentially private, that leaning a language is a matter of associating words with, or ostensibly defining words by reference to, subjective experience (the 'given'), and that communication is a matter of stimulating a pattern of associations in the mind of the hearer qualitatively identical with what in the mind of the speaker is linked with multiple mutually supporting misconceptions about language, experiences and their identity, the mental and its relation to behaviour, self - knowledge and knowledge of the states of minds of others.

1   The idea that there can be such a thing as a private language is one manifestation of a tactic committed to what Wittgenstein called 'Augustine's picture of language' - pre - theoretical picture according to which the essential function of words is to name items in reality, that the link between word and world is affected by 'ostensive definition', and describe a state of affairs. Applied to the mental, this knows that what a psychological predicate such as 'pain' if one knows, is acquainted with, what it stands for - a sensation one has. The word 'pain' is linked to the sensation it names by way of private ostensive definition, which is affected by concentration (the subjective analogue of pointing) on the sensation and undertaking to use the word of that sensation. First - person present tense psychological utterances, such as 'I have a pain' are conceived to be descriptions which the speaker, as it was, reads off the facts which are private accessibility to him.

2. Experiences are conceived to be privately owned and inalienable - no on else can have my pain, but not numerically, identical with mine. They are also thought to be epistemically private - only I really know that what I have is a pain, others can at best only believe or surmise that I am in pain.

3. Avowals of experience are expressions of self - knowledge. When I have an experience, e.g., a pain, I am conscious or aware that I have by introspection (conceived as a faculty of inner sense). Consequently, I have direct or immediate knowledge of my subjective experience. Since no one else can have what I have, or peer into my mind, my access is privileged. I know, and an certain, that I have a certain experience whenever I have it, for I cannot doubt that this, which I now have, in a pain.

4. One cannot gain introspective access to the experience of others, so one can obtain only indirect knowledge or belief about them. They are hidden behind the observable, behaviour,

inaccessible to direct observation, and inferred either analogically. Whereby, this argument is intended to establish our right to believe in the existence and nature of other minds, it admits it is possible that the objects we call persons are, other than themselves, mindless automata, but claims that we nonetheless, have sufficient reason for supposing this not to be the case. There is more evidence that they are not mindless automata than they are.

The real criticism of the argument is du e to Wittgenstein (1953). It is that the argument assumes that we at least understand the claims that there are subjects of experience other than ourselves, who enjoy experiences which are like ours but not ours: It only asks what reason we have to suppose that claim true. But if the argument does indeed express the ground of our right to believe in the existence of others, it is impossible to explain how we are able to achieve that understanding. So if there is a place for argument from analogy, the problem of other minds - the real, hard problem, which is how we acquire a conception of another mind - is insoluble. The argument is either redundant or worse.

Even so, the inference to the best explanation is claimed by many to be a legitimate form of non - deductive reasoning, which provides an important alternative to both deduction and enumerative induction. Indeed, some would claim that it is only through reasoning to the best explanation that one can justify beliefs about the external world, the past, theoretical entities in science, and even the future. Consider belief about the external world and assume that we know what we do about the external world through our knowledge of the subjective and fleeting sensations. It seems obvious that we cannot deduce any truths about the existence of physical objects from truths describing the character of our sensations. But either can we observe a correlation between sensations and something other than sensations since by hypothesis all we ever nave to rely on ultimately is knowledge of our sensations. Nevertheless, we may be able to posit physical objects as the best explanation for the character and order of our sensations. In the same way, various hypotheses about the past, might best be explained by present memory: Theoretical postulates in physics might best explain phenomena in the macro - world. And it is even possible that our access to the future to explain past observations. But what exactly is the form of an inference to the best explanation? However, if we are to distinguish between legitimate and illegitimate reasoning to the best explanation it would seem that we need a more sophisticated model of the argument form. It would seem that in reasoning to an explanation we need 'criteria' for choosing between alternative explanation. If reasoning to the best explanation is to constitute a genuine alterative to inductive reasoning, it is important that these criteria not be implicit premises which will convert our argument into an inductive argument

However, in evaluating the claim that inference to best explanation constitutes a legitimate and independent argument form, one must explore the question of whether it is a contingent fact that at least most phenomena have explanations and that explanations that satisfy a given criterion, simplicity, for example, is more likely to be correct and writers of texts, if the universe structure in such a way that simply, powerful, familiar explanations were usually the correct explanation. It is difficult to avoid the conclusion that this is true, but It would be an empirical fact about our universe discovered only a posterior. If the reasoning to the best explanation relies on such criteria, it seems that one cannot without circularity use reasoning to the best explanation to discover that the reliance on such criteria is safe. But if one has some independent was of discovering that simple, powerful, familiar explanations are more often correct, then why should we think that reasoning of the best explanation is an independent source of information about the world? Indeed, why should we not conclude that it would be more perspicuous to represent the reasoning this way. That is, simply an instance of familiar inductive reasoning.

5. The observable behaviour from which we thus infer consists of bare bodily movements caused by inner mental events. The outer (behaviour) are not logically connected with the inner (the mental). Hence, the mental are essentially private, known 'strictu sensu', only to its owner, and the private and subjective is better known than the public.

The resultant picture leads first to scepticism then, ineluctably to 'solipsism'. Since pretence and deceit are always logically possible, one can never be sure whether another person is really having the experience behaviourally appears to be having. But worse, if a given psychological predicate 'this' (which I have no one else could logically have - since experience is inalienable), then any other subjects of experience. Similar scepticism about defining samples of the primitive terms of a language is private, then I cannot be sure that what you mean by 'red' or 'pain' is not quantitatively identical with what I mean by 'green' or 'pleasure'. And nothing can stop us frm concluding that all languages are private and strictly mutually unintelligible.

Philosophers had always been aware of the problematic nature of knowledge of other minds and of mutual intelligibly of speech of their favour red picture. It is a manifestation of Wittgenstein's genius to have launched his attack at the point which seemed incontestable - namely, not whether I can know of the experiences of others, but whether I can understand the 'private language' of another in attempted communication, but whether I can understand my own allegedly private language.

The functionalist thinks of 'mental states' and events as causally mediating between a subject's sensory inputs and that subject's ensuing behaviour that what makes a mental state the doctrine that what makes a mental state the type of state it is - a pain, a smell of violets, a belief that koalas are dangerous - is the functional relation it bears to the subject's perceptual stimuli it beards to the subject's perceptual stimuli, behavioural responses and other mental states. That's not to say, that, functionalism is one of the great 'isms' that have been offered as solutions to the mind/body problem. The cluster of questions that all of these 'isms' promise to answer can be expressed as: What is the ultimate nature of the mental? At the most overall level, what makes a mental state mental? At the more specific level that has been the focus in recent years: What do thoughts have in common in virtue of which they are thoughts? That is, what makes a thought a thought? What makes a pain a pain? Cartesian Dualism said the ultimate nature of the mental of the mental was said the ultimate nature of the mental was to be found in a special mental substance. Behaviouralism identified mental states with behavioural disposition: Physicalism in its most influential version identifies mental states with brain states. Of course, the relevant physical state s are various sorts of neutral states. Our concepts of mental states such as thinking, and feeling are of course different from our concepts of neural states, of whatever.

Disaffected by Cartesian dualism and from the 'first - person' perspective of introspective psychology, the behaviouralists had claimed that there is nothing to the mind but the subject's behaviour and disposition to behave equally well against the behavioural betrayal, behaving just as pain - free human beings, would be the right sort of case. For example, for Rudolf to be in pain is for Rudolf to be either behaving in a wincing - groaning - and - favouring way or disposed to do so (in that not keeping him from doing so): It is nothing about Rudolf's putative inner life or any episode taking place within him.

Though behaviourism avoided a number of nasty objects to dualism (notably Descartes' admitted problem of mind - body interaction), some theorists were uneasy, they felt that it its total repudiation of the inner, behaviourism was leaving out something real and important. U.T. Place spoke of an 'intractable residue' of conscious mental items that bear no clear relations to behaviour of any particular sort. And it seems perfectly possible for two people to differ psychologically despite total similarity of they're actual and counter - factual behaviour, as in a Lockean case of 'inverted spectrum': For that matter, a creature might exhibit all the appropriate stimulus - response relations and lack mentation entirely.

For such reasons, Place and the Cambridge - born Australian philosopher J.J.C. Smart proposed a middle way, the 'identity theory', which allowed that at least some mental states and events are genuinely inner and genuinely episodic after all: They are not to be identified with outward behaviour or even with hypothetical disposition to behave. But, contrary to dualism, the episodic mental items are not ghostly or non - physical either. Rather, they are neurophysiological of an experience that seems to resist 'reduction' in terms of behaviour. Although 'pain' obviously has behavioural consequences, being unpleasant, disruptive and sometimes overwhelming, there is also something more than behaviour, something 'that it is like' to be in pain, and there is all the difference in the world between pain behaviour accompanied by pain and the same behaviour without pain. Theories identifying pain with neural events subserving it have been attacked, e.g., Kripke, on the grounds that while a genuine metaphysical identity y should be necessarily true, the association between pain and any such events would be contingent.

Nonetheless, the American philosopher's Hilary Putnam (1926-) and American philosopher of mind Alan Jerry Fodor (1935-), pointed out a presumptuous implication of the identity theory understood as a theory of types or kinds of mental items: That a mental type such s pain has always and everywhere the neurophysiological characterization initially assigned to it. For example, if the identity theorist identified pain itself with the firing of c - fibres, it followed that a creature of any species (earthly or science - fiction) could be in pain only if that creature had c - fibres and they were firing. However, such a constraint on the biology of any being capable of feeling pain is both gratuitous and indefensible: Why should we suppose that any organism must be made of the same chemical materials as us in order to have what can be accurately recognized pain? The identity theorists had overreacted to the behaviourists' difficulties and focussed too narrowly on the specifics of biological humans actual inner states, and in doing so, they had fallen into species chauvinism.

Fodor and Putnam advocated the obvious correction: What was important, were no t being c-fibres (per se) that were firing, but what the c-fibres were doing, what their firing contributed to the operation of the organism as a whole? The role of the c-fibres could have been preformed by any mechanically suitable component s long as that role was performed, the psychological containment for which the organism would have been unaffected. Thus, to be in pain is not per se, to have c - fibres that are firing, but merely to be in some state or other, of whatever biochemical description that play the same functional role as did that plays the same in the human beings the firing of c-fibres in the human being. we may continue to maintain that pain 'tokens', individual instances of pain occurring in particular subjects at particular neurophysiological states of these subjects at those times, throughout which the states that happed to be playing the appropriate roles: This is the thesis of 'token identity' or 'token physicalism'. But pan itself (the kind, universal or type) can be identified only with something mor e abstract: th e caudal or functional role that c - fibres share with their potential replacements or surrogates. Mental state - and identified not with neurophysiological types but with more abstract functional roles, as specified by 'stare - tokens' relations to the organism's inputs, outputs and other psychological states.

Functionalism has in itself the distinct souses for which Putnam and Fodor saw mental states in terms of an empirical computational theory of the mind, also, Smart's 'topic neutral' analyses led Armstrong and Lewis to a functional analysis of mental concepts. While Wittgenstein's idea of meaning as use led to a version of functionalism as a theory of meaning, further developed by Wilfrid Sellars (1912 - 89) and later Harman.

One motivation behind functionalism can be appreciated by attention to artefact concepts like 'carburettor' and biological concepts like 'kidney'. What it is for something to be a carburettor is for it to mix fuel and air in an internal combustion engine, and carburettor is a functional concept. In the case of 'kidney', the scientific concept is functional - defined in terms of a role in filtering the blood and maintaining certain chemical balances.

The kind of function relevant to the mind can be introduced through the parity - detecting automaton, wherefore according to functionalism, all there is to being in pain is having to say 'ouch', wonder whether you are ill, and so forth. Because mental states in this regard, entail for its method for defining automaton states is supposed to work for mental states as well. Mental states can be totally characterized in terms that involve only logico-mathematical language and terms for input signals and behavioural outputs. Thus, functionalism satisfied one of the desiderata of behaviourism, characterized the mental in entirely non - mental language.

Suppose we have a theory of mental states that specify all the causal relations among the stats, sensory inputs and behavioural outputs. Focussing on pain as a sample, mental state, it might say, among other things, that sitting on a tack causes pain an that pain causes anxiety and saying 'ouch'. Agreeing for the sake of the example, to go along with this moronic theory, functionalism would then say that could define 'pain' as follows: Bing in pain - being in the first of two states, the first of which is causes by sitting on tacks, and which in turn cases the other state and emitting 'ouch'. More symbolically:

Being in pain = Being an x such that

P Q[sitting on a tack cause s P and P

cause's both Q and emitting 'ouch; and

x is in P]

623

More generally, if T is a psychological theory with 'n' mental terms of which the seventeenth is 'pain', we can define 'pain' relative to T as follows (the 'F1' . . . 'Fn' are variables that replace the 'n' mental terms):

Being in pain = Being an x such that

F1 . . . Fn[T(F1 . . . Fn) & x is in F17]

The existentially quantified part of the right - hand side before the '&' is the Ramsey's sentence of the theory 'T'. In this way, functionalism characterizes the mental in non - mental terms, in terms that involve quantification over realization of mental states but no explicit mention of them: Thus, functionalism characterizes the mental in terms of structures that are tacked down to reality only at the inputs and outputs.

The psychological theory 'T' just mentioned can be either originating based on factual information or direct sense experiences as an empirical value for psychological theory or else a commonsense 'folk' theory, and the resulting functionalisms are very different. In the former case, which is named 'psychofunctionalism'. The functional definitions are supposed to fix the extensions of mental terms. In the latter case, conceptual functionalism, the functional definitions are aimed at capturing our ordinary mental concepts. (This distinction shows an ambiguity in the original question of what the ultimate nature of the mental is.) The idea of psychofunctionalism is that the scientific nature of the mental consists not in anything biological, but in something 'organizational', analogous to computational structure. Conceptual functionalism, by contrast, can be thought of as a development of logical Behaviouralism. Logical behaviouralisms thought that pain was a disposition to pan behaviour. But as the Polemical British Catholic logician and moral philosopher Thomas Peter Geach (1916 -) and the influential American philosopher and teacher Milton Roderick Chisholm (1916 - 99) pointed out, what counts as pain behaviour depends on the agent's belief and desires. Conceptual functionalism avoid this problem by defining each mental state in terms of its contribution to dispositions to behave - and have other mental states.

The functional characterization is given to assume a psychological theory with a finite number of mental state terms. In the case of monadic states like pain, the sensation of red, and so forth. It does seem a theoretical option to simply list the states and the=ir relations to other states, inputs and outputs. But for a number of reasons, this is not a sensible theoretical option for belief - states, desire - states, and other propositional - attitude states. For on thing, the list would be too long to be represented without combinational methods. Indeed, there is arguably no upper bound on the number of propositions anyone which could in principle be an object of thought. For another thing, there are systematic relations among belies: For example, the belief that 'John loves Mary'. Ann the belief that 'Mary loves John'. These belief - states represent the same objects as related to each other in converse ways. But a theory of the nature of beliefs can hardly just leave out such an important feature of them. We cannot treat 'believes-that -grass-is-green', 'believes-that-grass-is-green], and so forth, as unrelated', as unrelated primitive predicates. So we will need a more sophisticated theory, one that involves some sort of combinatorial apparatus. The most promising candidates are those that treat belief as a relation. But a relation to what? There are two distinct issues at hand. One issue is how to formulate the functional theory, for which our acquiring of knowledge - that acquires knowledge - how, abilities to imagine and recognize, however, the knowledge acquired can appear in embedded as contextually represented. For example, reason commits that if this is what it is like to see red, then this similarity of what it is like to see orange, least of mention, that knowledge has the same problem as to infer that non - cognitive analysis of ethical language have in explaining the logical behaviour of ethical predicates. For a suggestion in terms of a correspondence between the logical relations between sentences and the inferential relations among mental states. A second issue is that types of states could possibly realize the relational propositional attitude states. Fodor (1987) has stressed the systematicity of propositional attitudes and further points out that the beliefs whose contents are systematically related exhibit th e following sort of empirical relation: If one is capable of believing that Mary loves John, one is also capable of believing that John love Mary. Jerry Fodor argues that only a language of thought in the brain could explain this fact.

Jerry Alan Fodor (1935-), an American philosopher of mind who is well known for a resolute realism about the nature of mental functioning. Taking the analogy between thought and computation seditiously. Fodor believes that mental representations should be conceived as individual states with their own identities and structure, like formulae transformed by processes of computation or those of the 'Holist' such as Donald Herbert Davidson (1917-2003) or, 'instrumentalists about mental ascriptions, such as Daniel Clement Dennett (1952). In recent years he has become a vocal critic of some of the aspirations of cognitive science, literaturizing such books as 'Language of Thought' (1975, 'The Modularity of Mind (1983), 'Psychosemantics (1987), 'The Elm and the Expert(1994), 'Concepts: Where Cognitive Science went Wrong' (1998), and 'Hume Variations '(2003).

Purposively, 'Folk psychology' is primarily 'intentional explanation': It's the idea that people's behaviour can be explained b yy reference to the contents of their beliefs and desires. Correspondingly, the method - logical issue is whether intentional explanation can be co - opted to make science out of. Similar questions might be asked about the scientific potential of other folk - psychological concepts (consciousness for example), but, what make s intentional explanation problematic is that they presuppose that there are intentional states. What makes intentional states problematic is that they exhibit a pair of properties assembled in the concept of 'intentionality', in its current use the expression 'intentionality refers to that property of the mind by which it is directed at, about, or of objects and stat es of affairs in the world. Intentionality, so defined, includes such mental phenomena as belief, desire, intention, hope, fear, memory, hate, lust, disgust, and memory as well as perception and intentional action, however, there is in remaining that of:

(1) Intentional states have causal powers. Thoughts (more precisely, having of thoughts) make things happen: Typically, thoughts make behaviour happen. Self - pit y can make one weep, as can onions.

(2) Intentional states are semantically evaluable, beliefs, for example, area about how things are and are therefore true or false depending on whether things are the way that they are believed to be. Consider, by contrast, tables, chairs, onions, and the cat's being on the mat. Though they all have causal powers they are not about anything and are therefore not evaluable as true or false.

If there is to be an intentional science, there must be semantically evaluable things that have causal powers. Moreover, there must be laws about such things, including, in particular, laws that relate beliefs and desires to one another and to actions. If there are no intentional laws, then there is no intentional science. Perhaps, scientific explanation is not always explanation by law subsumption, but surely if often is, and there is no obvious reason why an intentional science should be exceptional in this respect. Moreover, one of the best reasons for supposing that common sense is right about there being intentional states is precisely that there seem to be many reliable intentional generalizations for such states to fall under. It is for us to assume that many of the truisms of folk psychology either articulate intentional laws or come pretty close doing so.

So, for example, it is a truism of folk psychology that rote repetition facilitates recall. (Moreover, and most generally, repetition improves performance 'How do you get to Carnegie Hall'?) This generalization relates the content to what you learn to the content of what you say to yourself while you are learning it: So, what it expresses, is, 'prima facie', a lawful causal relation between types of intentional states. Real psychology y has lots more to say on this topic, but it is, nonetheless, much more of the same. To a first approximation, repetition does causally facilitate recall, and that it does is lawful.

There are, to put it mildly, many other case of such reliable intentional causal generalizations. There are also many, many kinds of folk psychological generalizations about 'correlations' among intentional states, and these to are plausible candidates for flushing out as intentional laws. For example that anyone who knows what 7 + 5 is also to know what 7+ 6 is: That anyone who knows what 'John love's Mary' who knows what 'Mary' loves 'John', and so forth.

Philosophical opinion about folk psychological intentional generalizations runs the gamut from 'there are not any that are really reliable' to. They are all platitudinously true, hence not empirical at all. Nevertheless, suffice to say, that the necessity of 'if $7 + 5 = 12$ then $7 + 6 = 13$' is quite compatible with the 'contingency' of 'if someone knows that $7 + 5 = 12$, then he knows that $7 + 6 = 13$: And, then, part of the question 'how can there be an intentional science' is 'how can there be an intentional practice of law'?

Let us assume most generally, that laws support counterfactuals and are confirmed by their instances. Further, to assume that every law is standardized for its basic or not. Basic laws are exceptionless or intractably statistical. The only basic laws are laws of basic physics.

All Non - basic laws, including the laws of all the Non - basic sciences, including, in particular, the intentional laws of psychology, are 'c[eteris] p[aribus] laws: They hold only 'all else being equal'. There is - anyhow. There ought to be that a whole department of the philosophy of science devoted to the construal of cp laws: To making clear, for instances, how they can be explanatory, how they can support counterfactuals, how they can subsume the singular causal truths that instance them . . . and so forth. Omitting only these issues in what gives presence to the future, is, because they do not belong to philosophical psychology as such. If the laws of intentional psychology is a special, I, e., Non - basic science. Not because it is an intentional science.

There is a further quite general property that distinguishes cp laws from basic ones: Non - basic laws want mechanisms for their implementation. Suppose, for a working example, that some special science states that being 'F' causes ,'s to be 'G'. (Being irradiated by sunlight causes plants to photo - synthesize, as for being freely suspended near the earth's surface causes bodies to fall with uniform accelerating, and so on.) Then it is a constraint on this generalization's being lawful that 'How does, being 'F' cause's χ's to be 'G'? There must be an answer, this is, however, if we are continued to suppose that one of the ways special science; laws are different from basic laws. A basic law says that 'F's causes (or are), if there were, perhaps that aby explaining how, or why, or by what F's cause G's, the law would have not been basic but derived.

Typically - though variably - the mechanism that implements a special science law is defined over the micro - structure of the thing that satisfy the law. The answer to 'how does. Sunlight make plants photo - synthesize'? Its function implicates the chemical structure of plants: The answer to 'how does freezing make water solid'? This question surely implicates the molecular structure of waters' foundational elements, and so forth. In consequence, theories about how a law is implemented usually draw on or upon the vocabularies of two, or more levels of explanation.

If you are specially interested in the peculiarities of aggregates of matter at the $L1^{th}$ level (in plants, or minds, or mountains, as it might be) then you are likely to be specially interested in implementing mechanisms at the L -1th level (the 'immediately' mechanisms): This is because the characteristics of L - level laws can often be explained by the characteristics of their L - 1th level implementations. You can learn a lot about plants qua plants by studying their chemical composition. You learn correspondingly less by studying their subatomic constituents, though, no doubt, laws about plants are implemented, eventually, sub - atomically. The question thus arises of what mechanisms might immediately implement the intentional laws of psychology with that accounting for their characteristic features.

Intentional laws subsume causal interactions among mental processes, that much is truistic. But, in this context, something substantive, something that a theory of the implementation of intentional laws will account for. The causal processes that intentional states enter into have a tendency to preserve their semantic properties. For example, thinking true thoughts are so, that an inclining inclination to casse one to think more thoughts that are also true. This is not small matter: The very rationality of thought depends on such fact, in that ewe can consider or place them for interpretations as that true thoughts that ((P Q) and (P)) makes receptive to cause true thought that 'Q'.

A good deal has happened in psychology - notably since the Viennese founder of psychoanalysis, Sigmund Freud (1856 - 1939) - has consisted of finding new and surprising cases where mental processes are semantically coherent under intentional characterizations. Freud made his reputation by showing that this was true even much of the detritus of behaviours, dreams, verbal slips and the like, even to free or word association and ink - blob coloured identification cards (the Rorschach test). Even so, it turns out that psychology of normal mental processes is largely a grist for the same normative intention. For example, it turns out to be theoretically revealing to construe perceptual processes as inferences that take specifications of proximal stimulations as premises and yield specifications, and that are reliably truth preserving in ecologically normal circumstances. The psychology of learning cries out for analogous treatment, e.g., for treatment as a process of hypothesis formation and ratifying confirmation.

Intentional states, as or common - sense understands them, have both causal and semantic properties and that the combination appears to be unprecedented: Propositions are semantically evaluable, but they are abstract objects and have no casual powers. Onions are concrete particulars and have casual powers, however, they are not semantically evaluable. Intentional states seem to be unique in combining the two that is what so many philosophers have against them.

Suppose, once, again, that 'the cat is on the mat'. On the one hand, the thing as stated about the cat on the mat, is a concrete particular in good standing and it has, qua material object, an open - ended galaxy of causal powers. (It reflects light in ways that are essential to its legibility; It exerts a small but in particular detectable gravitational effect upon the moon, and whatever. On the other hand, what stands concrete is about something and is therefore semantically evaluable: It's true if and only if there is a cat where it says that there is. So, then, the inscription of 'the cat is on the mat,' has both content and causal powers, and so does my thought that the cat is on the mat.

At this point, we are asked of how many words are there in the sentence. 'The cat is on the mat'? There are, of course, at least two answers to this question, precisely because one can count either word types, of which there are five, or individual occurrences - known as tokens - of which there are six. Moreover, depending on how one chooses to think of word types, another answer is possible. Since the sentence contains definite articles, noun, a proposition and a verb, there are four grammatically different types of word in the sentence.

The type/token distinction, understood as a distinction between sorts of thing and instances, is commonly applied to mental phenomena. For example, one can think of pain in the type way as when we say that we have experienced burning pain many times: Or, in the token way, as when we speak of the burning pain currently being suffered. The type/token distinction for mental states and events becomes important in the context of attempts to describe the relationship between mental and physical phenomena. In particular, the identity theory asserts that mental states are physical states, and this raises the question whether the identity in question is of types or tokens.

Appreciably, if mental states are identical with physical states, presumably the relevant physical states are various sorts of neural state. Our concept of mental states such as thinking, sensing, and feeling and, and, of course, are different from our concepts of neural states, of whatever sort. Still, that is no problem for the identity theory. As J.J. Smart (1962) who first argued for the identity theory, and, emphasizes the requisite identity does not depend on our concepts of mental states or the meaning of mental terminology. For 'a' to be identical with 'b', both 'a' and 'b' must have exactly the same properties, however, the terms 'a' and 'b' need not mean the same. The principle of the indiscernibility of identical states that if 'a' is identical with 'b'. Then every property that 'a' has 'b' has, and vice versa. This is sometimes known as Leibniz's law.

However, the problem does seem to arise about the properties of mental states. Suppose pain is identical with a certain firing of c - fibres. Although a particular pain is the very same state as a neural firing, we identify that state in two different ways: As a pain and synaptic-firing. The state will therefore have certain properties in virtue of which we identify it as neural firing, the properties in virtue of which we identify it as a pain will be mental properties, whereas those in virtue of which we identify it as neural firing will be physical property. This has, as it seems, as too many to lead to a kind of duality, at which the level of the properties of mental states. Even so, if we reject a dualism of substances and take people simply to be physical organism, those organisms still have both mental and physical states.

The problem just sketched about mental properties is widely thought to be most pressing for sensations, since the painful quality of pains and the red quality of visualization in sensations that seem to be irretrievably non - physical. So even if mental states are all identicals with physical states, these states appear to have properties that are not physical. And if mental states do actually have non - physical properties, the identity of mental with physical states would not sustain the thoroughgoing mind - body materialism.

A more sophisticated reply to the difficultly about mental properties is due independently to the forth - right Australian 'materialist and together with J.J.C. Smart, the leading Australian philosophers of the second half of the twentieth century. D.M. Armstrong (1926 -) and the American philosopher David Lewis (1941 - 2002), who argue that for a state to be a particular sort of intentional state or sensation is for that state to bear characteristic causal relations to other particular occurrences. The properties in virtue of which we identify states as thoughts or sensations will still be neutral as between being mental and physical, since anything can bear a causal relations to anything else. But causal connections have a better chance than simplify in some unspecified respect of capturing the distinguishing properties of sensations and thoughts.

Early identity theorists insisted that the identity between mental and bodily events was contingent, meaning simply that the relevant identity statements were not conceptual truths. That leaves open the question of whether such identities would be necessarily true on other construals of necessity.

American logician and philosopher, Saul Aaron Kripke (1940-) made his early reputation as a logical prodigy, especially through the work on the completeness of systems of modal logic. The three classic papers are 'A Completeness Theorem in Modal Logic' (1959, 'Journal of Symbolic Logic') 'Semantical Analysis of Modal Logic' (1963, Zeltschrift fur Mathematische Logik und Grundlagen der Mathematik) and 'Semantical Considerations on Modal Logic (1963, Acta Philosohica Fennica). In Naming and Necessity' (1980), Kripke gave the classic modern treatment of the topic of reference, to clarify the distinction between names and 'definite descriptions, and opening the door to many subsequent attempts to a better understanding of a notion of reference in terms of a causal link between the use of a term and an original episode of attaching a name to a subject. His Wittgenstein on Rules and Private Language (1983) also proved seminal, putting the rule - following considerations at the centre of Wittgenstein studies, and arguing that the private language argument is an application of them. Kripke has also written influential work on the theory of truth and the solution of the 'semantic paradoxes'.

Nonetheless, Kripke (1980) has argued that such identities would have to be necessarily true if they were true at all. Some terms refer to things contingently, in that those terms would have referred to different things had circumstances been relevantly different. Kripke's example is 'The first Post - master General of the us of 'A', which, in a different situation, would have referred to somebody other than Benjamin Franklin. Kripke calls these term's non - rigid designators. Other terms refer to things necessarily, since no circumstances are possible in which they would refer to anything else, these terms are rigid designators.

If the term 'a' and 'b' refer to the same thing and both determine that thing necessarily, the identity statement 'a = b' is necessarily true. Kripke maintains that the term 'pain' and the term for the various brain states all determine the states they refer to necessarily: No circumstances are possible in which these terms would refer to different things. So, if pain were identical d with some particular brain state. But be necessarily identical with that state. Yet, Kripke argues that pain cannot be necessarily identical with any brain state, since the tie between pains and brain states plainly seems contingent. He concludes that they cannot be identical at all.

Kripke notes that our intuition about whether an identity is contingent can mislead us. Heat is necessarily identical with mean molecular kinetic energy: No circumstances are possible in which they are not identical. Still, it may at first sight appear that heat could have been identical with some other phenomena, but it appears that this way, Kripke argues only because we pick out heat by our sensation of heat, which bears only a contingent - bonding to mean molecular kinetic energy. It is the sensation of heat that actually seems to be connected contingently with mean molecular kinetic energy, not with mean molecular kinetic energy, not the physical heat itself.

Kripke insists, however, that such reasoning cannot disarm our intuitive sense that pain is connected only contingently with brain states. This is, because for a state to be pain is necessity for it to be felt as pain, unlike heat, in the case of pain there is no difference between the state itself and how that state is felt, and intuitions about the one are perforce intuitions about the other one are perforce intuitions about the other.

Kripke's assumption and the term 'pain' is open to question. As Lewis notes. One need not hold that 'pain' determines the same state in all possible situations indeed, the causal theory explicitly allows that it may not. And if it does not, it may be that pains and brain states are contingently identicals. But there is also a problem about some substantive assumption Kripke makes about the nature of pains, namely, those pains are necessarily felt as pains. First impression notwithstanding, there is reason to think not. There are times when we are not aware of our pains, for example, when we are suitably distracted, so the relationship between pains and our being aware of them may not be contingent after all, just as the relationship between physical heat and our sensation of heat is. And that would disarm the intuitions that pain is connected only contingently with brain states.

Kripke's argument focuses on pains and other sensations, which, because they have qualitative properties, are frequently held to cause the greater of problems for the identity theory. The American moral and political theorist Thomas Nagel (1937 -) traces to general difficulty for the identity theory to the consciousness of mental states. A mental state's being conscious, he urges, that there is something it is like to be in that state. And to understand that, we must adopt the point of view of the kind of creature that is in the state. But an account of something is objective, he insists, only insofar as it is independents of any particular type of point of view. Since consciousness is inextricably tied to points of view, no objective account of it is possible. And that conscious states cannot be identical with bodily states.

The viewpoint of a creature is central to what that creature's conscious states are like, because different kinds of crenatures have conscious states with different kinds of qualitative property. However, the qualitative properties of a creature's conscious states depend, in an objective way, on that creature's perceptual apparatus. we cannot always predict what anther creature's conscious states are like, just as we cannot always extrapolate from microscopic to macroscopic properties, at least without having a suitable theory that covers those properties. But what a creature's conscious states like depends in an objective way on its bodily endowment, which is itself objective. So, these considerations give us no reason to think that those conscious states are like is not also an objective matter.

'If a sensation is not conscious, there is nothing it's like to have it. So Nagel's idea that what it is like to have sensations is central to their nature suggests that sensations cannot occur without being conscious. And that in turn, seems to threaten their objectivity. If sensations must be conscious, perhaps they have no nature independently of how we ae aware of them, and thus no objective nature. Nonetheless, only conscious sensations seem to cause problems of the independent theory.

The notion of subjectivity, as Nagel again, see, is the notion of a point of view, what psychologists call a 'constructionist theory of mind'. Undoubtedly, this notion is clearly tied to the notion of essential subjectivity. This kind of subjectivity is constituted by an awareness of the world's being experienced differently by different subjects of experience. (It is thus possible to see how the privacy of phenomenal experience might be easily confused with the kind of privacy inherent in a point of view.)

Point-of-view subjectivity seems to take time to develop. The developmental evidence suggests that even toddlers are abl e to understand others as being subjects of experience. For instance, as a very early age, we begin ascribing mental states to other things - generally, to those same things to which we ascribe 'eating'. And at quite an early age we can say what others would see from where they are standing. We early on demonstrate an understanding that the information available is different from different perceiver. It is in these perceptual senses that we first ascribe the point-of-view- subjectivity.

Nonetheless, some experiments seem to show that the point - of - view subjectivity then ascribes to others is limited. A popular, and influential series of experiments by Wimmer and Perner (1983) is usually taken to illustrate these limitations (though there are disagreements about the interpretations, as such.) Two children - Dick and Jane - watch as an experimenter puts a box of candy somewhere, such as in a cookie jar, which is opaque. Jane leaves the room. Dick is asked where Jane will look for the candies, and he correctly answers. 'In the cookie jar'. The experimenter, in dick's view, then takes the candy out of the cookie jar and puts it in another opaque place, a drawer, ay. When Dick is asked where to look for the candy, he says quite correctly. 'In the drawer'. When asked where Jane will look for the candy when she returns. But Dick answers. 'In the drawer'. Dick ascribes to Jane, not the point - of - view subjectivity she is likely ton have, but the one that fits the facts. Dick is unable to ascribe to Jane belief - his ascription is 'reality driven - and his inability demonstrates that Dick does not as yet have a fully developed point - of - view subjectivity.

At around the age of four, children in Dick's position do ascribe the like point - of - view subjectivity to children in Jane's position ('Jane will look in the cookie jar'): But, even so, a fully developed notion of a point - of - view subjectivity is not yet attained. Suppose that Dick and Jane are shown a dog under a tree, but only Dick is shown the dog's arriving there by chasing a boy up the tree. If Dick is asked to describe, what Jane, who he knows not to have seen the dog under the tree. Dick will display a more fully developed point - of - view subjectivity only those description will not entail the preliminaries that only he witnessed. It turns out that four - year - olds are restricted by the age's limitation, however, only when children are six to seven do they succeed.

Yet, even when successful in these cases' children's point - of - view subjectivity is reality - driven. Ascribing a point - of - view, subjectivity to others is still in terms relative to information available. Only in our teens do we seem capable of understanding that others can view the world differently from ourselves, even when given access to the same information. Only then do we seem to become aware of the subjectivity of the knowing procedure itself: Interring the 'facts' can be coloured by one's knowing procedure and history. There are no 'merely' objective facts.

Thus, there is evidence that we ascribe a more and more subjective point of view to others: from the point - of - view subjectivity we ascribe being completely reality - drive, to the possibility that others have insufficient information, to they're having merely different information, and finally, to their understanding the same information differently. This developmental picture seems insufficient familiar to philosophers - and yet well worth our thinking about and critically evaluating.

The following questions all need answering does the apparent fact that our point - of - view subjectivity ascribed to others develop over time, becoming more and more of the 'private' notions, shed any light on the sort of subjectivity we ascribe to our own self? Do our self - ascriptions of subjectivity themselves become more and more 'private', metre and more removed both from the subjectivity of others and from the objective world? If so, what is the philosophical importance of these facts? At the last, this developmental history shows that disentangling our self from the world we live in is a complicate matter.

Based in the fundament of reasonableness, it seems plausibility that we share of our inherented perception of the world, that 'self - realization as 'actualized' of an 'undivided whole', drudgingly we march through the corpses to times generations in that we are founded of the last two decades. Here we have been of a period of extraordinary change, especially in psychology. Cognitive psychology, which focuses on higher mental processes like reasoning, decision masking, problem solving, language processing and higher-level visual processing, has become - perhaps - the dominant paradigm among experimental psychologists, while behaviouristically oriented approaches have gradually fallen into disfavour. Largely as a result of this paradigm shift, the level of interaction between the disciplines of philosophy and psychology has increased dramatically.

Nevertheless, developmental psychology was for a time dominated by the ideas of the Swiss psychologist and pioneer of learning theory, Jean Piaget (1896-1980), whose primary concern was a theory of cognitive developments (his own term was 'genetic epistemology). What is more, like modern - day cognitive psychologists, Piaget was interested in the mental representations and processes that underlie cognitive skills. However, Piaget's genetic epistemology y never co-existed happily with cognitive psychology, though Piaget's idea that reasoning is based in an internalized version of predicate calculus has influenced research into adult thinking and reasoning. One reason for the lack of declining side by side interactions between genetic epistemology and cognitive psychology was that, as cognitive psychology began to attain prominence, developmental psychologists were starting to question Piaget's ideas. Many of his empirical claims about the abilities, or more accurately the inabilities, of children of various ages were discovered to be contaminated by his unorthodox, and in retrospect unsatisfactory, empirical methods. And many of his theoretical ideas were seen to be vague, or uninterpretable, or inconsistent, however.

More than one of the central goals of thee philosophy of science is to provide explicit and systematic accounts of the theories and explanatory strategies exploited in th e sciences. Another common goal is to construct philosophically illuminating analysis or explanations of central theoretical concepts invoked in one or another science. In the philosophy of biology, for example, there is a rich literature aimed at understanding teleological explanations, and there has been a great deal of work on the structure of evolutionary theory on the structure of evolutionary theory and on such crucial concepts as fitness and biological function. The philosophy of physics is another are a in which studies of this sort have been actively pursued. In undertaking this work, philosophers need not (and typically do not) assume that there is anything wrong with the science the y are studying. Their goal simply to provide e accounts of the theories, concepts, and explanatory strategies that scientists are using - accounts th at are more explicit, systematic and philosophically sophisticated that an offered rather rough - and - ready accounts offered by scientists themselves.

Cognitive psychology is in many was a curious and puzzling science. Many of the theorists put forward by cognitive psychologists make use of a family of 'intentional' concepts - like believing that 'p', desiring that 'q', and representing 'r' - which do not appear in the physical or biological sciences, and these intentional concepts play a crucial role in many of the explanations offered by these theories.

If a person 'X' thinks that 'p', desires that 'p', believes that 'p'. Is angry at 'p' and so forth, then he or she is described as having a propositional attitude too 'p?'. The term suggests that these aspects of mental life are well thought of in terms of a relation to a 'proposition' and this is not universally agreeing. It suggests that knowing what someone believes, and so on, is a matter of identifying an abstract object of their thought, than understanding his or her orientation toward a greater extent of worldly objects.

Once, again, the directness or 'aboutness' of many, if not all, conscious states have side by side their summing 'intentionality'. The term was used by the scholastics, but belief thoughts, wishes, dreams, and desires are about things. Equally, we use to express these beliefs and other mental states are about things. The problem of intentionality is that of understanding the relation obtaining between a mental state, or its expression, and the things it is about. A number of peculiarities attend this relation. First, If I am in some relation to a chair, for instance by sitting on it, then both it and I am in some relation to a chair, that is, by sitting on it, then both it and I must exist. But while mostly one thinks about things that exist, sometimes (although this way of putting it has its problems) one has beliefs, hopes, and fears about things that do not, as when the child expects Santa Claus, and the adult fears snakes. Secondly, if I sit on the chair, and the chair is the oldest antique chair in all of Toronto, then I am on the oldest antique chair in the city of Toronto. But if I plan to avoid the mad axeman, and the mad axeman is in fact my friendly postal - carrier. I do not therefore plan to avoid my friendly postal - carrier. The extension of such is the predicate, is the class of objects that is described: The extension of 'red' is the class of red things. The intension is the principle under which it picks them out, or in other words the condition a thing must satisfy to be truly described by the predicate. Two predicates '. . . are a rational animal. '. . . is a naturally feathered biped might pick out the same class but they do so by a different condition? If the notions are extended to other items, then the extension of a sentence is its truth - value, and its intension a thought or proposition: And the extension of a singular term is the object referred to by it, if it so refers, and its intension is the concept by of which the object is picked out. A sentence puts a predicate on other predicate or term with the same extension can be substituted without it being possible that the truth - value changes: If John is a rational animal and we substitute the coexistence 'is a naturally feathered biped', then 'John is a naturally featherless biped', other context, such as 'Mary believes that John is a rational animal', may not allow the substitution, and are called 'intensional context'.`

What remains of a distinction between the context into which referring expressions can be put. A contest is referentially transparent if any two terms referring to the same thing can be substituted in a 'salva veritate', i.e., without altering the truth or falsity of what is aid. A context is referentially opaque when this is not so. Thus, if the number of the planets is nine, then the number of planets is odd, and has the same truth - value as 'nine is odd': Whereas, 'necessarily the number of planets is odd' or 'x knows that the number of planets is odd' need not have the same truth - value as 'necessarily nine is odd have the same truth - value as 'necessarily nine in odd' or 'x knows that nine is odd'. So while' . . . in odd' provides a transparent context, 'necessarily . . . is odd' and 'x knows that . . . is odd' do not.

Here, in a point, is the view that the terms in which we think of some area are sufficiently infected with error for it be better to abandon them than to continue to try to give coherence theories of their use. Eliminativism should be distinguished from scepticism which claims that we cannot know the truth about some area: Eliminativism claims that there is no truth there to be known, in the terms with which we currently think. An eliminativist about theology simply councils abandoning the terms or discourse of theology, and that will include abandoning worries about the extent of theological knowledge. Eliminativist in the philosophy of mind council abandoning the whole network of terms mind, consciousness' self, Qualia that usher in the problems of mind and body. Sometimes the argument for doing this is that we should wait for a supposed future e understanding of ourselves, based on cognitive science and better than our current mental descriptions provide, something it is supposed that physicalism shows that no mental description could possibly be true.

It seems, nonetheless, that of a widespread view that the concept is indispensable, we must declare seriously that science be that it cannot deal with the central feature of the mind or explain how serious science may include intentionality. One approach in which we communicate fears and beliefs have a two - faced aspect, involving both the object referred to, and the mod e of presentation under which they are thought of. we can see the mind as essentially directed onto existent things, and extensionally relate to them. Intentionality then becomes a feature of language, than a metaphysical or ontological peculiarity of the mental world.

While cognitive psychologists occasionally say a bit about the nature of intentional concepts and the explanations that exploit them, their comments are rarely systematic or philosophically illuminating. Thus, it is hardly surprising that many philosophers have seen cognitive psychology as fertile ground for the sort of careful descriptive work that is done in the philosophy of biology and the philosophy of physics. Jerry Fodor's 'Language of Thought' (1975) was a pioneering study in this genre, one that continues to have a major impact on the field.

The relation between language and thought is philosophy's chicken - or - egg problem. Language and thought are evidently importantly related, but how exactly are they related? Does language come first and make thought possible or vice versa? Or are they counter - balanced and parallel with each making the other possible?

When the question is stated this of such generality, however, no unqualified answer is possible. In some respect language is prior, in other respects thought is prior. For example, it is arguable that a language is an abstract pairing of expressions and meanings, a function, in the set - theatric sense, in that, this makes sense of the fact that Esperanto is a language no one speaks, and it explains why it is that, while it is a contingent fact that 'La neige est blanche' that snow is white among the French-speaking peoples. It is a necessary truth that it that in French and English are abstract objects in this sense, then they exist whether or not anyone speaks them: They even exist in possible worlds in which there are no thinkers. In this respect, then, language, as well as such notions as meaning and truth in a language, is prior to thought.

But even if languages are construed as abstractive expression - meaning pairing, they are construed what was as abstractions from actual linguistic practice - from the use of language in linguistic communicative behaviour - and there remains a clear sense in which language is dependent on thought. The sequence of marks, 'Point Peelie is the most southern point of Canada's geographical boundaries', among us that Point Peelie is the most southern lactation that hosts thousands of migrating species. Had our linguistic practice been different, Point Peelie is a home for migrating species and an attraction of hundreds of tourists, that in fact, that the province of Ontario is a home and a legionary resting point for thousands of migrating species, have nothing at all among us. Plainly that Point Peelie is Canada's most southern location in bordering between Canada and the Unites State of America. Nonetheless, Point Peelie is special to Canada has something to do with the belief and intentions underlying our use of words and structure that compose the sentence of Canada's most southern point and yet nearest point in bordering of the United States. More generally, it is a platitude that the semantic features that marks and sounds have a population of tourist and migrating species are at least partly determined by the attitudinal values for which this is the platitude, of course, which says that meaning depends, partially, on the use in communicative behaviours. So, here, is one clear sense in which language is dependent on thought: Thought is required to imbue marks and sounds with the somantic features they have as to host of populations.

The sense in which language does depend on thought can be wedded to the sense in which language does not depend on thought in the following way. we can say, that a sequence of marks or sounds (or, whatever) 'ò' 'q' in a language 'L', construed as a function from expressions onto meaning, iff L(ò) = q. This notion of meaning - in - a - language, like the notion of a language, is a mere set - theoretic notion that is independent of thought in that it presupposes nothing about the propositional attitude of language users: 'ò' can mean 'q' in 'L' even if 'L' has never very been used? But then, we can say that 'ò' also 'q' in a population 'P'. The question of moment then becomes: What relation must a population 'P' bear to a language 'L' in order for it to be the case that 'L' is a language of 'P', a language member s of 'P' actually speak? In whatever the answer to this question is, this much seems right: In order for a language to be a language of a population of speakers, those speakers must produce sentences of the language in their communicative behaviour. Since such behaviour is intentional, we know that the notion of a language's being the language of a population of speakers presupposes the notion of thought. And since that notion presupposes the notion of thought, we also know that the same is true of the correct account of the semantic features expression have in populations of speakers.

This is a pretty thin result, not on likely to be disputed, and the difficult question remain. we know that there is some relation 'R' such that a adaptive 'L' is used by a population 'P' iff 'L' bears 'R' to 'P'. Let us call this reflation, whatever it turns out to be, the 'actual - language relation'. we know that to explain the semantic features expressions have among those who are apt to produce those expressions, and we know that any account of the relation must require language users to have certain propositional attitudes. But how exactly is the actual language relation to be explained in terms of the propositional attitudes of language users? And what sort of dependence might those propositional attitude in turn have on language or on the semantic factures that are fixed by the actual - language relation? Further, what of the relation of language to thought, before turning to the relation of thought to language.

All must agree that the actual - language relation, and with it the semantic features linguistic items have among speakers, is at least, partly determined by the propositional attitudes of language users. This, however, leaves plenty of room for philosophers to disagree both about the extent of the determination and the nature of the determining propositional attitude. At one end of the determination spectrum, we have those who hold that the actual - language relation is wholly definable in terms on non - semantic propositional attitudes. This position in logical space is most taken as occupied by the programme, sometimes called intention - based semantics, of the English philosopher of language Paul Herbert Grice (1913 - 1988), introducing the important concept of an 'implicature' into the philosophy of language, arguing that not everything that is said is direct evidence for the meaning of some term, since many factors my determine the appropriateness of remarks independently of whether they are actually true. The point, however, undermines excessive attention to the niceties in conversation as reliable indicators of meaning, a methodology characteristic of 'linguistic philosophy'. In a number of elegant papers which identities is with a complex of sentences which it is uttered. The psychological is thus used to explain the semantic, and the question of whether this is the correct priority has prompted considerable subsequent discussion.

The foundational notion in this enterprise is a certain notion of 'speaker - semantics'. It is the species of communicative behaviour reported when we say, for example, that in uttering 'II pleut'. Pierre meant that it was raining, or that in waving her hand, the Queen meant that you were to leave the room. Intention - based semantics seeks to define this notion of speaker meaning wholly in terms of communicators' audience - directed intentions and without recourse to any semantic notions. Then it seeks to define the actual - language relation in terms of the now - defined notion of speaker meaning, together with certain ancillary notions such as that of a conventional regularity or practice, themselves defined wholly in terms of non-semantic propositional attitudes. The definition in terms of speaker meaning of other agent-semantic notions, such as the notions of an illocutionary act, and this, is part of the intention-based semantics programme.

Some philosophers object to intention - based semantics because they think it precludes a dependence of thought on the communicative use of language. This is a mistake, in that if intention - based semantics definitions are given a strong reductionist reading, as saying that public - language semantic properties (i.e., those semantic properties that supervene on use in communicative behaviour) just are psychological properties, it might still be that one could not have propositional attitudes unless one had mastery of a public - language, insofar as the concept of supervenience has seen increasing service in philosophy of mind. The thesis that the mental is supervenient on the physical - roughly, the claim that the mental character of a thing is wholly determine d by its physical nature - has played a key role in the formulation of some influential positions on the mind - bod y problem. In particular, versions of non - reductive physicalism. Mind - body supervenience has also been invoked in arguments for or against certain specific claims about the mental, and has been used to devise solutions to some central problems about the mind - for example, the problem of mental causation - such that the psychological level of description carries with it a mode of explanation which 'has no echo in physical theory'.

The 'content as to infer about mental events, states or processes with content include seeing that the door is shut: Believing you are being followed, and calculating the square root of 2. What centrally distinguishes states, events, or processes - are basic to simply being states - with content is that they involve reference to objects, properties or relations. A mental state with content can fail to refer, but there always exists a specific condition for a state with content to refer to certain things. When the state has a correctness or fulfilment condition, its correctness is determined by whether its referents have the properties the content specifies for them. It leaves open the possibility that unconscious states, as well as conscious states, have content. It equally allows the states identified by an empirical, computational psychology to have content. A correct philosophical understanding of this general notion of content is fundamental not only to the philosophy of mind and psychology, but also to the theory of knowledge and to metaphysics.

There is a long - standing tradition that emphasizes that the reason - giving relation is a logical or conceptual one. One way of bringing out the nature of this conceptual link is by the construction of reasoning, linking the agent's reason - providing states with the states for which they provide reasons. This reasoning is easiest to reconstruct in the case of reason for belief where the contents of the reason - providing beliefs inductively or deductively support the content of the rationalized belief. For example, I believe my colleague is in her room now, and my reasons are (1) she usually has a meeting in her room at 9:30 on Mondays and (2) it is to accept it as true, and it is relative to the objective of reaching truth that the rationalizing relations between contents are set for belief. They must be such that the truth of the premises makes likely the truth of the conclusion.

The causal explanatorial approach to reason - giving explanations also requires an account of the intentional content of our psychological states, which makes it possible for such content to be doing such work. It also provides a motivation for the reduction of intentional characterization as to extensional ones, in an attempt to fit such intentional causality into a fundamentally materialist world picture. The very nature of the reason - giving relation, however, can be seen to render such reductive projects unrealizable. This, therefore, leaves causal theorists with the task of linking intentional and non - intentional levels of description in such a way as to accommodate intentional causality, without over - either determination or a miraculous coincidence of prediction from within distinct causally explanatorial frameworks.

The idea that mentality is physically realized is integral to the 'functionalist' conception of mentality, and this commits most functionalists to mind - body supervenience in one form or another. As a theory of mind, supervenience of the mental - in the form of strong supervenience, or at least global supervenience - is arguably a minimum commitment of physicalism. But can we think of the thesis of mind - body supervenience itself as a theory of the mind - body relation - that is, as a solution to the mind - body problem?

A supervenience claim consists of covariance and a claim of dependence e (leaving aside the controversial claim of non - reducibility). This that the thesis that the mental supervenience on the physical amounts to the conjunction of the two claims (1) strong or global supervenience, and (2) the mental depends on the physical. However, the fact that the thesis says nothing about just what kind of dependence is involved in mind - body supervenience. When you compare the supervenience thesis with the standard positions on the mind-body problem, you are struck by what the supervenience thesis does not say. For each of the classic mind - body theories has something to say, not necessarily anything veery plausible, about the kind of dependence that characterizes the mind - body relationship. According to epiphenomenalism, for example, the dependence is one of causal dependence is one of casual dependence: On logical behaviourism, dependence is rooted in meaning dependence, or definability: On the standard type physicalism, the dependence is one that is involved in the dependence of macro - properties and son forth. Even Wilhelm Gottfried Leibniz (1646-1716) and Nicolas Malebranche (1638 - 1715) had something to say about this: The observed property convariation is due not to a direct dependancy relation between mind and body but rather to divine and interventions. That is, mind-body convariation was explained in terms of their dependence on a third factor-a sort of 'common cause' explanation.

It would seem that any serious theory addressing the mind - body problem must say something illuminating about the nature of psychophysical dependence, or why, contrary to common belief, there is no dependence. However, there is reason to think that 'supervenient dependence' does not signify a special type of dependence reflation. This is evident when we reflect on the varieties of ways in which we could explain the supervenience relation holds in a given case. For example, consider the supervenience of the moral on the descriptive the ethical naturalist will explain this on the basis of definability: The ethical intuitionist will say that the supervenience, and also the dependence, seems the brute fact that you discern through moral intuition. And the prescriptivist will attribute the supervenience to some form of consistency requirement on the language of evaluating and prescription. And distinct from all of these is Mereological supervenience, namely the supervenience of properties of a whole on properties and relations of its parts. What all this shows is that there is no single type of dependence relation common to all cases of supervenience: Supervenience holds in different cases for different reasons, and does not represent a type of dependence that can be put alongside causal dependence, meaning dependence, Mereological dependence and so forth.

If this is right, the supervenience thesis concerning the mental does not constitute an explanatory account of the mind - body relation, on a par with the classic alternatives on the mind-body problem. It is merely the claim that the mental covaried in a systematic way with the physical, an that this is due to a certain dependence relation yet to be specified and explained. In this sense, the supervenience thesis states the mind - bod y problem than offering a solution to it.

There seems to be a promising strategy for turning the supervenience thesis into a more substantive theory of mind, and it is this: To explicate mind - body supervenience as a special case of Mereological supervenience - that is, the dependence of the properties of a whole on the properties and relations characterizing its proper parts. Mereological dependence does seem to be a special form of dependence that is metaphysical and highly important. If one takes this approach, one would have to explain psychological properties as macroproperties of a whole organism that covary, in appropriate ways, with its microproperties, i.e., the way its constituents, tissue, and do on, are organized and function. This more specific supervenience thesis may well be a serious theory of the mind - body relation that can compete with the classic options in the field.

Previously, our considerations had fallen to arrange in making progress in the betterment of an understanding, fixed on or upon the alternatives as to be taken, accepted or adopted, even to bring into being by mental or physical selection, among alternates that generally are in agreement. These are minded in the reappearance of confronting or agreeing with solutions precedently recognized. That is of saying, whether or not this is plausible (that is a separate question), it would be no more logically puzzling than the idea that one could not have any propositional attitude unless one had one's with certain sorts of contents. Tyler Burge's insight is partly determined by the meanings of one's words in one's linguistic community. Burge (1979) is perfectly consistent with any intention - based semantics, reduction of the semantic to the psychological. Nevertheless, there is reason to be sceptical of the intention - based semantic programme. First, no intention - based semantic theorist has succeeded in stating a sufficient condition for more difficult task of starting a necessary - and - sufficient condition. And is a plausible explanation of this failure is that what typically makes an utterance an act of speaker meaning is the speaker's intention to be meaning or saying something, where the concept of meaning or saying used in the content of the intention is irreducibly semantic. Second, whether or not an intention - based semantic way of accounting for the actual - language relation in terms of speaker meaning. The essence of the intention - based semantic approach is that sentences used as conventional devices for making known a speaker's communicative understanding is an inferential process wherein a hearer perceives an utterance and, thanks to being party to relevant conventions or practices, infers the speaker's communicative intentions. Yet it appears that this inferential model is subject to insuperable epistemological difficulties, and. Third, there is no pressing reason to think that the semantic needs to be definable in terms of the psychological. Many intention - based semantic theorists have been motivated by a strong version of physicalism which requires the reduction of all intentional properties (i.e., all semantic and propositional - attitude properties) to physical or at least topic - neutral, or functional, properties, for it is plausible that there could be no reduction to the semantic and the psychological to the physical without a prior reduction of the semantic to the psychological. But it is arguable that such a strong version of physicalism is not what is required in order to fit the intentional into the natural order.

What is more, in the dependence of thought on language for which this claim is that propositional attitudes are relations to linguistic items which obtain, at least, partially, by virtue of the content those items have among language users. Thus, position does not imply that believers have to be language users, but it does make language an essential ingredient in the concept of belief. The position is motivated by two considerations (a) The supposition that believing is a relation to things that believing is a relation to things believed, for which of things have truth values and stand in logical relations to one another, and (b) The desires not to take things believed to be propositions - abstract things believed to be propositions - abstract, mind - and essentially the truth conditions that have. Now the tenet (a) is well motivated: The relational construal of propositional attitude s is probably the best way to account forms the quantitative in, 'Harvey believes something nasty about you'. But there are probable mistakes with taking linguistic items, rather than propositions, as the objects of belief In the first place, If Harvey believes that Flounders snore' is represented along the lines that of ('Harvey, but flounder snore'), then one could know the truth expressed by the sentience about Harvey without knowing the content of his beliefs: For one could know that he stands in the belief relation to 'flounders snore' without knowing its content. This is unacceptable, as in the second place, if Harvey believes that flounders snore, then what he believes that flounders snore, then what he believes - the reference of 'that flounders snore' - is that flounders snore. But what is this thing that flounders snore? well, it is abstract, in that it has no spatial location. It is mind and language independent, in that it exists in possible worlds for which there are neither thinkers nor speakers: and, necessarily, it is true if flounders snore. In short, it is a proposition - an abstract mind, and language - independent thing that has a truth condition and has essentially the truth condition it has.

A more plausible way that thought depend s on language is suggested by the topical thesis that we think in a 'language of thought'. On one reading, this is nothing more than the vague idea that the neural states that realize our thoughts 'have elements and structure in a way that is analogous to the way in which sentences have elements and structure'. Nonetheless, we can get a more literal rendering by relating it to the abstract conception of languages already recommended. On this conception, a language is a function from 'expressions' - sequences of marks or sounds or neural states or whatever - onto meaning, for which meanings will include the propositions of our propositional altitudes relations relate us to. we could then read the language of though t hypothesis as the claim that having propositional altitudes require s standing in a certain relation to a language whose expressions are neural state. There would now be more than one 'actualized - language relations. The one earlier of mention, the one discussed earlier might be better called the 'public - language relation'. Since the abstract notion of a language ha been so weakly construed. It is hard to see how the minimal language - of - thought proposal just sketched could fail to be true. At the same time, it has been given no interesting work to do. In trying to give it more interesting work, further dependencies of thought on language might come into play. For example, it has been claimed that the language of thought of a claim that the language of thought of a public - language user is the public language she uses: Her neural sentences are related to her spoken and written sentences in something like the way the written sentences are related to her spoken sentences. For another example, I that it might be claimed that even if one's language of thought is something like the way her written sentences are related to he r spoken sentences. For example, it might be claimed that even if one's language of thought is distinct from one's public language, the language - of thought relations makes presuppositions about the public-language relations in way that make the content of one's words in one's public language community.

Tyler Burge, has in fact shown that there is a sense for which though t content is dependent on the meanings of words in one's linguistic communications. Alfred's use of 'arthritis' is fairly standard, except that he is under the misconception that arthritis is not confined to the joints, he also applies the word to rheumatoid ailments not in the joints. Noticing an ailment in his thigh that is symptomatically like the disease in his hands and ankles, he says, to his doctor, 'I have arthritis in the thigh'. Here Alfred is expressing his false belief that he has arthritis in the thigh. But now consider a counterfactual situation that differs in just one respect (and, whatever it entails): Alfred's use of 'arthritis' is the correct use in his linguistic community. In this situation, Alfred would be expressing a true belief when he says 'I have arthritis in the thigh'. Since the proposition he believes is true while the proposition that he has arthritis in the thigh is false, he believes some other proposition. This shows that standing in the belief relation to a proposition can be partly determined by the meanings of words on one's public language. The Burge phenomenon seems real, but it would be nice to have a deep explanation of why thought content should be dependent on language in this way.

Finally, there is the old question of whether, or to what extent, a creature who does not understand a natural language can have thoughts. Now it seems pretty compelling that higher mammals and humans raised without language have their behaviour controlled by mental state that are sufficiently like our beliefs, desires, and intentions to share those labels. It also seems easy to imagine non - communicating creatures who have sophisticated mental lives (they build weapons, dams, bridges, have clever hunting devices, and so on). At the same time, ascription of particular contents to non-language-using creatures typically seem exercises in loose speaking (does the dog really believe that there is a bone in the yard?), and it is no accident that, as a matter of fact, creatures who do not understand a natural language have at best primitive mental lives. There is no accepting the primitive mental lives of animals account for their failure to master natural language, but the better explanation may be Chomsky's faculty unique to our species. As regards the inevitably primitive mental life of another wise normal human, raised without language, this might simply be due to the ignorance and lack of intellectual stimulation such a person would be doomed to. On the other hand, it might also be that higher thought requirements of a neural language with structures comparable to that of a natural language, and that such neural language ss are somehow acquired as the ascription of content to the propositional - attitude states of language less creatures is a difficult topic that needs more attention. It is possible of our ascriptions of propositional content, we will realize that these ascriptions are egocentrically based on a similarity to the language in which we express our beliefs. we might then learn that we have no principled basis for ascribing propositional content to a creature who does not speak something, or who does not have internal states with natural - language - like structure. It is somewhat surprising how little we know about thought's dependence on language.

The Language of Thought hypothesis has a compelling neatness about it. A thought is depicted as a structure of internal representational elements combined in a lawful way, and plays a certain functional role in an internal processing economy. So that the functionalist thinks of mental states and events as causally mediating between a subject's sensory inputs and that subjects ensuing behaviour. Functionalism itself is the stronger doctrine that what makes a mental state the type of state it is - a pain, a smell of violets, a belief that koalas are dangerous - is the functional relationist bears to the subject's perceptual stimuli, behavioural responses, and other mental states.

The representational theory of the mind arises with the recognition that thoughts have contents carried by mental representations.

Nonetheless, theorists seeking to account for the mind's activities have long sought analogues to the mind. In modern cognitive science, these analogues have provided the basses for simulation or modelling of cognitive performance seeing that cognitive psychology simulate one way of testings in a manner comparable to the mind, that offers support for the theory underlying the analogue upon which the simulation is based simulation, however, also serves a heuristic function, suggesting ways for which the mind might gainfully characteristically operate in physical terms. The problem is most obvious in the case of 'arbitrary' signs, like words, where it is clear that there is no connection between the physical properties of a word and what it denotes (the problem remains for Iconic representation). What kind of mental representation might support denotation and attribution if not linguistic representation? Perhaps, when thinking within the peculiarities that the mind and attributions thereof, being among the semantic properties of thoughts, are that 'thoughts' in having content, posses semantic properties, however, if thoughts denote and precisely attribute, sententialism may be best positioned to explain how this is possible.

Beliefs are true or false. If, as representationalism had it, beliefs are relations to mental representations, then beliefs must be relations to representations that have truth values among their semantic properties. Beliefs serve a function within the mental economy. They play a central part in reasoning and, thereby, contribute to the control of behaviour. To be rational, a set of beliefs, desires, and actions, also perceptions, intentions, decisions, must fit together in various ways. If they do not, in the extreme case they fail to constitute a mind at all - no rationality, no agent. This core notion of rationality in philosophy of mind thus concerns a cluster of personal identity conditions. That is, 'Holistic' coherence requirements on or upon the system of elements comprising a person's mind, related conception and epistemic or normative rationality are key linkages among the cognitive, as distinct rom qualitative mental stats. The main issue is characterizing these types of mental coherence.

Closely related to thought's systematicity is its productivity to have a virtual unbounded competence to think ever more complex novel thoughts having certain clear semantic ties to their less complex predecessor. Systems of mental representation apparently exhibit mental representation apparently exhibit the sort of productivity distinctive of spoken languages. Sententialism accommodates this fact by identifying the productive system of mental representation with a language of thought, the basic terms of which are subject to a productive grammar.

Possibly, in reasoning mental representations stand to one another just as do public sentences in valid 'formal derivations'. Reasoning would then preserve truth of belief by being the manipulation of truth - valued sentential representations according to rules so selectively sensitive to the syntactic properties of the representations as to respect and preserve their semantic properties. The sententialist hypothesis is thus that reasoning is formal inference. It is a process tuned primarily to the structure of mental sentences. Reasoners, then, are things very much like classical programmed computers. Thinking, according to sententialism, may then be like quoting. To quote an English sentence is to issue, in a certain way, a token of a given English sentence type: It is certainly not similarly to issue a token of every semantically equivalent type. Perhaps, thought is much the same. If to think is to token, a sentence in the language of thought, the sheer tokening of one mental sentence need not insure the tokening of another formally distinct equivalents, hence, thought's opacity.

Objections to the language of thought come from various quarters. Some will not tolerate any edition of representationalism, including Sententialism: Others endorse representationalism while denying that mental representations could involve anything like a language. Representationalism is launched by the assumption that psychological stat es ae relational, that being in psychological state minimally involves being related to something. But perhaps, psychological states are not at all relational. Verbalism begins by denying that expressions of psychological states are relational, infers that psychological states themselves are monadic and, thereby, opposes classical versions of representationalism, including Sententialism.

What all this is supposed to show, was that Chomsky and advances in computer science, the 1960s saw a rebirth of 'mentalistic' or 'cognitivist' approaches to psychology and the study of mind.

These philosophical accounts o cognitive theories and the concepts they invoke are generally much more explicit than the accounts provided by psychologists, and they inevitably smooth over some of the rough edges of scientists' actual practice. But if the account they give of cognitive theories diverges significantly from the theories that psychologists have just gotten it wrong. There is, however, a very different way in which philosophers have approached cognitive psychology. Rather than merely trying to characterize what cognitive psychology is actually doing, some philosophers try to say what it should and should not be doing. Their goal is not to explicate scientific practice, but to criticize and improve it. The most common target of this critical approach is the use of intentional concepts in cognitive psychology. Intentional notions have been criticized on various grounds. The two taken for our considerations are that they fail to supervene on the physiology of the cognitive agent, and that they cannot be 'naturalized'.

Perhaps, to an approach that is mos radical is the proposal that cognitive psychology should recast its theories and explanations in a way that does not appeal to intentional properties or 'syntactic' properties. Somewhat less radical is the suggestion that we can define a species of representation, which does supervene an organism's physiology, and that psychological explanations that appeal to ordinary ('wide') intentional properties can be replaced by explanations that invoke only their narrow counterparts. Nonetheless, many philosophers have urged that the problem lies in the argument, not in the way that cognitive psychology might be modified. However, many philosophers have urged that the problem lis in the argument, not in the way that cognitive psychology goes about its business. The most common critique of the argument focuses on the normative premise - the one that insists that psychological explanations ought not to appeal to 'wide' properties that fail to supervene on physiology. Why should it bot be that psychological explanations appeal to wide properties, the critics ask? : What exactly is wrong with psychological explanations invoking properties that do not supervene on physiology? Various answers have been proposed in the literature, though they typically end up invoking metaphysical principles that are less clear and less plausible than the normative thesis they are supposed to support.

Given to any psychological property that fails to supervene on physiology, it is trivial to characterize a narrow correlated property that does supervene. The extension of the correlate property includes all actual and possible objects in the extension of the original property, plus all actual and possible physiological duplicates of those objects. Theories originally stated in terms of wide psychological properties sated in terms of wide psychological properties can be recast in terms of their descriptive or explanatory power. It might be protested that when characterized in this way, narrow belief and narrow content are not really species of belief and content at all. Nevertheless, it is far from clear how this claim could be defended, or why we should care if it turns out to be right.

The worry about the 'naturalizability' of intentional properties is much harder to pin down. According to Fodor, the worry derives from a certain ontological intuition: That there is no place for intentional categories in a physicalistic view of the world, and thus, that the semantic and/or intentionality will prove permanently recalcitrant to integration in the natural order. If, however, intentional properties cannot be integrated into the natural order, then presumably they ought to be banished from serious scientific theorizing. Psychology should have no truck with them. Indeed, if intentional properties have no place in the natural order, then nothing in the natural world has intentional properties, and intentional states do not exist at all. So goes the worry. Unfortunately, neither Fodor nor anyone else has said anything very helpful about what is required to 'integrate' intentional properties into the natural order. There are, to be sure, various proposals to be found in the literature. But all of them seem to suffer from a fatal defect. On each account of what is required to naturalize a property or integrate it into the natural order, there are lots of perfectly respectable non - intentional scientific or common - sense properties that fail to meet the standards. Thus, all the proposals that have been made so far, end up being declined and thrown out.

Now, or course, the fact that no one has been able to give a plausible account of what is required to 'naturalize' the intentional may indicate nothing more than that their project is a difficult one. Perhaps with further work a more plausible account will be forthcoming. But one might also offer a very different diagnosis of the failure of all accounts of 'naturalizing' that have so far been offered. Perhaps the 'ontological intuition' that underlies the worry about integrating the intentional into the natural order is simply muddled. Perhaps, there is no coherent criterion of naturalization or naturalizability that all properties invoked in respectable science must meet, as, perhaps, that this diagnosis is the right one. Until those who are worried about the naturalizability of the intentional provide us with some plausible account of what is required of intentional categories if they are to find a place in 'a physicalistic view of the world'. Possibly we are justified in refusing to take their worry seriously.

Recently, John Searle (1992) has offered a new set of philosophical arguments aimed at showing that certain theories in cognitive psychology are profoundly wrong - headed. The theories that are the target of computational explanations of various psychological capacities - like the capacity to recognize grammatical sentences, or the capacity to judge which of two objects in one 's visual field is further away. Typically, these theories are set out in the form of a computer program - a set of rules for manipulating symbols - and the explanations offered for the exercise of the capacity in question is that people's brains are executing the program. The central claim in Searle' s critique is that being a symbol or a computational stat e is not an 'intrinsic' physical feature of a computer state or a brain state. Rather, being a symbol is an 'observer relative' feature. However, Searle maintains, only intrinsic properties of a system can play a role in causal explanations of how they work. Thus, appeal to symbolic or computational states of the brain could not possibly play a role in a 'casual account of cognition in knowledge'.

All of which, the above aforementioned surveyed, does so that implicate some of the philosophical arguments aimed at showing that cognitive psychology is confusing and in need of reform. My reaction to those arguments was none too sympathetic. In each case, it was maintained to the philological argument that is problematic, not the psychology it is criticizing.

It is fair to ask where we get the powerful inner code whose representational elements need only systematic construction to express, for example, the thought that cyclotrons are bigger and more than vast than black holes. Nonetheless, on this matter, the language of thought theorist has little to say. All that concept learning could be, assuming it is to be some kind of rational process and not due to mere physical maturation or a bump on the head. According to the language of thought theorist, is the trying out of combinations of existing representational elements to see if a given combination captures the sense (as evidenced in its use) of some new concept. The consequence is that concept learning, conceived as the expansion of our representational resources, simply does not happen. What happens instead is that we work with a fixed, innate repertoire of elements whose combination and construction must express any content we an ever learn to understand. And note that it is not the trivial claim that in some sense the resources a system starts with must set limits on what knowledge it can acquire. For these are limits which flow not, for example, from sheer physical size, number of neurons, connectivity of neurons, and so forth. But from a base class of genuinely representational elements. They are more like the limits that being restricted to the propositional calculus would place on the expressive power of a system than, say, the limits that having a certain amount of available memory storage would place on one.

But this picture of representational stasis in which all change consists in the redeployment of existing representational resources, is one that is fundamentally alien to much influential theorizing in developmental psychology. The prime example of a developmentalist who believed in a much stronger forms a much stronger form in genuine expansion of representational power at the very heart of a model of human development. In a similar vein, recent work in the field of connectivism seems to open up the possibility of putting well - specified models of strong representational change back into the centre of cognitive scientific endeavours.

Nonetheless, the understanding of how the underlying combinatoric code 'develops' the deep understanding of cognitive processes, than understanding the structure and use of the code itself (though, doubtless the projects would need to be pursued hand - in - hand).

The language of thought depicts thoughts as structures of concepts, for which in turn exist as elements (for any basic concept) or concatenations of elements (for the rest) in the inner code. The intentional states, as commonsense understands them, have both causal and semantic properties and that the combination appears to be unprecedented. However, a further problem about inferential role semantics is that it is, almost invariably, suicidally holistic. it seems, that, if externalism is right, then (some of) the intentional properties of thought are essentially 'extrinsic': They essentially involve mind-to-world relations. All and all, in assuming that the computational role of a mental representation is determined entirely by its intrinsic properties, such properties of its weigh t, shape, or electrical conductivity as it might be., hard to see how the extrinsic properties: Which is to say, that it is hard to see how there could be computationally sufficient conditions for being in an intentional state, for which is to say that it is hard to see how the immediate implementation of intentional laws could be computational.

However, there is little to be said about intrinsic relation s between basic representational items. Even bracketing the (difficult) question of which, if any words in our public language may express content s which have as their vehicles atomic items in the language of thought (an empirical question on which it is to assume that Fodor to be officially agnostic), the question of semantic relations between atomic items in the language of thought remains. Are there any such relations? And if so, in what do they consist? Two thought s are depicted as semantically related just in casse they share elements themselves (like the words of public language on which they are modelled) seem to stand in splendid isolation from one another. An advantage of some connectionist approaches lies precisely in their ability to address questions of the interrelation of basic representational elements (in act, activation vectors) by representing such items as location s in a kind of semantic space. In such a space related contents are always expressed by related representational elements. The connectionist's conception of significant structure thus goes much deeper than the Fodorian's. For the connectionist representations need never be arbitrary. Even the most basic representational items will bear non - accidental relations of similarity and difference to one another. The Fodorian, having reached representational bedrock, must explicitly construct any such further relations. They do not come for free as a consequence ee of using an integrated representational space. Whether this is a bad thing or a goo one will depend, of course, on what kind of facts we need to explain. But it is to suspect that representational atomism may turn out to be a conceptual economy that a science of the mind cannot afford.

The approach for ascribing contents must deal with the point that it seems metaphysically possible for here to be something that in actual and counterfactual circumstances behaves as if it enjoys states with content, when in fact it does not. If the possibility is not denied, this approach must add at least that the states with content causally interact in various ways with one - another, and also causally produce intentional action. For most causal theories, however, the radical separation of the causal and rationalizing role of reason - giving explanations is unsatisfactory. For such theorists, where we can legitimately point to an agent's reasons to explain a certain belief or action, then those features of the agent's intentional states that render the belief or action reasonable must be causally relevant in explaining how the agent came to believe or act in a way which they rationalize. One way of putting this requirement is that reason - giving states not only cause, but also causally explain their explananda.

On most accounts of causation an acceptance of the causal explanatory role of reason - giving connections requires empirical causal laws employing intentional vocabulary. It is arguments against the possibility of such laws that have, however, been fundamental for those opposing a causal explanatorial view of reasons. What is centrally at issue in these debates is the status of the generalizations linking intentional states to each other, and to ensuing intentional acts. An example of such a generalization would be, 'If a person desires 'X', believes 'A' would be a way of promoting 'X', is able to 'A' and has no conflicting desires than she will do 'A'. For many theorists such generalizations are between desire, belief and action. Grasping the truth of such a generalization is required to grasp the nature of the intentional states concerned. For some theorists the a priori elements within such generalization s as empirical laws. That, however, seems too quick, for it would similarly rule out any generalizations in the physical sciences that contain a priori elements, as a consequence of the implicit definition of their theoretical kinds in a causal explanation theory. Causal theorists, including functionalist in philosophy of mind, can claim that it is just such implicit definition that accounts for th a priori status of our intentional generalizations.

The causal explanatory approach to reason - giving explanations also requires an account of the intentional content of our psychological states, which makes it possible for such content to be doing such work. It also provides a motivation for the reduction of intentional characteristics to extensional ones, on an attempt to fit intentional causality into a fundamentally materialist world picture. The very nature of the reason - giving relation, however, can be seen to render such reductive projects unrealizable. This, therefore leaves causal theorists with the task of linking intentional and non - intentional levels of description in such a way as to accommodate intentional causality, without over - either determination or a miraculous coincidence of prediction from within distinct causally explanatorial frameworks.

The existence of such causal links could well be written into the minimal core of rational traditions required for the ascription of the contents in question. Yet, it is one thing to agree that the ascription of content involves a species of rational intelligibility. It is another to provide an explanation of this fact. There are competing explanations. One treatment regards rational intelligibility as ultimately dependent on or upon what we find intelligible, or on what we could come to find intelligible in suitable circumstances. This is an analogue of classical treatments of secondary qualities, and as such is a form of subjectivism about content. An alternative position regards the particular conditions for correct ascription of given contents as more fundamental. This alternative states that interpretation must respect these particular conditions. In the case of conceptual contents, this alternative could be developed in tandem with the view that concepts are individuated by the conditions for possessing them. These possession conditions would then function as constraints upon correct interpretation. If such a theorist also assigns references to concepts in such a way that the minimal rational transition are also always truth - preserving, he will also have succeeded in explaining why such transition are correct. Under an approach that treats conditions for attribution as fundamental, intelligibility need not be treated as a subjective property. There may be concepts we could never grasp because of our intellectual limitations, as there will be concepts that members of other species could not grasp. Such concepts have their possession conditions, but some thinkers could not satisfy those conditions.

Ascribing states with content to an actual person has to proceed simultaneously with attribution of a wide range of non - rational states and capacities. In general, we cannot understand a person's reasons for acting as he does without knowing the array of emotions and sensations to which he is subject: What he remembers and what he forgets, and how he reasons beyond the confines of minimal rationality. Even the content - involving perceptual states, which play a fundamental role in individuating content, cannot be understood purely in terms relating to minimal rationality. A perception of the world as being a certain way is not (and could not be) under a subject's rational control. Though it is true and important that perceptions give reasons for forming beliefs, the beliefs for which they fundamentally provide reasons - observational beliefs about the environment - have contents which can only be elucidated by referencing back to perceptual experience. In this respect (as in others) perceptual states differ from those beliefs and desires that are individuated by mentioning what they provide reasons for judging or doing: For frequently these latter judgements and actions can be individuated without reference back to the states that provide reasons for them.

What is the significance for theories of content of the fact that it is almost certainly adaptive for members of a species to have a system of states with representational contents which are capable of influencing their actions appropriately? According to teleological theories of content, a constitutive account of content - one which says what it is for a state to have a given content - must make use of the notion of natural function and teleology. The intuitive idea is that for a belief state to have a given content 'p' is for the belief - forming mechanisms which produced it to have the function b(perhaps derivatively) of producing that state only when it is the case that 'p'. One issue this approach must tackle is whether it is really capable of associating with states the classical, realistic, verification - transcendent contents which pre - theoretically, we attribute to hem. It is not clear that a content's holding unknowably can influence the replication of belief-forming mechanics. Bu t even if content itself proves to resist elucidation in terms of natural function and selection. It is still a very attractive view that selection must be mentioned in an account of what associate ss something - such as sentence - with a particular content, even though that content itself may be individuated by other .

Contents are normally specified by 'that . . .' clauses, and it is natural to suppose that a content has the same kind of sequential and hierarchical structure as the sentence that specifies it. This supposition would be widely accepted for conceptual content. It is, however, a substantive thesis that all content is conceptual. One way of treating one sort of perceptual content is to regard the content as determined by a spatial type, the type under which the region of space around the perceiver's must fall if the experience with that content is to represent the environment correctly. The type involves a specification of surfaces and features in the environment, and their distances are directed from the perceiver's body as origin. Such contents lack any sentence - like structure at all. Supporters of the view that all content is conceptual will argue that the legitimacy of using these spatial type in giving the content of experience does not undermine the thesis that all content is conceptual. Such supporters will say that the spatial type is just a way of capturing what can equally be captured by conceptual components such as 'that distance', or 'that direction', where these demonstratives are made available by the perception in question. Friends of non - conceptual content will respond that

648

these demonstratives themselves cannot be elucidated without mentioning the spatial types for which lack sentence - like structure.

The actions made rational by content - involving states are actions individuated in part by reference to the agent's relations to things and properties in his environment. Wanting to see a particular movie and believing that, that building over thee is a cinema showing it makes rational the action of walking in the direction of that building. Similarly, for the fundamental casse of a subject who has knowledge about his environment, a crucial factor in making rational the formations of particular attitude is the way the world is around him. One may expect, the n, that any theory that links the attribution of contents to states with rational intelligibility will be commit to the thesis that the content of a person's states depends in part on his relations to the world outside him. We call this thesis the thesis of externalism about content.

Externalism about content should steer a middle course. On the one had, it should not ignore the truism that the relations of rational intelligibility involve not things and properties in the world, but the way they are presented as being - an externalist should use some version of F)'s notion of mode of presentation. On the other hand, the externalist for whom considerations of rational intelligibility are pertinent to the individuation of content is likely to insist that we cannot dispense with the notion of something in the world - being presented in a certain way. If we dispense with the notion of something external bing presented in a certain way, we are in danger of regarding attributions of content as having no consequence for how an individual relates to his environment, in a way that is quite contrary to our intuitive understanding of rational intelligibility.

Externalism comes in more and fewer extreme versions. Consider a mind of a thinker who sees or perceives of a particular pear, and thinks a thought that the pear is ripe, where the demonstrative way of thinking of the pear expressed by 'that pear' is made available to him by his perceiving the pear. Some philosophers have held that the thinker would be employed of thinking were he perceiving a different perceptually based way of thinking were he perceiving a different pear. But Externalism need not be committed to this. In the perceptual state that makes available the way on thinking pear is presented as being in a particular distance, and as having certain properties. A position will still be Externalist if it holds that what is involved in the pear's being so presented is the collective role of these components of content in making intelligible in various circumstances the subject's relations to environmental directions distance and properties of object. This can be held without committed to the object - dependence of the way of thinking expressed by 'that pear'. This less strenuous form of Externalism must, though, address the epistemological arguments offered in favour of the more extreme versions, to the effect that only they are sufficiently world - involving.

The apparent dependence of the content of belief on factors external to the subject can be formulated as a failure of Supervenience of belief content upon facts about what is the case within the boundaries of the subject's body. To claim that such Supervenience fails is to make a model claim: That there can be two persons the same in respect of their internal physical states (and so in respect to those of their dispositions that are independent of content - involving states), who nevertheless differ in respect of which beliefs they have. Hilary Putnam (1926 -), the American philosopher of science, who became more prominent in his writing about 'Reason, Truth, and History' (1981) marked of a subtle position that he call's internal realism, initially related to a n ideal limit theory of truth, and apparently maintaining affinities with verificationism, but in subsequent work more closely aligned with minimalism. Putnam's concern in the later period has largely been to deny any serious asymmetry between truth and knowledge as obtained in moral s, and even Theology.

Nonetheless, in the case of content - involving perceptual states. It is a much more delicate matter to argue for the failure of Supervenience. The fundamental reason for this is answerable not only to factors ion the input side-what in certain fundamental cases causing the subject to be in the perceptual state - but also to factors on the perceptual state - but also to factors on the output side - what the perceptual state is capable of helping to explain amongst the subject's actions. If differences in perceptual content always involve differences in bodily - described actions in suitable counter - factual circumstances, and if these different actions always will after all be Supervenience of content - involving perceptual states on internal states. But if this should turn ut to be so, that is not a refutation of Externalism for perceptual contents. A different reaction to this situation of dependence ads one of Supervenience is in some cases too strong. A better is given by a constitutive claim: That what makes a state have the content it does are certain of its complex relations to external states of affairs. This can be held without commitment to the model separability of certain internal states from content - involving perceptual states.

Attractive as Externalism about content ma be, it has been vigorously contested notably by the American philosopher of mind Jerry Alan Fodor (1935-), who is known for a resolute realism about the nature of mental functioning. Taking the analogy between thought and computation seriously, Fodor believes that mental representations should be conceived as individual states with their own identities and structure, like formulae transformed by processes of computation or thought. His views are frequently contrasted with those of 'Holist' such as Herbert Donald Davidson (1917 - 2003), although Davidson is a defender of the doctrines of the 'indeterminacy' of radical translation and the 'inscrutability' of reference, his approach has seemed to many to offer some hope of identifying meaning as a respectable notion, even within a broadly 'extensional' approach to language. Davidson is also known for rejection of the idea of a 'conceptual scheme', thought of as something peculiar to one language or in one way of looking at the world, arguing that where the possibility of translation stops so does the coherence of the idea that there is anything to translate. Nevertheless, Fodor (1981) endorses the importance of explanation by content - involving states, but holds that content must be narrow, constituted by internal properties of an individual.

One influential motivation for narrow content is a doctrine about explanation that molecule - for - molecule counter - parts must have the same causal powers. Externalists have replied that the attributions of content - involving states presuppose some normal background or context for the subject of the states, and that content - involving explanations commonly take the presupposed background for granted. Molecular counter - parts can have different presuppose d backgrounds, and their content - involving states may correspondingly differ. Presupposition of a background of external relations in which something stands is found in other sciences outside those that employ the notion of content, including astronomy and geology.

A more specific concern of those sympathetic to narrow content is that when content is externally individuated, the explanatorial principles postulated in which content - involving states feature will be a priori in some way that is illegitimate. For instance, it appears to be a priori that behaviour is intentional under some description involving the concept 'water' will be explained by mental states that have the externally individuated concept about 'water' in their content. The Externalist about content will have a twofold response. First, explanations in which content - involving states are implicated will also include explanations of the subject's standing in a particular relation to the stuff water itself, and for many such relations, it is in no way a priori that the thinker's so standing has a psychological explanation at all. Some such cases will be fundamental to the ascription of Externalist content on treatments that tie such content to the rational intelligibility of actions relationally characterized. Second, there are other cases in which the identification of a theoretically postulated state in terms of its relations generates a priori truths, quite consistently with that state playing a role in explanation. It arguably is phenotypical characteristic, then it plays a causal role in the production of that characteristic in members of the species in question. Far from being incompatible with a claim about explanation, the characterization of genes that would make this a priori also requires genes to have a certain casual explanatory role.

Of anything, it is the friend of narrow content who has difficulty accommodating the nature content are fit to explain bodily movements in environment - involving terms. But we note, that the characteristic explananda of content - involving states, such as walking towards the cinema, are characterized in environment - involving terms. How is the theorist of narrow content to accommodate this fact? He may say, that we merely need to add a description of the context of the bodily movement, which ensures that the movement is in fact a movement toward the cinema. But mental property of an event to an explanation of that event does not give one an explanation of the event's having that environmental property, let alone a content - involving explanation of the fact. The bodily movement may also be a walking in the direction of Moscow, but it does not follow that we have a rationally intelligible explanation of the event as a walking in the direction of Moscow. Perhaps the theorist of narrow content would at this point add further relational proprieties of the internal states of such a kind that when his explanation is fully supplemented, it sustains the same counter - factuals and predications as does the explanation that mentions externally individuated content. But such a fully supplemented explanation is not really in competition with the Externalist's account. It begins to appear that if such extensive supplementation is adequate to capture the relational explananda it is also sufficient to ensure that the subject is in states with externally individuated contents. This problem, however, affects not only treatments of content as narrow, but any attempt to reduce explanation by content - involving states to explanation by neurophysiological states.

One of the tasks of a sub - personal computational psychology is to explain how individuals come to have beliefs, desires, perceptions and other personal - level content - involving properties. If the content of personal - level states is externally individuated, then the contents mentioned in the sub - personal psychology that is explanatory of those personal states must also be externally individuated. One cannot fully explain the presence of an externally individuated state by citing only states that are internally individuated. On an Externalist conception of sub - personal psychology, a content - involving computation commonly consists in the explanation of some externally individuated states by other externally individuated states.

This view of sub - personal content has, though, to be reconciled with the fact that the first states in an organism involved in the explanation - retinal states in the case of humana - are not externally individuated. The reconciliation is affected by the presupposed normal background, whose importance to the understanding of content we have already emphasized. An internally individuated state, when taken together with a presupposed external background, can explain the occurrence of an externally individuated state.

An Externalist approach to sub - personal content also has the virtue of providing a satisfying explanation of why certain personal - level states are reliably correct in normal circumstances. If the sub - personal computations that cause the subject to be in such states are reliably correct, and the final commutation is of the content of the personal - level state, then the personal-level state will be reliably correct. A similar point applies to reliable errors, too, of course. In either case, the attribution of correctness condition to the sub - personal state is essentially to the explanation.

Externalism generates its own set of issues that need resolution, notably in the epistemology of attributions. A content - involving state may be externally individuated, but a thinker does not need to check on his relations to his environment to know the content of his beliefs, desires, and perceptions. How can this be? A thinker's judgements about his beliefs are rationally responsive to his own conscious beliefs. It is a first step to note that a thinker's beliefs about his own beliefs will then inherit certain sensitivities to his environment that are present in his original (first - order) beliefs. But this is only the first step, for many important questions remain. How can there be conscious externally individuated states at all? Is it legitimate to infer from the content of one's states to certain general facts about one's environment, and if so, how, and under what circumstances?

Ascription of attitudes to others also needs further work on the Externalist treatment. In order knowledgeably to ascribe a particular content - involving attitude to another person, we certainly do not need to have explicit knowledge e of the external relations required for correct attribution of the attitude. How then do we manage it? Do we have tacit knowledge of the relation on which content depends, or do we in some way take our own case as primary, and think of the relations as whatever underlies certain of our own content - involving states? In the latter, in what wider view of other - ascription should this point be embedded? Resolution of these issues, like so much else in the theory of content, should provide us with some understanding of the conception each one has of himself as one mind amongst many, interacting with a common world which provides the anchor for the ascription of content.

There seems to have the quality of being an understandably comprehensive characteristic as 'thought', attributes the features of 'intentionality' or 'content': In thinking, as one thinks about certain things, and one thinks certain things about those things - one entertains propositions that maintain a position as promptly categorized for the states of affairs. Nearly all the interesting properties of thoughts depend upon their 'content': Their being coherent or incoherent, disturbing or reassuring, revolutionary or banal, connected logically or illogically to other thoughts. It is thus, hard to see why we would bother to talk of thought at all unless we were also prepared to recognize the intentionality of thought. So we are naturally curious about the nature of content: We want to understand what makes it possible, what constitutes it, what it stems from. To have a theory of thought is to have a theory of its content.

Four issues have dominated recent thinking about the content of thought, each may be construed as a question about what thought depends on, and about the consequences of its so depending (or not depending). These potential dependencies concern: (1) The world outside of the thinker himself, (2) language, (3) logical truth (4) consciousness. In each casse the question is whether intentionality is essentially or accidentally related to the items mentioned: Does it exist, that is, only by courtesy of the dependence of thought on the aid items? And this question determining what the intrinsic nature of thought is.

Thoughts are obviously about things in the world, but it is a further question whether they could exist and have the content they do whether or not their putative objects themselves exist. Is what I think intrinsically dependent on or upon the world in which I happen to think it? This question was given impetus and definition by a thought experiment due to Hilary Putnam, concerning a planet called 'twin earth'. On twin earth there live thinkers who are duplicates of us in all internal respects but whose surrounding environment contain different kinds of natural objects. The suggestion then is that what these thinkers refer to and think about is individuality dependent upon their actual environment, so that where we think about cats when we say 'cat' they think about that word - the different species that actually sits on their mats and so on. The key point is that since it is impossible to individuate natural kinds like cats solely by reference to the way they strike the people who think about them cannot be a function simply of internal properties of the thinker. The content, here, is relational in nature, is fixed by external facts as they bear upon the thinker. Much the same point can be made by considering repeated demonstrative reference to distinct particular objects: What I refer to when I say 'that bomb', of different bombs, depends on or upon the particular bomb in front of me and cannot be deduced from what is going on inside me. Context contributes to content.

Inspired by such examples, many philosophers have adopted an 'Externalist' view of thought content: Thoughts are not antonymous states of the individual, capable of transcending the contingent facts of the surrounding world. One is therefore not free to think whatever one's liking, as it was, whether or not the world beyond cooperates in containing suitable referents for those thoughts. And this conclusion has generated a number of consequential questions. Can we know our thoughts with special authority, given that they are thus hostage to external circumstances? How do thoughts cause other thoughts and behaviour, given that they are not identical with an internal states we are in? What kind of explanation are we giving when we cite thoughts? Can there be a science of thought if content does not generalize across environments? These questions have received many different answers, and, of course, not everyone agrees that thought has the kind of world - dependence claimed. Nonetheless, what has not been considered carefully enough, is the scope of the Externalist thesis - whether it applies to all forms of thought, all concepts. For unless this questions be answered affirmatively we cannot rule out the possibility that though in general depends on there being some thought that is purely internally determined, so that the externally fixed thoughts are a secondary phenomenon. What about thoughts concerning one's present sensory experience, or logical thoughts or ethical thought? Could there, indeed, be a thinker for whom internalism was generally correct? Is external individuation the rule or the exception? And might it take the rule or the exception? And might it take different forms in different cases?

Since words are also about things, it is natural to ask how their intentionality is connected to that of thoughts. Two views have been advocated: One view takes thought content to be self - subsisting relative to linguistic content, with the latter dependent upon the former: the other view takes thought comment to be derivative upon linguistic content, so that there can be no thought without a bedrock of language. Thus, arise controversies about whether animals really think, being non - speakers, or computers really use language., being non - thinkers. All such question depend critically upon what one is to mean by 'language'. Some hold that spoken language is unnecessary for thought but that there must be an inner language in order for thought to be possible, while others reject the very idea of an inner language, preferring to suspend thought from outer speech. However, it is not entirely clear what it amounts to assert (or deny)that there is an inner language of thought. If it merely that concepts (thought constituents) are structured in such a way as to be isomorphic with spoken language, then the claim is trivially true, given some natural assumptions. But if it that concepts just are 'syntactic' items orchestrated into springs of the same, then the claim is acceptable only in so far as syntax is an adequate basis for meaning - which, on the face of it, it is not. Concepts no doubt have combinatorial powers compactable to those of words, but the question is whether anything else can plausible be meant by the hypothesis of an inner language.

On the other hand, it appears undeniable that spoken language does have autonomous intentionality, but instead derives its meaning from the thought of speakers - though language may augment one's conceptual capacities. So thought cannot postdate spoken language. The truth seems to be that in human psychology speech and thought are interdependent in many ways, but there is no conceptual necessity about this. The only 'language' on which thought essentially depends itself: Thought indeed, depends upon there being insoluble concepts that can join with others to produce complete propositional statements. But this is merely to draw attention to a property any system of concepts must have: It is not to say what concepts are or how they succeed in moving between thoughts as they so. Appeals to language at this point, are apt to flounder on circularity, since words take on the power of concepts only insofar as they express them. Thus, there seems little philosophical illumination to be got from making thought depend on or upon language.

This third dependency question is prompted by the reflection that, while people are no doubt often irrational, woefully so, there seems to be sme kind of intrinsic limit to their unreason. Even the sloppiest thinker will not infer anything from anything. To do so is a sign of madness The question then is what grounds this apparent concession to logical prescription. Whereby, the hold of logic over thought? For the dependence there can seem puzzling: Why should the natural causal processes relations of logic, I am free to flout the moral law to any degree I desire, but my freedom to think unreasonably appears to encounter an obstacle in the requirement of logic? My thoughts are sensitive to logical truth in somewhat the way they are sensitive to the world surrounding me: They have not the independence of what lies outside my will or self that I fondly imagined. I may try to reason contrary to modus ponens, but my efforts will be systematically frustrated. Pure logic takes possession of my reasoning processes and steers them according to its own indicates, variably, of course, but in a systematic way that seems perplexing.

One view of tis is that ascriptions of thought are not attempts to map a realm of independent causal relations, which might then conceivably come apart from logical relations, but are rather just a useful method of summing up people's behaviours. Another view insists that we must acknowledge that thought is not a natural phenomenon in the way merely, and physical facts are: Thoughts are inherently normative in their nature, so that logical relations constitute their inner essence. Thought incorporates logic in somewhat the way Externalists say it incorporates the world. Accordingly, the study of thought cannot be a natural science in the way the study of (say) chemistry compounds is. Whether this view is acceptable, depends upon whether we can make sense of the idea that transitions in nature, such as reasoning appear to be, can also be transitions in logical space, i.e., be confined by the structure of that space. What must be thought, in such that this combination n of features is possible. Put differently, what is it for logical truth to be self - evident?

This dependency question has been studied less intensively than the previous three. The question is whether intentionality ids dependent on or upon consciousness for its very existence, and if so why. Could our thoughts have the very content they now have if we were not to be consciousness beings at all? Unfortunately, it is difficult to see how to mount an argument in either direction. On one hand, it can hardly be an accident that our thoughts are conscious and that this content is reflected in the intrinsic condition of our state of consciousness: It is not as if consciousness leaves off where thought content begins - as it does with, say, the neural basis of thought. Yet, on the other hand, it is by no clear what it is about consciousness that links it to intentionality in this way. Much of the trouble here stems from our exceedingly poor understanding of the nature of consciousness could arise from grain tissue (the mind - body problem), so that we fill to grasp the manner in which conscious states bear meaning. Perhaps content is fixed by extra - conscious properties and relations and only subsequently shows up in consciousness, as various naturalistic reductive accounts would suggest; Or perhaps, consciousness itself plays a more enabling role, allowing meaning to come into the word, hard as this may be to penetrate. In some ways the question is analogous to, say, the properties of 'pain': Is the aversive property of pain, causing avoidance behaviour and so forth, essentially independent of the conscious state of feeling, or is it that pain, could only have its aversion function in virtue of the conscious feedings? This is part of the more general question of the epiphenomenal character of consciousness: Is conscious awareness just a dispensable accompaniment of some mental feature - such as content or causal power - or is it that consciousness is structurally involved in the very determination of the feature? It is only too easy to feel pulled in both directions on this question, neither alterative being utterly felicitous. Some theorists, suspect that our uncertainty over such questions stems from a constitutional limitation to human understanding. We just cannot develop the necessary theoretical tools which to provide answers to these questions, so we may not in principle be able to make any progress with the issue of whether thought depends upon consciousness and why. Certainly our present understanding falls far short of providing us with any clear route into the question.

It is extremely tempting to picture thought as some kind of inscription in a mental medium and of reasoning as a temporal sequence of such inscriptions. On this picture all that a particulars thought requires in order to exist is that the medium in question should be impressed with the right inscription. This makes thought independent of anything else. On some views the medium is conceived as consciousness itself, so that thought depends on consciousness as writing depends on paper and ink. But ever since Wittgenstein wrote, we have seen that this conception of thought has to be mistaken, in particular of intentionality. The definitive characteristics of thought cannot be captured within this model. Thus, it cannot make room for the idea of intrinsic worlds - dependence. Since any interiorized inscription would be individualatively independent of items outside the putative medium of thought, but can it be made to square with the dependence of thought on logical pattens, since the medium could be configured in any way permitted by its intrinsic nature, within regard for logical truth - as sentences can be written down in any old order one likes. And it misconstrues the relation between thought and consciousness, since content cannot consist in marks on the surface of consciousness, so to speak. States of consciousness do contain particular meanings but not as a page contains sentences: The medium conception of the relation between content and consciousness is thus deeply mistaken. The only way to make meaning enter internally into consciousness is to deny that it as a medium for meaning to be expressed. However, it is marked and noted as the difficulty to form an adequate conception of how consciousness does carry content - one puzzle being how the external determinants of content find their way into the fabric of consciousness.

Only the alleged dependence of thought upon language fits the Na ve tempting inscriptional picture, but as we have attested to, this idea tends to crumble under examination. The indicated conclusion seems to be that we simply do not posses a conception of thought that makes its real nature theoretically comprehensible: Which is to say, that we have no adequate conception of mind? Once we form a conception of thought that makes it seem unmysterious as with the inscriptional picture. It turns out to have no room for content as it presents itself: While building in a content as it is leaves' us with no clear picture of what could have such content. Thought is 'real', then, if and only if it is mysterious.

In the philosophy of mind 'epiphenomenalism' that while there exist mental events, states of consciousness, and experience, they have themselves no causal powers, and produce no effect on the physical world. The analogy sometimes used is that of the whistle on the engine that makes the sound (corresponding to experiences), but plays no part in making the machinery move. Epiphenomenalism is a drastic solution to the major difficulties the existence of mind with the fact that according to physics itself only a physical event can cause another physical event an epiphenomenalism may accept one - way causation, whereby physical events produce mental events, or may prefer some kind of parallelism, avoiding causation either between mind and body or between body and mind. And yet, Occasionalism considers the view that reserves causal efficacy to the action of God. Events in the world merely form occasions on which God acts so as to bring about the events normally accompanying them, and thought of as their effects. Although, the position is associated especially with the French Cartesian philosopher Nicolas Malebranche (1638 - 1715), inheriting the Cartesian view that pure sensation has no representative power, and so adds the doctrine that knowledge of objects requires other representative ideas that are somehow surrogates for external objects. These are archetypes of ideas of objects as they exist in the mind of God, so that 'we see all things in God'. In the philosophy of mind, the difficulty to seeing how mind and body can interact suggests that we ought instead to think of hem as two systems running in parallel. When I stub my toe, this does so cause pain, but there is a harmony between the mental and the physical (perhaps due yo God) that ensures that there will be a simultaneous pain, when I form an intention and then act, the same benevolence ensures that my action is appropriated to my intention. The theory has never been wildly popular, and many philosophers would say that it was the result of a misconceived 'Cartesian dualism'. Nonetheless, a major problem for epiphenomenalism is that if mental events have no causal relationship it is not clear that they can be objects of memory, or even awareness.

The metaphor used by the founder of revolutionary communism, Karl Marx (1805-1900) and the German social philosopher and collaborator of Marx, Friedrich Engels (1820 - 95), to characterize the relation between the economic organization of society, which is its base, an the political, legal, and cultural organizations and social consciousness of a society, which is the super-structure. The sum total of the relations of production of material life conditions the social political, and intellectual life process in general. The way in which the base determines of much debate with writers from Engels onwards concerned to distance themselves from that the metaphor might suggest. It has also in production are not merely economic, but involve political and ideological relations. The view that all causal power is centred in the base, with everything in the super - structure merely epiphenomenal. Is sometimes called economicism? The problems are strikingly similar to those that are arising when the mental is regarded as supervenience upon the physical, and it is then disputed whether this takes all causal power away from mental properties.

Just the same, for if, as the causal theory of action implies, intentional action requires that a desire for something and a belief about how to obtain what one desires play a causal role in producing behaviour, then, if epiphenomenalism is true, we cannot perform intentional actions. Nonetheless, in describing events that happen does not of itself permit us to talk of rationality and intention, which are the categories we may apply if we conceive of them as actions. Ewe think of ourselves not only passively, as creatures within which things happen, but actively, as creatures that make things happen. Understanding this distinction gives rise to major problems concerning the nature of agency, of the causation of bodily events by mental events, and of understanding the 'will' and 'free will'. Other problems in the theory of action include drawing the distinction between the structures involved when we do one thing 'by' doing another thing. Even the placing and dating of action can give ruse to puzzles, as one day and in one place, and the victim then dies on another day and in another place. Where and when did the murder take place? The notion of applicability inherits all the problems of 'intentionality'. The specific problems it raises include characterizing the difference between doing something accidentally and doing it intentionally. The suggestion that the difference lies in a preceding act of mind or volition is not very happy, since one may automatically do what is nevertheless intensional, for example, putting one's foot forwards while walking. Conversely, unless the formation of a volition is intentional, and thus raises the same questions, the presence of a violation might be unintentional or beyond one's control. Intentions are more finely grained than movements, one set of movements may both be answering the question and starting a war, yet the one may be intentional and the other not.

However, according to the traditional doctrine of epiphenomenalism, things are not as they seem: In reality, mental phenomena can have no causal effects: They are casually inert, causally impotent. Only physical phenomena are casually efficacious. Mental phenomena are caused by physical phenomena, but they cannot cause anything. In short, mental phenomena are epiphenomenal.

The epiphenomenalist claims that mental phenomena seem to be causes only because there are regularities that involve types (or kinds) of mental phenomena. For example, instances of a certain mental type 'M', e.g., trying to raise one's arm might tend to be followed by instances of a physical type 'P', e.g., one's arms rising. To infer that instances of 'M' tend to cause instances of 'P' would be, however, to commit the fallacy of post hoc, ergo propter hoc. Instances of 'M' cannot cause instances of 'P': Such causal transactions are casually impossible. P - typ e events tend to be followed by M - type events because instances of such events are dual - effects of common physical causes, not because such instances causally interact. Mental events and states can figure in the web of causal relations only as effects, never as causes.

Epiphenomenalism is a truly stunning doctrine. If it is true, then no pain could ever be a cause of our wincing, nor could something's looking red to us ever be a cause of our thinking that it is red. A nagging headache could never be a cause of a bad mood. Moreover, if the causal theory of memory is correct, then, given epiphenomenalism, we could never remember our prior thoughts, or an emotion we once felt, or a toothache we once had, or having heard someone say something, or having seen something: For such mental states and events could not be causes of memories. Furthermore, epiphenomenalism is arguably incompatible with the possibility of intentional action. For if, s the casual theory of action implies, intentional action requires that a desire for something and a belief about how to obtain what one desires lay a causal role in producing behaviour, then, if epiphenomenalism is true, we cannot perform intentional actions. As it strands, to accommodate this point - most obviously, specifying the circumstances in which belief - desire explanations are to be deployed. However, matter are not as simple as the seem. Ion the functionalist theory, beliefs are casual functions from desires to action. This creates a problem, because all of the different modes of psychological explanation appeal to states that fulfill a similar causal function from desire to action. Of course, it is open to a defender of the functionalist approach to say that it is strictly called for beliefs, and not, for example, innate releasing mechanisms, that interact with desires in a way that generates actions. Nonetheless, this sort of response is of limited effectiveness unless some sort of reason - giving for distinguishing between a state of hunger and a desire for food. It is no use, in that it is simply to describe desires as functions from belief to actions.

Of course, to say the functionalist theory of belief needs to be expanded is not to say that it needs to be expanded along non - functionalist lines. Nothing that has been said out the possibility that a correct and adequate account of what distinguishes beliefs from non - intentional psychological states can be given purely in terms of respective functional roles. The core of the functionalist theory of self - reference is the thought that agents can have subjective beliefs that do not involve any internal representation of the self, linguistic or non - linguistic. It is in virtue of this that the functionalist theory claim to be able to dissolve such the paradox. The problem that has emerged, however, is that it remains unclear whether those putative subjective beliefs really are beliefs. Its thesis, according to which all cases of action to be explained in terms of belief - desire psychology have to be explained through the attribution of beliefs. The thesis is clearly at work as causally given to the utility conditions, and hence truth conditions, of the belief that causes the hungry creature facing food to eat what I in front of him - thus, determining the content of the belief to be. 'There is food in front of me', or 'I am facing food'. The problem, however, is that it is not clear that this is warranted. Chances would explain by the animal would eat what is in front of it. Nonetheless, the animal of difference, does implicate different thoughts, only one of which is a purely directive genuine thought.

Now, the content of the belief that the functionalist theory demands that we ascribe to an animal facing food is 'I am facing food now' or 'There is food in front of me now'. These are, it seems clear, structured thoughts, so too, for that matter, is the indexical thought 'There is food here now'. The crucial point, however, is that the casual function from desires to actions, which, in itself, is all that a subjective belief is, would be equally well served by the unstructured thought 'Food'.

At the heart of the reason - giving relation is a normative claim. An agent has a reason for believing, acting and so forth. If, given here to other psychological states this belief/action is justified or appropriate. Displaying someone's reasons consist in making clear this justificatory link. Paradigmatically, the psychological states that prove an agent with logical states that provide an agent with treason are intentional states individuated in terms of their propositional content. There is a long tradition that emphasizes that the reason - giving relation is a logical or conceptual representation. In the case of reason for actions the premises of any reasoning are provided by intentional states other than belief.

Notice that we cannot then, assert that epiphenomenalism is true, if it is, since an assertion is an intentional speech act. Still further, if epiphenomenalism is true, then our sense that we are enabled is true, then our sense that we are agents who can act on our intentions and carry out our purposes is illusory. We are actually passive bystanders, never the agent in no relevant sense is what happens up to us. Our sense of partial causal control over our exert no causal control over even the direction of our attention. Finally, suppose that reasoning is a causal process. Then, if epiphenomenalism is true, we never reason: For there are no mental causal processes. While one thought may follow anther, one thought never leads to another. Indeed, while thoughts may occur, we do not engage in the activity of thinking. How, the, could we make inferences that commit the fallacy of post hoc, ergo propter hoc, or make any inferences at all for that matter?

Neurophysiological research began to develop in earnest during the latter half of the nineteenth century. It seemed to find no mental influence on what happens in the brain. While it was recognized that neurophysiological events do not by themselves casually determine other neurophysiological events, there seemed to be no 'gaps' in neurophysiological causal mechanisms that could be filled by mental occurrences. Neurophysiological appeared to have no need of the hypothesis that there are mental events. (Here and hereafter, unless indicated otherwise, 'events' in the broadest sense will include states as well as changes.) This 'no gap' line of argument led some theorists to deny that mental events have any casual effects. They reasoned as follows: If mental events have any effects, among their effects would be neurophysiological ones: Mental events have no neurophysiological effects: Thus, mental events have no effect at all. The relationship between mental phenomena and neurophysiological mechanisms is likened to that between the steam - whistle which accompanies the working of a locomotive engine and the mechanisms of the engine, just as the steam - whistle which accompanies the working of a locomotive engine and the mechanisms of the engine: just as the steam - whistle is an effect of the operations of the mechanisms but has no casual influence on those operations, so too mental phenomena are effects of the workings of neurophysiological mechanisms, but have no causal influence on their operations. (The analogy quickly breaks down, as steam whistles have casual effects but the epiphenomenalist alleges that mental phenomenons have no causal effects at all.)

An early response to this 'no gap' line of argument was that mental events (and states) are not changes in (and states of) an immaterial Cartesian substance e, they are, rather changes in (and states of) the brain. While mental properties or kinds are not neurophysiological properties or kinds, nevertheless, particular mental events are neurophysiological events. According to the view in question, a given events can be an instance of both a neurophysiological type and a mental type, and thus be both a mental event and a neurophysiological event. (Compare the fact that an object might be an instance of more than one kind of object: For example, an object might be both a stone and a paper-weight.) It was held, moreover, that mental events have causal effects because they are neurophysiological events with causal effects. This response presupposes that causation is an 'extensional' relation between particular events that if two events are causally related, they are so related however they are typed (or described). Given that assumption is today widely held. And given that the causal relation is extensional, if particular mental events are indeed, neurophysiological events are causes, and epiphenomenalism is thus false.

This response to the 'no gap' argument, however, prompts a concern about the relevance of mental properties or kinds to causal relations. And in 1925 C.D. Broad tells us that the view that mental events are epiphenomenal is the view that mental events either (a) do not function at all as causal - factors, or hat (b) if they do, they do so in virtue of their physiological characteristics and not in virtue of their mental characteristics. If particular mental events are physiological events with causal effects, then mental events function as case - factors: They are causes, however, the question still remains whether mental events are causes in virtue of their mental characteristics., yet, neurophysiological occurrences without postulating mental characteristics. This prompts the concern that even if mental events are causes, they may be causes in virtue of their physiological characteristics. But not in virtue of their mental characteristics.

This concern presupposes, of course, that events are causes in virtue of certain of their characteristics or properties. But it is today fairly widely held that when two events are causally related, they are so related in virtue of something about each. Indeed, theories of causation assume that if two events 'x' and 'y' are causally related, and two other events 'a' and 'b' are not, then there must be some difference between 'x' and 'y' and 'a' and 'b' in virtue of which 'x' and 'y' are. But 'a' and 'b' are not, causally related. And they attempt to say what that difference is: That is, they attempt to say what it is about causally related events in virtue of which they are so related. For example, according to so - called 'nomic subsumption views of causation', causally related events will be so related in virtue of falling under types (or in virtue of having properties) that figure in a 'causal law'. It should be noted that the assumption that casually related events are so related in virtue of something about each is compatible with the assumption that the causal relation is an 'extensional' relationship between particular events. The weighs - less - than relation is an extensional relation between particular objects: If O weighs less than O*, then O and O* are so related, have them of a typed (or characterized, or described, nevertheless, if O weighs less than O*, then that is so in virtue of something about each, namely their weights and the fact that the weight of one is less than the weight of the other. Examples are readily multiplied. Extensional relations between particulars typically hold in virtue of something about the particular. It is, nonetheless, that we will grant that when two events are causally related, they are so related in virtue of something about each.

Invoking the distinction between types and tokens, and using the term 'physical', rather than the more specific term 'physiological'. Of the following are two broad distinctions of epiphenomenalism:

Token Epiphenomenalism: Mental events cannot cause anything.

> Type Epiphenomenalism: No event can cause anything in virtue of falling under a mental type.

So in saying. That property epiphenomenalism is the thesis that no event can cause anything in virtue of having a mental property. The conjunction of token epiphenomenalism and the claim those physical events cause mental events is, that, of course, the traditional doctrine of epiphenomenalism, as characterized earlier. on epiphenomenalism implies type epiphenomenalism, for if an event could cause something in virtue of falling under a mental type, then an event could be both epiphenomenalism would be false. Thus, if mental events cannot be causes, then events cannot be causes in virtue of falling under mental types. The denial of token epiphenomenalism does not, however, imply the denial of type epiphenomenalism, if a mental event can be a physical event that has causal effects. For, if so, then token epiphenomenalism may still be true. For it may be that events cannot be causes in virtue of falling under mental types. Mental events may be causes in virtue of falling under mental types. Thus, even if token epiphenomenalism is false, the question remains whether type epiphenomenalism is.

Suppose, for the sake of argument, that type epiphenomenalism is true. Why would that be a concern if mental events are physical events with causal effects? In our assumption that the causal relation is extensional, it could be true, consist with type epiphenomenalism, that pains cause winces, that desires cause behaviour, that perceptual experience cause beliefs and mental states cause memories, and that reasoning processes are causal processes. Nevertheless, while perhaps not as disturbing a doctrine as token epiphenomenalism, type epiphenomenalism can, upon reflection, seen disturbing enough.

Notice to begin with that 'in virtue of' expresses an explanatory relationship. In so doing, that 'in virtue of' is arguably a near synonym of the more common locution 'because of'. But, in any case, the following seems true so as to be adequate: An event causes a G - event in virtue of being an F - event if and only if it causes a G - event because of being an F - event.'In virtue of' implies 'because of', and in the case in question at least the implication seems to go in the other direction as well. Suffice it to note that were type epiphenomenalism consistent with its being the case that an event could have a certain effect because of falling under a certain mental type, then we would, indeed be owed an explanation of why it should be of any concern if type epiphenomenalism is true. We will, however, assume that type epiphenomenalism is inconsistent with that. We will assume that type epiphenomenalism could be reformulated as: No event can cause anything because of falling under a mental type. (And we will assume that property epiphenomenalism can be reformulated thus: No event can cause anything because of having a mental property.) To say that 'a' causes 'b' in virtue of being 'F' is too say that 'a' causes 'b' because of being 'F'; that is, it is to say that it is because 'a' is 'F' that it causes 'b'. So, understood, type epiphenomenalism is a disturbing doctrine indeed.

If type epiphenomenalism is true, then it could never be the case that circumstances are such that it is because some event or states is a sharp pain, or a desire to flee, or a belief that danger is near, that it has a certain sort of effect. It could never be the case that it is because some state in a desire of 'X' (impress someone) and another is a belief that one can 'X' by doing 'Y' (standing on one's head) that the states jointly result in one's doing 'Y' (standing on one's head). If type (property) epiphenomenalism is true, then nothing has any causal powers whatever in virtue of (because of) being an instance of a mental type, then, never be the case of a certain mental type that a state has the causal power in certain circumstances to provide some effect. For example, it could never the case that it is in virtue of being an urge to scratch (or a belief that danger is near) that a state has the causal power in certain circumstances to produce scratching behaviour (or fleeing behaviour) if type - epiphenomenalism is true, then the mental qua mental, so to speak, is casually impotent. That may very well seem disturbing enough.

What reason is there, however, for holding type epiphenomenalism? Even if neurophysiology does not need to postulate types of mental events, perhaps the science of psychology does. Note that physics has no need to postulate types of neurophysiological events: But that may well not lead one tp doubt that an event can have effects in virtue of being (say) a neuron firing. Moreover, mental types figure in our every day, casual explanations of behaviour, intentional action, memory, and reasoning. What reason is there, then, for holding that events cannot have effects in virtue of being instances of mental types? This question naturally leads to the more general question of which event types are such that events have effects in virtue of falling under them. This more general question is best addressed after considering a 'no gap' line of argument that has emerged in recent years.

Current physics includes quantum mechanics, a theory which appears able, in principle, to explain how chemical processes unfold in terms of the mechanics of sub - atomic particles. Molecular biology seems able, in principle, to explain how the physiological operations of systems in living things in terms of biochemical pathways, long chains of chemical reactions. On the evidence, biological organisms are complex physical objects, made up of molecular particles (there are noo entelechies or élan vital). Since we are all biological organisms, the movements of our bodies and of their minute parts, including the chemicals in our brains, and so forth, are causally determined, too whatsoever subatomic particles and fields. Such considerations have inspired a lin e of argument that only events within the domain of physics are causes.

Before presenting the argument, let us make some terminological stipulations: Let us henceforth use 'physical events' (states) and 'physical property' in as strict and narrow sense to mean, respectfully, a type of event (state) physics (or, by some improved version of current physics). Event if they figure in laws of physics. Finally, by 'a physical event (states) we will mean an even (state) that falls under a physical type. Only events within the domain of (current) physics (or, some improved eversion of current physics) count as physical in this strict and narrow sense.

Consider, then:

The Token - Exclusion Thesis

Only physical events can have causal effects (i.e., as a matter of causal necessity, only physical events have casual effects).

The premises of the basis argument for the token - exclusion thesis are:

Physical Caudal Closure

Only physical events can cause physical events.

Causation by way of Physical Effects

As a matter of at least, casual necessity, an event is a cause of another event if and only if it is a cause of some physical event?

These principles jointly imply the exclusion thesis. The principle of causation through physical effects is supported on the empirical grounds that every event occurs within space-time, and by the principle that an event is a cause of an event that occurs within a given region of space-time if and only if it is a cause of some physical event that occurs within that region of space-time. The following claim is offered in support of physical closure:

Physical causal Determination,

For any (caused) physical event, 'P', there is a chain of entirely physical events leading to 'P', each link of which casually determines its successor.

(A qualification: If strict determinism is not true, then each link will determine the objective probability of its successor.) Physics is such that there is compelling empirical reason to believe that physical causal determination hold. Every physical event will have a sufficient physical cause. More precisely, there will be a deterministic casual chan of physical events leading to any physical event, 'P'. Butt such links there will be, and such physical causal chains are entirely 'gap-less'. Now, to be sure, physical casual determination does not physical causal closure, the former, but not the latter, is consistent with non - physical events causing physical events. However, a standard epiphenomenalist response to this is that such non - physical events would be, without exception, over - determining causes of physical events, and it is ad hoc are over - determining non - physical events. Nonctheless, a standard epiphenomenalist response of this is that such non - physical events would be, without exception, over - determining causes of physical events, and it is ad hoc to maintain that non - physical events are over - determining causes of physical events.

Are mental events within the domain of physics? Perhaps, like objects, events can fall under many different types or kinds. We noted earlier that a given object might, for instance, be both a stone and a paper wight, however, we understand how a stone could be a paper - wight, but how, for instance could an event of subatomic particles and fields be a mental event? Suffice e it to note for a moment that if mental events are not within the domain of physics, then if the token - exclusion thesis is true, no mental event can ever cause anything: Token epiphenomenalism is true.

One might reject the token - exclusion thesis, however, on the grounds that, typical events within the domains of the special sciences - chemistry, the life sciences, and so on - are not within the domain of physics, but nevertheless have causal effects. One might maintain that neuron firing, for instance, cause either neuron firing, even though neurophysiological events are not within the domain of physics. Rejecting the token - exclusion either, however, requires arguing either that physical causal closure is false or that the principle of causation by way of physical effects is.

But one response to the 'no - gap' argument from physics is to reject physical casual closure. Recall that physical causal determination is consistent with non-physical events being over-determining causes of physical events. One might concede that it would be ad hoc to maintain that a non - physical event, 'N', is an over-determining cause of a physical event 'P', and that 'N' causes 'P' in a way that is independent of the causation of 'P' by other physical events. Nonetheless, 'N' can be a cause of another event, that 'N' can cause a physical event 'P' in a way that is dependent upon P's being caused by physical events. Again, one might argue that physical events 'underlie' non - physical events, and that a non - physical event 'N' can be a cause of anther event 'X' (physical or non - physical), in virtue of the physical event that 'underlie' 'N' being a cause of 'X'.

Another response is to deny the principle of causation through physical effects. Physical causal closure is consistent with non - physical events. One might concede physical causal closure but deny the principle of causation by way of physical effects, and argue that nonphysical events cause other nonphysical events without causing physical events. This would not require denying that (1) Physical events invariably 'underlie' nonphysical events or that (2) Whenever a nonphysical event causes another nonphysical event, some physical event that underlies the first event causes a physical event that underlies the second. Clams of both tenets (1) and (2) do not imply the principle of causation through physical effects. Moreover, from the fac t that a physical event 'P', causes another physical event 'P*'. It may not allow that 'P' causes every non - physical event that 'P*' underlies. That may not follow it the physical events that underlie nonphysical events casually suffice for those nonphysical events. It would follow from that, which for every nonphysical event, there is a causally sufficient physical event. But it may be denied that causal sufficiency suffices for causation: It may be argued that there are further constraints on causation that can fail to be met by an event that causally suffices for another. Moreover, it ma be argued that given the further constraints, non - physical events are the causes of nonphysical events.

However, the most common response to the 'no - gap' argument from physics is to concede it, ad thus to embrace its conclusion, the token - exclusion these, but to maintain the doctrine of 'token physicalism', the doctrine that every event (state) is within the domain of physics. If special science events and mental events are within the domain of physics, then they can be causes consistent with the token - exclusion thesis.

Now whether special science events and mental events are within the domain of physics depends, in part, on the nature of events, and that is a highly controversial topic about which there is nothing approaching a received view. The topic raises deep issues that are beyond the scope of this essay, yet the issues concerning the 'essence' of events and the relationship between causation and causal explanation, are in any case, . . . suffice it to note here that it is believed that the sme fundamental issues concerning the causal efficacy of the mental arise for all the leading theories of the 'relata' of casual relation. The issues just 'pop - up' in different places. However, that cannot be argued at this time, and it will not be for us to be assumed.

Since the token physicalism response to the no - gap argument from physics is the most popular response, is that special science events, and even mental events, are within the domain of physics. Of course, if mental events are within the domain of physics then, token epiphenomenalism can be false even if the token - exclusion is true: For mental events may be physical events which have causal effects.

Nevertheless, concerns about the causal relevance of mental properties and event types would remain. Indeed, token physicalism together with a fairly uncontroversial assumption, naturally leads to the question of whether events can be causes only in virtue of falling under types postulated by physics. The assumption is that physics postulates a system of event types that has the following features:

Physical Causal Comprehensiveness:

When two physical events are causally related, they are so related in virtue of falling under physical types. That thesis naturally invites the question of whether the following is true:

The Type - Exclusion Thesis:

An event can cause something only in virtue of falling under a physical type, i.e., a type postulated by physics. The type-exclusion thesis offer's one would - be answer to our earlier question of which effects types are such that events have effects in virtue of falling under them. If the answer is the correct one, it may, however, be in fact (if it is correct) that special science events and mental events are within the domain of physics will be cold comfort. For type physicalism, the thesis that every event type is a physical type, seems false. Mental types seem not to be physical types in our strict and narrow sense. No mental type, it seems, is necessarily coextensive (i.e., coextensive in every 'possible world') with any type postulated by physics. Given that, and given the type - exclusion thesis, type epiphenomenalism is true. However, typical special science types also fail to be necessarily coextensive with any physical types, and thus typical special science types fail to be physical types. Indeed, we individuate the sciences in part by the event (state) types they postulate. Given that typical special science types are not physical types (in our strict sense), then typical special science types are not such that even can have causal effects in virtue of falling under them.

Besides, a neuron firing is not a type of event postulated by physics, given the type exclusion thesis, no event could ever have any causal effects in virtue of being a firing of a causal effect. The neurophysiological qua neurophysiological is causally impotent. Moreover, if things have casual powers only in virtue of their physical properties, then an HIV virus, qua HIV virus, does not have the causal power to contribute to depressing the immune system: For being an HIV virus is not a physical property (in our strict sense). Similarly, for the same reason the SALK vaccine, qua SALK vaccine, would not have the causal power to contribute to producing an immunity to polio. Furthermore, if, as it seems, phenotype properties are not physical properties, phenotypic properties do not endow organisms with casual powers conducive to survival. Having hands, for instance, could never endow anything with casual powers conducive to survival since it could never endow anything with any causal powers whatsoever. But how, then, could phenotypic properties be units of natural selection? And if, as it seems, genotypes are not physical types, then, given the type exclusion thesis, genes do not have the causal power, qua genotypes, to transmit the genetic bases for phenotypes. How, then, could the role of genotypes as units of heredity be a causal role? There seem to be ample grounds for scepticism that any reason for holding the type - exclusion thesis could outweigh our reasons for rejecting it.

We noted that the thesis of universal physical causal comprehensiveness or 'upc-comprehensiveness' for short, invites the question of whether the type-exclusion thesis is true. But does upc- comprehensiveness while rejecting the type - exclusion thesis?

Notice that there is a crucial one-word difference between the two theses: The exclusion thesis contains the word 'only' in front of 'in virtue of', while thesis of upc - comprehensiveness does not. This difference is relevant because 'in virtue of' does not imply 'only in virtue of', I am a brother in virtue of being a male with a sister, but I am also a brother in virtue of being a male with a brother, and, of course, being a male with a brother, and conversely. Likewise, I live in the province of Ontario in virtue of living in the city of Toronto, but it is also true that I live in Canada in virtue of living in the County of York. Moreover, in the general case, if something 'x' bears a relation 'R', to something 'y' in virtue of x's being 'F' and y's being 'G'. Suppose that 'χ' weighs less than 'y' in virtue of χ's weighing lbs., and y's weighing lbs. Then, it is also true that 'χ' weighs less than 'y' in virtue of χ's weighing under lbs., and y's weighing over lbs. And something can, of course, weigh under lbs., without weighing lbs. To repeat, 'in virtue of' does not imply 'only in virtue of'.

Why, then, think that upc-comprehensiveness implies the type - exclusion thesis? The fact that two events are causally related in virtue of falling under physical types does not seem to exclude the possibility that they are also causally related in virtue of falling under nonphysical types, in virtue of the one being (say) a firing of a certain other neuron, or in virtue of one being a secretion of enzymes and the other being a breakdown of amino acids. Notice that the thesis of upc-comprehensiveness implies that whenever an event is an effect of another, it is so in virtue of falling under a physical type. But the thesis does not seem to imply that whenever an event vis an effect of another, it is so only in virtue of falling under a physical type. Upc - comprehensiveness seems consistent with events being effects in virtue of falling under nonphysical types. Similarly, the thesis seems consistent with events being causes in virtue of falling under non - physical types.

Nevertheless, an explanation is called for how events could be causes in virtue of falling under nonphysical types if upc-comprehensiveness is true. The most common strategy for offering such an explanation involves maintaining there is a dependence-determination relationship between non

- physical types and physical types. Upc-comprehensiveness, together with the claim that instances of nonphysical event types are causes or effects, implies that, as a matter of causal necessity, whenever an event falls under a non - physical event type, if falls under some physical type or other. The instantiation of nonphysical types by an event thus depends, as a matter of causal necessity, on the instantiation of some or other physical event type by the event. It is held that nonphysical types in physical context: Although as given nonphysical type might be 'realizable' by more than one physical type. The occurrence o a physical type in a physical context in some sense determines the occurrence of any nonphysical type that it 'realizes'.

Recall the considerations that inspired the 'no gap' arguments from physics: Quantum mechanics seems able, in principle, to explain how chemical processes unfold in terms of the mechanics of subatomic particles: Molecular biology seems able, in principle, to explain how the physiological operations of systems in living things occur in terms of biochemical pathways, long chains of chemical reactions. Types of subatomic causal processes 'implement' types of chemical processes. Many in the cognitive science community hold that computational processes implement that mental processes, and that computational processes are implemented, in turn, by neurophysiological processes.

The Oxford English Dictionary gives the everyday meaning of 'cognition' as 'the action or faculty of knowing'. The philosophical meaning is the same, but with the qualification that it is to be 'taken in its widest sense, including sensation, perception, conception, and volition'. Given the historical link between psychology and philosophy, it is not surprising that 'cognitive' in 'cognitive psychology' has something like this broader sense, than the everyday one. Nevertheless, the semantics of 'cognitive psychology', like that of many adjective - noun combinations, is not entirely transparent. Cognitive psychology is a branch of psychology, and its subject matter approximates to the psychological study that are largely historical, its scope is not exactly what one would predict.

Many cognitive psychologists have little interest in philosophical issues, as cognitive scientists are, in general, more receptive. Fodor, because of his early involvement in sentence processing research, is taken seriously by many psycholinguistics. His modularity thesis is directly relevant to questions about the interplay of different types of knowledge in language understanding. His innateness hypothesis, however, is generally regarded as unhelpful, and his prescription that cognitive psychology is primarily ignored. Dennett's recent work on consciousness treats a topic that is highly controversial, but his detailed discussion of psychological research findings has enhanced his credibility among psychologists. Overall, psychologists are happy to get on with their work without philosophers telling them about their 'mistakes'.

The hypotheses driving most of modern cognitive science is simple to state - the mind is a computer. What are the consequences for the philosophy of mind? This question acquires heightened interest and complexity from new forms of computation employed in recent cognitive theory.

Cognitive science has traditionally been based on or upon symbolic computation systems: Systems of rules for manipulating structures built up of tokens of different symbol types. (This classical kind of computation is a direct outgrowth of mathematical logic.) Since the mid-1980s, however, cognitive theory has increasingly employed connectionist computation: The spread of numerical activation across units - the view that one of the most impressive and plausible ways of modelling cognitive processes in by of a connectionist, or parallel distributed processing computer architecture. In such a system data is input into a number of cells as one level, or hidden units, which in turn delivers an output.

Such a system can be 'trained' by adjusting the weights a hidden unit accords to each signal from an earlier cell. The' training' is accomplished by 'back propagation of error', meaning that if the output is incorrect the network makers the minimum adjustment necessary to correct it. Such systems prove capable of producing differentiated responses of great subtly. For example, a system may be able to task as input written English, and deliver as output phonetically accurate speech. Proponents of the approach also, point pout that networks have a certain resemblance to the layers of cells that make up a human brain, and that like us. But unlike conventional computing programs, networks degrade gracefully, in the sense that with local damage they go blurry rather than crashed altogether. Controversy has concerned the extent to which the differentiated responses made by networks deserve to be called recognitions, and the extent to which non-recognizable cognitive function, including linguistic and computational ones, are well approached in these terms.

Some terminology will prove useful: that is, for which we are to stipulate that an event type 'T' is a casual type if and only if there is, at least one type T*, such that something can case a T* in virtue of being a 'T'. And by saying that an event type is realizable by physical event types or physical properties. For that of which is least causally possible for the event to be realized by a physical event type. Given that nonphysical causal types must be realizable by physical types, and given that mental types are nonphysical types, there are two ways that mental types might to be causal. First, mental types may fail to be realizable by physical types. Second, mental types might be realizable by physical types but fail to meet some further condition for being causal types. Reasons of both sorts can be found in the literature on mental causation for denting that any mental types are causal. However, there has been much attention paid to reasons for the first sort in this casse of phenomenal mental types (pain states, visual states, and so forth). And there has been much attention to reasons of the second sort in the case of intentional mental states (i.e., beliefs that P, desires that Q, intentions that R, and so on).

Notice that intentional states figure in explanations of intentional actions not in virtue of their intentional mode (whether they are beliefs or desires, and so on) but also in virtue of their contents, i.e., what is believed, or desired, and so forth. For example, what causally explains someone's doing 'A' (standing on his head) is that the person wants to 'X' (impress someone) and believes that by doing 'A' he will 'X'. The contents of the belief and desire (what is believed and what is desired) sem essential to the causal explanation of the agent's doing 'A'. Similarly, we often causally explain why someone came to believe that 'P' by citing the fact that the individual came to believe that 'Q' and inferred 'P' from 'Q'. In such cases, the contents of the states in question are essential to the explanation. This is not, of course, to say that contents themselves are causally efficacious, contents are not among the relata of causal relations. The point is, however, that we characterize states when giving such explanations not only as being as having intentional modes, but also as having certain contents: We type states for having certain contents, we type states for the purpose of such explanations in terms of their intentional modes and their contents. We might call intentional state types that might include content properties 'conceptual intentional state types', but to avoid prolixity, let us call them 'intentional state types' for short: Thus, for present purposes, b y 'intentional state types' we will mean types such as the belief that 'P; the desire that 'Q', and so on, and not types such as belief, desire and the like, and not types such as belief, desire, and so forth.

Although it was no part of American philosopher Hilary Putnam, who in 1981 marked a departure from scientific realism in favour of a subtle position that he called internal realism, initially related to an ideal limit theory of truth and apparently maintaining affinities with verification, but in subsequent work more closely aligned with 'minimalism', Putnam's concepts in the later period has largely to deny any serious asymmetry between truth and knowledge as it is obtained in natural science, and as it is obtained in morals and even theology. Still, purposively of raising concerns about whether ideational states are causal, the well-known 'twin earth' thought experiment have prompted such concerns. These thought-experiments are fairly widely held to show alike in every intrinsic physical respect can have intentional states with different contents. If they show that, then intentional state type fail to supervene on intrinsic physical state types. The reason is that with contents an individual's beliefs, desires, and the like, have, depends, in part, on extrinsic, contextual factors. Given that, the concern has been raised toast states cannot have effects in virtue of falling under intentional state types.

One concern seems to be that state cannot have effects in virtue of falling under intentional state types because individuals who are in all and only the same intrinsic states must have all and only the same causal powers. In response to that concern, it might be pointed out that causal power ss often depend on context. Consider weight. The weight of objects do not supervene on their intrinsic properties: Two objects can be exactly alike in every intrinsic respect (and thus have the same mass) yet have different weights. Weight depends, in part on extrinsic, contextual factors. Nonetheless, it seems true that an object can make a scale read 10lbs in virtue of weighing 10lbs. Thus, objects which are in exactly the am e type of intrinsic states may have different causal powers due to differences in their circumstances.

It should be noted, however, that on some leading 'externalist' theories of content, content, unlike weight, depends on a historical context, such as a certain set of content-involving states is for attribution of those states to make the subject as rationally intelligible as possible, in the circumstances. Call such as theory of content 'historical - externalist theories'. On one leading historical-externalist theory, the content of a state depends on the learning history of the individual on another. It depends on the selection history of the species of which the individual is a member. Historical - externalist theories prompt a concern that states cannot have causal effects in virtue of falling under intentional state types. Causal state types, it might be claimed, are never such that their tokens must have a certain causal ancestry. But, if so, then, if the right account of content is a historical - externalist account, then intentional types are not casual types. Some historical - externalists appear to concede this line of argument, and thus to effects in virtue of falling under intentional state types. However, explain how intentional-externalists attempt to explain how intentional types can be casual, even though their tokens must have appropriated causal ancestries. This issue is hotly debated, and remains unresolved.

Finally, by noting, why it is controversial, whether phenomenal state types can be realized by physical state types. Phenomenal state types are such that it is like something for a subject to be in them: It is, for instance, like something to have a throbbing pain. It has been argued that phenomenal state types are, for that reason, subjective to fully understand what it is to be in them. One must be able to take up is to be in them, one must be able to take up a certain experiential point of view. For, it is claimed, an essential aspect of what it is to be in a phenomenal state is what it is like to be in a phenomenal state is what it is like to be in the state, only by tasking up certain experiential point of view can one understand that aspect (in our strict and narrow sense) are paradigms' objective state, i.e., non - subjective states. The issue arises, then, as to whether phenomenal state types can be realized by physical types. How could an objective state realize a subjective one? This issue too is hotly debated, and remains unresolved. Suffice it to say, that only physical types and types realizable by physical types and types realizable by physical types are causal, and if phenomenal types are neither, then nothing can have any causal effects, so, then, in virtue of falling under a phenomenal type. Thus, it could never be the case, for example, that a state causally results in a bad mood in virtue of being a throbbing pain.

Philosophical theories are unlike scientific ones, scientific theories ask questions in circumstances where there are agreed - upon methods for answering the question and where the answers themselves are generally agreed upon. Philosophical theory: They attempt to model the known data to be seen from a new perspectives, a perspective that promotes the development of genuine scientific theory. Philosophical theories are, thus, proto-theories, as such, they are useful precisely in areas where no large-scale scientific theory exist. At present, which is exactly the state psychology it is in. Philosophy of mind, is to be a kind of propaedeutics to a psychological science. What is clear is that at the moment no universally accepted paradigm for a scientific psychological science exists. It is exactly in this kind of circumstance for a scientific psychology exists. It is exactly in this kind of circumstance that the philosophers of mind in the present context is to consider the empirical data available and to ry to form a generalized, coherent way of looking at those data tat will guide further empirical research, i.e., philosophers can provide a highly schematized model that will structure that research. And the resulting research will, in turn, help bring about refinements of the schematized theory, with the ultimate hope being that a closer, viable, scientific theory, one wherein investigators agree on the question and on the methods to be used to answer them, and will emerge. In these respects, philosophical theories of mind, though concerned with current empirical data, are too general in respect of the data to be scientific theories. Moreover, philosophical theories aimed primarily at a body of accepted data. As such, philosophical theories merely give as 'picture' of those data. Scientific theories not only have to deal with the given data but also have to make predictions, in that can be gleaned from the theory together with accepted data. This removal go unknown data is what forms the empirical basis of a scientific theory and allows it to be justified in a way quite distinct from the way in which philosophical theories are justified. Philosophical theories are only schemata, coherent pictus of the accepted data, only pointers toward empirical theory, and as the history of philosophy makers manifest, usually unsuccessful one - though I think this lack of success is any kind of a fault, these are different tasks.

In the philosophy of science, a generalization or set of generalizations purportedly making reference to unobservable entities, e.g., atoms, genes, quarks, unconscious wishes, and so forth. The ideal gas law, for example, refers only to such observables as pressure, temperature and volume and their properties. Although an older usage suggests lack of adequate evidence in support thereof ('merely a theory'), current philosophical usage does not carry that connotation. Einstein's special theory of relativity, for example, is considered extremely well founded.

There are two main views on the nature of theories. According to the 'received view' theories are partially interpreted axiomatic systems, according to the semantic view a theory is a collection of models.

The axiomatization or axiomatics belongs of a theory that usually emerges as a body of (supposed) truths that are not neatly organized, making the theory difficult to survey or study as a whole. The axiomatic method is an idea for organizing a theory: One tries to select from among the supposed truths a small number from which all the others can be seen to be deductively inferrable. This make the theory rather more tractable since, in a sense, all the truths are contained in those few. In a theory so organised, the few truths from which all others are deductively inferred are called 'axioms'. David Hilbert had argued that, just as algebraic and differential equations and physical precesses, could themselves be made mathematical objects, so axiomatic theories, like algebraic and differential equations, which are of representing physical processes and mathematical structures, could be made objects of mathematical investigation.

Wherein, a credibility programme of a speech given in 1900, the mathematician David Hilbert (1862-1943) identified 23 outstanding problems in mathematics. The first was the 'continuum hypothesis'. The second was the problem of the consistency of mathematics. This evolved into a programme of formalizing mathematic-reasoning, with the aim of giving meta - mathematical proofs of its consistency. (Clearly there is no hope of providing a relative consistency proof of classical mathematics, by giving a 'model' in some other domain. Any domain large and complex enough to provide a model would be raising the same doubts.) The programme was effectively ended by Kurt Gödel (1906 - 78), whose theorem of 1931, which showed that any system of arithmetic would need to make logical and mathematical assumptions at least as strong as arithmetic itself, and hence be just as much prey to hidden inconsistencies.

In the tradition (as in Leibniz, 1704), many philosophers had the conviction that all truths, or all truths about a particular domain, followed from a few principles. These principles were taken to be either metaphysically prior or epistemologically prior or both. In the first sense, they were taken to be entities of such a nature that what exist is 'caused' by them. When the principles were taken as epistemically prior, that is, as axioms, they were taken to be either epistemically privileged, e.g., self - evident, not needing to be demonstrated, or again, inclusive 'or', to be such that all truths do in need follow from them, in at least, by deductive inferences. Gödel (1984) showed - in the spirit of Hilbert, treating axiomatic theories as themselves mathematical objects - that mathematics, and even a small part of mathematics, elementary number theory, could not be axiomatized that more precisely, any class of axioms which is such that we could effectively decide, of that class, would be too small to capture all of the truths.

'Philosophy is to be replaced by the logic of science - that is to say, by the logical analysis of the concepts and sentences of the sciences, for the logic of science is nothing other than the logical syntax of the language of science', has a very specific meaning. The background was provided by

Hilbert's reduction of mathematics to purposes of philosophical analysis, any scientific theory could ideally be reconstructed as an axiomatic system formulated within the framework of Russell' s logic. Further analysis of a particular theory could then proceed a the logical investigation of its ideal logical reconstruction. Claims about theories in general were couched as claims about such logical systems.

In both Hilbert's geometry and Russell's logic had an attempt made to distinguish between logical and non - logical terms. Thus the symbol '&' might be used to indicate the logical relationship of conjunction between two statements, while 'P' is supposed to stand for a non - logical predicate. As in the case of geometry, the idea was that underlying any scientific theory is a purely formal logical structure captured in a set of axioms formulated in the appropriated formal language. A theory of geometry, for example, might include an axiom stating that for ant two distinct P's (points), 'p' and 'q', there exist a number 'L' (Line) such that O(p, I) and O(q, I), where 'O' is a two-place relationship between P's and L's (p lies on I). Such axioms, taken all together, were said to provide an implicit definition of the meaning of the non - logical predicates. In whatever of all the P's and L's might be, they must satisfy the formal relationships given by the axioms.

The logical empiricists were not primarily logicians: They were empiricists first. From an empiricist point of view, it is not enough that the non - logical terms of a theory be implicitly defined: They also require an empirical interpretation. This was provided by the 'correspondence rules' which explicitly linked some of the nonlogical terms of a theory with terms whose meaning was presumed to be given directly through 'experience' or 'observation'. The simplest sort of correspondence rule would be one that takes the application of an observationally meaningful term, such as 'dissolve', as being both necessary and sufficient for the applicability of a theoretical term, such as 'soluble'. Such a correspondence rule would provide a complete empirical interpretation of the theoretical term.

A definitive formulation of the classical view was provided by the German logical positivist Rudolf Carnap (1891-1970), who divided the nonlogical vocabulary of theories into theoretical and observational components. The observational terms were presumed to be given a complete empirical interpretation, which left the theoretical terms with only an indirect empirical interpretation provided by their implicit definition within an axiom system in which some of the terms possessed a complete empirical interpretation.

Among the issues generated by Carnap's formulation was the viability of 'the theory - observation distinction', of course, one could always arbitrarily designate some subset of non - logical terms as belonging to the observational vocabulary, but that would compromise the relevance of the philological analysis for an understanding of the original scientific theory. But what could be the philosophical basis for drawing the distinction? Take the predicate 'spherical', for example. Anyone can observe that a billiard ball is spherical. But what about the moon, on the one hand, or an invisible speck of sand, on the other. Is the application of the term? For which the 'spherical' in these objects are 'observational'?

Another problem was more formal, as did, that Craig's theorem seem to show that a theory reconstructed in the recommendations fashioned could be re - axiomatized in such a way as to dispense with all theoretical terms, while retaining all logical consequences involving only observational terms. Craig's theorem continues as a theorem in mathematical logic, held to have implications in the philosophy of science. The logician William Craig at Berkeley showed how, if we partition the vocabulary of a formal system (say, into the 'T' or theoretical terms, and the 'O' or observational terms) then if there is a fully 'formalized system' 'T' with some set 'S' of consequences containing only 'O' terms, there is also a system 'O' containing only the 'O' vocabulary but strong enough to give the same set 'S' of consequences. The theorem is a purely formal one, in that 'T' and 'O' simply separate formulae into the preferred ones, containing as nonlogical terms only one kind of vocabulary, and the objects. The theorem might encourage the thought that the theoretical terms of a scientific theory are in principle dispensable, since the same consequences can be derived without them.

However, Craig's actual procedure gives no effective way of dispensing with theoretical terms in advance, i.e., in the actual process of thinking about and designing the premises from which the set 'S' follows. In this sense 'O' remains parasitic upon its parent 'T'. Thus, as far as the 'empirical' content of a theory is concerned, it seems that we can do without the theoretical terms. Carnap's version of the classical vew seemed to imply a form of instrumentionalism, a problem which Carl Gustav Hemel (1905 - 97) christened 'the theoretician's dilemma'.

In the late 1940s, the Dutch philosopher and logician Evert Beth published an alternative formalism for the philosophical analysis of scientific theories. He drew inspiration from the work of Alfred Tarski, who studied first biology and then mathematics. In logic he studied with Kotarinski, Lukasiewicz and Lesniewski publishing a succession of papers from 1923 onwards. Yet he worked on decidable and undecidable axiomatic systems, and in the course in his mathematical career he published over 300 papers and books, on topics ranging from set theory to geometry and algebra. And also, drew further inspiration from Rudolf Carnap, the German logical positivist who left Vienna to become a professor at Prague in 1931, and felt Nazism to become professor in Chicago in 1935. He subsequently worked at Los Angeles from 1952 to 1951. All the same, Evert Beth drew inspirations from von Neumann's work on the foundations of quantum mechanics. Twenty years later, Beth's emigrant who left Holland around the time Beth's and van Fraassen. Here we are consider the comprehensibility in following the explication for which as preconditions between the 'syntactic' approach of the classical view and the 'semantic' approach of Beth and van Fraassen, and further consider the following simple geometrical theory as van Fraassen in 1989, presented first in the form of:

A1: For any two lines, at most one point lies on both.

A2: For any two points, exactly one line lies on both.

A3: On every line are at least two points.

Note first, that these axioms are stated in more or less everyday language. On the classical view one would have first to reconstruct these axioms in some appropriate formal language, thus introducing quantifiers and other logical symbols. And one would have to attach appropriate correspondence rules. Contrary to common connotations of the word 'semantic', the semantic approach down - plays concerns with language as such. Any language will do, so long as it is clear enough to make reliable discriminations between the objects which satisfy the axiom and those which do not. The concern is not so much with what can be deduced from their axioms, valid deduction being matter of syntax alone. Rather, the focus is on 'satisfaction', what satisfies the axioms - a semantic notion. These objects are, in the technical, logical sense of the term, models of the axioms. So, on the semantic approach, the focus shifts from the axiom as linguistic entities, to the models, which are non - linguistic entities.

It is not enough to be in possession of a general interpretation for the terms used to characterize the models, one must also be able to identify particular instances - for example, a particular nail in a particular board. In real science must effort and sophisticated equipment may be required to make the required identification, for example, of a star as a white dwarf or of a formation in the ocean floor as a transformed fault. On a semantic approach, these complex processes of interpretation and identification, while essential in being able t use a theory, have no place within the theory itself. This is inn sharp contrast to the classical view, which has the very awkward consequence that various innovations in instrumenting itself. The semantic approach better captures the scientist's own understanding of the difference between theory and instrumentation.

On the classical view the question 'What is a scientific theory'? Receives a straightforward answer. A theory is (1) a set of uninterrupted axioms in a specific formal language plus (2) a set of correspondence rules that provide a partial empirical interpretation in terms of observable entities and processes. A theory is thus true if and only if the interpreted axioms are all true. To obtain a similarly straightforward answer a little differently. Return to the axiom for placements as considered as free - standing statements. The definition could be formulated as follows: Any set of points and lines constitute a seven-pointed geometry is not even a candidate for truth or falsity, one can hardly identify a theory with a definition. But claims to the effect that various things satisfy the definition may be true or false of the world. Call these claims theoretical hypotheses. So we may say that, on the semantic approach, a theory consists of (1) a theoretical definition plus (2) a number of theoretical hypotheses. The theory may be said to be true just in case all its associated theoretical hypotheses are true.

Adopting a semantic approach to theories still leaves wide latitude in the choice of specific techniques for formulating particular scientific theories. Following Beth, van Fraassen adopts a 'state space' representation which closely mirrors techniques developed in theoretical physics during the nineteenth century - techniques were carried over into the developments of quantum and relativistic mechanics. The technique can be illustrated most simply for classical mechanics.

Consider a simple harmonic oscillator, which consists of a mass constrained to move in one dimension subject to a linear restoring force - a weight bouncing gently while from a spring provides a rough example of such a system. Let 'x' represent the single spatial dimension, 't' the time., 'p' the momentum, 'k' the strength of the restoring force, ands 'm' the mass. Then a linear harmonic oscillator may be 'defined' as a system which satisfies the following differential equation of motion:

$$dx/dt = DH/Dp. \quad Dp/dt = - DH/Dx, \text{ where } H = (k/2)x2 + (1/2m)p2$$

The Hamiltonian, 'H', represents the sun of the kinetic and potential energy of the system. The state of the system at any instant of time is a point in a two - dimensional position - momentum space. The history of any such system is this state space is given by an ellipse, as in time, the system repeatedly traces by revealing the ellipse onto the 'x' axis covering classical mechanics. It remains to be any real world system, such as a bouncing spring, satisfies this definition.

Other advocates of a semantic approach defer from the Beth - van Fraassen point of view in the type of formalism they would employ in reconstructing actual scientific theories. One influential approach derives from the word of Pattrick Suppes during he 1950s and 1960s, some of which inspired Suppes was by the logician J.C.C. Mckinsey and Alfred Tarski. In its original form. Suppes's view was that theoretical definition should be formulated in the language of set theory. Suppes's approach, as developed by his student Joseph Sneed (1971), has been adopted widely in Europe, and particularly in Germany, by the late Wolfgang Stegmüller (1976) and his students. Frederick Suppe has shares features of both the state space and the set - theoretical approaches.

Most of those who have developed 'semantic' alternatives to the classical 'syntactic' approach to the nature of scientific theories were inspired by the goal of reconstructing scientific theories - a goal shared by advocates of the classical view. Many philosophers of science now question whether there is any point in treating philosophical reconstructions as scientific theories. Rather, insofar as the philosophy of science focuses on theories at all, it is the scientific versions, in their own terms, that should be of primary concern. But many now argue that the major concern should be directed toward the whole practice of science, in which theories are but a part. In these latter pursuits what is needed is not a technical framework for reconstructing scientific theories, but merely a general imperative framework for talking about required theories and their various roles in the practice of science. This becomes especially important when considering science such as biology, in which mathematical models play less of a role than in physics.

At this point, at which there are strong reasons for adopting a generalized model-based understanding of scientific theories which makes no commitments to any particular formalism - for example, state spaces or set-theoretical predicates. In fact, one can even drop the distinction between 'syntactic' and 'semantic' as a leftover from an old debate. The important distinction is between an account of theories that takes models as fundamental versus that takes statements, particularly laws, as fundamental. A major argument for a model-based approach is that just given. There seem in fact to be few, if any, universal statements that might even plausibly be true, let alone known to be true, and thus available to play the role which laws have been thought to play in the classical account of theories, rather, what have often been taken to be universal generalisations should be interpreted as parts of definitions. Again, it may be helpful to introduce explicitly the notion of an idealized, theoretical model, an abstract entity which answers s precisely to the correspondence theoretical definition. Theoretical models thus provide, though only by something of which theoretical definitions may be true. This makes it possible to interpret much of scientific' theoretical discourse as being about theoretical models than directly about the world. What have traditionally been interpreted as laws of nature thus out to be merely statements describing the behaviour of theoretical models?

If one adopts such a generalized model-based understanding of scientific theories, one must characterize the relationship between theoretical models and real systems. Van Fraassen (1980) suggests that it should be one of isomorphism. But the same considerations that count against there being true laws in the classical sense also count against there being anything in the real world strictly isomorphic in any theoretical model, or even isomorphic to an 'empirical' sub-model. What is needed is a weaker notion of similarity, for which it must be specified both in which respect the theoretical model and the real system are similar, and to what degree. These specifications, however, like the interpretation of terms used in characterizing the model and the identification of relevant aspects of real systems, are not part of the model itself. They are part of a complex practice in which models are constructed and tested against the world in an attempt to determine how well they 'fit'.

Divorced from its formal background, a model-based understanding of theories is easily incorporated into a general framework of naturalism in the philosophy of science. It is particularly well-suited to a cognitive approach to science. Today the idea of a cognitive approach to the study of science something quite different - indeed, something antithetical to the earlier meaning. A 'cognitive approach' is now taken to be one that focuses on the cognitive structures and processes exhibited in the activities of individual scenists. The general nature of these structures and processes is the subject matter of the newly emerging cognitive science. A cognitive approach to the study of science appeals to specific features of such structures and processes to explain the model and choices of individual scientists. It is assumed that to explain the overall progress of science one must ultimately also appeal to social factors and social approaches, but not one in which the cognitive excludes the social. Both are required for an adequate understanding of science as the product of human activities.

What is excluded by the newer cognitive approach to the study of science is any appeal to a special definition of rationality which would make rationality a categorical or transcendent feature of science. Of course, scientists have goals, both individual and collective, and they employ more or less effective for achieving these goals. So one may invoke an 'instrumental' or 'hypothetical' notion of rationality in explaining the success or failure of various scientific enterprise. But what is it at issue is just the effectiveness of various goal-directed activities, not rationality in any more exalted sense which could provide a demarcation criterion distinguishing science from other activities, sch as business or warfare. What distinguishes science is its particular goals and methods, not any special form of rationality. A cognitive approach to the study of science, then, is a species of naturalism in the philosophy of science.

Naturalism in the philosophy of science, and philosophy generally, is more an overall approach to the subject than a set of specific doctrines. In philosophy it may be characterized only by the most general ontological and epistemological principles, and then more by what it opposes than by what it proposes.

Besides ontological naturalisms and epistemological type naturalism, it seems that its most probably the single most important contributor to naturalism in the past century was Charles Robert Darwin (1809-82), who, while not a philosopher, naturalist is both in the philosophical and the biological sense of the term. In 'The Descent of Man' (1871) Darwin made clear the implications of natural selection for humans, including both their biology and psychology, thus undercutting forms of anti - naturalism which appealed not only to extra-natural vital forces in biology, but to human freedom, values, morality, and so forth. These supposed indicators of the extra - natural are all, for Darwin, merely products of natural selection.

All and all, among advocates of a cognitive approach there is near unanimity in rejecting the logical positivist leal of scientific knowledge as being represented in the form of an interpreted, axiomatic system. But there the unanimity ends. Many employ a 'mental models' approach derived from the work of Johnson - Laird (1983). Others favour 'production rules' if this, infer that, a long usage for which the continuance by researchers in computer science and artificial intelligence, while some appeal to neural network representations.

The logical positivist are notorious for having restricted the philosophical study of science to the 'context of justification', thus relegating questions of discovery and conceptual change to empirical psychology. A cognitive approach to the study of science naturally embraces these issues as of central concern. Again, there are differences. The pioneering treatment, inspired by the work of Herbert Simon, who employed techniques from computer science and artificial intelligence to generate scientific laws from finite data. These methods have now been generalized in various directions, while appeals to study of analogical reasoning in cognitive psychology, while Gooding (1990) develops a cognitive model of experimental procedure. Both Nersessian and Gooding combine cognitive with historical methods, yielding what Neressian calls a 'cognitive-historical' approach. Most advocates of a cognitive approach to conceptual change are insistent that a proper cognitive understanding of conceptual change avoids the problem of incommensurability between old and new theories.

No one employing a cognitive approach to the study of science thinks that there could be an inductive logic which would pick out the uniquely rational choice among rival hypotheses. But some, such as Thagard (1991) think it possible to construct an algorithm that could be run on a computer which would show which of two theories is best. Others seek to model such judgements as decisions by individual scientists, whose various personal, professional, and social interests are necessarily reflected in the decision process. Here, it is important to see how experimental design and the result of experiments may influence individual decisions as to which theory best represents the real world.

The major differences in approach among those who share a general cognitive approach to the study of science reflect differences in cognitive science itself. At present, 'cognitive science' is not a unified field of study, but an amalgam of parts of several previously existing fields, especially artificial intelligence, cognitive psychology, and cognitive neuroscience. Linguistic, anthropology, and philosophy also contribute. Which particular approach a person takes has typically been determined more by developing a cognitive approach may depend on looking past specific disciplinary differences and focussing on those cognitive aspects of science where the need for further understanding is greatest.

Broadly, the problem of scientific change is to give an account of how scientific theories, proposition, concepts, and/or activities alter over the corpuses of times generations. Must such changes be accepted as brute products of guesses, blind conjectures, and genius? Or are there rules according to which at least some new ideas are introduced and ultimately accepted or rejected? Would such rules be codifiable into coherent systems, a theory of 'the scientific method'? Are they more like rules of thumb, subject to exceptions whose character may not be specifiable, not necessarily leading to desired results? Do these supposed rules themselves change over time? If so, do they change in the light of the same factors as more substantive scientific beliefs, or independently of such factors? Does science 'progress'? And if so, is its goal the attainment of truth, or a simple or coherent account (true or not) of experience, or something else?

Controversy exists about what a theory of scientific change should be a theory of the change 'of'. Philosophers long assumed that the fundamental objects of study of study are the acceptance or rejection of individual belief or propositions, change of concepts, positions, and theories being derivative from that. More recently, some have maintained that the fundamental units of change are theories or larger coherent bodies of scientific belief, or concepts or problems. Again, the kinds of causal factors which an adequate theory of scientific change should consider are far from evident. Among the various factors said to be relevant are observational data: The accepted background of theory, higher-level methodological constraints, psychological, sociological, religious, meta - physical, or aesthetic factors influencing decisions made by scientists about what to accept and what to do.

These issues affect the very delineation of the field of philosophy of science, in what ways, if any, does it, in its search for a theory of scientific change, differ from and rely on other areas, particularly the history and sociology of science? One traditional view was that those others are not relevant at all, at least in any fundamental way. Even if they are, exactly how do they relate to the interest peculiar to the philosophy of science? In defining their subject many philosophers have distinguished maltsters internal to scientific development - ones relevant to the discovery and/or justification of scientific claims - from ones external thereto - psychological, sociological, religious, metaphysical, and so forth, not directly relevant but frequently having a causal influence. A line of demarcation is thus drawn between science and non - science, and simultaneously between philosophy of science, concerned with the internal factors which function as reasons (or count as reasoning), and other disciplines, to which the external, nonrational factors are relegated.

This array of issues is closely related to that of whether a proper theory of scientific change is normative or descriptive. Is philosophy of science confined in description of what scientific cases be described with complete accuracy as it is descriptive, to what extent must scientific cases be described with compete accuracy? Can the theory of internal factors be a 'rational reconstruction' a retelling that partially distorts what actually happened in order to bring out the essential reasoning involved?

Or should a theory of scientific change be normative, prescribing how science ought to proceed? Should it counsel scientists about how to improve their procedures? Or would it be presumptuous of philosophers to advise them about how to do what they would it be presumptuous of philosophers to advise them about how to do what they are far better prepared to do? Most advocates of normative philosophy of science agree that their theories are accountable somehow to the actual conduct of science. Perhaps philosophy should clarify what is done in the best science: But can what qualifies as 'best science' be specified without bias? Feyerabend objects to taking certain developments as paradigmatic of good science. With others, he accepts the 'Pessimistic induction' according to which, since all past theories have proved incorrect, present ones can be expected to do so also, what we consider good science, eve n the methodological rules we rely on, may be rejected in the future.

Much discussion of scientific change since Handon centres on the distinction between context of discovery and justification. The distinction is usually ascribed to the philosopher of science and probability theorist Hand Reichenbach (1891-1953) and, as generally interpreted, reflective attitude of the logical empiricist movement and of the philosopher of science Raimund Karl Popper (1902-1994) who overturns the traditional attempts to found scientific method in the support that experience gives in suitably formed generalizations and theories. Stressing the difficulty, the problem of 'induction' put in front of any such method. Popper substitutes an epistemology that starts with the hold, imaginative formation of hypotheses. These face the tribunal of experience, which has the power to falsify, but not to confirm them. A hypotheses that survives the ordeal of attempted refutation between science and metaphysics, that an unambiguously refuted law statement may enjoy a high degree of this kind of 'confirmation', where can be provisionally accepted as 'corroborated', but never assigned a probability.

The promise of a 'logic' of discovery, in the sense of a set of algorithmic, content - neutral rules of reasoning distinct from justification, remains unfulfilled. Upholding the distinction between discovery and justification, but claiming nonetheless that discovery is philosophically relevant, many recent writers propose that discovery is a matter of a 'methodology', 'rationale', or 'heuristic;' rather than a 'logic'. That is, only a loose body of strategies or rules of thumb - still formulable discoveries, there is content of scientific belief - which one has some reason to hope will lead to the discovery of a hypothesis.

In the enthusiasm over the problem of scientific change in the 1960s nd 1970s, the most influential theories were based on holistic viewpoints within which scientific 'traditions' or 'communities' allegedly worked. The American philosopher of science Samuel Thomas Kuhn (1922-96) suggested that the defining characteristic of a scientific tradition is its 'commitment' to a shared 'paradigm'. A paradigm is 'the source of the methods, problem - field, and standards of solution accepted by any mature scientific community at any given time'. Normal science, the working out of the paradigm, gives way to scientific revolution when 'anomalies' in it precipitate a crisis leading to adoptions of a new paradigms. Besides many studies contending that Kuhn's model fails for some particular historical case, three major criticisms of Kuhn's view are as follows. First, ambiguities exist in his notion of a paradigm. Thus a paradigm includes a cluster of components, including 'conceptual, theoretical, instrumental, and methodological' communities: It involves more than is capturable in a single theory, or even in words. Second, how can a paradigm fall, since it determine s what count as facts, problems, and anomalies? Third, since what counts as a 'reason' is paradigm - dependent, there remains of no paradigmatic reason for accepting a new paradigm upon the failure of an older one.

Such radical relativism is exacerbated by the 'incommensurability' thesis shared by Kuhn (1962) and Feyerabend (1975), are, even so, that, Feyerabend's differences with Kuhn can be reduced to two basic ones. The first is that Feyerabend's variety of incommensurability is more global and cannot be localized in the vicinity of a single problematic term or even a cluster of terms. That is, Feyerabend holds that fundamental changes of theory lead to changes in the meaning of all the terms in a particular theory. The other significant difference concerns the reasons for incommensurability. Whereas Kuhn thinks that incommensurability stems from specific transactional difficulties involving problematic terms, as Feyerabend's variety of incommensurability seems to result from a kin d of extreme holism about the nature of meaning itself. Feyerabend is more consistent than Kuhn in giving a linguistic characterization of incommensurability, and there seems to be more continuity in his usage over time. He generally frames the incommensurability claim in term's of language, but the precis e reasons he cites for incommensurability are different from Kuhn's. One of Feyerabend's most detailed attempts to illustrate the concept of incommensurability involves the medieval European impetus theory and Newtonian classical mechanics. He claims that 'the concept of impetus, as fixed by the usage established in the impetus theory, cannot be defined in a reasonable way within Newton's theory'.

Yet, on several occasions Feyerabend explains the reasons for incommensurability by saying that there are certain 'universal rules' or 'principles of construction' which govern the terms of one theory and which are violated by the other theory. Since the second theory violates such rules, any attempt to state the claims of that theory in terms of the first will be rendered futile. 'We have a point of view (theory, framework, cosmos, modes of representation) whose elements (concepts, facts, picture) are built up in accordance e with certain principles of construction. The principle s involve e something ;like a 'closure', there are things that cannot be said, or 'discovered', without violating the principles (which does not mean contradicting them). Stating such terms as 'universal' he states: 'Let us call a discovery, or a statement, or an attitude incommensurable with the cosmos (the theory, the framework) if it suspends some of its universal principles'. As an example, of this phenomena, consider two theories, 'T' and T*, where 'T' is classical celestial mechanics, including the space-time framework, and 'T' is general relativity theory. Such principles as the absence of an upper limit for velocity, governing all the terms in celestial mechanics, and these terms cannot be expressed at once such principles are violated, as they will be by general relativity theory. Even so, the meaning of terms is paradigm - dependent, so that a paradigm tradition is 'not only incompatible but often actually incommensurable with that which has gone before'. Different paradigms cannot even be compared, for both standards of comparison and meaning are paradigm-dependent.

Response to incommensurability have been profuse in the philosophy of science, and only a small fractions can be sampled at this point, however, two main trends may be distinguished. The first denies some aspects of the claim, and suggests a method of forging a linguistic comparison among theories, while the second, though not necessarily accepting the claim of linguistic incommensurability proceeds to develop other ways of comparing scientific theories.

Inn the first camp are those who have argued that at least one component of meaning is unaffected by untranslatability: Namely, reference, Israel Scheller (1982) enunciates this influential idea in responses to incommensurability, but he does not supply a theory of reference to demonstrate how the reference of terms from different theories can be compared. Later writers seem to be aware of the need for a full-blown theory of reference to make this response successful. Hilary Putnam (1975) argues that the causal theory of reference can be used to give an account of the meaning of natural kind terms, and suggests that the same can be done for scientific terms in general, but the causal theory was first proposed as a theory of reference for proper names, and there are serious problems with the attempt to apply it to science. An entirely different language response to the incommensurability claim is found in the American philosopher Herbert Donald Davidson (1917-2003), where the construction takes place within a generally 'holistic' theory of knowledge and meaning. A radial interpreter can tell when a subject holds a sentence term and using the principle of 'charity' ends up making an assignment of truth conditions to individual sentences, although Davidson is a defender of the doctrine of the 'indeterminacy' of radical translations and the reusability 'of reference, his approach has seemed to many to offer some hope of identifying meaning as an extensional approach to language. Davidson is also known for rejection of the idea of a conceptual scheme, thought of s something peculiar to one language or one way of looking at the world.

The second kind of response to incommensurability proceeds to look or non - linguistic ways of making a comparison between scientific theories. Among these responses one can distinguish two main approaches. One approach advocates expressing theories in model-theoretic terms, thus espousing a mathematical mode of comparisons. This position has been advocated by writers such as Joseph Sneed and Wolfgang Stegmüller, who have shown how to discern certain structural similarities among theories in mathematical physics. But the methods of this 'structural approach' do not seem applicable to any but the most highly mathematized scientific theories. Moreover, some advocate of this approach have claimed that it lends support to a model-theoretic analogue of Kuhn's incommensurability claim. Another trend which has scientific theories to be entities in the minds or brains of scientists, and regard them as amendable to the techniques of recent cognitive science, proponents include Paul Churchlands, Ronald Gierre, and Paul Thagard. Thagard's (1992) s perhaps the most sustained cognitive attempt to rely to incommensurability. He uses techniques derived from the connectionist research program in artificial intelligence, but relies crucially from a linguistic mode of representing scientific theories without articulating the theory of meaning presupposed. Interestingly, neither cognitivist who urges acing connectionist methods to represent scientific theories. Churchlands (1992), argues that connectionist models vindicate Feyerabend's version of incommensurability.

The issue of incommensurability remains a live one. It does not arise just for a logical empiricist account of scientific theories, but for any account that allows for the linguistic representation of theories. Discussions of linguistic meaning cannot be banished from the philosophical analysis of science, simply because language figures prominently in the daily work of science itself, and its place is not about to be taken over by any other representational medium. Therefore, the challenge facing anyone who holds that the scientific enterprise sometimes requires us to mk e a point-by-point linguistic comparison of rival theories is to respond to the specific semantic problem raised by Kuhn and Feyerabend. However, if one does not think that such a piecemeal comparison of theories is necessary, then the challenge is tp articulate another way of putting scientific theories in the balance and weighing them against one-another.

The state of science at any given time is characterized, in part at least, by the theories that are 'accepted' at that time. Presently, accepted theories include quantum theory, the general theory of relativity, and the modern synthesis of Darwin and Mendel, as well as lower level (but still clearly theoretical) assertions such as that DNA has a double helical structure, that the hydrogen atom contains a single electron and so firth. What precisely involves the accepting of a theory?

The commonsense answer might appear to be that given by the scientific realist, to accept a theory, at root, to believe it to be true for at any rate, 'approximately' or 'essentially' true. Not surprising, the state of theoretical science at any time is in fact too complex to be captured fully by any such single notion.

For one thing, theories are often firmly accepted while being explicitly recognized to be idealizations. The use of idealizations raises as number of problems for the philosopher of science. One such problem is that of confirmation. On the deductive nomological model of scientific theories, which command virtually universal assent in the eighteenth and nineteenth centuries, is that confirming evidence for a hypothesis of evidence which increases its probability. Nonetheless, presumably, if it could be shown that any such hypothesis is sufficiently well confirmed by the evidence, then that would be grounds for accepting it. If, then, it could be shown that observational evidence could confirm such transcendent hypotheses at all, then that would go some way to solving the problem of induction. Nevertheless, thinkers as diverse in their outlook as Edmund Husserl and Albert Einstein have pointed to idealizations as the hall - mark of modern science.

Once, again, theories may be accepted, not be regarded as idealizations, and yet be known not to be strictly true - for scientific, rather than abstruse philosophical, reasons. For example, quantum theory and relativity theory were uncontroversially listed as among those presently accepted in science. Yet, it is known that the two theories, yet relativity requires all theories are not quantized, yet quantum theory say that fundamentally everything is. It is acknowledged that what is needed is a synthesis of the two theories, a synthesis which cannot of course (in view of their logical incommutability) leave both theories, as presently understood, fully intact, (This synthesis is supposed to be supplied by quantum field theory, but it is not yet known how to articulate that theory fully) none of this, that the present quantum and relativistic theories regarded as having an authentically conjectural character. Instead, the attitude seems to be that they are bound to survive in modified form as limited cases in the unifying theory of the future - this is why a synthesis is consciously sought.

In addition, there are theories that are regarded as actively conjectured while nonetheless being accepted in some sense: It is implicitly allowed that these theories might not live on as approximations or limiting cases in further sciences, though they are certainly the best accounts we presently have of their related range of phenomena. This used to be (perhaps still is) the general view of the theory of quarks, few would put these on a par with electrons, say, but all regard them as more than simply interesting possibilities.

Finally, the phenomenon of change in accepted theory during the development of science must be taken into account: But from the beginning, the distance between idealization and the actual practice of science was evident. Karl Raimund Popper (1902-1994), the philosopher of science, was to note, that an element of decision is required in determining what constitute a 'good' observation, a question of this sort, which leads to an examination of the relationship between observation and theory, has prompted philosophers of science to raise a series of more specific questions. What reasoning was in fact used to make inferences about light waves, which cannot be observed from diffraction patterns that can be? Was such reasoning legitimate? Are they to be construed as postulating entities just as real as water waves only much smaller? Or should the wave theory be understood non realistically as an instrumental device for organizing the predicting observable optical phenomena such as the reflection, refraction, and diffraction of light? Such questions presuppose that here is a clear distinction between what can and cannot be observed. Is such a distinction clear? If so, how is it to be drawn? As, these issues are among the central ones raised by philosophers of science about theory that postulates unobservable entities

Reasoning begins in the 'context of justification', as this is accomplished by deriving conclusions deductively from the assumptions of the theory. Among these conclusions at least some will describe states of affairs capable of being establish ed as true or false by observations. If these observational conclusions turns out to be true, the theory is shown to be empirically supported or probable. On a weaker version due to Karl Popper (1959), the theory is said to be 'corroborated', meaning simply that it has been subjected to test and has not been falsified. Should any of the observational conclusions turn out to be false, the theory is refuted, and must be modified or replaced. So a hypothetico - deductivist can postulate any unobservable entities or events he or she wishes in the theory, so long as all the observational conclusions of the theory are true.

However, against the then generally accepted view that the empirical science are distinguished by their use of an inductive method. Popper's 1934 book had tackled two main problems: That of demarcating science from non - science (including pseudo - science and metaphysics), and the problem of induction. Again, Popper proposed a falsifications criterion of demarcation: Science advances unverifiable theories and tries to falsify them by deducing predictive consequences and by putting the more improbable of these to searching experimental tests. Surviving such testing provided no inductive support for the theory, which remain a conjecture, and may be overthrown subsequently. Popper's answer to the Scottish philosopher, historian and essayist David Hume (1711 - 76), was that he was quite right about the invalidity of inductive inference, but that this does not matter, because these play no role in science, in that the problem of induction drops out.

Then, is a scientific hypothesis to be tested against protocol statements, such that the basic statements in the logical positivist analysis of knowledge, thought as reporting the unvanishing and pre - theoretical deliverance of experience: What it is like here, now, for me. The central controversy concerned whether it was legitimate to couch them in terms of public objects and their qualities or whether a less theoretical committing, purely phenomenal content could be found. The former option makes it hards to regard then as truly basic, whereas the latter option,makes it difficult to see how they can be incorporated into objectives science. The controversy is often thought to have been closed in favour of a public version by the 'private language' argument. Difficulties at this point led the logical positivist to abandon the notion of an epistemological foundation altogether, and to flirt with the 'coherence theory' of truth', it is widely accepted that trying to make the connection between thought and experience through basic sentences depends on an untenable 'myth of the given'.

Popper advocated a strictly non-psychological reading of the empirical basis of science. He required 'basic' statements to report events that are 'observable' only in that they involve relative position and movement of macroscopic physical bodies in certain space-time regions, and which are relatively easy to tests. Perceptual experience was denied an epistemological role (though allowed a causal one),: Basic statements are accepted as a result of a convention or agreement between scientific observers. Should such an agreement break down, the disputed basic statements would need to be tested against further statements that are still more 'basic' and even easier to test.

But there is an easy general result as well: Assuming that a theory is any deductively closed set of sentences as assuming, with the empiricist, that the language in which these sentences are expressed has two sorts of predates (observational and theoretical) and, finally, assuming that the entailment of the evidence is the only constraint on empirical adequacy, then there are always indefinitely many different theories which are equally empirically adequate as any given theory. Take a theory as the deductive closure of some set of sentences in a language in which the two sets of predicates are differentiated: Consider the restriction of 'T' to quantifier - free sentences expressed purely in the observational vocabulary, then any conservative extension of that restricted set of T's consequences back into the full vocabulary is a 'theory' co-empirically adequate with - entailing the same singular observational statement as - 'T'. Unless very special conditions apply (conditions which do not apply to any real scientific theory), then some of these empirically equivalent theories will formally contradict 'T'. (A similarly straightforward demonstration works for the currently a fashionable account of theories as set of models.)

Many of the problems concerning scientific change have been clarified, and many new answers suggested. Nevertheless, concepts central to it (like 'paradigm', 'core', 'problem', constraint', 'verisimilitude') still remain formulated in highly general, even programmatic ways. Many devastating criticisms of the doctrine based of them have not been answered satisfactory.

Problems centrally important for the analysis of scientific change have been neglected, there are, for instance, lingering echoes of logical empiricism in claims that the methods and goals of science are unchanging, and thus are independent of scientific change itself, or that if they do change, they do so for reasons independent of those involved in substantive scientific change itself. By their very nature, such approaches fail to address the change that actually occur in science. For example, even supposing that science ultimately seeks the general and unalterable goal of 'truth' or 'verisimilitude', that injunction itself gives guidance as to what scenists should seek or others should go about seeking it. More specific goals do provide guidance, and, as the tradition from technological mechanistic to gauge - theoretic goals illustrate, those goals are often altered in light of discoveries about what is achieved, or about what kinds of theories are promising. A theory of scientific change should account for these kinds of goal changes, and for how, once accepted, they alter the rest of the patterns of scientific reasoning and change, including ways in which mor general goals and methods may be preconceived.

Traditionally, philosophy has concerned itself with relations between propositions which are specifically relevant to one-another in form or content. So viewed, philosophical explanation of scientific change should appeal to factors which are clearly more scientifically relevant in their content to the specific direction of new scientific research and conclusions than are social factors whose overt relevance lies elsewhere. However, in recent years many writers, especially in the 'strong programme' in the sociology of science have maintained that all purported 'rational' practices must be assimilated to social influences.

Such claims are excessive. Despite allegations that even what is counted as evidence is a matter of mere negotiated agreement, many consider that the last word has not been said on the idea tat there is in some deeply important sense a 'given', inn experience in terms of which we can, at least partially, judge theories. Again, studies continue to document the role of reasonably accepted prior beliefs ('background information') which can help guide those and other judgements. Even if we can no longer naively affirm the sufficiency of 'internal' givens and background scientific information to account for what science should and can be, and certainly for what it is often in human practice, neither should we take the criticisms of it nor granted, accepting that scientific change is explainable only by appeal to external factors.

Equally, we cannot accept too readily the assumption (another logical empiricist legacy) that our task is to explain science and its evolution by appeal to meta - scientific rules or goals, or metaphysical principles, arrived at in the light of purely philosophical analysis, and altered (if at all) by factors independent of substantive science. For such trans-scientific analysis, even while claiming to explain 'what science is', do so in terms 'external' to the processes bty which science actually changes.

Externalist claims are premature: Not enough is yet understood about the roles of indisputable scientific consecrations in shaping scientific change, including changes of method and goals. Even if we ultimately cannot accept the traditional 'internalist' approach in philosophy of science, as philosophers concerned with the form and content of reasoning we must determine accurately how far it can be carried. For that task, historical and contemporary case studies are necessary but insufficient: Too often the positive implications of such studies are left unclear, and their too hasty assumption is often that whatever lessons are generated therefrom apply equally to later science. Larger lessons need to be a systematic account integrating the revealed patterns of scientific reasoning and the ways they are altered into a coherent interpretation of the knowledge-seeking enterprise - a theory of scientific change. Whether such efforts are successful or not, it only through attempting to give sch a coherent account in scientific terms, or through understanding our failure ton do so, that it will be possible to assess precisely the extent to which trans. - scientific factors (meta-scientific, social, or otherwise) must be included in accounts of scientific change.

That for being on one side, it is noticeable that the modifications for which of changes have conversely been revealed as a quality specific or identifying to those of something that makes or sets apart the unstretching obligation for ones approaching the problem. That it has echoed over times generations in making different or become different, to transforming substitution for or among its own time of change. Finding in the resulting grains of residue that history has amazed a gradual change of attitudinal values for which times changes in 1925, where the old quantum mechanics of Planck, Einstein, and Bohr was replaced by the new (matrix) quantum mechanics of Born, Heisenberg, Jordan, and Dirac. In 1926 Schrödinger developed wave mechanics, which proved to be equivalent to matrix mechanics in the sense that they ked to the same energy levels. Dirac and Jordan joined the two theories into one transformation quantum theory. In 1932 von Neumann presented his Hilbert space formations of quantum mechanics and proved a representation theorem showing that sequences in transformation theory were isomorphic notions of theory identity are involved, as theory individuation of theoretical equivalence and empirical equivalences.

What determines whether theories T1 and T2, are instances of the same theory or distinct theories? By construing scientific theories as partially interpreted syntactical axiom system TC, positivism made specific of the axiomatization individuating factures of the theory. Thus, different choices of axioms T or alternations in the correspondence rules - say, to accommodate a new measurement procedure - resulting in a new scientific meaning of the theorized descriptive terms 'ô'. Thus, significant alternations in the axiomatization would result not only in a new theory T'C' but one with changed meaning ô'. Kuhn and Feyerabend maintained that the resulting change could make TC and T'C' non-comparable, or 'incommensurable'. Attempts to explore individuation issues for theories through the medium of meanings change or incommensurability proved unsuccessful and have been largely abandoned.

Individuation of theories in actual scientific practice is at odds with the positivistic analyses. For example, difference equation, differential equations, and Hamiltonian versions of classical mechanics, are all formulations of one theory, though they differ in how fully they characterize classical mechanics. It follows that syntactical specifics of theory formulation cannot be undeviating features, which is to say that scientific theories are not linguistic entities. Rather, theories must be some sort of extra-linguistic structure which can be referred to through th medium of alterative and even in equivalent formulations (as with classical mechanics). Also, the various experimental designs, and so forth, incorporated into positivistic correspondence rules cannot be individuating features of theories. For improved instrumentation or experimental technique does not automatically produce a new theory. Accommodating these individuation features was a main motivation for the semantic conception of theories where theories are state spaces or other extra-linguistic structures standing in mapping relations to phenomena.

Scientific theories undergo developments, are refined, and change. Both syntactic and semantic analysis of theories concentrate on theories at mature stages of development, and it is an open question either approach adequately individuates theories undergoing active development.

Under what circumstances are two theories equivalent? On syntactical approaches, axiomatizations T1 and T2 having a common definitional extension would be sufficient Robinson's theorem which says that T1 and T2 must have a model in common t be compatible. They will be equivalent if theory have precisely the same (or equivalent) sets of models. On the semantic conception the theories will be two distinct sets of structures (models) M1 and M2. The theories will be equivalent just in case we can prove a representation theorem showing that M1 and M2 are isomorphic (structurally equivalent). In this way von Neumann showed that trandformations quantum theory and the Hilbert Space formulation were equivalent.

Many philosophers contend that only part of the structure or content of theories is descriptive of empirical reality. Under what circumstances are two theories identical or equivalent in empirical content? Positivists viewed theories as having a separable observational or empirical component, 'O', which could be described in a theory-neutral observation language, That if O1 and O2 be the observational content of two theories. The two theories are empirically equivalent just in case O1 and O2 meet appropriate requirements for theoretical equivalence. The notion of a theory of a theory-independent observation language was challenged by the view that observation and empirical facts were 'theory-dependant'. Thus, syntactically equivalent O1 and O2 might not be empirically equivalent.

If one adopts such a generalized model-based understanding of scientific theories, one must characterize the relationship between theoretical models and real systems, van Fraassen (1980) suggests that it should be one of isomorphism. But the same considerations that count against there being true laws in the classical sense also count against there being isomorphic to an 'empirical' sub model. What is needed is a weaker notion of similarity, for which it must be specified both in which respects the theoretical model and the real system are similar, and to what degree,. These specifications, however, like the interpretation of terms used in characterizing the model and the identification of relevant aspects of real systems, are not part of the model itself. They are part of a complex practice in which,models are constructed and tested against the world in an attempt to determine how well they 'fit'.

However, according to this view, the alternate approaches of the 'analogical conception of theories' are historically changed entities, and consist essentially of hypothetic modes or analogues of reality, not primarily on formal systems. Theoretically models of data, and the real world are related in complex networks of analogy, which are continually being modified as new data are obtained and new models develop. Analogies with familiar entices and events introduce in language. Inferences within theories and from theory to data and predictions, are metaphysical principles of the 'analogy of nature', a principle that is weaker than the usual assumptions of 'natural kinds' or 'unnatural laws'. It has been suggested that a suitable philosophical model for the difficult concept of 'analogy' may be found on artificial learning systems such as parallel distributive processing.

In van Fraassen's version of the semantic conception, a theory formulation 'T' is given a semantic interpretation in terms of a logical space into which lots of models can be mapped. (This presupposes his theory of semi-interpreted language), for which the Cambridge mathematician and philosopher, Ramsey Frank Plumpton (1903-30) made important contributions to mathematical logic, probability theory, the philosophy of science. He showed how the distinction between the semantic paradoxes, such as that of the 'Liar, and Russell's paradox made unnecessary the ramified type theory of 'Principia Mathematica', and the resulting axiom of reducibility. In the philosophy of language Ramsey was one of the first thinkers to accept a 'Redundancy theory of truth', which he combined with radial views of the function of many kinds of propositions, neither generalizations, nor causal propositions, nor those treating probability or ethics, describe facts, but each has a different specific function in our intellectual economy. Ramsey was one of the earliest commentators on the early works of Ludwig Wittgenstein)1889-1951), and his continuing friendship with the latter led Wittgenstein 's return to Cambridge and to philosophy in 1929.

Ramset advocated, that a sentence generated by taking all the sentences affirmed in a scientific theory that use some term, e.g., 'quark', replacing the term by a variable, and existentially quantifying into the result. Instead of saying that quarks have such-and-such properties, the Ramsey sentence says that there is something that has those properties. If the process is repeated for all of a group of the theoretical terms, the sentence gives the 'topic-neutral' 'structure of the theory', but removes any implication that we know what the terms so treated denote. It leaves open the possibility of identifying the theoretical item with whatever it is that best fits the description provided. However, it was pointed out by the Cambridge mathematician Newman, that if the process is carried out for all except the 'Löwenheim-Skolem theorem', the result will be interpretable in any domain of sufficient cardinally, and the content of the theory may reasonably be felt to have been lost. Whereby a theory is empirically adequate if the actual world 'A' is among those models. Two theories are empirically equivalent if for each model 'M' of T1 there is a model M' of T2 such that all empirical substructures of 'M' are isomorphic to empirical substructures of M', and vice versa with T1 and T2 exchanged. In this sense wave and matrix mechanics are equivalent. Van Fraassen assumes an observability/ non- observability distinction in presenting his empirical adequacy account, but this notion and his formal account can be divorced from such distinction, the generalized empirical adequacy notion being applicable wherever not all of the structure in a theory (e.g., the dimensionality of the state space) corresponds to reality.

For all this as to their accounts of the underlaying idea is that two theories are empirically equivalent if those sub-portions of the theories making empirically ascertainable claims are consistent (in the sense of Robinson's theorem) and assert the same facts, Suppe's quasi-realistic version of the semantic conception (which employs no observability/ non- observability distinction) maintains that theories with variables 'v' purport to describe only the world 'A' would be if phenomena were isolated from all influences other than 'v', Thus, the theory structure 'M' stands in a counterfactual mapping relation to the actual world 'A'. Typically, neither 'A' nor its 'v' portion will be among the sub-models of empirically true theory 'M', so true theories will not be empirically adequate. Further, two closely related theories M1 and M2 with variables, v1 and v2 could both be counter-factually true of the actual world without their formulations having extensional models in common (violating Robinson's theorem), Thus, issues of empirical equivalence largely become pre-empted by questions of empirical truth for the theories.

Philosophical theories are unlike scientific ones. Scientific theories ask questions in circumstances where there are agreed-upon methods for answering the questions and where the answers themselves are generally greed upon. Philosophical theories, in contrast, are forerunners of scientific theory: They attempt to model the known data in ways that allow those data to be seen from a new perceptive that promotes the development of genuine scientific theory. Philosophical theories are, thus, proto-theories. As such, they are useful precisely in areas where no large-scale scientific theory exists. At present that is exactly the state psychology is in. However, philosophy, is exactly in this kind of circumstance, in that the philosopher can be helpful to empirical scientist, the task for philosophers of mind in the present context is to consider the empirical data available and to try form a generalisation, in the coherent way of looking at those data that will guide further empirical research,. I.e., philosophers can provide a highly schematized model that will structure that research. And the resulting research will, in turn, help bring about refinements of the schematized theory, with the ultimate hope being that a closely honed, viable, scientific theory (one wherein investigators agree on the question and on the methods to be used to answer them) will emerge. In these respects, philosophical theories of mind, though concerned with current empirical data, are too general in respect to the data to be scientific theories. Moreover, philosophical theories are aimed primarily at a body of accepted data. Scientific theories not only have to deal with the given data bit also have to make predictions (and postdictions) about unknown data, prediction (and postdiction) that can be gleaned from the theory together with accepted data. This removal to unknown data is what forms the empirical basis of a scientific theory and allows it to be justified in a way quite distinct from the way in which philosophical theories are justified. Philosophical theories are only schemata, coherent, 'pictures' of the accepted data, only pointed toward empirical theory (and as the history of philosophy makes manifest, usually unsuccessful ones -though it is wise to think that this task of success is any kind of fault: These are difficult tasks.

The emerging of evidence, and emphasized that a Cartesian theory of mind is rooted in scientific results and evidence, and that science presuppose an existent external world. That is, that scientific explanation of the world have ontological commitment that a philosophic theory, compatible with and underlying that scientific theory, can and should avoid.

The upshot of these facts is that scientific explanation - psychological ones in the present case - from a Cartesian standpoint will resemble many externalist ones almost exactly. Suppose for the case in point (scientific explanation of concept possession and acquisition) that something like Jerry Alan Fodor, who believes that mental representations should be conceived as individual states with their own identities and structure-like formulae transformed to posses of computation or thought. Nonetheless, Fodor's casual account of concepts were most nearly correct among externalist alternatives. Then Cartesian might provide an account that would take in and utilize many of the same data that Fodor's account relies on. That is, as regards psychology, there will be few differences, and perhaps no obvious ons between Cartesian science and externalist science. Cartesian science, can even allow for interactions among individuals to matter in concept, acquisitions and possession. So Cartesianism can even allow for much of the motivation behind Anti-Individualist Externalism. That is, the dispute between internalism and Externalism is not so much about what the empirical data are, but about how to interpret those data so as to understand the determinants of contents in 'concepts' and thoughts. Even so, much psychology of concept acquisition can be practised without ever considering this high-level problem of what makes content 'content'. The scientists are more interested in the empirical determinants of how concepts and thoughts, with this content, come into being. But these sorts of questions are prior to, and independents of, the question of what makes content 'content'. And on the scientific questions, the ones present-day empirical scientist are trying to answer, the internalism/externalism dispute will matter little, and internalists and externalists can agree on similar scientific solutions to these scientific disputes.

Moreover, given that the grounds for Cartesianism are themselves based in science, philosophical theories like scepticism are almost certainly false, least of mention, the odds of something like a natural clone existing are staggeringly small, vanishingly close to impossible. As Wittgenstein (1969) also pointed out, if our scientific beliefs are totally misguided, enormous numbers of our other beliefs would have to be surrounded - to the point of speechlessness. Nevertheless, there is no god reason for surrounding our beliefs, for scepticism to be right, belief after belief would have to be peeled away and rejected. The American philosopher of mind, Daniel Clement Dennett (1942-) his conception of our understanding of each other, in terms of taking up an 'intentional state', which is useful for prediction and explanation, has been widely discussed. These concerns of whether we can usefully take up the stance towards inanimate things, and whether the account does sufficient justice to the real existence of mental states. Dennett has also been a major force in illustrating how the philosophy of mind needs to be informed by work in surrounding sciences. Dennett gives a good account of the vast knowledge that would be required to cause a brain in a vat to have perceptual-like experience. Moreover, there is excellent reason to think that scepticism is false, and none - except for its very possibility - to think it true. So, once, again, the rational position for Cartesianism will involve a commitment to a scientific account of concept acquisition and possession that depends on interactions between an organism and its environment, including its interactions with like organisms, and also including built-in genetic neural constraints on its concept-forming abilities. That is, in its rejection of scepticism, Cartesian science will look simply like science.

All and all, Cartesianism allows for much greater distance between the way the world is and our concepts of it than do many sorts of externalism. Science, from a Cartesian point of view can be seen as an attempt to discover concepts that better fit on the world than do our pre-scientific ones. We should not expect the content of our pre-scientific concepts of the world to determine by the way physics now tells us the world really is (excerpt causally determined - we can no longer, but no less,. Expect our concepts to closely match the world than to closely match each other's.

These consequences of Cartesianism matter for psychology and not merely for philosophy. Nor is it likely that these two constituents constitute the only consequences of Cartesianism relevant to present scientific interests. But even if there are no others, these are important enough in their own right to make the time spent defending Cartesianism worthwhile - even for present-day psychology.

If Cartesian accounts of the world allow for causal and other interactions between environment and organism, then it should come as no surprise that Cartesian accounts of concept acquisition do so likewise. So when it comes to spatial, temporal, and object concepts, Cartesian driven scientific accounts will not be much different, if at all different, from externalist-driven ones. A Cartesian does not need to claim that it is technically possible that brains in a vat or disembodied minds, say, could posses these concepts, only that it is metaphysically possible. The data and theories mentioned are representative, in relevant respects, of every relevant scientific theory that, least of mention, its trying to understand our possession of spatial, temporal, and object concepts.

All the same, they are close to being right: Ascriptions of propositional-attitude states are deeply theoretical, but the states themselves are accessed apperceptively, i.e., we have direct non-inferential access to them, despite arguments with Dennett and Fodor that propositional-attitude ascriptions are a matter of theory, deep differences exist between views that differences that affect one's view of psychology. And the role given to apperception (that its role is appointed) is a key to understanding those differences, these claims need careful explication and elucidation. Soon, the consistency of 'directly' apprehended, yet theoretical' should be more apparent.

Within this intermittent interval, we are occupied of some spatial peculiarity in a particular point of space and time. However, to further our investigations, the role of phenomena in perception will again be that phenomena, on a broader scale, that play a different and lesser role than might be thought.

When it comes to role of phenomenal states in perception, there are three major possibilities (1) Phenomenal properties are 'read off' in making perceptual judgements. This view holds that perception is itself non-cognitive: An experiencing of phenomenal properties. Any cognitive act is post-perceptual and derived from the perceiver's 'read off,' the phenomenal properties perceived. Call this position the 'read off' 'position. (2) Phenomenal properties are not 'read off'. They are non-cognitive causes of perception, which is a cognitive state - a judgement. While phenomenal states are not themselves perceptions, not even necessary to perception, they - at least sometimes - play an integral, causal role in perception and so cannot be completely discounted in explaining perception itself. Call this position the 'causal position'. (3) Phenomenal properties are merely epiphenomenal of perceptual processes. While phenomena may not be epiphenomenal altogether (for instance, they,may be causes of thoughts about themselves), they play no 'read off' of casual role in perception itself. Like the causal position, this view, the 'epiphenomenal position' regards perceptual as a cognitive state.

In its most extreme form, the 'read off' position holds that perception has both an inner component and an outer component, the inner being a representation of the outer. The inner component, which is something like a photograph or portrait (however distorted), I a phenomenon, directly accessible only to the person whose phenomenon it is. Phenomena have properties such as colour and shape: And because of the projective nature of the inner properties, phenomena represents the outer word to us, bringing us to believe that like properties exist out there as well as in us. To distinguish the ways in which properties such as colour or shape are instantiated, or else to distinguish the kinds of things that possess the properties, the inner instance of the property is labelled a 'phenomenal' property: The outer, a real (or, eternal) property. Because red exists phenomenally, we ascribe red to objects out in the world, and because square exists phenomenally, we ascribe square to objects in the world. We name the phenomenal colour 'red', and that also provides the name of the red colour. And similarly for 'square', hold the position in the 'phenomenal view'. Apparently, this view has often come under attack. A philosopher influenced by objects, there are only phenomenal states. And if phenomena are not objects, then properties such as colour and shape are not ascribable to them. Colour and shape, it is concluded, are properties only of real objects.

We begin with the philosophical reflection about the mind that begins with a curious, even unsettling fact: The mind and its contents - sensations, thoughts, emotions, and so forth -seem to be radically unlike anything physical. Such that we are to consider the following:

The mind is a conscious subject: It states have phenomenal feel,. There is something that is like to be in pain, say, or to imagine eating a strawberry. But what is physical lacks such feel. We may project phenomenal qualities, such as colours, on to physical objects. But the physical realm is in itself phenomenally lifeless.

The mind's contents lack spatial location. A thought, for example, may be about a spatially located object, e.g., the Rogers Centre, but the thought itself is not located anywhere. By contrast, occupants of the physical world are necessarily located in space.

Some mental states are representational, the have intentionality. Now it is true that parts of the physical world, such is representational because we bestow meaning on it. It is due to our semantic conventions that the words on this page stand for something, so the intentionality of the physical is in this way derived. But the mental has origin - that is, inderived - intentionality. My thought about the Rogers Centre is in itself about something in a way that no physical representation is.

These (alleged differences are all metaphysical, they point to a fundamental difference in nature between the mental and physical. The mind-body divide can also be drawn epistemologically: We know about the mental - at least our own minds - in a way that is quite different from the way we know about the physical: For example:

Our primary of discovering truths about the physical world is perception (sight, hearing, and so forth). But our primary of discovering truths about our own mental states is introspection. And whatever exact knowledge of the mental than outward perception gives us of the physical.

Our knowledge of our own minds is more secure than our knowledge of the external physical world. While you may have doubts about whether you are reading a book right now - perhaps you are hallucinating or dreaming - you cannot doubt that it seems to you as if you are reading a book, that your mind contains this sort of appearance. Mental states are 'self-illuminating' in a way that no physical states are.

The mental is private, your own mental states are uniquely your own, directly accessible only to yourself. But the physical world is public, in principle, it is equally accessible to everyone.

And so, a metaphysical distinction was drawn in antiquity between qualities which really belong to objects in the world and qualities which only appear to belong to them, or which human beings only believe to belong to them, because of the effects those objects produce in human beings, typically through the sense organs. Objects must posses some quality or other in order to produce their effects, so the view is not that there are no qualifies to impute certain qualities to them. It is only that some of the qualities which are imputed to objects, e.g., colour, sweetness, bitterness and so on, are not possessed by those objects knowledge of nature is knowledge of what qualifies objects actually have, and of how they bring about their effects. To claim such knowledge is to impute certain qualities to objects, the richer one's knowledge the more such qualities will be imputed,. But when the imputation is true, or amounts to knowledge, the qualifies are not merely imputed they are also in fact present in the object. The metaphysical view holds that those are the only qualities which objects really have. The rest of our conception of the world has a human source.

This is so far not distinction between two kinds of qualities ('primary' and 'secondary') which objects possess, or between qualities which are imputed to objects and qualities which are not, but qualifies which objects really have and qualities which are merely imputed to them which they do not in fact possess. It is a claim about what is really so.

Descartes found nothing but confusion in the attempt even to impute to objects those very effects which they bring about though the senses. The 'sensations' caused people's minds by the qualities of bodies which affect them could not themselves be in the external objects. Nor does it make sense to suppose that bodies could in some way 'resemble' those sensory effects. For Descartes the essence of body is extensions, so no quality that is not a mode of extension could possibly belong to the body at all. Colours, odours, sounds, and so forth, are on his view nothing but sensations. 'When we say that we perceive colours in objects, this is really just the same as saying in objects, this is really just the same as saying that we perceive something in the objects whose nature we do not know, but which produce in us a certain very clear and vivid sensation which we call the sensation of colour.' if ewe try to think of colours as something real outside our minds 'there is no way of understanding what sort of things they are'.

This again is not a distinction between two kinds of qualities which belong to bodies. It is a distinction between qualities which belong and bodies (all of which are modes of extensions such as shape, position, motion, and so on, and what we unreflectively and confusedly come to think are qualities of bodies.

The term 'secondary quality' appears to have been coined by the Irish scientist Robert Boyle (1623-92),whose 'corpuscular philosophy' was shard by the English philosopher John Locke (1632-1704). But it is no easy to say what either he or Locke meant by the term. They were not consistent in their use of it. Locke, like Boyle, distinguished an object's qualities from the powers it has to produce effects. It has the powers only in virtue of possessing some 'primary; or 'real' qualities. The effects it is capable of producing occur either in other bodies or in minds. If in minds, the effects are 'ideas', e.g., of colour or sweetness or bitterness, or of roundness or squareness. These ideas in turn are employed in thoughts to the effect that the object in question is, e.g., coloured or sweet or bitter or round or square or moving. We have such thoughts, according to Locke, by thinking that the object in question 'resembles' the idea we have in the mind. Additionally. Locke identifies 'secondary qualities' as 'such qualities which in truth are nothing in the objects themselves but powers to produce various sensations within us by their 'primary qualities'. This can be taken in at least two ways. It could,mean that, in addition to its 'primary qualities all there really is in an object we call coloured, sweet or bitter, and o forth, is its power to produce ideas of colour, sweetness. Or bitterness, and so on, in us by virtue of the operations of those 'primary' or 'real' qualities. That is compatible with the earlier view that colour, sweetness, bitterness, and so on, are not real in objects. Or it could (and does seem to) mean that 'secondary qualities' such as colour. Sweetness, bitterness, and so forth, are themselves nothing more than certain powers which objects have to effect us in certain ways. But such powers, on Locke's view, really do belong to objects endowed with the appropriate 'primary' or 'real' properties. To identify 'secondary qualities' with such powers in this way ailed imply that such 'secondary qualities' as colour, sweetness, bitterness, and so on, since they are nothing but powers, really do belong to or exist in objects after all. Imputations of colour, sweetness, and so on, to objects would then be true, not false or confused, as on those earlier views.

A distinction drawn in this way between 'primary' and 'secondary' qualities would not be a distinction between qualities which objects really poses and qualities which only mistakenly or confusedly think they possess. Nor would it even be a distinction between two kinds of qualities, strictly speaking. It would be a distinction between qualities and certain kinds of powers, both of which reality belong to objects. But Locke confusedly sometimes calls both of them 'qualities'. This is Locke's way of saying what really belongs to the objects around us: Only them is so, only mistakenly imputed 'secondary' qualities to objects, but in the 'secondary' qualities to objects, but in the case of the 'primary' qualities the imputations are true. But that is inconsistent with the idea that the 'secondary' qualities which we impute are nothing but powers, since the imputations would then be imputations of certain powers, and so would be true of all objects with the appropriate 'primary' or 'real' qualities.

The Irish idealist George Berkeley (1685-1753), published De Motu ('On Motion') attacking Newton's philosophy of space, a topic he returned to much later in 'The Analyst' of 1734. However, with the consequent shrinking of reality down to a world of minds and their own sensations or 'ideas' that they are of the impossibility of 'inert senseless matter' and the merit of a scheme based on a pervading all-wise providence whose production is the conceptual order, the world of ideas, that makes up our lives, runs through all Berkeley's writings. What he saw and emphasized is its great rigour was the impossibility of bridging the gap opened by the Cartesian target is the comfortable, commonsense view of mind as entirely different from matter, yet in satisfactory contact with a material world about which it can know a great deal. Berkeley deploys many of the arguments of ancient 'scepticism' and others found as Malebranche and the French philosopher Pierre Bayle (1647-1706), to undermine this synthesis, showing that once the separation of mind from the material world is as complete as Descartes makes it, the hope of knowing or understanding anything about the supposed external world quite vanishes. Unlike Cartesian scepticism, which stresses the bare possibility of things not being as we take them to be. Berkeley urges the actual inconsistencies in the conceptual scheme left by Cartesianism, which entrap such thinkers as Locke and arguably commonsense itself. His way out is not to advocates scepticism, which he consistently regards with extreme repugnance, but to reform the relation between mind and the world so that contact is re-established. Unfortunately, this introduces subjective idealism, in which the subject apprehends as the world is just the relationship between the subjects's own mental states plus an uneasy relationship with archetypes of the subject's idea in the mind of God in promoting his system Berkeley makes brilliant use of the scepticism problems that will bedevil alternatives, as well as of the problems faced by particular elements of the conceptual scheme. He opposes problems of causation, substance, perception and understanding, although his system had proved incredible to virtually all subsequent philosophers, as importance lies in the challenge, it offers in a commonsense that vaguely hopes that these notions fit together in a satisfactory way.

Berkeley objected to Locke that it is nonsense to speak of a 'resemblance' between an idea and an object, just as Descartes had ridiculed the idea that a sensation could resemble the object that causes it. 'An idea can be like nothing but an idea', Berkeley says, that is a general refection of Locke's account of how we are able to think of things existing independently of the mind. If it is correct if it works as much against what Locke says of our ideas of 'primary' qualities as it does against that he says of our ideas of 'secondary' qualitites

Boyle speaks of the 'texture' of a body whose minute corpuscles are arranged in a certain way. It is in virtue of possessing that 'texture' that the body is 'disposed' or has the power to produce ideas of certain kinds in perceivers, even if no on is perceiving it at the moment. This has tempted some philosophers in recent years to identify 'secondary' qualities, not with the powers which objects have to affect us in certain ways, but with the quality 'bases' of those causal powers. The colour or sweetness or bitterness and so forth of an object would then be some real but possibly unknown castled of the object which is responsible for the specific effect it has on us. This agin, would imply that 'secondary' qualitites, so understood are really in objects. And it would have the consequence that 'secondary' qualities are true qualities, not just powers. But it would seem to have no room for a distinction between 'secondary' and 'primary' or 'real' qualities of bodies. The 'bases' of all the causal powers of objects are to be understood in terms of their 'primary' or 'real' qualities.

Another strategy is to show that the qualities said to be perceived or thought about in such cases are really qualities that do belong to objects after all. This can take the form of arguing that, e.g., the word 'coloured' just the same as 'has the power to produce perceptions of colour in human beings, or meant the same as 'has that quality which produces perceptions of colour in human beings, or that the same as the physical term, whatever it is, which denotes that quality which in fact produces perceptions of colour in human beings. These perceptions of colour in human beings. These are all these about the meanings of terms for allegedly 'secondary' qualities. Or it might be held only that a so-called 'secondary' quality term in fact denotes the very same quality as is denoted the very same quality as denoted by some purely physical term. This would simply identify the very quality in question with some physical quality or power and not of two different qualities, but only one, without holding the terms that denote it must have the same that the term that denote it must have the same meaning. In either case it would have the consequences that when we see coloured, what we see, or what seems as to believe and to belong to the object, is that very physical quality or power which colour is said to be. Then, again, it would leave no distinction between qualities which really belong to ones object and qualities which are only mistaken or confusedly imputing them.

Philosophical issues about 'perception' tend to be issues specifically about sense-perception. In English (and the same is true of comparable terms in many other languages) the term 'perception' has a wider connotation than anything has to do with the senses and sense-organs, though it generally involves the idea of what may imply, if only the idea of what may imply. If only in a metaphorical sense, a point of view. Thus it is now increasingly common for news-commentators, for example, to speak of people's perception of a certain set of events,.even though those people have not been wittinesses of them. In one sense, however, there is nothing new about this: Inn seventeenth and eighteenth-centuries philosophical usage, words for perception were used with a much wider coverage than sense-perception alone. It is, however, sense-perception that typically raised the largest and most obvious philosophical problems.

Such problems may be said to fal into two categories. There are, first, the epistemological problem s about the role of sense-perception in connection with the acquisition and possession of knowledge of the world around us. These problems; -does perception give us knowledge of the so-called 'external world? How and to what extent? - have become dominant in epistemology since Descartes because of his invocation of the method of doubt, although they undoubtedly existed in philosophers' minds in one way or another-before that. In early and middle twentieth-century Anglo-Saxon philosophy such problems centred on the question whether there are firm data provided by the senses-so-called sense-data - and if so what is the relation of such sense-data to so-called material objects. Such problems are not essentially problems for the philosophers of mind, although certain answers to questions about perception which undoubtedly belong to the philosophy of mind can certainly add to epistemological difficulties. If perception is assimilated, for example, to sensation, there is an obvious temptation to think that perception we are restricted at any rate initially, to the contents of our minds.

The second category of problems under the heading of the philosophy of mind - are thus in a sense prior to the problems that exercised many empiricists in the first half of this century. They are problems about how perception is to be construed and how it related to a number of other aspects of the mind's functioning - sensations, concepts and other things involved in our understandings of things, belief and judgement, the imagination, our action in relation to the world around us, and the causal processes involved inn thee physics, biology and psychology of perception. Some of the last were central to the considerations that Aristotle raised about perception in his De Anima.

There is, but with availing circumstances that no simple, agreed-upon definition of consciousness exists. Attempted definitions tend to be tautological (for example, consciousness defined as awareness) or merely descriptive (for example, consciousness described as sensations, thoughts, or feelings). Despite this problem of definition, the subject of consciousness has had a remarkable history. At one time the primary subject matter of psychology, consciousness as an area of study suffered an almost total demise, later reemerging to become a topic of current interest.

Most of the philosophical discussions of consciousness arose from the mind-body issues posed by the French philosopher and mathematician René Descartes in the 17th century. Descartes asked: Is the mind, or consciousness, independent of matter? Is consciousness extended (physical) or unextended (nonphysical)? Is consciousness determinative, or is it determined? English philosophers such as John Locke equated consciousness with physical sensations and the information they provide, whereas European philosophers such as Gottfried Wilhelm Leibniz and Immanuel Kant gave a more central and active role to consciousness.

The philosopher who most directly influenced subsequent exploration of the subject of consciousness was the 19th-century German educator Johann Friedrich Herbart, who wrote that ideas had quality and intensity and that they may inhibit or facilitate one another. Thus, ideas may pass from 'states of reality' (consciousness) to 'states of tendency' (unconsciousness), with the dividing line between the two states being described as the threshold of consciousness. This formulation of Herbart clearly presages the development, by the German psychologist and physiologist Gustav Theodor Fechner, of the psychophysical measurement of sensation thresholds, and the later development by Sigmund Freud of the concept of the unconscious.

The experimental analysis of consciousness dates from 1879, when the German psychologist Wilhelm Max Kundt started his research laboratory. For Kundt, the task of psychology was the study of the structure of consciousness, which extended well beyond sensations and included feelings, images, memory, attention, duration, and movement. Because early interest focussed on the content and dynamics of consciousness, it is not surprising that the central methodology of such studies was introspection; that is, subjects reported on the mental contents of their own consciousness. This introspective approach was developed most fully by the American psychologist Edward Bradford Titchener at Cornell University. Setting his task as that of describing the structure of the mind, Titchener attempted to detail, from introspective self-reports, the dimensions of the elements of consciousness. For example, taste was 'dimensionalized' into four basic categories: sweet, sour, salt, and bitter. This approach was known as structuralism.

By the 1920s, however, a remarkable revolution had occurred in psychology that was to essentially remove considerations of consciousness from psychological research for some 50 years: Behaviourism captured the field of psychology. The main initiator of this movement was the American psychologist John Broadus Watson. In a 1913 article, Watson stated, 'I believe that we can write a psychology and never use the terms consciousness, mental states, mind . . . imagery and the like.' Psychologists then turned almost exclusively to behaviour, as described in terms of stimulus and response, and consciousness was totally bypassed as a subject. A survey of eight leading introductory psychology texts published between 1930 and the 1950s found no mention of the topic of consciousness in five texts, and in two it was treated as a historical curiosity.

Beginning in the late 1950s, however, interest in the subject of consciousness returned, specifically in those subjects and techniques relating to altered states of consciousness: sleep and dreams, meditation, biofeedback, hypnosis, and drug-induced states. Much of the surge in sleep and dream research was directly fuelled by a discovery relevant to the nature of consciousness. A physiological indicator of the dream state was found: At roughly 90-minute intervals, the eyes of sleepers were observed to move rapidly, and at the same time the sleepers' brain waves would show a pattern resembling the waking state. When people were awakened during these periods of rapid eye movement, they almost always reported dreams, whereas if awakened at other times they did not. This and other research clearly indicated that sleep, once considered a passive state, were instead an active state of consciousness.

During the 1960s, an increased search for 'higher levels' of consciousness through meditation resulted in a growing interest in the practices of Zen Buddhism and Yoga from Eastern cultures. A full flowering of this movement in the United States was seen in the development of training programs, such as Transcendental Meditation, that were self-directed procedures of physical relaxation and focussed attention. Biofeedback techniques also were developed to bring body systems involving factors such as blood pressure or temperature under voluntary control by providing feedback from the body, so that subjects could learn to control their responses. For example, researchers found that persons could control their brain-wave patterns to some extent, particularly the so-called alpha rhythms generally associated with a relaxed, meditative state. This finding was especially relevant to those interested in consciousness and meditation, and a number of 'alpha training programs emerged.

Another subject that led to increased interest in altered states of consciousness was hypnosis, which involves a transfer of conscious control from the subject to another person. Hypnotism has had a long and intricate history in medicine and folklore and has been intensively studied by psychologists. Much has become known about the hypnotic state, relative to individual suggestibility and personality traits; the subject has now largely been demythologized, and the limitations of the hypnotic state are fairly well known. Despite the increasing use of hypnosis, however, much remains to be learned about this unusual state of focussed attention.

Finally, many people in the 1960s experimented with the psychoactive drugs known as hallucinogens, which produce disorders of consciousness. The most prominent of these drugs are lysergic acid diethylamide, or LSD; mescaline, and psilocybin; the latter two have long been associated with religious ceremonies in various cultures. LSD, because of its radical thought-modifying properties, was initially explored for its so-called mind-expanding potential and for its psychotomimetic As the concept of a direct, simple linkage between environment and behaviour became unsatisfactory in recent decades, the interest in altered states of consciousness may be taken as a visible sign of renewed interest in the topic of consciousness. That persons are active and intervening participants in their behaviour has become increasingly clear. Environments, rewards, and punishments are not simply defined by their physical character. Memories are organized, not simply stored. An entirely new area called cognitive psychology has emerged that centres on these concerns. In the study of children, increased attention is being paid to how they understand, or perceive, the world at different ages. In the field of animal behaviour, researchers increasingly emphasize the inherent characteristics resulting from the way a species has been shaped to respond adaptively to the environment. Humanistic psychologists, with a concern for self-actualization and growth, have emerged after a long period of silence. Throughout the development of clinical and industrial psychology, the conscious states of persons in terms of their current feelings and thoughts were of obvious importance. The role of consciousness, however, was often de-emphasized in favour of unconscious needs and motivations. Trends can be seen, however, toward a new emphasis on the nature of states of consciousness.

Conscious mental process of evoking ideas or images of objects, events, relations, attributes, or processes never before experienced or perceived. Imagination, perception (the conscious integration of sensory impressions of external objects and events), and memory (the mental evocation of previous experiences) is essentially similar mental processes. This is particularly true when their content consists of sensory images. Psychologists occasionally distinguish between imagination that is passive or reproductive, by which mental images originally perceived by the senses are elicited, and imagination that is active, constructive, or creative, by which the mind produces images of events or objects that are either insecurely related or unrelated to past and present reality. At one time the term imagination included the reviving or 'recollecting' processes (memory), as well as the process of creating mental images (imagination). The present stricter definition of imagination excludes and contrasts with that of memory, as the concept of forming something new contrasts with that of reviving something old.

When an imagined and a real perception are simultaneous, the imagined perception may be confused with or even mistaken for the true perception. One objectively measurable example of this phenomenon is synesthesia, an experience in which the stimulation of one sense elicits a perception that ordinarily would be elicited had another sense been stimulated, as when a loud noise registers as a blinding light, or vice versa. Events and objects that apparently are perceived in dreams are examples of imaginative exercises that are neither verifiable nor repeatable. In all these psychological events, imagination takes over the functions of perception. The most extreme examples of this kind of confusion are the hallucinations suffered by victims of severe mental disorders. When a genuine perception is assumed by the individual to be an imagined one, an opposite error is said to exist; this rare occurrence can be induced in the laboratory under experimental conditions, as in one well-known case in which subjects are asked to imagine a scene or object on a screen, upon which, unknown to them, the same scene or object has been dimly projected. The subject almost invariably believes that the projected image is the product of his or her own imagination even when it does not correspond exactly with the imagined perception.

The faculties of Cognition are the associations that apply of an act or the process of knowing. Cognition includes attention, perception, memory, reasoning, judgment, imagining, thinking, and speech. Attempts to explain the way in which cognition works are as old as philosophy itself; the term, in fact, comes from the writings of Plato and Aristotle. With the advent of psychology as a discipline separates from philosophy, cognition has been investigated from several viewpoints.

An entire field - cognitive psychology - has arisen since the 1950s. It studies cognition mainly from the standpoint of information handling. Parallels are stressed between the functions of the human brain and the computer concepts such as the coding, storing, retrieving, and buffering of information. The actual physiology of cognition is of little interest to cognitive psychologists, but their theoretical models of cognition have deepened understanding of memory, psycholinguistics, and the development of intelligence.

Social psychologists since the mid-1960s have written extensively on the topic of cognitive consistency - that is to say, the tendency of a person's beliefs and actions to be logically consistent with one another. When cognitive dissonance, or the lack of such consistency, arises, the person unconsciously seeks to restore consistency by changing his or her behaviour, beliefs, or perceptions. The manner in which a particular individual classifies cognitions in order to impose order has been termed cognitive style.

In philosophy, a system of thought that emphasizes the role of reason in obtaining knowledge, in contrast to empiricism, which emphasizes the role of experience, especially sense perception. Rationalism has appeared in some form in nearly every stage of Western philosophy, but it is primarily identified with the tradition stemming from the 17th-century French philosopher and scientist René Descartes. Descartes believed that geometry represented the ideal for all sciences and philosophy. He held that by of reason alone, certain universal, self-evident truths could be discovered, from which the remaining content of philosophy and the sciences could be deductively derived. He assumed that these self-evident truths were innate, not derived from sense experience. This type of rationalism was developed by other European philosophers, such as the Dutch philosopher Baruch Spinoza and the German philosopher and mathematician Gottfried Wilhelm Leibniz. It was opposed, however, by British philosophers of the empiricist tradition, such as John Locke, who believed that all ideas are derived from the senses.

Epistemological rationalism has been applied to other fields of philosophical inquiry. Rationalism in ethics is the claim that certain primary moral ideas are innate in humankind and that such first moral principles are self-evident to the rational faculty. Rationalism in the philosophy of religion is the claim that the fundamental principles of religion are innate or self-evident and that revelation is not necessary (see Deism). Since the end of the 1800s, however, rationalism has played chiefly an antireligious role in theology.

Epistemology, as a branch of collectable knowledge in philosophy that addresses the philosophical problems surrounding the theory of knowledge. Epistemology is concerned with the definition of knowledge and related concepts, the sources and criteria of knowledge, the kinds of knowledge possible and the degree to which each is certain, and the exact relation between the one who knows and the object known.

In the 5th century Bc, the Greek Sophists questioned the possibility of reliable and objective knowledge. Thus, a leading Sophist, Gorgias, argued that nothing really exists, that if anything did exist it could not be known, and that if knowledge were possible, it could not be communicated. Another prominent Sophist, Protagoras, maintained that no person's opinions can be said to be more correct than another's, because each is the sole judge of his or her own experience. Plato, following his illustrious teacher Socrates, tried to answer the Sophists by postulating the existence of a world of unchanging and invisible forms, or ideas, about which it is possible to have exact and certain knowledge. The things one sees and touches, they maintained, are imperfect copies of the pure forms studied in mathematics and philosophy. Accordingly, only the abstract reasoning of these disciplines yields genuine knowledge, whereas reliance on sense perception produces vague and inconsistent opinions. They concluded that philosophical contemplation of the unseen world of forms is the highest goal of human life.

Aristotle followed Plato in regarding abstract knowledge as superior to any other, but disagreed with him as to the proper method of achieving it. Aristotle maintained that almost all knowledge is derived from experience. Knowledge is gained either directly, by abstracting the defining traits of a species, or indirectly, by deducing new facts from those already known, in accordance with the rules of logic. Careful observation and strict adherence to the rules of logic, which were first set down in systematic form by Aristotle, would help guard against the pitfalls the Sophists had exposed. The Stoic and Epicurean schools agreed with Aristotle that knowledge originates in sense perception, but against both Aristotle and Plato they maintained that philosophy is to be valued as a practical guide to life, rather than as an end in itself.

After many centuries of declining interest in rational and scientific knowledge, the Scholastic philosopher Saint Thomas Aquinas and other philosophers of the Middle Ages helped to restore confidence in reason and experience, blending rational methods with faith into a unified system of beliefs. Aquinas followed Aristotle in regarding perception as the starting point and logic as the intellectual procedure for arriving at reliable knowledge of nature, but he considered faith in scriptural authority as the main source of religious belief.

From the 17th to the late 19th century, the main issue in epistemology was reasoning versus sense perception in acquiring knowledge. For the rationalists, of whom the French philosopher René Descartes, the Dutch philosopher Baruch Spinoza, and the German philosopher Gottfried Wilhelm Leibniz were the leaders, the main source and final test of knowledge was deductive reasoning based on self-evident principles, or axioms. For the empiricists, beginning with the English philosophers Francis Bacon and John Locke, the main source and final test of knowledge was sense perception.

Bacon inaugurated the new era of modern science by criticizing the medieval reliance on tradition and authority and also by setting down new rules of scientific method, including the first set of rules of inductive logic ever formulated. Locke attacked the rationalist belief that the principles of knowledge are intuitively self-evident, arguing that all knowledge is derived from experience, either from experience of the external world, which stamps sensations on the mind, or from internal experience, in which the mind reflects on its own activities. Human knowledge of external physical objects, he claimed, is always subject to the errors of the senses, and he concluded that one cannot have absolutely certain knowledge of the physical world.

The Irish philosopher George Berkeley agreed with Locke that knowledge comes through ideas, but he denied Locke's belief that a distinction can be made between ideas and objects. The British philosopher David Hume continued the empiricist tradition, but he did not accept Berkeley's conclusion that knowledge was of ideas only. He divided all knowledge into two kinds: knowledge of relations of ideas - that is, the knowledge found in mathematics and logic, which is exact and certain but provide no information about the world; and knowledge of matters of fact - that is, the knowledge derived from sense perception. Hume argued that most knowledge of matters of fact depends upon cause and effect, and since no logical connection exists between any given cause and its effect, one cannot hope to know any future matter of fact with certainty. Thus, the most reliable laws of science might not remain true—a conclusion that had a revolutionary impact on philosophy.

The German philosopher Immanuel Kant tried to solve the crisis precipitated by Locke and brought to a climax by Hume; his proposed solution combined elements of rationalism with elements of empiricism. He agreed with the rationalists that one can have exact and certain knowledge, but he followed the empiricists in holding that such knowledge is more informative about the structure of thought than about the world outside of thought. He distinguished three kinds of knowledge: analytical a priori, which is exact and certain but uninformative, because it makes clear only what is contained in definitions; synthetic a posteriori, which conveys information about the world learned from experience, but is subject to the errors of the senses; and synthetic a priori, which is discovered by pure intuition and is both exact and certain, for it expresses the necessary conditions that the mind imposes on all objects of experience. Mathematics and philosophy, according to Kant, provide this last. Since the time of Kant, one of the most frequently argued questions in philosophy has been whether or not such a thing as synthetic a priori knowledge really exists.

During the 19th century, the German philosopher Georg Wilhelm Friedrich Hegel revived the rationalist claim that absolutely certain knowledge of reality can be obtained by equating the processes of thought, of nature, and of history. Hegel inspired an interest in history and a historical approach to knowledge that was further emphasized by Herbert Spencer in Britain and by the German school of historicism. Spencer and the French philosopher Auguste Comte brought attention to the importance of sociology as a branch of knowledge, and both extended the principles of empiricism to the study of society.

The American school of pragmatism, founded by the philosophers Charles Sanders Peirce, William James, and John Dewey at the turn of this century, carried empiricism further by maintaining that knowledge is an instrument of action and that all beliefs should be judged by their usefulness as rules for predicting experiences.

In the early 20th century, epistemological problems were discussed thoroughly, and subtle shades of difference grew into rival schools of thought. Special attention was given to the relation between the act of perceiving something, the object directly perceived, and the thing that can be said to be known as a result of the perception. The phenomenalists contended that the objects of knowledge are the same as the objects perceived. The neorealists argued that one has direct perceptions of physical objects or parts of physical objects, rather than of one's own mental states. The critical realists took a middle position, holding that although one perceives only sensory data such as Colour and sounds, these stand for physical objects and provide knowledge thereof.

A method for dealing with the problem of clarifying the relation between the act of knowing and the object known was developed by the German philosopher Edmund Husserl. He outlined an elaborate procedure that he called phenomenology, by which one is said to be able to distinguish the way things appear to be from the way one thinks they really are, thus gaining a more precise understanding of the conceptual foundations of knowledge.

During the second quarter of the 20th century, two schools of thought emerged, each indebted to the Austrian philosopher Ludwig Wittgenstein. The first of these schools, logical empiricism, or logical positivism, had its origins in Vienna, Austria, but it soon spread to England and the United States. The logical empiricists insisted that there is only one kind of knowledge: scientific knowledge; that any valid knowledge claim must be verifiable in experience; and hence that much that had passed for philosophy was neither true nor false but literally meaningless. Finally, following Hume and Kant, a clear distinction must be maintained between analytic and synthetic statements. The so-called verifiability criterion of meaning has undergone changes as a result of discussions among the logical empiricists themselves, as well as their critics, but has not been discarded. More recently, the sharp distinction between the analytic and the synthetic has been attacked by a number of philosophers, chiefly by American philosopher W.V.O. Quine, whose overall approach is in the pragmatic tradition.

The latter of these recent schools of thought, generally referred to as linguistic analysis, or ordinary language philosophy, seem to break with traditional epistemology. The linguistic analysts undertake to examine the actual way key epistemological terms are used - terms such as knowledge, perception, and probability -and to formulate definitive rules for their use in order to avoid verbal confusion. British philosopher John Langshaw Austin argued, for example, that to say a statement was true added nothing to the statement except a promise by the speaker or writer. Austin does not consider truth a quality or property attaching to statements or utterances.

The Common-Sense school, was the philosophical movement that originated in Scotland in the 18th century and spread abroad, particularly to France and the United States. The movement is also known as Scottish realism or the Scottish School. Its founder was Scottish philosopher Thomas Reid, who described the movement's basic tenets in these words:

'If there are certain principles, as I think there are, which the constitution of our nature leads us to believe, and which we are under a necessity to take for granted in the common concerns of life, without being able to give a reason for them -these are what we call the principles of common sense. What is manifestly contrary to them, is what we call absurd'.

Commonsense philosophy was a rejection of the 'ideal system' that originated in the writings of 17th-century French philosopher René Descartes. The idea of this system culminated in the 18th century in the skepticism of British philosophers John Locke and David Hume, who questioned the objective value of reasoning based on the senses. The Common-Sense School maintained that ordinary human experience was all that was needed for proving (1) the existence of the self, (2) the existence of real objects directly perceived, and (3) certain so-called first principles, upon which sound morality and religious beliefs may be established.

Basically sound and appealing in its purpose and intent, commonsense philosophy was weak in the technical development of its leading ideas. It aimed, in the words of Scottish philosopher Sir William Hamilton, to 'appeal from the heretical conclusions of particular philosophies to the catholic principles of all philosophy.' But the Common-Sense School often fell into the error of dogmatically asserting the prejudices of untutored opinion against careful and competent criticism

The most notable achievement of the Common-Sense school was a realistic treatment of sense perception by its founder Thomas Reid, although even this treatment contained confusions and contradictions. Other proponents of the school included Reid's contemporaries James Oswald and James Beattie, who exalted the philosophically untrained common man as a court of last appeal on moral and religious questions. Scottish theologian George Campbell used his sermons to reply to Hume's famous essay Of Miracles (1748). Sir William Hamilton, a historian of ideas who introduced the works of German philosopher Immanuel Kant to Scottish readers, also regarded himself as a member of the Common-Sense school.

The principles of the Common-Sense School became a force in France through the translation of Reid's works into French and their subsequent adoption by Victor Cousin, founder of the modern school of eclecticism. In the United States, the Common-Sense school formed the dominant academic philosophy throughout the middle decades of the 19th century. James McCosh carried commonsense philosophy with him when he left Scotland in 1868 to become president of Princeton University in Princeton, New Jersey. Noah Porter taught commonsense realism to generations of students at Yale University in New Haven, Connecticut, during the mid-and late 1800s. The Common-Sense school declined, both in the academic world and in literary circles, before the rise of the idealistic tradition that stemmed from Kant and Hegel. The school's influence on philosophy has resurfaced in the 20th-century realistic schools of thought.

The philosophical debate that had led to conclusions useful to the architects of classical physics can be briefly summarized, such when Thale's fellow Milesian Anaximander claimed that the first substance, although indeterminate, manifested itself in a conflict of oppositions between hot and cold, moist and dry. The idea of nature as a self-regulating balance of forces was subsequently elaborated upon by Heraclitus (d. after 480 BC), who asserted that the fundamental substance is strife between opposites, which is itself the unity of the whole. It is, said Heraclitus, the tension between opposites that keeps the whole from simply 'passing away.'

Parmenides of Elea (B.c. 515 BC) argued in turn that the unifying substance is unique and static being. This led to a conclusion about the relationship between ordinary language and external reality that was later incorporated into the view of the relationship between mathematical language and physical reality. Since thinking or naming involves the presence of something, said Parmenides, thought and language must be dependent upon the existence of objects outside the human intellect. Presuming a one-to-one correspondence between word and idea and actual existing things, Parmenides concluded that our ability to think or speak of a thing at various times implies that it exists at all times. Hence the indivisible One does not change, and all perceived change is an illusion.

These assumptions emerged in roughly the form in which they would be used by the creators of classical physics in the thought of the atomists. Leucippus: 450-420 BC and Democritus (460-c. 370 BC). They reconciled the two dominant and seemingly antithetical concepts of the fundamental character of being-Becoming (Heraclitus) and unchanging Being (Parmenides)-in a remarkable simple and direct way. Being, they said, is present in the invariable substance of the atoms that, through blending and separation, make up the thing of changing or becoming worlds.

The last remaining feature of what would become the paradigm for the first scientific revolution in the seventeenth century is attributed to Pythagoras (Bc. 570 Bc). Like Parmenides, Pythagoras also held that the perceived world is illusory and that there is an exact correspondence between ideas and aspects of external reality. Pythagoras, however, had a different conception of the character of the idea that showed this correspondence. The truth about the fundamental character of the unified and unifying substance, which could be uncovered through reason and contemplation, is, he claimed, mathematical in form.

Pythagoras established and was the cental figure in a school of philosophy, religion and mathematics; He was apparently viewed by his followers as semi-divine. For his followers the regular solids (symmetrical three-dimensional forms in which all sides are the same regular polygons) and whole numbers became revered essences of sacred ideas. In contrast with ordinary language, the language of mathematics and geometric forms seemed closed, precise and pure. Providing one understood the axioms and notations, and the meaning conveyed was invariant from one mind to another. The Pythagoreans felt that the language empowered the mind to leap beyond the confusion of sense experience into the realm of immutable and eternal essences. This mystical insight made Pythagoras the figure from antiquity most revered by the creators of classical physics, and it continues to have great appeal for contemporary physicists as they struggle with the epistemological implications of the quantum mechanical description of nature.

Yet, least of mention, progress was made in mathematics, and to a lesser extent in physics, from the time of classical Greek philosophy to the seventeenth century in Europe. In Baghdad, for example, from about A.D. 750 to A.D. 1000, substantial advancement was made in medicine and chemistry, and the relics of Greek science were translated into Arabic, digested, and preserved. Eventually these relics reentered Europe via the Arabic kingdom of Spain and Sicily, and the work of figures like Aristotle universities of France, Italy, and England during the Middle Ages.

For much of this period the Church provided the institutions, like the reaching orders, needed for the rehabilitation of philosophy. But the social, political and an intellectual climate in Europe was not ripe for a revolution in scientific thought until the seventeenth century. Until later in time, lest as far into the nineteenth century, the works of the new class of intellectuals we called scientists, whom of which were more avocations than vocation, and the word scientist do not appear in English until around 1840.

Copernicus (1473-1543) would have been described by his contemporaries as an administrator, a diplomat, an avid student of economics and classical literature, and most notable, a highly honoured and placed church dignitaries. Although we named a revolution after him, his devoutly conservative man did not set out to create one. The placement of the Sun at the centre of the universe, which seemed right and necessary to Copernicus, was not a result of making careful astronomical observations. In fact, he made very few observations in the course of developing his theory, and then only to ascertain if his prior conclusions seemed correct. The Copernican system was also not any more useful in making astrological calculations than the accepted model and was, in some ways, much more difficult to implement. What, then, was his motivation for creating the model and his reasons for presuming that the model was correct?

Copernicus felt that the placement of the Sun at the centre of the universe made sense because he viewed the Sun as the symbol of the presence of a supremely intelligent and intelligible God in a man-centred world. He was apparently led to this conclusion in part because the Pythagoreans believed that fire exists at the centre of the cosmos, and Copernicus identified this fire with the fireball of the Sun. the only support that Copernicus could offer for the greater efficacy of his model was that it represented a simpler and more mathematical harmonious model of the sort that the Creator would obviously prefer. The language used by Copernicus in 'The Revolution of Heavenly Orbs,' illustrates the religious dimension of his scientific thought: 'In the midst of all the sun reposes, unmoving. Who, indeed, in this most beautiful temple would place the light-giver in any other part than from where it can illumine all other parts?'

The belief that the mind of God as Divine Architect permeates the working of nature was the guiding principle of the scientific thought of Johannes Kepler (or Keppler, 1571-1630). For this reason, most modern physicists would probably feel some discomfort in reading Kepler's original manuscripts. Physics and metaphysics, astronomy and astrology, geometry and theology commingle with an intensity that might offend those who practice science in the modern sense of that word. Physical laws, wrote Kepler, 'lie within the power of understanding of the human mind; God wanted us to perceive them when he created us of His own image, in order . . . that we may take part in His own thoughts. Our knowledge of numbers and quantities is the same as that of God's, at least insofar as we can understand something of it in this mortal life.'

Believing, like Newton after him, in the literal truth of the words of the Bible, Kepler concluded that the word of God is also transcribed in the immediacy of observable nature. Kepler's discovery that the motions of the planets around the Sun were elliptical, as opposed perfecting circles, may have made the universe seem a less perfect creation of God on ordinary language. For Kepler, however, the new model placed the Sun, which he also viewed as the emblem of a divine agency, more at the centre of mathematically harmonious universes than the Copernican system allowed. Communing with the perfect mind of God requires as Kepler put it 'knowledge of numbers and quantity.'

What is more, was, suggested Einstein, belief in the word of God as it is revealed in biblical literature that allowed him to dwell in a 'religious paradise of youth' and to shield himself from the harsh realities of social and political life. In an effort to recover that inner sense of security that was lost after exposure to scientific knowledge, or to become free once again of the 'merely personal', he committed himself to understanding the 'extra-personal world within the frame of given possibilities', or as seems obvious, to the study of physics. Although the existence of God as described in the Bible may have been in doubt, the qualities of mind that the architects of classical physics associated with this God were not. This is clear in the comments from which Einstein uses of mathematics, . . . 'Nature is the realization of the simplest conceivable mathematical ideas. I am convinced that we can discover, by of purely mathematical construction, those concepts and those lawful connections between them that furnish the key to the understanding of natural phenomena. Experience remains, of course, the sole criteria of physical utility of a mathematical construction. But the creative principle resides in mathematics. In a certain sense, therefore, I hold it true that pure thought can grasp reality, as the ancients dreamed.'

This article of faith, first articulated by Kepler, that 'nature is the realization of the simplest conceivable mathematical ideas' allowed for Einstein to posit the first major law of modern physics much as it allows Galileo to posit the first major law of classical physics. During which time, when the special and then the general theories of relativity had not been confirmed by experiment and many established physicists viewed them as at least minor heresies, Einstein remained entirely confident of their predictions. Ilse Rosenthal-Schneider, who visited Einstein shortly after Eddington's eclipse expedition confirmed a prediction of the general theory (1919), described Einstein's response to this news: When I was giving expression to my joy that the results coincided with his calculations, he said quite unmoved, 'But I knew the theory was correct,' and when I asked, what if there had been no confirmation of his prediction, he countered: 'Then I would have been sorry for the dear Lord -the theory is correct.'

Einstein was not given to making sarcastic or sardonic comments, particularly on matters of religion. These unguarded responses testify to his profound conviction that the language of mathematics allows the human mind access to immaterial and immutable truths existing outside of the mind that conceived them. Although Einstein's belief was far more secular than Galileo's, it retained the same essential ingredients.

What continued in the twenty-three-year-long debate between Einstein and Bohr, least of mention? The primary article drawing upon its faith that contends with those opposing to the merits or limits of a physical theory, at the heart of this debate was the fundamental question, 'What is the relationship between the mathematical forms in the human mind called physical theory and physical reality?' Einstein did not believe in a God who spoke in tongues of flame from the mountaintop in ordinary language, and he could not sustain belief in the anthropomorphic God of the West. There is also no suggestion that he embraced ontological monism, or the conception of Being featured in Eastern religious systems, like Taoism, Hinduism, and Buddhism. The closest that Einstein apparently came to affirming the existence of the 'extra-personal' in the universe was a 'cosmic religious feeling', which he closely associated with the classical view of scientific epistemology.

The doctrine that Einstein fought to preserve seemed the natural inheritance of physics until the advent of quantum mechanics. Although the mind that constructs reality might be evolving fictions that are not necessarily true or necessary in social and political life, there was, Einstein felt, a way of knowing, purged of deceptions and lies. He was convinced that knowledge of physical reality in physical theory mirrors the preexistent and immutable realm of physical laws. And as Einstein consistently made clear, this knowledge mitigates loneliness and inculcates a sense of order and reason in a cosmos that might appear otherwise bereft of meaning and purpose.

What most disturbed Einstein about quantum mechanics was the fact that this physical theory might not, in experiment or even in principle, mirrors precisely the structure of physical reality. There is, for all the reasons we seem attested of, in that an inherent uncertainty in measurement made, . . . a quantum mechanical process reflects of a pursuit that quantum theory in itself and its contributive dynamic functionalities that there lay the attribution of a completeness of a quantum mechanical theory. Einstein's fearing that it would force us to recognize that this inherent uncertainty applied to all of physics, and, therefore, the ontological bridge between mathematical theory and physical reality - does not exist. And this would mean, as Bohr was among the first to realize, that we must profoundly revive the epistemological foundations of modern science.

The world view of classical physics allowed the physicist to assume that communion with the essences of physical reality via mathematical laws and associated theories was possible, but it made no other provisions for the knowing mind. In our new situation, the status of the knowing mind seems quite different. Modern physics distributively contributed its view toward the universe as an unbroken, undissectable and undivided dynamic whole. 'There can hardly be a sharper contrast,' said Melic Capek, 'than that between the everlasting atoms of classical physics and the vanishing 'particles' of modern physics as Stapp put it: 'Each atom turns out to be nothing but the potentialities in the behaviour pattern of others. What we find, therefore, are not elementary space-time realities, but rather a web of relationships in which no part can stand alone, every part derives its meaning and existence only from its place within the whole"

The characteristics of particles and quanta are not isolatable, given particle-wave dualism and the incessant exchange of quanta within matter-energy fields. Matter cannot be dissected from the omnipresent sea of energy, nor can we in theory or in fact observe matter from the outside. As Heisenberg put it decades ago, 'the cosmos appears to be a complicated tissue of events, in which connection of different kinds alternate or overlay or combine and thereby determine the texture of the whole. This that a pure reductionist approach to understanding physical reality, which was the goal of classical physics, is no longer appropriate.

While the formalism of quantum physics predicts that correlations between particles over space-like separated regions are possible, it can say nothing about what this strange new relationship between parts (quanta) and whole (cosmos) was by an outside formalism. This does not, however, prevent us from considering the implications in philosophical terms, as the philosopher of science Errol Harris noted in thinking about the special character of wholeness in modern physics, a unity without internal content is a blank or empty set and is not recognizable as a whole. A collection of merely externally related parts does not constitute a whole in that the parts will not be 'mutually adaptive and complementary to one and another.'

Wholeness requires a complementary relationship between unity and differences and is governed by a principle of organization determining the interrelationship between parts. This organizing principle must be universal to a genuine whole and implicit in all parts that constitute the whole, even though the whole is exemplified only in its parts. This principle of order, Harris continued, 'is nothing really in and of itself. It is the way parts are organized and not another constituent addition to those that constitute the totality.'

In a genuine whole, the relationship between the constituent parts must be 'internal or immanent' in the parts, as opposed to a mere spurious whole in which parts appear to disclose wholeness due to relationships that are external to the parts. The collection of parts that would allegedly constitute the whole in classical physics is an example of a spurious whole. Parts constitute a genuine whole when the universal principle of order is inside the parts and thereby adjusts each to all that they interlock and become mutually complementary. This not only describes the character of the whole revealed in both relativity theory and quantum mechanics. It is also consistent with the manner in which we have begun to understand the relation between parts and whole in modern biology.

Modern physics also reveals, claims Harris, a complementary relationship between the differences between parts that constituted contentual representations that the universal ordering principle that is immanent in each of the parts. While the whole cannot be finally disclosed in the analysis of the parts, the study of the differences between parts provides insights into the dynamic structure of the whole present in each of the parts. The part can never, nonetheless, be finally isolated from the web of relationships that disclose the interconnections with the whole, and any attempt to do so results in ambiguity.

Much of the ambiguity in attempted to explain the character of wholes in both physics and biology derives from the assumption that order exists between or outside parts. But order in complementary relationships between differences and sameness in any physical event is never external to that event -the connections are immanent in the event. From this perspective, the addition of non-locality to this picture of the dynamic whole is not surprising. The relationship between part, as quantum event apparent in observation or measurement, and the undissectable whole, revealed but not described by the instantaneous, and the undissectable whole, revealed but described by the instantaneous correlations between measurements in space-like separated regions, is another extension of the part-whole complementarity to modern physics.

If the universe is a seamlessly interactive system that evolves to a higher level of complexity, and if the lawful regularities of this universe are emergent properties of this system, we can assume that the cosmos is a singular point of significance as a whole that evinces of the 'progressive principal order' of complementary relations its parts. Given that this whole exists in some sense within all parts (quanta), one can then argue that it operates in self-reflective fashion and is the ground for all emergent complexities. Since human consciousness evinces self-reflective awareness in the human brain and since this brain, like all physical phenomena can be viewed as an emergent property of the whole, it is reasonable to conclude, in philosophical terms at least, that the universe is conscious.

But since the actual character of this seamless whole cannot be represented or reduced to its parts, it lies, quite literally beyond all human representations or descriptions. If one chooses to believe that the universe be a self-reflective and self-organizing whole, this lends no support whatsoever to conceptions of design, meaning, purpose, intent, or plan associated with any mytho-religious or cultural heritage. However, If one does not accept this view of the universe, there is nothing in the scientific descriptions of nature that can be used to refute this position. On the other hand, it is no longer possible to argue that a profound sense of unity with the whole, which has long been understood as the foundation of religious experience, which can be dismissed, undermined or invalidated with appeals to scientific knowledge.

While we have consistently tried to distinguish between scientific knowledge and philosophical speculation based on this knowledge -there is no empirically valid causal linkage between the former and the latter. Those who wish to dismiss the speculative assumptions as its basis to be drawn the obvious freedom of which is firmly grounded in scientific theory and experiments there is, however, in the scientific description of nature, the belief in radical Cartesian division between mind and world sanctioned by classical physics. Seemingly clear, that this separation between mind and world was a macro-level illusion fostered by limited awarenesses of the actual character of physical reality and by mathematical idealization that were extended beyond the realm of their applicability.

Thus, the grounds for objecting to quantum theory, the lack of a one-to-one correspondence between every element of the physical theory and the physical reality it describes, may seem justifiable and reasonable in strictly scientific terms. After all, the completeness of all previous physical theories was measured against the criterion with enormous success. Since it was this success that gave physics the reputation of being able to disclose physical reality with magnificent exactitude, perhaps a more comprehensive quantum theory will emerge to insist on these requirements.

All indications are, however, that no future theory can circumvent quantum indeterminancy, and the success of quantum theory in co-ordinating our experience with nature is eloquent testimony to this conclusion. As Bohr realized, the fact that we live in a quantum universe in which the quantum of action is a given or an unavoidable reality requires a very different criterion for determining the completeness or physical theory. The new measure for a complete physical theory is that it unambiguously confirms our ability to co-ordinate more experience with physical reality.

If a theory does so and continues to do so, which is certainly the case with quantum physics, then the theory must be deemed complete. Quantum physics not only works exceedingly well, it is, in these terms, the most accurate physical theory that has ever existed. When we consider that this physics allows us to predict and measure quantities like the magnetic moment of electrons to the fifteenth decimal place, we realize that accuracy per se is not the real issue. The real issue, as Bohr rightly intuited, is that this complete physical theory effectively undermines the privileged relationship in classical physics between 'theory' and 'physical reality'.

If the universe is a seamlessly interactive system that evolves to higher levels of complex and complicating regularities of which ae lawfully emergent in property of systems, we can assume that the cosmos is a single significant whole that evinces progressive order in complementary relations to its parts. Given that this whole exists in some sense within all parts (quanta), one can then argue that in operates in self-reflective fashion and is the ground from all emergent plexuities. Since human consciousness evinces self-reflective awareness in te human brain (well protected between the cranium walls) and since this brain, like all physical phenomena, can b viewed as an emergent property of the whole, it is unreasonable to conclude, in philosophical terms at least, that the universe is conscious.

Nevertheless, since the actual character of this seamless whole cannot be represented or reduced to its parts, it lies, quite laterally, beyond all human representation or descriptions. If one chooses to believe that the universe be a self-reflective and self-organizing whole, this lends no support whatsoever to conceptual representation of design, meaning, purpose, intent, or plan associated with mytho-religious or cultural heritage. However, if one does not accept this view of the universe, there is noting in the scientific description of nature that can be used to refute this position. On the other hand, it is no longer possible to argue that a profound sense of unity with the whole, which has long been understood as foundation of religious experiences, but can be dismissed, undermined, or invalidated with appeals to scientific knowledge.

While we have consistently tried to distinguish between scientific knowledge and philosophical speculation based on this of what is obtainable, let us be quite clear on one point - there is no empirically valid causal linkage between the former and the latter. Those who wish to dismiss the speculative base on which is obviously free to do as done. However, there is another conclusion to be drawn, in that is firmly grounded in scientific theory and experiment there is no basis in the scientific descriptions of nature for believing in the radical Cartesian division between mind and world sanctioned by classical physics. Clearly, his radical separation between mind and world was a macro-level illusion fostered by limited awareness of the actual character of physical reality nd by mathematical idealizations extended beyond the realms of their applicability.

Nevertheless, the philosophical implications might prove in themselves as a criterial motive in debative consideration to how our proposed new understanding of the relationship between parts and wholes in physical reality might affect the manner in which we deal with some major real-world problems. This will issue to demonstrate why a timely resolution of these problems is critically dependent on a renewed dialogue between members of the cultures of human-social scientists and scientist-engineers. We will also argue that the resolution of these problems could be dependent on a renewed dialogue between science and religion.

As many scholars have demonstrated, the classical paradigm in physics has greatly influenced and conditioned our understanding and management of human systems in economic and political realities. Virtually all models of these realities treat human systems as if they consist of atomized units or parts that interact with one another in terms of laws or forces external to or between the parts. These systems are also viewed as hermetic or closed and, thus, its discreteness, separateness and distinction.

Consider, for example, how the classical paradigm influenced or thinking about economic reality. In the eighteenth and nineteenth centuries, the founders of classical economics -figures like Adam Smith, David Ricardo, and Thomas Malthus conceived of the economy as a closed system in which intersections between parts (consumer, produces, distributors, and so forth) are controlled by forces external to the parts (supply and demand). The central legitimating principle of free market economics, formulated by Adam Smith, is that lawful or law-like forces external to the individual units function as an invisible hand. This invisible hand, said Smith, frees the units to pursue their best interests, moves the economy forward, and in general legislates the behaviour of parts in the best vantages of the whole. (The resemblance between the invisible hand and Newton's universal law of gravity and between the relations of parts and wholes in classical economics and classical physics should be transparent.)

After roughly 1830, economists shifted the focus to the properties of the invisible hand in the interactions between pats using mathematical models. Within these models, the behaviour of pats in the economy is assumed to be analogous to the awful interactions between pats in classical mechanics. It is, therefore, not surprising that differential calculus was employed to represent economic change in a virtual world in terms of small or marginal shifts in consumption or production. The assumption was that the mathematical description of marginal shifts n the complex web of exchanges between parts (atomized units and quantities) and whole (closed economy) could reveal the lawful, or law-like, machinations of the closed economic system.

These models later became one of the fundamentals for microeconomics. Microeconomics seek to describe interactions between parts in exact quantifiable measures-such as marginal cost, marginal revenue, marginal utility, and growth of total revenue as indexed against individual units of output. In analogy with classical mechanics, the quantities are viewed as initial conditions that can serve to explain subsequent interactions between parts in the closed system in something like deterministic terms. The combination of classical macro-analysis with micro-analysis resulted in what Thorstein Veblen in 1900 termed neoclassical economics-the model for understanding economic reality that is widely used today

Beginning in the 1939s, the challenge became to subsume the understanding of the interactions between parts in closed economic systems with more sophisticated mathematical models using devices like linear programming, game theory, and new statistical techniques. In spite of the growing mathematical sophistication, these models are based on the same assumptions from classical physics featured in previous neoclassical economic theory-with one exception. They also appeal to the assumption that systems exist in equilibrium or in perturbations from equilibria, and they seek to describe the state of the closed economic system in these terms.

One could argue that the fact that our economic models are assumptions from classical mechanics is not a problem by appealing to the two-domain distinction between micro-level macro-level processes expatiated upon earlier. Since classical mechanic serves us well in our dealings with macro-level phenomena in situations where the speed of light is so large and the quantum of action is so small as to be safely ignored for practical purposes, economic theories based on assumptions from classical mechanics should serve us well in dealing with the macro-level behaviour of economic systems.

The obvious problem, . . . acceded peripherally, . . . nature is relucent to operate in accordance with these assumptions, in that the biosphere, the interaction between parts be intimately related to the hole, no collection of arts is isolated from the whole, and the ability of the whole to regulate the relative abundance of atmospheric gases suggests that the whole of the biota appear to display emergent properties that are more than the sum of its parts. What the current ecological crisis reveals in the abstract virtual world of neoclassical economic theory. The real economies are all human activities associated with the production, distribution, and exchange of tangible goods and commodities and the consumption and use of natural resources, such as arable land and water.

Although expanding economic systems in the really economy ae obviously embedded in a web of relationships with the entire biosphere, our measure of healthy economic systems disguises this fact very nicely. Consider, for example, the healthy economic system written in 1996 by Frederick Hu, head of the competitive research team for the World Economic Forum - short of military conquest, economic growth is the only viable for a country to sustain increases in natural living standards . . . An economy is internationally competitive if it performs strongly in three general areas: Abundant productive inputs from capital, labour, infrastructure and technology, optimal economic policies such as low taxes, little interference, free trade and sound market institutions. Such as the rule of law and protection of property rights.

The prescription for medium-term growth of economies ion countries like Russia, Brazil, and China may seem utterly pragmatic and quite sound. But the virtual economy described is a closed and hermetically sealed system in which the invisible hand of economic forces allegedly results in a health growth economy if impediments to its operation are removed or minimized. It is, of course, often trued that such prescriptions can have the desired results in terms of increases in living standards, and Russia, Brazil and China are seeking to implement them in various ways.

In the real economy, however, these systems are clearly not closed or hermetically sealed: Russia uses carbon-based fuels in production facilities that produce large amounts of carbon dioxide and other gases that contribute to global warming: Brazil is in the process of destroying a rain forest that is critical to species diversity and the maintenance of a relative abundance of atmospheric gases that regulate Earth temperature, and China is seeking to build a first-world economy based on highly polluting old-world industrial plants that burn soft coal. Not to forget, . . . the victual economic systems that the world now seems to regard as the best example of the benefits that can be derived form the workings of the invisible hand, that of the United States, operates in the real economy as one of the primary contributors to the ecological crisis.

The philosophical debate that had led to conclusions useful to the architects of classical physics can be briefly summarized, such when Thale's fellow Milesian Anaximander claimed that the first substance, although indeterminate, manifested itself in a conflict of oppositions between hot and cold, moist and dry. The idea of nature as a self-regulating balance of forces was subsequently elaborated upon by Heraclitus (d. after 480 BC), who asserted that the fundamental substance is strife between opposites, which is itself the unity of the whole. It is, said Heraclitus, the tension between opposites that keeps the whole from simply 'passing away.'

Parmenides of Elea (B.c. 515 BC) argued in turn that the unifying substance is unique and static being. This led to a conclusion about the relationship between ordinary language and external reality that was later incorporated into the view of the relationship between mathematical language and physical reality. Since thinking or naming involves the presence of something, said Parmenides, thought and language must be dependent upon the existence of objects outside the human intellect. Presuming a one-to-one correspondence between word and idea and actual existing things, Parmenides concluded that our ability to think or speak of a thing at various times implies that it exists at all times. Hence the indivisible One does not change, and all perceived change is an illusion.

These assumptions emerged in roughly the form in which they would be used by the creators of classical physics in the thought of the atomists. Leucippus: ?lL. 450-420 BC and Democritus (460-c. 370 BC). They reconciled the two dominant and seemingly antithetical concepts of the fundamental character of being-Becoming (Heraclitus) and unchanging Being (Parmenides)-in a remarkable simple and direct way. Being, they said, is present in the invariable substance of the atoms that, through blending and separation, make up the thing of changing or becoming worlds.

The last remaining feature of what would become the paradigm for the first scientific revolution in the seventeenth century is attributed to Pythagoras (Bc. 570 Bc). Like Parmenides, Pythagoras also held that the perceived world is illusory and that there is an exact correspondence between ideas and aspects of external reality. Pythagoras, however, had a different conception of the character of the idea that showed this correspondence. The truth about the fundamental character of the unified and unifying substance, which could be uncovered through reason and contemplation, is, he claimed, mathematical in form.

Pythagoras established and was the cental figure in a school of philosophy, religion and mathematics; He was apparently viewed by his followers as semi-divine. For his followers the regular solids (symmetrical three-dimensional forms in which all sides are the same regular polygons) and whole numbers became revered essences of sacred ideas. In contrast with ordinary language, the language of mathematics and geometric forms seemed closed, precise and pure. Providing one understood the axioms and notations, and the meaning conveyed was invariant from one mind to another. The Pythagoreans felt that the language empowered the mind to leap beyond the confusion of sense experience into the realm of immutable and eternal essences. This mystical insight made Pythagoras the figure from antiquity most revered by the creators of classical physics, and it continues to have great appeal for contemporary physicists as they struggle with the epistemological implications of the quantum mechanical description of nature.

Yet, least of mention, progress was made in mathematics, and to a lesser extent in physics, from the time of classical Greek philosophy to the seventeenth century in Europe. In Baghdad, for example, from about A.D. 750 to A.D. 1000, substantial advancement was made in medicine and chemistry, and the relics of Greek science were translated into Arabic, digested, and preserved. Eventually these relics reentered Europe via the Arabic kingdom of Spain and Sicily, and the work of figures like Aristotle universities of France, Italy, and England during the Middle Ages.

For much of this period the Church provided the institutions, like the reaching orders, needed for the rehabilitation of philosophy. But the social, political and an intellectual climate in Europe was not ripe for a revolution in scientific thought until the seventeenth century. Until later in time, lest as far into the nineteenth century, the works of the new class of intellectuals we called scientists, whom of which were more avocations than vocation, and the word scientist do not appear in English until around 1840.

Copernicus (1473-1543) would have been described by his contemporaries as an administrator, a diplomat, an avid student of economics and classical literature, and most notable, a highly honoured and placed church dignitaries. Although we named a revolution after him, his devoutly conservative man did not set out to create one. The placement of the Sun at the centre of the universe, which seemed right and necessary to Copernicus, was not a result of making careful astronomical observations. In fact, he made very few observations in the course of developing his theory, and then only to ascertain if his prior conclusions seemed correct. The Copernican system was also not any more useful in making astrological calculations than the accepted model and was, in some ways, much more difficult to implement. What, then, was his motivation for creating the model and his reasons for presuming that the model was correct?

Copernicus felt that the placement of the Sun at the centre of the universe made sense because he viewed the Sun as the symbol of the presence of a supremely intelligent and intelligible God in a man-centred world. He was apparently led to this conclusion in part because the Pythagoreans believed that fire exists at the centre of the cosmos, and Copernicus identified this fire with the fireball of the Sun. the only support that Copernicus could offer for the greater efficacy of his model was that it represented a simpler and more mathematical harmonious model of the sort that the Creator would obviously prefer. The language used by Copernicus in 'The Revolution of Heavenly Orbs,' illustrates the religious dimension of his scientific thought: 'In the midst of all the sun reposes, unmoving. Who, indeed, in this most beautiful temple would place the light-giver in any other part than from where it can illumine all other parts?'

The belief that the mind of God as Divine Architect permeates the working of nature was the guiding principle of the scientific thought of Johannes Kepler (or, Keppler 1571-1630). For this reason, most modern physicists would probably feel some discomfort in reading Kepler's original manuscripts. Physics and metaphysics, astronomy and astrology, geometry and theology commingle with an intensity that might offend those who practice science in the modern sense of that word. Physical laws, wrote Kepler, 'lie within the power of understanding of the human mind; God wanted us to perceive them when he created us of His own image, in order . . . that we may take part in His own thoughts. Our knowledge of numbers and quantities is the same as that of God's, at least insofar as we can understand something of it in this mortal life.'

Believing, like Newton after him, in the literal truth of the words of the Bible, Kepler concluded that the word of God is also transcribed in the immediacy of observable nature. Kepler's discovery that the motions of the planets around the Sun were elliptical, as opposed perfecting circles, may have made the universe seem a less perfect creation of God on ordinary language. For Kepler, however, the new model placed the Sun, which he also viewed as the emblem of a divine agency, more at the centre of mathematically harmonious universes than the Copernican system allowed. Communing with the perfect mind of God requires as Kepler put it 'knowledge of numbers and quantity.'

Since Galileo did not use, or even refer to, the planetary laws of Kepler when those laws would have made his defence of the heliocentric universe more credible, his attachment to the god-like circle was probably a more deeply rooted aesthetic and religious ideal. But it was Galileo, even more than Newton, who was responsible for formulating the scientific idealism that quantum mechanics now force us to abandon. In 'Dialogue Concerning the Two Great Systems of the World,' Galileo said about the following about the followers of Pythagoras: 'I know perfectly well that the Pythagoreans had the highest esteem for the science of number and that Plato himself admired the human intellect and believed that it participates in divinity solely because it is able to understand the nature of numbers. And I myself am inclined to make the same judgement.'

This article of faith-mathematical and geometrical ideas mirror precisely the essences of physical reality was the basis for the first scientific law of this new science, a constant describing the acceleration of bodies in free fall, could not be confirmed by experiment. The experiments conducted by Galileo in which balls of different sizes and weights were rolled simultaneously down an inclined plane did not, as he frankly admitted, their precise results. And since a vacuum pumps had not yet been invented, there was simply no way that Galileo could subject his law to rigorous experimental proof in the seventeenth century. Galileo believed in the absolute validity of this law in the absence of experimental proof because he also believed that movement could be subjected absolutely to the law of number. What Galileo asserted, as the French historian of science Alexander Koyré put it, was 'that the real are in its essence, geometrical and, consequently, subject to rigorous determination and measurement.'

The popular image of Isaac Newton (1642-1727) is that of a supremely rational and dispassionate empirical thinker. Newton, like Einstein, had the ability to concentrate unswervingly on complex theoretical problems until they yielded a solution. But what most consumed his restless intellect were not the laws of physics. In addition to believing, like Galileo that the essences of physical reality could be read in the language of mathematics, Newton also believed, with perhaps even greater intensity than Kepler, in the literal truths of the Bible.

While Hertz made this statement without having to contend with the implications of quantum mechanics, the feeling, the described remains the most enticing aspects sets of physics.

That elegant mathematical formulae provide a framework for understanding the origins and tranformations of a cosmos of enormous age and dimensions are a staggering discovery for bidding physicists. Professors of physics do not, of course, tell their students that the study of physical laws in an act of communion with thee perfect mind of God or that these laws have an independent existence outside the minds that discover them. The business of becoming a physicist typically begins, however, with the study of classical or Newtonian dynamics, and this training provides considerable covert reinforcement of the feeling that Hertz described, perhaps, the best way to examine the legacy of the dialogue between science and religion in the debate over the implications of quantum non-locality is to examine the source of Einstein's objections tp quantum epistemology in more personal terms. Einstein apparently lost faith in the God portrayed in biblical literature in early adolescence. But, as appropriated, . . . the 'Autobiographical Notes' give to suggest that there were aspects that carry over into his understanding of the foundation for scientific knowledge, . . . 'Thus I came despite the fact I was th e son of an ent it y irriligious [Jewish] Breden heritage that I was the son of an entirely irreligious [Jewish] Breeden heritage, which is deeply held of its religiosity, which, however, found an abrupt end at the age of 12. Though the reading of popular scientific books I soon reached the conviction that much in the stories of the Bible could not be true. The consequence waw a positively frantic [orgy] of freethinking coupled with the impression that youth is intentionally being deceived by the stat through lies that it was a crushing impression. Suspicion against every kind of authority grew out of this experience. . . . It was clear to me that the religious paradise of youth, which was thus lost, was a first attempt ti free myself from the chains of the 'merely personal'. . . . The mental grasp of this extra-personal world within the frame of the given possibilities swam as highest aim half consciously and half unconsciously before the mind's eye.'

What is more, was, suggested Einstein, belief in the word of God as it is revealed in biblical literature that allowed him to dwell in a 'religious paradise of youth' and to shield himself from the harsh realities of social and political life. In an effort to recover that inner sense of security that was lost after exposure to scientific knowledge, or to become free once again of the 'merely personal', he committed himself to understanding the 'extra-personal world within the frame of given possibilities', or as seems obvious, to the study of physics. Although the existence of God as described in the Bible may have been in doubt, the qualities of mind that the architects of classical physics associated with this God were not. This is clear in the comments from which Einstein uses of mathematics, . . . 'Nature is the realization of the simplest conceivable mathematical ideas. I am convinced that we can discover, by of purely mathematical construction, those concepts and those lawful connections between them that furnish the key to the understanding of natural phenomena. Experience remains, of course, the sole criteria of physical utility of a mathematical construction. But the creative principle resides in mathematics. In a certain sense, therefore, I hold it true that pure thought can grasp reality, as the ancients dreamed.'

This article of faith, first articulated by Kepler, that 'nature is the realization of the simplest conceivable mathematical ideas' allowed for Einstein to posit the first major law of modern physics much as it allows Galileo to posit the first major law of classical physics. During which time, when the special and then the general theories of relativity had not been confirmed by experiment and many established physicists viewed them as at least minor heresies, Einstein remained entirely confident of their predictions. Ilse Rosenthal-Schneider, who visited Einstein shortly after Eddington's eclipse expedition confirmed a prediction of the general theory (1919), described Einstein's response to this news: When I was giving expression to my joy that the results coincided with his calculations, he said quite unmoved, 'But I knew the theory was correct,' and when I asked, what if there had been no confirmation of his prediction, he countered: 'Then I would have been sorry for the dear Lord -the theory is correct.'

Einstein was not given to making sarcastic or sardonic comments, particularly on matters of religion. These unguarded responses testify to his profound conviction that the language of mathematics allows the human mind access to immaterial and immutable truths existing outside of the mind that conceived them. Although Einstein's belief was far more secular than Galileo's, it retained the same essential ingredients.

What continued in the twenty-three-year-long debate between Einstein and Bohr, least of mention? The primary article drawing upon its faith that contends with those opposing to the merits or limits of a physical theory, at the heart of this debate was the fundamental question, 'What is the relationship between the mathematical forms in the human mind called physical theory and physical reality?' Einstein did not believe in a God who spoke in tongues of flame from the mountaintop in ordinary language, and he could not sustain belief in the anthropomorphic God of the West. There is also no suggestion that he embraced ontological monism, or the conception of Being featured in Eastern religious systems, like Taoism, Hinduism, and Buddhism. The closest that Einstein apparently came to affirming the existence of the 'extra-personal' in the universe was a 'cosmic religious feeling', which he closely associated with the classical view of scientific epistemology.

The doctrine that Einstein fought to preserve seemed the natural inheritance of physics until the advent of quantum mechanics. Although the mind that constructs reality might be evolving fictions that are not necessarily true or necessary in social and political life, there was, Einstein felt, a way of knowing, purged of deceptions and lies. He was convinced that knowledge of physical reality in physical theory mirrors the preexistent and immutable realm of physical laws. And as Einstein consistently made clear, this knowledge mitigates loneliness and inculcates a sense of order and reason in a cosmos that might appear otherwise bereft of meaning and purpose.

What most disturbed Einstein about quantum mechanics was the fact that this physical theory might not, in experiment or even in principle, mirrors precisely the structure of physical reality. There is, for all the reasons we seem attested of, in that an inherent uncertainty in measurement made, . . . a quantum mechanical process reflects of a pursuit that quantum theory in itself and its contributive dynamic functionalities that there lay the attribution of a completeness of a quantum mechanical theory. Einstein's fearing that it would force us to recognize that this inherent uncertainty applied to all of physics, and, therefore, the ontological bridge between mathematical theory and physical reality -does not exist. And this would mean, as Bohr was among the first to realize, that we must profoundly revive the epistemological foundations of modern science.

The world view of classical physics allowed the physicist to assume that communion with the essences of physical reality via mathematical laws and associated theories was possible, but it made no other provisions for the knowing mind. In our new situation, the status of the knowing mind seems quite different. Modern physics distributively contributed its view toward the universe as an unbroken, undissectable and undivided dynamic whole. 'There can hardly be a sharper contrast,' said Melic Capek, 'than that between the everlasting atoms of classical physics and the vanishing 'particles' of modern physics as Stapp put it: 'Each atom turns out to be nothing but the potentialities in the behaviour pattern of others. What we find, therefore, are not elementary space-time realities, but rather a web of relationships in which no part can stand alone, every part derives its meaning and existence only from its place within the whole"

The characteristics of particles and quanta are not isolatable, given particle-wave dualism and the incessant exchange of quanta within matter-energy fields. Matter cannot be dissected from the omnipresent sea of energy, nor can we in theory or in fact observe matter from the outside. As Heisenberg put it decades ago, 'the cosmos appears to be a complicated tissue of events, in which connection of different kinds alternate or overlay or combine and thereby determine the texture of the whole. This that a pure reductionist approach to understanding physical reality, which was the goal of classical physics, is no longer appropriate.

While the formalism of quantum physics predicts that correlations between particles over space-like separated regions are possible, it can say nothing about what this strange new relationship between parts (quanta) and whole (cosmos) was by an outside formalism. This does not, however, prevent us from considering the implications in philosophical terms, as the philosopher of science Errol Harris noted in thinking about the special character of wholeness in modern physics, a unity without internal content is a blank or empty set and is not recognizable as a whole. A collection of merely externally related parts does not constitute a whole in that the parts will not be 'mutually adaptive and complementary to one and another.'

Wholeness requires a complementary relationship between unity and differences and is governed by a principle of organization determining the interrelationship between parts. This organizing principle must be universal to a genuine whole and implicit in all parts that constitute the whole, even though the whole is exemplified only in its parts. This principle of order, Harris continued, 'is nothing really in and of itself. It is the way parts are organized and not another constituent addition to those that constitute the totality.'

In a genuine whole, the relationship between the constituent parts must be 'internal or immanent' in the parts, as opposed to a mere spurious whole in which parts appear to disclose wholeness due to relationships that are external to the parts. The collection of parts that would allegedly constitute the whole in classical physics is an example of a spurious whole. Parts constitute a genuine whole when the universal principle of order is inside the parts and thereby adjusts each to all that they interlock and become mutually complementary. This not only describes the character of the whole revealed in both relativity theory and quantum mechanics. It is also consistent with the manner in which we have begun to understand the relation between parts and whole in modern biology.

But since the actual character of this seamless whole cannot be represented or reduced to its parts, it lies, quite literally beyond all human representations or descriptions. If one chooses to believe that the universe be a self-reflective and self-organizing whole, this lends no support whatsoever to conceptions of design, meaning, purpose, intent, or plan associated with any mytho-religious or cultural heritage. However, If one does not accept this view of the universe, there is nothing in the scientific descriptions of nature that can be used to refute this position. On the other hand, it is no longer possible to argue that a profound sense of unity with the whole, which has long been understood as the foundation of religious experience, which can be dismissed, undermined or invalidated with appeals to scientific knowledge.

In quantum physics, however, the hydrogen atom cannot be visualized with such macro-level analogies. The orbit of the electron is not a circle, in which a planet-like object moves, and each orbit is described in terms of a probability distribution for finding the electron in an average position corresponding to each orbit as opposed to an actual position. Without observation or measurement, the electron could be in some sense anywhere or everywhere within the probability distribution, also, the space between probability distributions is not empty, it is infused with energetic vibrations capable of manifesting itself as the befitting quanta.

`The energy levels manifest at certain distances because the traditions between orbits occurs in terms of precise units of Planck's constant. If any attentive effects to comply with or measure where the particle-like aspect of the electron is, in that the existence of Planck's constant will always prevent us from knowing precisely all the properties of that electron that we might presume to be they're in the absence of measurement. Also, the two-split experiment, as our presence as observers and what we choose to measure or observe are inextricably linked to the results obtained. Since all complex molecules are built from simpler atoms, what is to be done, is that liken to the hydrogen atom, of which case applies generally to all material substances.

The grounds for objecting to quantum theory, the lack of a one-to-one correspondence between every element of the physical theory and the physical reality it describes, may seem justifiable and reasonable in strict scientific terms. After all, the completeness of all previous physical theories was measured against that criterion with enormous success. Since it was this success that gave physicists the reputation of being able to disclose physical reality with magnificent exactitude, perhaps a more complex quantum theory will emerge by continuing to insist on this requirement.

All indications are, however, that no future theory can circumvent quantum indeterminacy, and the success of quantum theory in co-ordinating our experience with nature is eloquent testimony to this conclusion. As Bohr realized, the fact that we live in a quantum universe in which the quantum of action is a given or an unavoidable reality requires a very different criterion for determining the completeness of physical theory. The new measure for a complete physical theory is that it unambiguously confirms our ability to co-ordinate more experience with physical reality.

If a theory does so and continues to do so, which is certainly the case with quantum physics, then the theory must be deemed complete. Quantum physics not only works exceedingly well, it is, in these terms, the most accurate physical theory that has ever existed. When we consider that this physics allows us to predict and measure quantities like the magnetic moment of electrons to the fifteenth decimal place, we realize that accuracy perse is not the real issue. The real issue, as Bohr rightly intuited, is that this complete physical theory effectively undermines the privileged relationships in classical physics between physical theory and physical reality. Another measure of success in physical theory is also met by quantum physics -eloquence and simplicity. The quantum recipe for computing probabilities given by the wave function is straightforward and can be successfully employed by any undergraduate physics student. Take the square of the wave amplitude and compute the probability of what can be measured or observed with a certain value. Yet there is a profound difference between the recipe for calculating quantum probabilities and the recipe for calculating probabilities in classical physics.

In quantum physics, one calculates the probability of an event that can happen in alternative ways by adding the wave functions, and then taking the square of the amplitude. In the two-split experiment, for example, the electron is described by one wave function if it goes through one slit and by another wave function if it goes through the other slit. In order to compute the probability of where the electron is going to end on the screen, we add the two wave functions, compute the obsolete value of their sum, and square it. Although the recipe in classical probability theory seems similar, it is quite different. In classical physics, one would simply add the probabilities of the two alternative ways and let it go at that. That classical procedure does not work here because we are not dealing with classical atoms in quantum physics additional terms arise when the wave functions are added, and the probability is computed in a process known as the 'superposition principle'. That the superposition principle can be illustrated with an analogy from simple mathematics. Add two numbers and then take the square of their sum, as opposed to just adding the squares of the two numbers. Obviously, $(2 + 3)2$ is not equal to $22 + 32$. The former is 25, and the latter are 13. In the language of quantum probability theory:

$$| \Psi + \Psi |2 | \Psi |2 + \Psi | \Psi2 |2$$

Where Ø1 and Ø2 are the individual wave functions on the left-hand side, the superposition principle results in extra terms that cannot be found on the right-handed side the left-hand faction of the above relation is the way a quantum physicists would compute probabilities and the right-hand side is the classical analogue. In quantum theory, the right-hand side is realized when we know, for example, which slit through which the electron went. Heisenberg was among the first to compute what would happen in an instance like this. The extra superposition terms contained in the left-hand side of the above relation would not be there, and the peculiar wave-like interference pattern would disappear. The observed pattern on the final screen would, therefore, be what one would expect if electrons were behaving like bullets, and the final probability would be the sum of the individual probabilities. But when we know which slit the electron went through, this interaction with the system causes the interference pattern to disappear.

In order to give a full account of quantum recipes for computing probabilities, one g-has to examine what would happen in events that are compounded. Compound events are events that can be broken down into a series of steps, or events that consist of a number of things happening independently the recipe here calls for multiplying the individual wave functions, and then following the usual quantum recipe of taking the square of the amplitude.

The quantum recipe is $| \Psi1 \cdot \Psi2 |2$, and, in this case, it would be the same if we multiplied the individual probabilities, as one would in classical theory. Thus the recipes of computing results in quantum theory and classical physics can be totally different from quantum superposition effects are completely non-classical, and there is no mathematical justification to why the quantum recipes work. What justifies the use of quantum probability theory is the same thing that justifies the use of quantum physics -it has allowed us in countless experiments to vastly extend our ability to co-ordinate experience with nature.

The view of probability in the nineteenth century was greatly conditioned and reinforced by classical assumptions about the relationships between physical theory and physical reality. In this century, physicists developed sophisticated statistics to deal with large ensembles of particles before the actual character of these particles was understood. Classical statistics, developed primarily by James C. Maxwell and Ludwig Boltzmann, was used to account for the behaviour of a molecule in a gas and to predict the average speed of a gas molecule in terms of the temperature of the gas.

The presumption was that the statistical average were workable approximations those subsequent physical theories, or better experimental techniques, would disclose with precision and certainty. Since nothing was known about quantum systems, and since quantum indeterminacy is small when dealing with macro-level effects, this presumption was quite reasonable. We know, however, that quantum mechanical effects are present in the behaviour of gasses and that the choice to ignore them is merely a matter of convincing in getting workable or practical resulted. It is, therefore, no longer possible to assume that the statistical averages are merely higher-level approximations for a more exact description.

Perhaps the acknowledged defence of the classical conception of the relationship between physical theory ands physical reality is the celebrated animal introduced by the Austrian physicist Erin Schrödinger (1887-1961) in 1935, in a 'thought experiment' showing the strange nature of the world of quantum mechanics. The cat is thought of as locked in a box with a capsule of cyanide, which will break if a Geiger counter triggers. This will happen if an atom in a radioactive substance in the box decays, and there is a chance of 50% of such an event within an hour. Otherwise, the cat is alive. The problem is that the system is in an indeterminate state. The wave function of the entire system is a 'superposition' of states, fully described by the probabilities of events occurring when it is eventually measured, and therefore 'contains equal parts of the living and dead cat'. When we look and see we will find either a breathing cat or a dead cat, but if it is only as we look that the wave packet collapses, quantum mechanic forces us to say that before we looked it was not true that the cat was dead and not true that it was alive, the thought experiment makes vivid the difficulty of conceiving of quantum indeterminacies when these are translated to the familiar world of everyday objects.

The 'electron,' is a stable elementary particle having a negative charge, 'e', equal to:

$$1.602\ 189\ 25 \times 10\text{-}19\ C$$

And a rest mass, m0 equal to;

$$9.109\ 389\ 7 \times 10\text{-}31\ kg$$
$$\text{equivalent to } 0.511\ 0034\ MeV\ /\ c2$$

It has a spin of ½ and obeys Fermi-Dirac Statistics. As it does not have strong interactions, it is classified as a 'lepton'.

The discovery of the electron was reported in 1897 by Sir J.J. Thomson, following his work on the rays from the cold cathode of a gas-discharge tube, it was soon established that particles with the same charge and mass were obtained from numerous substances by the 'photoelectric effect', 'thermionic emission' and 'beta decay'. Thus, the electron was found to be part of all atoms, molecules, and crystals.

Free electrons are studied in a vacuum or a gas at low pressure, whereby beams are emitted from hot filaments or cold cathodes and are subject to 'focussing', so that the particles in which an electron beam in, for example, a cathode-ray tube, where in principal methods as (i) Electrostatic focussing, the beam is made to converge by the action of electrostatic fields between two or more electrodes at different potentials. The electrodes are commonly cylinders coaxial with the electron tube, and the whole assembly forms an electrostatic electron lens. The focussing effect is usually controlled by varying the potential of one of the electrodes, called the focussing electrode. (ii) Electromagnetic focussing, by way that the beam is made to converge by the action of a magnetic field that is produced by the passage of direct current, through a focussing coil. The latter are commonly a coil of short axial length mounted so as to surround the electron tube and to be coaxial with it.

The force FE on an electron or magnetic field of strength E is given by FE = Ee and is in the direction of the field. On moving through a potential difference V, the electron acquires a kinetic energy eV, hence it is possible to obtain beams of electrons of accurately known kinetic energy. In a magnetic field of magnetic flux density 'B', an electron with speed 'v' is subject to a force, FB = Bev sin è, where è is the angle between 'B' and 'v'. This force acts at right angles to the plane containing 'B' and 'v'.

The mass of any particle increases with speed according to the theory of relativity. If an electron is accelerated from rest through 5kV, the mass is 1% greater than it is at rest. Thus, accountably, must be taken of relativity for calculations on electrons with quite moderate energies.

According to 'wave mechanics' a particle with momentum 'mv' exhibits' diffraction and interference phenomena, similar to a wave with wavelength ë = h/mv, where 'h' is the Planck constant. For electrons accelerated through a few hundred volts, this gives wavelengths rather less than typical interatomic spacing in crystals. Hence, a crystal can act as a diffraction grating for electron beams.

Owing to the fact that electrons are associated with a wavelength ë given by ë = h/mv, where 'h' is the Planck constant and (mv) the momentum of the electron, a beam of electrons suffers diffraction in its passage through crystalline material, similar to that experienced by a beam of X-rays. The diffraction pattern depends on the spacing of the crystal planes, and the phenomenon can be employed to investigate the structure of surface and other films, and under suitable conditions exhibit the characteristics of a wave of the wavelength given by the equation ë = h/mv, which is the basis of wave mechanics. A set of waves that represent the behaviour, under appropriate conditions, of a particle, e.g., its diffraction by a crystal lattice, that is given the 'de Broglie equation.' They are sometimes regarded as waves of probability, since the square of their amplitude at a given point represents the probability of finding the particle in unit volume at that point.

The first experiment to demonstrate 'electron diffraction', and hence the wavelike nature of particles. A narrow pencil of electrons from a hot filament cathode was projected 'in vacua' onto a nickel crystal. The experiment showed the existence of a definite diffracted beam at one particular angle, which depended on the velocity of the electrons, assuming this to be the Bragg angle, stating that the structure of a crystal can be determined from a set of interference patterns found at various angles from the different crystal faces, least of mention, the wavelength of the electrons was calculated and found to be in agreement with the 'de Broglie equation.'

At kinetic energies less than a few electro-volts, electrons undergo elastic collision with atoms and molecules, simply because of the large ratio of the masses and the conservation of momentum, only an extremely small trandfer of kinetic energy occurs. Thus, the electrons are deflected but not slowed down appreciably. At slightly higher energies collisions are inelastic. Molecules may be dissociated, and atoms and molecules may be excited or ionized. Thus it is the least energy that causes an ionization

A A+ + e

Where the [ion] and the electron are far enough apart for their electrostatic interaction to be negligible and no extra kinetic energy removed is that in the outermost orbit, i.e., the level strongly bound electrons. It is also possible to consider removal of electrons from inner orbits, in which their binding energy is greater. As an excited particle or recombining, ions emit electromagnetic radiation mostly in the visible or ultraviolet.

For electron energies of the order of several GeV upwards, X-rays are generated. Electrons of high kinetic energy travel considerable distances through matter, leaving a trail of positive ions and free electrons. The energy is mostly lost in small increments (about 30 eV) with only an occasional major interaction causing X-ray emissions. The range increases at higher energies.

The positron - the antiparticle of the electron, i.e., an elementary particle with electron mass and positive charge equal to that of the electron. According to the relativistic wave mechanics of Dirac, space contains a continuum of electrons in states of negative energy. These states are normally unobservable, but if sufficient energy can be given, an electron may be raised into a state of positive energy and suggested itself observably. The vacant state of negativity behaves as a positive particle of positive energy, which is observed as a positron.

The simultaneous formation of a positron and an electron from a photon is called 'pair production', and occurs when the annihilation of gamma-ray photons with an energy of 1.02 MeV passes close to an atomic nucleus, whereby the interaction between the particle and its antiparticle disappear and photons or other elementary particles or antiparticles are so created, as accorded to energy and momentum conservation.

At low energies, an electron and a positron annihilate to produce electromagnetic radiation. Usually the particles have little kinetic energy or momentum in the laboratory system before interaction, hence the total energy of the radiation is nearly $2m_0c_2$, where m_0 is the rest mass of an electron. In nearly all cases two photons are generated. Each of 0.511 MeV, in almost exactly opposite directions to conserve momentum. Occasionally, three photons are emitted all in the same plane. Electron-positron annihilation at high energies has been extensively studied in particle accelerators. Generally the annihilation results in the production of a quark, and an antiquark, fort example, e+ e ì+ ì or a charged lepton plus an antilepton (e+e ì+ì). The quarks and antiquarks do not appear as free particles but convert into several hadrons, which can be detected experimentally. As the energy available in the electron-positron interaction increases, quarks and leptons of progressively larger rest mass can be produced. In addition, striking resonances are present, which appear as large increases in the rate at which annihilations occur at particular energies. The I / PSI particle and similar resonances containing an antiquark are produced at an energy of about 3 GeV, for example, giving rise to abundant production of charmed hadrons. Bottom (b) quark production occurs at greater energies than about 10 GeV. A resonance at an energy of about 90 GeV, due to the production of the Z0 gauge boson involved in weak interaction is currently under intensive study at the LEP and SLC e+ e colliders. Colliders are the machines for increasing the kinetic energy of charged particles or ions, such as protons or electrons, by accelerating them in an electric field. A magnetic field is used to maintain the particles in the desired direction. The particle can travel in a straight, spiral, or circular paths. At present, the highest energies are obtained in the proton synchrotron.

The Super Proton Synchrotron at CERN (Geneva) accelerates protons to 450 GeV. It can also cause proton-antiproton collisions with total kinetic energy, in centre-of-mass co-ordinates of 620 GeV. In the USA the Fermi National Acceleration Laboratory proton synchrotron gives protons and antiprotons of 800 GeV, permitting collisions with total kinetic energy of 1600 GeV. The Large Electron Positron (LEP) system at CERN accelerates particles to 60 GeV.

All the aforementioned devices are designed to produce collisions between particles travelling in opposite directions. This gives effectively very much higher energies available for interaction than our possible targets. High-energy nuclear reaction occurs when the particles, either moving in a stationary target collide. The particles created in these reactions are detected by sensitive equipment close to the collision site. New particles, including the tauon, W, and Z particles and requiring enormous energies for their creation, have been detected and their properties determined.

While, still, a 'nucleon' and 'anti-nucleon' annihilating at low energy, produce about half a dozen pions, which may be neutral or charged. By definition, mesons are both hadrons and bosons, justly as the pion and kaon are mesons. Mesons have a substructure composed of a quark and an antiquark bound together by the exchange of particles known as Gluons.

The conjugate particle or antiparticle that corresponds with another particle of identical mass and spin, but has such quantum numbers as charge (Q), baryon number (B), strangeness (S), charms © and Isospin (I3) of equal magnitude but opposite sign. Examples of a particle and its antiparticle include the electron and positron, proton and antiproton, the positive and negatively charged pions, and the 'up' quark and 'up' antiquark. The antiparticle corresponding to a particle with the symbol 'a' is usually denoted. When a particle and its antiparticle are identical, as with the photon and neutral pion, this is called a 'self-conjugate particle'.

The critical potential to excitation energy required to change am atom or molecule from one quantum state to another of higher energy, is equal to the difference in energy of the states and is usually the difference in energy between the ground state of the atom and a specified excited state. Which the state of a system, such as an atom or molecule, when it has a higher energy than its ground state.

The ground state contributes the state of a system with the lowest energy. An isolated body will remain indefinitely in it, such that it is possible for a system to have possession of two or more ground states, of equal energy but with different sets of quantum numbers. In the case of atomic hydrogen there are two states for which the quantum numbers n, I, and m are 1, 0, and 0 respectively, while the spin may be $+ \frac{1}{2}$ with respect to a defined direction. An allowed wave function of an electron in an atom obtained by a solution of the 'Schrödinger wave equation' in which a hydrogen atom, for example, the electron moves in the electrostatic field of the nucleus and its potential energy is e2 / r, where 'e' is the electron charge and 'r' its distance from the nucleus. A precise orbit cannot be considered as in Bohr's theory of the atom, but the behaviour of the electron is described by its wave function, Ψ

which is a mathematical function of its position with respect to the nucleus. The significance of the wave function is that $|\Psi|2$ dt is the probability of locating the electron in the element of volume dt.

Solution of Schrödinger's equation for the hydrogen atom shows that the electron can only have certain allowed wave functions (eigenfunctions). Each of these corresponds to a probability distribution in space given by the manner in which $|\Psi|2$ varies with position. They also have an associated value of the energy 'E'. These allowed wave functions, or orbitals, are characterized by three quantum numbers similar to those characterized the allowed orbits in the earlier quantum theory of the atom: 'n', the 'principal quantum number, can have values of 1, 2, 3, and so forth the orbital with n =1 has the lowest energy. The states of the electron with n = 1, 2, 3, and so forth, are called 'shells' and designate the K, L, M shells, and so forth 'I', the 'azimuthal quantum numbers', which for a given value of 'n' can have values of 0, 1, 2, . . . (n 1) An electron in the 'I' shell of an atom with n = 2 can occupy two sub-shells of different energy corresponding to I = 0, I = 1, and I = 2. Orbitals with I = 0, 1, 2 and 3 are called s, p, d, and ¦ orbitals respectively. The significance of one quantum number is that it gives the angular momentum of the electron. The orbital angular momentum of an electron is given by:

$$?[I(I + 1)(h/2\eth).$$

'm', the 'magnetic quantum number, which for a given value of I can have values, L (I-1), . . ., 0, . . . (I-1), I. Thus, for a 'p' orbital for orbits with m = 1, 0, and 1. These orbitals, with the same values of 'n' and 'I' but different 'm' values, have the same energy. The significance of this quantum number is that it indicates the number of different levels that would be produced if the atom were subjected to an external magnetic field.

Printed in the United States
By Bookmasters